UNDERSTANDING BIOLOGY
for Advanced Level
Third Edition

Glenn Toole

Vice Principal, Pendleton Sixth Form College, Salford

Susan Toole

Head of Biology, The Hulme Grammar School for Girls, Oldham
and Former Team Leader and Examiner for A-Level Biology

STANLEY THORNES PUBLISHERS LTD.

Originally published in 1987 by Hutchinson Education

Second edition published in 1991
Third edition published in 1995 by
Stanley Thornes (Publishers) Ltd
Ellenborough House
Wellington Street
CHELTENHAM GL50 1YW

96   97   98   99   00   /   10   9   8   7   6   5   4   3

A catalogue record of this book is available from the British
Library.

ISBN 0 7487 1718 8

Related titles
Skills in Advanced Biology by Wilbert Garvin

   Volume 1: Dealing with data (0 8595 0588 X)
   Worksheet masters (0 8595 0589 8)

   Volume 2: Observing, recording and interpreting (0 8595 0817 X)
   Teacher's supplement (0 7487 0043 9)

   Volume 3: Investigating (0 7487 2048 0)

Artwork by Tech-Set Ltd, Geoff Jones, Annabel Milne,
Angela Lumley, Mark Dunn, Tim Smith and David Oliver
Typeset by Tech-Set Ltd, Gateshead, Tyne & Wear
Printed and bound in Italy

Acknowledgements

To Annabel Milne for permission to reproduce artwork on
pp. 54, 55, 59, 263, 266, 267, 394, 396, 473, 474, 503, 524.

To the following for permission to reproduce photographs:

Ardea London Ltd p. 170 (Jean-Paul Ferrero); Biological
Sciences Review pp. 307, 518; Biophoto Associates pp. 49,
54, 56, 58, 59 (top, bottom), 62, 64, 76, 80 (top, bottom), 85,
86, 89, 90 (top, bottom), 93 (middle), 133, 141, 149, 188, 198,
213, 219, 264 (top left, bottom left), 362 (bottom), 397, 411,
446 (bottom), 447, 454 (bottom), 459 (top), 494, 495, 514 (top,
bottom), 557 (bottom), 565, 608, 618; Biopol p. 604; Bruce
Coleman Limited, pp. 170, 202 (Jane Burton), 170 (Hans
Reinhard), 203 (top, bottom) (Kim Taylor), 317, 359 (bottom)
(John Murray), 366 (Dr Norman Myers), 381 (bottom)
(Adrian Davies), 437, 493 (top) (Dr M P Kahl), 519 (Rocco
Longo); Culham/Harwell Photographic Group, AEA
Technology, p. 4; Dr David Thornton, Dept. of Biol. Sciences,
University of Salford, p. 281; Ecoscene, pp. 380 (Andrew D
R Brown), 381 (top) (Nick Hawkes); FLPA, pp. 8 (D P Wilson),
65 (J C Allen), 87 (Mark Newman), 361 (top) (D T Grewcock),
497 (top) (Leonard Lee Rue); Gene Cox, pp. 1, 86, 225, 304
(top), 401; GSF Picture Library, pp. 268, 337, 364 (D Hoffman),
392, 603 (W Pierdon); Heather Angel, pp. 93 (top), 250 (all),
353; Holt Studios International, pp. 12, 362 (top), 570, 609,
376, 378 (bottom) (Nigel Cattlin), 329 (Inga Spence), 378
(Richard Anthony), 594 (bottom) (Inga Spence); ICCE, p.
326 (Alain Compost); J & S Professional Photography, p. 594
(top) (Jim Lowe); Martyn F Chillmaid, pp. 342, 536, 548, 560;
Oxford Scientific Films, pp. 93 (bottom), 205 (J K Burras),
223 (London Scientific Films), 339 (K G Vock, Okapia), 345,
361 (bottom) (Kim Westerskov), 381 (middle) (Kathie
Atkinson), 493 (bottom) (Norbert Rosing), 497 (bottom) (M
Austerman, Animals Animals), 550 (G I Bernard), 562 (David
Thompson), 577 (Colin Milkins); Panos Pictures, pp. 337
(Rob Cousins), 359 (top) (J Hartley), 602 (Sean Sprague);
Picturepoint, 340; Ralston Photography, 599; Science Photo
Library, pp. 99, 334, 459 (bottom), 542 (bottom), 23 (Richard
Kirby), 49 (David Scharf), 77, 531 (CNRI), 124 (Simon
Fraser/RVI, Newcastle-upon-Tyne), 126 (Philippe
Plailly/Eurelios), 127 (David Parker), 133, 429 (Biophoto
Associates), 145 (Moredun Animal Health), 221 (top,
bottom), 445 (Dr Jeremy Burgess), 221 (middle), 389 (Claude
Nuridsany and Marie Perennou), 229 (bottom), 473, 505
(top, bottom) (Prof. P Motta, Dept. of Anatomy, University
'La Sapienza', Rome), 269, 589 (James Holmes, Celltech
Ltd), 288 (Damien Lovegrove), 309, (J C Revy), 368, 373
(Simon Fraser), 375 (US Dept of Energy), 387 (ASA
Thoresen), 388 (M I Walker), 390 (John Mead), 395, 416
(Martin Dohrn), 399 (top) (Astrid and Hanns-Frieder
Michler), 399 (bottom) (Manfred Kage), 406, 547 (top) (Bill
Longcore), 423 (Professors P M Motta and S Correr), 427
(Eamonn McNulty), 487 (Nancy Kedersha, UCLA), 524, 527
(Dr Don Fawcett), 538 (Dr Colin Chumbley), 542 (top) (Tim
Beddow), 542 (middle) (Catherine Pouedras), 547 (bottom)
(Adam Hart-Davies), 559 (bottom) (Secchi-Lecaque,Roussel-
UCLAF, CNRI), 607 (A B Dowsett); Science PicturesLimited,
pp. 84, 92, 186, 214 (top, middle, bottom), 219, 228, 229 (top),
237, 263, 264 (right), 293, 294, 300, 301, 302, 304 (middle,
bottom), 405, 424, 439, 440 (top, bottom), 446 (top), 448 (left,
right), 451, 454 (top), 460, 500, 503, 526, 557 (top), 559 (top);
The Environmental Picture Library, p. 257 (Robert Brook);
University of Oxford, pp. 71, 73; Wildlife Matters,
p. 370.

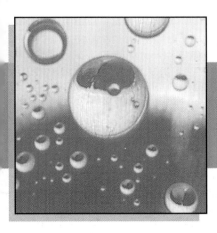

Contents

Part II — The Continuity of Life

Part IV   Transport and Exchange Mechanisms

Part V   Coordination, Response and Control

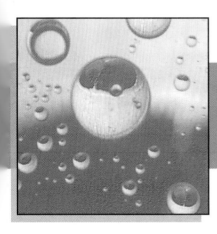

Preface

In keeping with science in general, A- and AS-level Biology syllabuses have undergone radical changes over the past few years. The move towards modular courses and the introduction of a larger number of optional topics has expanded the scope of biological knowledge required from a course textbook. In addition, many students need to fulfil a requirement of their syllabus for an extended practical investigation or a project during their course. All biology syllabuses however have a Core of common topics building on the knowledge and understanding of Science in the National Curriculum. This book gives a clear and comprehensive coverage of these areas and aims to make the difficult transition from Double Science (GCSE) to A/AS-level Biology as easy and rewarding as possible.

This book is intended primarily for students taking A- and AS-level Biology, but in addition, the text will also be valuable to students studying Human and Social Biology, the Scottish Higher Grade examination, GNVQ Science (Advanced) or the International Baccalaureat. Most of the material for these courses is covered by this book.

Detailed analysis of all the major new A- and AS-level Biology syllabuses has been undertaken to ensure that all common compulsory topics are covered. In addition, the popular options such as Biotechnology are included as separate chapters, while others like Health and Disease are integrated into the text throughout.

The style and accessibility of the book, with its many clear diagrams and summary tables have been retained, but the content has been extensively revised with much new work and some rearrangement of existing material to bring it fully in line with the new requirements. Full colour has been used throughout, not only to make the book unusually attractive but also to improve understanding by making the diagrams clearer and access easier.

New features have been developed for the Third Edition. They include Notebooks; short explanatory sections on key underpinning topics of biology and other sciences such as pH and mole calculations. Notebooks are designed to cover basic principles required for the course.

Applications have always been a strong feature of this book. They have been expanded and highlighted in the Third Edition and include numerous topics of applied biology such as pacemakers, gene therapy, biosensors and washing powders as well as health topics like cancer, allergies, Alzheimer's disease and eating disorders.

Project suggestions are provided because most new syllabus require students to undertake an independent practical investigation on a topic of their choice. The suggestions, included at relevant points in the text, are designed as starting points for enquiry and to stimulate innate curiosity. Did You Know? offers fascinating and often startling facts about the biological world; while introducing a light-hearted note, they also provide worthwhile factual information about the topic being studied.

All nomenclature in this book is consistent with the recommendations made by the Institute of Biology and the Association for Science Education. The sample questions of the Examining Boards at the end of each chapter have been extensively updated with recent examples.

Overall the book is intended to provide a clear, highly readable text that is sufficiently detailed to satisfy the requirements of all major syllabuses but without unnecessary detail which can cloud the underlying issues. Biology is an exciting subject and we hope that this book will provide not only the information needed for examination success but also the stimulus to investigate and to enquire further.

We would like to express our gratitude to Wilbert Garvin for compiling the project suggestions and to Adrian Wheaton, Malcolm Tomlin and all those at Stanley Thornes (Publishers) Ltd without whose patient encouragement and hard-working efficiency the quality of this new edition could not have been achieved.

Glenn and Susan Toole
1995

Transverse section through kidney tubules to show cellular organization (*opposite*)

Part I
LEVELS OF ORGANIZATION

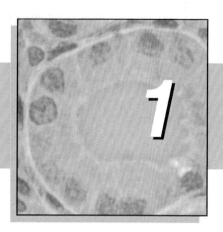

1

Size and complexity

Biology covers a wide field of information over a considerable size range. On the one hand it involves the movements of electrons in photosynthesis and on the other the migrations of individuals around the earth. Within this range it is possible to recognize seven levels of organization, each of which forms the basis of the next. The most fundamental unit is the **atom**; atoms group to form **molecules**, which in turn, may be organized into **cells**. Cells are grouped into **tissues** which collectively form **organs**, which form **organisms**. A group of organisms of a single species may form a **population**.

TABLE 1.1 **Metric units**

Units of size			
1 kilometre	(km)	=	1000 (10^3) metres
1 metre	(m)		
1 centimetre	(cm)	=	1/100 (10^{-2}) metre
1 millimetre	(mm)	=	1/1000 (10^{-3}) metre
1 micrometre (micron)	(μm)	=	1/1 000 000 (10^{-6}) metre
1 nanometre	(nm)	=	1/1 000 000 000 (10^{-9}) metre
1 picometre	(pm)	=	1/1 000 000 000 000 (10^{-12}) metre

1.1   Atomic organization

Atoms are the smallest unit of a chemical element which can exist independently. They comprise a nucleus which contains positively charged particles called **protons**, the number of which is referred to as the **atomic number**. For each proton there is a particle of equal negative charge called an **electron**, so the atom has no overall charge. The electrons are not within the nucleus, but orbit in fixed quantum shells around it (see Fig. 1.1). There is a fixed limit to the number of electrons in any one shell. There may be up to seven such shells each with its own energy level; electrons in the shells nearest the nucleus have the least energy. The addition of energy, e.g. in the form of heat or light, may promote an electron to a higher energy level within a shell. Such an electron almost immediately returns to its original level, releasing its newly absorbed energy as it does so. This electron movement is important biologically in processes such as photosynthesis (Chapter 14).

2

	HYDROGEN	CARBON	NITROGEN	OXYGEN
Atomic nucleus				
Proton (positively charged)		6p 6n	7p 7n	8p 8n
Electron (negatively charged) in a 3-dimentional orbit around the nucleus				
Electron shell				
Atomic number	1	6	7	8
Number of protons	1	6	7	8
Number of neutrons	0	6	7	8
Relative atomic mass	1	12	14	16
Number of electrons				
First quantum shell	1	2	2	2
Second quantum shell	–	4	5	6
Total	1	6	7	8

Fig. 1.1 Atomic structure of four commonly occurring biological elements

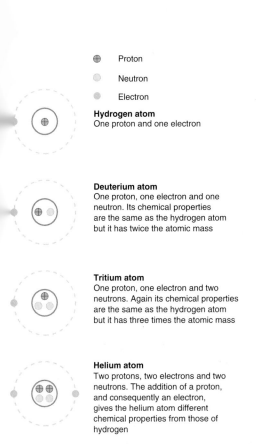

⊕ Proton

◉ Neutron

● Electron

Hydrogen atom
One proton and one electron

Deuterium atom
One proton, one electron and one neutron. Its chemical properties are the same as the hydrogen atom but it has twice the atomic mass

Tritium atom
One proton, one electron and two neutrons. Again its chemical properties are the same as the hydrogen atom but it has three times the atomic mass

Helium atom
Two protons, two electrons and two neutrons. The addition of a proton, and consequently an electron, gives the helium atom different chemical properties from those of hydrogen

Fig. 1.2 Atomic structure of the atoms of hydrogen, deuterium, tritium and helium

The nucleus of the atom also contains particles called **neutrons** which have no charge. Protons and neutrons contribute to the mass of an atom, but electrons have such a comparatively small mass that their contribution is negligible. However, the number of electrons determines the chemical properties of an atom. (See Fig. 1.2.)

1.1.1 Ions

As we have seen, atoms do not have any overall charge because the number of protons is always the same as the number of electrons and both have equal, but opposite, charges. If an atom loses or gains electrons it becomes an **ion**. The addition of electrons produces a negative ion while the loss of electrons gives rise to a positive ion. The loss of an electron is called **oxidation**, while the gain of an electron is called **reduction**. The atom losing an electron is said to be oxidized, while that gaining an electron is said to be reduced. The loss of an electron from a hydrogen atom, for instance, would leave a hydrogen ion, comprising just a single proton. Having an overall positive charge it is written as H^+. Where an atom, e.g. calcium, loses two electrons its overall charge is more positive and it is written Ca^{2+}. The process is similar where atoms gain electrons, except that the overall charge is negative, e.g. Cl^-. Ions may comprise more than one type of atom, e.g. the sulphate ion is formed from one sulphur and four oxygen atoms, with the addition of two electrons, SO_4^{2-}.

1.1.2 Isotopes

The properties of an element are determined by the number of protons and hence electrons it possesses. If protons (positively charged) are added to an element, then an equivalent number of electrons (negatively charged) must be added to maintain an overall neutral charge. The properties of the element would th change – indeed it now becomes a new element. For example, can be seen from Fig. 1.1 that the addition of one proton, one electron and one neutron to the carbon atom, transforms it int nitrogen atom.

If however, a neutron (not charged) is added, there is no ne for an additional electron and so its properties remain the sam As neutrons have mass, the element is heavier. Elements whic have the same chemical properties as the normal element, but have a different mass, are called **isotopes**. Hydrogen normally comprises one proton and one electron and consequently has a atomic mass of one. The addition of a neutron doubles the atomic mass to two, without altering the element's chemical properties. This isotope is called **deuterium**. Similarly, the addition of a further neutron forms the isotope **tritium**, which has an atomic mass of three (Fig. 1.2).

Isotopes can be traced by various means, even when incorporated in living matter. This makes them exceedingly useful in tracing the route of certain elements in a variety of biological processes.

NOTEBOOK

Using isotopes as tracers

Autoradiograph of labelled leaves

Isotopes are varieties of atoms which differ in their mass. They are usually taken up and used in biological systems in the same way as the 'normal' form of the element, but they can be detected because they have different properties. Isotopes have been used to study photosynthesis, respiration, DNA replication and protein synthesis. Isotopes such as ^{15}C and ^{13}C are not radioactive and so do not decay but they can be detected using a mass spectrometer or a nuclear magnetic resonance (NMR) spectrometer.

To use a mass spectrometer the sample to be studied is vaporized is such a way that the molecules become charged. They then pass through a magnetic field which deflects them and the machine records the abundance of each ion with a particular charge : mass ratio. Isotopes with an uneven number of protons or electrons spin, like spinning bar magnets. A NMR spectrometer detects each type of spinning nucleus.

Radioactive isotopes can be used in a different way to follow biological processes. For example, when studying photosynthesis leaves may be exposed to $^{14}CO_2$ instead of $^{12}CO_2$. The 'labelled' carbon is incorporated into the carbohydrate produced and can be detected using autoradiography. This technique relies on the ability of radioisotopes to 'fog' photographic film as they emit radiation. When the process is combined with chromatography (see Section 14.3) it is possible to identify which individual compounds have taken up the radioactive carbon. If an accurate measure of the radioactivity in a sample is required a scintillation counter can be used.

APPLICATION

Radioisotopes in medicine

Radioisotopes can be used both for diagnosis and treatment of disease as well as for research into possible causes. A particularly important isomer used to study lung and heart complaints is ^{99m}technetium. It has a half life of only 6 hours and so decays very rapidly. It can be used as an aerosol, in very low concentrations, to show up available air spaces in patients' lungs. It may be used to label red blood cells so that the distribution of blood within the spaces of the heart or in deep veins can be shown. This is useful if there is a possibility of blood clots having formed.

The radiation emitted by a radioisotope can be used to destroy damaged tissue. For example ^{137}caesium is inserted in a sealed probe to destroy cancerous cells in the cervix. ^{131}Iodine is taken up selectively by the thyroid gland and can be used in carefully calculated doses to destroy a specific amount of that gland.

^{90}Yttrium in a silicate injection kills synovial tissues which are eroding the ends of the bones in sufferers of rheumatoid and osteo arthritis.

1.2   Molecular organization

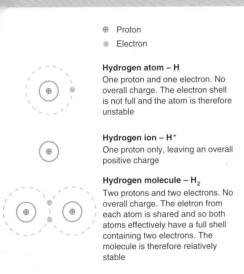

⊕ Proton
● Electron

Hydrogen atom – H
One proton and one electron. No overall charge. The electron shell is not full and the atom is therefore unstable

Hydrogen ion – H⁺
One proton only, leaving an overall positive charge

Hydrogen molecule – H₂
Two protons and two electrons. No overall charge. The eletron from each atom is shared and so both atoms effectively have a full shell containing two electrons. The molecule is therefore relatively stable

Fig. 1.3 Atomic structure of a hydrogen atom, a hydrogen ion and a hydrogen molecule

We have seen that the electron shells around an atom may each contain a maximum number of electrons. The shell nearest the nucleus may possess a maximum of two electrons and the next shell a maximum of eight. An atom is most stable, i.e. least reactive, when its outer electron shell contains the maximum possible number of electrons. For example, helium, with a full complement of two electrons in its outer shell, is inert. In a hydrogen atom, the electron shell has a single electron and so the atom is unstable. If two hydrogen atoms share their electrons they form a hydrogen **molecule**, which is more stable. The two atoms are effectively combined and the molecule is written as H$_2$. The sharing of electrons in order to produce stable molecules is called **covalent bonding**.

The oxygen atom contains eight protons and eight neutrons in the nucleus with eight orbiting electrons. The inner quantum shell contains its maximum of two electrons, leaving six electrons in the second shell (Fig. 1.1). As this second shell may contain up to eight electrons, it requires two electrons to complete the shell and become stable. It may therefore combine with two hydrogen atoms by sharing electrons to form a water molecule (Fig. 1.4). In this way the outer shells of the oxygen atom and both hydrogen atoms are completed and a relatively stable molecule is formed.

Carbon with its six electrons (Fig. 1.1) has an inner shell containing two, leaving four in the outer shell. It requires four

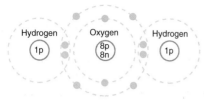

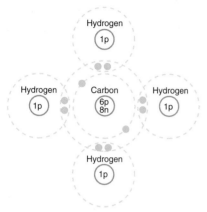

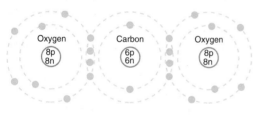

Water molecule (H_2O)
The oxygen atom shares 2 electrons with each hydrogen atom. Both molecules thereby complete their outer shell - the hydrogen atom with 2 electrons, the oxygen atom with 8

Methane molecule (CH_4)
The carbon atom shares 2 electrons with each hydrogen atom. Each hydrogen atom thus completes its outer shell with 2 electrons, while the carbon atom completes its outer shell with 8

Carbon dioxide molecule (CO_2)
The carbon atom shares 4 electrons with each oxygen atom. All three atoms thereby complete their outer shells with 8 electrons

Fig. 1.4 Atomic models of the molecules of water, methane and carbon dioxide

TABLE 1.2 **Relative abundance by weight of elements in humans compared to the earth's crust**

Element	Human	Earth's crust
Oxygen	63.0	46.5
Carbon	19.5	0.1
Hydrogen	9.5	0.2
Nitrogen	5.0	0.0001
Phosphorus	0.5	1.5

more electrons to fill this shell. It may therefore combine with four hydrogen atoms each of which shares its single electron. This molecule is called methane CH_4 (Fig. 1.4). It may also combine with two oxygen atoms, each of which shares two electrons. This molecule is carbon dioxide (Fig. 1.4).

When an atom, e.g. hydrogen, requires one electron to complete its outer shell it is said to have a **combining power (valency)** of one. Oxygen, which requires two electrons to complete its outer shell, has a combining power of two. Likewise nitrogen has a combining power of three and carbon of four.

When two atoms share a single electron, the bond is referred to as a **single bond** and is written with a single line, e.g. the hydrogen molecule is H—H and water may be represented as H—O—H. If two atoms share two electrons a **double bond** is formed. It is represented by a double line, e.g. carbon dioxide may be written as O=C=O. To form stable molecules, hydrogen must therefore have a single bond; oxygen two bonds (either two singles or one double); nitrogen must have three bonds (either three singles, or one single and one double); and carbon must have four bonds. It should now be apparent that these four atoms can combine in a number of different ways to form a variety of molecules. This partly explains the abundance of these elements in living organisms although some are relatively rare in the earth's crust (Table 1.2).

Carbon in particular can be seen to be almost 200 times more abundant in living organisms than in the earth's crust. Why should this be so? In the first place, carbon with its combining power of four can form molecules with a wide variety of other elements such as hydrogen, oxygen, nitrogen, sulphur, phosphorus and chlorine. This versatility allows great diversity in living organisms. More importantly, carbon can form long chains linked by single, double and triple bonds. These chains may be thousands of carbon atoms long. Such large molecules are essential to living organisms, not least as structural components. Furthermore, these chains have great stability – another essential feature. Carbon compounds may also form rings. These rings and chains may be combined with each other to give giant molecules of almost infinite variety. Examples of the size, diversity and complexity of carbon molecules can be found among the three major groups of biological compounds: carbohydrates, fats and proteins.

These are discussed in more detail in Chapter 2.

1.2.1 Ionic bonding

In addition to forming covalent bonds through the sharing of electrons, atoms may stabilize themselves by losing or gaining electrons to form ions. The loss of an electron (oxidation) leaves the atom positively charged (oxidized). The gain of an electron (reduction) leaves the atom negatively charged (reduced). Oppositely charged atoms attract one another forming **ionic bonds**. Sodium, for example, tends to lose an electron forming a Na^+ ion; chlorine tends to gain an electron forming a Cl^- ion. These two oppositely charged ions form ionic bonds and form sodium chloride (common salt).

1.2.2 Hydrogen bonds

The electrons in a molecule do not distribute themselves evenly but tend to group at one position. This region will consequently be more negative than the rest of the molecule. The molecule is said to be **polarized**. The negative region of such a molecule will be attracted to the positive region of a similarly polarized molecule. A weak electrostatic bond between the two is formed. In biological systems this type of bond is frequently a hydrogen bond. These bonds are weak individually, but collectively form important forces which alter the physical properties of molecules. Water forms hydrogen bonds which, as we shall see in Chapter 22, significantly affect its properties and hence its biological importance.

1.3 Cellular organization

In 1665, Robert Hooke, using a compound microscope, discovered that cork was composed of numerous small units. He called these units, **cells**. In the years which followed, Hooke and other researchers discovered that many other types of material were similarly composed of cells. By 1838, the amount of plant material shown to be composed of cells convinced Matthias Schleiden, a German botanist, that all plants were made up of cells. The following year, Theodor Schwann reached the same conclusion about the organization of animals. Their joint findings became known as the **cell theory**. It was of considerable biological significance as it suggested a common denominator for all living matter and so unified the nature of organisms. The theory makes the cell the fundamental unit of structure and function in living organisms. Hooke had originally thought the cell to be hollow, and that the wall represented the living portion. It soon became clear that cells were far from hollow. With the development of better light microscopes, first the nucleus and then organelles such as the chloroplasts became visible. One hundred years after Schleiden and Schwann put forward the cell theory, the development of the **electron microscope** revolutionized our understanding of cell structure. With its ability to magnify up to 500 times more than the light microscope, the electron microscope revealed the fine structure of cells including many new organelles. This detail is called the **ultrastructure** of the cell. The complexity of cellular structure so revealed led to the emergence of a new field of biology, **cytology** – the study of cell ultrastructure. This shows that while organisms are very diverse in their structures and cells vary considerably in size and shape, there is nevertheless a remarkable similarity in their basic structure and organization. This structure and organization is studied in Chapter 4.

1.4   Colonial organization

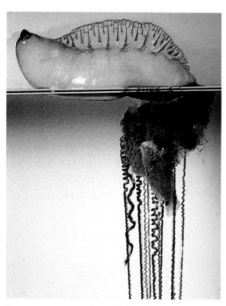

Portuguese man-of-war (*Physalia*) eating fish

The first colonies may have arisen when individual unicells failed to separate after cell division. Within colonies each cell i capable of carrying out all the essential life processes. Indeed, separated from the colony, any cell is capable of surviving independently. The only advantage of a colonial grouping is tl the size of the unit probably deters some predators and thus increases the group's survival prospects.

If one cell in a colony should lose the ability to carry out a vital process, it could only survive by relying on other cells in the colony to perform the process on its behalf. The loss of one function, however, might permit the cell to perform one or othe of its functions more efficiently, because the energy and resources required by the missing function could be directed towards the remaining ones. In this way, the individual cells within a colony could have become different from one another both structure and function, a process known as **differentiatior** Further changes of this type would finally result in cells performing a single function. This is known as **specialization**. Clearly specialization must be organized in such a way that all essential functions are still performed by the colony as a whole. With increasing specialization, and the consequent loss of more and more functions, any cell becomes increasingly dependent or others in the colony for its survival. This **interdependence** of cells must be highly organized. Groups of cells must be coordinated so that the colony carries out its activities efficiently Such coordination between the different cells is called **integration**. Once the cells become so dependent on each other that they are no longer capable of surviving independently, then the structure is no longer a colony but a **multicellular organism**.

1.5   Tissue organization

A tissue is a group of similar cells, along with any intercellular substance, which performs a particular function. Some cells, e.g. unicellular protozoans and algae, perform all functions which are essential to life. It is impossible for such cells to be efficient at all functions, because each function requires a different type of cellular organization. Whereas one function might require the cell to be long and thin, another might require it to be spherical. One function might require many mitochondria, another, very few. Acid conditions might suit one activity but not another. No one cell can possibly provide the optimum conditions for all activities. For this reason, cells are specialized to perform one, or at most a few, functions. To increase efficiency, cells performing the same functions are grouped together into a tissue. The study of tissues is called **histology**. Some organisms, e.g. cnidarians, are at the tissue level of organization. Their physiological activities are performed by tissues rather than organs.

1.6 Organ level of organization

An organ is a structural and functional unit of a plant or animal. It comprises a number of tissues which are coordinated to perform a variety of functions, although one major function often predominates. The majority of plants and animals are composed of organs. Most organs do not function independently but in groups called **organ systems**. A typical organ system is the digestive system which comprises organs such as the stomach, duodenum, ileum, liver and pancreas. Certain organs may belong to more than one system. The pancreas, for example, forms part of the **endocrine (hormone)** system as well as the digestive system, because it produces the hormones insulin and glucagon, as well as the digestive enzymes amylase and trypsinogen.

1.7 Social level of organization

A **population** is a number of individuals of the same species which occupy a particular area at the same time. In itself, a population is not a level of organization as no organization exists between the individual members. In some species, however, the individuals do exhibit some organization in which they cooperate for their mutual benefit. Such a population is more accurately termed a **society**. It differs from a colony (although the term is often used) in that the individuals are not physically connected to one another, but totally separate. As with a colony, the individuals can survive independently of others in the society, although usually somewhat less successfully. Unlike most colonies, there is considerable coordination between the society members and communication forms an integral part of their organization. Societies may exist simply because there is safety in numbers, e.g. schools of fish. They may enable more successful hunting, as in wolves, or aid the successful rearing of young, as in baboons. In insects, however, the degree of organization is considerable. There is **division of labour** which leads to differentiation of individuals in order to perform specialized functions. In a bee society for instance, the queen is the only fertile female and has a purely reproductive role. The drones (males) also function reproductively while the workers (sterile females) perform a variety of tasks such as collecting food, feeding the larvae and guarding and cleaning the hive. Complex societies can readily be compared to an organism with its organs each specialized for a major function. Some account of the organization of a bee colony is given in Section 27.7.5.

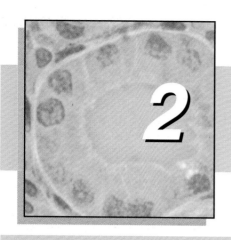

2 Molecular organization

2.1 Inorganic ions

Water is the most important inorganic molecule in biology and its chemical structure and properties are described in Chapter 2. Dissolved in the water within living organisms are a large number of inorganic ions. Typically they constitute about 1% of an organism by weight, but they are nonetheless essential. They are divided into two groups: the **macronutrients** or **major elements** which are needed in very small quantities, and the **micronutrients** or **trace elements** which are needed in minute amounts (a few parts per million). Although the elements mostly fall into the same category for plants and animals, there are a few exceptions. Chlorine, for example, is a major element in animals but a trace element in plants. In addition to the essential elements listed in Table 2.1 (on this and the next page), some organisms also have specific requirements such as vanadium, chromium and silicon.

TABLE 2.1 **Inorganic ions and their functions in plants and animals**

Macronutrients/ main elements	Functions	Notes
Nitrate NO_3^- Ammonium NH_4^+	Nitrogen is a component of amino acids, proteins, vitamins, coenzymes, nucleotides and chlorophyll. Some hormones contain nitrogen, e.g. auxins in plants and insulin in animals	A deficiency of nitrogen in plants causes chlorosis (yellowing of leaves) and stunted growth
Phosphate PO_4^{3-} Orthophosphate $H_2PO_4^-$	A component of nucleotides, ATP and some proteins. Used in the phosphorylation of sugars in respiration. A major constituent of bone and teeth. A component of cell membranes in the form of phospholipids	Deficiency of phosphates in plants leads to stunted growth, especially of roots, and the formation of dull, dark green leaves. In animals, deficiency can result in a form of bone malformation called rickets
Sulphate SO_4^{2-}	Sulphur is a component of some proteins and certain coenzymes, e.g. acetyl coenzyme A	Sulphur forms important bridges between the polypeptide chains of some proteins, giving them their tertiary structure. A deficiency in plants causes chlorosis and poor root development
Potassium K^+	Helps to maintain the electrical, osmotic and anion/cation balance across cell membranes. Assists active transport of certain materials across the cell membrane. Necessary for protein synthesis and is a co-factor in photosynthesis and respiration. A constituent of sap vacuoles in plants and so helps to maintain turgidity	Potassium plays an important role in the transmission of nerve impulses. A deficiency in plants leads to yellow-edged leaves and premature death

cont.

TABLE 2.1 *cont.*

Macronutrients/ main elements	Functions	Notes
Calcium Ca^{2+}	In plants, calcium pectate is a major component of the middle lamella of cell walls and is therefore necessary for their proper development. It also aids the translocation of carbohydrates and amino acids. In animals, it is the main constituent of bones, teeth and shells. Needed for the clotting of blood and the contraction of muscle	In plants, deficiency causes the death of growing points and hence stunted growth. In animals, deficiency leads to rickets and delay in the clotting of blood
Sodium Na^+	Helps to maintain the electrical, osmotic and anion/cation balance across cell membranes. Assists active transport of certain materials across the cell membrane. A constituent of the sap vacuole in plants and so helps maintain turgidity	In animals, it is necessary for the functioning of the kidney, nerves and muscles; deficiency may cause muscular cramps. Sodium is so common in soils that deficiency in plants is rare. Sodium ions have much the same function as potassium ions and may be exchanged for them
Chlorine Cl^-	Helps to maintain the electrical, osmotic and anion/cation balance across cell membranes. Needed for the formation of hydrochloric acid in gastric juice. Assists in the transport of carbon dioxide by blood (chloride shift)	In animals, deficiency may cause muscular cramps. Its widespread availability in soils makes deficiency in plants practically unknown
Magnesium Mg^{2+}	A constituent of chlorophyll. An activator for some enzymes, e.g. ATPase. A component of bone and teeth	Deficiency in plants leads to chlorosis
Iron Fe^{2+} or Fe^{3+}	A constituent of electron carriers, e.g. cytochromes, needed in respiration and photosynthesis. A constituent of certain enzymes, e.g. dehydrogenases, decarboxylases, peroxidases and catalase. Required in the synthesis of chlorophyll. Forms part of the haem group in respiratory pigments such as haemoglobin, haemoerythrin, myoglobin and chlorocruorin	Deficiency in plants leads to chlorosis and in animals to anaemia
Micronutrients/ trace elements		
Manganese Mn^{2+}	An activator of certain enzymes e.g. phosphatases. A growth factor in bone development	Deficiency in plants produces leaves mottled with grey and in animals, bone deformations
Copper Cu^{2+}	A constituent of some enzymes, e.g. cytochrome oxidase and tyrosinase. A component of the respiratory pigment haemocyanin	Deficiency in plants causes young shoots to die back at an early stage
Iodine I^-	A constituent of the hormone thyroxine, which controls metabolism in animals	Iodine is not required by higher plants. Deficiency in humans causes cretinism in children and goitre in adults; in some other vertebrates it is essential for metamorphic changes
Cobalt Co^{2+}	Constituent of vitamin B_{12}, which is important in the synthesis of RNA, nucleoprotein and red blood cells	Deficiency in animals causes pernicious anaemia
Zinc Zn^{2+}	An activator of certain enzymes, e.g. carbonic anhydrase. Required in plants for leaf formation, the synthesis of indole acetic acid (auxin) and anaerobic respiration (alcoholic fermentation)	Carbonic anhydrase is important in the transport of carbon dioxide in vertebrate blood. Deficiency in plants produces malformed, and sometimes mottled, leaves
Molybdenum Mo^{4+} or Mo^{5+}	Required by plants for the reduction of nitrate to nitrite in the formation of amino acids. Essential for nitrogen fixation by prokaryotes	Deficiency produces a reduction in crop yield. Not vital in most animals
Boron BO_3^{3+} or B_4O^{2+}	Required for the uptake of Ca^{2+} by roots. Aids the germination of pollen grains and mitotic division in meristems	Boron is not required by animals. Deficiency in plants causes death of young shoots and abnormal growth. May cause specific diseases such as 'internal cork' of apples and 'heart rot' of beet and celery
Fluorine F^-	A component of teeth and bones	Not required by most plants. Associates with calcium to form calcium fluoride which strengthens teeth and helps prevent decay

Chlorosis (thistle) – healthy plants shown for comparison

2.2 Carbohydrates

Carbohydrates comprise a large group of organic compounds which contain carbon, hydrogen and oxygen and which are either aldehydes or ketones. The word carbohydrate suggests that these organic compounds are hydrates of carbon. Their general formula is $C_x(H_2O)_y$. The word carbohydrate is convenient rather than exact, because while most examples do conform to the formula, e.g. glucose —$C_6H_{12}O_6$, sucrose $C_{12}H_{22}O_{11}$, a few do not, e.g. deoxyribose —$C_5H_{10}O_4$. Carbohydrates are divided into three groups: the **monosaccharides** ('single-sugars'), the **disaccharides** ('double-sugars') and the **polysaccharides** ('many-sugars').

The functions of carbohydrates, although variable, are in the main concerned with storage and liberation of energy. A few, such as cellulose, have structural roles. A full list of individual carbohydrates and their functions is given in Table 2.3 on page 19.

2.3 Monosaccharides

Monosaccharides are a group of sweet, soluble crystalline molecules of relatively low molecular mass. They are named with the suffix -ose. Monosaccharides contain either an aldehyde group (—CHO), in which case they are called **aldoses** or **aldo-sugars**, or they contain a ketone group (C = O), in which case they are termed **ketoses** or **keto-sugars**. The general formula for a monosaccharide is $(CH_2O)_n$. Where n = 3, the sugar is called a **triose** sugar, n = 5, a **pentose** sugar, and n = 6, a **hexose** sugar. Table 2.2 classifies some of the more important monosaccharides.

TABLE 2.2 **Classification of monosaccharides**

	Trioses $(C_3H_6O_3)$	Pentoses $(C_5H_{10}O_5)$	Hexoses $(C_6H_{12}O_6)$
Aldoses (—CHO) (Aldo-sugars)	Glyceraldehyde	Ribose Arabinose Xylose	Glucose Galactose Mannose
Ketoses (C=O) (Keto-sugars)	Dihydroxyacetone	Ribulose Xylulose	Fructose Sorbose

2.3.1 Structure of monosaccharides

Probably the best known monosaccharide, glucose, has the formula $C_6H_{12}O_6$. All but one of the six carbon atoms possesses an hydroxyl group (—OH). The remaining carbon atom forms part of the aldehyde group. Glucose may be represented by a straight chain of six carbon atoms. These are numbered beginning at the carbon of the aldehyde group. Glucose in common with other hexoses and pentoses easily forms stable ring structures. At any one time most molecules exist as rings rather than a chain. In the case of glucose, carbon atom number 1 may combine with the oxygen atom on carbon 5. This forms a six-sided structure known as a **pyranose** ring. In the case of fructose, it is carbon atom number 2 which links with the oxygen on carbon atom 5. This forms a five-sided structure called a **furanose** ring (Fig. 2.1). Both glucose and fructose can exist in both pyranose and furanose forms.

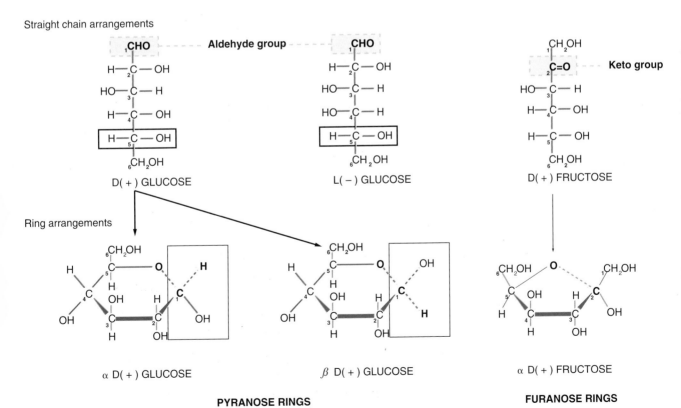

Fig. 2.1 Structure of various isomers of glucose and fructose

Glucose, in common with most carbohydrates, can exist as a number of **isomers** (they possess the same molecular formula but differ in the arrangement of their atoms). One type of isomerism, called **sterioisomerism**, occurs when the same atoms or groups, are joined together but differ in their arrangement in space. One form of sterioisomerism, called **optical isomerism**, results in isomers which can rotate the plane of polarized light (light which is vibrating in one plane only). The isomer which rotates the plane of polarized light to the right is called the **dextro(D or +) form**; the isomer rotating it to the left is called the **laevo(L or −) form**. (By present convention, however, the D and L forms are named by different criteria, regardless of the

---

NOTEBOOK

The mole

The mole is the scientific unit for the amount of a substance and is expressed as the symbol – **mol**. One mole of any substance contains the same number of particles (atoms, molecules or ions). This number is known as **Avogadro's constant** and is equal to 6.023×10^{23}. To give you some idea of the vast size of this number it is equal to the total human population of one hundred million million worlds identical to ours!

Different atoms (and therefore molecules and ions) have different masses. Chemists use the atomic weight of carbon, set at 12, as a standard against which to compare the weight of other atoms. Thus hydrogen which has a mass one twelfth that of carbon is given the mass of 1. These are known as relative atomic masses. The relative atomic mass of an element in grams always contains a mole of its atoms (i.e. 6.023×10^{23} atoms). The same is true of the relative **molecular** mass of a molecule. Thus:

> a mole of hydrogen atoms (H) has a mass of 1 g
> a mole of hydrogen molecules (H_2) has a mass of 2 g
> a mole of oxygen atoms (O) has a mass of 16 g
> a mole of oxygen molecules (O_2) has a mass of 32 g
> a mole of water molecules (H_2O) has a mass of 18 g.

To find out the number of moles in a given mass of a substance we simply divide the mass (in grams) by the mass of one mole,

> e.g. in 90 g of water there are $\frac{90}{18}$ moles

> $= 5$ moles (or $5 \times 6.023 \times 10^{23}$ molecules)

When dealing with gases, we use volume rather than weight to measure amounts. A mole of any gas at standard temperature and pressure (0 °C and 1 atmosphere) occupies $22.4 \, dm^3$ (litres). At room temperature (20 °C) this volume expands to $24 \, dm^3$. That all gases, regardless of the mass of the molecules they comprise, should occupy the same volume may seem surprising. In a gas however, the molecules are so far apart, that the size of the molecule itself is unimportant in terms of the volume occupied. Imagine several balls bouncing around inside a large hall – they could all be fitted in whether they were golf balls, tennis balls or footballs.

The concentration of a solution can be expressed in moles. A $1 \, mol \, dm^{-3}$ solution (1M solution) contains 1 mol in each dm^3 of the

direction in which they rotate polarized light.) While the chemical and physical properties of the two forms are the same, many enzymes will only act on one type. There would seem to be no reason why one form should be preferred to another, and yet almost all naturally occurring carbohydrates are of the D(+) form. It must be assumed that at an early stage in evolution the D(+) form was selected by chance and the consequent development of enzymes specific to this type ensured that all subsequent development was based on this form. Both D(+) and L(−) forms of glucose are shown in Fig. 2.1. The D(+) and L(−) forms of glucose arise because the relevant carbon atom has four different groups attached to it. This is called an **asymmetric**

solution. In other words to make up a 1 mol dm^{-3} (1 M) solution of a substance we add the relative molecular mass in grams of that substance to 1 dm^3 (litre) of water. In the case of sucrose ($C_{12}H_{22}O_{11}$) this is:

Molecule	Number in sucrose	Relative atomic mass (g)	Total mass (g)
CARBON	12	12	$12 \times 12 = 144$
HYDROGEN	22	1	$22 \times 1 = 22$
OXYGEN	11	16	$11 \times 16 = 176$
			Total 342 g

Hence we dissolve 342 g of sucrose in 1 dm^3 of water

To convert moles into number of molecules, mass or volume and vice versa simply follow the scheme below.

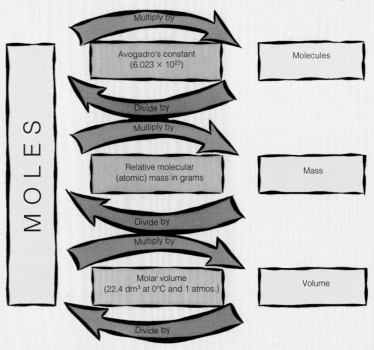

Conversion to and from moles

carbon atom. Another asymmetric carbon atom arises when glucose forms a ring structure. This gives rise to two further isomers, the α-**form** and the β-**form**. Both types occur natural[ly] and, as we shall see later, result in considerable biological differences when they form polymers. Fig. 2.1 again illustrate[s] both types.

2.4 Disaccharides

Monosaccharides may combine together in pairs to give a **disaccharide** (double-sugar). The union involves the loss of a single water molecule and is therefore a **condensation reaction**. The addition of water, under suitable conditions, is necessary i[f] the disaccharide is to be split into its constituent monosaccharide[s]. This is called **hydrolysis** 'water-breakdown' or, more accuratel[y] 'breakdown *by* water'. The bond which is formed is called a **glycosidic bond**. It is usually formed between carbon atom 1 o[f] one monosaccharide and carbon atom 4 of the other, hence it is called a 1–4 glycosidic bond (see Fig. 2.2). Any two monosaccharides may be linked in this way to form a disacchari[de] of which maltose, sucrose and lactose are the most common.

Disaccharides, like monosaccharides, are sweet, soluble and crystalline. Maltose and lactose are reducing sugars, whereas sucrose is a non-reducing sugar. The significance of this is considered in Section 2.5.5.

maltose (malt sugar) = glucose + glucose

sucrose (cane sugar) = glucose + fructose

lactose (milk sugar) = glucose + galactose

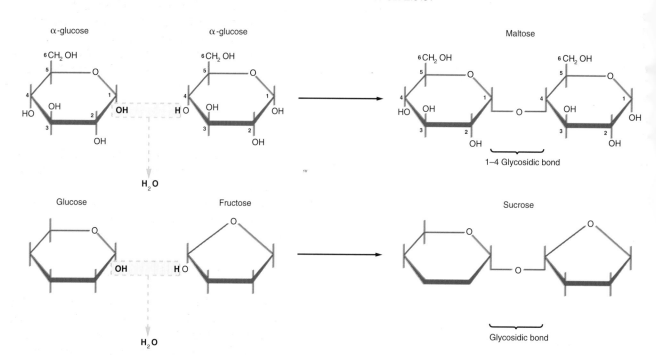

The removal of water (condensation) from the two hydroxyl groups (—OH) on carbons 1 and 4 of the respective glucose molecules, forms a maltose molecule. Some carbon and hydrogen atoms have been omitted for simplicity.
Sucrose is formed by a condensation reaction between one glucose and one fructose molecule. The process shown is much simplified.

Fig. 2.2 Formation of maltose and sucrose

2.5 Polysaccharides

In the same way that two monosaccharides may combine in pairs to give a disaccharide, many monosaccharides may combine by condensation reactions to give a **polysaccharide**. The number of monosaccharides which combine is variable and the chain produced can be branched or unbranched. The chains may be folded, thus making them compact and therefore ideal for storage. The size of the molecule makes them insoluble – another feature which suits them for storage as they exert no osmotic influence and do not easily diffuse out of the cell. Upon hydrolysis, polysaccharides can be converted to their constituent monosaccharides ready for use as respiratory substrates. Starch and glycogen are examples of storage polysaccharides. Not all polysaccharides are used for storage; cellulose, for example, is a structural polysaccharide giving strength and support to cell walls.

2.5.1 Starch

Starch is a polysaccharide which is found in most parts of the plant in the form of small granules. It is a reserve food formed from any excess glucose produced during photosynthesis. It is common in the seeds of some plants, e.g. maize, where it forms the food supply for germination. Indirectly these starch stores form an important food supply for animals.

Starch is a mixture of two substances: amylose and amylopectin. Starches differ slightly from one plant species to the next, but on the whole they comprise 20% amylose, 79% amylopectin, and 1% of other substances such as phosphates and fatty acids. A comparison of amylose and amylopectin is given in Fig. 2.3.

2.5.2 Glycogen

Glycogen is the major polysaccharide storage material in animals and fungi and is often called 'animal starch'. It is stored mainly in the liver and muscles. Like starch it is made up of α-glucose molecules and exists as granules. It is similar to amylopectin in structure but it has shorter chains (10–20 glucose units) and is more highly branched.

2.5.3 Cellulose

Cellulose typically comprises up to 50% of a plant cell wall, and in cotton it makes up 90%. It is a polymer of around 10 000 β-glucose molecules forming a long unbranched chain. Many chains run parallel to each other and have cross linkages between them (Fig. 2.4). These help to give cellulose its considerable stability which makes it a valuable structural material. The stability of cellulose makes it difficult to digest and therefore not such a valuable food source to animals, which only rarely produce cellulose-digesting enzymes. Some, however, have formed symbiotic relationships with organisms which can digest cellulose. To these organisms it is the major component of their

arch and glycogen are degraded in the digestive tract by -amylase, β-amylase and amylo-α (1→6)-glucosidase.

-amylase is an endoglucosidase which randomly hydrolyses α- (1→4) linkages of the side chains of glycogen and amylopectin. It can cleave either side of a branch point except in very highly branched regions.

β-amylase, an exoglycosidase, sequentially removes β-maltose from the ends of the outer branches but stops cleavage before any branch points are reached.

The structures remaining after hydrolysis by α- and β-amylase are called limit dextrins and comprise about three dozen glucose residues.

Amylo-α-(1→6)-glucosidase, the debranching enzyme, catalyses the hydrolysis of the α- (1→6) glycosidic bonds of the limit dextrins, thereby permitting further breakdown by α- and β-amylase.

PROJECT

1. Put scrapings from round and wrinkled pea seeds on a microscope slide.

2. Stain with iodine/potassium iodide solution and cover with a cover slip.

3. Compare the starch grains from both types of seed.

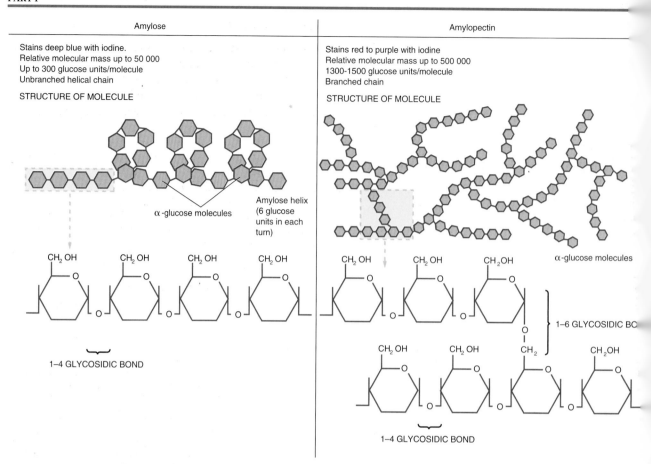

Amylose	Amylopectin
Stains deep blue with iodine. Relative molecular mass up to 50 000 Up to 300 glucose units/molecule Unbranched helical chain	Stains red to purple with iodine Relative molecular mass up to 500 000 1300-1500 glucose units/molecule Branched chain

Fig. 2.3. Comparison of the properties and structures of amylose and amylopectin

diet. Cellulose's structural strength has long been recognized by humans. Cotton is used in the manufacture of fabrics. Rayon is produced from cellulose extracted from wood and its remarkable tensile strength makes it especially useful in the manufacture of tyre cords. Cellophane, used in packaging, and celluloid, used in photographic film, are also cellulose derivatives. Paper is perhaps the best known cellulose product.

Being composed of β-glucose units, the chain, unlike that of starch, has adjacent glucose molecules rotated by 180°. This allows hydrogen bonds to be formed between the hydroxyl (—OH) groups on adjacent parallel chains which help to give cellulose its structural stability.

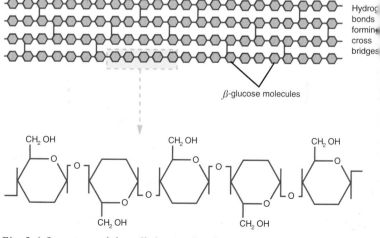

Fig. 2.4 Structure of the cellulose molecule

2.5.4 Other polysaccharides

Chitin – Chemically and structurally chitin resembles cellulose. It differs in possessing an acetyl-amino group (NH.OCCH$_3$) instead of one of the hydroxyl (—OH) groups. Like cellulose it has a structural function and is a major component of the exoskeleton of insects and crustacea. It is also found in fungal cell walls.

E 2.3   **Carbohydrates and their functions**

oup of bohydrates	Name of carbohydrate	Type/composition	Function
nosaccharides ioses $_3H_6O_3$)	Glyceraldehyde	Aldose sugar	The phosphorylated form is the first formed sugar in photosynthesis, and as such may be used as a respiratory substrate or be converted to starch for storage. It is an intermediate in glycolysis
	Dihydroxyacetone	Ketose sugar	Respiratory substrate. Intermediate in glycolysis
ntoses $_5H_{10}O_5$)	Ribose/deoxyribose	Aldose sugars	Makes up part of nucleotides and as such gives structural support to the nucleic acids RNA and DNA. Constituent of hydrogen carriers such as NAD, NADP and FAD. Constituent of ATP
	Ribulose	Ketose sugar	Carbon dioxide acceptor in photosynthesis
lexoses $_6H_{12}O_6$)	Glucose	Aldose sugar	Major respiratory substrate in plants and animals. Synthesis of disaccharides and polysaccharides. Constituent of nectar
	Galactose	Aldose sugar	Respiratory substrate. Synthesis of lactose
	Mannose	Aldose sugar	Respiratory substrate
	Fructose	Ketose sugar	Respiratory substrate. Synthesis of inulin. Constituent of nectar. Sweetens fruits to attract animals to aid seed dispersal
Disaccharides	Sucrose	Glucose + fructose	Respiratory substrate. Form in which most carbohydrate is transported in plants. Storage material in some plants, e.g. *Allium* (onion)
	Lactose	Glucose + galactose	Respiratory substrate. Mammalian milk contains 5% lactose, therefore major carbohydrate source for sucklings
	Maltose	Glucose + glucose	Respiratory substrate
Polysaccharides	Amylose } starch Amylopectin	Unbranched chain of α-glucose with 1–4 glycosidic links + branched chain of α-glucose units with 1–4 and 1–6 glycosidic links	Major storage carbohydrate in plants
	Glycogen	Highly branched short chains of α-glucose units with 1–4 glycosidic links	Major storage carbohydrate in animals and fungi
	Cellulose	Unbranched chain of β-glucose units with 1–4 glycosidic links + cross bridges	Gives structural support to cell walls
	Inulin	Unbranched chain of fructose with 1–2 glycosidic links	Major storage carbohydrate in some plants, e.g. Jerusalem artichoke; *Dahlia*
	Chitin	Unbranched chain of β-acetylglucosamine units with 1–4 glycosidic links	Constituent of the exoskeleton of insects and crustacea

Inulin – This is a polymer of fructose found as a storage carbohydrate in some plants, e.g. *Dahlia* root tubers.

Mucopolysaccharides – This group includes **hyaluronic acid**, which forms part of the matrix of vertebrate connective tissue. It is found in cartilage, bones, the vitreous humour of the eye and synovial fluid. The anticoagulant **heparin** is also a member of this group of polysaccharides.

2.5.5 Reducing and non-reducing sugars

All monosaccharides, whether aldo- or keto-sugars, are capable of reducing copper (II) sulphate in Benedict's reagent to copper (I) oxide. When monosaccharides combine to form disaccharides this reducing ability is often retained with the result that sugars such as lactose and maltose, although disaccharides, are still reducing sugars. In a few cases, however, the formation of a disaccharide results in the loss of this reducing ability. This is true of the formation of sucrose which is therefore a non-reducing sugar.

TABLE 2.4 **Relationship between amount of reducing sugar and colour of precipitate on boiling with Benedict's reagent**

Amount of reducing sugar	Colour of solution and precipitate
No reducing sugar	Blue
Increasing quantity of reducing sugar	Green
	Yellow
	Brown
	Red

2.6   Lipids

Lipids are a large and varied group of organic compounds. Like carbohydrates, they contain carbon, hydrogen and oxygen, although the proportion of oxygen is much smaller in lipids. They are insoluble in water but dissolve readily in organic solvents such as acetone, alcohols and others. They are of two types: fats and oils. There is no basic difference between these two; fats are simply solid at room temperatures (10–20°C) whereas oils are liquid. The chemistry of lipids is very varied but they are all esters of **fatty acids** and an alcohol, of which **glycerol** is by far the most abundant. Glycerol has three hydroxyl (—OH) groups and each may combine with a separate fatty acid, forming a **triglyceride** (Fig. 2.5). It is a condensation reaction and thus hydrolysis of the triglyceride will again yield glycerol and three fatty acids.

The three triglycerides may all be the same, thereby forming a simple triglyceride, or they may be different in which case a mixed triglyceride is produced. In either case it is a condensation reaction.

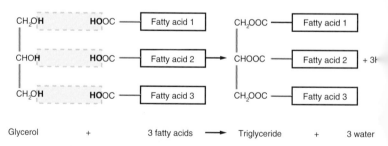

Fig. 2.5 Formation of a triglyceride

2.6.1 Fatty acids

As most naturally occurring lipids contain the same alcohol, namely glycerol, it is the nature of the fatty acids which determines the characteristics of any particular fat. All fatty acids contain a carboxyl group (—COOH). The remainder of the molecule is a hydrocarbon chain of varying length (examples are given in Table 2.5). This chain may possess one or more double bonds in which case it is said to be **unsaturated**. If, however, it possesses no double bonds it is said to be **saturated**.

It can be seen from Table 2.5 that the hydrocarbon chains may be very long. Within the fat they form long 'tails' which extend from the glycerol molecule. These 'tails' are **hydrophobic**, i.e. they repel water.

TABLE 2.5 **Nature and occurrence of some fatty acids**

Name of fatty acid	General formula	Saturated/ unsaturated	Occurrence
Butyric	C_3H_7COOH	Saturated	Butter fat
Linoleic	$C_{17}H_{31}COOH$	Unsaturated	Linseed oil
Oleic	$C_{17}H_{33}COOH$	Unsaturated	All fats
Palmitic	$C_{15}H_{31}COOH$	Saturated	Animal and vegetable fats
Stearic	$C_{17}H_{35}COOH$	Saturated	Animal and vegetable fats
Arachidic	$C_{19}H_{39}COOH$	Saturated	Peanut oil
Cerotic	$C_{25}H_{51}COOH$	Saturated	Wool oil

2.6.2 Phospholipids

Phospholipids are lipids in which one of the fatty acid groups is replaced by phosphoric acid (H_3PO_4) (Fig. 2.6). The phosphoric acid is **hydrophilic** (attracts water) in contrast to the remainder of the molecule which is **hydrophobic** (repels water). Having one end of the phospholipid attracting water while the other end repels it affects its role in the cell membrane.

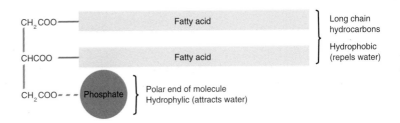

Fig. 2.6 Structure of a phospholipid

2.6.3 Waxes

Waxes are formed by combination with an alcohol other than glycerol. This alcohol is much larger than glycerol, and therefore waxes have a more complex chemical structure. Their main rôle is in waterproofing plants and animals, although they form storage compounds in a few organisms, e.g. castor oil and in fish.

21

2.6.4 Functions of lipids

1. **An energy source** – Upon breakdown they yield $38\,kJ\,g^{-1}$ energy. This compares favourably with carbohydrates which yield $17\,kJ\,g^{-1}$.

2. **Storage** – On account of their high energy yield upon break down, they make excellent energy stores. For the equivalent amount of energy stored they possess less than half the mass carbohydrate. This makes them especially useful for animals where locomotion requires mass to be kept to a minimum. In plants they are useful in seeds where dispersal by wind or insects makes small mass a necessity. This explains the abundance of oils extracted from seeds and fruits, e.g. olive, linseed, castor, peanut, coconut and sunflower. Their insolubi is another advantage, as they are not easily dissolved out of ce

3. **Insulation** – Fats conduct heat only slowly and so are usefu insulators. If fat is to be stored because of its concentrated ener supply, it may as well be put to a secondary use. In endotherm animals, such as mammals, it is stored beneath the skin (subcutaneous fat) where it helps to retain body heat. In aquati mammals, such as whales, seals and manatees, hair is ineffecti as an insulator because it cannot trap water in the same way as it can air. These animals therefore have extremely thick subcutaneous fat, called blubber, which forms an effective insulator.

4. **Protection** – Another secondary use to which stored fat is pu is as a packing material around delicate organs. Fat surroundin the kidneys, for instance, helps to protect them from physical damage.

5. **Waterproofing** – Terrestrial plants and animals have a need to conserve water. Animal skins produce oil secretions, e.g. from the sebaceous glands in mammals, which waterproof the body. Oils also coat the fur, helping to repel water which would otherwise wet it and reduce its effectiveness as an insulator. Birds spread oil over their feathers, from a special gland near the cloaca, for the same purpose. Insects have a waxy cuticle to prevent evaporative loss in the same way that plant leaves have one to reduce transpiration.

6. **Cell membranes** – Phospholipids are major components of the cell membrane and contribute to many of its properties (see Section 4.2.2).

7. **Other functions** – Lipids perform a host of miscellaneous functions in different organisms. For example, plant scents are fatty acids (or their derivatives) and so aid the attraction of insects for pollination. Bees use wax in constructing their honeycombs.

2.6.5 Steroids

Steroids are related to lipids, and **cholesterol** is perhaps the best known. It is found in animals where it is important in the synthesis of steroid hormones, such as oestrogen and cortisone. Other important steroids include vitamin D and bile acids.

Cholesterol

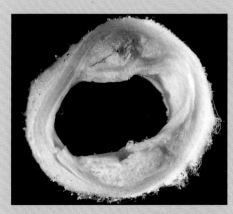

Cholesterol

Cholesterol is a lipid containing four rings of carbon and hydrogen atoms with a branched side chain. A single hydroxyl group (OH^-) gives the molecule a small charge as the result of ionization. Cholesterol is very hydrophobic.

Most cholesterol in the body is found in the membranes of cells. The plasma membrane has the most, almost one cholesterol molecule for each phospholipid. In internal membranes the ratio is closer to $1:10$.

Cholesterol is essential for the functioning of plasma membranes where it plays two main roles. Firstly it limits the uncontrolled leakage of small molecules (water and ions) in and out of the plasma membrane. The cell can thus control the passage of solutes and ions using specialized membrane proteins and without wasting energy counteracts their leakage. Cholesterol is an important constituent of myelin and helps to prevent the outward flow of ions which would 'short circuit' the movement of nerve impulses along the axon. The second role of cholesterol in membranes is to pull together the fatty acid chains in phospholipids, restricting their movement, but not making them solid. Cholesterol is also used by the liver for making bile salts and, in small quantities, is used to make steroids in the ovaries, testes and adrenal glands.

In total the body contains a pool of about 120–150 g of cholesterol which is maintained by biosynthesis in the liver and intestine and by ingestion of meat, seafood, eggs and dairy produce. Vegans take in no cholesterol but most other diets result in an intake of approximately 0.5 g, the body making a further 0.5 g, per day. Cholesterol is lost from the body mainly as bile salts, but also as bile, in cells from the lining of the intestine and a tiny percentage as steroid hormones in the urine.

Cholesterol is insoluble in water but can be carried in the blood plasma in the form of lipoproteins. The balance of these lipoproteins is usually maintained by special receptors in the liver cells but saturated fats in the diet decrease their activity and hence lead to a rise in plasma cholesterol. Deposits of crystalline cholesterol and droplets of cholesterol esters can cause thickening of the artery walls (atherosclerosis). This can lead to heart attacks (from blocking of coronary arteries), strokes (brain arteries blocked) or blockages of arteries in the legs. Atherosclerosis may follow damage caused to the artery walls by high blood pressure and smoking. Smoking considerably decreases the concentration of the antioxidant vitamins E and C in the blood resulting in the oxidation of some lipoproteins.

The products of oxidation are often toxic to the cells of the artery and cause them to behave abnormally. The damaged cells release substances which cause the blood to clot and the artery to contract. Macrophages which degrade the oxidized lipoproteins are unable to deal with the cholesterol it carries. Eventually they fill with cholesterol and die, depositing the cholesterol back into the artery.

TS human aorta with a fatty atheroma partially obstructing the interior

2.7 Proteins

Proteins are organic compounds of large molecular mass (up 40 000 000 for some viral proteins but more typically several thousand, e.g. haemoglobin = 64 500). They are not truly solul in water, but form colloidal suspensions (the nature of colloid dealt with in Section 22.1.5). In addition to carbon, hydrogen oxygen, they always contain nitrogen, usually sulphur and sometimes phosphorus. Whereas there are relatively few carbohydrates and fats, the number of proteins is almost limitless. A simple bacterium such as *Escherichia coli* has arour 800, and humans have over 10 000. They are specific to each species. Glucose is glucose in whatever organism it occurs, bu proteins vary from one species to another. Indeed, it is the proteins rather than the fats or carbohydrates which determine the characteristics of a species. Proteins are rarely stored in organisms, except in eggs or seeds where they are used to form the new tissue. The word protein (from the Greek) means 'of fi importance' and was coined by a Dutch chemist, Mulder, because he thought they played a fundamental rôle in cells. We now know that proteins form the structural basis of all living cells and that Mulder's judgement was sound.

PROJECT

Breakfast cereals have labels on the outside of the packet indicating the amounts of the various ingredients

Use your knowledge of the various food tests to find out if the claims on the labels are correct.

2.7.1 Amino acids

Amino acids are a group of over a hundred chemicals of which around twenty commonly occur in proteins. They always contain a basic group, the amino group ($-NH_2$) and an acid group, the carboxyl group ($-COOH$). (See Fig. 2.7.) Most amino acids have one of each group and are therefore neutral, but a fev have more amino groups than carboxyl ones (basic amino acids, while others have more carboxyl than amino groups (acidic amino acids). With the exception of glycine, all amino acids have an asymmetric carbon atom and therefore exhibit optical isomerism, having both D(+) and L(−) forms. Whereas all naturally occurring carbohydrates are of the D(+) form, all naturally occurring amino acids are of the L(−) form. Amino acids are soluble in water where they form ions. These ions are formed by the loss of a hydrogen atom from the carboxyl group, making it negatively charged. This hydrogen atom associates with the amino group, making it positively charged. The ion is therefore **dipolar** – having a positive and a negative pole. Such ions are called **zwitterions** (see Fig. 2.8). Amino acids therefore have both acidic and basic properties, i.e. they are **amphoteric**. Being amphoteric means that amino acids act as **buffer solutions**. A buffer solution is one which resists the tendency to alter its pH even when small amounts of acid or alkali are added to it. Such a property is essential in biological systems where any sudden change in pH could adversely affect the performance of enzymes.

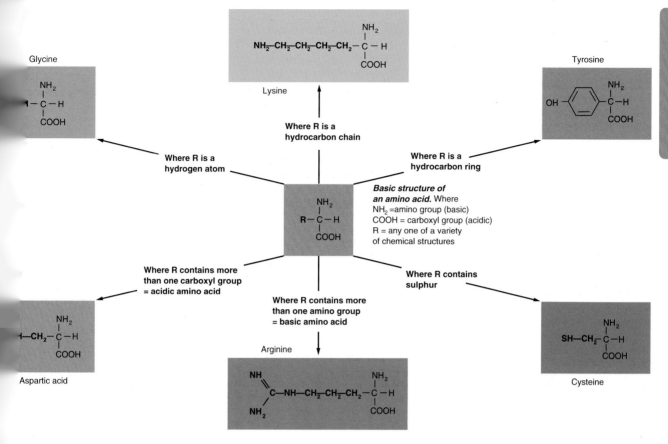

2.7 Structure of a range of amino acids

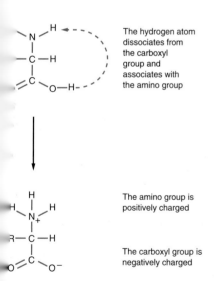

g. 2.8 Zwitterion formation in amino acids

2.7.2 Formation of polypeptides

We have seen that monosaccharides may be linked to form disaccharides and polysaccharides by the loss of water (condensation reaction). Similarly, fats are formed from condensation reactions between fatty acids and glycerol. The formation of polypeptides follows the same pattern. A condensation reaction occurs between the amino group of one amino acid and the carboxyl group of another, to form a **dipeptide** (see Fig. 2.9, on page 28). Further combinations of this type extend the length of the chain to form a **polypeptide** (see Figs. 2.10 and 2.11).

A polypeptide usually contains many hundreds of amino acids. Polypeptides may be linked by forces such as disulphide bridges to give proteins comprising thousands of amino acids.

2.7.3 Structure of polypeptides

The chains of amino acids which make up a polypeptide have a specific three-dimensional shape (see Fig. 2.12). This shape is important in the functioning of proteins, especially enzymes. The shape of a polypeptide molecule is due to four types of bonding which occur between various amino acids in the chain.

The first type of bond is called a **disulphide bond**. It arises between sulphur-containing groups on any two cysteine molecules. These bonds may arise between cysteine molecules in the same amino acid chain (intrachain) or between molecules in different chains (interchain).

The second type of bond is the **ionic bond**. We have seen amino acids form zwitterions (Section 2.7.1) which have NH and COO⁻ groups. The formation of peptide bonds when making a polypeptide means that the COOH and NH₂ grou are not available to form ions. In the case of acidic amino ac however, there are additional COOH groups which may ior to give COO⁻ groups. In the same way, basic amino acids m still retain NH_3^+ groups even when combined into the struct

NOTEBOOK

Electrophoresis

Electrophoresis is a technique used to separate molecules of different electrical charge. Under the influence of an electrical fie **anions** (negatively charged ions) will move towards the **anode** (positive electrode) while **cations** (positively charged ions) are attracted to the cathode (negative electrode).

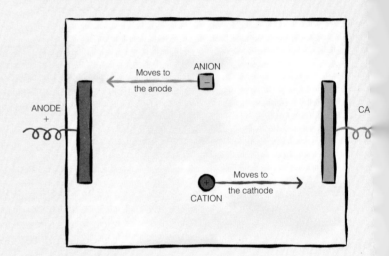

Two factors affect the speed with which charged molecules mo towards an electrode:

1. The amount of charge – the greater the charge the faster the molecule moves.
2. The size of the molecule – small molecules move faster than larger ones with the same charge.

Amino acids and proteins are **amphoteric** (have both basic and acidic properties) because they are **zwitterions** (have positively ar negatively charged groups).

The amount of positive or negative charge is affected by pH. Each molecule has a specific pH at which the total positive charge is exactly equal to the total negative charge, i.e. it is electrically neutral and has no tendency to move to either the anode or cathode of an electric field. This is known as the **isoelectric point** At higher pH protein and amino acid molecules become more negatively charged while at lower pH they become more positively charged.

of a polypeptide. In addition NH_3^+ and COO^- can occur at the ends of a polypeptide chain. Any of these available NH_3^+ and COO^- groups may form ionic bonds which help to give a polypeptide molecule its particular shape. These ionic bonds are weak and may be broken by alterations in the pH of the medium around the polypeptide.

The third type of bond is the **hydrogen bond**. This occurs between certain hydrogen atoms and certain oxygen atoms

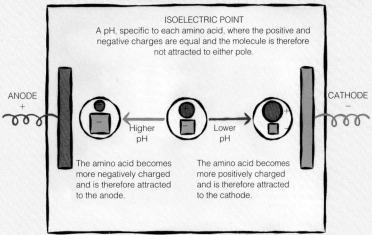

The molecules being separated have to be supported in an appropriate medium such as paper or a thin layer of gel.

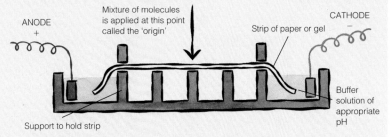

ypical apparatus for carrying out electrophoresis

Either end of a strip of the medium is dipped in a small reservoir of buffer solution of the appropriate pH. Each reservoir also contains an electrode.

The electrical field is applied for a specific period of time and then the position of the molecules is determined by adding a suitable stain to colour them. The molecules are separated according to their charges; the negatively charged ones moving to the anode with the most negatively charged ones moving furthest. The positively charged ones move to the cathode and again the more positive they are the closer they get to the cathode.

It is possible to treat the mixture being separated in such a way that all the molecules are equally negatively charged. These can then be loaded at the cathode end of the apparatus and will be attracted to the anode. The distance they travel in a given time will then depend not on their charge but their size, the smaller molecules moving further than the larger ones.

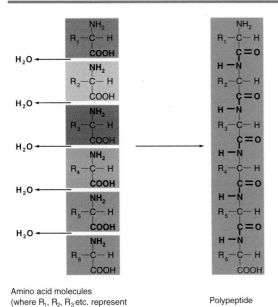

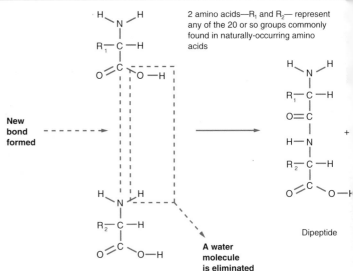

Amino acid molecules
(where R₁, R₂, R₃ etc. represent
any of the 20 or so groups
found in naturally-occurring amino
acids)

Polypeptide
(part of)

Fig. 2.10 Formation of a polypeptide

Fig. 2.9 Formation of a dipeptide

A simplified representation of a polypeptide chain to show three types of bonding responsible for shaping the chain. In practice the polypeptide chains are longer, contain more of these three types of bond and have a three dimensional shape.

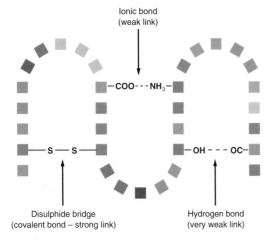

Fig. 2.11 Types of bond in a polypeptide chain

within the polypeptide chain. The hydrogen atoms have a small positive charge on them (electropositive) and the oxygen atoms small negative charge (electronegative). The two charged atoms are attracted together and form a hydrogen bond. While each bond is very weak, the sheer number of bonds means that they play a considerable rôle in the shape and stability of a polypeptide molecule.

The fourth type is **hydrophobic interactions** which are interactions between non-polar R groups. These cause the protein to fold as hydrophobic side groups are shielded from water.

2.7.4 Fibrous proteins

The fibrous proteins have a primary structure of regular repetitive sequences. They form long chains which may run parallel to one another, being linked by cross bridges. They are very stable molecules and have structural rôles within organisms. Collagen is a good example. It is a common constituent of animal connective tissue, especially in structures requiring physical strength, e.g. tendons. It has a primary structure which is largely a repeat of the tripeptide sequence, glycine – proline – alanine, and forms a long unbranched chain. Three such chains are wound into a triple helix, with cross bridges linking them to each other and providing additional structural support. (Compare the repeating glucose units, parallel chains and cross links of the structural carbohydrate cellulose.)

2.7.5 Globular proteins

In contrast to fibrous proteins, the globular proteins have highly irregular sequences of amino acids in their polypeptide chains. Their shape is also different, being compact globules. If a fibrous protein is likened to a series of strands of string twisted into a rope, then a globular protein can be thought of as the same string rolled into a ball. These molecules are far less stable and have metabolic rôles within organisms. All enzymes are globular proteins. Globular and fibrous proteins are compared in Table 2.6, on page 30.

Hair perming

The protein keratin, which makes up human hair, has a high percentage of the amino acid cysteine. The disulphide bridges formed between cysteine molecules are largely responsible for the shape of the hair. Hair is straight or curly because the keratin contains disulphide linkages that enable the molecules to hold their particular shapes. When hair is permed it is first treated with a reducing agent that breaks some of the —S—S— bonds.

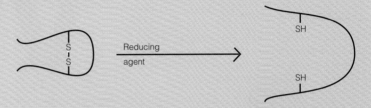

This allows the molecules to become more flexible and the hair is then set into the desired shape, using rollers or curlers. An oxidizing agent is then added which reverses the above reaction, forming new disulphide bonds, which now hold the molecules together in the desired positions. The straightening of curly hair is done in the same way. 'Perms' are not of course truly permanent. The hair keeps growing and the new hair has the same disulphide linkages as the original hair.

2.7.6 Conjugated proteins

Many proteins incorporate other chemicals within their structure. These proteins are called **conjugated proteins** and the non-protein part is referred to as the **prosthetic group**. The prosthetic group plays a vital rôle in the functioning of the protein. Some examples are given in Table 2.7 (see page 31).

2.12 Fine structure of the fibrous protein collagen

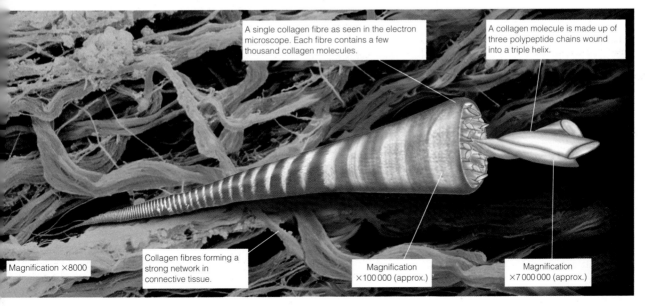

A single collagen fibre as seen in the electron microscope. Each fibre contains a few thousand collagen molecules.

A collagen molecule is made up of three polypeptide chains wound into a triple helix.

Magnification ×8000

Collagen fibres forming a strong network in connective tissue.

Magnification ×100 000 (approx.)

Magnification ×7 000 000 (approx.)

(a) *The primary structure of a protein is the sequence of amino acids found in its chains. This sequence determines its properties and shape. Following the elucida amino acid sequence of the hormone insulin, by Frederick Sanger in 1954, the pri structure of many other proteins is now know.*

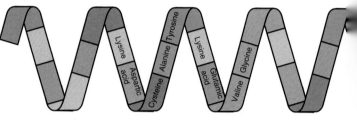

	Lysine	Aspartic acid	Cysteine	Alanine	Tyrosine	Lysine	Glutamic acid	Valine	Glycin

(b) *The secondary structure is the shape which the polypeptide chain forms as a re hydrogen bonding. This is most often a spiral known as the α-helix, although other configurations occur.*

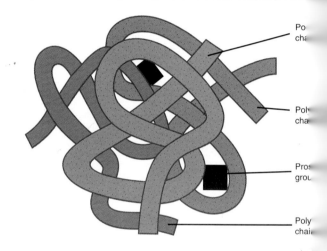

(c) *The tertiary structure is due to the bending and twisting of the polypeptide helix into a compact structure. All three types of bond, disulphide, ionic and hydrogen, contribute to the maintenance of the tertiary structure.*

(d) *The quarternary structure arises from the combination of a number of different polypeptide chains, and associated non-protein groups, into a large complex protein*

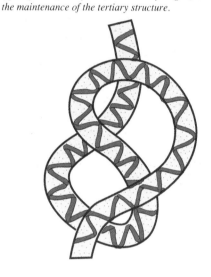

Fig. 2.13 Structure of proteins

TABLE 2.6   **Comparison of globular and fibrous proteins**

Fibrous proteins	Globular proteins
Repetitive regular sequences of amino acids	Irregular amino acid sequences
Actual sequences may vary slightly between two examples of the same protein	Sequence highly specific and never varies between two examples of the same protein
Polypeptide chains form long parallel strands	Polypeptide chains folded into a spherical shape
Length of chain may vary in two examples of the same protein	Length always identical in two examples of the same protein
Stable structure	Relatively unstable structure
Insoluble	Soluble – forms colloidal suspensions
Support and structural functions	Metabolic functions
Examples include collagen and keratin	Examples include all enzymes, some hormones (e.g. insulin) and haemoglobin

TABLE 2.7 **Examples of conjugated proteins**

Name of protein	Where found	Prosthetic group
Haemoglobin	Blood	Haem (contains iron)
Mucin	Saliva	Carbohydrate
Casein	Milk	Phosphoric acid
Cytochrome oxidase	Electron carrier pathway of cells	Copper
Nucleoprotein	Ribosomes	Nucleic acid

2.7.7 Denaturation of proteins

We have seen that the three-dimensional structure of a protein is, in part at least, due to fairly weak ionic and hydrogen bonds. Any agent which breaks these bonds will cause the three-dimensional shape to be changed. In many cases the globular proteins revert to a more fibrous form. This process is called **denaturation**. The actual sequence of amino acids is unaltered; only the overall shape of the molecule is changed. This is still sufficient to prevent the molecule from carrying out its usual functions within an organism.

Denaturation may be temporary or permanent and is due to a variety of factors as shown in Table 2.8.

E 2.8 **Factors causing protein denaturation**

ctor	Explanation	Example
at	Causes the atoms of the protein to vibrate more (increased kinetic energy), thus breaking hydrogen and ionic bonds	Coagulation of albumen (boiling eggs makes the white more fibrous and less soluble)
ids	Additional H^+ ions in acids combine with COO^- groups on amino acids and form COOH. Ionic bonds are hence broken	The souring of milk by acid (e.g. *Lactobacillus* bacterium produces lactic acid, lowering pH and causing it to denature the casein, making it insoluble and thus forming curds)
lkalis	Reduced number of H^+ ions causes NH_3^+ groups to lose H^+ ions and form NH_2. Ionic bonds are hence broken	
organic chemicals	The ions of heavy metals such as mercury and silver are highly electropositive. They combine with COO^- groups and disrupt ionic bonds. Similarly, highly electronegative ions, e.g. cyanide (CN^-), combine with NH_3^+ groups and disrupt ionic bonds	Many enzymes are inhibited by being denatured in the presence of certain ions, e.g. cytochrome oxidase (respiratory enzyme) is inhibited by cyanide
Organic chemicals	Organic solvents alter hydrogen bonding within a protein	Alcohol denatures certain bacterial proteins. This is what makes it useful for sterilization
Mechanical force	Physical movement may break hydrogen bonds	Stretching a hair breaks the hydrogen bonds in the keratin helix. The helix is extended and the hair stretches. If released, the hair returns to its normal length. If, however, it is wetted and then dried under tension, it keeps its new length – the basis of hair styling

TABLE 2.9 **Functions of proteins**

Vital activity	Protein example	Function
Nutrition	Digestive enzymes, e.g. trypsin amylase lipase	Catalyses the hydrolysis of proteins to polypeptides Catalyses the hydrolysis of starch to maltose Catalyses the hydrolysis of fats to fatty acids and glycerol
	Fibrous proteins in granal lamellae	Help to arrange chlorophyll molecules in a position to receive maximum am of light for photosynthesis
	Mucin	Assists trapping of food in filter feeders. Prevents autolysis. Lubricates gut
	Ovalbumin	Storage protein in egg white
	Casein	Storage protein in milk
Respiration and transport	Haemoglobin/haemoerythrin/ haemocyanin/chlorocruorin	Transport of oxygen
	Myoglobin	Stores oxygen in muscle
	Prothrombin/fibrinogen	Required for the clotting of blood
	Mucin	Keeps respiratory surface moist
	Antibodies	Essential to the defence of the body, e.g. against bacterial invasion
Growth	Hormones, e.g. thyroxine	Controls growth and metabolism
Excretion	Enzymes, e.g. urease; arginase	Catalyse reactions in ornithine cycle and therefore help in protein breakdow and urea formation
Support and movement	Actin/myosin	Needed for muscle contraction
	Ossein	Structural support in bone
	Collagen	Gives strength with flexibility in tendons and cartilage
	Elastin	Gives strength and elasticity to ligaments
	Keratin	Tough for protection, e.g. in scales, claws, nails, hooves, skin
	Sclerotin	Provides strength in insect exoskeleton
	Lipoproteins	Structural components of all cell membranes
Sensitivity and coordination	Hormones, e.g. insulin/glucagon ACTH vasopressin	Control blood sugar level Controls the activity of the adrenal cortex Controls blood pressure
	Rhodopsin/opsin	Visual pigments in the retina, sensitive to light
	Phytochromes	Plant pigments important in control of flowering, germination, etc.
Reproduction	Hormones, e.g. prolactin	Induces milk production in mammals
	Chromatin	Gives structural support to chromosomes
	Gluten	Storage protein in seeds – nourishes the embryo
	Keratin	Forms horns and antlers which may be used for sexual display

2.8 Nucleic acids

Like proteins, nucleic acids are informational macromolecules.
They are made up of chains of individual units called **nucleotide**
The structure of nucleic acids and their constituent nucleotides ar
closely related to their functions in heredity and protein synthesis
For this reason the details of their structure will be left until the
nature of the genetic code is discussed in Chapter 7.

2.9   Questions

,B,C and **D** are the structural formulae of four
ients.

A

B

C

D

(a) (i) Complete Table I.1 below to identify to
which class of nutrient **A**, **B**, **C** and **D**
each belong. (Specific names are **not**
required.)

(ii) State a principal function of each class of
nutrient in the body.

I.1

npound	Class of nutrient	Body function

(8 marks)

(b) Complete the equation below to show how
two molecules of **B** can react together to form
a dimer.

(2 marks)

(c) State the names of the smaller molecules
formed from **A** and **D** by hydrolysis during
digestion. (3 marks)

(d) Large molecules similar to **D** often exhibit
bonding between adjacent chains. In each
case, name, and show by diagram, **two** types
of bonding or interaction that occur between
such chains. (4 marks)

(e) The body also requires about 12 inorganic
nutrients. State the names of **two** of these
inorganic nutrients together with a body
function for each. (4 marks)
(Total 21 marks)

UCLES (Modular) June 1992 (Biochemistry) No. 1

2. The table below refers to monosaccharides and
amino acids. If the statement is correct for that
substance put a tick (✔) in the appropriate box and if
the statement is incorrect put a cross (✗).

Statement	Mono-saccharides	Amino acids
Always contain nitrogen		
May be polymerized into macromolecules		
Released by complete hydrolysis of nucleic acids		
Insoluble in water		
May be linked by glycosidic bonds		
Released by complete hydrolysis of cellulose		
May be broken down and used in the TCA cycle (Krebs cycle)		
Always contain carbon, hydrogen and oxygen		

(Total 8 marks)

ULEAC June 1990, Paper I, No. 3

3. The diagrams below show the structural formulae
of two amino acids, P and Q.

Amino acid P   Amino acid Q

(a) Name **two** elements, other than carbon,
hydrogen and oxygen, which may be present
in groups R_1 and R_2. (2 marks)

(b) P and Q may be linked during protein
synthesis. In this reaction certain atoms from
P and Q combine to form new molecules.

(i) On the diagram, draw a circle around the
atoms that are removed when P and Q
are linked together. (1 mark)

(ii) Draw a line connecting the atoms in P
and Q that are bonded together. (1 mark)

(iii) Name the bond formed by this reaction.

(1 mark)

(c) Explain how the different properties of groups such as R_1 and R_2 are important in the structure and functioning of proteins.

(4 marks)

(Total 9 marks)

ULEAC June 1991, Paper I, No. 11

4. Complete the numbered lines in the table below which presents the name, natural location(s) and a biological function of each of four important polysaccharides.

Name	Location(s)	Function
1........................	(i) potato tuber 2.(ii).....................	3..
Glycogen	4.(i)...................... 5.(ii).....................	6..
Cellulose	7...........................	8..
Chitin	9...........................	10......................................

(Total 10 marks)

NISEAC June 1993, Paper II, No. 2

5. *(a)* The table below lists three biologically important metallic elements. Complete the table with the names of **TWO** compounds containing calcium, **THREE** of iron and **ONE** of magnesium. In the third column **BRIEFLY** state **ONE** function of **EACH** compound.

Element	Compound	Function of compound
Calcium	1. 2.	
Iron	1. 2. 3.	
Magnesium	1.	

(12 marks)

(b) Hydrogencarbonate ions in solution have buffering role.

(i) What is the function of a buffer in a living system? *(1 m*

(ii) Name **TWO** mammalian body fluids which hydrogencarbonate ions have a buffering role. *(2 m*

(Total 15 m

NISEAC June 1992, Paper II, N

6. *(a)* (i) Describe the general structure of an amino acid.

(ii) With the aid of a diagram, show the results when **two** amino acids are join together.

(iii) Why are amino acids described as *amphoteric*? *(6 m*

(b) What happens in the human body if certa amino acids are:

(i) present in excess,

(ii) in short supply?

(c) Proteins may be described as having:

(i) primary structure,

(ii) secondary structure,

(iii) tertiary structure,

(iv) quaternary structure.

Explain the meaning of these terms when applied to proteins. *(10 ma*

(d) Give an account of how a protein molecul such as an enzyme, is synthesized in a cel

(Total 30 ma

Oxford June 1994, Paper 2, N

See *Further questions* pp. 614–20, question 7.

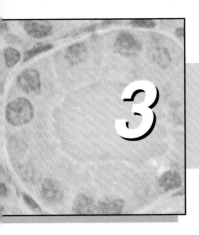

3 Enzymes

Until recently it was thought that all biological catalysts were enzymes. We now know that other substances may carry out catalytic functions in living organisms. **Abzymes** are antibodies with catalytic properties and **ribozymes** are molecules of RNA which act catalytically on themselves. Most biological catalysts however are globular proteins known as enzymes. A catalyst is a substance which alters the rate of a chemical reaction without itself undergoing a permanent change. As they are not altered by the reactions they catalyze, enzymes can be used over and over again. They are therefore effective in very small amounts. Enzymes cannot cause reactions to occur, but only speed up ones which would otherwise take place extremely slowly. The word 'enzyme' means 'in yeast', and was used because they were first discovered by Eduard Buchner in an extract of yeast.

3.1 Enzyme structure and function

Enzymes are complex three-dimensional globular proteins, some of which have other associated molecules. While the enzyme molecule is normally larger than the substrate molecule it acts upon, only a small part of the enzyme molecule actually comes into contact with the substrate. This region is called the **active site**. Only a few of the amino acids of the enzyme molecule make up the active site. These so-called **catalytic amino acids** are often some distance apart in the protein chain but are brought into close proximity by the folding of that chain (see Fig. 3.1).

...talytic amino acids A, B and C, although some ...ce apart in the chain, are close together when the ...: is folded.

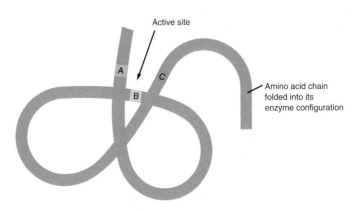

...1 Catalytic amino acids forming the ...e site

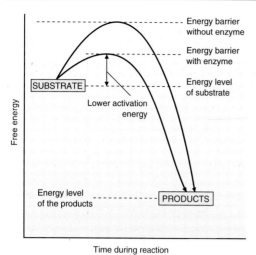

Fig. 3.2 How enzymes lower the activation energy

3.1.1 Enzymes and activation energy

Before a reaction can take place it must overcome an energy barrier by exceeding its activation energy. Enzymes operate lowering this activation energy and thus permit the reaction occur more readily (Fig. 3.2). As heat is often the source of activation energy, enzymes often dispense with the need for heat and so allow reactions to take place at lower temperatu Many reactions which would not ordinarily occur at the temperature of an organism do so readily in the presence of enzymes.

3.1.2 Mechanism of enzyme action

Enzymes are thought to operate on a **lock and key mechani** In the same way that a key fits a lock very precisely, so the substrate fits accurately into the active site of the enzyme molecule. The two molecules form a temporary structure ca the **enzyme-substrate complex**. The products have a differe shape from the substrate and so, once formed, they escape fr the active site, leaving it free to become attached to another substrate molecule. The sequence is summarized in Fig. 3.3.

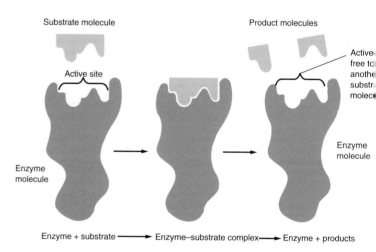

Fig. 3.3 Mechanism of enzyme action

Modern interpretations of the lock and key mechanism suggest that in the presence of the substrate the active site change in order to suit the substrate's shape. The enzyme is flexible and moulds to fit the substrate molecule in the sam way that clothing is flexible and can mould itself to fit the shape of the wearer. The enzyme initially has a binding configuration which attracts the substrate. On binding to th enzyme, the substrate disturbs the shape of the enzyme anc causes it to assume a new configuration. It is this new configuration which is catalytically active and which in turr affects the shape of the substrate thus lowering its activatio energy. This is referred to as an **induced fit** of the substrate the enzyme.

3.2 Properties of enzymes

The properties of enzymes can be explained in relation to the lock and key mechanism of enzyme action, and the theory of induced fit.

3.2.1 Specificity

All enzymes operate only on specific substrates. Just as a key has a specific shape and therefore fits only complementary locks, so only substrates of a particular shape will fit the active site of an enzyme. Some locks are highly specific and can only be opened with a single key. Others are opened by a number of similar keys; yet others may be opened by many different keys. In the same way, some enzymes will act only on one particular isomer. Others act only on similar molecules; yet others will break a particular chemical linkage, wherever it occurs.

3.2.2 Reversibility

Chemical reactions are reversible, and equations are therefore often represented by two arrows to indicate this reversibility. (See opposite.)

At any one moment the reaction (shown left) may be proceeding predominantly in one direction. If, however, the conditions are changed, the direction may be reversed. It may be that the reaction proceeds from left to right in acid conditions, but in alkaline conditions it goes from right to left. In time, reactions reach a point where the reactants and the product are in **equilibrium** with one another. Enzymes catalyze the forward and reverse reactions equally. They do not therefore alter the equilibrium itself, only the speed at which it is reached. Carbonic anhydrase is an enzyme which catalyzes a reaction in either direction depending on the conditions at the time. In respiring tissues where there is much carbon dioxide it converts carbon dioxide and water into carbonic acid. In the lungs, however, the removal of carbon dioxide by diffusion means a low concentration of carbon dioxide, and hence the carbonic acid breaks down into carbon dioxide and water. Both reactions are catalyzed by carbonic anhydrase, as shown opposite.

$$A + B \rightleftharpoons C + D$$

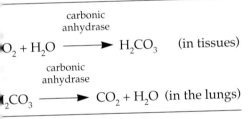

$$\overset{\text{carbonic}}{\underset{\text{anhydrase}}{}}$$
$$CO_2 + H_2O \longrightarrow H_2CO_3 \quad \text{(in tissues)}$$

$$\overset{\text{carbonic}}{\underset{\text{anhydrase}}{}}$$
$$H_2CO_3 \longrightarrow CO_2 + H_2O \quad \text{(in the lungs)}$$

3.2.3 Enzyme concentration

The active site of an enzyme may be used again and again. Enzymes therefore work efficiently at very low concentrations. The number of substrate molecules which an enzyme can act upon in a given time is called its **turnover number**. This varies from many millions of substrate molecules each minute, in the case of catalase, to a few hundred per minute for slow acting enzymes. Provided the temperature and other conditions are suitable for the reaction, and provided there are excess substrate molecules, the rate of a reaction is directly proportional to the enzyme concentration. If the amount of substrate is restricted it may limit the rate of reaction. The addition of further enzyme cannot increase the rate and the graph therefore tails off (Fig. 3.4).

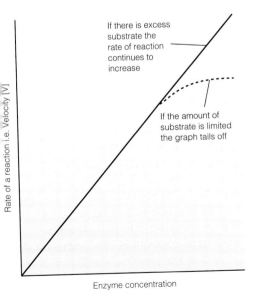

If there is excess substrate the rate of reaction continues to increase

If the amount of substrate is limited the graph tails off

Rate of a reaction i.e. Velocity [V]

Enzyme concentration

Fig. 3.4 Graph to show the effect of enzyme concentration on the rate of an enzyme-controlled reaction

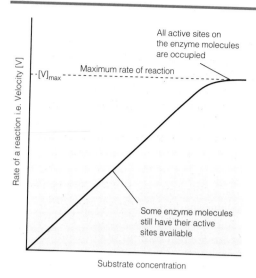

Fig. 3.5 Graph to show the effect of substrate concentration on the rate of an enzyme-controlled reaction

3.2.4 Substrate concentration

For a given amount of enzyme, the rate of an enzyme-controll reaction increases with an increase in substrate concentration up to a point. At low substrate concentrations, the active sites the enzyme molecules are not all used – there simply are not enough substrate molecules to occupy them all. As the substra concentration is increased, more and more sites come into use point is reached, however, where all sites are being used; increasing the substrate concentration cannot therefore increas the rate of reaction, as the amount of enzyme is the limiting factor. At this point the graph tails off (Fig. 3.5).

3.2.5 Temperature

An increase in temperature affects the rate of an enzyme-controlled reaction in two ways:

1. As the temperature increases, the kinetic energy of the substrate and enzyme molecules increases and so they move faster. The faster these molecules move, the more often they collide with one another and the greater the rate of reaction.

2. As the temperature increases, the more the atoms which ma up the enzyme molecules vibrate. This breaks the hydrogen bonds and other forces which hold the molecules in their precis shape. The three-dimensional shape of the enzyme molecules is altered to such an extent that their active sites no longer fit the substrate. The enzyme is said to be **denatured** and loses its catalytic properties. (See Section 2.7.7.)

The actual effect of temperature on the rate of reaction is the combined influence of these two factors and is illustrated in Fig. 3.6.

The optimum temperature for an enzyme varies considerably Many arctic and alpine plants have enzymes which function efficiently at temperatures around 10 °C, whereas those in algae inhabiting some hot springs continue to function at temperature around 80 °C. For many enzymes the optimum temperature lies around 40 °C and denaturation occurs at about 60 °C.

3.2.6 pH

The precise three-dimensional molecular shape which is vital to the functioning of enzymes is partly the result of hydrogen bonding. These bonds may be broken by the concentration of hydrogen ions (H^+) present. pH is a measure of hydrogen ion concentration. It is measured on a scale of 1–14, with pH 7 being the neutral point. A pH less than 7 is acid, one greater than 7 is alkaline.

By breaking the hydrogen bonds which give enzyme molecules their shape, any change in pH can effectively denature enzymes. Each enzyme works best at a particular pH, and deviations from this optimum may result in denaturation. Fig. 3.7 illustrates the different pH optima of four enzymes.

3.2.7 Inhibition

The rate of enzyme-controlled reactions may be decreased by the presence of inhibitors. They are of two types: **reversible inhibitors** and **non-reversible inhibitors**.

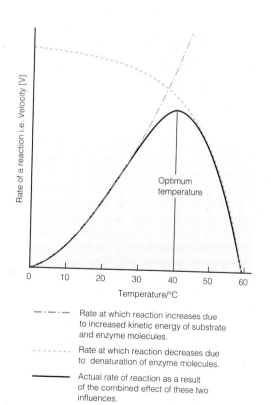

- — · — · — Rate at which reaction increases due to increased kinetic energy of substrate and enzyme molecules.

- – – – – – Rate at which reaction decreases due to denaturation of enzyme molecules.

- ———— Actual rate of reaction as a result of the combined effect of these two influences.

Fig. 3.6 Graph to show the effect of temperature on the rate of an enzyme-controlled reaction

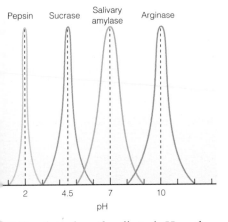

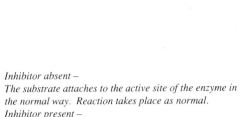

3.7 Graph to show the effect of pH on the *f reaction of four different enzymes*

Reversible inhibitors

The effect of this type of inhibitor is temporary and causes no permanent damage to the enzyme because the association of the inhibitor with the enzyme is a loose one and it can easily be removed. Removal of the inhibitor restores the activity of the enzyme to normal. There are two types: **competitive** and **non-competitive**.

Competitive inhibitors compete with the substrate for the active sites of enzyme molecules. The inhibitor may have a structure which permits it to combine with the active site. While it remains bound to the active site, it prevents substrate molecules occupying them and so reduces the rate of the reaction. The same quantity of product is formed, because the substrate continues to use any enzyme molecules which are unaffected by the inhibitor. It does, however, take longer to make the products. If the concentration of the substrate is increased, less inhibition occurs. This is because, as the substrate and inhibitor are in direct competition, the greater the proportion of substrate molecules the greater their chance of finding the active sites, leaving fewer to be occupied by the inhibitor.

Malonic acid is a competitive inhibitor. It competes with succinate for the active sites of succinic dehydrogenase, an important enzyme in the Krebs cycle (Section 16.3).

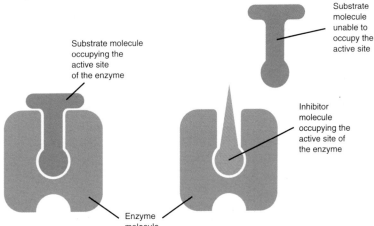

3.8 Competitive inhibition

Non-competitive inhibitors do not attach themselves to the active site of the enzyme, but elsewhere on the enzyme molecule. They nevertheless alter the shape of the enzyme molecule in such a way that the active site can no longer properly accommodate the substrate. As the substrate and inhibitor molecules attach to different parts of the enzyme they are not

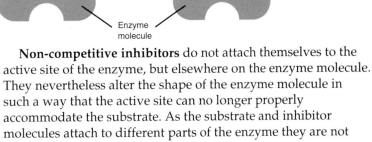

Inhibitor absent –
The substrate attaches to the active site of the enzyme in the normal way. Reaction takes place as normal.
Inhibitor present –
The inhibitor prevents the normal enzyme – substrate complex being formed. The reaction rate is reduced.

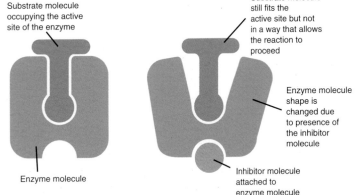

3.9 Non-competitive inhibition

39

PROJECT

Many different investigations into enzyme activity can be carried out using the digestion of starch by amylase, staining with iodine/potassium iodide solution, and a colorimeter, or using starch gels with wells, for example

(a) substrate concentration
(b) enzyme concentration
(c) temperature
(d) pH.

competing for the same sites. An increase in substrate concentration will not therefore reduce the effect of the inhib

Cyanide is a non-competitive inhibitor. It attaches itself to copper prosthetic group of cytochrome oxidase, thereby inhibiting respiration (Section 16.4).

Non-reversible inhibitors
Non-reversible inhibitors leave the enzyme permanently damaged and so unable to carry out its catalytic function. He metal ions such as mercury (Hg^{2+}) and silver (Ag^+) cause disulphide bonds to break. These bonds help to maintain the shape of the enzyme molecule. Once broken the enzyme molecule's structure becomes irreversibly altered with the permanent loss of its catalytic properties.

NOTEBOOK

Why pH?

The term was first used by the Danish biochemist S.P.L. Sörenson when researching into the best conditions for brewing beer. Acidity is the result of free hydrogen ions (H^+) in a solution. The concentration is often very low, however. Vinegar, for example, typically has a concentration of 0.001 mol dm^{-3}. This is a rather long-winded way of expressing acid and base strength, especially when 1 M sodium hydroxide has a hydrogen ion concentration of 0.000 000 000 000 01 mol^{-3}. Sörenson appreciated that 0.001 can be written as 10^{-3} and 0.000 000 000 000 01 as 10^{-14}. He then simply ignored the 10 and the minus sign to give values of 3 and 1 respectively. pH is therefore the negative power (p) of the hydrogen ion concentration (H).

H$^+$ concentration in mol dm^{-3}	10^{-14}	10^{-13}	10^{-12}	10^{-11}	10^{-10}	10^{-9}	10^{-8}	10^{-7}	10^{-6}	10^{-5}	10^{-4}	10^{-3}	10^{-2}	10^{-1}
pH	14	13	12	11	10	9	8	7	6	5	4	3	2	1

3.3 Enzyme cofactors

A **cofactor** is a non-protein substance which is essential for some enzymes to function efficiently. There are three types: **activators**, **coenzymes** and **prosthetic groups**.

3.3.1 Activators

Activators are substances which are necessary for the functioning of certain enzymes. The enzyme thrombokinase, which converts prothrombin into thrombin during blood clotting, is activated by calcium (Ca^{2+}) ions. In the same way

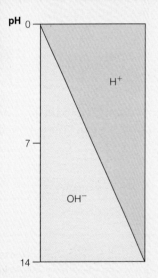

Relative concentrations of H^+ and OH^- ions at different pHs.

Why is water neutral at pH 7?

Some water molecules are always dissociated into hydrogen (H^+) and hydroxyl (OH^-) ions

$$H_2O \rightleftharpoons H^+ = OH^-$$

At 25 °C 1 dm^3 of water contains 10^{-7} moles of H^+ ions and therefore has a pH of 7. It follows from the equation above that there will also be 10^{-7} moles of OH^- ions. As the concentration of H^+ ions increases, that of OH^- ions decreases correspondingly. For example, where the concentration of H^+ ions is 10^{-5} mol dm^{-3} that of OH^- ions is 10^{-9} mol dm^{-3}. The two concentrations multiplied always give a value of 10^{-14}. Hence the pH scale is 0–14. At pH 0 almost all the ions are H^+ whereas at pH 14 they are almost entirely OH^-.

The pH scale is not linear but is logarithmic, based on a factor of 10. pH 6 = 0.000 001 mol dm^{-3} of H^+ and pH 5 = 0.000 01 mol dm^{-3} of H^+, i.e. pH 5 is 10 times more acidic than pH 6. In the same way pH 3 is 10 times more acidic than pH 4 and 100 times more so than pH 5.

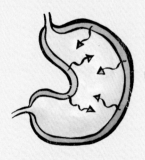

The lower the pH the greater the concen- tration of H^+

GASTRIC JUICE pH 1

Sweat is 1000× more acid than water

SWEAT pH 4

Water at pH 7 has H^+ concentration of 0.000 000 1 (10^{-7})

WATER pH 7

salivary amylase requires the presence of chloride (Cl^-) ions before it will efficiently convert starch into maltose. It is possib that these activators assist in forming the enzyme-substrate complex by moulding either the enzyme or substrate molecule into a more suitable shape.

3.3.2 Coenzymes

Coenzymes are non-protein organic substances which are essential to the efficient functioning of some enzymes, but are not themselves bound to the enzyme. Many coenzymes are derived from vitamins, e.g. **nicotinamide adenine dinucleotid (NAD)** is derived from nicotinic acid, a member of the vitamin complex. NAD acts as a coenzyme to dehydrogenases by actin; as a hydrogen acceptor.

3.3.3 Prosthetic groups

Like coenzymes, prosthetic groups are organic molecules, but unlike them they are bound to the enzyme itself. Perhaps the best known prosthetic group is **haem**. It is a ring-shaped organi molecule with iron at its centre. Apart from its rôle as an oxyge carrier in haemoglobin, it is also the prosthetic group of the electron carrier cytochrome and of the enzyme catalase.

3.4   Classification of enzymes

Enzymes are classified into six groups according to the type of reaction they catalyse. Table 3.1 summarizes this internationally accepted classification.

TABLE 3.1   **The classification of enzymes**

Enzyme group	Type of reaction catalysed	Enzyme examples
1. Oxidoreductases	Transfer of O and H atoms between substances, i.e. all oxidation-reduction reactions	Dehydrogenases Oxidases
2. Transferases	Transfer of a chemical group from one substance to another	Transaminases Phosphorylases
3. Hydrolases	Hydrolysis reactions	Peptidases Lipases Phosphatases
4. Lyases	Addition or removal of a chemical group other than by hydrolysis	Decarboxylases
5. Isomerases	The rearrangement of groups within a molecule	Isomerases Mutases
6. Ligases	Formation of bonds between two molecules using energy derived from the breakdown of ATP	Synthetases

Each enzyme is given two names:

A **systematic** name, based on the six classification groups. These names are often long and complicated.

A **trivial** name which is shorter and easier to use.
   The trivial names are derived by following three procedures:

1. Start with the name of the substrate upon which the enzyme acts, e.g. succinate.

2. Add the name of the type of reaction which it catalyses, e.g. dehydrogenation.

3. Convert the end of the last word to an -ase suffix, e.g. dehydrogenase.

The example above gives succinic dehydrogenase. Another example would be DNA polymerase. This enzyme catalyses the formation (and breakdown) of the nucleic acid DNA by polymerization. Some of the commercial uses of enzymes are considered in Sections 30.3.5 and 30.7.

3.5   Control of metabolic pathways

With many hundreds of reactions taking place in any single cell it is clear that a very structured system of control of metabolic pathways is essential. If the cell were merely a 'soup' of substrates, enzymes and products, the chances of particular reactants meeting would be small and the metabolic processes inefficient. In addition, different enzymes need different conditions, e.g. a particular pH, and it would be impossible to provide these in such an unstructured 'soup'. Cells contain organelles, and enzymes are often bound to these inner membranes in a precise order. This increases the chances of them coming into contact with their appropriate substrates, and leads to efficiency. The organelles may also have varying conditions to suit the specific enzymes they contain. By controlling these conditions, and the enzymes available, the cell can control the metabolic pathways within it.

Cells also make use of the enzyme's own properties to exercise control over metabolic pathways. The end-product of a pathway may inhibit the enzyme at the start **(end product inhibition)**.

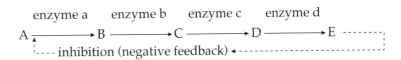

In the example above, the product E acts as an inhibitor to enzyme a. If the level of product E falls, this inhibition is reduced, and so more A is converted to B, and subsequently more E is produced. If the level of E rises above normal, inhibition of enzyme a increases and so the level of E is reduced. In this way homeostatic control of E is achieved, more details of which are given in Section 25.1. The mechanism is termed **negative feedback** because the information from the end of the pathway which is fed back to the start has a negative effect, i.e. a high concentration of E reduces its own production rate.

These forms of inhibition are, for obvious reasons, reversible, i.e. they do not permanently damage the enzymes. They frequently affect the nature of an enzyme's active site by binding with the enzyme at some point on the molecule remote from the active site. Such effects are termed **allosteric** and refer to the ability of the enzyme to have more than one shape. One shape renders the enzyme active, another renders it inactive.

3.6   Questions

1. (a) Summarize the characteristic properties of enzymes. *(13 marks)*
  (b) Discuss enzyme inhibition. *(10 marks)*
  (Total 23 marks)

  UCLES June 1992, Paper II, No. 6

2. (a) What is an enzyme? With reference to **one named** example, outline the way in which an enzyme functions. *(8 marks)*
  (b) Explain how each of the following affects the rate of an enzyme-controlled reaction.
    (i) temperature
    (ii) enzyme concentration
    (iii) substrate concentration *(15 marks)*
    (Total 23 marks)

  UCLES (Modular) March 1993, Foundation, No. 1

3. The following are all types of enzyme:

  A oxidoreductase
  B transferase
  C hydrolase

Using each letter once, more than once or not at all, identify the type of enzyme that would catalyse the reactions below:
  (a) ATP + glucose → ADP + glucose 6-phosphate; *(1 mark)*
  (b) sucrose + water → glucose + fructose; *(1 mark)*
  (c) aspartate   + α-ketoglutarate →
    (amino acid)   (carboxylic acid)

    glutamate   + oxaloacetate.
    (amino acid)   (carboxylic acid) *(1 mark)*
    (Total 3 marks)

  AEB June 1991, Paper I, No. 13

4. 'Enzymes are specific organic catalysts which lower the activation energy of metabolic reactions. They operate either intracellularly or extracellularly and in some cases function only in the presence of co-factors. Their reactions occur only within controlled physiological limits, outside which denaturation will occur.'
  (a) Explain fully what is meant by
    (i) specific organic catalysts,
    (ii) activation energy,
    (iii) controlled physiological limits. *(6 marks)*
  (b) Describe fully the process of denaturation, explaining why it affects enzyme activity. *(5 marks)*
  (c) Explain what is meant by a co-factor. Give **two** examples and state their functions. *(6 marks)*

  (d) Name **three** types of intracellular enzyme indicating the reactions that they carry ou[t] *(3 ma[rks])*
  (Total 20 ma[rks])

  WJEC June 1993, Paper A1, N[o.]

5. (a) What is meant by the term *enzyme*? *(2 ma[rks])*
  (b) The graph below shows the effect of substrate concentration on the rate of an enzyme-catalysed reaction.

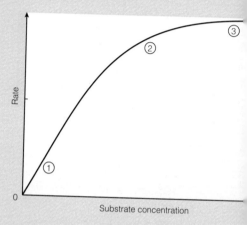

  Interpret the three regions of the graph labelled 1, 2, 3. *(3 mark[s])*
  (c) Enzymatic reactions normally have a temperature coefficient (Q_{10}) of 2 between 0 and 40 °C. What does this mean? *(1 mar[k])*
  (d) At higher temperatures the enzyme is rapidly *denatured*.
    (i) What has happened? *(1 mar[k])*
    (ii) How does this affect enzyme action? *(2 mark[s])*
  (e) Name **two** factors, in addition to substrate concentration and temperature, which affect the rate of enzyme activity. *(2 mark[s])*
  (f) Give an example of **one** enzyme which acts *intracellularly* and state its function. *(2 mark[s])*
  (g) Give an example of **one** enzyme which acts *extracellularly* and state its function. *(2 marks[)]*
  (h) Distinguish between *anabolic* and *catabolic* reactions. *(2 marks[)]*
  (i) Give a brief account of the way anabolic and catabolic reactions may be linked in a cell. *(3 marks[)]*
  (Total 20 marks[)]

  O&C SEB June 1991, Paper I, No. 6

The diagram shows a complex formed between an ~~enzyme~~ and its substrate.

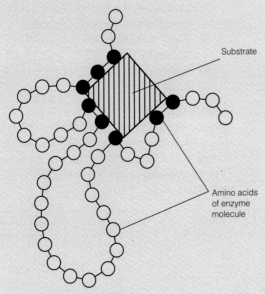

Substrate

Amino acids
of enzyme
molecule

(a) What is represented by the amino acids that have been coloured black? *(1 mark)*

(b) Use information in the diagram and your own knowledge to explain why:

(i) a restriction enzyme used in genetic engineering can only cut a length of DNA at a specific base sequence; *(2 marks)*

(ii) cyanide prevents the action of the enzyme cytochrome oxidase by acting as a non-competitive inhibitor. *(2 marks)*

(Total 5 marks)

AEB June 1993, Paper I, No. 11

The graph shows the effect of increasing substrate ~~concentration~~ on the rate of an enzyme catalysed ~~reaction~~ at a temperature of 35 °C and a constant pH.

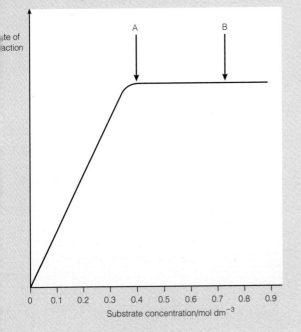

~~Rate of~~ ~~reaction~~

0   0.1   0.2   0.3   0.4   0.5   0.6   0.7   0.8   0.9
Substrate concentration/mol dm^{-3}

(a) Explain the shape of the curve between points **A** and **B**. *(2 marks)*

(b) Sketch on the graph a curve to illustrate the effect of increasing substrate concentration on the rate of reaction at a temperature of 25 °C. *(2 marks)*

(c) Explain the following observations.

(i) At a substrate concentration of 0.7 mol dm^{-3} an equal volume of competitive inhibitor at a concentration of 0.05 mol dm^{-3} was added. The rate of reaction remained the same. *(1 mark)*

(ii) At a substrate concentration of 0.7 mol dm^{-3} an equal volume of non-competitive inhibitor at a concentration of 0.05 mol dm^{-3} was added. The rate of reaction was reduced dramatically. *(1 mark)*

(Total 6 marks)

AEB November 1993, Paper I, No. 14

8. Catalase is an enzyme found in liver cells. Its activity may be studied by measuring the release of oxygen from hydrogen peroxide.

$$2H_2O_2 \xrightarrow{\text{Catalase}} 2H_2O + O_2$$
Hydrogen   Water   Oxygen
peroxide

(a) Hydrogen peroxide was provided as a 5% solution. Complete the table below to show the volumes of this solution and of water that would be necessary to make 10 cm^3 of each of the solutions of the final concentrations shown.

Volume of 5% hydrogen peroxide/cm^3						
Volume of water/cm^3						
Final concentrations of hydrogen peroxide/per cent	0	1	2	3	4	5

(1 mark)

A liver extract was first made up by homogenizing and filtering 50 g of pig's liver. 1 cm^3 of the extract was mixed with 99 cm^3 of ice cold buffer solution at pH 7 and centrifuged for 5 minutes. The supernatant was removed and 0.5 cm^3 of it was made to 500 cm^3 with buffer solution to produce the catalase solution used in experiments.

(b) In this procedure, give the reasons for:

(i) using a buffer solution rather than water,

(ii) using a buffer solution that was ice cold. *(2 marks)*

(c) By how much was the original liver extract diluted to obtain the catalase solution used in experiments? *(1 mark)*

The diagram below shows the apparatus used to investigate the activity of the catalase solution.

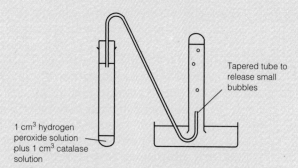

Tapered tube to release small bubbles

1 cm³ hydrogen peroxide solution plus 1 cm³ catalase solution

In one experiment, 1 cm³ of catalase solution was added in turn to 1 cm³ of each of the different hydrogen peroxide solutions. The rate of reaction was measured 10 seconds after mixing by counting the number of bubbles produced during the next 15 seconds. The following results were obtained.

Concentration of hydrogen peroxide/ per cent	0	1	2	3	4	5
Number of bubbles produced in 15 seconds	0	7	12	16	18	19

(d) Account for the nature of the results obtained. (2 marks)

(e) The experiment described is open to several criticisms. Explain, in detail, **two** ways in which it might be improved. (4 marks)

(Total 10 marks)

NEAB June 1993, Paper IIA, No. 6

9. The enzyme phenol oxidase is often released when plant cells are disrupted and leads to the oxidation of colourless phenols into coloured products. Samples of an extract containing phenol oxidase were subjected to various treatments, then mixed with a solution of phenols buffered at pH 7 and incubated at 35 °C for 10 minutes. The pre-treatments and results are shown below.

Tube	Pre-treatment of enzyme extract	Colour of extract after incubation with phenol
A	None	Intense brown
B	Incubated with protease for 10 minutes	Colourless
C	Mixed with trichloroacetic acid for 5 minutes	Colourless
D	Mixed with mercuric chloride for 5 minutes	Very light yellow

(a) Assuming these experiments were all appropriately standardized, what do you conclude from the results of each experiment

(b) How would you have discovered
 (i) if any **non-enzymic** oxidation of phenol occurs in Tube A?
 (ii) if **enzymic** oxidation of phenols occurs Tube D? (3 mar

In some further experiments, samples of the enzyme extract were mixed with different substrate concentrations, with or without the presence of a standard amount of a chemical, PTU (phenylthiourea). The results are shown in the table below.

Concentration of substrate/mM	Initial rate with PTU present/units	Initial rate without PTU present/units
0.5	2.4	4.2
1.0	4.1	6.3
1.5	5.1	7.1
2.0	5.5	7.6
2.5	5.5	7.6

(c) Plot these results, using the same axes, on squared paper. (3 mar

(d) Phenylthiourea (PTU) binds to copper atom
 (i) Suggest a hypothesis which might reasonably explain how PTU inhibits th action of phenol oxidase.
 (ii) Explain how, from the information provided, it is possible to determine that the inhibition caused by PTU is non-competitive rather than competitive. (3 mar

(Total 13 mark

NEAB June 1994, Paper IIA, No

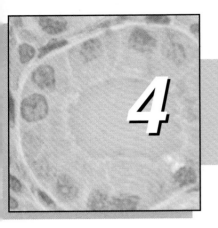

4   Cellular organization

The cell is the fundamental unit of life. All organisms, whatever their type or size, are composed of cells. The modern theory of cellular organization states that:

1. All living organisms are composed of cells.

2. All new cells are derived from other cells.

3. Cells contain the hereditary material of an organism which is passed from parent to daughter cells.

4. All metabolic processes take place within cells.

4.1   Cytology – the study of cells

All cells are self-contained and more or less self-sufficient units. They are surrounded by a cell membrane and have a nucleus, or a nuclear area, at some stage of their existence. They show remarkable diversity, both in structure and function. They are basically spherical in shape, although they show some variation where they are modified to suit their function. In size they normally range from 10–30 μm.

TABLE 4.1   Comparison of prokaryotic and eukaryotic cells

Prokaryotic cells	Eukaryotic cells
No distinct nucleus; only diffuse area(s) of nucleoplasm with no nuclear membrane	A distinct, membrane-bounded nucleus
No chromosomes – circular strands of DNA	Chromosomes present on which DNA is located
No membrane-bounded organelles such as chloroplasts and mitochondria	Chloroplasts and mitochondria may be present
Ribosomes are smaller	Ribosomes are larger
Flagella (if present) lack internal 9 + 2 fibril arrangement	Flagella have 9 + 2 internal fibril arrangement
No mitosis or meiosis occurs	Mitosis and/or meiosis occurs

4.1.1 The structure of prokaryotic cells

Prokaryotic cells (*pro* – 'before', *karyo* – 'nucleus') were probably the first forms of life on earth. Their hereditary material, DNA, is not enclosed within a nuclear membrane. This absence of a true nucleus only occurs in two groups, the bacteria and the blue-green bacteria (Section 5.3). There are no membrane-bounded organelles within a prokaryotic cell, the structure of which is shown on page 51.

4.1.2 Structure of the eukaryotic cell

Eukaryotic cells (*Eu* – 'true', *karyo* – 'nucleus') probably arose a little over 1000 million years ago, nearly 2500 million years after their prokaryotic ancestors. The development of eukaryotic cells from prokaryotic ones involved considerable changes, as can be seen from Table 4.1. The essential change was the development of membrane-bounded organelles, such as mitochondria and

NOTEBOOK

Microscopy

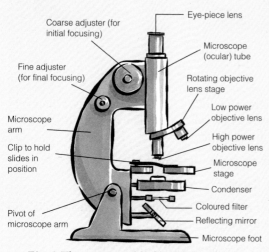

Fig. 1 The compound light microscope

Labels:
- Coarse adjuster (for initial focusing)
- Fine adjuster (for final focusing)
- Microscope arm
- Clip to hold slides in position
- Pivot of microscope arm
- Eye-piece lens
- Microscope (ocular) tube
- Rotating objective lens stage
- Low power objective lens
- High power objective lens
- Microscope stage
- Condenser
- Coloured filter
- Reflecting mirror
- Microscope foot

Have you ever wished you could see that little bit better – perhaps read what someone else is writing from a distance or to recognize who exactly it is in the crowd at a football match? How frustrating can be when you can't quite make out the print of the newspaper the person opposite you on the train. This is how early scientists must have felt when they strained their eyes to decipher the detail structure of organisms. How they must have rejoiced at the development of first the glass lens and then the simple light microscope.

The light microscope opened up a new world of structural detail for the biologist, revealing the variety of cell forms making up organisms. In time however, their curiosity again became thwarted as the limitations of the light microscope prevented them observing the fine detail within cells. The problem is that the wavelength of light limits the light microscope to distinguishing objects which are 0.2 μm or further from each other. The problem could only be overcome by using a form of radiation which had a wavelength less than that of light. So in 1933 the electron microscope was developed. This instrument works on the same principles as the light microscope except that instead of light rays, with their wavelength in the order of 500 nm, a beam of electrons of wavelengths 0.005 is used. This means that the electron microscope can magnify objects up to 500 000 times compared to the best light microscope which magnify only around 2000 times.

Whereas the light microscope uses glass lenses to focus the light rays, the electron beam of the electron microscope is focused by means of powerful electromagnets. The image produced by the electron microscope cannot be detected directly by the naked eye. Instead, the electron beam is directed on to a screen from which black and white photographs, called **photoelectronmicrographs** can be taken. A comparison of the radiation pathways in light and electron microscopes is given in Fig. 2.

There are two main types of electron microscope. In the **transmission electron microscope** (TEM), a beam of electrons passed through thin, specially prepared slices of material. As the molecules in air would absorb the electrons, a vacuum has to be created within the instrument. Where electrons are absorbed by the material, and do not therefore reach the screen, the image is dark. Such areas are said to be **electron dense**. Where the electrons penetrate, the screen appears bright. These areas are termed **electron transparent**. As electrons have a very small mass, they not easily penetrate materials and so sections need to be exceedingly thin. This sectioning creates a flat image and the natural contouring of a specimen cannot be seen. To overcome the problem, the **scanning electron microscope** (SEM) was developed. In this instrument a fine beam of electrons is passed to and fro across the specimen, beginning at one end and working across to the other. The specimen scatters many electrons, while others are absorbed. Low energy secondary electrons may be emitted by the specimen. The scattered electrons and the low energy secondary ones are amplified and transmitted to a screen. The resultant image shows holes and depressions as dark areas and ridges and extensions of the surface as bright areas. In this way the natural contouring of the material may be observed.

Comparison of advantages and disadvantages of the light and electron microscopes

LIGHT MICROSCOPE	ELECTRON MICROSCOPE
Advantages	**Disadvantages**
Cheap to purchase and operate	Expensive to purchase and operate
Small and portable – can be used almost anywhere	Very large and must be operated in special rooms
Unaffected by magnetic fields	Affected by magnetic fields
Preparation of material is relatively quick and simple, requiring only a little expertise	Preparation of material is lengthy and requires considerable expertise and sometimes complex equipment
Material rarely distorted by preparation	Preparation of material may distort it
Natural colour of the material can be observed	All images are in black and white
Disadvantages	**Advantages**
Magnifies objects up to 2000×	Magnifies objects over 500 000×
The depth of field is restricted	It is possible to investigate a greater depth of field

Electron microscope

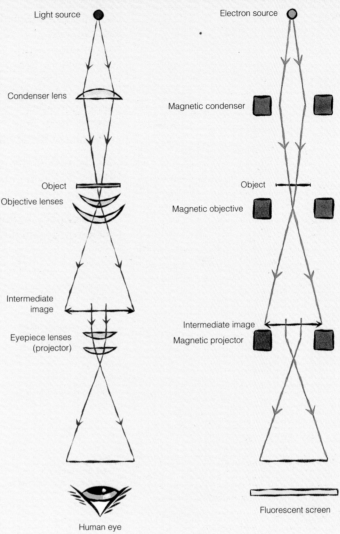

Light source

Condenser lens

Object
Objective lenses

Intermediate
image

Eyepiece lenses
(projector)

Human eye

Electron source

Magnetic condenser

Object
Magnetic objective

Intermediate image
Magnetic projector

Fluorescent screen

Fig. 2 Comparison of radiation pathways in light and electron microscopes

The problem with both these forms of electron microscope is that complex preparation techniques coupled with the need for a high vacuum means that not only must the material being observed be dehydrated and therefore dead, it is frequently considerably distorted; what you see may be very different from the original material. In response to this a new generation of electron microscopes has been developed – the environmental scanning electron microscope (ESEM). These microscopes allow the material on view to be kept at a low vacuum while the region around the electron gun is at a high vacuum. This is achieved by separating the microscope column into a series of chambers each with its own pressure. There is only a minute hole between chambers – wide enough to allow the tiny electron beam through. A special low voltage detector which can operate at low vacuums is used to detect the scattered electrons, secondary electrons and X-rays which provide the image.

False-colour scanning electron micrograph of a fruit fly, *Drosophila sp.*, laying an egg. (Magnification ×80)

49

chloroplasts, within the outer plasma membrane of the cell. The presence of membrane-bounded organelles confers four advantages:

1. Many metabolic processes involve enzymes being embedded in a membrane. As cells become larger, the proportion of membrane area to cell volume is reduced. This proportion is increased by the presence of organelle membranes.

2. Containing enzymes for a particular metabolic pathway within organelles means that the products of one reaction will always be in close proximity to the next enzyme in the sequence. The rate of metabolic reactions will thereby be increased.

3. The rate of any metabolic pathway inside an organelle can be controlled by regulating the rate at which the membrane surrounding the organelle allows the first reactant to enter.

4. Potentially harmful reactants and/or enzymes can be isolated inside an organelle so they won't damage the rest of the cell.

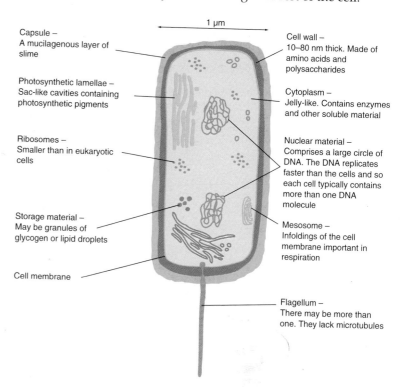

Figure 4.1 Structure of the prokaryotic cell, e.g. a generalized bacterial cell

It is possible that such organelles arose as separate prokaryotic cells which developed a symbiotic relationship with larger prokaryotic ones. This would explain the existence of one membraned structure inside another, the ability of mitochondria and chloroplasts to divide themselves (self-replication) and the presence of DNA within these two organelles. Alternatively, the organelles may have arisen by invaginations of the plasma membrane which became 'pinched off' to give a separate membrane-bounded structure within the main cell. Although many variations of the eukaryotic cell exist, there are two main types, the plant cell and the animal cell. (See Figs. 4.2 and 4.3 on the pages 52–3.)

4.1.3 Differences between plant and animal cells

The major differences between plant and animal cells are given in Table 4.2.

TABLE 4.2 **Differences between plant and animal cells**

Plant cells	Animal cells
Tough, slightly elastic cellulose cell wall present (in addition to the cell membrane)	Cell wall absent – only a membrane surrounds the cell
Pits and plasmodesmata present in the cell wall	No cell wall and therefore no pits or plasmodesmata
Middle lamella join cell walls of adjacent cells	Middle lamella absent – cells are joined by intercellular cement
Plastids, e.g. chloroplasts and leucoplasts, present in large numbers	Plastids absent
Mature cells normally have a large single, central vacuole filled with cell sap	Vacuoles, e.g. contractile vacuoles, if present, are small and scattered throughout the cell
Tonoplast present around vacuole	Tonoplast absent
Cytoplasm normally confined to a thin layer at the edge of the cell	Cytoplasm present throughout the cell
Nucleus at edge of the cell	Nucleus anywhere in the cell but often central
Lysosomes not normally present	Lysosomes almost always present
Centrioles absent in higher plants	Centrioles present
Cilia and flagella absent in higher plants	Cilia or flagella often present
Starch grains used for storage	Glycogen granules used for storage
Only some cells are capable of division	Almost all cells are capable of division
Few secretions are produced	A wide variety of secretions are produced

4.2 Cell ultrastructure

4.2.1 Cytoplasmic matrix

All the cell organelles are contained within a cytoplasmic matrix, sometimes called the **hyaloplasm** or **cytosol**. It is an aqueous material which is a solution or colloidal suspension of many vital cellular chemicals. These include simple ions such as sodium, phosphates and chlorides, organic molecules such as amino acids, ATP and nucleotides, and storage material such as oil

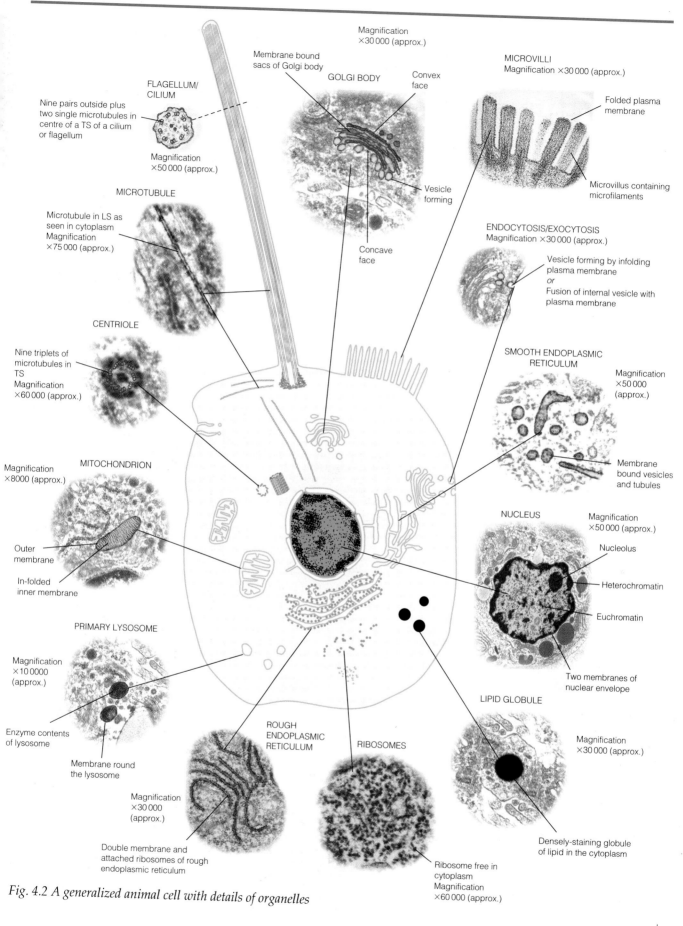

Magnification
×30 000 (approx.)

Membrane bound
sacs of Golgi body

GOLGI BODY

Convex
face

MICROVILLI
Magnification ×30 000 (approx.)

Folded plasma
membrane

FLAGELLUM/
CILIUM

Nine pairs outside plus
two single microtubules in
centre of a TS of a cilium
or flagellum

Magnification
×50 000 (approx.)

Vesicle
forming

Concave
face

Microvillus containing
microfilaments

MICROTUBULE

Microtubule in LS as
seen in cytoplasm
Magnification
×75 000 (approx.)

ENDOCYTOSIS/EXOCYTOSIS
Magnification ×30 000 (approx.)

Vesicle forming by infolding
plasma membrane
or
Fusion of internal vesicle with
plasma membrane

CENTRIOLE

Nine triplets of
microtubules in
TS
Magnification
×60 000 (approx.)

SMOOTH ENDOPLASMIC
RETICULUM

Magnification
×50 000
(approx.)

Membrane
bound vesicles
and tubules

MITOCHONDRION

Magnification
×8000 (approx.)

NUCLEUS

Magnification
×50 000 (approx.)

Nucleolus

Heterochromatin

Outer
membrane

Euchromatin

In-folded
inner membrane

PRIMARY LYSOSOME

Magnification
×10 0000
(approx.)

Two membranes of
nuclear envelope

LIPID GLOBULE

Enzyme contents
of lysosome

ROUGH
ENDOPLASMIC
RETICULUM

RIBOSOMES

Magnification
×30 000 (approx.)

Membrane round
the lysosome

Magnification
×30 000
(approx.)

Densely-staining globule
of lipid in the cytoplasm

Double membrane and
attached ribosomes of rough
endoplasmic reticulum

Ribosome free in
cytoplasm
Magnification
×60 000 (approx.)

Fig. 4.2 A generalized animal cell with details of organelles

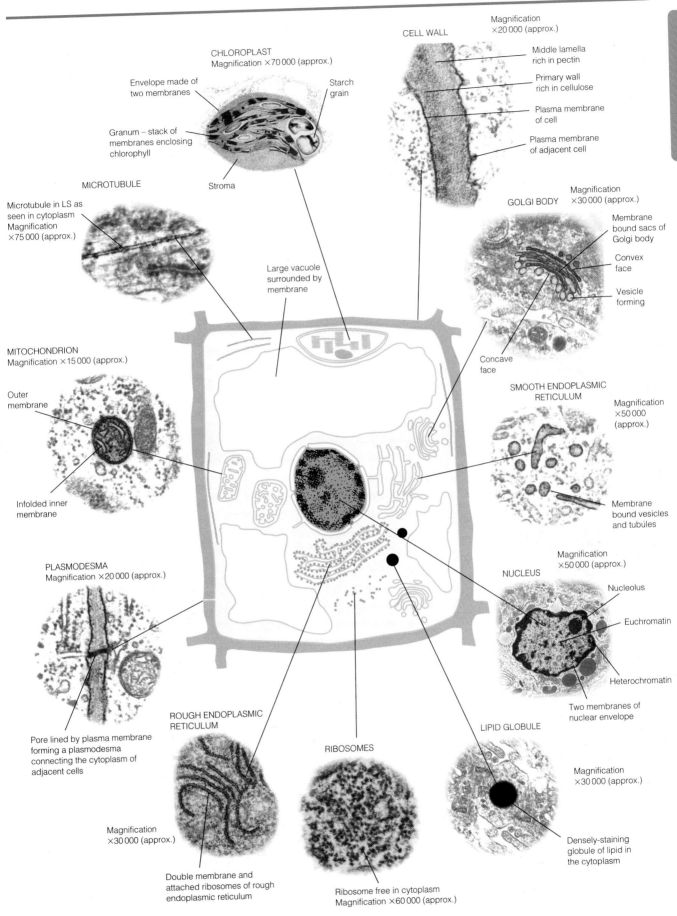

CHLOROPLAST
Magnification ×70 000 (approx.)

Envelope made of
two membranes

Starch
grain

Granum – stack of
membranes enclosing
chlorophyll

Stroma

CELL WALL

Magnification
×20 000 (approx.)

Middle lamella
rich in pectin

Primary wall
rich in cellulose

Plasma membrane
of cell

Plasma membrane
of adjacent cell

MICROTUBULE

Microtubule in LS as
seen in cytoplasm
Magnification
×75 000 (approx.)

Large vacuole
surrounded by
membrane

GOLGI BODY

Magnification
×30 000 (approx.)

Membrane
bound sacs of
Golgi body

Convex
face

Vesicle
forming

Concave
face

MITOCHONDRION
Magnification ×15 000 (approx.)

Outer
membrane

Infolded inner
membrane

**SMOOTH ENDOPLASMIC
RETICULUM**

Magnification
×50 000
(approx.)

Membrane
bound vesicles
and tubules

Magnification
×50 000 (approx.)

NUCLEUS

Nucleolus

Euchromatin

Heterochromatin

Two membranes of
nuclear envelope

PLASMODESMA
Magnification ×20 000 (approx.)

Pore lined by plasma membrane
forming a plasmodesma
connecting the cytoplasm of
adjacent cells

**ROUGH ENDOPLASMIC
RETICULUM**

Magnification
×30 000
(approx.)

Double membrane and
attached ribosomes of rough
endoplasmic reticulum

RIBOSOMES

Ribosome free in cytoplasm
Magnification ×60 000 (approx.)

LIPID GLOBULE

Magnification
×30 000 (approx.)

Densely-staining
globule of lipid in
the cytoplasm

Fig. 4.3 A generalized plant cell with detail of organelles

droplets. Many important biochemical processes, including glycolysis, occur within the cytoplasm. It is not static but capable of mass flow, which is called **cytoplasmic streaming**.

4.2.2 Cell membrane

The cell membrane's main function is to serve as a boundary between the cell and its environment. It is not, however, inert but a functional organelle. It may permanently exclude certain substances from the cell while permanently retaining others. Some substances may pass freely in and out through the membrane. Yet others may be excluded at one moment only to pass freely across the membrane on another occasion. On account of the membrane's ability to permit different substances to pass across it at different rates, it is said to be **partially permeable**.

There is little dispute that the cell membrane is made up almost entirely of two chemical groups – proteins and phospholipids. In 1972, J. J. Singer and G. L. Nicholson suggested a structure for the cell membrane. There is a bimolecular phospholipid layer with inwardly directed hydrophobic tails and a variety of protein molecules with an irregular arrangement (Fig. 4.4, on the next page). Some proteins occur on the surface of

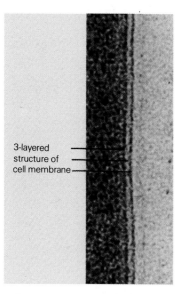

3-layered structure of cell membrane

Cell membrane (EM) (×250 000)

NOTEBOOK

Cell fractionation and centrifugation

If you shake up particles of different sizes within a liquid such as water and leave them to settle, they separate out – the largest and heaviest at the bottom and the smallest and lightest at the top. The same principle can be applied to separate out the various components of cells. Dividing the cell into its parts (or fractions) is called **cell fractionation** and is achieved by the process of **centrifugation** using a **centrifuge**.

A centrifuge is a machine which can spin tubes containing liquid suspensions at a very high speed. The effect is to exert a force on the contents of the tube similar to, but much greater than, that of gravity. The faster the speed and the longer the time for which the tubes are spun, the greater the force. At slower speeds (less force) the larger fragments collect at the bottom of the tube and the smaller ones remain in suspension in the liquid near the top of the tube – **supernatant liquid**. If the larger fragments are removed and the supernatant recentrifuged at a faster speed (more force), the larger of these smaller fragments will collect at the bottom. By continuing in this way, smaller and smaller fragments may be recovered. As the size of any organelle is relatively constant, each organelle will tend to separate from the supernatant at a specific speed of rotation. If the suspension of cell fragments is spun at a slower speed than that required to separate out a particular organelle, all larger fragments and organelles can be collected and discarded. Spinning the supernatant at the appropriate speed will now cause a new fraction to be collected. This fraction will be a relatively pure sample of the required organelles. Since the process involves centrifuging at different speeds, it is called **differential centrifugation**.

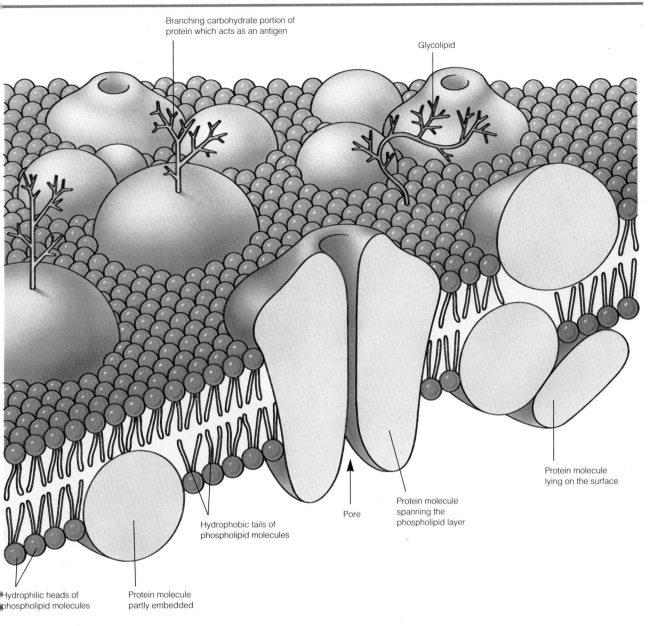

Branching carbohydrate portion of protein which acts as an antigen

Glycolipid

Hydrophobic tails of phospholipid molecules

Pore

Protein molecule spanning the phospholipid layer

Protein molecule lying on the surface

Hydrophilic heads of phospholipid molecules

Protein molecule partly embedded

Fig. 4.4 The fluid-mosaic model of the cell membrane

the phospholipid layer (**peripheral** or **extrinsic proteins**) while others extend into it (**integral** or **intrinsic proteins**) and some even extend completely across (**transmembrane proteins**). Viewed from the surface, the proteins are dotted throughout the phospholipid layer in a mosaic arrangement. Other research suggests that the phospholipid layer is capable of much movement, i.e. is fluid. It was these facts which gave rise to its name, the **fluid-mosaic model**. Also present in the membrane is cholesterol which interacts with the phospholipids to make the membrane less fluid.

The proteins in the membrane have a number of functions. Apart from giving structural support they are very specific, varying from cell to cell. It is this specificity which allows cells to be recognized by other agents in the body, e.g. enzymes, hormones and antibodies. In the fluid-mosaic model it is thought probable that the proteins also assist the active transport of materials across the membrane.

MEMBRANOUS ORGANELLES

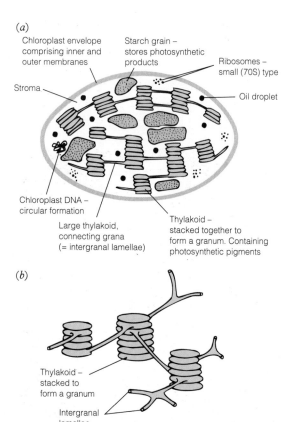

Nucleus (EM) (×12 000 approx.)

4.2.3 The nucleus

When viewed under a microscope, the most prominent feature of a cell is the nucleus. While its shape, size, position and chemical composition vary from cell to cell, its functions are always the same, namely, to control the cell's activity and to retain the organism's hereditary material, the chromosomes. It is bounded by a double membrane, the **nuclear envelope**, the outer membrane being continuous with the endoplasmic reticulum and often having ribosomes on its surface. The inner membrane has three proteins on its surface which act as anchoring points for chromosomes during interphase (see Section 8.2). It possesses many large pores (typically 3000 per nucleus) 40–100 nm in diameter, which permit the passage of large molecules, such as RNA, between it and the cytoplasm. The cytoplasm-like material within the nucleus is called **nucleoplasm**. It contains **chromatin** which is made up of coils of DNA bound to proteins. During division the chromatin condenses to form the chromosomes but these are rarely, if ever, visible in a non-dividing cell. The denser, more darkly staining areas of chromatin are called **heterochromatin**.

Within the nucleus are one or two small spherical bodies, each called a **nucleolus**. They are not distinct organelles as they are not bounded by a membrane. They manufacture ribosomal RNA, a substance in which they are especially rich, and assemble ribosomes.

The functions of a nucleus are:

1. To contain the genetic material of a cell in the form of chromosomes.

2. To act as a control centre for the activities of a cell.

3. To carry the instructions for the synthesis of proteins in the nuclear DNA.

4. To be involved in the production of ribosomes and RNA.

5. In cell division.

4.2.4 The chloroplast

Chloroplasts belong to a larger group of organelles known as **plastids**. In higher plants most chloroplasts are 5–10 μm long and are bounded by a double membrane, the **chloroplast envelope**, about 30 nm thick. While the outer membrane has a similar structure to the plasma membrane, the inner one is folded into a series of lamellae and is highly selective in what it allows in and out of the chloroplast.

Within the chloroplast envelope are two distinct regions. The **stroma** is a colourless, gelatinous matrix in which are embedded structures rather like stacks of coins in appearance. These are the **grana**. Each granum, and there may be around fifty in a chloroplast, is made up of between two and a hundred closed flattened sacs called **thylakoids**. Within these are located the photosynthetic pigments such as chlorophyll, details of which are given in Section 14.2.2. Some thylakoids have tubular extensions which interconnect adjacent grana (Fig. 4.5).

Fig. 4.5 Structure of the chloroplasts

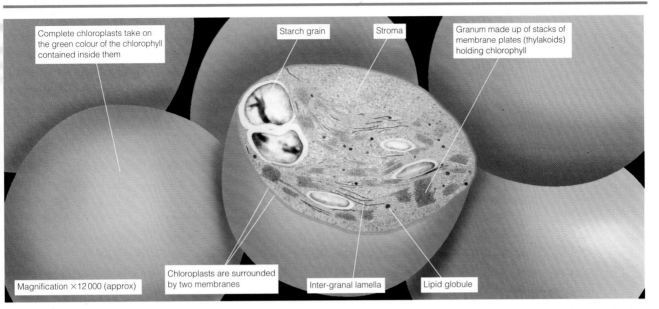

Fig. 4.6 Chloroplasts

Also present within the stroma are a series of starch grains which act as temporary stores for the products of photosynthesis. A number of smaller granules within the stroma readily take up osmium salts during the preparation of material for the electron microscope. They are called **osmiophilic granules** (*osmio* – 'osmium', *philo* – 'liking') but their function is not yet clear. A small amount of DNA is always present within the stroma, as are oil droplets.

4.2.5 The mitochondrion

Mitochondria are found within the cytoplasm of all eukaryotic cells, although in highly specialized cells such as mature red blood cells they may be absent. They range in shape from spherical to highly elongated and are typically 5 μm in length and 0.2 μm across. They are bounded by a double membrane, the outer of which controls the entry and exit of chemicals. The inner membrane is folded inwards, giving rise to extensions called

Fig. 4.7 Mitochondria

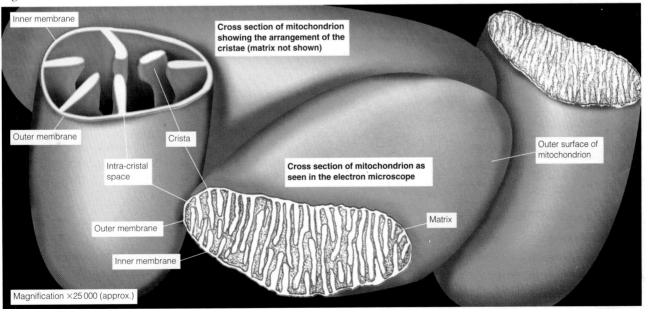

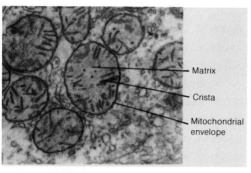

Mitochondrion (EM) (×15 000 approx.)

(a) Stereogram of the mitochondrion

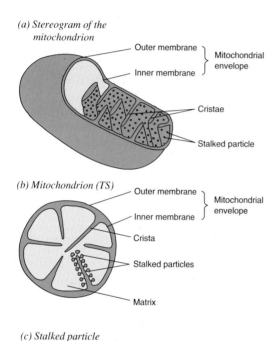

(b) Mitochondrion (TS)

(c) Stalked particle

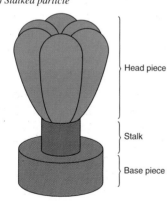

Fig. 4.8 Structure of a mitochondrion

cristae, some of which extend across the entire organelle. They function to increase the surface area on which respiratory processes take place. The surface of these cristae has stalked granules along its length (Fig. 4.8).

The remainder of the mitochondrion is the **matrix**. It is a semi-rigid material containing protein, lipids and traces of DNA. Electron-dense granules of 25 nm diameter also occur.

Mitochondria function as sites for certain stages of respiration, details of which are given in Section 16.4.1. The number of mitochondria in a cell therefore varies with its metabolic activity. Highly active cells may possess up to 1000. Similarly the number of cristae increases in metabolically active cells, giving weight to the proposition that respiratory enzymes are located on them.

4.2.6 Endoplasmic reticulum

The endoplasmic reticulum (ER) is an elaborate system of membranes found throughout the cell, forming a cytoplasmic skeleton. It is an extension of the outer nuclear membrane with which it is continuous. The membranes form a series of sheets which enclose flattened sacs called **cisternae** (Fig. 4.9 on page 59). Its structure varies from cell to cell and can probably change its nature rapidly; the membranes of the ER may be loosely organized or tightly packed. Where the membranes are lined with ribosomes they are called **rough endoplasmic reticulum**. The rough ER is concerned with protein synthesis (Section 7.6) and is consequently most abundant in those cells which are rapidly growing or secrete enzymes. In the same way, damage to a cell often results in increased formation of ER in order to produce the proteins necessary for the cell's repair. Where the membranes lack ribosomes they are called **smooth endoplasmic reticulum**. The smooth ER is concerned with lipid synthesis and is consequently most abundant in those cells producing lipid-related secretions, e.g. the sebaceous glands of mammalian skin and cells secreting steroids.

The functions of the ER may thus be summarized as:

1. Providing a large surface area for chemical reactions.

2. Providing a pathway for the transport of materials through the cell.

3. Producing proteins, especially enzymes (rough ER).

4. Producing lipids and steroids (smooth ER).

5. Collecting and storing synthesized material.

6. Providing a structural skeleton to maintain cellular shape (e.g. the smooth ER of a rod cell from the retina of the eye).

4.2.7 Golgi apparatus (dictyosome)

The Golgi apparatus, named after its discoverer Camillo Golgi, has a similar structure to the smooth endoplasmic reticulum but is more compact. It is composed of stacks of flattened sacs made of membranes. The sacs are fluid-filled and pinch off smaller

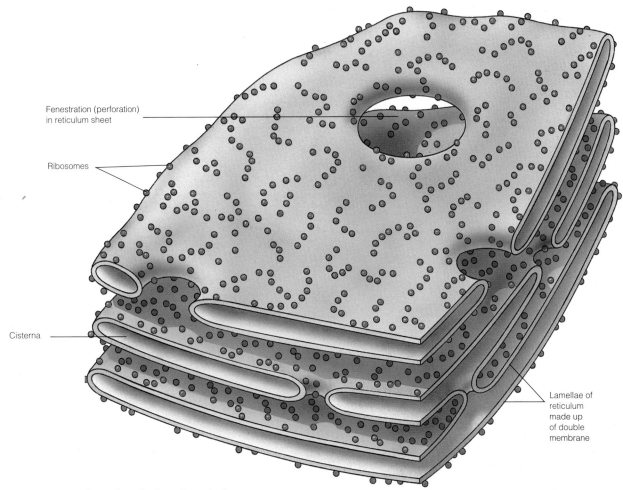

Fenestration (perforation) in reticulum sheet

Ribosomes

Cisterna

Lamellae of reticulum made up of double membrane

Fig. 4.9 Structure of rough endoplasmic reticulum

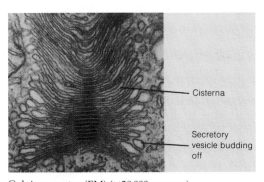

Cisterna

Secretory vesicle budding off

Golgi apparatus (EM) (×30 000 approx.)

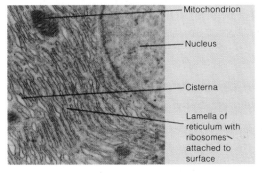

Mitochondrion

Nucleus

Cisterna

Lamella of reticulum with ribosomes attached to surface

Endoplasmic reticulum (EM) (×9000 approx.)

membranous sacs, called **vesicles**, at their ends. There is normally only one Golgi apparatus in each animal cell but in plant cells there may be a large number of stacks known as **dictyosomes**. Its position and size varies from cell to cell but it is well developed in secretory cells and neurones and is small in muscle cells. All proteins produced by the endoplasmic reticulum are passed through the Golgi apparatus in a strict sequence. They pass first through the cis-Golgi network which returns to the ER any proteins wrongly exported by it. They then pass through the stack of cisternae which modify the proteins and lipids undergoing transport and add labels which allow them to be identified and sorted at the next stage, the trans Golgi network. Here the proteins and lipids are sorted and sent to their final destinations. In general the Golgi acts as the cell's post office, receiving, sorting and delivering proteins and lipids. More specifically its functions include:

1. Producing glyco-proteins such as mucin required in secretions, by adding the carbohydrate part to the protein.

2. Producing secretory enzymes, e.g. the digestive enzymes of the pancreas.

3. Secreting carbohydrates such as those involved in the production of new cell walls.

4. Transporting and storing lipids.

5. Forming lysosomes as described in Section 4.2.8.

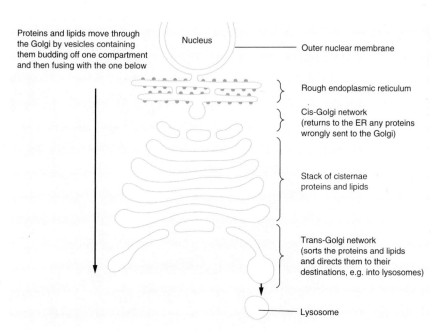

Fig. 4.10 The Golgi apparatus and its relationship to the nucleus, endoplasmic reticulum and lysosomes

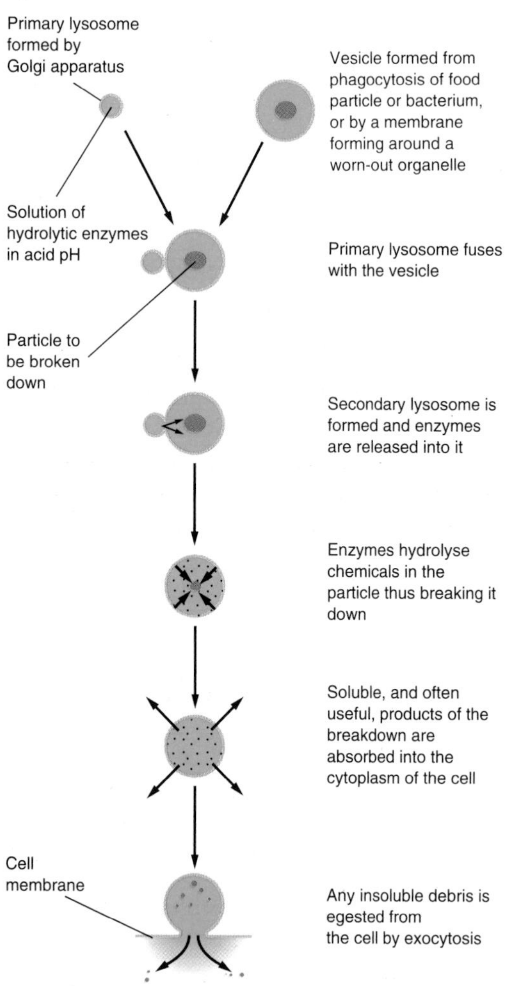

4.2.8 Lysosomes

Lysosomes (*lysis* – 'splitting', *soma* – 'body') are spherical bodies, some 0.1 to 1.0 μm in diameter. They contain around 50 enzymes, mostly hydrolases, in acid solution. They isolate these enzymes from the remainder of the cell and by so doing prevent them from acting upon other chemicals and organelles within the cell.

The functions of lysosomes are:

1. To digest material which the cell consumes from the environment. In the case of white blood cells, this may be bacteria or other harmful material. In Protozoa, it is the food which has been consumed by phagocytosis. In either case the material is broken down within the lysosome, useful chemicals are absorbed into the cytoplasm and any debris is egested by the cell by exocytosis (Fig. 4.11).

2. To digest parts of the cell, such as worn-out organelles, in a similar way to that described in **1**. This is known as **autophagy**. After the death of the cell they are responsible for its complete breakdown, a process called **autolysis** (*auto* – 'self', *lysis* – 'splitting').

3. To release their enzymes outside the cell (**exocytosis**) in order to break down other cells, e.g. in the reabsorption of tadpole tails during metamorphosis.

In view of their functions, it is hardly surprising that lysosomes are especially abundant in secretory cells and in phagocytic white blood cells.

Fig. 4.11 The functioning of a lysosome

4.2.9 Microbodies (peroxisomes)

Microbodies are small spherical membrane-bounded bodies which are between 0.5 and 1.5 μm in diameter. Apart from being slightly granular, they have no internal structure. They contain a number of metabolically important enzymes, in particular the enzyme catalase, which catalyses the breakdown of hydrogen peroxide. Hence these microbodies are sometimes called peroxisomes.

Hydrogen peroxide is a potentially toxic by-product of many biochemical reactions within organisms. Peroxisomes containing catalase are therefore particularly numerous in actively metabolizing cells like those of the liver.

$$2H_2O_2 \xrightarrow{\text{catalase}} 2H_2O + O_2$$

Hydrogen peroxide   Water   Oxygen

4.2.10 Vacuoles

A fluid-filled sac bounded by a single membrane may be termed a vacuole. Within mature plant cells there is usually one large central vacuole. The single membrane around it is called the **tonoplast**. A plant vacuole contains a solution of mineral salts, sugars, amino acids, wastes (e.g. tannins) and sometimes also pigments such as **anthocyanins**.

Plant vacuoles serve a variety of functions:

1. The sugars and amino acids may act as a temporary food store.

2. The anthocyanins are of various colours and so may colour petals to attract pollinating insects, or fruits to attract animals for dispersal.

3. They act as temporary stores for organic wastes, such as tannins. These may accumulate in the vacuoles of leaf cells and are removed when the leaves fall.

4. They occasionally contain hydrolytic enzymes and so perform functions similar to those of lysosomes (Section 4.2.8).

5. They support herbaceous plants, and herbaceous parts of woody plants by providing an osmotic system which creates a pressure potential (Section 22.3).

In animal cells, vacuoles are much smaller but may occur in larger numbers. Common types include food vacuoles, phagocytic vacuoles and contractile vacuoles. The latter are important in the osmoregulation of certain protozoans (Section 23.3.1).

PROJECT

1. Use eye-piece and stage micrometers to measure the thicknesses of various coloured hairs from the heads of your colleagues.

2. Examine a selection of light and electron micrographs of similar structures. List the similarities and differences.

3. (a) Examine electron micrographs of a selection of animal and plant cells where the magnifications are given.
   (b) Measure and calculate the mean sizes of a variety of organelles, e.g. nuclei, mitochondria, etc.
   (c) Are there any differences in the mean sizes of the organelles in animal and plant cells?

NON-MEMBRANOUS STRUCTURES

4.2.11 Ribosomes

Ribosomes are small cytoplasmic granules found in all cells. They are around 20 nm in diameter in eukaryotic cells (80S type) but slightly smaller in prokaryotic ones (70S type). They may occur in groups called **polysomes** and may be associated with endoplasmic reticulum or occur freely within the cytoplasm. Despite their small size, their enormous numbers mean that they can account for up to 20% of the mass of a cell.

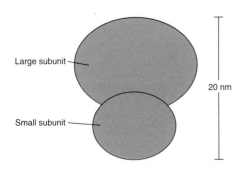

Large subunit

Small subunit

20 nm

Fig. 4.12 Structure of a ribosome

Ribosomes are made up of one large and one small sub-unit and comprise RNA known as **ribosomal RNA** and protein. They are important in the synthesis of proteins where they move along messenger RNA in succession (see Section 7.6).

4.2.12 Storage granules

Every cell contains a limited store of food energy. This store may be in the form of soluble material such as the sugar found in the vacuoles of plant cells. It may also occur in insoluble form, as grains or granules, within cells or organelles.

Starch grains occur within chloroplasts and the cytoplasm of plant cells. Starch may also be stored in specialized leucoplasts called amyloplasts. **Glycogen granules** occur throughout the cytoplasm of animal cells. They store animal starch or glycogen. **Oil or lipid droplets** are found within the cytoplasm of both plant and animal cells.

4.2.13 Microtubules

Microtubules occur widely throughout eukaryotic cells but are not found in prokaryotic ones. They are slender, unbranched tubes 24 nm in diameter and up to several microns in length. They are made of two similar proteins **alpha- and beta-tubulin**, each of which comprises 450 amino acids. The arrangement of these proteins is shown in Fig. 4.13.

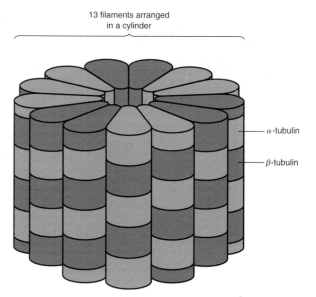

13 filaments arranged in a cylinder

α-tubulin

β-tubulin

Fig. 4.13 Arrangement of alpha- and beta-tubulin within a microtubule

The functions of microtubules are:

1. To provide an internal skeleton (**cytoskeleton**) for cells and so help determine their shape.

2. To aid transport within cells by providing routes along which materials move.

3. To form a framework along which the cellulose cell wall of plants is laid down.

4. As major components of cilia and flagella where they are grouped in a very precise way and contribute to their movement.

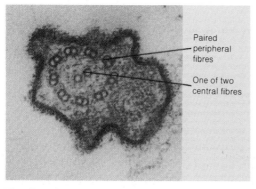

Paired peripheral fibres

One of two central fibres

Flagellum (EM) (×92 000 approx.)

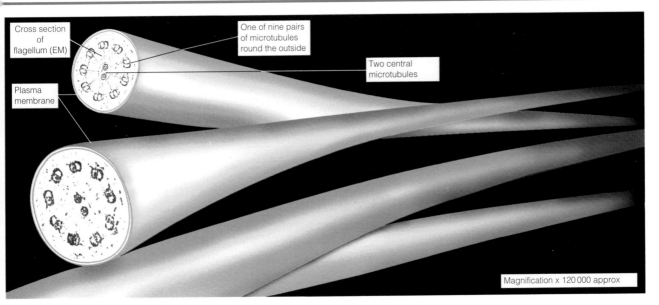

Fig. 4.14 Flagella

5. In the spindle during cell division and within the centrioles from which the spindle is formed. Here they help to draw chromosomes or chromatids to opposite poles (see Chapter 8).

4.2.14 Cilia and flagella

Cilia and flagella are almost identical, except that cilia are usually shorter and more numerous. Both are around $0.2\,\mu m$ in diameter; cilia are up to $25\,\mu m$ long whereas flagella may be $1000\,\mu m$ long. They are found in a limited number of cells but are nevertheless of great importance. The structure of a cilium is shown in Fig. 4.15. They function to either move an entire organism, e.g. cilia on the protozoan *Paramecium*, or to move material within an organism, e.g. the cilia lining the respiratory tract move mucus towards the throat. In the human respiratory tract there are around 200 cilia (each $7\,\mu m$ long) on each epithelial cell, giving a density of 10^9 cilia per cm^2.

4.2.15 Centrioles

Centrioles have the same basic structure as the basal bodies of cilia. They are hollow cylinders about $0.2\,\mu m$ in diameter. They arise in a distinct region of the cytoplasm known as the **centrosome**. It contains two centrioles. At cell division they migrate to opposite poles of the cell where they synthesize the microtubules of the spindle. Despite the absence of centrioles, the cells of higher plants do form spindles.

4.2.16 Microfilaments

Microfilaments are very thin strands about 6 nm in diameter. They are usually made up of the protein actin although a smaller proportion are of myosin. As these are the two proteins involved in muscle contraction, it seems probable that microfilaments play a rôle in movement within cells and possibly of the cells as a whole in some cases.

(a) LS Basal region of a cilium

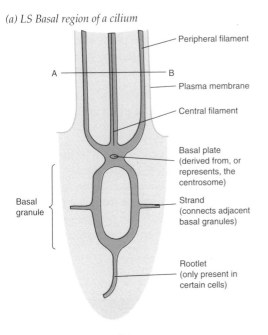

(b) TS Cilium (Section A/B)

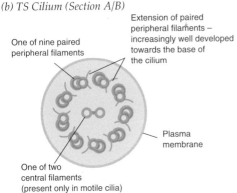

Fig. 4.15 Structure of a cilium

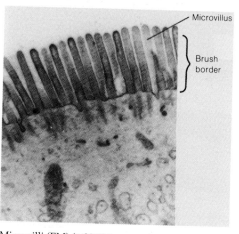

Microvilli (EM) (×23 000 approx.)

4.2.17 Microvilli

Microvilli are tiny finger-like projections about 0.6 μm in length on the membranes of certain cells, such as those of the intestinal epithelium and the kidney tubule. They should not be confused with the much larger villi which are multicellular structures. Microvilli massed together appear similar to the bristles of a brush, hence the term **brush border** given to the edge of cells bearing microvilli. Actin filaments within the microvilli allow them to contract, which, along with their large surface area, facilitates absorption.

4.2.18 Cellulose cell wall

A cell wall is a characteristic feature of plant cells. It consists of cellulose microfibrils embedded in an amorphous polysaccharide matrix. The structure and properties of cellulose are discussed in Section 2.5.3 and the detailed structure of a cellulose microfibril is given in Fig. 4.16. The matrix is usually composed of polysaccharides, e.g. pectin or lignin. The microfibrils may be regular or irregular in arrangement.

The main functions of the cell wall are:

1. To provide support in herbaceous plants. As water enters the cell osmotically the cell wall resists expansion and an internal pressure is created which provides turgidity for the plant.

2. To give direct support to the cell and the plant as a whole by providing mechanical strength. The strength may be increased by the presence of lignin in the matrix between the cellulose fibres.

3. To permit the movement of water through and along it and so contribute to the movement of water in the plant as a whole, in particular in the cortex of the root.

4. In some cell walls the presence of cutin, suberin or lignin in the matrix makes the cells less permeable to substances. Lignin helps to keep the water within the xylem and cutin in the epidermis of leaves prevents water being lost from the plant. Suberin in root endodermal cells prevents movement of water across them, thus concentrating its movement through special passage cells.

Fig. 4.16 Structure of a cellulose microfibril

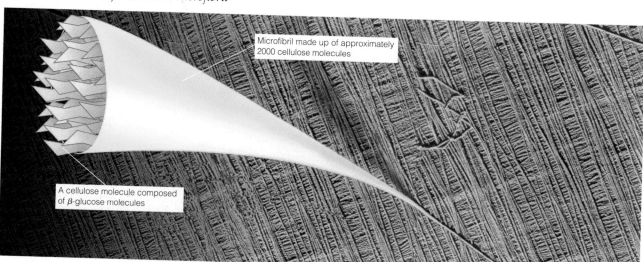

Microfibril made up of approximately 2000 cellulose molecules

A cellulose molecule composed of β-glucose molecules

Cellulose and paper

Cotton plants ready for harvesting

Cellulose is a polysaccharide made up of several thousand glucose units linked together in long chains by β-1,4 links. As the major constituent of plant cell walls, cellulose is a readily available and renewable raw material. Although its most important use is in the manufacture of paper, cellulose is also the main component of many other household goods. The seed hairs of the cotton plant (*Gossypium*) are almost pure cellulose and their natural twist makes them easy to spin for use in a variety of textiles from clothes to curtains. Other products using cellulose are derived from various plants, e.g. linen from flax (*Linum usitatissimum*) and rattan furniture from the stems of an Asian climbing plant.

Cellulose can be processed to form an even wider range of products, such as, thickeners in paints, stabilizers in foods and cosmetics, cellophane, adhesive tape and materials for dialysis membranes.

Paper making began in China in about AD 100 using hemp and flax but it was not until the eighteenth century that wood began to be used as a source of paper-making fibre. The properties of any paper depend on the plant fibres from which it is derived as well as the processing techniques. The cellulose in fibres provides the necessary characteristics of chemical stability, flexibility, high tensile strength and good bonding ability.

The first stage in paper making is pulping. When only mechanical processes are used to heat and grind the wood chips a high yield of relatively low grade pulp (suitable for newspaper) is produced. Chemical processes can also be employed to remove the lignin so that a better quality paper is produced which does not discolour. However the processes also remove some of the cellulose and a low yield is produced. Sometimes pure cellulose pulps are required, for example as a starting point for viscose or cellophane production.

High quality pulp, such as that produced from cotton, is used for the long-lasting paper required for bank notes or for specialist paper like filter paper. It is about seven or eight times more expensive than low grade pulp.

In Britain every person uses about 163 kg of paper per year but since 1981 over 50% of the fibre used by our paper industry has come from waste paper. Much of the rest comes from European forests which are carefully managed plantations, actually increasing in size by 3% per year. In recent years techniques have improved so that any grade of paper can be produced from waste fibre but it is still difficult to remove the inks and have a white product without the use of potentially harmful chemicals. Further recycling of waste paper would avoid the possible polluting effects of burning it (thus increasing atmospheric carbon dioxide levels) or burying it in landfill sites.

5. The arrangement of the cellulose fibrils in the cell wall can determine the pattern of growth and hence the overall shape of a cell.

6. Occasionally cell walls act as food reserves.

4.3   Movement in and out of cells

The various organelles and structures within a cell require a variety of substances in order to carry out their functions. In turn they form products, some useful and some wastes. Most of these substances must pass in and out of the cell. This they do by **diffusion, osmosis, active transport, phagocytosis** and **pinocytosis**.

4.3.1 Diffusion

Diffusion is the process by which a substance moves from a region of high concentration of that substance to a region of low concentration of the same substance. Diffusion occurs because the molecules of which substances are made are in random motion (kinetic theory). The process is explained in Fig. 4.17.

The rate of diffusion depends upon:

1. The concentration gradient – The greater the difference in concentration between two regions of a substance the greater the rate of diffusion. Organisms must therefore maintain a fresh supply of a substance to be absorbed by creating a stream over the diffusion surface. Equally, the substance, once absorbed, must be rapidly transported away.

2. The distance over which diffusion takes place – The shorter the distance between two regions of different concentration the greater the rate of diffusion. The rate is proportional to the reciprocal of the square of the distance (inverse square law). Any structure in an organism across which diffusion regularly takes place must therefore be thin. Cell membranes for example are only 7.5 nm thick and even epithelial layers such as those lining the alveoli of the lungs are as thin as 0.3 μm across.

3. The area over which diffusion takes place – The larger the surface area the greater the rate of diffusion. Diffusion surfaces frequently have structures for increasing their surface area and hence the rate at which they exchange materials. These structures include villi and microvilli.

4. The nature of any structure across which diffusion occurs – Diffusion frequently takes place across epithelial layers or cell membranes. Variations in their structure may affect diffusion. For example, the greater the number and size of pores in cell membranes the greater the rate of diffusion.

5. The size and nature of the diffusing molecule – Small molecules diffuse faster than large ones. Fat-soluble ones diffuse more rapidly through cell membranes than water-soluble ones.

4.3.2 Facilitated diffusion

This special form of diffusion allows more rapid exchange. It may involve channels within a membrane which make diffusion of specific substances easier. These channels form water-filled connections across the lipid bilayer which allow water-soluble substances to move across. They are important therefore in transporting ions. The channels are selective in that they will open or close in response to certain signals such as a change in voltage or the binding of another molecule. In this way the cell can control the entry and exit of molecules and ions.

An alternative form of facilitated diffusion involves different protein molecules in the membrane called **carrier proteins**. These bind molecules to them and then change shape as a result of this binding in such a way that the molecules are released to the inside of the membrane (Fig. 4.18).

1. *If 10 particles occupying the left-hand side of a closed vessel are in random motion, they will collide with each other and the sides of the vessel. Some particles from the left-hand side move to the right, but initially there are no available particles to move in the opposite direction, so the movement is in one direction only. There is a large concentration gradient and diffusion is rapid.*

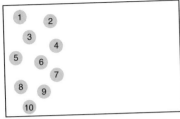

2. *After a short time the particles (still in random motion) have spread themselves more evenly. Particles can now move from right to left as well as left to right. However with a higher concentration of particles (7) on the left than on the right (3) there is a greater probability of a particle moving to the right than in the reverse direction. There is a smaller concentration gradient and diffusion is slower.*

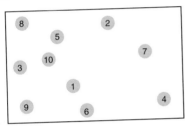

3. *Some time later, the particles will be evenly distributed throughout the vessel and the concentrations will be equal on each side. The system is in equilibrium. The particles are not however static but remain in random motion. With equal concentrations on each side, the probability of a particle moving from left to right is equal to the probability of one moving in the opposite direction. There is no concentration gradient and no net diffusion.*

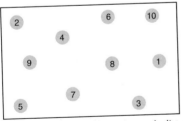

4. *At a later stage the particles remain evenly distributed and will continue to do so. Although the number of particles on each side remains the same, individual particles are continuously changing position. This situation is called* **dynamic equilibrium**.

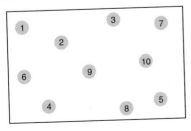

Figure 4.17 Diffusion

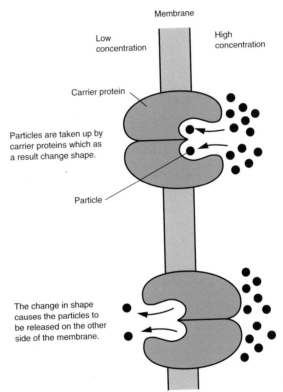

Fig. 4.18 Facilitated diffusion by carrier proteins

In all cases facilitated diffusion does not involve the use of energy (i.e. it is passive) and hence material is moved along a concentration gradient (i.e. from high to low concentration).

4.3.3 Osmosis

Osmosis is a special form of diffusion which involves the movement of solvent molecules. The solvent in biological systems is invariably water. Most cell membranes are permeable to water and certain solutes only. Such membranes are termed **partially permeable**. Osmosis in living organisms can therefore be defined as: **the passage of water from a region where it is highly concentrated to a region where its concentration is lower, through a partially permeable membrane**. The process is explained in Fig. 4.19 on page 68.

If a solution is separated from its pure solvent, as in Fig 4.19, the pressure which must be applied to stop water entering that solution, and so prevent osmosis, is called the **osmotic pressure**. The more concentrated a solution the greater is its osmotic pressure. This is a hypothetical situation and, as a solution does not actually exert a pressure under normal circumstances, the term 'osmotic potential' is preferred. As the osmotic potential is in effect the potential of a solution to pull water into it, it always has a negative value. A more concentrated solution therefore has a more positive osmotic pressure but a more negative osmotic potential.

Osmosis not only occurs when a solution is separated from its pure solvent by a partially permeable membrane but also arises when such a membrane separates two solutions of different concentrations. In this case water moves from the more dilute, or **hypotonic**, solution, to the more concentrated, or **hypertonic**, solution. When a dynamic equilibrium is established and both solutions are of equal concentration they are said to be **isotonic**. The above terms should only be applied to animal cells. The osmotic relationships of plant cells should be described in terms of water potential (Section 22.3).

Consider Fig. 4.20 on the next page. Initially the water molecules on the right of the partially permeable membrane collide with the membrane more often than those on the left, which are to some extent impeded by the glucose molecules. In other words the water on the right has a greater potential energy than that on the left of the membrane. The greater the number of collisions the water molecules make on the membrane, the greater the pressure on it. This pressure is called the **water potential** and is represented by the greek letter psi (Ψ).

Under standard conditions of temperature and pressure (25 °C and 100 kPa) pure water is designated a water potential of zero. The addition of solute to pure water lowers its water potential because the solute molecules impede the water molecules, reducing the number of collisions they make with the membrane. It therefore exerts less pressure and has a lower water potential. Given that pure water has a water potential of zero, all solutions therefore have a lower one i.e. they have negative water potentials. The more concentrated a solution the more negative is its water potential. Water will diffuse from a region of less negative (higher) water potential to one of more negative (lower) water potential.

4.3.4 Active transport

Diffusion and osmosis are passive processes, i.e. they occur without the expenditure of energy. Some molecules are transported in and out of cells by active means, i.e. energy is required to drive the process.

The energy is necessary because molecules are transported against a **concentration gradient**, i.e. from a region of low concentration to one of a high concentration. It is thought that the process occurs through the proteins that span the membrane. These accept the molecule to be transported on one side of the membrane and, by a change in the structure of the protein, convey it to the other side (see Fig. 4.21). A good example of

---

1. *Both solvent (water) and solute (glucose) molecules are in random motion, but only solvent (water) molecules are able to cross the partially permeable membrane. This they do until their concentration is equal on both sides of the membrane.*

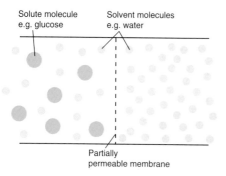

Solute molecule e.g. glucose

Solvent molecules e.g. water

Partially permeable membrane

2. *Once the water molecules are evenly distributed, in theory a dynamic equilibrium should be established. However, the water molecules on the left of the membrane are impeded to some extent by the glucose molecules from crossing the membrane. With no glucose present on the right of the membrane, water molecules move more easily to the left than in reverse direction.*

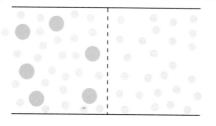

3. *A situation is reached whereby additional water molecules accumulate on the left of the membrane, until their greater concentration offsets the blocking effect of the glucose. The probability of water molecules moving in either direction is the same, and a dynamic equilibrium is established.*

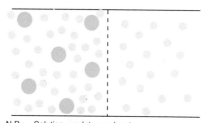

N.B.　Solution = solute + solvent
e.g.　Glucose solution = glucose powder + water

Fig. 4.19 Osmosis

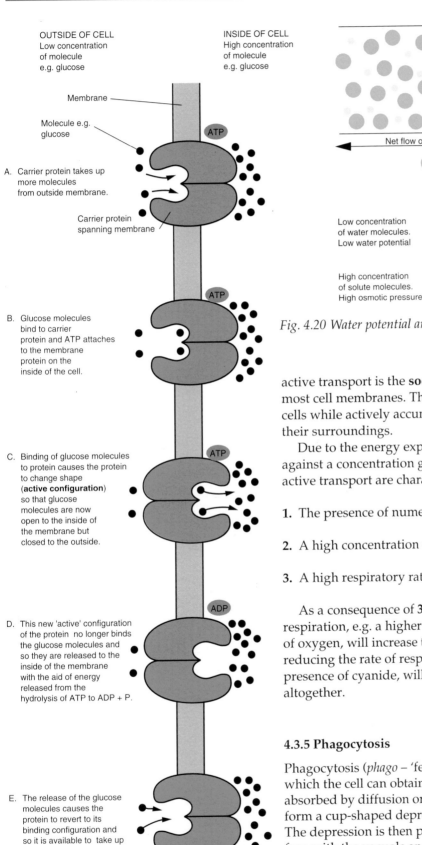

A. Carrier protein takes up more molecules from outside membrane.

B. Glucose molecules bind to carrier protein and ATP attaches to the membrane protein on the inside of the cell.

C. Binding of glucose molecules to protein causes the protein to change shape (**active configuration**) so that glucose molecules are now open to the inside of the membrane but closed to the outside.

D. This new 'active' configuration of the protein no longer binds the glucose molecules and so they are released to the inside of the membrane with the aid of energy released from the hydrolysis of ATP to ADP + P.

E. The release of the glucose molecules causes the protein to revert to its binding configuration and so it is available to take up more glucose molecules from the outside.

Fig. 4.21 Active transport

Partially permeable membrane

Net flow of molecules

Solute molecule, e.g. glucose
Water molecule

Low concentration of water molecules. Low water potential

High concentration of water molecules. High water potential

High concentration of solute molecules. High osmotic pressure

Low concentration of solute molecules. Low osmotic pressure

Fig. 4.20 Water potential and osmotic pressure

active transport is the **sodium-potassium pump** which exists in most cell membranes. This actively removes sodium ions from cells while actively accumulating potassium ions into them from their surroundings.

Due to the energy expenditure necessary to move molecules against a concentration gradient, cells and tissues carrying out active transport are characterized by:

1. The presence of numerous mitochondria.

2. A high concentration of ATP.

3. A high respiratory rate.

As a consequence of **3**, any factor which increases the rate of respiration, e.g. a higher temperature or increased concentration of oxygen, will increase the rate of active transport. Any factor reducing the rate of respiration or causing it to cease, e.g. the presence of cyanide, will cause active transport to slow or stop altogether.

4.3.5 Phagocytosis

Phagocytosis (*phago* – 'feeding', *cyto* – 'cell') is the process by which the cell can obtain particles which are too large to be absorbed by diffusion or active transport. The cell invaginates to form a cup-shaped depression in which the particle is contained. The depression is then pinched off to form a vacuole. Lysosomes fuse with the vacuole and their enzymes break down the particle, the useful contents of which may be absorbed (Fig. 4.22). The process only occurs in a few specialized cells (called **phagocytes**), such as white blood cells where harmful bacteria can be ingested, or *Amoeba* where it is a means of feeding.

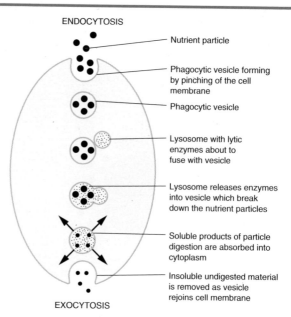

ENDOCYTOSIS

Nutrient particle

Phagocytic vesicle forming by pinching of the cell membrane

Phagocytic vesicle

Lysosome with lytic enzymes about to fuse with vesicle

Lysosome releases enzymes into vesicle which break down the nutrient particles

Soluble products of particle digestion are absorbed into cytoplasm

Insoluble undigested material is removed as vesicle rejoins cell membrane

EXOCYTOSIS

Fig. 4.22 Endocytosis and exocytosis

4.3.6 Pinocytosis

Pinocytosis or 'cell drinking' is very similar to phagocytosis except that the vesicles produced, called **pinocytic vesicles**, are smaller. The process is used for the intake of liquids rather than solids. Even smaller vesicles, called **micropinocytic vesicles**, may be pinched off in the same way.

Both pinocytosis and phagocytosis are methods by which materials are taken into the cell in bulk. This process is called **endocytosis**. By contrast, the reverse process, in which materials are removed from cells in bulk, is called **exocytosis** (Fig. 4.22).

4.4   Questions

1. (a) In the study of cells, what are the limitations of (i), the light microscope and (ii) the electron microscope?   *(4 marks)*
    (b) Show the structure of the cell membrane (plasmalemma) by means of a fully-labelled diagram.   *(4 marks)*
    (c) Describe the functions of the various chemical components of the cell membrane.   *(8 marks)*
    (d) Distinguish between pinocytosis and phagocytosis. Describe examples of **each** in living cells.   *(4 marks)*
    (Total 20 marks)

    NEAB June 1993, Paper IB, No. 4

2. (a) Describe the structure of
        (i) the cell wall, and
        (ii) the cell membrane in plant cells.
    Include in your answer the arrangement of the chemical constituents and explain how they affect the functions of the two structures.   *(12 marks)*
    (b) Discuss the role and distribution of membranes in
        (i)   mitochondria,
        (ii) endoplasmic reticulum,
        (iii) nucleus.   *(8 marks)*
    (Total 20 marks)

    NEAB June 1991, Paper IB, No. 6

3. The diagram shows a fluid-mosaic model of a cell membrane.

    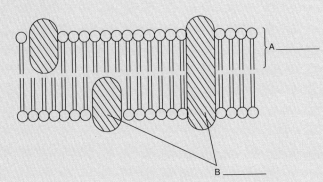

    (a) Label the molecules **A** and **B** on the diagram.   *(2 marks)*
    (b) Why is the model described as being *fluid*?   *(1 mark)*
    (c) Give **two** functions in the membrane of the molecules labelled **B**.   *(2 marks)*
    (Total 5 marks)

    AEB June 1990, Paper I, No. 8

4. The table below refers to a bacterial cell, a liver cell and a palisade mesophyll cell and structures which may be found in them.

    If the feature is present, place a tick (✓) in the appropriate box and if the feature is absent, place a cross (✗) in the appropriate box.

Feature	Bacterial cell	Liver cell	Palisade cell
Nuclear envelope			
Cell wall			
Glycogen granules			
Microvilli			
Chloroplasts			

    (Total 5 marks)

    ULEAC June 1992, Paper I, No. 8

5. Study the electron-micrograph below and then answer the following questions.

    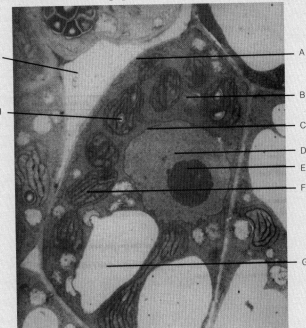

    (a) Give **four** reasons why the cell shown can be identified as a plant cell.   *(4 marks)*
    (b) Name structures **B, C, D, E, G** and **I**.   *(3 marks)*
    (c) For each of parts **E, F** and **H**, briefly explain **one** function.   *(3 marks)*
    (d) Suggest a name for this type of plant cell and state (with a reason) whether you consider it to be a young or old (mature) cell.   *(2 marks)*
    (Total 12 marks)

    Oxford June 1991, Paper I, No. 1

6. The diagram shows the stages involved in separating liver-cell components.

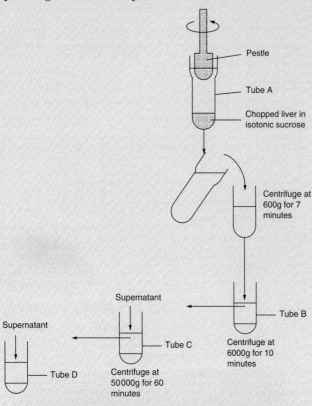

(a) Why is the tissue placed in an isotonic solution? (2 marks)

(b) The rotating pestle is lowered into Tube A. What is the purpose of this? (1 mark)

(c) The sediment in tube **C** would be rich in mitochondria. Give **one** way in which you could determine that this mitochondrial fraction was:

(i) pure; (ii) metabolically active. (2 marks)

(d) Name **one** organelle that would be found in the sediment in Tube **D**. (1 mark)
(Total 6 marks)

AEB June 1992, Paper I, No. 6

7. The figure is a cutaway diagram of a mitochondrion from a liver cell.

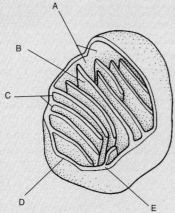

(a) Name the parts labelled **A, B, C, D** and **E**. (5 marks)

(b) Suggest why liver cells contain large numbers of mitochondria. (1 mark)

(c) (i) The matrix contains many hundreds of oxidative enzymes. Name **two** substances which enter the mitochondria and are oxidized.

(ii) Name **three other** substances which enter the mitochondria. (5 marks)

(d) Mitochondria contain DNA and ribosomes. State the significance of their presence. (4 marks)

(e) (i) Where does oxidative phosphorylation occur in mitochondria?

(ii) Explain briefly what is meant by oxidative phosphorylation. (5 marks)
(Total 20 marks)

UCLES June 1992, Paper II, No. 1

8. The diagram shows part of an animal cell and is based on a series of electron micrographs.

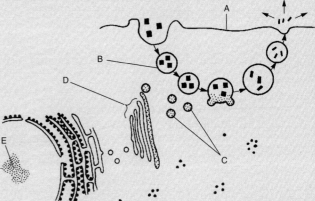

(a) (i) Name the structures labelled **A, B, C, D, E**. (5 marks)

(ii) Label with the letter **F** a structure where the protein contents of **D** are synthesized. (1 mark)

(iii) What is the part played by **E** in the synthesis of this protein? (2 marks)

(iv) Describe **two** functions **D** may have in cells of organisms. (2 marks)

(b) (i) Name the process illustrated in the diagram which results in large particles entering the cell. (1 mark)

(ii) This process is common in some types of white blood cell. Suggest **one** reason why this cell activity is important to the body. (1 mark)

(iii) Name **one** group of organisms which feed using the process illustrated in the diagram. (1 mark)
(Total 13 marks)

WJEC June 1993, Paper A2, No. 2

9. The electron micrograph below shows part of a palisade cell of a leaf.

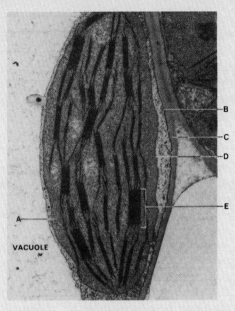

Magnification ×16 500 (approx)

(a) Name the parts labelled **A**, **B**, **C** and **D**.
(4 marks)

(b) Name **one** carbohydrate present in structure **C**.
(1 mark)

(c) Given that the magnification is ×40 000, calculate the actual length of structure **E**. Show your working.
(3 marks)
(Total 8 marks)

ULEAC June 1993, Paper I, No. 2

10. Outline the functions of each of the following cell organelles.

(a) Golgi apparatus (2 marks)
(b) Microtubules (2 marks)
(c) Ribosomes (2 marks)
(d) Nucleolus (2 marks)
(Total 8 marks)

ULEAC June 1991, Paper I, No. 6

11. (a) Outline the methods of preparation of a sample of tissue for examination with the electron microscope. Explain why each stage in the preparation is necessary. (8 marks)

(b) Explain briefly how an image of cell ultrastructure is formed by the electron microscope. (4 marks)

(c) Discuss the advantages and disadvantages of transmission electron microscopy compared to light microscopy. (6 marks)
(Total 18 marks)

UCLES June 1994, Paper 2, No. 6

12. Peptic cells from the lining of the mammalian stomach secrete the enzyme precursor pepsinogen. Some of these cells were isolated and maintained in a culture solution containing radioactively labelled amino acids. Samples of the cells were taken at regular intervals and prepared for electron microscopy. Below is a drawing from an electron micrograph of a peptic cell. The time taken for radioactivity to be detected in the various cell organelles viewed under the electron microscope is shown on the left of the drawing.

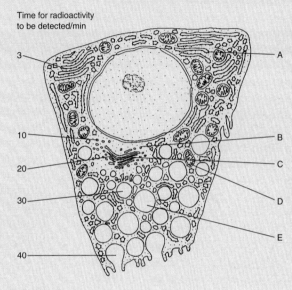

(a) Name the organelles **A** to **D**. (4 marks)

(b) Outline the sequence of events that result in the detection, at different times, of radioactivity in the organelles labelled **A** to **C**. (6 marks)

(c) Describe briefly the role of the nucleus in the synthesis of pepsinogen. (3 marks)

(d) Suggest how the secretory material in the organelles labelled **E** passes out of the cell. (3 marks)

(e) Explain why relatively large numbers of the organelle **D** are required in secretory cells. (3 marks)
(Total 19 marks)

UCLES June 1994, (AS) Paper 1, No. 1

See *Further questions* pp. 614–20, questions 16, 18.

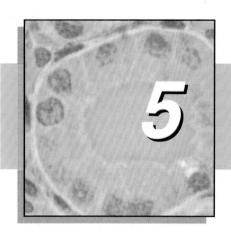

5 Classification of organisms

5.1 Principles of classification

Before any study can be made of living organisms it is necessary to devise a scheme whereby the enormous diversity of them can be organized into manageable groups. This grouping of organisms is known as **classification** and the study of biological classification is called **taxonomy**. A good universal system of classification aids communication between scientists and allows information about a particular organism to be found more readily. There is no 'correct' scheme of classification since organisms form a continuum and any division of them into groups has been devised solely for human convenience.

During the eighteenth century, the Swedish botanist Linnaeus devised a scheme of classification which has become widely accepted. In this scheme organisms are grouped together according to their basic similarities. Relationships are based on homologous rather than analogous characteristics. **Homologous** characters are ones which have a fundamental similarity of origin, structure and position, regardless of their function in the adult. **Analogous** characters are ones which have a similar function in the adult but which are not homologous, i.e. they do not have the same origin. For example, wings of butterflies and birds are both used for flight but their origins are not similar. Classification based on homology is called **natural classification**. It now embraces biochemical and chromosome studies as well as the morphology and anatomy used by Linnaeus. A successful natural classification should reflect the true evolutionary relationships of organisms.

5.1.1 Taxonomic ranks

It is convenient to distinguish large groups of organisms from smaller subgroups and a series of rank names has been devised to identify the different levels within this hierarchy. The rank names used today are largely derived from those used by Linnaeus over 200 years ago. The largest groups are known as **phyla** and the organisms in each phylum have a body plan radically different from organisms in any other phylum. Diversity within each phylum allows it to be divided into **classes**. Each class is divided into **orders** of organisms which have additional features in common. Each order is divided into **families** and at this level differences are less obvious. Each family is divided into **genera** and each genus into **species**.

TABLE 5.1 **Classification of three organisms**

Rank	Cabbage white butterfly	Human	Sweet pea
Phylum	Arthropoda	Chordata	Angiospermae
Class	Insecta	Mammalia	Dicotyledoneae
Order	Lepidoptera	Primates	Rosales
Family	Pieridae	Hominidae	Papilionaceae
Genus	*Pieris*	*Homo*	*Lathyrus*
Species	*brassica*	*sapiens*	*odoratus*

With the gradual acceptance that all species arose by adaptation of existing forms, the basis of this hierarchy became evolutionary. Species are groups that have diverged most recently, genera somewhat earlier and so on up the taxonomic ranks.

Every organism is given a scientific name according to an internationally agreed system of nomenclature, first devised by Linnaeus. The name is always in Latin and is in two parts. The first name indicates the genus and is written with an initial capital letter; the second name indicates the species and is written with a small initial letter. These names are always distinguished in text by italics or underlining. This system of naming organisms is known as **binomial nomenclature**.

Table 5.1 shows the use of rank names in classifying a cabbage white butterfly, a human and a sweet pea. Only the obligate ranks of classification to which every organism must be assigned have been shown in the table. However, a taxonomist may use a large number of additional categories within this scheme as shown below:

kingdom, subkingdom, grade, **Phylum**, subphylum, superclass, **Class**, subclass, infraclass, superorder, **Order**, suborder, infraorder, superfamily, **Family**, subfamily, tribe, **Genus**, subgenus, **Species**, subspecies, variety.

Living organisms are divided into 5 kingdoms:

Prokaryotae, **Fungi**, **Protoctista**, **Plantae** and **Animalia**. It is difficult to fit viruses into this scheme of classification because they are on the border of living and non-living. For this reason they are dealt with separately.

5.2   Viruses

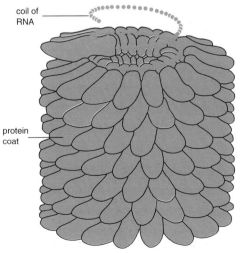

Fig. 5.1 Simplified diagram of tobacco mosaic virus

coil of RNA

protein coat

Viruses are smaller than bacteria, ranging in size from about 20 nm to 300 nm. They cannot be seen through a light microscope and pass through filters which retain bacteria. Many can be crystallized and they can only multiply inside living cells. They do, however, contain nucleic acids such as DNA or RNA and must therefore be considered as being on the border between living and non-living. They are not classified with any other living organisms. They are made up of a nucleic acid core surrounded by a coat of protein; outside cells these inert particles are known as **virions**. Most viruses found in animal cells and those attacking bacteria (known as **bacteriophages**) have the nucleic acid DNA. Other animal viruses and plant viruses contain RNA. Electron microscopy and X-ray diffraction have shown viruses to be a variety of shapes such as spherical, e.g. poliomyelitis, straight rods, e.g. tobacco mosaic virus (TMV), or flexible rods, e.g. potato virus X. Bacteriophages have a distinct 'head' and 'tail'.

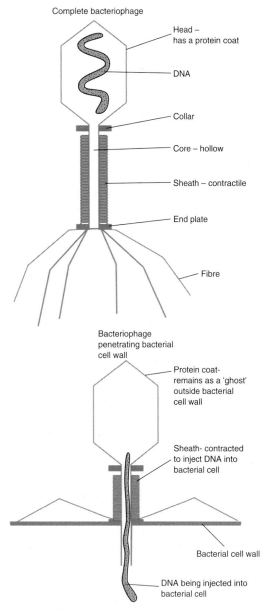

Complete bacteriophage

Head –
has a protein coat

DNA

Collar

Core – hollow

Sheath – contractile

End plate

Fibre

Bacteriophage
penetrating bacterial
cell wall

Protein coat-
remains as a 'ghost'
outside bacterial
cell wall

Sheath- contracted
to inject DNA into
bacterial cell

Bacterial cell wall

DNA being injected into
bacterial cell

Fig. 5.2 Structure of a bacteriophage

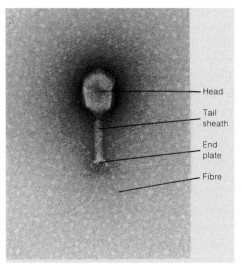

Head

Tail
sheath

End
plate

Fibre

A bacteriophage (EM) (×108 000)

5.2.1 Transmission of viruses

Two viruses that have been widely studied are the **tobacco mosaic virus (TMV)**, which attacks tomato, blackcurrant, potato and orchid as well as tobacco itself, and the **T$_2$ phage**, a bacteriophage, which infects *Escherichia coli*. Tobacco mosaic virus is rod-shaped with a length of about 300 nm and a diameter of 15 nm. It comprises 94% protein and 6% RNA, the nucleic acid determining its characteristics. TMV is very infectious, being carried on seed coats, by grasshoppers and by mechanical means. The only effective way to limit its effect is to maintain virus-free stock.

T$_2$ phage is tadpole-shaped, the head having a diameter of about 70 nm and the tail a length of about 0.2 μm. The cycle of infection of this bacteriophage has been particularly well studied and is explained in Fig. 5.3.

T$_2$ phage immediately kills the bacterium it enters and is therefore known as a **virulent phage**. In **temperate phages** the process is much less rapid and the host and phage may exist together for many generations. Host DNA may become incorporated in the viral DNA, and this DNA is carried to the next host, thereby resulting in new characteristics. This process of **transduction** is an important method by which antibiotic resistance spreads throughout a population of bacteria.

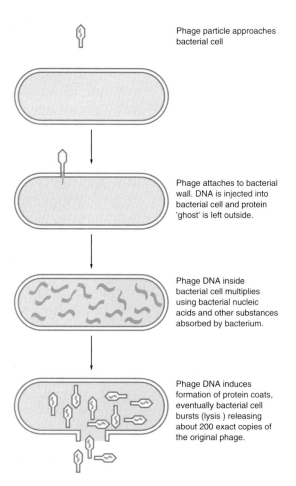

Phage particle approaches bacterial cell

Phage attaches to bacterial wall. DNA is injected into bacterial cell and protein 'ghost' is left outside.

Phage DNA inside bacterial cell multiplies using bacterial nucleic acids and other substances absorbed by bacterium.

Phage DNA induces formation of protein coats, eventually bacterial cell bursts (lysis) releasing about 200 exact copies of the original phage.

Fig. 5.3 Life cycle of a virulent phage (e.g. T$_2$ phage)

5.2.2 Retroviruses

Probably the best known retrovirus is the Human Immunodeficiency virus (HIV) which causes AIDS (Acquired Immune Deficiency Syndrome), further details of which are given in Section 21.3.4.

The genetic information in a retrovirus is RNA. While many viruses possess RNA, retroviruses are different in that they can use it to synthesize DNA. This is a reversal of the usual genetic process in which RNA is made from DNA and the reason retroviruses are so called (*retro* = behind or backwards).

In 1970 the enzyme capable of synthesizing DNA from RNA was discovered and given the name **reverse transcriptase** (as it catalyses the opposite process to transcriptase which synthesizes RNA from DNA). The discovery of this enzyme, more details of which are given in Section 7.8.2, has considerable importance for genetic engineering.

The DNA form of the retrovirus genes is called the **provirus** and is significant in that it can be incorporated into the host's DNA. Here it may remain latent for long periods before the DNA of the provirus is again expressed and new viral RNA produced. During this time any division of the host cell results in the proviral DNA being duplicated as well. In this way the number of potential retroviruses can proliferate considerably. This explains why individuals infected with the HIV virus often display no symptoms for many years before suddenly developing full-blown AIDS.

When incorporated into the host DNA the provirus is capable of activating the host genes in its immediate vicinity. Where these genes are concerned with cell division or growth, and are 'switched off' at the time, their activation by the provirus can result in a malignant growth known as **cancer**. The RNA produced by these newly activated genes may become packaged inside new retrovirus particles being assembled inside the host cell. This RNA may then be delivered, along with the retroviral RNA, to the next cell the virus infects. This new cell will then become potentially cancerous.

Host genes which have been acquired by retroviruses in this way are called **oncogenes** (*oncos* = tumour). Very few human cancers are caused by retroviruses in this way but research into them has led to the discovery of similar genes found in human chromosomes. These genes can be activated by chemicals or forms of radiation rather than viruses, and their investigation has already helped to prevent some cancers and may, in time, provide a cure.

Retroviruses can cause diseases other than cancer, but most are harmless. Some proviral DNA has become such an integral part of the host-cell DNA that it is passed on from one generation to the next via the gametes and is, in effect, part of the host's genetic make-up. Such a virus is referred to as an **endogenous** virus.

5.2.3 Economic importance of viruses

Viruses cause a variety of infectious diseases in humans, other animals and plants. The symptoms shown by plants may be localized or distributed throughout the plant. The same virus may have quite different effects in different hosts and these symptoms may be influenced by environmental conditions.

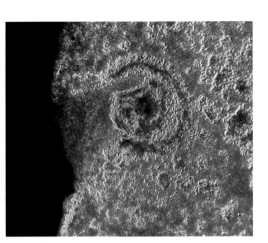

False-colour transmission electron micrograph of a Human Immunodeficiency Virus (HIV) – shown in red – infecting a T-lymphocyte

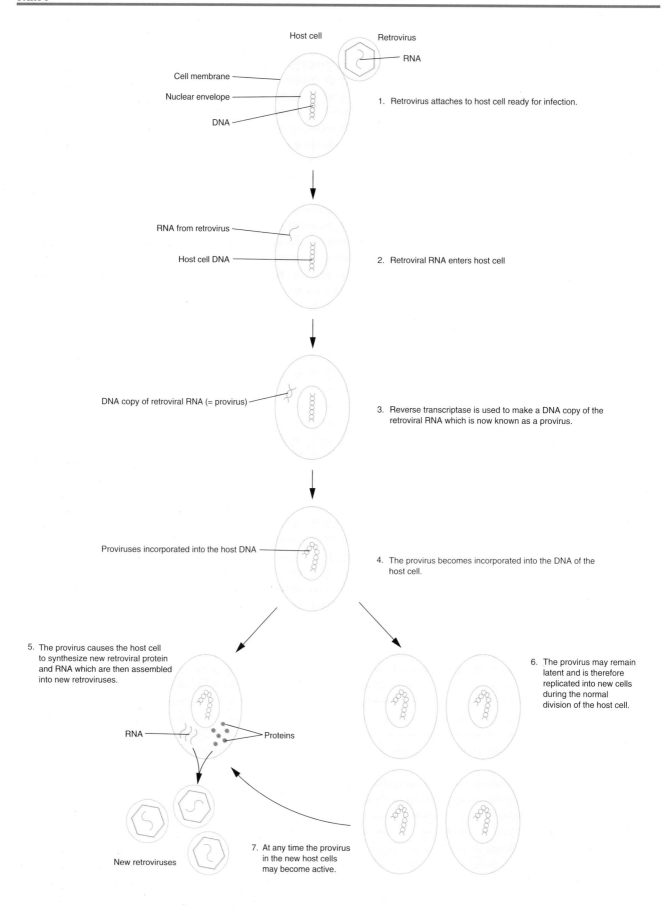

Fig. 5.4 Life cycle of a retrovirus

Viral diseases are often difficult to treat because antibiotics cannot be used. Vaccines may be produced but these are not always effective because one virus may exist in a variety of forms. Methods of control therefore depend primarily on prevention, such as breeding resistant species, removal of the source of infection and the protection of susceptible plants and animals.

Viral diseases of plants include those caused by TMV, potato virus X, barley yellow dwarf virus and turnip yellow mosaic virus.

Retroviruses cause a number of diseases including a degenerative brain disease in sheep, anaemia in cattle and some cancers, but by far the most important one is AIDS caused by the HIV retrovirus. It may, however, prove possible to use retroviruses to cure diseases by utilizing them to insert useful genes into cells where particular genes are defective. Inherited diseases such as phenylketonuria and thalassaemia are the most likely to be cured by this means.

5.3   Prokaryotae

The cyanobacteria (blue-green bacteria) and bacteria which comprise the Prokaryotae are the only living prokaryotic organisms. As such they are the living organisms which most closely resemble the first forms of life. The differences between prokaryotic and eukaryotic cells are given in Section 4.1.2.

No Prokaryotae are truly multicellular although the blue-green bacteria are commonly found in filaments and clusters. This is either because their cell walls fail to separate completely at cell division or because they are held together by a mucilagenous sheath. Most blue-green bacteria can photosynthesize and many are capable of nitrogen fixation. They are important colonizers of bare land and were probably among the first organisms to evolve.

5.3.1 Bacteria

Bacteria are the smallest cellular organisms and are the most abundant.

Fig. 5.5 shows the structure of a typical bacterial cell. Such cells may vary in the nature of the cell wall. In some forms the glyco-protein is supplemented by large molecules of lipopolysaccharide. Cells which lack the lipopolysaccharide combine with dyes like gentian violet and are said to be **gram positive**. Those with the lipopolysaccharide are not stained by gentian violet and are said to be **gram negative**. Gram positive bacteria are more susceptible to both antibiotics and lysozyme than are gram negative ones. Bacteria may be coated with a slime capsule which is thought to interfere with phagocytosis by the white blood cells. Bacteria are generally distinguished from each other by their shape. Spherical ones are known as **cocci** (singular – coccus), rod-shaped as **bacilli** (singular – bacillus) and spiral ones as **spirilla** (singular – spirillum).

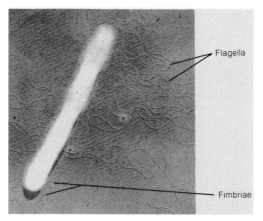

A rod bacterium showing flagella (EM) (×1000 approx.)

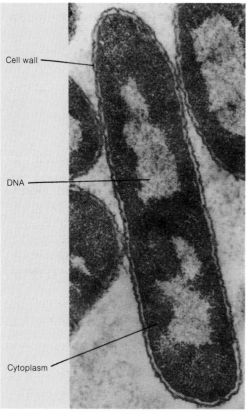

E. coli (EM) (×38 000 approx.)

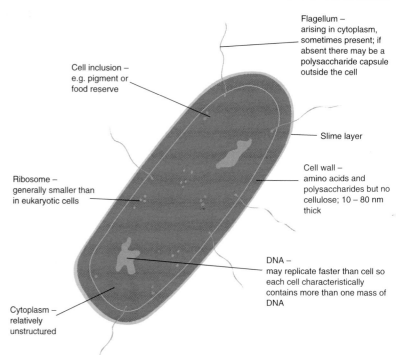

Fig. 5.5 Generalized bacterial cell

Cocci may stick together in chains – **streptococcus**, or in clusters – **staphylococcus**. Bacteria show considerable diversity in their metabolism. The majority are heterotrophic and most of these are saprobionts. They are responsible, with the Fungi, for decaying and recycling organic material in the soil. Others are parasitic, some causing disease but many having little effect on their host. Numerous gut bacteria have a symbiotic relationship with their host, for example helping to digest the cellulose ingested by ruminants.

Other bacteria are autotrophic. **Photosynthetic bacteria** are anaerobic and often use sulphur compounds as electron donors rather than the water used by higher plants.

$$CO_2 + 2H_2S \xrightarrow{\text{light}} (CH_2O) + H_2O + 2S$$

Some bacteria derive their energy from inorganic molecules such as ammonia, nitrite, sulphur or hydrogen sulphide. These are the **chemosynthetic bacteria**, some of which are essential links in the nitrogen cycle. For example, one group oxidizes ammonia or ammonium compounds to nitrites and energy, and another oxidizes nitrites to nitrates and energy.

$$2NH_4 + 3O_2 \longrightarrow 2NO_2^- + 4H^+ + 2H_2O + \text{energy} \ (\textit{Nitrosomonas})$$

$$2NO_2^- + O_2 \longrightarrow 2NO_3^- + \text{energy} \ (\textit{Nitrobacter})$$

Bacteria reproduce by binary fission, one cell being capable of giving rise to over 4×10^{21} cells in 24 hours. Under certain circumstances conjugation occurs and new combinations of genetic material result. Bacteria may also produce thick-walled spores which are highly resistant, often surviving drought and extremes of temperature.

5.3.2 Economic importance of bacteria

It is easy to think of all bacteria as pathogens but it is important to remember that many are beneficial to humans. These benefits include:

1. The breakdown of plant and animal remains and the recycling of nitrogen, carbon and phosphorus.

2. Symbiotic relationships with other organisms. For example supplying vitamin K and some of the vitamin B complex in humans, breaking down cellulose in herbivores.

3. Food production, e.g. some cheeses, yoghurts, vinegar.

4. Manufacturing processes, e.g. making soap powders, tanning leather and retting flax to make linen.

5. They are easily cultured and may be used for research, particularly in genetics. They are also used for making antibiotics, amino acids, enzymes and SCP (single cell protein).

Further details of how humans exploit these beneficial uses of bacteria are given in Chapter 30.

Detrimental effects of bacteria include deterioration of stored food and damage to buried metal pipes caused by sulphuric acid production by *Thiobacillus* and *Desulphovibrio*.

5.4 Fungi

The Fungi are a large group of organisms composed of about 80 000 named species. For many years they were classified with the plants but are now recognized as a separate kingdom. This separation is based on the presence of the polysaccharide chitin found in their cell walls, rather than the cellulose present in plant cell walls. Their bodies are usually a **mycelium** of thread-like multinucleate **hyphae** without distinct cell boundaries. The Fungi lack chlorophyll and are therefore unable to photosynthesize. They feed heterotrophically, generally as saprobionts or parasites, and details of these methods of feeding will be found in Chapter 15.

Within this kingdom there are three phyla, Zygomycota, Ascomycota and Basidiomycota (see Table 5.2).

5.4.1 Economic importance of fungi

Many fungi are beneficial to humans. Examples include:

1. Decomposition of sewage and organic material in the soil.

2. Production of antibiotics, notably from *Penicillium* and *Aspergillus*.

3. Production of alcohol for drinking and industry.

4. Production of other foods. Citric acid for lemonade is produced by the fermentation of glucose by *Aspergillus*. Yeasts are used in bread production and the food yeast *Candida utilis* has been investigated as a source of single cell protein (SCP).

5. Experimental use, especially for genetic investigations.

PROJECT

The fungus *Rhytisma acerinum* grows on sycamore leaves where it forms 'tar spots'

1. Select a particular sycamore tree when the leaves first appear in the spring and which are developing tar spots.

2. From this time until the leaves fall in the autumn, map the distribution of the fungus in the tree at regular intervals to determine the spread of *Rhytisma*.

TABLE 5.2 **Classification of the fungi**

KINGDOM FUNGI	No chlorophyll; do not photosynthesize Heterotrophic Cell walls contain chitin rather than cellulose Body usually a mycelium Carbohydrate stored as glycogen Reproduce by means of spores without flagella		
Zygomycota	**Ascomycota**		**Basidiomycota**
No septa in hyphae; large branched mycelium formed	Septa in hyphae		Septa in hyphae; large 3-dimensional structures often formed
Asexual reproduction by sporangia producing spores or by conidia	Asexual reproduction by conidia		Asexual reproduction unusual but spores formed
Conjugation gives rise to a zygospore	Sexual reproduction by ascospores forming in an ascus		Sexual reproduction by formation of basidiospores outside basidia
e.g. Mucor – pin mould *Rhizopus* – bread mould (See Fig. 5.6)	*e.g. Saccharomyces* – yeast *Erysiphe* – powdery mildew *Aspergillus* and *Penicillium* – saprophytic moulds (See Fig. 5.7)		*e.g. Agaricus campestris* – field mushroom *Coprinus* – ink cap toadstool (See Fig. 5.9)

Many fungi are also harmful to humans, causing decomposition of stored foods and deterioration of natural materials such as leather and wood. Fungi more commonly cause disease in plants than in animals but some of the plants infected are of great economic importance to humans. Powdery mildew, caused by *Erysiphe graminae*, causes serious damage to cereal crops.

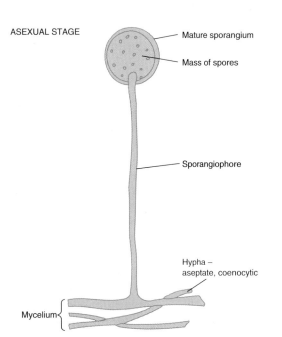

ASEXUAL STAGE

- Mature sporangium
- Mass of spores
- Sporangiophore
- Hypha – aseptate, coenocytic
- Mycelium

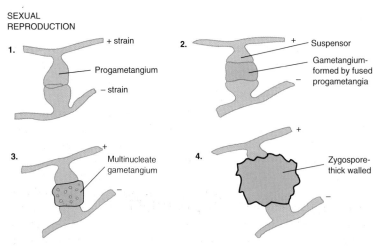

SEXUAL REPRODUCTION

1. + strain
   - Progametangium
   - − strain
2. +
   - Suspensor
   - Gametangium- formed by fused progametangia
   −
3. +
   - Multinucleate gametangium
   −
4. +
   - Zygospore- thick walled
   −

Fig. 5.6 Mucor

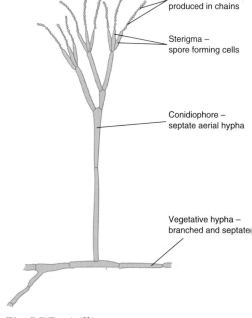

- Conidia – produced in chains
- Sterigma – spore forming cells
- Conidiophore – septate aerial hypha
- Vegetative hypha – branched and septate

Fig. 5.7 Penicillium

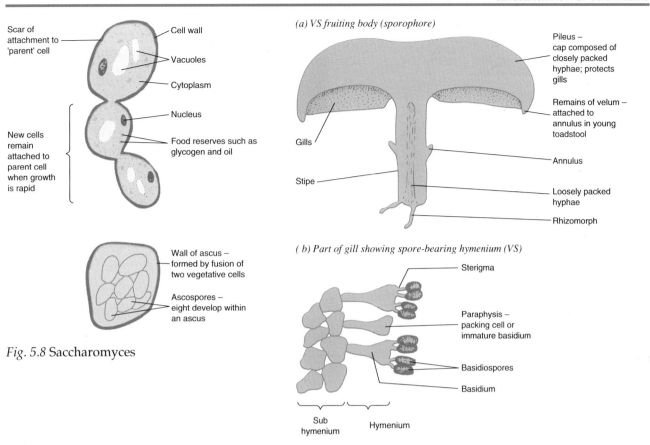

Scar of attachment to 'parent' cell

Cell wall

Vacuoles

Cytoplasm

Nucleus

New cells remain attached to parent cell when growth is rapid

Food reserves such as glycogen and oil

(a) VS fruiting body (sporophore)

Pileus – cap composed of closely packed hyphae; protects gills

Remains of velum – attached to annulus in young toadstool

Gills

Stipe

Annulus

Loosely packed hyphae

Rhizomorph

Wall of ascus – formed by fusion of two vegetative cells

Ascospores – eight develop within an ascus

(b) Part of gill showing spore-bearing hymenium (VS)

Sterigma

Paraphysis – packing cell or immature basidium

Basidiospores

Basidium

Sub hymenium

Hymenium

Fig. 5.8 Saccharomyces

Fig. 5.9 Agaricus

5.5 Protoctista

The kingdom Protoctista is made up of single celled eukaryotic organisms or organisms which are assemblages of similar cells. Apart from this one common feature the kingdom is very varied and includes all nucleated algae, all protozoa and slime moulds. In this section the algae will be used to illustrate the group.

5.5.1 Algae

This is a collective name for a varied group of phyla with no one diagnostic feature. They are normally aquatic or live in damp terrestrial habitats. Sub-divisions are mainly associated with biochemical differences related to photosynthesis.

The **Chlorophyta** are green algae which range in form from unicells such as *Chlamydomonas* and *Chlorella* through colonies like *Volvox* and filaments like *Spirogyra* to delicate thalloid genera like *Ulva*. They contain the same photosynthetic pigments as higher plants but the chloroplasts which contain them vary. *Chlamydomonas* has a single bowl-shaped chloroplast and that of *Spirogyra* is spiral. Both have starch deposits called **pyrenoids**. *Chlamydomonas* also has a light-sensitive spot and will swim, by means of flagella, towards the light. Both genera are capable of asexual and sexual reproduction.

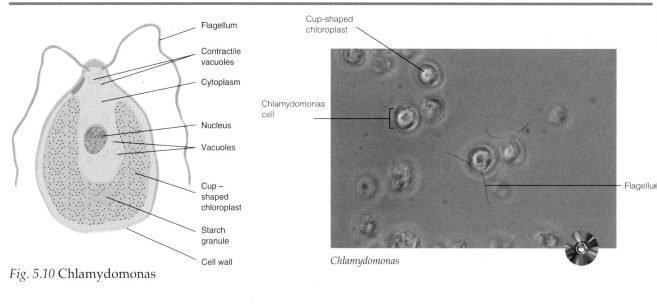

Fig. 5.10 Chlamydomonas

Chlamydomonas

Fig. 5.11 Spirogyra

The Phaeophyta are a phylum which shows great diversity in structure and method of reproduction. It includes all the larger seaweeds as well as small, branched filamentous ones such as *Ectocarpus*. Genera like *Fucus* are well adapted for life in the intertidal zone where they are frequently buffeted by waves and may be exposed at low tide. Both asexual and sexual reproduction are shown, although the latter is unusual in *Fucus*.

TABLE 5.3   **Classification of the algae**

ALGAE	No stems, roots or leaves No sclerenchyma No vascular tissue No archegonia Other photosynthetic pigments, in addition to chlorophyll a	
Chlorophyta (green algae)	**Phaeophyta (brown algae)**	
Chlorophyll a and b present	Chlorophyll a and c, xanthophylls (e.g. fucoxanthin)	
Food reserve is starch	Food reserves include mannitol and laminarin	
Cellulose cell walls	Cell wall includes alginic acid	
Unicellular, filamentous or thalloid	No unicellular forms	
Mostly fresh-water	Almost entirely marine	

e.g. *Chlamydomonas*
   Chlorella
   Pleurococcus
   Spirogyra
	e.g. *Ascophyllum*
	   Fucus	

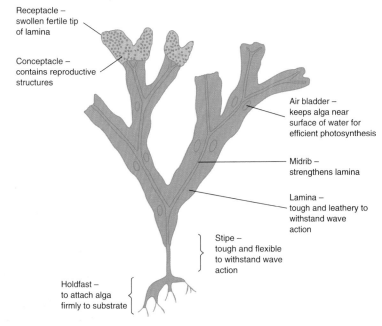

Fig. 5.12 Fucus

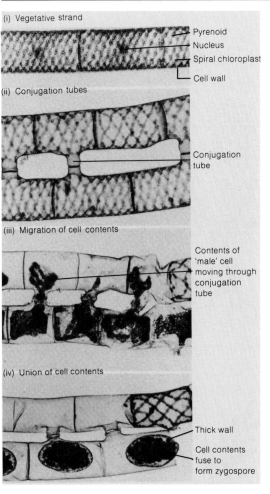

(i) Vegetative strand

— Pyrenoid
— Nucleus
— Spiral chloroplast
— Cell wall

(ii) Conjugation tubes

Conjugation tube

(iii) Migration of cell contents

Contents of 'male' cell moving through conjugation tube

(iv) Union of cell contents

— Thick wall

Cell contents fuse to form zygospore

Spirogyra (×500 approx.) – showing conjugation

5.5.2 Economic importance of algae

At least half the carbon fixation of the earth is carried out by algae in the surface layers of oceans. This primary production is at the base of all aquatic food chains. These algae are also responsible for half the oxygen released by plants into the atmosphere.

Algae can be used in some parts of the world as a direct food source for humans and they may be used as fertilizers on coastal farms. Unicellular green algae such as *Chlorella* are easy to cultivate and can be used as a source of a single cell protein (SCP) for human and animal consumption.

Green algae provide oxygen for the aerobic bacteria which break down sewage.

Derivatives of alginic acid found in the cell walls of many brown algae are non-toxic and readily form gels. These alginates are used as thickeners in many products including ice cream, hand cream, polish, medicine, paint, ceramic glazes and confectionery.

Excessive numbers of algae may develop in bodies of water following pollution by fertilizers or other chemicals. These 'blooms' cause the water to smell and taste unpleasant and may lead to oxygen depletion and the death of fish (Section 18.5.3).

5.6 Plantae

The organisms included in this kingdom are made up of more than one eukaryotic cell, have cell walls containing cellulose and photosynthesize using chlorophyll as the main pigment.

5.6.1 Bryophyta

The mosses and liverworts which make up the Bryophyta are small plants generally found in moist terrestrial habitats. They have no roots and no vascular tissue. They all show alternation of generations in which the sporophyte and gametophyte are almost equally conspicuous, although the sporophyte is attached to, and dependent on, the gametophyte throughout its life. Although it is thought that bryophytes arose from green algae and colonized land over 400 million years ago, they are still very dependent on water for their existence.

5.6.2 Filicinophyta (ferns)

Ferns have large leaves called fronds which are coiled in bud. Most living ferns are quite small and have no direct economic importance to humans, although they are significant

Funaria with capsules

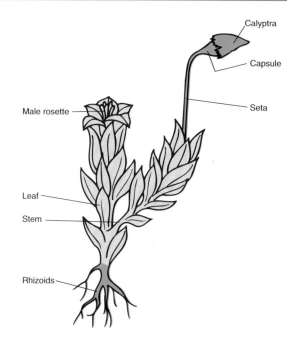

Fig. 5.13 External features typical of a moss

Dryopteris

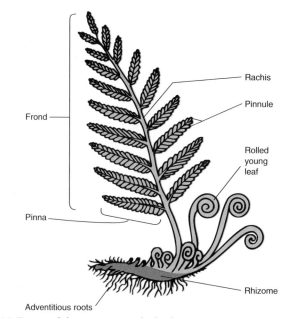

Fig. 5.14 External features typical of a fern

groundcover plants in moist areas. The larger ferns which formed the dominant terrestrial vegetation for about 70 million years from the Devonian to the Permian periods contributed greatly to the coal measures now so useful to humans.

5.6.3 Angiospermophyta

Angiosperms form the dominant terrestrial vegetation today. They are found in a wide range of habitats and have even re-established themselves in fresh water and the sea. Their evolution has closely paralleled that of the insects on which many species depend for pollination. They are extremely well suited to life on land both in their morphology, e.g. efficient water-carrying xylem vessels, and in their reproduction, e.g. seeds enclosed in an ovary.

The two angiosperm classes, **monocotyledoneae** and **dicotyledoneae**, differ in a number of respects, the most significant of which are shown in Table 5.4.

TABLE 5.4 **Comparison of monocotyledoneae and dicotyledoneae**

Monocotyledoneae	Dicotyledoneae
Embryo has one cotyledon	Embryo has two cotyledons
Narrow leaf with parallel venation	Broad leaf with net-like venation
Scattered vascular bundles in stem	Ring of vascular bundles in stem
Rarely vascular cambium present and normally no secondary growth	Vascular cambium present which can lead to secondary growth
Many xylem groups in root	Few xylem groups in root
Flower parts usually in threes	Flower parts usually in fours or fives
Calyx and corolla not easily distinguishable	Usually distinct calyx and corolla
Often wind pollinated	Often insect pollinated
e.g. *Avena* – oats *Iris* *Triticum* – wheat *Lilium* – lily	e.g. *Ranunculus* – buttercup *Lamium* – nettle *Cheiranthus* – wallflower *Bellis* – daisy

5.7   Animalia (animals)

The organisms included in this kingdom are non-photosynthetic multicellular organisms with nervous coordination.

5.7.1 Phylum Cnidaria

Animals in this phylum are all aquatic and predominantly marine. They are **diploblastic** with a body wall composed of two cell layers, an inner **gastrodermis** and an outer **epidermis**. Between these lies a jelly-like **mesogloea** which may contain cells derived from the other two layers. The body is organized around a central cavity, the **gastrovascular cavity**, which has one opening serving as both mouth and anus. This opening is surrounded by tentacles bearing specialized stinging cells or **nematocysts**.

The whole body is radially symmetrical and occurs in two main forms: a jellyfish-like medusoid phase and a hydroid or polyp phase. When the same organism can exist in a number of morphologically distinct forms it is said to show **polymorphism**. Some species have both medusoid and polyp phases in their life history, others only one.

Did you know?

Ninety-nine per cent of all the animal species that have ever lived are now extinct.

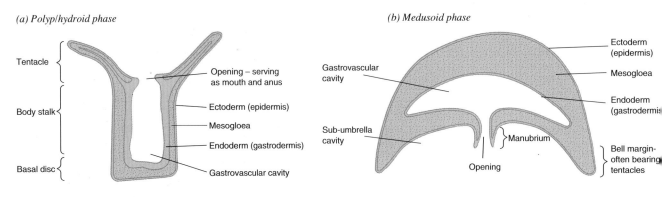

(a) Polyp/hydroid phase

Tentacle

Body stalk

Basal disc

Opening – serving as mouth and anus

Ectoderm (epidermis)

Mesogloea

Endoderm (gastrodermis)

Gastrovascular cavity

(b) Medusoid phase

Gastrovascular cavity

Sub-umbrella cavity

Manubrium

Opening

Ectoderm (epidermis)

Mesogloea

Endoderm (gastrodermis)

Bell margin – often bearing tentacles

Fig. 5.15 A comparison of hydroid and medusoid phases

TABLE 5.5 **Classification of the Cnidaria**

PHYLUM CNIDARIA		Diploblastic Single body cavity with one opening surrounded by tentacles Radial symmetry Polymorphism with free-swimming medusoid and/or sedentary polyps Nematocysts Planula larva	
Class	**Hydrozoa**	**Scyphozoa**	**Anthozoa**
	Dominant polyp	Reduced polyp	Only polyp
	Reduced medusa	Dominant medusa	No medusa
	No mesenteries (divisions in gastrovascular cavity)	Mesenteries in young polyp	Large mesenteries
Examples	*Obelia* – colonial; marine *Hydra* – solitary; freshwater *Physalia* – Portuguese man-of-war	*Aurelia* – solitary; free-swimming marine jellyfish	*Actinia* – solitary; marine sea anemone *Corallium* – colonial; marine coral

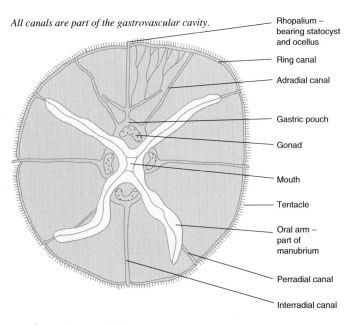

All canals are part of the gastrovascular cavity.

Rhopalium – bearing statocyst and ocellus

Ring canal

Adradial canal

Gastric pouch

Gonad

Mouth

Tentacle

Oral arm – part of manubrium

Perradial canal

Interradial canal

Fig. 5.16 Aurelia – oral view

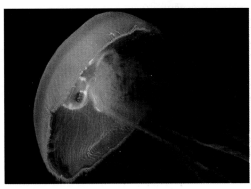

Aurelia aurita (jellyfish)

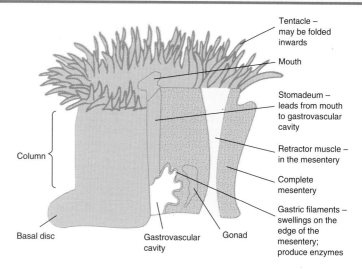

Fig. 5.17 Actinia – *cut to show part of the internal structure*

5.7.2 Phylum Platyhelminthes

The Platyhelminthes are a group of flatworms with a definite head region and bilateral symmetry. They are **triploblastic** with a body wall composed of an outer epidermis and an inner gastrodermis separated by a relatively undifferentiated region of mesoderm called the **mesenchyme**. The phylum contains around 14 000 different species many of which are parasitic and of great economic importance.

The excretory and reproductive systems are very diffuse, and cover the digestive system.

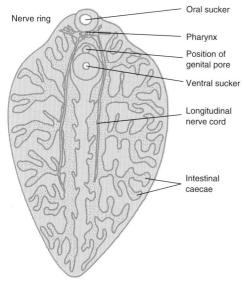

ig. 5.18 Fasciola

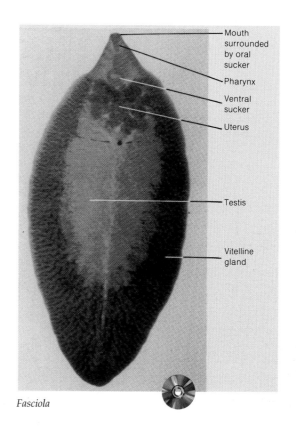

Fasciola

TABLE 5.6 **The parasitic flatworms**

Class	Parasite	Primary host	Intermediate host	Harm caused to primary host
Trematoda	*Fasciola hepatica* – liver-fluke	Sheep and cattle	Snails	Liver rot
	Clonorchis sinensis – Chinese liver-fluke	Humans	Aquatic snails and freshwater fish	Damage to liver where they feed on blood; large numbers block bile ducts
	Schistosoma – blood-fluke	Humans	Freshwater snails	Bilharzia (schistosomiasis); damage to lungs and liver and localized swellings; hepatitis
Cestoda	*Taenia solium* – pork tapeworm	Humans	Pig	Anaemia; diarrhoea; loss of weight; intestinal pains, heavy infestations may block gut and its associated ducts
	Diphyllobothrium latum – fish tapeworm	Humans and carnivores	Copepod (crustacean) and freshwater fish	
	Echinococcus granulosus	Humans, sheep and cattle	Dog	Hydatid cysts – 70% in liver, 20% in lungs and rest elsewhere

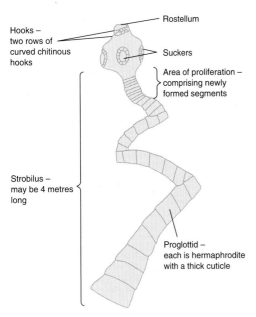

Fig. 5.19 Taenia

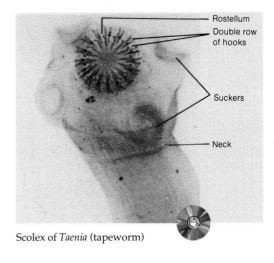

Scolex of *Taenia* (tapeworm)

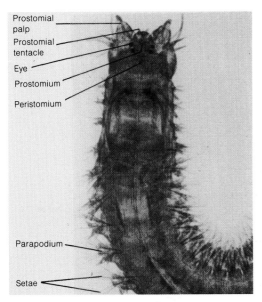

Nereis

5.7.3 Phylum Annelida

The annelids or 'true worms' are coelomate animals showing metameric segmentation. There are about 9000 species living in the sea, fresh water or moist soil.

Economically the most significant annelids are the earthworms which contribute to soil formation and improvement in the following ways:

1. Tunnels improve aeration and drainage.

2. Dead vegetation is pulled into the soil where decay by saprobionts takes place.

3. Mixing of soil layers.

4. Addition of organic matter by excretion and death.

5. Secretions of gut neutralize acid soils.

6. Improving tilth by passing soil through gut.

Annelids in general contribute to food chains and leeches used to be of medical importance. None of the parasites cause major infestations of humans or of domesticated animals.

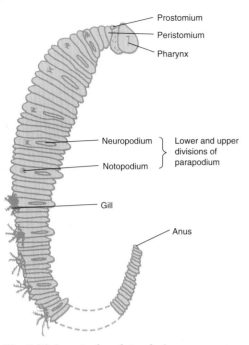

Fig. 5.20 Arenicola – *lateral view*

TABLE 5.7   **Classification of the Annelida**

PHYLUM ANNELIDA		Metameric segmentation Non-chitinous cuticle Chaetae (bristles)	
Class	**Polychaeta**	**Oligochaeta**	**Hirudinea**
	Parapodia	No parapodia	No parapodia
	Many chaetae	Few chaetae	No chaetae
	Outer rings correspond to inner septa	Outer rings correspond to inner septa	Outer rings more numerous than inner septa
	Distinct head	No distinct head	No distinct head
	No suckers	No suckers	Suckers (ectoparasite)
	Separate sexes	Hermaphrodite	Hermaphrodite
	Larvae	No larvae	No larvae
Examples	*Nereis* – ragworm *Arenicola* – lugworm (See Fig. 5.20)	*Lumbricus* – earthworm (See Fig. 5.21)	*Hirudo* – leech (See Fig. 5.22)

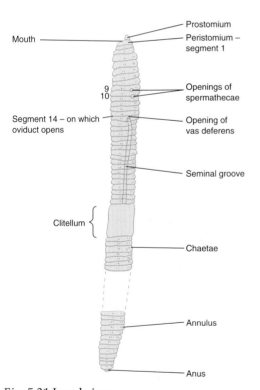

Fig. 5.21 Lumbricus

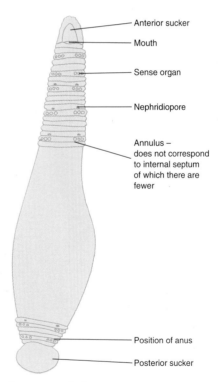

Fig. 5.22 Hirudo – *ventral view*

5.7.4 Phylum Arthropoda

Arthropods make up about three quarters of living animal species. They have adapted to live successfully in both aquatic and terrestrial habitats and may be free-living or parasitic. **Trilobites** are an extinct group of arthropods, nearly 4000 species of which have been described from fossils. They became extinct at the end of the Palaeozoic era.

Because of the size and importance of this phylum it will be studied a class at a time.

TABLE 5.8   **Phylum Arthropoda**

PHYLUM ARTHROPODA	Exoskeleton, mainly comprising a chitinous cuticle Jointed appendages Dorsal heart and open blood system Growth in stages after moulting **(ecdysis)**

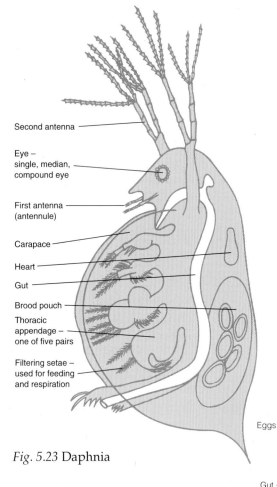

Second antenna

Eye –
single, median,
compound eye

First antenna
(antennule)

Carapace

Heart

Gut

Brood pouch

Thoracic
appendage –
one of five pairs

Filtering setae –
used for feeding
and respiration

Fig. 5.23 Daphnia

Crustacea
There are about 26 000 species known. They are extremely abundant and many are relatively large.

TABLE 5.9   **Classification of the Crustacea**

Superclass Crustacea	Cephalothorax – formed by fusion of head and thorax Strong exoskeleton – impregnated with calcium carbonate Two pairs of antennae Three pairs of mouthparts	
Class	**Branchiopoda**	**Malacostraca**
	One pair of compound eyes which may be fused to form a single eye	One pair of stalked compound eyes
	Thoracic appendages with bristles for filter feeding	Eight pairs of thoracic appendages for walking and feeding
	No abdomen	Abdomen with appendages for swimming
	Body enclosed in carapace of two pieces	Carapace covers thorax
Examples	*Daphnia* – water flea (See Fig. 5.23)	*Carcinus* – crab *Astacus* – crayfish *Oniscus* – woodlouse *Leander* – prawn

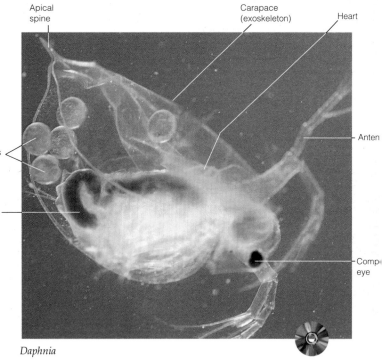

Apical spine

Carapace (exoskeleton)

Heart

Anten

Eggs

Gut

Comp
eye

Daphnia

Insecta
The Insecta comprise more than 750 000 species which show great diversity. They are extremely successful terrestrial animals in terms of numbers of species, individuals and habitats.

There are a few insects such as springtails and silverfish which do not develop wings. The remainder may be subdivided as shown in Table 5.11.

TABLE 5.10   **The Insecta**

Class Insecta	Body comprises head, thorax and abdomen Three pairs of thoracic legs Two pairs of thoracic wings (except Apterygota) – sometimes modified One pair of antennae One pair of compound eyes Respiration by tracheae

Cockroach

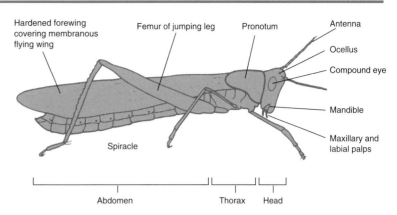

Fig. 5.24 Locust

TABLE 5.11 **The winged insects**

	Exopterygota	Endopterygota
	Wings develop externally Metamorphosis incomplete (**hemimetabolous**) Egg → nymphal stages → adult	Wings develop internally Metamorphosis complete (**holometabolous**) Egg → larval stages → pupa → adult
Order	Orthoptera, e.g. *Locusta* – locust (See Fig. 5.24) Odonata, e.g. *Libellula* – dragonfly Dictyoptera, e.g. *Periplaneta* – cockroach	Lepidoptera, e.g. *Pieris* – cabbage white butterfly Diptera, e.g. *Musca* – house-fly Hymenoptera, e.g. *Apis* – honey-bee

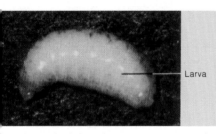

Apis (honey-bee), brood stages

Apis adults – Queen (*left*), drone (*middle*) and worker (*right*)

5.7.5 Phylum Mollusca

This phylum is the second largest in the animal kingdom, comprising about 100 000 living species. There is also a very long fossil record of molluscs stretching back to the Pre-Cambrian period. Most living species are marine and examples include octopus and squid as well as members of the classes shown in Table 5.12.

TABLE 5.12   **Classification of the Mollusca**

PHYLUM MOLLUSCA	Body divided into head, muscular foot and visceral mass Mantle may secrete a calcareous shell Bilateral symmetry, but torsion may lead to asymmetry	
Class	**Gastropoda**	**Pelycopoda (Bivalves)**
	Asymmetrical (due to torsion of visceral mass)	Bilateral symmetry
	Single shell, usually coiled	Shell in two valves
	Well developed head with tentacles and eyes	Reduced head and no tentacles
	Radula (rasping tongue) for feeding	Gills with cilia used for filter feeding
Examples	*Littorina* – winkle *Limax* – slug *Helix* – snail (See Fig. 5.25)	*Mytilus* – mussel *Anodonta* – freshwater mussel (See Fig. 5.26)

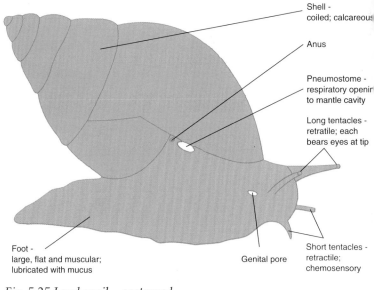

Fig. 5.25 Land snail – gastropod

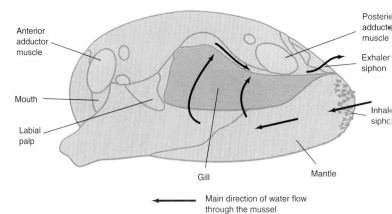

Fig. 5.26 Anodonta – viewed from left side with left valve and one gill removed

5.7.6 Phylum Chordata

This phylum includes the vertebrates which have evolved during the past 500 million years to become the dominant animals of land, sea and air. In number of species and individuals they do not rival the arthropods but their biomass and ecological dominance are much greater.

TABLE 5.13 **The chordata**

PHYLUM CHORDATA	Gill-slits present in pharnyx Post-anal tail, at some stage in development Notochord Dorsal, tubular nerve cord
SUB-PHYLUM VERTEBRATA (Craniata)	Well developed head with brain encased in cranium Vertebral column replaces notochord

PROJECT

Design a simple dichotomous key for identifying common animals that you found, for example:

(a) in leaf litter
(b) on a rocky shore
(c) under stones
(d) in a fluorescent lamp cover, etc.

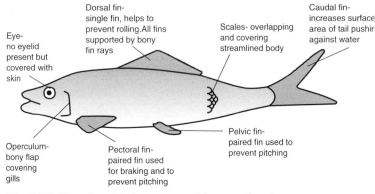

Fig. 5.27 Class Osteichthyes, genus Clupea – herring

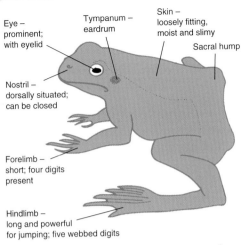

Eye – prominent; with eyelid

Tympanum – eardrum

Skin – loosely fitting, moist and slimy

Sacral hump

Nostril – dorsally situated; can be closed

Forelimb – short; four digits present

Hindlimb – long and powerful for jumping; five webbed digits

Fig. 5.28 Class Amphibia, genus Rana *– frog*

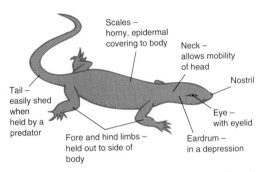

Scales – horny, epidermal covering to body

Neck – allows mobility of head

Nostril

Tail – easily shed when held by a predator

Fore and hind limbs – held out to side of body

Eye – with eyelid

Eardrum – in a depression

Fig. 5.29 Class Reptilia, genus Lacerta *– lizard*

TABLE 5.14   **Classification of the vertebrata**

Class	Characteristics	Examples		
Chondrichthyes	Cartilaginous endoskeleton No operculum Heterocercal tail fin No swim bladder	*Scyliorhinus* – dogfish *Raja* – ray		
Osteichthyes	Bony endoskeleton Bony scales Operculum over gills Homocercal tail fin Swim bladder present	*Clupea* – herring *Gasterosteus* – stickleback *Salmo* – salmon		
Amphibia	No scales Tympanum (eardrum) visible Lungs in adult Aquatic larvae Metamorphosis	*Rana* – frog *Bufo* – toad *Triturus* – newt		
Reptilia	Dry skin with horny scales Teeth – all the same type (homodont) Eggs with yolk and leathery shell No gills No larval stages	*Lacerta* – lizard *Natrix* – grass snake *Chelonia* – turtle *Crocodilus* – crocodile		
Aves	Endothermic (warm-blooded) Feathers Beak (no teeth) Forelimbs modified into wings Air sacs in light bones	*Columba* – pigeon		
Mammalia	Endothermic Hair Sweat and sebaceous glands Mammary glands Pinna (external ear) Heterodont (different types of teeth) Diaphragm	Order	Insectivora Carnivora Cetacea Chiroptera Rodentia Primates	*Talpa* – mole *Canis* – dog *Delphinus* – dolphin *Desmodus* – vampire bat *Rattus* – rat *Pan* – chimpanzee

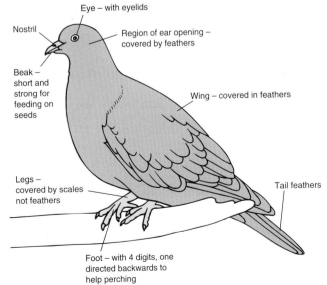

Nostril

Eye – with eyelids

Region of ear opening – covered by feathers

Beak – short and strong for feeding on seeds

Wing – covered in feathers

Legs – covered by scales not feathers

Tail feathers

Foot – with 4 digits, one directed backwards to help perching

Fig. 5.30 Class Aves, genus Columba *– pigeon*

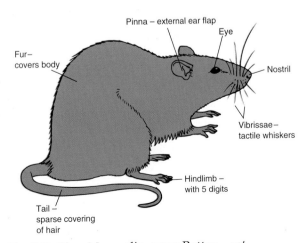

Pinna – external ear flap

Eye

Fur – covers body

Nostril

Vibrissae – tactile whiskers

Tail – sparse covering of hair

Hindlimb – with 5 digits

Fig. 5.31 Class Mammalia, genus Rattus *– rat*

5.8   Questions

1. Three organisms, a clam, a moss and an earthworm, are illustrated below. The completed table will give the name of the organism, the major group, a subgroup and a characteristic of the subgroup for each organism. Complete the table by filling in the blank sections.

Organism	Major group	Subgroup	Characteristic of subgroup
	Bryophyta		
		Chaetopoda/ Oligochaeta	
			Shell in two parts

(Total 6 marks)

ULEAC January 1994, Paper I, No. 1

2. *(a)* Define the term *taxon*.   *(2 marks)*
(b) The following terms, used in classification of organisms, have each been allotted a letter:

Class – C; Family – F; Genus – G; Phylum – P; Order – O; Species – S

In the table below mark, by means of a tick (✓) in the space provided, that sequence which you consider to be the correct hierarchy used in classification.

(✓)

1	G	C	F	P	S	O	
2	C	O	F	G	P	S	
3	P	F	C	O	S	G	
4	P	C	O	F	G	S	
5	C	P	O	G	S	F	
6	C	P	G	O	F	S	

(1 mark)

(c) Using the same letter labelling, assign an appropriate letter to the following:
  (i)   Bryophyta
  (ii) a named fish (e.g. a salmon *Salmo salar*)
  (iii) earthworms (Lumbricidae)
  (iv) Chordata
  (v)   Insecta
  (vi) A snail, e.g. *Cepea*

(6 marks)
(Total 9 marks)

O&C SEB June 1993, Paper 2/3, No. 1

3. The animals shown in the drawings appear radially symmetrical.

Animal A   Animal B

(a) What is meant by *radially symmetrical*?
(1 mark)
(b) Give **two** advantages of radial symmetry to these organisms.   *(2 marks)*
(c) Animal **A** belongs to the phylum Cnidaria ('Coelenterata') and animal **B** to the phylum Annelida. Give **two** characteristics which would enable you to classify each animal correctly.   *(2 marks)*
(Total 5 marks)

AEB June 1990, Paper I, No. 1

4. The diagram below shows a virus attached to a bacterial cell.
(a) Name the component parts where indicated.
(2 marks)

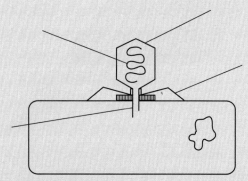

(b) What type of virus is shown? *(1 mark)*

(c) What type of bacterium is indicated? *(1 mark)*

(d) What process has occurred immediately prior to the stage shown in the diagram? *(1 mark)*

(e) Briefly explain what **two** processes will next occur in sequence. *(2 marks)*

(f) (i) Name **two** diseases of human beings caused by viruses. *(1 mark)*

  (ii) Name **one** disease of plants caused by viruses. *(1 mark)*

(g) Viruses affecting plant cells can be spread by insect vectors (e.g. aphids or greenfly). Briefly suggest why such insects are usually necessary if plants are to be infected. *(1 mark)*

(h) One recent 'A' level Biology textbook states, rather dogmatically, that: 'Viruses are the smallest living organisms….' Another book states: '… many biologists have found it impossible to classify viruses as either living or nonliving. The terms simply have no meaning when applied to them (viruses).'

Give (i) one argument in favour of viruses being classified as living organisms, and (ii) one argument in favour of viruses being classified as nonliving.

(2 marks)
(Total 12 marks)

Oxford June 1991, Paper I, No. 3

5. A bacteriophage is a virus which attacks cells of bacteria. The following diagrams represent the life cycle of a bacteriophage.

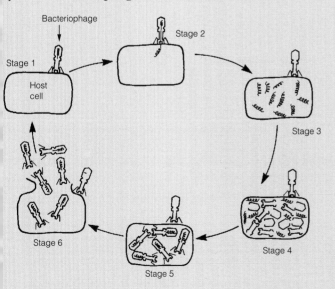

(a) Name the substance which is being introduced into the host cell by the bacteriophage in stage 2. *(1 mark)*

(b) Describe what is happening during stage 3 and stage 4. *(2 marks)*

(c) What term is used to describe the splitting open of the host cell to release new viruses? *(1 mark)*

(Total 4 marks)

SEB (revised) May 1991, Higher Grade Paper I, No. 9

6. The table below refers to some major groups of animals. For each group, state **two** external features which are characteristic of that group.

Group	Characteristic external features	
Cnidarians	1	
	2	
Annelids	1	
	2	
Arthropods	1	
	2	
Molluscs	1	
	2	
Chordates	1	
	2	

(Total 10 marks)

ULEAC June 1993, Paper I, No. 9

7. For each of the descriptions listed below, identify the taxonomic group described.

(a) Unsegmented, soft-bodied animals with calcareous shells *(1 mark)*

(b) Plants with a conspicuous sporophyte generation which has complex leaves bearing sporangia in clusters *(1 mark)*

(c) Radially symmetrical animals that are composed of two cellular layers *(1 mark)*

(d) Vertebrate animals which have aquatic larvae and land-living adults *(1 mark)*

(Total 4 marks)

ULEAC January 1993, Paper I, No. 3

Computer representation of part of a DNA molecule (*opposite*)

Part II

THE CONTINUITY OF LIFE

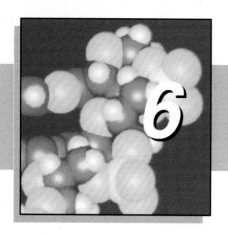

6 Inheritance in context

The obvious similarities between children and their parents, or sometimes their grandparents, have long been recognized. Despite many attempts to explain this phenomenon, it is only in recent years that our knowledge of the process of heredity has enabled us to understand the mechanism fully.

The fact that, in some species, both male and female are needed to produce offspring was realized from early times. The rôle each sex played was, however, a matter of argument. Aristotle believed that the male's semen was composed of an incomplete blend of ingredients which upon mixing with the menstrual fluid of the female, gained form and power and became the new organism. Apart from minor refinements this belief was generally accepted until the seventeenth century. When Anton van Leeuwenhoek observed sperm in human semen, the idea arose that these contained a miniature human. When these sperm were introduced into the female, one would implant in the womb and develop there, the female's rôle being nothing more than a convenient incubator. Around this period Regnier de Graaf discovered in ovaries what was later to be called the Graafian follicle. This was thought by another group of scientists to contain the miniature human, the sperm simply acting as a stimulus for its development.

The problem with both these beliefs was that it could easily be observed that any offspring tended to show characteristics of both parents rather than just one. This led, in the last century, to the idea that both parents contributed hereditary characteristics and the offspring was merely an intermediate blend of both. While closer to present thinking, it too had one flaw. Logically the offspring of a cross between a red flower and a white flower should have pink flowers and the children of a tall father and a short mother should be of medium height. It took the rediscovery of the work of Mendel at the beginning of this century to provide what is now an accepted explanation. Both parents do provide hereditary material within the sperm and ovum. The offspring therefore has two sets of genetic information – one from the mother and one from the father. For any individual characteristic, e.g. eye colour, only one of the two factors expresses itself. An individual with one factor for blue eyes and one for brown eyes will always have brown eyes. A few characters do show an intermediate state between two contrasting factors, but this is relatively rare. Only one of each pair of factors will be present in any one gamete.

Alongside these changes in the last century, another development took place. It was originally believed that new species arose spontaneously in some manner. By the end of the

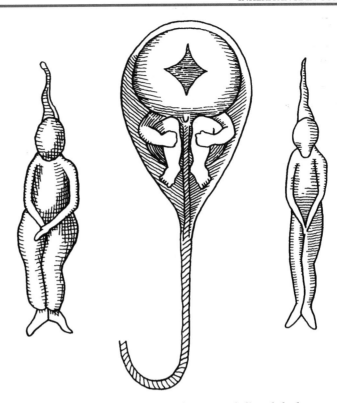

Fig. 6.1 Observers in the seventeenth century believed the human sperm contained a tiny copy of the parent – they thought they could see these down the microscope and called them homunculi

century it was more or less accepted that they were formed by adaptation of existing forms. Natural selection is considered to be the mechanism by which these changes arise and it depends upon there being much variety among individuals of a species. Without this variety and consequent selection of the types best suited to the present conditions, species could not adapt and evolve to meet the changing demands of the environment.

If this theory of evolution is accepted, then the process of inheritance must permit variety to occur. At the same time, if the offspring are to be supplied with the same genetic information as the parents, the genetic material must be extremely stable. This stability is especially important to ensure that favourable characteristics are passed on from one generation to the next. This then is the paradox of inheritance – how to reconcile the genetic stability needed to preserve useful characteristics with the genetic variability necessary for evolution. To satisfy both requirements it is necessary to have hereditary units which are in themselves exceedingly stable, which can be reassorted in an almost infinite variety of ways. The idea can be likened to a pack of playing cards. The cards themselves are stable, fixed units, but the number of different possible combinations in a typical hand of thirteen cards is immense. Imagine how much greater are the possible combinations of the thousands of hereditary units in a typical organism.

A summary of the historical events which contributed to our current understanding of heredity is given in Table 6.1.

TABLE 6.1   **Historical review of events leading to present-day knowledge of reproduction and heredity**

Name	Date	Observation/discovery	Name	Date	Observation/discovery
Aristotle	384–322 BC	Mixing of male semen and female semen (menstrual fluid) was like blending two sets of ingredients which gave 'life'	Morgan	early 1900s	Pioneered use of *Drosophila* in genetics experiments and described linkage
General scientific belief	Up to 17th century	Simple organisms arose spontaneously out of non-living material	Garrod	1908	Postulated mutations as sources of certain hereditary diseases
van Leeuwenhoek	1677	Discovered sperm – it was generally believed that these contained miniature organisms which only developed when introduced into a female	Johannsen	1909	Coined term 'gene' as hereditary unit
			Janssens	1909	Observed chiasmata and crossing over
			Sturtevant	1913	Mapped genes on chromosomes of *Drosophila*
de Graaf	1670s	Described the ovarian follicle (later called Graafian follicle)	Muller	1920s	Observed mutagenic effect of X-rays
Lamarck	1809	Proposed theory of evolution based on inheritance of acquired characteristics	Oparin	1923	Suggested theory of origin of life
			Griffith	1928	Produced evidence suggesting that a chemical 'transforming principle' was responsible for carrying genetic information
Darwin	1859	*On the Origin of Species by Means of Natural Selection* published	Beadle and Tatum	1941	Produced evidence supporting the one gene, one enzyme hypothesis
Pasteur	1864	Experimentally disproved the theory of spontaneous generation	Avery, McCarty and McCleod	1944	Showed nucleic acid to be the chemical which carried genetic information
Mendel	1865	Experiments on the genetics of peas and formulation of his two laws	Hershey and Chase	1952	Showed DNA to be the hereditary material
Hertwig	1875	Witnessed fusion of nuclei during fertilization	Watson and Crick	1953	Formulated the detailed structure of DNA
Flemming	1882	Described all stages of mitosis	Kornberg	1956	Produced DNA copies from single DNA template using DNA polymerase
de Vries	1900	Rediscovery of the significance of Mendel's 1865 experiment			
Sutton	1902	Observed pairing of homologous chromosomes during meiosis and suggested these carried genetic information	Meselsohn and Stahl	1959	Described mechanism of semi-conservative replication in DNA
			Jacob and Monod	1961	Postulated existence of mRNA in theory on control of protein synthesis

7 DNA and the genetic code

7.1 Evidence that the nucleus contains the hereditary material

The universal occurrence of a nucleus at some stage of the life cycle of cells suggests that it performs an essential rôle. The functions of the nucleus are listed in Section 4.2.3. The fundamental rôle of the nucleus in determining the features of a cell was established by Hämmerling. Working with individual cells is normally a difficult task, not least because of their small size. Hämmerling, however, used unusually large single-celled algae belonging to the genus *Acetabularia*. Each cell is up to 5 cm in length, making the sectioning of it relatively easy.

Fig. 7.1, over the page, gives a summary of the experiments using two species of *Acetabularia*, which show the nucleus to contain the hereditary material. The experiments are based on those of Hämmerling although they incorporate some refinements made possible by modern techniques.

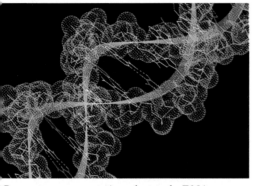

Computer representation of part of a DNA molecule

7.2 Evidence that DNA is the hereditary material

7.2.1 Chromosome analysis

With the nucleus having been shown to contain the hereditary material, attention focussed on determining the precise nature of this material. As **chromosomes** only become visible during cell division, it was hardly surprising that they quickly attracted attention. Chromosomes were shown to be made up of protein and DNA. Of the two, protein was thought a more likely candidate as it was known to be a complex molecule existing in an almost infinite number of forms – a necessary characteristic of a material which must carry an immense diversity of information. Later work showed this not to be the case and research centred on the DNA.

7.2.2 Metabolic stability of DNA

Any material which is responsible for transferring information from one generation to another must be extremely stable. If it were altered to any extent imperfect copies would be made. Unlike protein, DNA shows remarkable metabolic stability. If DNA is labelled with a radioactive isotope it can be shown that its rate of disappearance from the DNA is very slow. This suggests that, once formed, a DNA molecule undergoes little if any alteration.

METHOD	RESULTS	CONCLUSION	EXPLANATION IN LIGHT OF PRESENT KNOWLEDGE
Experiment 1 *A. mediterranea* is cut into two approximately equal halves	The portion without the nucleus degenerates The portion with the nucleus regenerates a new cap of the same type	The information for the regeneration of the cap is contained in, or produced by, the lower portion which contains the nucleus	The DNA in the nucleus produces mRNA which enters the cytoplasm where it provides the instructions for the formation of the enzymes needed in the production of a new cap. In the absence of a nucleus, the upper portion cannot do this
Experiment 2 *A. mediterranea* is cut to isolate the stalk section which does not contain the nucleus	A new cap is regenerated from the stalk section	The information on how to regenerate the cap is present in the stalk	As the nucleus produces a constant supply of mRNA there is sufficient in the cytoplasm of the stalk to provide instructions on how to form the enzymes necessary for the regeneration of the cap
Experiment 3 The regenerated cap from the previous experiment is again removed	The stalk does not regenerate a cap for a second time	The information on how to regenerate the cap, which is contained in the stalk, must be used up and so cannot effect a second regeneration	The mRNA is broken down once its rôle in regenerating the cap is complete. It is therefore not available for the cap to be generated a second time. In the absence of a nucleus, there is no new source of mRNA
Experiment 4 The stalk of *A. mediterranea* is grafted onto the base portion (which contains the nucleus) of *A. crenulata* – a species possessing a different shaped cap	The cap of *A.crenulata* is regenerated	The influence on cap regeneration of the base portion (with nucleus) is greater than the influence of the stalk portion (without the nucleus)	With the nucleus of *A. crenulata* present a constant supply of mRNA is available to regenerate this type of cap. The mRNA from *A.mediterranea* is limited to that present in the stalk when it was separated from its nucleus. The influence of the mRNA from *A. crenulata* is therefore greater
Experiment 5 The nucleus from a decapitated *A. crenulata* is removed and replaced with a transplanted nucleus from *A. mediterranea*	The cap regenerated is of the *A. mediterranea* type	As the only part of *A. mediterranea* which is present is the nucleus, it alone must contain the instruction on how to regenerate the cap	The situation similar to that in experiment 4 except that it is the mRNA of *A. mediterranea* which is present in greater quantities, because its nucleus is present and forms a constant supply of mRNA

Fig. 7.1 Summary of experiments to show that the nucleus contains hereditary material

7.2.3 Constancy of DNA within a cell

Almost all the DNA of a cell is associated with the chromosomes in the nucleus. Small amounts do occur in cytoplasmic organelles such as mitochondria, but this represents a small proportion of the total. Analysis shows that the amount of DNA remains constant for all cells within a species except for the gametes, which have almost exactly half the usual quantity. Prior to cell division the amount of DNA per cell doubles. This is shared equally between the two daughter cells which therefore have the usual quantity. These changes are consistent with those expected of hereditary material which is being transmitted from cell to cell during division.

7.2.4 Correlation between mutagens and their effects on DNA

Mutagens are agents which cause **mutations** in living organisms. A mutation is an alteration to an organism's characteristics which is inherited. Many agents are known mutagens; they include X-rays, nitrous acid and various dyes. It can be shown that these mutagens all alter the structure of DNA in some way. A typical example is ultra-violet light of wavelength 260 nm. It both causes mutations and alters the structure of the pyrimidine bases of which DNA is made. This suggests that it is this alteration of DNA which is the source of the mutation and DNA must therefore be the hereditary material.

7.2.5 Experiments on bacterial transformation

The most convincing evidence for the genetic rôle of DNA was provided by Griffith in 1928. He experimented on the bacterium *Pneumococcus* which causes pneumonia. It exists in two forms:

1. The harmful form – a virulent (disease-causing) type which has a gelatin coat. When grown on agar it produces **shiny, smooth** colonies and is therefore known as the S-strain.

2. The safe form – a non-virulent (does not cause disease) type which does not have a gelatin coat. When grown on agar it produces **dull, rough** colonies and is therefore known as the R-strain.

Griffith's experiments may be summarized thus:

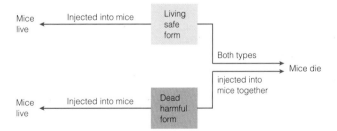

The living safe form and dead harmful form, while not causing pneumonia when injected separately, did so when injected together. The resulting dead mice were found to contain living harmful forms of *Pneumococcus*. If one discounts the improbable explanation that the dead harmful forms have been resurrected, how then could the living safe forms suddenly have acquired the ability to form a gelatin coat, produce smooth colonies and cause pneumonia? It is possible that the safe form had mutated into the harmful form, but this is unlikely. Furthermore, the experiment can be repeated with similar results and the likelihood of the same mutation arising each and every time is so improbable that it can be discounted.

If pneumonia is caused by some toxin produced by *Pneumococcus*, then the harmful type must have the ability to produce it, whereas the safe type does not. The explanation could therefore be that the dead harmful type has the information on how to make the toxin but, being dead, is unable to manufacture it. The safe type, being alive, is potentially able to make the toxin but lacks the information on how to go about it. If then the recipe for the toxin can in some way be transferred from the dead harmful to the living safe variety, the toxin can be manufactured and pneumonia will result. As the substance was able to transform one strain of *Pneumococcus* into another, it became known as the **transforming principle**.

7.2.6 Experiments to identify the transforming principle

The identity of the transforming principle was determined by Avery, McCarty and McCleod in 1944. In a series of experiments they isolated and purified different substances from the dead harmful types of *Pneumococcus*. In turn they tested the ability of each to transform living safe types into harmful ones. Purified DNA was shown to be capable of bringing about transformation, and this ability ceased when the enzyme which breaks down DNA (deoxyribonuclease) was added.

7.2.7 Transduction experiments

In 1952 Hershey and Chase performed a series of experiments involving the bacterium *Escherichia coli* and a bacteriophage (T_2 phage) which attacks it. (Details of a phage life cycle are given in Section 5.2.1.) T_2 phage transfers to *E. coli* the necessary hereditary material needed to make it manufacture new T_2 phage viruses. As the T_2 phage virus is composed of just DNA and protein, one or the other must constitute the hereditary material. Hershey and Chase carefully labelled the protein of one phage sample with radioactive sulphur (^{35}S) and the DNA of another phage sample with radioactive phosphorus (^{32}P). They then separately introduced each sample into a culture of *E. coli* bacteria. At a critical stage, when the viruses had transferred their hereditary material into the bacterial cells, the two organisms were separated mechanically and each culture of bacteria was examined for radioactivity. The culture injected with radioactive DNA contained radioactive bacteria, while that injected with radioactive protein did not. The evidence was conclusive: DNA was the hereditary material; but if further proof were needed this was provided by electron microscope studies which actually traced the movement of DNA from viruses into bacterial cells.

7.3 Nucleic acids

Adenosine monophosphate (adenylic acid)

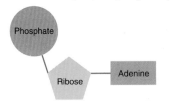

Fig. 7.2 Structure of a typical nucleotide

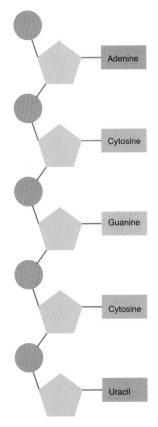

Fig. 7.3 Structure of section of polynucleotide, e.g. RNA

7.3.1 Structure of nucleotides

Individual nucleotides comprise three parts:

1. **Phosphoric acid** (phosphate H_3PO_4). This has the same structure in all nucleotides.

2. **Pentose sugar.** Two types occur, ribose ($C_5H_{10}O_5$) and deoxyribose ($C_5H_{10}O_4$).

3. **Organic base.** There are five different bases which are divided into two groups, described on the next page.

 (*a*) **Pyrimidines** – these are single rings each with six sides. Examples found in nucleic acids are: **cytosine**, **thymine** and **uracil**.

 (*b*) **Purines** – these are double rings comprising a six-sided and a five-sided ring. Two examples are found in nucleic acids: **adenine** and **guanine**.

The three components are combined by condensation reactions to give a nucleotide, the structure of which is shown in Fig. 7.2. By a similar condensation reaction between the sugar and phosphate groups of two nucleotides, a **dinucleotide** is formed. Continued condensation reactions lead to the formation of a **polynucleotide** (Fig. 7.3).

The main function of nucleotides is the formation of the nucleic acids **RNA** and **DNA** which play vital rôles in protein synthesis and heredity. In addition they form part of other metabolically important molecules. Table 7.1 gives some examples.

TABLE 7.1 **Biologically important molecules containing nucleotides, and their functions**

Molecule	Abbreviation	Function
Deoxyribonucleic acid	DNA	Contains the genetic information of cells
Ribonucleic acid	RNA	All three types play a vital rôle in protein synthesis
Adenosine monophosphate Adenosine diphosphate Adenosine triphosphate	AMP ADP ATP	Coenzymes important in making energy available to cells for metabolic activities, osmotic work, muscular contractions, etc.
Nicotinamide adenine dinucleotide Flavine adenine dinucleotide	NAD FAD	Electron (hydrogen) carriers important in respiration in transferring hydrogen atoms from the Krebs cycle along the respiratory chain
Nicotinamide adenine dinucleotide phosphate	NADP	Electron (hydrogen) carrier important in photosynthesis for accepting electrons from the chlorophyll molecule and making them available for the photolysis of water
Coenzyme A	CoA	Coenzyme important in respiration in combining with pyruvate to form acetyl coenzyme A and transferring the acetyl group into the Krebs cycle

NAME OF MOLECULE	CHEMICAL STRUCTURE	REPRESENTATIVE SHAPE
Phosphate	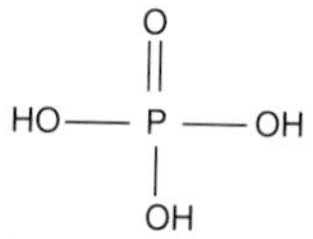	
Ribose	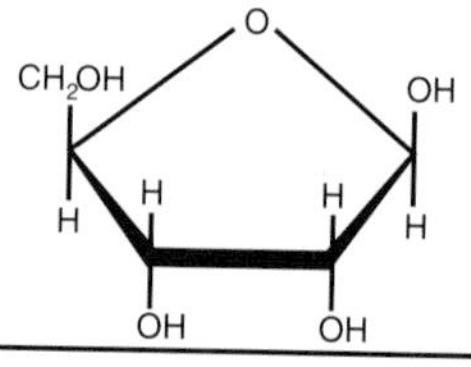	
Deoxyribose	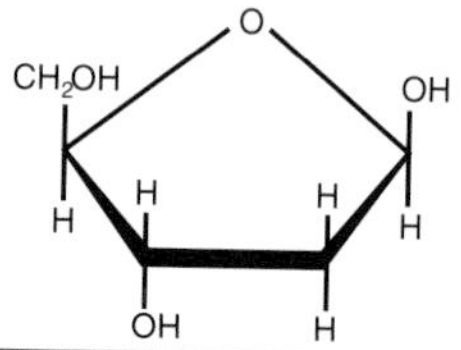	
Adenine (a purine)	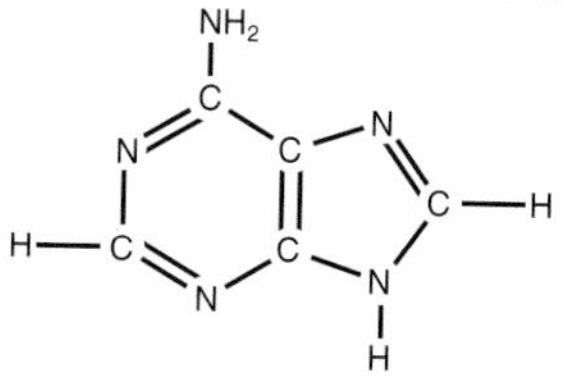	 Adenine
Guanine (a purine)	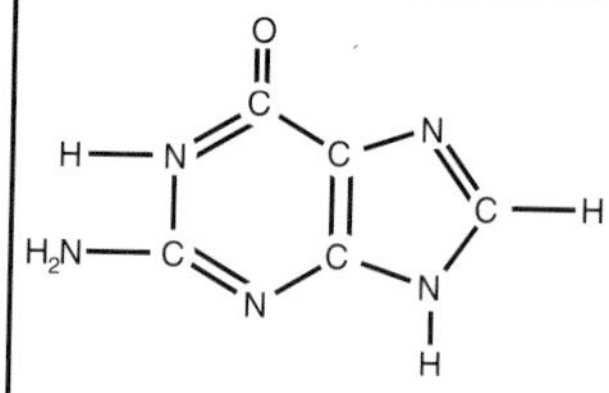	 Guanine
Cytosine (a pyrimidine)	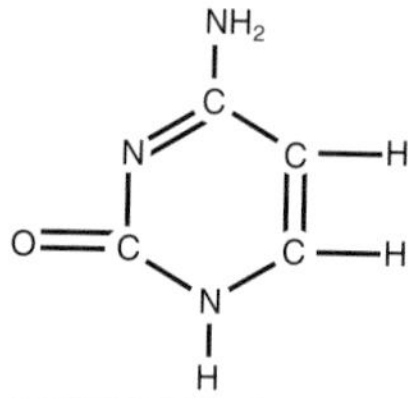	Cytosine
Thymine (a pyrimidine)	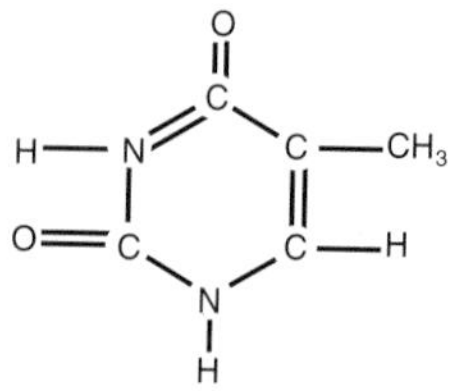	Thymine
Uracil (a pyrimidine)	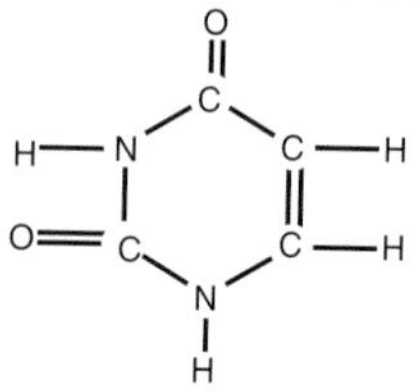	Uracil

Fig. 7.4 Structure of molecules in a nucleotide

7.3.2 Ribonucleic acid (RNA)

RNA is a single-stranded polymer of nucleotides where the pentose sugar is always ribose and the organic bases are adenine, guanine, cytosine and uracil. Its basic structure is given in Fig. 7.3). There are three types of RNA found in cells, all of which are involved in protein synthesis.

Ribosomal RNA (rRNA) is a large, complex molecule made up of both double and single helices. Although it is manufactured by the DNA of the nucleus, it is found in the cytoplasm where it makes up more than half the mass of the ribosomes. It comprises more than half the mass of the total RNA of a cell and its base sequence is similar in all organisms.

Transfer RNA (tRNA) is a small molecule (about eighty nucleotides) comprising a single strand. Again it is manufactured by nuclear DNA. It makes up 10–15% of the cell's RNA and all types are fundamentally similar. It forms a clover-leaf shape (Fig. 7.5), with one end of the chain ending in a cytosine–cytosine–adenine sequence. It is at this point that an amino acid attaches itself. There are at least twenty types of tRNA, each one carrying a different amino acid. At an intermediate point along the chain is an important sequence of three bases, called the **anticodon**. These line up alongside the appropriate codon on the mRNA during protein synthesis (Section 7.6).

Messenger RNA (mRNA) is a long single-stranded molecule, of up to thousands of nucleotides, which is formed into a helix. Manufactured in the nucleus, it is a mirror copy of part of one strand of the DNA helix. There is hence an immense variety of types. It enters the cytoplasm where it associates with the ribosomes and acts as a template for protein synthesis (Section 7.6). It makes up less than 5% of the total cellular RNA. It is easily and quickly broken down, sometimes existing for only a matter of minutes.

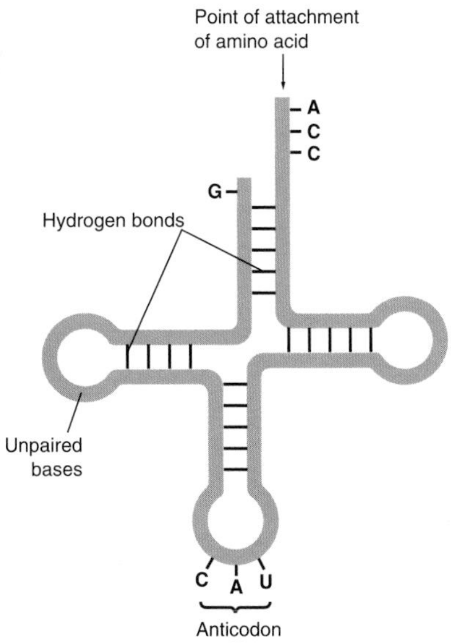

7.3.3 Deoxyribonucleic acid (DNA)

DNA is a double-stranded polymer of nucleotides where the pentose sugar is always deoxyribose and the organic bases are adenine, guanine, cytosine and thymine, but never uracil. Each of these polynucleotide chains is extremely long and may contain many million nucleotide units.

By the early 1950s, information on DNA from a variety of sources had been collected, but no molecular structure had been agreed. The available facts about DNA included:

1. It is a very long, thin molecule made up of nucleotides.

2. It contains four organic bases: adenine, guanine, cytosine and thymine.

3. The amount of guanine is usually equal to that of cytosine.

4. The amount of adenine is usually equal to that of thymine.

5. It is probably in the form of a helix whose shape is maintained by hydrogen bonding.

Using the accumulated evidence, James Watson and Francis Crick in 1953 suggested a molecular structure which proved to be one of the greatest milestones in biology. They postulated a

Fig. 7.5 Structure of transfer RNA

Did you know?

Each cell in the body contains about two metres of DNA. If all the DNA in all the cells of a single human were stretched out it would reach to the moon and back 8000 times.

TABLE 7.2 **Differences between RNA and DNA**

RNA	DNA
Single polynucleotide chain	Double polynucleotide chain
Smaller molecular mass (20 000–2 000 000)	Larger molecular mass (100 000–150 000 000)
May have a single or double helix	Always a double helix
Pentose sugar is ribose	Pentose sugar is deoxyribose
Organic bases present are adenine, guanine, cytosine and uracil	Organic bases present are adenine, guanine, cytosine and thymine
Ratio of adenine and uracil to cytosine and guanine varies	Ratio of adenine and thymine to cytosine and guanine is one
Manufactured in the nucleus but found throughout the cell	Found almost entirely in the nucleus
Amount varies from cell to cell (and within a cell according to metabolic activity)	Amount is constant for all cells of a species (except gametes and spores)
Chemically less stable	Chemically very stable
May be temporary – existing for short periods only	Permanent
Three basic forms: messenger, transfer and ribosomal RNA	Only one basic form, but with an almost infinite variety within that form

double helix of two nucleotide strands, each strand being linked to the other by pairs of organic bases which are themselves joined by hydrogen bonds. The pairings are always cytosine with guanine and adenine with thymine. This was not only consistent with the known ratio of the bases in the molecule, but also allowed for an identical separation of the strands throughout the molecule, a fact shown to be the case from X-ray diffraction patterns. As the purines, adenine and guanine, are double ringed structures (Fig. 7.4) they form much longer links if paired together than the two single ringed pyrimidines, cytosine and thymine. Only by pairing one purine with one pyrimidine can a consistent separation of three rings' width be achieved. In effect, the structure is like a ladder where the deoxyribose and phosphate units form the uprights and the organic base pairings form the rungs. However, this is no ordinary ladder; instead it is twisted into a helix so that each upright winds around the other. The two chains that form the uprights run in opposite directions, i.e. are **antiparallel**. The structure of DNA is shown in Figs. 7.6 and 7.7.

The structure postulated both fitted the known facts about DNA and was consistent with its biological rôle. Its extreme length (around 2.5 billion base pairs in a typical mammalian cell) permitted a very long sequence of bases which could be almost infinitely various, thus providing an immense store of genetic information. In addition its structure allowed for its replication. The separation of the two strands would result in each half attracting its complementary nucleotide to itself. The subsequent joining of these nucleotides would form two identical DNA double helices. This fitted the observation that DNA content doubles prior to cell division. Each double helix could then enter one of the daughter cells and so restore the normal quantity of DNA.

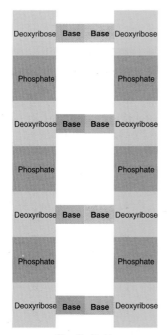

Simplified ladder

DNA structure may be likened to a ladder where alternating phosphate and deoxyribose molecules make up the 'uprights' and pairs of organic bases comprise the 'rungs'.

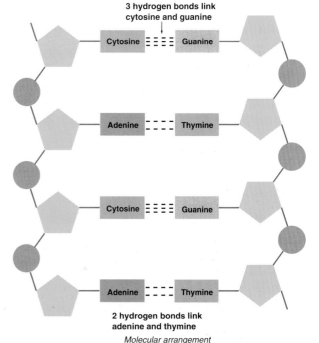

Molecular arrangement

Note the base pairings are always cytosine–guanine and adenine–thymine. This ensures a standard 'rung' length. Note also that the 'uprights' run in the opposite direction to each other (i.e. are antiparallel).

Fig. 7.6 Basic structure of DNA

7.3.4 Differences between RNA and DNA

Despite the obvious similarities between these two nucleic acids, a number of differences exist and these are listed in Table 7.2 on the previous page.

7.4 DNA replication

The uprights are composed of deoxyribose–phosphate molecules, the rungs of pairs of bases

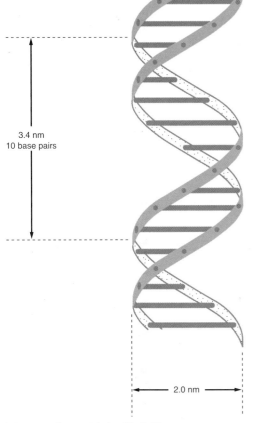

3.4 nm
10 base pairs

2.0 nm

Fig. 7.7 The DNA double helix structure

The Watson–Crick model for DNA allows for a relatively simple method by which the molecule can make exact copies of itself, something which must occur if genetic information is to be transmitted from cell to cell and from generation to generation. Replication is controlled by the enzyme DNA polymerase and an illustrated description is given in Fig. 7.9 on the next page.

Evidence for **semi-conservative replication** came from experiments by Meselsohn and Stahl. They grew successive generations of *Escherichia coli* in a medium where all the available nitrogen was in the form of the isotope ^{15}N (heavy nitrogen). In time, all the nitrogen in the DNA of *E. coli* was of the heavy nitrogen type. As DNA contains much nitrogen, the molecular weight of this DNA was measurably greater than that of DNA with normal nitrogen (^{14}N).

The *E. coli* containing the heavy DNA were then transferred into a medium containing normal nitrogen (^{14}N). Any new DNA produced would need to use this normal nitrogen in its manufacture. The question was, would the new DNA all be of the light type (contain only ^{14}N) or would it, as the semi-conservative replication theory suggests, be made up of one original strand of heavy DNA and one new strand of light DNA? In the latter case its weight would be intermediate between the heavy and light types. To answer this they allowed *E. coli* to divide once and collected all the first generation cells. The DNA from them was then extracted and its relative weight determined by special techniques involving centrifugation with caesium chloride. As the results depicted in Fig. 7.10 show, the weight was indeed intermediate between the heavy and light DNA types, thus confirming the semi-conservative replication theory.

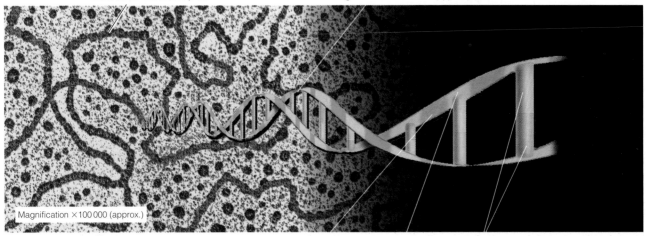

Electron micrograph of part of the long DNA molecule with associated protein molecules

Drawing of the DNA double helix without associated protein. Magnification $\times 13\,000\,000$ (approx.)

Magnification $\times 100\,000$ (approx.)

Fig. 7.8 Deoxyribonucleic acid

Phosphate   Deoxyribose sugar   Complementary base pair

1. *A representative portion of DNA, which is about to undergo replication, is shown.*

2. *DNA polymerase causes the two strands of the DNA to separate.*

3. *The DNA polymerase completes the splitting of the strand. Meanwhile free nucleotides are attracted to their complementary bases.*

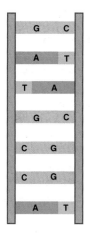

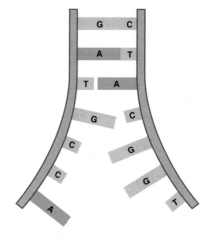

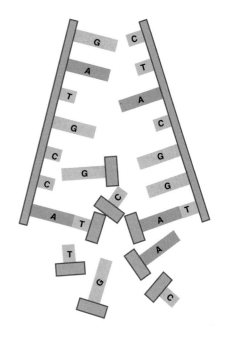

4. *Once the nucleotides are lined up they join together (bottom 3 nucleotides). The remaining unpaired bases continue to attract their complementary nucleotides.*

5. *Finally all the nucleotides are joined to form a complete polynucleotide chain. In this way two identical strands of DNA are formed. As each strand retains half of the original DNA material, this method of replication is called the semi-conservative method.*

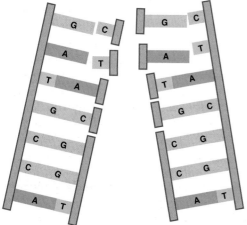

Fig 7.9 The replication of DNA

If a second generation of *E. coli* is grown from the first generation it is found to comprise half light and half intermediate weight DNA. Can you explain this?

Analysis shows that the replication of DNA takes place during interphase, shortly before cell division. Thus when the chromatids appear during prophase each has a double helix of DNA.

7.5 The genetic code

DNA extracted from *E. coli* grown in a medium containing normal nitrogen (^{14}N)

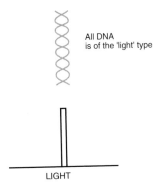

All DNA is of the 'light' type

LIGHT

DNA extracted from *E. coli* grown in a medium containing heavy nitrogen (^{15}N) and then transferred to a medium containing normal nitrogen (^{14}N)

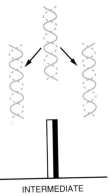

INTERMEDIATE

DNA extracted from *E. coli* grown in a medium containing heavy nitrogen (^{15}N)

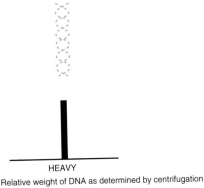

HEAVY

Relative weight of DNA as determined by centrifugation

Fig. 7.10 Interpretation of Meselsohn–Stahl experiments on semi-conservative replication of DNA

Once the structure of DNA had been elucidated and its mechanism of replication discovered, one important question remained: how exactly are the genetic instructions stored on the DNA in such a way that they can be used to mastermind the construction of new cells and organisms? Most chemicals within cells are similar regardless of the type of cell or species of organism. It is in their proteins and DNA that cells and organisms differ. It seems a reasonable starting point, therefore, to assume that the DNA in some way provides a 'code' for an organism's proteins. Moreover, most chemicals in cells are manufactured with the aid of enzymes, and all enzymes are proteins. Therefore by determining which enzymes are produced, the DNA can determine an organism's characteristics. Every species possesses different DNA and hence produces different enzymes. The DNA of different species differs not in the chemicals which it comprises, but in the sequence of base pairs along its length. This sequence must be a code that determines which proteins are manufactured.

Proteins show almost infinite variety. This variety likewise depends upon a sequence, in this instance the sequence of amino acids in the protein (Section 2.7). There are just twenty amino acids which regularly occur in proteins, and each must presumably have its own code of bases on the DNA. With only four different bases present in DNA, if each coded for a different amino acid, only four different amino acids could be coded for. Using a pair of bases, sixteen different codes are possible – still inadequate. A triplet code of bases produces sixty-four codes, more than enough to satisfy the requirements of twenty amino acids. This is called the **triplet code**.

The next problem was to determine the precise codon for each amino acid. Nirenberg devised a series of experiments towards the end of the 1950s which allowed him to break the code. He synthesized mRNA which had a triplet of bases repeated many times, e.g. GUA, GUA, GUA etc. He prepared test tubes which contained cell-free extracts of *E. coli*, i.e. they possessed all the necessary biochemical requirements for protein synthesis. Twenty tubes were set up, each with a different radioactively labelled amino acid. His synthesized mRNA was added to each tube and the presence of a polypeptide was looked for. Only in the test tube containing valine was a polypeptide found, indicating that GUA codes for valine. By repeating the process for all sixty-four possible combinations of bases, Nirenberg was able to determine which amino acid each coded for.

In some cases only the first two bases of the codon are relevant. Valine for instance is coded for by GU*, where * can be any of the four bases. Some amino acids have up to six codons. Arginine, for example, has CGU, CGC, CGA, CGG, AGA and AGG. At the other extreme, methionine, with AUG, and tryptophan, with UGG, have only one codon each. As there is more than one triplet for most amino acids it is called a **degenerate code** (a term derived from cybernetics). There are three codons UAA, UAG and UGA which are not amino acid codes. These are **stop** or **nonsense codons** and their importance is discussed in Section 7.6.4

All the codons are **universal**, i.e. they are precisely the same for all organisms.

The code is also **non-overlapping** in that each triplet is read separately. For example, CUGAGCUAG is read as CUG–AGC–UAG and not CUG–UGA–GAG–AGC etc., where each triplet overlaps the previous one, in this case by two bases. Overlapping would allow more information to be provided by a given base sequence, but it limits flexibility. Some viruses, with limited amounts of DNA, may use overlapping codes, but this is very rare.

7.6 Protein synthesis

If the triplet code on the DNA molecule determines the sequence of amino acids in a given protein, how exactly is the information transferred from the DNA, and how is the protein assembled? There are four main stages in the formation of a protein:

1. Synthesis of amino acids.

2. Transcription (formation of mRNA).

3. Amino acid activation.

4. Translation.

7.6.1 Synthesis of amino acids

In plants, the formation of amino acids occurs in mitochondria and chloroplasts in a series of stages:

(a) absorption of nitrates from the soil (Section 22.7.1);

(b) reduction of these nitrates to the amino group (NH_2);

(c) combination of these amino groups with a carbohydrate skeleton (e.g. α-ketoglutarate from Krebs cycle);

(d) transfer of the amino groups from one carbohydrate skeleton to another by a process called **transamination**. In this way all twenty amino acids can be formed.

Animals usually obtain their supply from the food they ingest, although they have some capacity to synthesize their own amino acids (called **non-essential amino acids**). The remaining nine – **essential amino acids** – must be provided in the diet.

7.6.2 Transcription (formation of messenger RNA)

Transcription is the process by which a complementary mRNA copy is made of the specific region (=**cistron**) of the DNA molecule which codes for a polypeptide (about 17 base pairs). A specific region of the DNA molecule, called a cistron, unwinds. This unwinding is the result of hydrogen bonds between base pairs in the DNA double helix being broken. This exposes the bases along each strand. Each base along one strand attracts its complementary RNA nucleotide, i.e. a free guanine base on the DNA will attract an RNA nucleotide with a cytosine base. It should be remembered, however, that uracil, and not thymine, is attracted to adenine (Fig. 7.11).

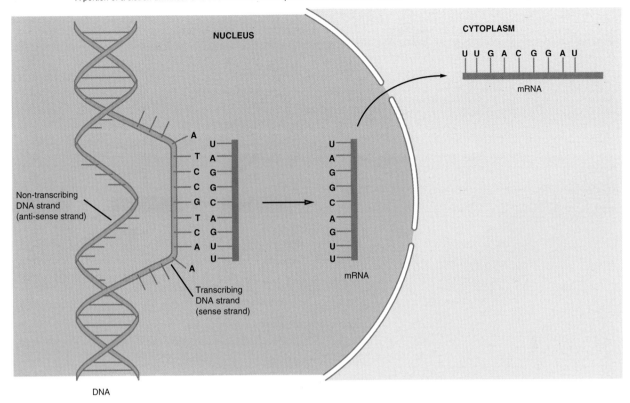

A portion of a cistron unwinds. One strand acts as a template for the formation of mRNA

Fig. 7.11 Transcription

The enzyme **RNA polymerase** moves along the DNA adding one complementary RNA nucleotide at a time to the newly unwound portion of DNA. The region of base pairing between the DNA and the RNA is only around 12 base pairs at any one time as the DNA helix reforms behind the RNA polymerase. The DNA thus acts as a **template** against which mRNA is constructed. A number of mRNA molecules may be formed before the RNA polymerase leaves the DNA, which closes up reforming its double helix. Being too large to diffuse across the nuclear membrane, the mRNA leaves instead through the nuclear pores. In the cytoplasm it is attracted to the ribosomes. Along the mRNA is a sequence of triplet codes which have been determined by the DNA. Each triplet is called a **codon**.

7.6.3 Amino acid activation

Activation is the process by which amino acids combine with tRNA using energy from ATP. Fig. 7.5 shows the structure of a tRNA molecule. Each type of tRNA binds with a specific amino acid which means there must be at least twenty types of tRNA. Each type differs, among other things, in the composition of a triplet of bases called the **anticodon**. What all tRNA molecules have in common is a free end which terminates in the triplet CCA. It is to this free end that the individual amino acids become attached, although how each specific amino acid is specified is not known (Fig. 7.12). The tRNA molecules with attached amino acids now move towards the ribosomes.

115

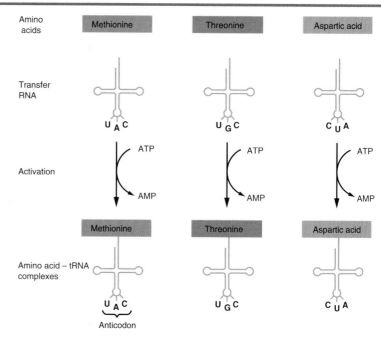

Fig. 7.12 Activation

7.6.4 Translation

Translation is the means by which a specific sequence of amino acids is formed in accordance with the codons on the mRNA. A group of ribosomes becomes attached to the mRNA to form a structure called a **polysome**. The complementary anticodon of a tRNA–amino acid complex is attracted to the first codon on the mRNA. The second codon likewise attracts its complementary

Many ribosomes may move along the mRNA at the same time, thus forming many identical polypeptides.

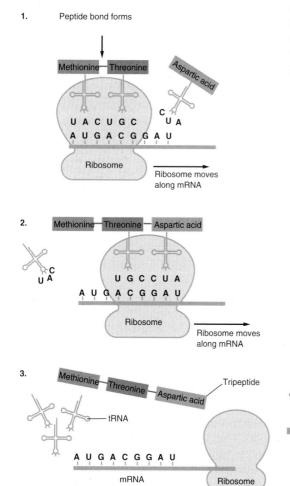

Fig. 7.13(a) Translation

Fig. 7.13(b) Polypeptide formation

anticodon. The ribosome acts as a framework which holds the mRNA and tRNA amino acid complex together until the two amino acids form a peptide bond between each other. Once they have combined, the ribosome will move along the mRNA to hold the next codon–anticodon complex together until the third amino acid is linked with the second. In this way a polypeptide chain is assembled, by the addition of one amino acid at a time. Second and subsequent ribosomes may pass along the mRNA immediately behind the first. In this way many identical polypeptides are produced simultaneously.

Once each amino acid is linked, the tRNA which carried it to the mRNA is released back into the cytoplasm. It is again free to combine with its specific amino acid. The ribosome continues along the mRNA until it reaches one of the nonsense codes (Section 7.5) at which point the polypeptide is cast off. The process of translation is summarized in Fig. 7.13, on the previous page.

The polypeptides so formed must now be assembled into proteins. This may involve the spiralling of the polypeptide to give a secondary structure, its folding to give a tertiary structure and its combination with other polypeptides and/or prosthetic groups to give a quaternary structure (see Fig. 2.13).

7.7   Gene expression and control

The part of the DNA molecule which specifies a polypeptide is termed a **cistron** by the molecular biologist. The geneticist, however, terms a similar functional unit of DNA a **gene**. Experiments have confirmed the theory that **one gene specifies one polypeptide**. Some polypeptides are required continually by a cell and these must therefore be produced continuously. Others are only required in certain circumstances and need not be produced all the time, indeed it would be needlessly wasteful to do so. How then are these genes switched on or off? Jacob and Monod studied the problem using *E. coli*. They considered two different methods of control:

1. Enzyme induction – Some genes are only switched on when the enzymes they code for are needed. *E. coli* normally respires glucose, but if grown on a medium containing lactose it will start to produce two enzymes, one enabling the lactose to be absorbed and the other to allow the lactose to be respired. The presence of the lactose in some way switches on the appropriate genes needed to produce these enzymes.

2. Enzyme repression – Some genes are normally switched on but may be switched off in special circumstances. *E. coli* manufactures its own amino acid, tryptophan, using a group of enzymes called tryptophan synthetase. If, however, tryptophan is added to the growing medium, it is absorbed by the bacterium, which consequently stops the production of tryptophan synthetase. The presence of tryptophan must switch off the appropriate gene.

Jacob and Monod put forward the concept of an **operon** – a group of adjacent genes which act together. The operon has two main sections:

1. The structural genes – These are the genes responsible for the production of the polypeptides which make up an enzyme or group of enzymes.

2. The operator gene – The gene which regulates the structural genes, in effect switching them on or off.

A third gene, the **regulator gene**, is involved. This is not part of the operon and may be some distance from it on the DNA. The regulator gene codes for a protein called the **repressor**. This repressor can bind with the operator gene and prevent it switching on the structural genes, i.e. it represses the production of those particular enzymes.

Enzyme induction and repression can both be explained in terms of this system, differing only in their effects on the repressor. In the case of enzyme induction, the inducer combines with the repressor in such a way that it prevents it binding with the operator gene. In the absence of the repressor, the operator gene switches on the structural genes which begin mRNA production which in turn synthesizes the relevant enzymes. To take the earlier example, lactose combines with the repressor and prevents it binding with the operator gene, which thus switches on the structural genes. The result is the manufacture of the enzymes which allow the lactose to be absorbed and respired.

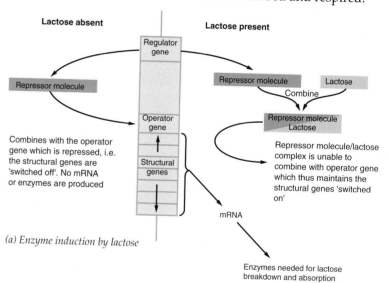

(a) Enzyme induction by lactose

Fig. 7.14 (a) Jacob–Monod hypothesis of control of enzyme synthesis

In the case of enzyme repression, the **co-repressor molecule**, as it is called, combines with the repressor. Together they bind with the operator gene and cause it to switch off the structural genes. These cease to produce mRNA and enzyme production stops. To use the earlier example, tryptophan combines with the repressor allowing it in turn to bind with the operator gene. The operator gene, being repressed, switches off the structural genes and the production of tryptophan synthetase ceases.

This system of control helps to maintain a steady state within cells despite fluctuations in the supply of materials. As such it is an example of **cellular homeostasis** (Section 25.6).

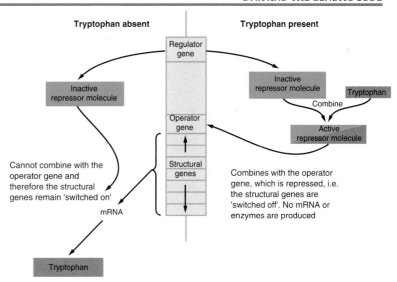

(b) Enzyme repression by tryptophan

Fig. 7.14(b) Jacob–Monod hypothesis of control of enzyme synthesis

7.8   Genetic engineering

Perhaps the most significant scientific advance in recent years has been the development of technology which allows genes to be manipulated, altered and transferred from organism to organism – even to transform DNA itself. This has enabled us to use rapidly reproducing organisms such as bacteria as chemical factories producing useful, often life-saving, substances. The list of these substances expands almost daily and includes hormones, antibiotics, interferon and vitamins. Details of the production of some of these chemicals are given in Chapter 30.

7.8.1 Recombinant DNA technology

It has been known that a number of human diseases are the result of individuals being unable to produce for themselves chemicals which have a metabolic rôle. Many such chemicals, e.g. insulin and thyroxine, are proteins and therefore the product of a specific portion of DNA. The treatment of such deficiencies had previously been to extract the missing chemical from either an animal or human donor. This has presented problems. While the animal extracts may function effectively, subtle chemical differences in their composition have been detected by the human immune system, which has responded by producing antibodies which destroy the extract. Even chemically compatible extracts from human donors present a risk of infection from other diseases, as the transmission of the HIV virus to haemophiliacs illustrates only too well. Whether from animals or humans, the cost of such extracts is considerable.

It follows that there are advantages in producing large quantities of 'pure' chemicals from non-human sources. As a result, methods have been devised for isolating the portion of human DNA responsible for the production of insulin and combining it with bacterial DNA in such a way that the microorganism will continually produce the substance. This DNA, which results from the combination of fragments from two different organisms, is called **recombinant DNA**.

7.8.2 Techniques used to manipulate DNA

The manipulation of DNA involves three main techniques, each using a specific enzyme or group of enzymes:

1. Cutting of DNA into small sections using restriction endonucleases

These enzymes are used to cut DNA between specific base sequences which the enzyme recognizes. For example, Hae III nuclease recognizes a four base-pair sequence and cuts it as shown by the arrow:

$$G \quad G \uparrow C \quad C \left.\right\} \quad \text{four complementary base-pairs}$$
$$C \quad C \downarrow G \quad G \qquad \text{on DNA double helix}$$

The Hind III nuclease however recognizes a six base-pair sequence, cutting it as shown by the arrow below:

$$A \uparrow A \quad G \quad C \quad T \quad T \left.\right\} \quad \text{six complementary base-}$$
$$T \quad T \quad C \quad G \quad A \downarrow A \qquad \text{pairs on DNA double helix}$$

As any sequence of four base-pairs is likely to occur more frequently than a six base-pair sequence, the nucleases recognizing four base-pairs cut DNA into smaller sections than those recognizing six base-pairs. The latter group are, however, more useful as the longer sections they produce are more likely to contain an intact gene.

2. Production of copies of DNA using either plasmids or reverse transcriptase

In bacterial cells there are small circular loops of DNA called **plasmids**. Plasmids are distinct from the larger circular portions of DNA which make up the bacterial chromosome. Bacteria replicate their plasmid DNA so that a single cell contains many copies. If a portion of DNA from, say, a human cell, is inserted into a plasmid and it is reintroduced into the bacterial cell, replication of the plasmid will result in up to 200 identical copies of the human DNA being made. A population of bacteria containing this human DNA can now be grown to provide a permanent source of it. By repeating the process for other DNA portions, a complete library of human DNA can be maintained. Geneticists can then select as required, any gene (a portion of DNA) they require for further investigation or use, in much the same way as a book is selected from a conventional library. Selection is, however, more complex requiring the use of DNA probes or specific antibodies. This collection of genetic information is called a **genome library**.

A second method of duplicating particular portions of DNA is appropriate where the protein for which it codes is synthesized in a specific organ. Thyroxine, for example, is produced in the thyroid gland and therefore cells from this gland would be expected to contain a relatively large amount of messenger RNA which codes for thyroxine. Reverse transcriptase (Section 5.2.2) can be used to synthesize DNA, called **copy DNA (cDNA)**, from the mRNA in thyroid cells. A large proportion of the cDNA produced is likely to code for thyroxine and it can be isolated using the techniques described in Section 7.8.3.

3. Joining together portions of DNA using DNA ligase

The recombination of pieces of DNA, e.g. the addition of cDNA into bacterial plasmid DNA, is carried out with the aid of the enzyme **DNA ligase**.

7.8.3 Gene cloning

The techniques described in the previous section are utilized in the process of gene cloning in which multiple copies of a specific gene are produced which may then be used to manufacture large quantities of valuable products.

Manufacture involves the following stages:

1. Identification of the required gene.

2. Isolation of that gene.

3. Insertion of the gene into a vector.

4. Insertion of the vector into a host cell.

5. Multiplication of the host cell.

6. Synthesis of the required product by the host cell.

7. Separation of the product from the host cell.

8. Purification of the product.

Figure 7.15 on the next page illustrates how gene cloning is used in the production of insulin. The bacteria produced in this way can be grown in industrial fermenters using a specific nutrient medium under strictly controlled conditions. The bacteria may then be collected and the insulin extracted from them by suitable means. Alternatively, it is possible to engineer bacteria which secrete the insulin and this can be extracted from the medium which is periodically drawn off. Details of these, and other, fermentation techniques are given in Chapter 30.

7.8.4 Applications of genetic engineering

The techniques illustrated above may be utilized to manufacture a range of materials which can be used to treat diseases and disorders. In addition to insulin, human growth hormone is now produced by bacteria in sufficient quantity to allow all children in this country requiring it to be treated. Among other hormones being produced in this manner are erythropoietin, which controls red blood cell production and calcitonin which regulates the levels of calcium in the blood (Section 26.4). Much research is taking place into the production of antibiotics and vaccines through recombinant DNA technology and already interferon, a chemical produced in response to viral infection, is in production.

A form of abnormal haemoglobin is produced by a defective gene causing a disease called **thalassaemia**. It is likely that genetic engineering will provide a cure for the disease, by the transference of a normal gene for haemoglobin into patients afflicted by the disease.

The scope of recombinant DNA technology is not restricted to the field of medicine. In agriculture, it is now possible to transfer

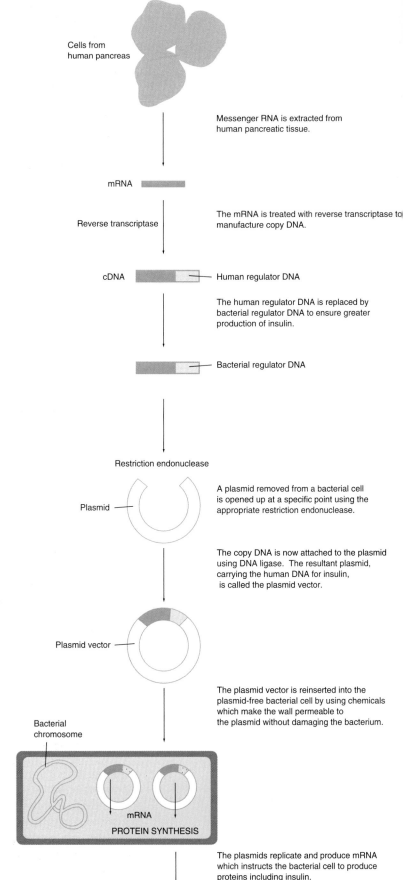

Cells from
human pancreas

Messenger RNA is extracted from
human pancreatic tissue.

mRNA

Reverse transcriptase

The mRNA is treated with reverse transcriptase to
manufacture copy DNA.

cDNA — Human regulator DNA

The human regulator DNA is replaced by
bacterial regulator DNA to ensure greater
production of insulin.

— Bacterial regulator DNA

Restriction endonuclease

Plasmid

A plasmid removed from a bacterial cell
is opened up at a specific point using the
appropriate restriction endonuclease.

The copy DNA is now attached to the plasmid
using DNA ligase. The resultant plasmid,
carrying the human DNA for insulin,
is called the plasmid vector.

Plasmid vector

The plasmid vector is reinserted into the
plasmid-free bacterial cell by using chemicals
which make the wall permeable to
the plasmid without damaging the bacterium.

Bacterial
chromosome

mRNA
PROTEIN SYNTHESIS

The plasmids replicate and produce mRNA
which instructs the bacterial cell to produce
proteins including insulin.

*Fig. 7.15 Use of plasmid vector in gene
cloning*

INSULIN

genes which produce toxins with insecticidal properties from bacteria to higher plants such as potatoes and cotton. In this way these plants have 'built-in' resistance to certain insect pests. The saving in time and money by not having to spray such crops regularly with insecticides is obvious, to say nothing of avoiding killing harmless or beneficial insect species which inevitably happens however carefully spraying is carried out. It may prove possible to transfer genes from nitrogen-fixing bacteria to cereal crops, to enable them to fix their own nitrogen. There would then be less need to apply expensive nitrogen fertilizers thus reducing the pollution problems of 'run-off' (Section 18.5.3).

There would seem to be no end of possibilities – transfer of genes conferring resistance to all manner of diseases, development of plants with more efficient rates of photosynthesis, the control of weeds and the development of oil-digesting bacteria to clear up oil spillages are just some of the potential uses of recombinant DNA technology. There are however ethical as well as practical problems to be overcome before many of these ideas can be brought to fruition. Some of these problems are discussed in the following section.

7.8.5 Implications of genetic engineering

The benefits of genetic engineering are obvious but it is not without its hazards. It is impossible to predict with complete accuracy what the ecological consequences might be of releasing genetically engineered organisms into the environment. It is always possible that the delicate balance that exists in any habitat may be irretrievably damaged by the introduction of organisms with new gene combinations. It is also possible that organisms designed for use in one environment may escape to others with harmful consequences. We know that viruses can transfer genes from one organism to another. Advantageous genes added to our domestic animals or crop plants may be transferred in this way to their competitors making them even greater potential dangers. The escape of a single pathogenic bacterium into a susceptible population could result in considerable damage to a species. Perhaps more sinister is the fear that the ability to manipulate genes could allow human characteristics and behaviour to be modified. In the wrong hands this could be used by individuals, groups or governments in order to achieve certain goals, control opposition or gain ultimate power.

Even without these dangers there are still ethical issues which arise from the development of recombinant DNA technology. Is it right to replace a 'defective' gene with a 'normal' one? Is the answer the same for a gene which causes the bearer pain, as it is where the gene has a merely cosmetic effect? Who decides what is 'defective' and what is 'normal'? A 'defective' gene may actually confer some other advantage, e.g. sickle-cell gene (Section 10.4.3). Is there a danger that we shall in time reduce the variety so essential to evolution, by the progressive removal of unwanted genes or, by combining genes from different species, are we actually increasing variety and favouring evolution? Where a gene probe detects a fatal abnormality, what criteria, if any, should be applied before deciding whether to carry out an abortion?

APPLICATION

Cystic fibrosis and gene therapy

One person in 2000 in Britain suffers from cystic fibrosis. CF patients produce mucus secretions which are too viscous. Thick, sticky mucus blocks the pancreatic duct and prevents pancreatic enzymes from reaching the duodenum and clogging up of the lungs leads to recurrent infections.

The long arm of chromosome 7 carries the CFTR (cystic fibrosis transmembrane conductance regulator) gene which codes for a protein which is essential for chloride transport. Everyone has two copies of this gene in every cell, one from the mother and one from the father. As long as one copy of the CFTR gene works correctly chloride transport is adequate but if two defective genes are present CF results.

Microbiologists have succeeded in isolating and cloning the CFTR gene and have found that 70% of its mutations consist of the same change to the normal DNA sequence. CFTR is a big gene, covering 250 000 base pairs of DNA and the commonest mutation is a small deletion in which three nucleotides are missing. After transcription and translation the CFTR gene lacks one amino acid, phenylalanine number 508, in the protein chain. People who have at least one chromosome bearing this mutation can be identified in the laboratory. This is done by using a blood sample or shed cells in a mouthwash to collect a DNA sample. DNA polymerase is used to make many copies of a 50 base-pair segment of both CFTR genes. The copies are then run on an electrophoretic gel (see Notebook page 26) in which small fragments move faster than big ones. If either of the CFTR genes has the three-base deletion it will move more quickly.

This test is quick and easy. A series of other, more expensive, tests would enable another 20% of CF carriers to be identified. But once a carrier for CF, or any other genetic disease, has been identified is it possible to replace the mutant gene with a normal version of the gene? Gene therapy is now becoming a realistic possibility, although there are many medical and ethical problems to be solved.

In **germ-line therapy** the approach could be to repair the gene in a fertilized egg so that the repaired gene would be

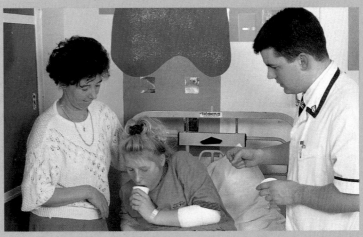

Cystic fibrosis patient coughing up mucus following treatment to loosen it

copied into each daughter cell at mitosis. In this way the mutant gene will not only have been eliminated from the person receiving treatment, but also from all his or her offspring. This raises the ethical question of whether we have the right to alter the genes of future generations and following the report of the Warnock committee such germ-line therapy is prohibited in Britain.

Research now centres on a different approach: **somatic cell gene therapy**. Although the mutated gene occurs in every cell of the patient's body, somatic cell gene therapy would target just the affected tissues, for example:

the lungs in cystic fibrosis;

the muscles in Duchenne muscular dystrophy;

blood cells or their precursors (reticulocytes in the bone marrow) in β – thalassaemia.

These tissues are fully differentiated and will eventually die so treatments may have to be repeated as the treated cells die and are replaced. It has not yet been possible to isolate and treat the undifferentiated stem cells which give rise to the mature tissues. It is relatively easy to introduce large molecules of DNA into a cell nucleus using a fine glass needle or by subjecting cells in a test tube to an electric pulse which causes temporary holes in the plasma membrane and induces them to take up DNA. DNA can also be carried into the cell by a suitable virus.

There are two types of genetic disease: those due to loss of function and those due to gain of function. Gain-of-function diseases are normally dominant and the mutated gene which causes them is doing something positively bad. Such a gene would need to be removed or neutralized and this is proving a very difficult task. On the other hand gene supplementation is providing a possible treatment for loss-of-function diseases and the approach looks promising for CF sufferers. It is hoped that an aerosol inhaler could be used to restore the missing gene to the lung epithelia. This would leave patients with their digestive problems but would solve the problem of congested and infected lungs.

An example of successful treatment by gene supplementation is in Severe Combined Immunodeficiency Disease (SCID). The gene coding for adenosine deaminase is mutated and homozygotes are unable to deaminate adenosine. This leads to the death of lymphocytes and sufferers have no immunity at all. Since 1990 two affected children have had some of their lymphocyte precursor cells infected with a special virus carrying the missing gene. The treatment is repeated every month or so to replace the lymphocytes as they die and the two children now attend normal schools.

For many Mendelian recessive disorders in which the damage is confined to an accessible tissue, gene therapy by gene supplementation looks a promising way forward. Even cancer, where mutations are confined to tumour cells, may be a candidate for gene therapy in the future.

APPLICATION

Transgenic animals

While it is possible to clone plants from any differentiated plant cell, the same is not true of animals. It is however possible to mix together cells from two different embryos, such as a sheep and a goat, to produce a chimaera, in this example called a geep.

The problem with chimaeras is that no-one can predict which cells will form which part of the animal. A more precise method of producing desired characteristics is to augment traditional animal breeding by genetic engineering to make transgenic animals.

For a brief time following the fusion of a sperm and an ovum the cell contains two pronuclei. Cloned DNA (Section 7.8.3) can be injected into one of these and in some cases it will integrate into one or more of the chromosomes. The manipulated offspring are then transferred to a foster mother and the resulting offspring screened for the presence of the introduced gene. A more recent approach which allows the gene to be positioned more precisely involves transferring the DNA into embryonic stem cells (cells from embryos prior to implantation). These cells can be grown indefinitely in a test tube and can be monitored to see if they produce the desired protein before they are injected into a normal embryo and transferred to a foster mother.

Genetic engineering may be used to improve animal health. For example, genes responsible for resistance to a particular disease could be introduced into otherwise vulnerable animals. Transgenic animals may also be used to produce rare and expensive proteins for use in human medicine. The genes coding for certain proteins may be expressed in a sheep's mammary glands and the protein recovered from the milk. Researchers have managed to insert the gene for a blood clotting protein known as factor IX alongside the regulator of the lactoglobulin gene which encodes for one of the proteins in sheep's milk. The transgenic ewes now produce factor IX in their milk and it can be used to treat one type of haemophilia. Some people suffer congenital emphysema (page 399) because they are unable to make a protein known as ATT (alpha-1-antitrypsin). This protein can also be extracted from the milk of transgenic ewes.

In transgenic animals the 'foreign' DNA has become stably integrated into the animal's own genome so that it can be passed from generation to generation, effectively giving rise to a new strain of the animal.

This transgenic ram has had a human gene incorporated into its DNA. This gene for the production of the protein alpha-1-antitrypsin (ATT) is inherited by the ram's offspring and the protein is secreted in the ewe's milk

It is inevitable that we shall remain inquisitive about the world in which we live and, in particular, about ourselves. Scientific research will therefore continue. The challenge is to develop regulations and safeguards within moral boundaries which permit genetic engineering to be used in a safe and effective way to the benefit of both individuals in particular and humans in general.

7.9 Genetic fingerprinting

The pattern of dermal ridges and furrows which constitute the fingerprint not only persist unchanged throughout our lives, but are also unique to each one of us (identical twins excepted). For this reason they have long been used to help solve crimes by comparing the fingerprint pattern of the suspect with the impressions left, as a result of the furrow's oily secretions, at the scene of the crime. To this well-tried and successful forensic technique has now been added another – **genetic fingerprinting**.

While having nothing to do with either fingers or printing, the technique is equally, if not more, successful in identifying individuals from 'information' they provide. This 'information' is contained in a spot of blood, a sample of skin, a few sperm – in fact almost any cell of the body.

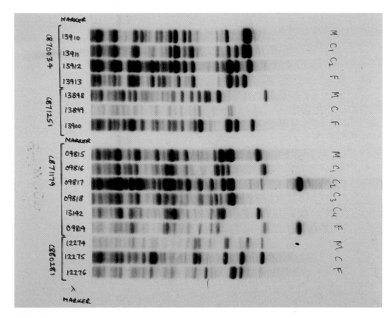

DNA fingerprint

The technique, developed by Alec Jeffreys of Leicester University, takes around six days and involves the following stages:

1. The DNA is separated from the sample.

2. Restriction endonucleases are used to cut the DNA into sections.

3. The DNA fragments are separated in an agarose gel using electrophoresis.

4. The fragments are transferred to a nitrocellulose (or nylon) membrane – a process called **Southern Blotting** after its inventor, Professor Southern.

5. Radioactive DNA probes (gene probes) are used to bind to specific portions of the fragments known as the core sequences.

6. The portions of the DNA not bound to the radioactive probes are washed off.

7. The remaining DNA still attached to the nylon membrane is placed next to a sheet of X-ray film.

8. The radioactive probes on this DNA expose the film, revealing a pattern of light and dark bands when it is developed. The pattern makes up the genetic fingerprint.

The patterns, like fingerprints, are unique to each individual (except identical twins) and remain unchanged throughout life. Unlike fingerprints, however, the pattern is inherited from both parents. The scope for genetic fingerprinting, therefore, extends beyond catching criminals, it can also be used in paternity suits for example (Section 9.5.3). To do this, white blood cells are taken from the mother and the possible father. From the pattern of bands of the child are subtracted those bands which correspond to the mother's bands. If the man is truly the parent, he must possess all the remaining bands in the child's genetic fingerprint.

As sperm contain DNA, they too can be used to provide a genetic fingerprint, leading to a remarkably accurate method of determining guilt, or otherwise, of an accused rapist. The method has also been successfully applied in immigration cases where the relationship of an immigrant to someone already resident in a country, is in dispute. Confirming the pedigree of animals, detecting some inherited diseases and monitoring bone-marrow transplants are other applications of the technique.

Despite the fact that we are, as yet, unclear as to what exactly the dark bands of the genetic fingerprint represent, the chances of two individuals (other than identical twins) having identical patterns is so small, that the technique is widely used and its results accepted as accurate.

7.10   Questions

1. (a) Explain the theoretical basis of genetic fingerprinting. *(10 marks)*

(b) Describe the practical procedures involved in making a genetic fingerprint. *(10 marks)*

(c) How can this technique be useful in forensic science? *(3 marks)*

(Total 23 marks)

UCLES (Modular) March 1992,
(Genetics and its applications), No. 1

2. The table below refers to DNA and RNA.

If the feature is correct, place a tick (✓) in the appropriate box and if the feature is incorrect, place a cross (✗) in the appropriate box.

Feature	DNA	RNA
Contains ribose		
Is single stranded		
Contains adenine, guanine and cytosine		
Contains thymine and uracil		
Always contains equal proportions of purines and pyrimidines		

(Total 5 marks)

ULEAC January 1994, Paper I, No. 3

3. The diagram shows some of the steps involved in the synthesis of insulin in a eukaryotic cell.

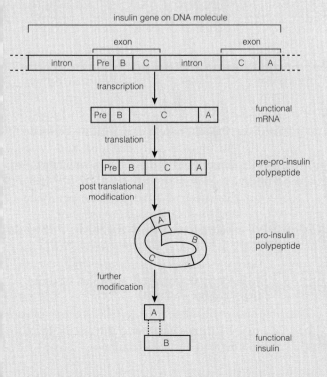

(a) Where exactly in the cell do the following processes take place?
  (i) Transcription *(1 mark)*
  (ii) Translation *(1 mark)*

(b) Suggest an organelle where modification of the pro-insulin polypeptide might occur. *(1 mark)*

(c) In genetic engineering, the information coding for insulin is extracted from the eukaryotic cell in the form of mRNA rather than DNA.
Suggest **two** reasons for this. *(2 marks)*

(d) Name the type of enzyme needed to convert the pro-insulin to its functional form. *(1 mark)*
(Total 6 marks)

AEB November 1993, Paper I, No. 17

4. The diagram shows a nucleotide from a molecule of DNA.

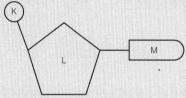

(a) Name the parts labelled K, L and M. *(1 mark)*

(b) Part of a DNA molecule has the sequence of bases shown.
Write the sequence of bases in the corresponding part of the molecule of mRNA synthesized on the DNA.

DNA molecule	A	T	C	G	C	G
mRNA molecule						

(1 mark)

(c) The percentage of bases in one strand of a DNA double helix is as follows
T = 40% and C = 22%
  (i) What is the percentage of bases A and G together in the same strand of DNA? *(1 mark)*
  (ii) What is the percentage of bases A and G together in the complementary strand of DNA? *(1 mark)*

(d) The following section of mRNA codes is for a peptide made of ten amino acids. The sequence of bases is shown below:

Code starts   Code ends
↓   ↓

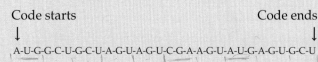

A-U-G-G-C-U-G-C-U-A-G-U-A-G-U-C-G-A-A-G-U-A-U-G-A-G-U-G-C-U

When the peptide was analysed, four types of amino acids were identified in the quantities shown in the table.

Amino acids		Number
Name (letter)		**Number**
Methionine	(M)	2
Alanine	(L)	3
Arginine	(R)	1
Serine	(S)	4

(i) How many consecutive bases in a codon specify a single amino acid?

(1 mark)

(ii) Work out the amino acid sequence in the peptide and write the correct sequence in the boxes provided, using the letters from the table.

M	L	L	S	S	R	S	M	S	L

(1 mark)
(Total 6 marks)

SEB (revised) May 1991, Higher Grade Paper I, No. 5

---

5. Read the following passage on genetic engineering and write on the dotted lines the most appropriate word or words to complete the passage.

The isolation of specific genes during a genetic engineering process involves forming eukaryotic DNA fragments. These fragments are formed using .... enzymes which make staggered cuts in the DNA within specific base sequences. This leaves single-stranded 'sticky ends' at each end. The same enzyme is used to open up a circular loop of bacterial DNA which acts as a .... for the eukaryotic DNA. The complementary sticky ends of the bacterial DNA are joined to the DNA fragment using another enzyme called .... DNA fragments can also be made from .... template. Reverse transcriptase is used to produce a single strand of DNA and the enzyme, ... , catalyses the formation of a double helix. Finally new DNA is introduced into host .... cells. These can then be cloned on an industrial scale and large amounts of protein harvested. An example of a protein currently manufactured using this technique is ....

(Total 7 marks)

ULEAC June 1992, Paper I, No. 7

6. (a) Read through the following passage and then fill in the blank spaces with the most appropriate biological term.

A molecule of DNA is composed of many polymerized units called nucleotides. Each nucleotide consists of a .... base joined to a .... sugar with a phosphate group. The DNA consists of two strands running parallel to each other and coiled in a double helix. The strands are held together by .... bonds which must be broken by the enzyme .... during replication. In RNA, the base .... is replaced by .... and the sugar present is .... Three varieties of RNA exist in cells. One sort, ... RNA, is found in high concentration in the specialized part of the nucleus called the .... , where it is probably synthesized. The pores seen in the nuclear envelope are probably important in allowing .... RNA to pass out to the .... situated on the endoplasmic reticulum. The third type of RNA is .... RNA and this combines with .... in the cytosol in order to locate them correctly during protein synthesis. (13 marks)

(b) Analysis of a sample of DNA extracted from a tissue showed that 38% of the bases were adenine. What percentage of the bases in the DNA would be guanine? Show how you arrive at your answer. (2 marks)

(Total 15 marks)

Oxford June 1992, Paper I, No. 5

---

7. The table shows the percentages of the bases in samples of DNA from different sources.

Source of DNA	Percentage of base present in sample			
	Adenine (A)	Guanine (G)	Cytosine (C)	Thymine (T)
Yeast	31.3	18.7	17.1	32.9
Wheat	27.3	22.7	22.8	27.1
Broad bean	29.7	20.6	20.1	29.6
Ox thymus gland	28.2	21.5	22.5	27.8
Ox spleen	27.9	22.7	22.1	27.3
Ox sperm	28.7	22.2	22.2	27.2

(a) Describe and explain the relation between:
(i) the total amount of adenine and the total amount of thymine in any one organism;
(2 marks)
(ii) the total amount of adenine + guanine and the total amount of cytosine + thymine in any one organism. (2 marks)

(b) Explain how it is possible to have two very different organisms such as ox and wheat with very similar proportions of bases in their respective DNA. (1 mark)

The table shows the genetic code. The position of each base in a mRNA triplet may be read from the table to give the abbreviated amino acid name.

1st position	2nd position				3rd position
	U	C	A	G	
U	Phe	Ser	Tyr	Cys	U
	Phe	Ser	Tyr	Cys	C
	Leu	Ser	STOP	STOP	A
	Leu	Ser	STOP	Trp	G
C	Leu	Pro	His	Arg	U
	Leu	Pro	His	Arg	C
	Leu	Pro	Gln	Arg	A
	Leu	Pro	Gln	Arg	G
A	Ile	Thr	Asn	Ser	U
	Ile	Thr	Asn	Ser	C
	Ile	Thr	Lys	Arg	A
	Met	Thr	Lys	Arg	G
G	Val	Ala	Asp	Gly	U
	Val	Ala	Asp	Gly	C
	Val	Ala	Glu	Gly	A
	Val	Ala	Glu	Gly	G

(c) Give the sequence of amino acids forming the polypeptide represented by the length of mRNA below. Read the mRNA from left to right.

U C C C C A C C G G U C U A A

(1 mark)

(d) Using examples from the table, explain what is meant by:
   (i) a degenerate code; *(2 marks)*
   (ii) a codon; *(2 marks)*
   (iii) its corresponding anticodon. *(2 marks)*

(e) errors may occur in copying DNA. Suggest **one** explanation for the fact that many of these have little or no effect on the polypeptide for which they code. *(2 marks)*

(f) Describe **four** differences between mitosis and meiosis. Write your answers in a suitable table. *(4 marks)*

The graph shows the amounts of DNA in a number of individual nuclei from a mouse testis.

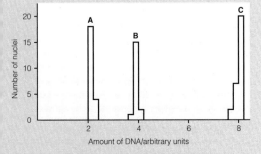

(g) (i) Which of the three groups of nuclei, A, B or C represents mature sperm cells? Give a reason for your answer. *(2 marks)*
   (ii) Explain, in terms of cell division, why there are three groups of nuclei.

   (3 marks)

(h) Suggest **one** reason why more valid results can be obtained from an investigation of this type if it is carried out on a testis rather than an ovary. *(1 mark)*

(Total 24 marks)

AEB November 1992, Paper II, No. 1

8. Cells of the bacterium *E. coli* were grown for many generations on a medium containing the heavy isotope of nitrogen, ^{15}N. This labelled all the DNA in the bacteria.

The cells were transferred to a medium containing ^{14}N and allowed to grow. During each generation of bacteria, the DNA replicates once. Samples of the bacteria were removed from the culture after one generation time and after two generation times. The DNA from each sample was extracted and centrifuged. As the DNA containing ^{15}N is slightly heavier than that containing ^{14}N, the relative amounts of DNA labelled with ^{14}N and ^{15}N can be determined.

The diagram shows two reference tubes and the results of this experiment.

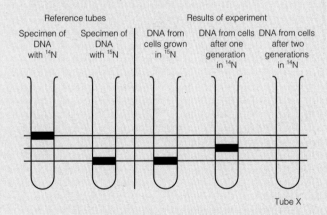

(a) Which part of the DNA molecule in the original culture would have been labelled with ^{15}N? *(1 mark)*

(b) Explain why the DNA occupies an intermediate position after one generation in the ^{14}N-containing medium. *(2 mark)*

(c) Complete the diagram to show the position of the band or bands of DNA after two generations in the ^{14}N-containing medium (Tube X). *(1 mark)*

(Total 4 marks)

AEB June 1991, Paper I, No. 17

131

9. (a) What is the name given to the process during which messenger RNA is synthesized?

(1 mark)

(b) List **three** ways in which messenger RNA differs in structure from DNA. *(3 marks)*

The table below presents the mRNA codons for some amino acids.

Amino acid	mRNA codon
Glutamic acid	GAU
Phenylalanine	UUC
Lysine	AAG
Proline	CCU
Threonine	ACC
Valine	GAA

(c) A mutation may result in a change in the sequence of amino acids. For example, in sickle cell anaemia a polypeptide of haemoglobin contains valine in place of glutamic acid. Using the information in the table above, describe precisely, with reference to the structure of the DNA molecule, the DNA mutation that has occurred in this example. *(2 marks)*

The diagram shows a stage in the synthesis of a part of a polypeptide.

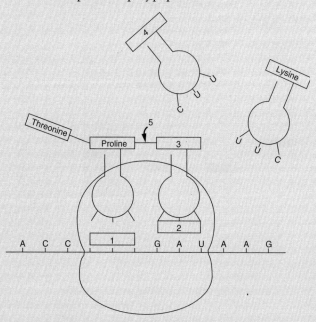

(d) Using the information in the table and the diagram determine:
  (i) the mRNA codon at **1**
  (ii) the tRNA anticodon at **2**
  (iii) its associated amino acid at **3**
  (iv) the amino acid at **4**
  (v) the type of bond formed at **5** *(5 marks)*

(e) Outline precisely what happens at the next stage in the elongation of the polypeptide shown. *(3 marks)*

(Total 14 marks)

NISEAC June 1992, Paper II, No. 5

10. The diagram below represents the molecular structure of part of a DNA molecule.

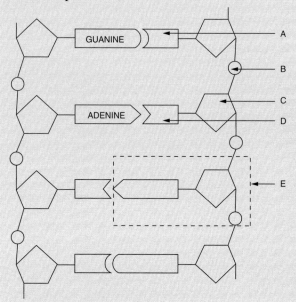

(a) Name **A–E**. *(5 marks)*
(b) State **three** differences between DNA and mRNA. *(3 marks)*
(c) (i) Distinguish between *transcription* and *translation*. *(4 marks)*
    (ii) Where do they take place? *(2 marks)*
(d) Why is the replication of DNA referred to as *semi-conservative*? *(2 marks)*
(e) Name **two** organelles, in addition to the nucleus, which contain DNA. *(2 marks)*
(f) With reference to DNA
   (i) define the term *gene*, *(2 marks)*
   (ii) define the term *gene mutation*. *(2 marks)*

(Total 22 marks)

Oxford June 1994, Paper 1, No. 1

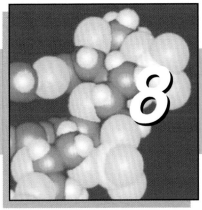

8

Cell division

Modern cell theory, as described in Chapter 4, states that 'all new cells are derived from other cells'. The process involved is **cell division**. All 10^{14} cells which comprise a human are derived, through cell division, from the single zygote formed by the fusion of two gametes. These gametes in turn were derived from the division of certain parental cells. It follows that all cells in all organisms have been formed from successive divisions of some original ancestral cell. The remarkable thing is that, while cells and organisms have diversified considerably over millions of years, the process of cell division has remained much the same.

There are two basic types:

Mitosis which results in all daughter cells having the same number of chromosomes as the parent.

Meiosis which results in the daughter cells having only half the number of chromosomes found in the parent cell.

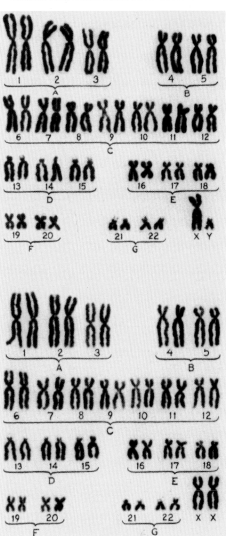

Human karyotypes (male and female)

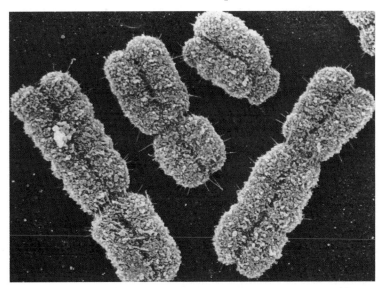

False-colour scanning electron micrograph (SEM) of a group of human chromosomes

8.1 Chromosomes

8.1.1 Chromosome structure

Chromosomes carry the hereditary material DNA (15%). In addition they are made up of protein (70%) and RNA (10%). Individual chromosomes are not visible in a non-dividing (resting) cell, but the chromosomal material can be seen,

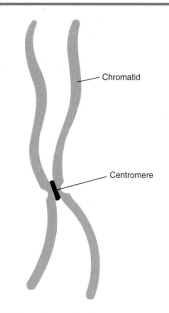

Fig. 8.1 Structure of a chromosome

especially if stained. This material is called **chromatin**. It is only at the onset of cell division that individual chromosomes become visible. They appear as long, thin threads between 0.25 μm and 50 μm in length. Each chromosome is seen to consist of two threads called **chromatids** joined at a point called the **centromere** (Fig. 8.1). Chromosomes vary in shape and size, both within and between species.

8.1.2 Chromosome number

The number of chromosomes varies from one species to another but is always the same for normal individuals of one species. Table 8.1 gives some idea of the range of chromosome number in different species. It can be seen that the numbers are not related to either the size of the organism or to its evolutionary status; indeed it is quite without significance.

Although the chromosome number of a cell varies from two to 300 or more, the majority of organisms have between ten and forty chromosomes in each of their cells. With well over one million different species, it follows that many share the same chromosome number, twenty-four being the most common.

TABLE 8.1 **The chromosome number of a range of species**

Species	Chromosome number
Certain roundworms	2
Crocus (*Crocus balansae*)	6
Fruit fly (*Drosophila melanogaster*)	8
Onion (*Allium cepa*)	16
Maize (*Zea mays*)	20
Locust (*Locusta migratoria*)	24
Lily (*Lilium longiflorum*)	24
Tomato (*Solanum lycopersicum*)	24
Cat (*Felis cattus*)	38
Mouse (*Mus musculus*)	40
Human (*Homo sapiens*)	46
Potato (*Solanum tuberosum*)	48
Horse (*Equus caballus*)	64
Dog (*Canis familiaris*)	78
Certain Protozoa	300+

8.2 Mitosis

Dividing cells undergo a regular pattern of events, known as the **cell cycle**. This cycle may be divided into two basic parts:

Interphase – when the cell undergoes a period of intense chemical activity. The amount of DNA is doubled during this

Nuclear division, or mitosis, typically occupies 5–10% of the total cycle. The cycle may take as little as 20 minutes in a bacterial cell, although it typically takes 8–24 hours.

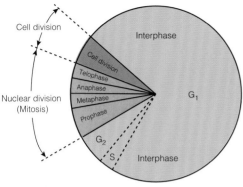

Fig. 8.2 The cell cycle

period. Interphase is divided into three phases: first growth phase (G_1) follows cell division and is the period where most cell organelles are synthesized and the cell grows rapidly. The synthesis (S) phase is next, during which DNA replication occurs and finally the second growth phase (G_2) when the centrioles replicate and energy stores increase.

Mitosis – when the nucleus is mechanically active as it divides.

In single-celled organisms and actively dividing cells, the cycle is continuous, i.e. the cells continue to divide regularly. In some cells, like those of the liver, division ceases after a certain time and only resumes if damaged or lost tissue needs replacing. In specialized tissues, such as nerves, division ceases completely once the cells are mature.

Interphase
Although often termed the **resting phase** because the chromosomes are not visible, interphase is in fact a period of considerable metabolic activity. It is during this phase that the DNA content of the cell is doubled. Duplication of the cell organelles also takes places at this time.

Prophase
The chromosomes become visible as long, thin tangled threads. Gradually they shorten and thicken, and each is seen to comprise two chromatids joined at the centromere. With the exception of higher plant cells which lack them, the centrioles migrate to opposite ends or **poles** of the cell. From each centriole, microtubules develop and form a star-shaped structure called an **aster**. Some of these microtubules, called **spindle fibres**, span the cell from pole to pole. Collectively they form the **spindle**. The nucleolus disappears and finally the nuclear envelope disintegrates, leaving the chromosomes within the cytoplasm of the cell.

Metaphase
The chromosomes arrange themselves at the centre or **equator** of the spindle, and become attached to certain spindle fibres at the centromere. Contraction of these fibres draws the individual chromatids slightly apart.

Anaphase
The centromeres split and further shortening of the spindle fibres causes the two chromatids of each chromosome to separate and migrate to opposite poles. The shortening of the spindle fibres is due to the progressive removal of the **tubulin** molecules of which they are made. The energy for this process is provided by mitochondria which are observed to collect around the spindle fibres.

Telophase
The chromatids reach their respective poles and a new nuclear envelope forms around each group. The chromatids uncoil and lengthen, thus becoming invisible again. The spindle fibres disintegrate and a nucleolus reforms in each new nucleus.

PROJECT

1. Examine longitudinal sections of bean root using a microscope. Locate the area just behind the tip of the root where mitosis is occurring.

2. Count the number of cells at interphase and at the various phases of mitosis and from this data determine the relative time taken for each phase.

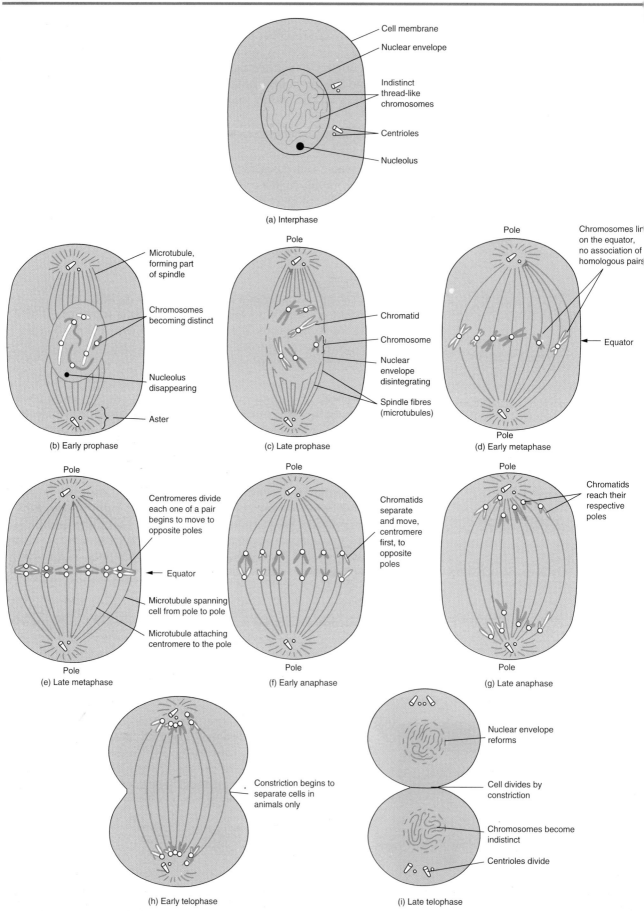

Cell membrane
Nuclear envelope
Indistinct thread-like chromosomes
Centrioles
Nucleolus

(a) Interphase

Microtubule, forming part of spindle
Chromosomes becoming distinct
Nucleolus disappearing
Aster

(b) Early prophase

Pole
Chromatid
Chromosome
Nuclear envelope disintegrating
Spindle fibres (microtubules)

(c) Late prophase

Pole
Chromosomes lie on the equator, no association of homologous pairs
Equator

(d) Early metaphase

Pole
Centromeres divide each one of a pair begins to move to opposite poles
Equator
Microtubule spanning cell from pole to pole
Microtubule attaching centromere to the pole
Pole

(e) Late metaphase

Pole
Chromatids separate and move, centromere first, to opposite poles
Pole

(f) Early anaphase

Pole
Chromatids reach their respective poles
Pole

(g) Late anaphase

Constriction begins to separate cells in animals only

(h) Early telophase

Nuclear envelope reforms
Cell divides by constriction
Chromosomes become indistinct
Centrioles divide

(i) Late telophase

Fig. 8.3 Stages of mitosis

136

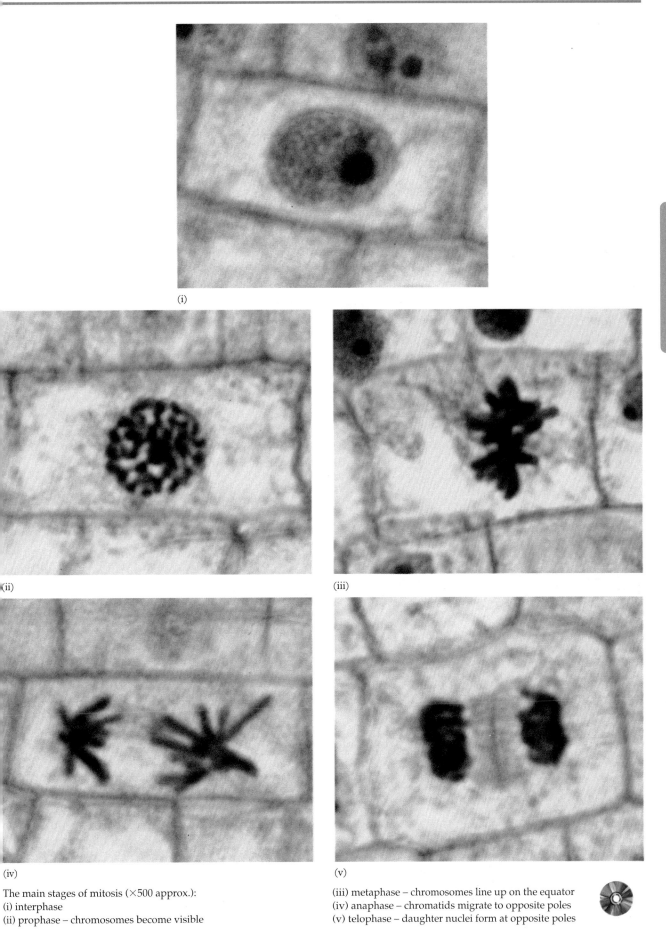

The main stages of mitosis (×500 approx.):
(i) interphase
(ii) prophase – chromosomes become visible
(iii) metaphase – chromosomes line up on the equator
(iv) anaphase – chromatids migrate to opposite poles
(v) telophase – daughter nuclei form at opposite poles

137

8.2.1 Differences between mitosis in plant and animal cells

The cells of higher plants lack centrioles and do not form asters as in animal cells. Spindle formation still takes place, however, and therefore the centrioles would seem not to be the centre of spindle synthesis.

In animal cells, cell division or **cytokinesis** occurs by the constriction of the centre of the parent cell from the outside inwards (Fig. 8.3h). In plant cells, however, the process occurs by the growth of a **cell plate** across the equator of the parent cell from the centre outwards. The plate is formed from the fusion of vesicles produced by the dictyosome. Cellulose is laid down on this plate to form the cell wall.

Whereas most animals cells are, if the need arises, capable of mitosis, only a specialized group of plant cells, called **meristematic cells**, are able to do so.

8.3 Meiosis (reduction division)

Meiosis involves one division of the chromosomes followed by two divisions of the nucleus and cell. The result is that the number of chromosomes in each cell is reduced by half. The **diploid (2n)** parent cell gives rise to **four haploid (n)** daughter cells. Meiosis occurs in the formation of gametes, sperm and ova, in animals, and in the production of spores in most plants.

Meiosis comprises two divisions:

1. First meiotic division – similar to mitosis except for a highly modified prophase stage.

2. Second meiotic division – a typically mitotic division.

The process is continuous but for convenience is divided into the same stages as mitosis. The symbols I and II indicate the first and second meiotic divisions respectively.

Interphase
The cell is in the non-dividing condition during which it replicates its DNA and organelles.

Prophase I
Organisms have two sets of chromosomes, one derived from each parent. Any two chromosomes which determine the same characteristics, e.g. eye colour, blood groups, etc., are called an **homologous pair**. Although each chromosome of a pair determines the same characteristics, they need not be identical. For instance, while one of the pair may code for blue eyes, the other may code for brown eyes.

Prophase I of meiosis is similar to prophase in mitosis, in that the chromosomes become visible, shorten and fatten, but differs in that they associate in their homologous pairs. They come together by a process termed **synapsis** and each pair is called a **bivalent**.

Each chromosome of the pair is seen to comprise two chromatids. These chromatids wrap around each other. The chromatids of the pair partially repel one another although they remain joined at certain points called **chiasmata** (singular – **chiasma**). It is at these points that chromatids may break and recombine with a different chromatid. This swapping of portions of chromatids is termed **crossing over**. The chromatids continue to repel one another although at this stage they still remain attached at the chiasmata. The nucleolus disappears and the nuclear envelope breaks down. Where present, the centrioles migrate to the poles and the spindle forms.

Metaphase I
The bivalents arrange themselves on the equator of the cell with each of a pair of homologous chromosomes orientated to opposite poles. This arrangement is completely random relative to the orientation of other bivalents. The genetic significance of this will be discussed later. The spindle fibres attached to the centromeres contract slightly, pulling the chromosomes apart as much as the chiasmata allow.

Anaphase I
The spindle fibres, which are attached to the centromeres, contract and pull the homologous chromosomes apart. One of each pair is pulled to one pole, its sister chromosome to the opposite one.

Telophase I
The chromosomes reach their opposite poles and a nuclear envelope forms around each group. In most cells the spindle fibres disappear and the chromatids uncoil. Cell division, or **cleavage**, may follow. The nucleus may enter interphase although no replication of the DNA takes place. In some cells this stage does not occur and the cell passes from anaphase I directly into prophase II.

Prophase II
In those cells where telophase and interphase take place, the nucleolus disappears and the nuclear envelope breaks down. Where centrioles are present these divide and move to opposite poles. The poles on this occasion are at right angles to the plane of the previous cell division and therefore the spindle fibres develop at right angles to the spindle axis of the first meiotic division.

Metaphase II
The chromosomes arrange themselves on the equator of the new spindle. The spindle fibres attach to the centromere of each chromosome.

Anaphase II
The centromeres divide and are pulled by the spindle fibres to opposite poles, carrying the chromatids with them.

Telophase II
Upon reaching their opposite poles, the chromatids unwind and become indistinct. The nuclear envelope and the nucleolus are reformed. The spindle disappears and the cells divide to give four cells, collectively called a **tetrad**.

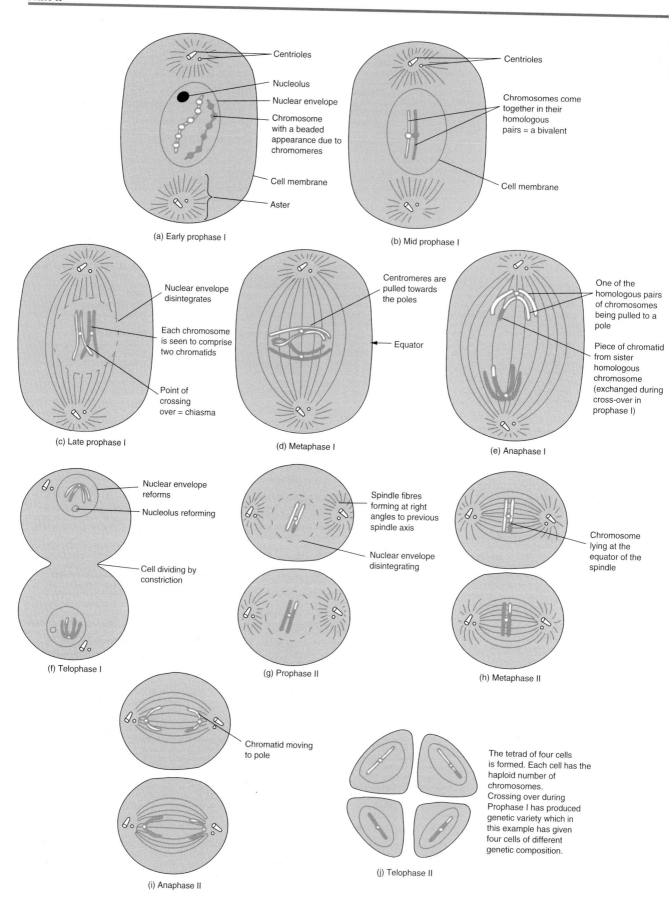

Fig. 8.4 Stages of meiosis (only one pair of chromosomes shown)

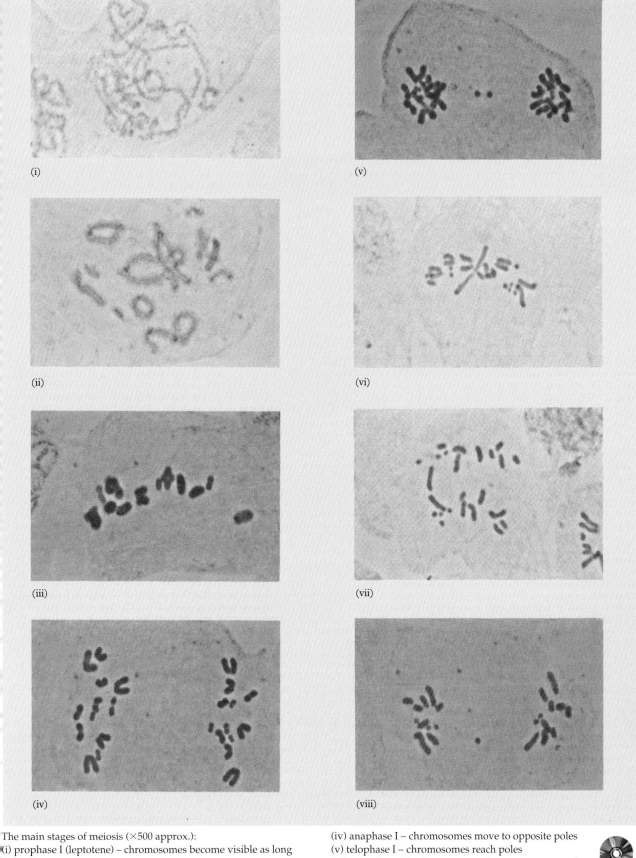

The main stages of meiosis (×500 approx.):

(i) prophase I (leptotene) – chromosomes become visible as long beaded structures

(ii) prophase I (diakinesis) – chiasmata and crossing over

(iii) metaphase I – homologous pairs of chromosomes line up on equator

(iv) anaphase I – chromosomes move to opposite poles

(v) telophase I – chromosomes reach poles

(vi) metaphase II – chromosomes line up on equator of daughter cell

(vii) anaphase II – chromatids move to opposite poles

(viii) telophase II – chromatids reach opposite poles

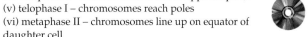

8.4 Comparison of mitosis and meiosis

The process of nuclear division is basically the same in mitosis and meiosis. The appearance and behaviour of the chromosomes is similar, but there are nevertheless some differences. These are listed in Table 8.2.

8.5 The significance of cell division

8.5.1 Significance of mitosis

The significance of mitosis is its ability to produce daughter cells which are exact copies of the parental cell. It is important in three ways.

(a) Growth

If a tissue is to extend by growth it is important that the new cells are identical to the existing cells. Cell division must therefore be by mitosis.

(b) Repair

Damaged cells must be replaced by exact copies of the originals if the repair is to return a tissue to its former condition. Mitosis is the means by which this is achieved.

(c) Asexual reproduction

If a species is successful in colonizing a particular habitat, there is little advantage, in the short term, in producing offspring which differ from the parents, because these may be less successful. It is better to establish quickly a colony of individuals which are similar to the parents. In simple animals and most plants this is achieved by mitotic divisions.

8.5.2 Significance of meiosis

The long-term survival of a species depends on its ability to adapt to a constantly changing environment. It should also be able to colonize a range of new environments. To achieve both these aims it is necessary for offspring to be different from their parents as well as different from each other.

There are three ways in which this variety is brought about with the aid of meiosis.

(a) Production and fusion of haploid gametes

Variety of offspring is increased by mixing the genotype of one parent with that of the other. This is the basis of the sexual process in organisms. It involves the production of special sex cells, called gametes, which fuse together to produce a new organism. Each gamete must contain half the number of chromosomes of the adult if the chromosome number is not to double at each generation. It is therefore essential that meiosis, which halves the number of chromosomes in daughter cells, occurs at some stage in the life cycle of a sexually reproducing organism. Meiosis is thus instrumental in permitting variety in organisms, and giving them the potential to evolve.

TABLE 8.2 **Differences between mitosis and meiosis**

Mitosis	Meiosis
A single division of the chromosomes and the nucleus	A single division of the chromosomes but a double division of the nucleus
The number of chromosomes remains the same	The number of chromosomes is halved
Homologous chromosomes do not associate	Homologous chromosomes associate to form bivalents in prophase I
Chiasmata are never formed	Chiasmata may be formed
Crossing over never occurs	Crossing over may occur
Daughter cells are identical to parent cells (in the absence of mutations)	Daughter cells are genetically different from parental ones
Two daughter cells are formed	Four daughter cells are formed, although in females only one is usually functional
Chromosomes shorten and thicken	Chromosomes coil but remain longer than in mitosis
Chromosomes form a single row at the equator of the spindle	Chromosomes form a double row at the equator of the spindle during metaphase I
Chromatids move to opposite poles	Chromosomes move to opposite poles during the first meiotic division

(b) The creation of genetic variety by the random distribution of chromosomes during metaphase I

When the pairs of homologous chromosomes arrange themselves on the equator of the spindle during metaphase I of meiosis, they do so randomly. Although each one of the pair determines the same general features, they differ in the detail of these features. The random distribution and consequent independent assortment of these chromosomes produces new genetic combinations. A simple example is shown in Fig. 8.5.

(c) The creation of genetic variety by crossing over between homologous chromosomes

During prophase I of meiosis, equivalent portions of homologous chromosomes may be exchanged. In this way new genetic combinations are produced and linked genes separated.

The variety which meiosis brings about is essential to the process of evolution. By providing a varied stock of individuals it permits the natural selection of those best suited to the existing conditions and so ensures that species constantly change and adapt when these conditions alter. This is the main significance of meiosis.

In arrangement 1, the two pairs of homologous chromosomes orientate themselves on the equator in such a way that the chromosome carrying the allele for brown eyes and the one carrying the allele for blood group A migrate to the same pole. The alleles for blue eyes and blood group B migrate to the opposite pole. Cell 1 therefore carries the alleles for brown eyes and blood group A while cell 2 carries the ones for blue eyes and blood group B.

In arrangement 2, the left hand homologous pair of chromosomes is shown orientated the opposite way around. As this orientation is random this arrangement is equally as likely as the first one. The result of this different arrangement is that cell 3 carries the alleles for blue eyes and blood group A whereas cell 4 carries ones for brown eyes and blood group B.

All four resultant cells are different from one another. With more homologous pairs the number of possible combinations becomes enormous. A human, with 23 such pairs, has the potential for $2^{23} = 8\,388\,608$ combinations.

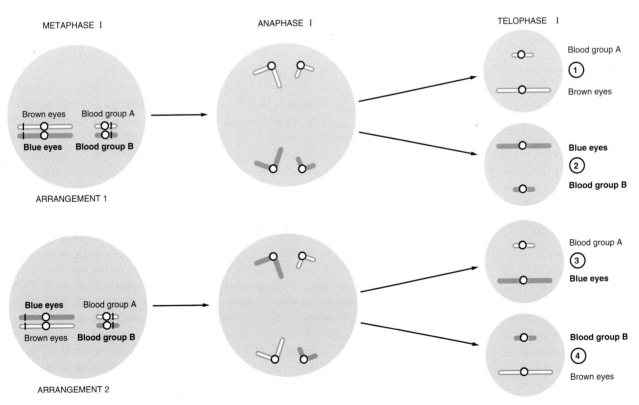

Fig. 8.5 Variety brought about by meiosis

APPLICATION

Cancer

Superficially cancer seems to be a range of diseases in which tumours arise at different sites, grow at different rates and may be benign or lethal. However, in all cancers a population of cells multiplies in an unregulated manner independent of normal control mechanisms. Most cancers are believed to arise from single cells which transform from a normal into a malignant state. If the abnormal cell is not recognized and destroyed by the immune system it multiplies into a small mass and then into a large primary tumour with its own blood supply. Primary tumours can be managed and the patients are often cured. The more usual causes of death are the effects of secondary tumours, or metastases, in distant organs. Transport within the lymph duct will give rise to metastases in lymph nodes adjacent to the primary tumour while transport within the blood may give rise to more distant secondaries.

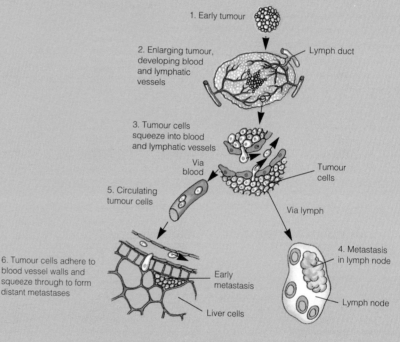

The early development and spread of secondary tumour

Cancers can occur at any age but overall incidence rises rapidly with age and they are a more prevalent cause of death in men than in women. Cancers which are clearly age-related are carcinomas of the gut, skin, urinary tract and some leukaemias. In breast cancer the incidence rises steeply up to menopause and then declines; cervical cancer peaks in the 30s.

It has been suggested that the cumulative effect of carcinogens in the environment may account for cancers whose incidence increases with age. The pattern of breast cancer suggests a hormonal link and cervical cancer correlates with sexual behaviour, age at which intercourse began and possibly the number of sexual partners.

As a cause of death today, cancer is second only to diseases of the circulatory system and in terms of working years lost it

APPLICATION continued

exceeds all other diseases. In the Western world the commonest forms of cancer are those of the lung, breast, skin, gut and prostate gland. In men the greatest killer is lung cancer while in women it is breast cancer. There are, however, considerable national variations. Skin cancer is 200 times more prevalent in parts of Australia than in India – possibly linked to the long exposure of Caucasian skin to sunlight. Stomach cancer is 30 times more common in Japan than in parts of Britain – possibly due to dietary or genetic effects.

The transformation of a normal cell into a cancerous one may be brought about by exposure to certain chemicals, by radiation or by the action of retroviruses (see Section 5.2.2). Chemical carcinogens fall into several classes which include polycyclic hydrocarbons, nitrosamines, aromatics, amines and others. The first of these classes is found in soot and cigarette smoke. It has been estimated that 25–30% of environmentally associated cancers can be attributed to smoking and it is now the main cause of lung cancer which kills over 40 000 people a year in Britain.

Carcinogenic transformation may also be brought about by exposure to radiation, the most damaging being short wavelengths – X-rays, gamma rays and some ultra-violet rays. Ultra-violet rays do not penetrate beyond superficial layers of the skin but they are responsible for a significant number of skin cancers.

Some cancers, including Burkitt's lymphoma and AIDS-related Kaposi's sarcoma have been linked to viruses, namely EB virus and HIV-I respectively, but in general it is now recognized that the relationship involves certain cancer-causing genes termed oncogenes rather than the viruses themselves.

If cancer cannot be prevented then early detection and diagnosis can at least improve the likelihood of a cure. Progress in this field has been rapid and now relies on specific biochemical and chemical reagents as well as sophisticated instruments such as ultrasound, computer tomography (CT) scanning and magnetic resonance imaging (MRI).

Once diagnosed, the choice of therapy will depend on the nature, site and extent of the tumours but may include surgery, radiotherapy, the use of drugs or immunotherapy. Surgery remains a major weapon in the fight against cancer although it is often used in association with radio- or chemotherapy. Radiotherapy has been used in cancer treatments for many years although many tumours respond poorly, especially those with low internal oxygen levels. There are many classes of anti-cancer drugs in use but the search still goes on for derivatives which retain their activity but have reduced toxicity.

It is generally accepted that the immune system plays an important role in preventing cancer and research is now centred on finding suitable molecules to stimulate the immune reaction. Such substances include interferons and interleukins, both produced by genetically engineered bacteria.

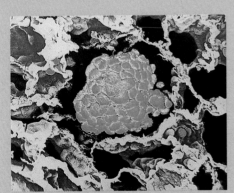

False colour scanning electron micrograph (SEM) of a bronchial carcinoma (lung tumour) filling an alveolus

8.6   Questions

1.   (a) Describe fully the similarities and differences
     between cell division by mitosis and cell
     division by meiosis.   (11 marks)
     (b) Explain the advantages and disadvantages
     for organisms in which meiosis takes place
     during the life cycle, compared with those
     that reproduce only by mitosis.   (9 marks)
                                          (Total 20 marks)

     WJEC June 1992, Paper A1, No. 2

2.   Write an essay on the behaviour of the chromosomes
during nuclear division.   (Total 25 marks)

     O&C SEB June 1993, Paper 2/3, No. 3

3.   Figures **A** and **B** show two stages of meiosis in a
diploid plant.

Figure A

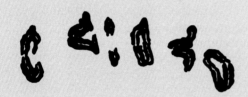

Figure B

(a) Give **one** piece of evidence from Figure **A** to
     support the fact that this cell is undergoing
     meiosis.   (1 mark)
(b) Giving **one** reason for your answer in each
     case, identify the stage of division shown in:
     (i) Figure **A**; (ii) Figure **B**.   (4 marks)
(c) What is the haploid chromosome number in
     this plant?   (1 mark)
                                          (Total 6 marks)

     AEB June 1991, Paper I, No. 16

4.   The diagram shows the metaphase I stage of
meiosis in a cell, where the diploid number is eight
(2n = 8).

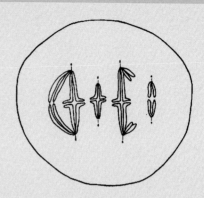

Draw a diagram to show the appearance of the same
cell during anaphase I. Label **two** structures which are
visible only when a cell is dividing.   (Total 5 marks)

     ULEAC January 1994, Paper I, No. 4

5.   Diagram A below shows the different phases
during the cell cycle of an animal cell. Diagram B
shows the quantity of DNA present during these
different phases. G_1 and G_2 represent growth phases,
separated by an intermediate phase S, in the cell cycle.

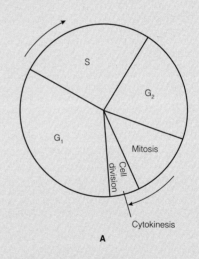

A

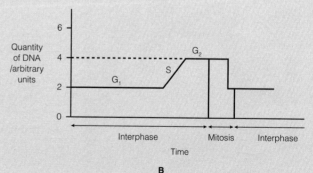

B

(a) Describe how the quantity of DNA in cells is
     increased during phase S, between growth
     phase G_1 and growth phase G_2, leading up to
     mitosis.   (2 marks)

(b) (i) What would be the quantity of DNA in arbitrary units in a cell at the end of mitosis? *(1 mark)*

(ii) Describe how the quantity of DNA is returned to this level. *(2 marks)*

(c) Name **one** metabolic process of a cell which stops when a nucleus divides. *(1 mark)*

(d) Indicate the importance of mitosis in the cell cycle. *(2 marks)*

(Total 8 marks)

ULEAC January 1993, Paper I, No. 2

Fig. 1 shows four animal cells in different stages of mitotic division.

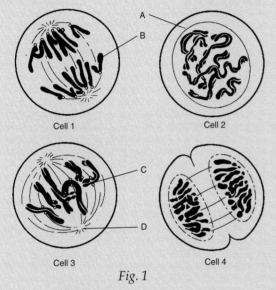

Fig. 1

(a) Name the structures labelled **A**, **B**, **C** and **D**. *(2 marks)*

(b) (i) Name the stages of division shown by cells 1 and 3.

(ii) Using the number given to each cell above, arrange the stages as they occur in the mitotic sequence. *(3 marks)*

(c) State precisely what is occurring in cell 4. *(3 marks)*

(d) State how mitosis maintains genetic stability in an organism. *(2 marks)*

(Total 10 marks)

UCLES June 1993, Paper II, No. 1

7. Fig. 1.1 shows an animal cell (**A**) undergoing meiosis.

(a) State the diploid number of chromosomes of the cell. *(1 mark)*

(b) Suggest where the cell could be found in a mammal. *(1 mark)*

(c) Name the stage of division shown at **B**, giving **one** reason for your choice. *(2 marks)*

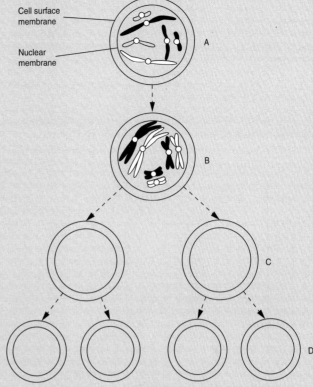

Fig. 1.1

(d) On Fig. 1.1, draw in the chromosomes that could be found in the cells at **C** and **D**. Be careful to distinguish the different chromosomes by shading and size, as given in the diagram. *(4 marks)*

(e) Explain **two** different ways in which meiosis contributes to genetic variation. State at which stage of division each occurs. *(6 marks)*

(Total 14 marks)

UCLES (Modular) June 1993,
Genetics and its Applications, No. 1

8. Fig. 2 represents the relative amounts of DNA per nucleus during several cell divisions in animal tissue.

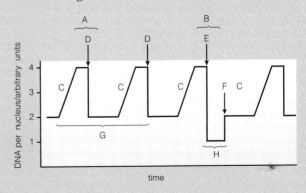

Fig. 2

(a) Name the type of nuclear division occurring at **A** and **B**. Give reasons for your answer. *(4 marks)*

147

(b) State what is occurring at the points marked
C, **D**, **E** and **F**. *(4 marks)*

(c) What type of cells are produced at **G** and **H**?
(2 marks)
(Total 10 marks)

UCLES June 1992, Paper II, No. 2

9. The graph shows the movement of chromosomes during mitosis. Curve **A** shows the mean distance between the centromeres of the chromosomes and the corresponding pole of the spindle.

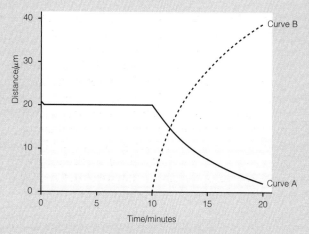

(a) What is represented by curve **B**? *(1 mark)*

(b) (i) At what time did anaphase begin?
(1 mark)

(ii) Explain how **one** piece of evidence from the graph supports your answer.
(2 marks)
(Total 4 marks)

AEB June 1992, Paper I, No. 8

10. (a) Diagram I shows a three dimensional view of one of the stages of mitosis in a typical animal cell. The chromosomes are shown lying in the same plane.

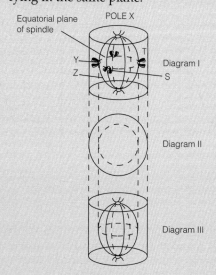

(i) Name the stage shown in the diagram **I**.
(1 mark)

(ii) Name the structure labelled **Y**. *(1 mark)*

(iii) What is the function of:
1. structure **Z**?
2. mitosis? *(2 marks)*

(iv) Diagram **II** shows the outline of the equator of the cell. Draw **all** the chromosomes in diagram **I** as they would appear looking down from pole **X**. *(4 marks)*

(v) Copy outline diagram **III** and draw the chromosomes labelled **S** and **T** as they would appear in the next stage of mitosis. *(2 marks)*

(b) The details of mitosis vary in different organisms. The diagram shows nuclear division in a unicellular organism.

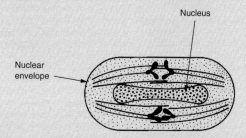

Use the information in the diagram to suggest **two** ways in which nuclear division in this species differs from that in an animal cell. *(2 marks)*
(Total 12 marks)

WJEC June 1994, Paper A2, No. 7

11. (a) What is:
(i) a bivalent; *(1 mark)*
(ii) a chiasma? *(1 mark)*

Diagrams **A**, **B** and **C** show the same stage in mitosis, meiosis I and meiosis II in a plant cell.

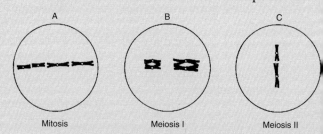

(b) Identify the stage shown giving your reason.
(1 mark)

(c) The cell in diagram **A** has 20 units of DNA. How many units of DNA would there be in a cell from this plant at the end of:
(i) mitosis; *(1 mark)*
(ii) meiosis? *(1 mark)*
(Total 5 marks)

AEB June 1994, Common Paper 1, No.16

9 Heredity and genetics

Gregor Mendel

It has been estimated that all the ova from which the present human population was derived could be contained within a five-litre vessel. All the sperm which fertilized these ova could be contained within a thimble. Put another way, all the information necessary to produce in excess of 5 000 000 000 different humans can be stored within this relatively tiny volume. Indeed, we now know that the genetic information itself occupies only a small proportion of the ovum. Imagine the space occupied if all the characteristics of every living human being were printed in book form (see 'Did You Know?' on p. 114). Whatever else it may be, the genetic information is remarkably condensed. In Chapter 7 we described the chemical nature of this genetic information, namely DNA. In this chapter we shall concern ourselves with the means by which it is transmitted from generation to generation.

Genetics is a fundamental and increasingly important branch of biology. Humans have unknowingly used genetic principles in the breeding of animals and plants for many thousands of years. An understanding of the underlying genetic principles is, however, fairly recent. In fact the term genetics was first used only at the beginning of this century. The understanding of the chemical foundations of heredity and genetics came even more recently with the discovery of the structure of DNA by Watson and Crick in 1953.

Observation of individuals of the same species shows them all to be recognizably similar. This is **heredity**. Closer inspection reveals minor differences by which each individual can be distinguished. This is **variation**.

The genetic composition of an organism is called the **genotype**. This often sets limits within which individual characteristics may vary. Such variation may be due to the effect of environmental influences, e.g. the genotype may determine a light-coloured skin, but the precise colour of any part of the skin will depend upon the extent to which it is exposed to sunlight. The **phenotype**, or set of characteristics, of an individual is therefore determined by the interaction between the genotype and the environment. Any change in the genotype is called a **mutation** and may be inherited. Any change in the phenotype only is called a **modification** and it is not inherited.

9.1 Mendel and the laws of inheritance

Gregor Mendel (1822–84) was an Austrian monk and teacher. He studied the process of heredity in selected features of the garden pea *Pisum sativum*. He was not the first scientist to study heredity, but he was the first to obtain sufficiently numerous, accurate and detailed data upon which sound scientific conclusions could be based. Partly by design and partly by luck, Mendel made a suitable choice of characteristics for study. He isolated pea plants which were **pure-breeding**. That is, when bred with each other, they produced consistently the same characteristics over many generations. He referred to each character as a **trait**. He chose traits which had two contrasting features, e.g. he chose stem length, which could be either long or short and flower colour which could be red or white. It must be remembered that Mendel began his ten-year-long experiments in 1856, when the nature of chromosomes and genes was yet to be discovered.

9.1.1 Monohybrid inheritance (Mendel's Law of Segregation)

Monohybrid inheritance refers to the inheritance of a single character only. One trait which Mendel studied was the shape of the seed produced by his pea plants. This showed two contrasting forms, round and wrinkled. When he crossed plants which were pure-breeding for round seed with ones pure-breeding for wrinkled seed, all the resulting plants produced round seed. The first generation of a cross is referred to as the **first filial generation (F_1)**. When individuals of the F_1 generation were intercrossed the resulting **second filial generation (F_2)** produced 7324 seeds, 5474 of which were round and 1850 wrinkled. This is a ratio of 2.96 : 1.

In all his crosses, Mendel found that one of the contrasting features of a pair was not represented in the F_1 generation. This feature reappeared in the F_2 generation where it was consistently outnumbered 3 to 1 by the contrasting feature.

The significance of these findings was that the F_1 seeds were not intermediate between the two parental types, i.e. partly wrinkled, partly smooth. This shows that there was no blending or mixing of the features. It also indicated that as only one of the features expressed itself in the F_1, this feature was **dominant** to the other. The feature which does not express itself in the F_1 is said to be **recessive**. In the example given, round is dominant and wrinkled is recessive.

In interpreting his results, Mendel concluded that the features were passed on from one generation to the next via the gametes. The parents he decided must possess two pieces of information about each character. However, only one of these pieces of information was found in an individual gamete. On the basis of this he formulated his first law, the **Law of Segregation**, which states:

The characteristics of an organism are determined by internal factors which occur in pairs. Only one of a pair of such factors can be represented in a single gamete.

We know that Mendel's 'factors' are specific portions of a chromosome called **genes**. We also know that the process which produces gametes with only one of each pair of factors is meiosis. On the basis of his results, Mendel had effectively predicted the existence of genes and meiosis.

9.1.2 Representing genetic crosses

Genetic crosses are usually represented in a form of shorthand. There is more than one system of this shorthand, but the following one has been adopted here because it is both quick and less liable to errors, especially under the pressures of an examination.

TABLE 9.1

Instruction	Reason/notes	Example [round and wrinkled seed]
Choose a single letter to represent each characteristic	An easy form of shorthand. In some conventional genetic crosses, e.g. in *Drosophila*, there are set symbols, some of which use two letters	—
Choose the first letter of one of the contrasting features	When more than one character is considered at one time such a logical choice means it is easy to identify which letter refers to which character	Choose either R (round) or W (wrinkled)
If possible, choose the letter in which the upper and lower case forms differ in shape as well as size	If the upper and lower case forms differ it is almost impossible to confuse them regardless of their size	Choose R, because the upper case form (R) differs in shape from the lower case form (r), whereas W and w differ only in size, and are more likely to be confused.
Let the upper case letter represent the dominant feature and the lower case letter the recessive one. Never use two different letters where one character is dominant. Always state clearly what feature each symbol represents	The dominant and recessive features can easily be identified. Do *not* use two different letters as this indicates incomplete dominance or codominance	Let R = round and r = wrinkled Do *not* use R for round and W for wrinkled
Represent the parents with the appropriate pairs of letters. Label them clearly as 'parents' and state their phenotypes	This makes it clear to the reader which the symbols refer to	Parents: Round seed RR v Wrinkled seed rr
State the gametes produced by each parent. Label them clearly, and encircle them. Indicate that meiosis has occurred	This explains why the gametes only possess one of the two parental factors. Encircling them reinforces the idea that they are separate	Gametes: meiosis ↓ Ⓡ meiosis ↓ Ⓡ
Use a type of chequerboard or matrix, called a **Punnett square**, to show the results of the random crossing of the gametes. Label male and female gametes even though this may not affect the results	This method is less liable to error than drawing lines between the gametes and the offspring. Labelling the sexes is a good habit to acquire – it has considerable relevance in certain types of crosses, e.g. sex-linked crosses	♂ gametes: ♀ gametes / Ⓡ Ⓡ; Ⓡ Rr Rr; Ⓡ Rr Rr
State the phenotype of each different genotype and indicate the numbers of each type. Always put the upper case (dominant) letter first when writing out the genotype.	Always putting the dominant feature first can reduce errors in cases where it is not possible to avoid using symbols with the upper and lower case letters of the same shape	All offspring are plants producing round seed (Rr)

NB Always carry out the above procedures in their entirety. Once you have practised a number of crosses, it is all too easy to miss out stages or explanations. Not only does this lead to errors, it often makes your explanations impossible for others to follow. *You* may understand what you are doing, but if the reader cannot follow it, it isn't much use, neither will it bring full credit in an examination.

151

9.1.3 Genetic representation of the monohybrid cross

Using the principles outlined in Table 9.1, the full genetic explanation of one of Mendel's experiments is shown below.

Let R = allele for round seed
    r = allele for wrinkled seed

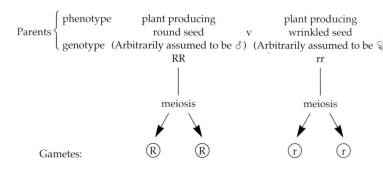

Parents
- phenotype — plant producing round seed v plant producing wrinkled seed
- genotype (Arbitrarily assumed to be ♂) (Arbitrarily assumed to be ♀)
  RR   rr

meiosis   meiosis

Gametes: R   R   r   r

		♂ gametes	
♀ gametes		R	R
F₁ generation:	r	Rr	Rr
	r	Rr	Rr

All offspring are plants producing round seed (Rr)

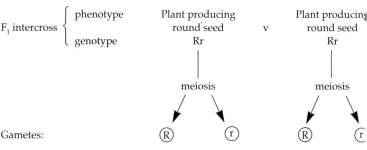

F₁ intercross
- phenotype — Plant producing round seed v Plant producing round seed
- genotype — Rr   Rr

meiosis   meiosis

Gametes: R   r   R   r

		♂ gametes	
♀ gametes		R	r
F₂ generation:	R	RR	Rr
	r	Rr	rr

3 plants producing round seeds (1 × RR + 2 × Rr)
1 plant producing wrinkled seeds (rr)

Mendel's actual results gave a ratio of 2.96 : 1, a very good approximation to the 3 : 1 ratio which the theory suggests should be achieved. Any discrepancy is due to statistical error. Such errors are inevitable. Imagine for instance tossing a coin ten times – it should in theory come down heads five times and tails five times. More often than not, some other ratio is achieved in practice. The actual results are rarely exactly the same as predicted by the theory. The larger the sample, the more nearly the results approximate to the theoretical value. This was an essential aspect of Mendel's experiments. Probably because he was trained partly as a mathematician, he appreciated the need to collect large numbers of offspring if he was to draw meaningful conclusions from his experiments.

The use of 'F_1 generation' should be limited to the offspring of homozygous parents. Similarly, 'F_2 generation' should refer only to the offspring of the F_1 generation. In all other cases 'offspring (1)' should replace 'F_1 generation', and 'offspring (2)' should replace 'F_2 generation'. The complete set of headings in order therefore will be:

Parents: phenotypes

Parents: genotypes

Gametes

Offspring (1) genotypes

Offspring (1) phenotypes

Gametes

Offspring (2) genotypes

Offspring (2) phenotypes

Whether the variation from an expected ratio is the result of statistical chance or not can be tested for mathematically using the chi-squared test. Details of this are given in Section 10.3.

9.1.4 Genes and alleles

A character such as the shape of the seed coat in peas is determined by a single gene. The gene is therefore the basic unit of inheritance. It is a region of the chromosome or, more specifically, a length of the DNA molecule, which has a particular function (see Section 7.7). Each gene may have two, or occasionally more, alternative forms. Each form of the gene is called an **allele**. The gene for the shape of the seed coat in peas has two alleles, one determining round shape, the other wrinkled. The position of a gene within a DNA molecule is called the **locus**. When two identical alleles occur together at the same locus on a chromosome, they are said to be **homozygous**, e.g. when two alleles for round seeds occur together (RR) they are said to be **homozygous dominant**. Similarly, the two alleles for wrinkled seeds (rr) are referred to as **homozygous recessive**. Where the two alleles differ (Rr) they are termed **heterozygous**.

9.1.5 Dihybrid inheritance (Mendel's Law of Independent Assortment)

Dihybrid inheritance refers to the simultaneous inheritance of two characters. In one of his experiments Mendel investigated the inheritance of seed shape (round v. wrinkled) and seed colour (green v. yellow) at the same time. He knew from his monohybrid crosses that round seeds were dominant to wrinkled ones and yellow seeds were dominant to green. He chose to cross plants with both dominant features (round and

---

PROJECT

1. Cross *Drosophila* of two types e.g. normal $\times$ vestigial wing.

2. Collect the offspring and cross them to get a second generation.

3. Count the numbers of normal and vestigial winged flies in the second generation.

4. Using the χ^2 test, determine if your numbers agree with the Mendelian ratio of 3:1.

yellow) with ones that were recessive for both (wrinkled and green). The F_1 generation yielded plants all of which produced round, yellow seeds – hardly surprising as these are the two dominant features.

Mendel planted the F_1 seeds, raised the plants and allowed them to self-pollinate. He then collected the seeds. Of the 556 seeds produced, the majority, 315, possessed the two dominant features – round and yellow. The smallest group, 32, possessed the two recessive features – wrinkled and green. The remaining 209 seeds were of types not previously found. They combined one dominant and one recessive feature; 108 were round (dominant) and green (recessive) and 101 were wrinkled (recessive) and yellow (dominant). At first inspection these results may appear to contradict those obtained in the monohybrid cross, but as Table 9.2 shows, the ratio of dominant to recessive for each feature is still $3:1$, as expected.

TABLE 9.2 **Results of Mendel's dihybrid cross**

Parents: round, yellow seeds v. wrinkled, green seeds

F_1 generation: all round, yellow seeds

F_2 generation:

		Seed shape			
		Round	Wrinkled	Total	Approx. ratio
Seed colour	Yellow	315	101	416	3 yellow
	Green	108	32	140	1 green
	Total	423	133		
	Approx. ratio	3 round	1 wrinkled		

Approx. ratio:

round, yellow (2 dominants)	round, green (dominant + recessive)	wrinkled, yellow (recessive + dominant)	wrinkled, green (2 recessives)
9 :	3 :	3 :	1

The significance of these findings was that as the features of seed shape and colour had each produced a $3:1$ ratio (dominant : recessive), the two features had behaved completely independently of one another. The presence of one had not affected the behaviour of the other. On the basis of these findings, Mendel formulated his second law, the Law of Independent Assortment, which states:

Each of a pair of contrasted characters may be combined with either of another pair.

With our present knowledge of genetics the law could now be rewritten as:

Each member of an allelic pair may combine randomly with either of another pair.

PROJECT

1. From Table 9.2 we find that:
   315 seeds were round and yellow
   108 were round and green
   101 were wrinkled and yellow
   32 were wrinkled and green.

   Using the χ^2 test, find out if these figures agree with the Mendelian ratio of $9:3:3:1$. Look up other numbers counted by Mendel and do the same.

2. Carry out dihybrid crosses, for example with *Drosophila*, count the F2 offspring and find out if the numbers agree with Mendel's $9:3:3:1$ ratio.

9.1.6 Genetic representation of the dihybrid cross

Using the principles outlined in Table 9.1, the full genetic explanation of this dihybrid cross is shown below.

Let R = allele for round seed
r = allele for wrinkled seed
G = allele for yellow seed
g = allele for green seed

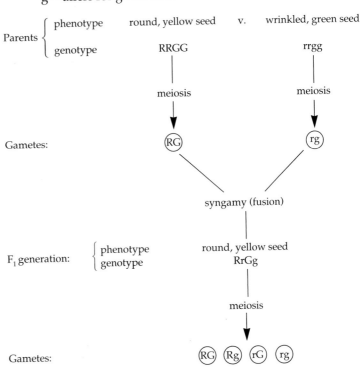

As the plants are self-pollinated the male and female gametes are of the same types. The offspring of this cross may therefore be represented in the following Punnett square.

♀ gametes	♂ gametes			
	RG	Rg	rG	rg
RG	RRGG	RRGg	RrGG	RrGg
Rg	RRGg	RRgg	RrGg	Rrgg
rG	RrGG	RrGg	rrGG	rrGg
rg	RrGg	Rrgg	rrGg	rrgg

In the following list, '−' represents either the dominant or recessive allele.

Total

R–G– = round, yellow seed   9 (315)

R–gg = round, green seed   3 (108)

rrG– = wrinkled, yellow seed   3 (101)

rrgg = wrinkled, green seed   1 (32)

Allowing for statistical error, Mendel's results (shown in brackets) were a reasonable approximation to the expected 9 : 3 : 3 : 1 ratio.

155

9.2   The test cross

One common genetic problem is that an organism which shows a dominant character can have two possible genotypes. For example, a plant producing seeds with round coats could either be homozygous dominant (RR) or heterozygous (Rr). The appearance of the seeds (phenotype) is identical in both cases. It is often necessary, however, to determine the genotype accurately. This may be achieved by crossing the organism of unknown genotype with one whose genotype is accurately known. One genotype which can be positively identified from its phenotype alone is one which shows the recessive feature. In the case of the seed coat, any pea seed with a wrinkled coat must have the genotype rr. By crossing the dominant character, the unknown genotype can be identified. To take the above example

Let R = allele for round seeds
      r = allele for wrinkled seeds

If the plant producing round seed has the genotype RR:

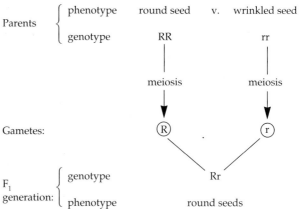

The only possible offspring are plants which produce round seeds

If the plant producing round seeds has the genotype Rr:

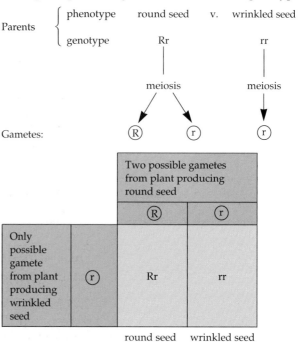

TABLE 9.3 Dihybrid backcross

Possible genotypes of plant producing round, yellow seeds	Possible gametes	Genotypes of offspring crossed with plant producing wrinkled, green seeds (gamete = rg)	Phenotype (type of seeds produced)
RRGG	RG	RrGg	All round and yellow
RrGG	RG	RrGg	$\frac{1}{2}$ round and yellow
	rG	rrGg	$\frac{1}{2}$ wrinkled and yellow
RRGg	RG	RrGg	$\frac{1}{2}$ round and yellow
	Rg	Rrgg	$\frac{1}{2}$ round and green
RrGg	RG	RrGg	$\frac{1}{4}$ round and yellow
	Rg	Rrgg	$\frac{1}{4}$ round and green
	rG	rrGg	$\frac{1}{4}$ wrinkled and yellow
	rg	rrgg	$\frac{1}{4}$ wrinkled and green

The offspring comprise equal numbers of plants producing round seeds and ones producing wrinkled seeds.

If some of the plants produce seeds with wrinkled coats (rr), then the unknown genotype must be Rr. An exact 1:1 ratio as above is not often achieved in practice, but this is unimportant as the presence of a single plant producing wrinkled seeds is proof enough (the possibility of a mutation must be discounted as it is highly unlikely and totally unpredictable). Such a plant with its rr genotype could only be produced if both parents donated an r gamete. The only way a plant which produces round seed can donate such a gamete is if it is heterozygous (Rr).

If all the offspring of our test cross were plants producing round seed, then no definite conclusions could be drawn, since both parental genotypes (Rr and RR) are capable of producing such offspring. However, provided a large number of offspring are produced, the absence of ones producing wrinkled seeds would strongly indicate that the unknown genotype was RR. Had it been Rr, half the offspring should have produced wrinkled seeds. While it would be theoretically possible for no wrinkled seeds to arise, this would be highly improbable where the sample was large.

It is possible to perform a dihybrid test cross. A plant which produces round, yellow seeds has four possible genotypes, namely: RRGG, RrGG, RRGg and RrGg. To determine the genotype of such a plant, it must be crossed with one producing wrinkled, green seeds. Such a plant has only one possible genotype, rrgg, and produces only one type of gamete, namely rg. The outcome of each of the crosses is shown in Table 9.3.

From the table it can be seen that the unknown genotypes can be identified from the results of the test cross as follows. If the offspring contain at least one plant producing wrinkled, yellow seeds, the unknown genotype is RrGG; if round, green seeds it is RRGg, and if wrinkled, green seeds it is RrGg. If the number of offspring is large and *all* produce round, yellow seeds it is highly probable that the genotype is RRGG.

9.3 Sex determination

In humans there are twenty-three pairs of chromosomes. Of these, twenty-two pairs are identical in both sexes. The twenty-third pair, however, is different in the male from the female. The twenty-two identical pairs are called **autosomes** whereas the twenty-third pair are referred to as **sex chromosomes** or **heterosomes**. In females, the two sex chromosomes are identical and are called **X chromosomes**. In males, an X chromosome is also present, but the other of the pair is smaller in size and called the **Y chromosome**. Unlike other features of an organism, sex is determined by chromosomes rather than genes.

Humans of the genotype XXY arise from time to time and are phenotypically male, while genotypes with just one X chromosome (XO) are phenotypically female. This suggests that it is the presence of the Y chromosome which makes a human male; in its absence the sex is female. How then does the Y chromosome determine maleness? The Y chromosome possesses several copies of a **testicular differentiating gene** which codes for the production of a substance which causes the undifferentiated gonads to become testes. In the absence of this gene and hence this substance, the gonads develop into ovaries.

It can be seen that in humans the female produces gametes which all contain an X chromosome and are therefore the same. She is called the **homogametic sex** ('same gametes'). The male, however, produces gametes of two genetic types: one which contains an X chromosome, the other a Y chromosome. The male is called the **heterogametic sex** ('different gametes').

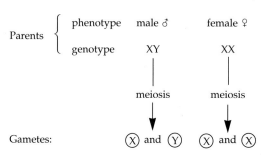

Offspring	♂ gametes	
♀ gametes	Ⓧ	Ⓨ
Ⓧ	XX	XY
Ⓧ	XX	XY

Sex ratio   1 female : 1 male

Sex determination differs in other organisms. In birds, most reptiles, some fish and all butterflies, the male is the homogametic sex (XX) and the female is the heterogametic sex (XY). In some insects, while the female is XX, the Y chromosome is absent in the male, which is therefore XO. In the fruit fly *Drosophila*, the female is XX and the male XY; however, the Y chromosome is not smaller, as in humans, but simply a different shape.

9.4   Linkage

For just twenty-three pairs of chromosomes to determine the many thousands of different human characteristics, it follows that each chromosome must possess many different genes. Any two genes which occur on the same chromosome are said to be **linked**. All the genes on a single chromosome form a **linkage group**.

Under normal circumstances, all the linked genes remain together during cell division and so pass into the gamete, and hence the offspring, together. They do not therefore segregate in accordance with Mendel's Law of Independent Assortment. Fig. 9.1 shows the different gametes produced if a pair of genes A and B are linked rather than on separate chromosomes.

If genes A and B occur on the same chromosome i.e. are linked

Only one homologous pair is needed to accommodate all four alleles

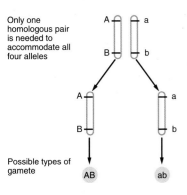

Possible types of gamete

If genes A and B occur on separate chromosomes i.e. are not linked

Two homologous pairs are needed to accommodate all four alleles

According to Mendel's Law of Independent Assortment, any one of a pair of contrasted characters may combine with any of another pair. There are thus four different possible types of gamete

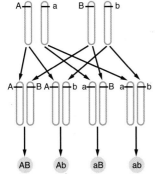

Fig. 9.1 Comparison of gametes produced by an organism heterozygous for two genes A and B, when they are linked and not linked.

9.4.1 Crossing over and recombination

It is known that genes for flower colour and fruit colour in tomatoes are on the same chromosome. Plants with yellow flowers bear red fruit, those with white flowers bear yellow fruit. If the two types are crossed, the following results are obtained.

Let R = allele for red fruit (dominant) and
r = allele for yellow fruit (recessive)
W = allele for yellow flowers (dominant) and
w = allele for white flowers (recessive)

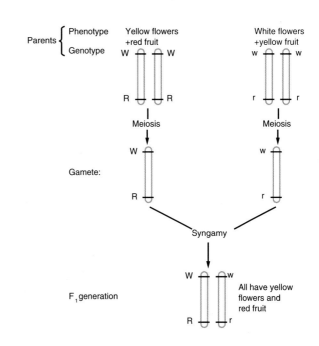

If the F_1 generation is intercrossed (i.e. self-pollinated), the following results would be expected:

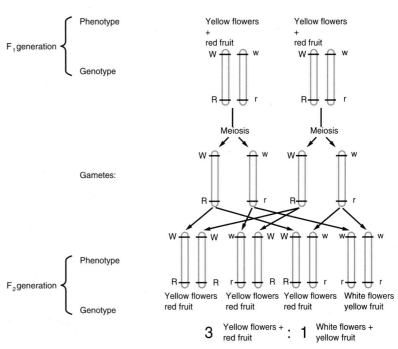

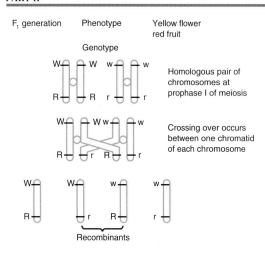

F₁ generation   Phenotype   Yellow flower red fruit

Genotype

Homologous pair of chromosomes at prophase I of meiosis

Crossing over occurs between one chromatid of each chromosome

Recombinants

If the plant is self-fertilized

F₂ generation:	♂ gametes			
♀ gametes	WR	Wr	wR	wr
WR	WWRR	WWRr	WwRR	WwRr
Wr	WWRr	WWrr	WwRr	Wwrr
wR	WwRR	WwRr	wwRR	wwRr
wr	WwRr	Wwrr	wwRr	wwrr

New combinations WWrr and Wwrr (yellow flowers, yellow fruit) wwRR and wwRr (white flowers, red fruit)

When the actual cross is performed, however, the following results are typical if 100 F₂ plants are produced:

Yellow flowers and red fruit	68
Yellow flowers and yellow fruit	7
White flowers and red fruit	7
White flowers and yellow fruit	18

What then is the explanation? Could it be that the two characters are not linked, but occur on separate chromosomes? If this were so, it would be a normal dihybrid cross, and a $9:3:3:1$ ratio should be found. For 100 plants this would mean a $56:19:19:6$ distribution. This is sufficiently different from the actual ratio of $68:7:7:18$ which was obtained for it to be discounted. For the answer, we have to go back to Section 8.3 and the events in prophase I of meiosis. During this stage portions of the chromatids of homologous chromosomes were exchanged in the process called crossing over. Could this be the explanation as to how the two unexpected phenotypes (yellow flowers/yellow fruit and white flowers/red fruit) came about? To find out, let us consider the same F₁ intercross as before but assume that in one parent crossing over took place and this plant was subsequently self-pollinated (see opposite).

The new combinations are thus the result of crossing over in prophase I of meiosis. These new combinations are called **recombinants**. As shown, this cross produces a $9:3:3:1$ ratio. However, in practice, crossing over will not always occur between the two genes. In some cases it may not occur at all; in others it may occur in such a way that the two genes are not separated. In these circumstances the only gametes are WR and wr. For this reason plants with yellow flowers and red fruit, and those with white flowers and yellow fruit, occur in greater numbers than expected.

9.4.2 Sex linkage

Sex linkage refers to the carrying of genes on the sex chromosomes. These genes determine body characters and have nothing to do with sex. The X chromosome carries many such genes, the Y chromosome has very few. Features linked on the Y chromosome will only arise in the heterogametic (XY) sex, i.e. males in mammals, females in birds. Features linked on the X chromosome may arise in either sex.

White eye colour is a sex-linked character in the fruit fly *Drosophila*. It is carried on the X chromosome and the male is the heterogametic sex. To represent sex-linked crosses, the same principles which were laid down in Table 9.1 should be followed. The letter representing each allele should, however, be attached to the letter X to indicate it is linked to it. No corresponding allele is found on the Y chromosomes, which therefore have no attached letter.

The expected results of a cross between a white-eyed male mutant and a **wild-type** (red-eyed) female are shown on the next page. 'Wild-type' is a term used to describe an organism as it normally occurs in nature. The **reciprocal cross** is also shown. A reciprocal cross is one where the same genetic features are used, but the sexes are reversed. In this case the reciprocal cross is between a white-eyed mutant female and a wild-type (red-eyed) male.

Red eyes are dominant over white eyes.

Therefore let R represent the allele for red eyes and
r represent the allele for white eyes

As the genes for eye colour are carried on the X chromosome the alleles are represented as X^R and X^r respectively. In *Drosophila* the male is the heterogametic sex (XY) and the female is the homogametic sex (XX).

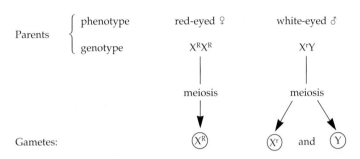

F$_1$ generation:

	$\male$ gametes	
$\female$ gametes	X^r	Y
X^R	$X^R X^r$	$X^R Y$

50% red-eyed $\female$ ($X^R X^r$)
50% red-eyed $\male$ ($X^R Y$)

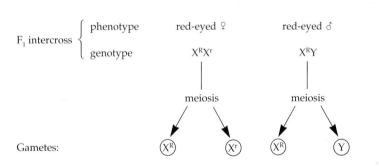

F$_2$ generation:

	$\male$ gametes	
$\female$ gametes	X^R	Y
X^R	$X^R X^R$	$X^R Y$
X^r	$X^R X^r$	$X^r Y$

50% red-eyed $\female$ ($X^R X^R$ and $X^R X^r$)
25% red-eyed $\male$ ($X^R Y$)
25% white-eyed $\male$ ($X^r Y$)

continued on next page

continued from previous page

Reciprocal cross

Parents { phenotype   white-eyed ♀   red-eyed ♂

genotype $X^r X^r$ $X^R Y$

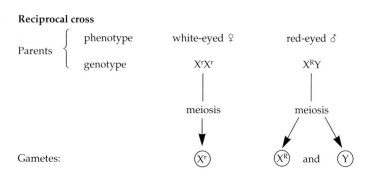

	♂ gametes	
♀ gametes	X^R	Y
X^r	$X^R X^r$	$X^r Y$

50% red-eyed ♀
50% white-eyed ♂

F_1 intercross { phenotype   red-eyed ♀   white-eyed ♂

genotype $X^R X^r$ $X^r Y$

Gametes: X^R and X^r X^r Y

F_2 generation:

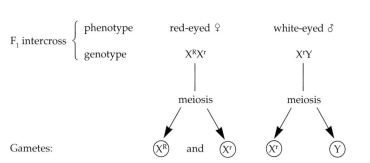

	♂ gametes	
♀ gametes	X^r	Y
X^R	$X^R X^r$	$X^R Y$
X^r	$X^r X^r$	$X^r Y$

25% red-eyed ♀ ($X^R X^r$)
25% red-eyed ♂ ($X^R Y$)
25% white-eyed ♂ ($X^r Y$)
25% white-eyed ♀ ($X^r X^r$)

Two well known sex-linked genes in humans are those causing haemophilia and red–green colour-blindness. Both are linked to the X chromosome and both occur almost exclusively in males. For the condition to arise in females requires the double recessive state and as the recessive allele is relatively rare in the population this is unlikely to occur. In females the recessive allele is normally masked by the appropriate dominant allele which occurs on the other X chromosome. These heterozygous females are not themselves affected but are capable of passing the recessive allele to their offspring. For this reason such females are termed **carriers**. When the recessive allele occurs in males it expresses itself because the Y chromosome cannot carry any corresponding dominant allele. The inheritance of red–green colour-blindness is illustrated below.

Normal sight is dominant over red–green colour-blindness. Therefore let B represent the allele for normal sight and b represent the allele for colour-blindness

As this gene is carried on the X chromosome, its alleles are represented as X^B and X^b respectively. In humans the male is the heterogametic sex (XY) and the female is the homogametic sex (XX).

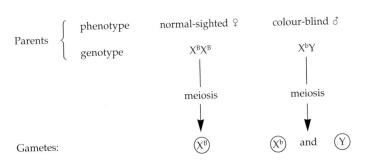

Parents
- phenotype: normal-sighted ♀   colour-blind ♂
- genotype: $X^B X^B$ $X^b Y$

meiosis   meiosis

Gametes: X^B X^b and Y

	♂ gametes	
♀ gametes	X^b	Y
X^B	$X^B X^b$	$X^B Y$

50% normal-sighted carrier ♀ ($X^B X^b$)
50% normal-sighted ♂ ($X^B Y$)

F_1 intercross
- phenotype: normal-sighted carrier ♀   normal-sighted ♂
- genotype: $X^B X^b$ $X^B Y$

meiosis   meiosis

Gametes: X^B X^b X^B Y

F_2 generation:

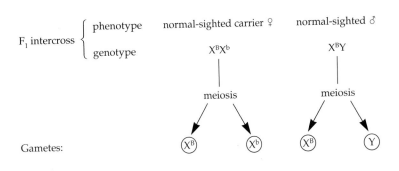

	♂ gametes	
♀ gametes	X^B	Y
X^B	$X^B X^B$	$X^B Y$
X^b	$X^B X^b$	$X^b Y$

25% normal-sighted ♀ ($X^B X^B$)   25% normal-sighted ♂ ($X^B Y$)
25% normal-sighted carrier ♀ ($X^B X^b$)   25% colour-blind ♂ ($X^b Y$)

PROJECT

The Ishihara test can be used to find out if someone is red-green colour-blind

1. Use this test to find out how much more common is the incidence of red–green colour-blindness in boys than in girls.

2. Are there any who are colour-blind in one eye and not in the other?

163

A study of the crosses reveals that the recessive gene causing colour-blindness is exchanged from one sex to the other at each generation. The father passes it to his daughters, who thus become carriers. The daughters in turn may pass it to their sons, who are thus colour-blind. This pattern of inheritance is perhaps more obvious when viewed another way. As the male is XY, his Y chromosome must have been inherited from his father as the mother does not possess a Y chromosome. The X chromosome and hence colour-blindness must therefore have been inherited from the mother. The colour-blind male can only donate his X chromosome to his daughters as it is bound to fuse with another X chromosome – the only type the mother produces. Colour-blind females can only arise from a cross between a carrier female and a colour-blind male. As both types are rare in the population, the chance of this happening is very small indeed. Even then, there is only a one in four chance of any single child of such a cross being a colour-blind female.

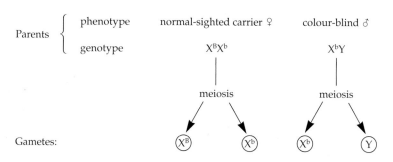

| Parents | phenotype | normal-sighted carrier ♀ | colour-blind ♂ |
| | genotype | $X^B X^b$ | $X^b Y$ |

Gametes: X^B X^b X^b Y

Offspring genotypes:

♀ gametes	♂ gametes	
	X^b	Y
X^B	$X^B X^b$	$X^B Y$
X^b	$X^b X^b$	$X^b Y$

25% normal-sighted carrier ♀ ($X^B X^b$)   25% colour-blind ♂ ($X^b Y$)
25% normal-sighted ♂ ($X^B Y$)   25% colour-blind ♀ ($X^b X^b$)

The inheritance of haemophilia follows a similar pattern to that of colour-blindness. Haemophilia is the inability of the blood to clot leading to slow and persistent bleeding, especially in the joints. Unlike colour-blindness it is potentially lethal. For this reason, the recessive allele causing it is even rarer in the population. Haemophiliac females are thus highly improbable, and in any case are unlikely to have children as the onset of menstruation at puberty is often fatal. Haemophilia is the result of an individual being unable to produce one of the many clotting factors, namely **factor 8** or **anti-haemophiliac globulin (AHG)** (see Application on p. 126). The extraction of this factor from donated blood now permits haemophiliacs to lead near-normal lives, although they still run the same risk of conveying the disease to their children.

Any mutant recessive allele, such as that causing haemophilia, is normally rapidly diluted among the many normal alleles in a population. Its expression is thus a rare event. If, however, there is close breeding between members of a family in which a

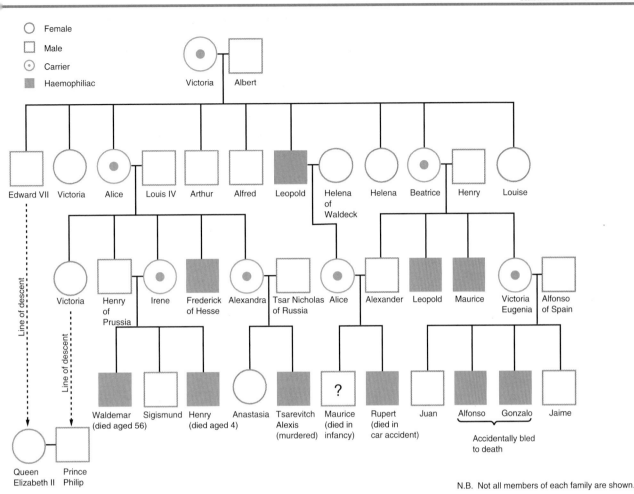

Fig. 9.2 Transmission of haemophilia from Queen Victoria

mutant gene exists, the chances of it expressing itself are enhanced. This accounts for the higher than normal occurrence of the haemophiliac gene among members of various European royal families. The origin of this particular gene can be traced back to England's Queen Victoria, who had a haemophiliac son, Leopold Duke of Albany. Prior to this there was no history of the gene among the royal family, although it existed elsewhere in the population. In order to marry someone of similar status, members of the European royal families were limited in their choice of partners and tended to marry within a relatively small circle. In effect, the gene pool was very restricted. As a result there was a disproportionately large number of haemophiliacs in these families. Fig. 9.2 traces the inheritance of this gene. The present English royal family is unaffected, as it is descended from Edward VII who did not inherit the haemophilia gene. The chart also illustrates another method of representing sex-linked crosses.

It is unusual to find dominant mutant genes linked to the X chromosome in humans, but one example is the congenital absence of incisor teeth. These conditions occur in both sexes but are more common in females as they have two X chromosomes. Genes linked to the Y chromosome are very rare, but hairy ear rims are an example.

9.5 Allelic interaction

PROJECT

Surveys can be carried out among your fellow students to find, for example:

1. The relationships between skin, hair and eye colour.

2. The proportions of tongue-rollers and non-rollers; ear-lobe types, etc.

3. If there are any correlations in laterality studies, for example, between handedness and
   (a) eye dominance
   (b) arm folding
   (c) volumes of hands
   (d) lengths of fingers, etc.

Up to now we have looked at inheritance in a straightforward way – the black and white of genetics so to speak. We now turn our attention to the less straightforward situations – the many shades of grey which exist when alleles interact in different ways.

9.5.1 Codominance

So far we have dealt with situations where one of the alleles of a gene is dominant and the other recessive. Sometimes however, alleles express themselves equally in the phenotype. Such a condition is called **codominance**.

One example of codominance occurs when a snapdragon (*Antirrhinum*) with red flowers is crossed with one with white flowers. All the F_1 generation produce flowers of intermediate colour, namely pink. The F_2 generation produces red, pink and white flowers in the ratio $1:2:1$. The cross may be represented by the procedure shown below:

N.B. Where codominance is involved, it is normal to use different letters to represent each allele, e.g. R to represent red flowers and W to represent white flowers. The use of upper and lower cases of one letter, e.g. R and r or W and w, would imply dominance and be confusing. It is also usual to assign the gene an upper case letter (in the above case C = colour) and use superscript upper case letters to designate the different alleles.

Let C^R = the allele for red flowers

C^W = the allele for white flowers

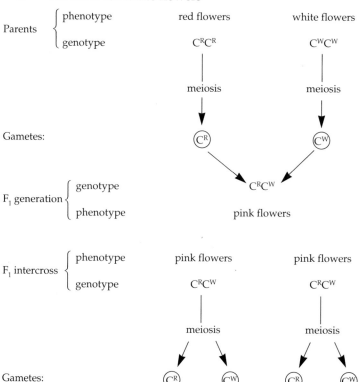

F_2 generation:

♂ gametes		
♀ gametes	C^R	C^W
C^R	C^RC^R	C^RC^W
C^W	C^RC^W	C^WC^W

25% red flowers (C^RC^R)
50% pink flowers (C^RC^W)
25% white flowers (C^WC^W)

9.5.2 Partial dominance

Sometimes both alleles express themselves in the phenotype, but one more so than another. This is an intermediate stage between complete dominance and codominance. There are many blends of partial dominance which lead to a wide range of intermediate varieties between two extremes.

9.5.3 Multiple alleles

All examples so far studied have involved a gene having two alternative alleles. We now look at a situation where a gene has more than two possible alleles.

In humans the inheritance of the ABO blood groups is determined by a gene I which has three different alleles. Any two of these can occur at a single locus at any one time.

Allele A causes production of antigen A on red blood cells.

Allele B causes production of antigen B on red blood cells.

Allele O causes no production of antigens on red blood cells.

Alleles A and B are codominant and allele O is recessive to both.

The transmission of these alleles occurs in normal Mendelian fashion.

A cross between an individual of group AB and one of group O therefore gives rise to individuals none of whom possess either parental blood group.

TABLE 9.4 **Possible genotypes of blood groups in the ABO system**

Blood group	Possible genotypes
A	I^AI^A or I^AI^O
B	I^BI^B or I^BI^O
AB	I^AI^B
O	I^OI^O

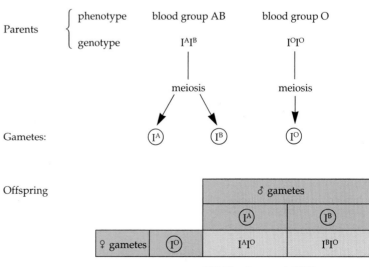

Parents
phenotype   blood group AB   blood group O
genotype I^AI^B I^OI^O

meiosis   meiosis

Gametes: I^A I^B I^O

Offspring

♂ gametes		
		I^A	I^B
♀ gametes	I^O	I^AI^O	I^BI^O

50% blood group A (I^AI^O)
50% blood group B (I^BI^O)

167

A cross between certain individuals of blood group A and certain individuals of blood group B may produce offspring with any one of the four blood groups.

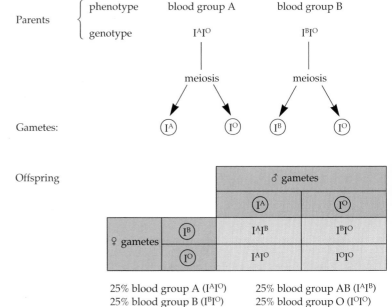

25% blood group A ($I^A I^O$) 25% blood group AB ($I^A I^B$)
25% blood group B ($I^B I^O$) 25% blood group O ($I^O I^O$)

Paternity suits
Although blood groups cannot prove who is the father of a child, it is possible to use their inheritance to show that an individual could not possibly be the father. Imagine a mother who is blood group B having a child of blood group O. She claims the father is a man whose blood group is found to be AB. As the child is group O its only possible genotype is $I^O I^O$. It must therefore have inherited one I^O allele from each parent. The mother, if $I^B I^O$, could donate such an allele. The man with blood group AB can only have the genotype $I^A I^B$. He is unable to donate an I^O allele and cannot therefore be the father.

TABLE 9.5 **Possible genotypes of rabbits with different coat colour**

Coat colour	Possible genotypes
Full	$C^F C^F$ or $C^F C^{CH}$ or $C^F C^H$ or $C^F C^A$
Chinchilla	$C^{CH} C^{CH}$ or $C^{CH} C^H$ or $C^{CH} C^A$
Himalayan	$C^H C^H$ or $C^H C^A$
Albino	$C^A C^A$

Dominance series
Coat colour in rabbits is determined by a gene C which has four possible alleles:

Allele C^F determines full coat colour and is dominant to
Allele C^{CH} which determines chinchilla coat and is in turn dominant to
Allele C^H which determines Himalayan coat and is in turn dominant to
Allele C^A which determines albino coat colour.

There is therefore a dominance series, and each type has a range of possible genotypes.

Inheritance is once again in normal Mendelian fashion.

9.5.4 Pleiotropy and lethal alleles

We have looked at situations so far where an allele determines one character. Sometimes however an allele may affect more than one character. Such an allele is termed **pleiotropic**. An example was given on page 124 where the allele for cystic fibrosis caused the production of especially viscous mucus. One effect was the blockage of the pancreatic duct leading to poor digestion as a result of the pancreatic enzymes not being able to enter the duodenum. These enzymes may therefore accumulate in the pancreas, so digesting its cells including the Islets of Langerhans which produce insulin. Unable to produce insulin the patient suffers diabetes – a second effect of the allele. The viscous mucus also blocks alveoli and bronchioles, leading to breathing problems – a third effect.

Sometimes one of the effects of a pleiotropic allele is lethal. In the case of the house mouse (*Mus musculus*) such an allele controls coat colour as well as some other characteristic which is lethal when present in the homozygous dominant condition.

Whenever two yellow mice are bred together the offspring are always in the ratio of 2 yellow to 1 agouti (grey). The expected ratio is 3 yellow to 1 agouti. Examination of pregnant yellow mice reveals that the homozygous yellow embryo always dies.

Let Y represent the dominant allele for yellow fur
and y represent the recessive allele for agouti fur

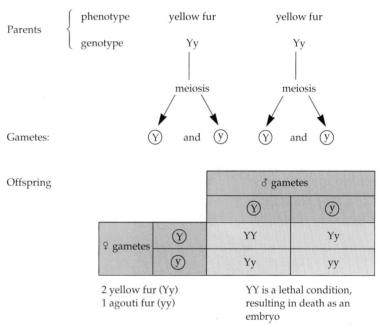

2 yellow fur (Yy)
1 agouti fur (yy)

YY is a lethal condition, resulting in death as an embryo

9.6 Gene interaction

We have examined how alleles at a single locus (i.e. the alleles of a single gene) may interact and we have called this 'allelic interaction'. Sometimes the alleles of more than one gene (i.e. at more than one locus) interact. This we call **gene interaction**, although in practice it is still the alleles of these genes which are influencing each other.

9.6.1 Simple interaction

This occurs where a group of genes or a **gene complex** act together to determine a single character. An example in humans

Rose-type comb

Walnut-type comb

Single-type comb

Fig. 9.3 Types of chicken comb

is the inheritance of skin pigmentation which is controlled by two genes A and B. An individual with the genotype AABB produces darkly pigmented skin whereas an individual with the genotype aabb has white unpigmented skin. A mating between these two types produces an intermediate skin colour (genotype AaBb). In the F_2 generation skin colour varies from dark (AABB) through dark brown (AABb or AaBB), half-coloured (AAbb or AaBb or aaBB), light brown (Aabb or aaBb) to white (aabb).

Gene complexes do not always result in the blending of features to produce intermediates; they can create entirely new features. Take the case of comb shape in poultry. When a pure-breeding fowl with a rose-type comb is crossed with a pure-breeding pea-type comb fowl, all the offspring produced have walnut-type combs. All three types are represented in the F_2 generation along with a fourth variety – the single comb, as shown below.

Let R represent the allele for rose-type comb and
  P represent the allele for pea-type comb

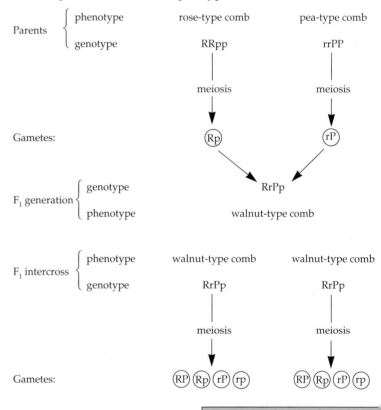

9 walnut-type comb (R–P–)
3 rose-type comb (R–pp)
3 pea-type comb (rrP–)
1 single-type comb (rrpp)

9.6.2 Epistasis

This arises when the allele of one gene suppresses or masks the action of another. An example occurs in mice where three genes determine coat colour. However the absence of a dominant allele at one of the loci results in no pigment being produced and the coat being albino. This occurs regardless of the genes present at the other loci, even if these produce normal coat colour. The gene at the third locus clearly suppresses the action of the others.

9.6.3 Polygenes

Many genes acting together are referred to as polygenes. Imagine a character determined by five genes, each gene having a dominant or recessive allele. An organism inheriting five dominant alleles will lie at one end of the spectrum and one with five recessive alleles will lie at the other. Between these extremes will be a continuum of types depending on the relative proportions of dominant and recessive alleles. Polygenes give rise to continuous variation, which is discussed further in Section 10.2.1.

9.7   Questions

1. When a female fruit fly with red eyes and grey body was crossed with a male with brown eyes and yellow body all of the 47 offspring had red eyes and grey bodies. These offspring were then used to set up a backcross in order to test the validity of Mendel's law of independent assortment of genes and the phenotypes of their offspring were carefully recorded.

(a) (i) Explain what is meant by independent assortment of genes.
   (ii) What evidence is there that these parents were pure breeding for these characteristics?
   (iii) Explain how a backcross would be set up. *(9 marks)*
(b) What phenotypes would you expect from interbreeding the F_1 if the two genes in question were located
   (i) on separate autosomes,
   (ii) on the same autosome,
   (iii) on the X chromosome?
   Explain your answers. *(10 marks)*
(c) Name **one** statistical test that could be applied to test the significance of these results. *(1 mark)*
   (Total 20 marks)

NEAB June 1991, Paper IB, No. 5

2. Red–green colour blindness is a sex-linked trait. The allele (c) for red–green colour blindness is recessive to the normal allele (C). The diagram below represents the inheritance of this characteristic.

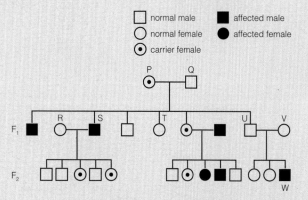

(i) With respect to vision, what are the phenotypes of the individuals labelled P, Q and S on the diagram? *(2 marks)*
(ii) With respect to the sex chromosomes, what are the genotypes of individuals P, Q and S? *(2 marks)*
(iii) State the genotypes, with respect to colour blindness, of the gametes of individuals R and S. *(2 marks)*

(iv) Individual W is colour blind. From the family tree, explain how this has occurred. *(1 mark)*
   (Total 7 marks)

SEB Human Biology Specimen 1992, Higher Grade Paper II, No. 3

3. Pure-breeding pea plants with grey seed coats and tall stems were crossed with plants with white seed coats and short stems. The offspring (F_1 generation) all had grey seed coats and tall stems.

(a) State suitable symbols for the alleles for colour of seed coat and height of stem. *(1 mark)*
(b) Explain, by means of a genetic diagram, the ratios of genotypes and phenotypes you would expect if the F_1 generation were self-fertilized. *(5 marks)*
(c) Explain how you could find out the genotype of a plant with grey seed coats and tall stems. *(4 marks)*
   (Total 10 marks)

UCLES (Modular) June 1993, (Foundation), No. 4

4. The inheritance of coat colour in cats is influenced by a gene which is present on the X chromosome but not on the Y. The allele for black coat colour can be represented by the symbol B and that for ginger coat colour by G. These alleles show incomplete dominance, and the hairs of heterozygous cats show bands of both black and ginger. Their coats are called tortoiseshell.

(a) Define the following terms:
   (i) Gene
   (ii) Allele
   (iii) Incomplete dominance. *(6 marks)*
(b) Using words and symbols from the passage, complete the following cross. *(7 marks)*

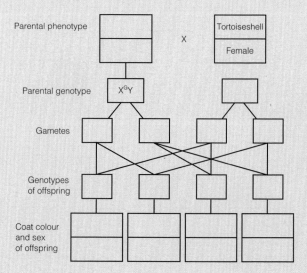

(c) If a black male were crossed with a tortoiseshell female, what proportion of the female offspring would you expect to have black coats? *(1 mark)*

(d) Cats may also have patches of white fur in combination with black, ginger or tortoiseshell. Give a reason for the inheritance of white patches being independent of the other colours. *(2 marks)*

(Total 16 marks)

WJEC June 1990, Paper A2, No. 6

. The diagram shows part of a family tree in which he inherited condition of phenylketonuria occurs.

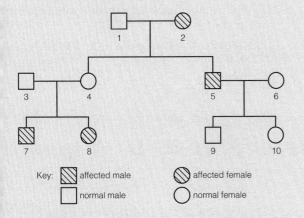

Key:
- ▨ affected male
- ◢ affected female
- ☐ normal male
- ○ normal female

(a) Identify and explain **one** piece of evidence from this family tree to show that the allele for phenylketonuria is recessive to the allele for the normal condition. *(2 marks)*

(b) Giving a reason for your answer in each case, identify **one** individual who must be:
  (i) heterozygous; *(1 mark)*
  (ii) homozygous. *(1 mark)*

(c) If individual 10 married a man who was heterozygous for the gene, what is the probability that their first child would be affected? *(1 mark)*

(Total 5 marks)

AEB November 1992, Paper I, No. 17

5. Read the following passage.

Cystic fibrosis affects people who inherit two copies of the cystic fibrosis allele, one from each parent. In the United States, approximately one thousand children with the condition are born each year and some 12 million people (about 1 in 20 of European descent) are carriers of this allele. It only affects the secretory epithelia which transport salt and water between the blood and outside world. No one has yet been able to identify the abnormal protein that produces the disorder. It seemed that the primary defect lay in the regulation of a protein that forms an ion channel in the cell-surface membrane. This channel normally transports chloride ions and, in people with cystic fibrosis, cannot be activated or 'opened' correctly. Consequent disruption of the transport of chloride ions and water causes a thick dehydrated mucus to build up, interfering with pancreatic secretion and, in the respiratory tract, creating an environment suitable for growth of bacteria.

Despite these findings, scientists could not be sure whether the defective protein was itself a channel or a separate protein that regulated the channel. One obvious approach, called 'forward' genetics, is to purify the chloride-channel protein and determine its sequence of amino acids and then to synthesize the corresponding base sequence so that the allele itself can be identified. This corresponding base sequence is radioactive and known as a probe.

Meanwhile, molecular geneticists investigated a small region of the seventh chromosome, which they knew to contain the cystic fibrosis allele. Without knowing anything about the precise sequence of bases that made up the allele, researchers were able to search the DNA in the right region until they isolated part of this allele. This is the principle of 'reverse' genetics.

From blood samples of people with cystic fibrosis and their families they extracted DNA. Because they did not know where the gene was, their approach was to look for markers – sequences of DNA physically near to it on the same chromosome. The closer a marker is to the cystic fibrosis allele, the more likely it is that both will be inherited together in future generations.

Adapted from: New Scientist, October 1989

Using information in the passage and your own knowledge, answer the following questions.

(a) Using examples from the passage to illustrate your answer, explain what is meant by:
  (i) a gene; *(2 marks)*
  (ii) an allele. *(2 marks)*

(b) (i) What is the evidence from the passage that cystic fibrosis is a recessive condition? *(1 mark)*

  (ii) What is the probability that a child with cystic fibrosis will be born to parents, both of whom are carriers for the condition? Explain how you arrived at your answer. *(2 marks)*

(c) Suggest how cystic fibrosis might interfere with the process of digestion. *(2 marks)*

(d) By means of a simple flow diagram, outline **one** hypothesis to account for the way in which dehydrated mucus is produced by people with cystic fibrosis. *(4 marks)*

(e) Outline the principle by which a probe (line 31) could be used to identify a particular allele. *(3 marks)*

(f) What is the difference between 'forward' genetics (line 25) and 'reverse' genetics (line 39)? *(3 marks)*

(g) From which cells in the blood sample (line 40) would the workers involved with 'reverse' genetics have isolated DNA? Give a reason for your answer. *(2 marks)*

(h) Explain why, 'The closer a marker is to the cystic fibrosis allele, the more likely it is that both will be inherited together in future generations' (lines 45–47). *(3 marks)*

(Total 24 marks)

AEB June 1991, Paper II, No. 6

7. In *Drosophila melanogaster*, kidney bean-shaped eye is recessive to round-shaped eye and orange eye colour is recessive to red eye colour.

A fly, homozygous for both round-shaped eye and red colour, was crossed with a fly having kidney bean-shaped orange coloured eyes. The offspring (F_1) were allowed to interbreed and the investigator expected to find a $9:3:3:1$ segregation for eye shape and colour in the F_2 generation. Instead, the following progeny were produced.

red colour, round-shaped eye	520
orange colour, kidney bean-shaped eye	180
red colour, kidney bean-shaped eye	54
orange colour, round eye-shaped eye	45

(a) Account for the basis for the expected $9:3:3:1$ segregation in the F_2 generation. *(6 marks)*

(b) Explain the observed result. *(6 marks)*

(Total 12 marks)

O&C SEB June 1991, Paper I, No. 3

8. Tomato plants show variation in their leaf shapes (described as cut-leaved and potato-leaved) and in their height (tall and dwarf). Two crosses were carried out as follows.

Cross 1

Cut-leaved tall tomato plants were crossed with potato-leaved dwarf tomato plants. The offspring from cross 1 were all cut-leaved tall plants.

Cross 2

The offspring from cross 1 were allowed to cross with each other and produced the following.

920 cut-leaved tall plants
291 potato-leaved tall plants
299 cut-leaved dwarf plants
101 potato-leaved dwarf plants

(a) (i) Using the symbols **A** and **a** to indicate alleles for leaf shape and **B** and **b** to indicate alleles for height, complete the following table.

Characteristic	Symbol
Potato-leaved	
Cut-leaved	
Tall	
Dwarf	

(1 mark)

(ii) How did you know which characteristics were dominant? *(1 mark)*

(b) (i) Complete the genetic diagram below for cross 1.

Parental phenotypes   Cut-leaved tall plant   ×   Potato-leaved dwarf plant
Parental genotypes   ....

Gametes   ....
Offspring genotype   ....
Offspring phenotype   Cut-leaved tall

(3 marks)

(ii) Complete the genetic diagram below for cross 2.

Phenotypes   Cut-leaved tall plant   ×   Cut-leaved tall plant
Genotypes   ....
Gametes   ....
   ....

(4 marks)

(iii) Give all the possible genotypes produced by the cut-leaved tall plants as a result of cross 2. *(2 marks)*

(iv) Comment on the differences between the $9:3:3:1$ ratio of phenotypes and the actual values obtained. *(2 marks)*

(c) Suggest how these ratios might be different if the genes controlling these characters were linked. *(2 marks)*

(Total 15 marks)

ULEAC June 1992, Paper I, No. 13

9. The position of the flower on the stem of the garden pea is governed by a gene with two alleles.

Flowers growing in the leaf axils are produced by the action of a dominant allele **A**, those growing only at the tip of the stem by its recessive allele **a**.

Coloured flowers are produced by a dominant allele **F** and white flowers by its recessive allele **f**. The two pairs of alleles are not genetically linked.

(a) A plant with coloured flowers in the leaf axils (heterozygous for both characters) is crossed with a pure breeding plant of the same phenotype.

(i) Draw a genetic diagram to represent this cross. Include in your diagram the ratio of genotypes produced in this cross. *(5 marks)*

(ii) What will be the appearance and position of the flowers produced? *(1 mark)*

(b) Describe how you could carry out an experiment to find out which of the offspring are homozygous for both characters. Give details of your procedure and explain your reasoning at each stage. *(5 marks)*

(c) (i) What do you understand by the term *autosomal linkage*? *(1 mark)*

(ii) A cross was made between two plants, one of which was homozygous dominant and the other was homozygous recessive for flower position and flower colour. The offspring of this cross were allowed to self-pollinate. What would you expect to be the appearance and ratio of the offspring of the second cross if the two characteristics were linked? *(2 marks)*

(iii) Suggest a reason why you may not obtain an exact ratio from such a cross. *(1 mark)*

ULEAC January 1993, Paper I, No. 13

10. The ABO blood group is governed by a set of three multiple alleles, I^A, I^B and I^O. I^A and I^B are codominant, I^O is recessive. Another blood group system is known as the MN system. The MN blood groups are governed by a pair of codominant alleles giving three possible blood groups, M, MN and N.

(a) A man of blood group B married a woman of unknown ABO blood group. They had three children. One of the children had blood group A, one had blood group AB and one had blood group O.

(i) State the genotypes of the parents and give an explanation for your answer. *(5 marks)*

(ii) Draw a genetic diagram to show the inheritance of ABO blood groups in this family. *(2 marks)*

(b) In another family, the man is accusing his wife of infidelity. Their first and second children, whom they both claim, are of blood groups O and AB respectively. The third child, whom the man disclaims, is blood group B.

Can this information be used to support the man's accusation that the third child is not his? Give the reasoning for your answer. *(4 marks)*

(c) Another test was carried out using the MN blood group system. In the test, it was found that the third child was blood group M. The man was blood group N.

Explain why this shows that the man is unlikely to be the child's father. *(4 marks)*

(Total 15 marks)

ULEAC June 1993, Paper I, No. 12

11. *This question requires a critical discussion of Mendel's laws of inheritance in the light of information concerning the inheritance of plumage colour in budgerigars. You are then asked to solve a genetics problem and outline the possible events leading to the evolution of Warfarin resistance in rats.*

SECTION A

Plumage colour in budgerigars is controlled by two pairs of alleles. One pair controls the general colour, blue being recessive to green. The density of colour is determined by a second pair of alleles,
5 independently inherited, which show incomplete dominance, i.e. the density of the colour of the heterozygote is intermediate between that of the two homozygotes. As a result of gene interaction between these two pairs of alleles there are six
10 possible phenotypes, described as shown in the table below:

Colour	Density of colour		
	Pale	Mid	Dark
Green	1. Light green	2. Mid green	3. Olive
Blue	4. Sky blue	5. Mid blue	6. Mauve

Answer the following questions which relate to the above information.

(a) (i) What is meant by 'independently inherited' (line 5)? *(1 mark)*

(ii) Explain the difference between 'incomplete dominance' (line 5) and 'gene interaction' (line 8). *(2 marks)*

(iii) Stated in modern terms, Mendel's findings led him to suggest:

1. that each characteristic of a diploid organism is controlled by a single pair of alleles;

2. that, if an organism has two unlike alleles for a characteristic, then one is expressed to the exclusion of the other;

3. that during breeding a gamete can contain only one allele for any locus, and which allele it contains is determined randomly.

How does information like that obtained from the study of plumage colour in budgerigars affect these Mendelian concepts of inheritance? *(2 marks)*

(b) Using the symbols G and g for colour and D and d for density, list the possible genotypes for each of the six phenotypes in the table above. *(3 marks)*

(c) Determine the expected proportions of both genotypes and phenotypes of the offspring of the following crosses:

   (i) sky blue cock × pure bred, light green hen.
   (ii) mid blue cock × heterozygous mid green hen. *(6 marks)*

(d) Several matings between 2 birds of different plumage but unknown genotype produced offspring which were all mid green. Determine two possible crosses which could have produced this outcome. For each cross, state the genotypes of the parents involved. *(2 marks)*

SECTION B

The rat poison Warfarin was introduced in the early 1950s. Warfarin-resistant rats were first discovered in 1958, and by 1972 over half of the rat population in some areas of Britain was resistant to Warfarin. Give a logical description of the probable mechanism by which Warfarin-resistant rat populations have evolved. *(4 marks)*

(Total 20 marks)

NISEAC June 1991, Paper III, No. 2

12. In the fruit fly (*Drosophila*), males are the heterogametic sex (XY). Two different crosses using these flies gave the following results.

	CROSS A		CROSS B
Parents	normal × 'cut' wing ♀ wing ♂	Parents	'cut' × normal wing ♀ wing ♂
F_1	all normal wing	F_1	normal wing ♀ 'cut' wing ♂
F_2	789 normal wing ♀ 391 normal wing ♂ 376 'cut' wing ♂	F_2	356 normal wing ♀ 339 'cut' wing ♀ 342 normal wing ♂ 333 'cut' wing ♂
N.B. ♂ = male ♀ = female			

(a) Given that the 'cut' wing characteristic is controlled by a single gene, explain how these results show that the 'cut' wing allele
   (i) is recessive.
   (ii) is carried on the X-chromosome, rather than on an autosome. *(2 marks)*

(b) Using **N** as the normal wing allele, and **n** as the allele for 'cut' wing, give the genotypes represented in Cross B.

   ♀ parent   ♂ parent
   ♀ F_1 ♂ F_1
   ♀ F_2 normal wing   ♂ F_2 normal wing
   ♀ F_2 'cut' wing   ♂ F_2 'cut' wing
                              (2 marks)

Another fruit fly mutant, caused by a change in a single gene, is known as bar-eye. Normal, round-eyed flies have nearly one thousand facets in their compound eyes. This number is greatly reduced in a bar-eyed individual, in which the eye appears as a narrow vertical strip.

   In another investigation, a virgin female from a homozygous stock with normal, round eyes was crossed with a bar-eyed male. All the female offspring had bar-eyes, and all the males had round eyes.

(c) What may be concluded about the nature of the bar-eyed allele from this information? *(2 marks)*

An F_2 generation was obtained from the previous investigation, but the eggs were separated into two batches immediately after laying, with each batch being maintained at a different temperature. Once the F_2 adults had emerged, the mean number of facets per eye for each phenotypic class was recorded. The results are shown in the table.

Phenotypic class in F_2 offspring		Mean number of facets per eye at the development temperature	
		15°C	30°C
Males	round-eyed	996	997
	bar-eyed	270	74
Females	round-eyed	997	996
	bar-eyed	214	40

(d) (i) Using this information, give **two** conclusions about the expression of the bar-eye allele.
   (ii) Suggest a hypothesis which might reasonably account for one of the conclusions given in (d)(i). *(4 marks)*

(Total 10 marks)

NEAB June 1994, Paper IIA, No. 6

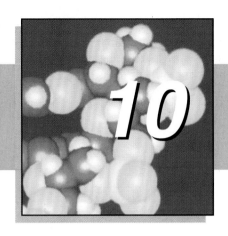

10 Genetic change and variation

Within any given population there are variations among individual organisms. It is this variation which forms the basis of the evolutionary theory of Darwin. There are two basic forms of variation: **continuous variation**, where the individuals in a population show a gradation from one extreme to the other, and **discontinuous (discrete) variation**, where there is a limited number of distinct forms within the population. Any study of variation inevitably involves the collection of large quantities of data.

10.1 Methods of recording variation

The investigation of variation within a population may involve recording the number of individuals which possess a particular feature, e.g. black fur. On the other hand, it may involve recording the number of individuals which fall within a set range of values, e.g. those weighing between 1 kg and 10 kg. What is being measured in both cases is the **frequency distribution**. This may be presented in a number of ways. To illustrate each method the same set of data is used throughout (Table 10.1), although it is not really suitable for some methods of presentation.

10.1.1 Table of data

Tabulation is the simplest means of presenting data. It is a useful method of recording information initially but is less useful for demonstrating the relationship between two variables.

10.1.2 Line graph

A graph typically has two axes, each of which measures a variable. One variable has fixed values which are selected by the experimenter. This is called the **independent variable**. The other variable is the measurement taken and as such is not selected by the experimenter. This is called the **dependent variable**. In Table 10.1, 'height' is the independent variable and 'frequency' the dependent variable. The values of the independent variable are plotted along the horizontal axis (also known as the x axis or abscissa) and the values of the dependent variable are plotted along the vertical axis (also known as the y axis or ordinate). The

TABLE 10.1 **Frequency of heights (measured to the nearest 2 cm) of a sample of humans**

Height/cm	Frequency
140	0
142	1
144	1
146	6
148	23
150	48
152	90
154	175
156	261
158	352
160	393
162	462
164	458
166	443
168	413
170	264
172	177
174	97
176	63
178	46
180	17
182	7
184	4
186	0
188	1
190	0

corresponding values of the two variables can be plotted as points on the graph known as **coordinates**. These points may then be joined to give a line or smooth curve, as in Fig. 10.1.

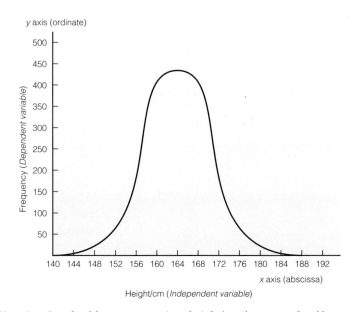

Fig. 10.1 Graph of frequency against height/cm for a sample of humans

10.1.3 Histogram

Axes are drawn in much the same way as for a line graph. The values for the independent variable on the *x* axis are, however, normally reduced, often by grouping the data into convenient classes. For example, the twenty-six values for height used on the line graph may be reduced to ten by grouping the heights into sets of 5 cm, e.g. 140–145, 145–150 etc. Instead of plotting points, vertical columns are drawn. The method is illustrated in Fig. 10.2.

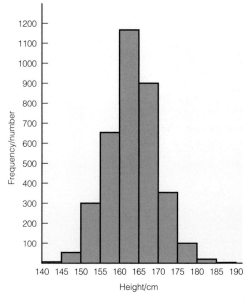

Fig. 10.2 Histogram showing height frequencies in a sample human population

10.1.4 Bar graph

This is similar to a histogram except that a non-numerical value is plotted on the *y* axis. Let us suppose the sample population is divided into non-numerical sets such as racial groups and sex. These can be plotted along the *y* axis, with average height being plotted along the *x* axis. The resultant bar graph is shown in Fig. 10.3.

10.1.5 Kite graph

This is a form of bar graph, which gives more detailed information on the frequency of a non-numerical variable. To take the information given in Fig. 10.3, it simply reveals the average height of each group and not the frequency at different heights. In a kite graph, the frequency of particular heights is plotted vertically for certain non-numerical variables, e.g. males and females. A kite graph is shown in Fig. 10.4.

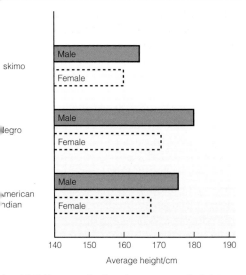

Fig. 10.3 Bar graph showing average height variation according to racial group and sex

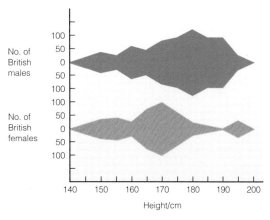

Fig. 10.4 Kite graphs to show the height frequency for British males and females in the population sample

10.1.6 Pie chart

Pie charts are a simple and clearly visible means of showing how a whole sample is divided up into specified parts. A circle (the pie) represents the whole and it is subdivided into different sized sections according to the relative proportions of each constituent part. To be effective the pie chart should not be divided into a large number of portions, nor should it be used when it is necessary to read off precise information from the chart. It is simply a means of giving an idea of relative proportions. Fig. 10.5 shows a pie chart depicting the proportions of our sample which fall within broad height ranges.

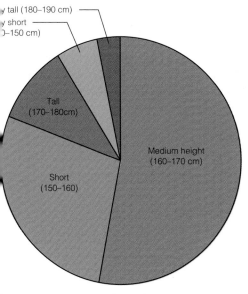

Fig. 10.5 Pie chart to show the relative proportions of five height categories in a population sample

10.2 Types of variation

10.2.1 Continuous variation

Certain characteristics within a population vary only very marginally between one individual and the next. This results in a gradation from one extreme to the other, called continuous variation. The height and weight of organisms are two

PROJECT

Determine the distribution of characteristics of your fellow students. The list is endless, but examples would be:

(a) heights and weights of males and females of different ages
(b) lengths of middle fingers
(c) sizes of feet
(d) lengths or widths of ears etc.

characteristics which show such a gradation. If a frequency distribution for such a characteristic is plotted, a bell-shaped graph similar to that in Fig. 10.1 is obtained. This is called a **normal distribution curve** or **Gaussian curve** (after the mathematician Fredrick Gauss). It is discussed further in Section 10.2.2.

Characteristics which show continuous variation are controlled not by one, but by the combined effect of a number of genes, called **polygenes**. Thus any character which results from the interaction of many genes is called a **polygenic character**. The effect of an individual gene is small, but their combined effect is marked. The random assortment of the genes during prophase I of meiosis ensures that individuals possess a range of genes from any polygenic complex. Where a group of genes all favouring the development of a tall individual combine, a very tall individual results. A combination of genes favouring small size results in a very short individual. These extremes are rare because it is probable that an individual will possess genes from both extremes. The combined effect of these genes produces individuals of intermediate height.

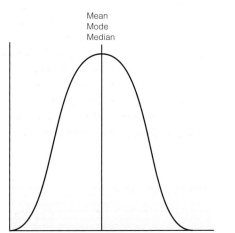

Mean
Mode
Median

Fig. 10.6 A normal distribution curve where the mean, mode and median have the same value

10.2.2 The normal distribution (Gaussian) curve

Fig. 10.6 shows a normal distribution curve; its bell-shape is typical for a feature which shows continuous variation, e.g. height in humans. The graph is symmetrical about a central value. Occasionally the curve is shifted slightly to one side. This is called a skewed distribution and is illustrated in Fig. 10.7. There are three main terms used in association with normal distribution curves, whether skewed or not. To illustrate these terms let us consider the values given in the table for the number of children in eleven different families.

The mean (arithmetic mean)
This is the average of a group of values. In our example opposite this is found by totalling the number of children in all families and dividing it by the number of families.

Total children in all families
$$= 0 + 1 + 1 + 1 + 2 + 2 + 3 + 4 + 6 + 6 + 7 = 33$$
Total number of families A–K = 11

Mean = 33 ÷ 11 = 3

The mode
This is the single value of a group which occurs most often. In our example more families have one child than any other number. The mode is therefore equal to 1.

The median
This is the central or middle value of a set of values. In our example the values are already arranged in ascending order of the number of children in each family. There are eleven families. The sixth family in the series (family F) is therefore the middle family of the group. There are five families (A–E) with the same number or fewer children, and five families (G–K) with more children. As family F has two children the median is 2.

Family	Number of children
A	0
B	1
C	1
D	1
E	2
F	2
G	3
H	4
I	6
J	6
K	7

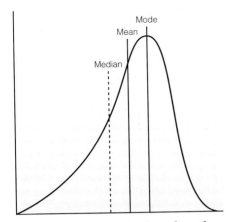

Fig. 10.7 A skewed distribution where the mean, mode and median have different values

Fig 10.6 shows a typical symmetrical normal distribution curve in which the mean and mode (and often the median) have the same value.

Fig 10.7 shows a skewed distribution in which the mean, mode and median all have different values.

The mean height of a sample population gives a good indication of its relative height compared to other sample populations. It does not, however, give any indication of the distribution of height within the sample. Indeed, the mean can be misleading. A population made up of individuals who were either 140 or 180 cm tall would have a mean of 160 cm, and yet no single individual would be anywhere near this height. It is therefore useful to have a value which gives an indication of the range of height either side of the mean. This value is called the **standard deviation (SD)**. It is calculated as follows:

$$SD = \sqrt{\frac{\Sigma d^2}{n}}$$

Σ = the sum of
d = difference between each value in the sample and the mean
n = the total number of values in the sample

How then does the standard deviation provide information on the range within a sample? Let us suppose the mean height of a sample human population is 170 cm and its standard deviation is ±10 cm. This means that over two thirds (68%) of the sample have heights which are within 10 cm of 170 cm, i.e. 68% of the sample have heights between 160 cm and 180 cm. Furthermore we can say that 95% of the sample lie within two standard deviations of the mean. In our example two standard deviations = 2 × 10 = 20 cm. In other words, 95% of the sample have heights between 150 cm and 190 cm. Fig. 10.8 illustrates these values.

10.2.3 Discontinuous (discrete) variation

Certain features of individuals in a population do not show a gradation between extremes but instead fall into a limited number of distinct forms. There are no intermediate types. For example, humans may be separated into distinct sets according to their blood groups. In the ABO system there are just four groups: A, B, AB and O. Unlike continuous variation, which is controlled by many genes (polygenes), a feature which exhibits discontinuous variation is normally controlled by a single gene. This gene may have two or more alleles. Features exhibiting discontinuous variation are normally represented on histograms, bar graphs or pie charts.

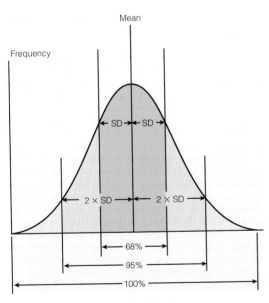

Fig. 10.8 The normal distribution curve showing the values for standard deviation

10.3   The chi-squared test

Imagine tossing a coin 100 times. It is reasonable to expect it to land heads on 50 occasions and tails on 50 occasions. In practice it would be unusual if these results were obtained (try it if you like!). If it lands heads 55 times and tails only 45 times, does this mean the coin is weighted or biased in some way, or is it purely a chance deviation from the expected result?

The **chi-squared test** is the means by which the statistical validity of results such as these can be tested. It measures the extent of any deviation between the expected and observed results. This measure of deviation is called the **chi-squared value** and is represented by the Greek letter chi, shown squared, i.e. χ^2. To calculate this value the following equation is used:

$$\chi^2 = \sum \frac{d^2}{x}$$

where $\sum$ = the sum of

d = difference between observed and expected results (the deviation)

x = the expected result

Using our example of the coin tossed 100 times, we can calculate the chi-squared value. We must first calculate the deviation from the expected number of times the coin should land heads:

Expected number of heads in 100 tosses of the coin $(x) = 50$

Actual number of heads in 100 tosses of the coin $= 55$

Deviation (d) $\overline{\quad 5 \quad}$

Therefore $\dfrac{d^2}{x} = \dfrac{5^2}{50} = \dfrac{25}{50} = 0.5$

We then make the same calculation for the coin landing tails.

Expected number of tails in 100 tosses of the coin $(x) = 50$

Actual number of tails in 100 tosses of the coin $= 45$

Deviation (d) $\overline{\quad 5 \quad}$

Therefore $\dfrac{d^2}{x} = \dfrac{5^2}{50} = \dfrac{25}{50} = 0.5$

The chi-squared value can now be calculated by adding these values:

Therefore $\chi^2 = 0.5 + 0.5 = 1.0$.

The whole calculation can be summarized thus:

$$\chi^2 = \sum \frac{d^2}{x}$$

$$\chi^2 = \left[\frac{\overset{\text{Heads}}{(55-50)^2}}{50} \right] + \left[\frac{\overset{\text{Tails}}{(50-45)^2}}{50} \right]$$

$$= \left[\frac{(5)^2}{50} \right] + \left[\frac{(5)^2}{50} \right]$$

$$= \frac{1}{2} + \frac{1}{2}$$

$$= 1.0$$

To find out whether this value is significant or not we need to use a chi-squared table, part of which is given in Table 10.2. Before trying to read these tables it is necessary to decide how

TABLE 10.2 **Part of a χ^2 table (based on Fisher)**

Degrees of freedom	Number of classes	χ^2							
1	2	0.00	0.10	0.45	1.32	2.71	3.84	5.41	6.64
2	3	0.02	0.58	1.39	2.77	4.61	5.99	7.82	9.21
3	4	0.12	1.21	2.37	4.11	6.25	7.82	9.84	11.34
4	5	0.30	1.92	3.36	5.39	7.78	9.49	11.67	13.28
5	6	0.55	2.67	4.35	6.63	9.24	11.07	13.39	15.09
Probability that deviation is due to chance alone		0.99 (99%)	0.75 (75%)	0.50 (50%)	0.25 (25%)	0.10 (10%)	0.05 (5%)	0.02 (2%)	0.01 (1%)

many **classes of results** there are in the investigation being carried out. In our case there are two classes of results, 'heads' and 'tails'. This corresponds to one degree of freedom. We now look along the row showing 2 classes (i.e. one degree of freedom) for our calculated value of 1.0. This lies between the values of 0.45 and 1.32 on the table. Looking down this column we see that this corresponds to a probability between 0.50 (50%) and 0.25 (25%). This means that the probability that chance alone could have produced the deviation is between 0.50 (50%) and 0.25 (25%). If this probability is greater than 0.05 (5%), the deviation is said to be **not significant**. In other words the deviation is due to chance. If the deviation is less than 0.05 (5%), the deviation is said to be **significant**. In other words, some factor other than chance is affecting the results. In our example the value is greater than 0.05 (5%) and so we assume the deviation is due to chance. Had we obtained 60 heads and 40 tails, a chi-squared value of slightly less than 0.05 (5%) would be obtained, in which case we would question the validity of the results and assume the coin might be weighted or biased in some way. This test is especially useful in genetic experiments.

In *Drosophila*, normal (wild-type) wings are dominant to vestigial wings. Suppose we cross two normal-winged individuals both believed to be heterozygous for this character. We should expect a 3 : 1 ratio of normal wings to vestigial wings. In practice, of 48 offspring produced, 30 have normal wings and 18 have vestigial wings. Is this close enough to a 3 : 1 ratio to justify the view that both parents were heterozygous?

Applying the chi-squared test:

	Normal wings	Vestigial wings
Expected number of *Drosophila* (x)	36	12
Actual number of *Drosophila*	30	18
Deviation (d)	6	6
	$d^2 = 36$	36

$$\chi^2 = \sum \frac{d^2}{x}$$
$$= \frac{36}{36} + \frac{36}{12}$$
$$= 1.0 + 3.0$$
$$= 4.0$$

With two classes of results (vestigial and normal wings) there is just one degree of freedom. Using the relevant row on the chi-squared table we find that the value of 4.0 lies between 3.84 and 5.41, i.e. 0.05 (5%) and 0.02 (2%), which means that the possibility that the deviation is due to chance is less than 5%. The deviation is therefore significant and we cannot assume the parents are heterozygous.

In another experiment domestic fowl with walnut combs were crossed with each other. The expected offspring ratio of comb types was 9 walnut, 3 rose, 3 pea and 1 single. In the event, the 160 offspring produced 93 walnut combs, 24 rose combs, 36 pea combs and 7 single combs. Applying the chi-squared test:

	Walnut	Rose	Pea	Single
Expected number of comb types (x)	90	30	30	10
Actual number of comb types	93	24	36	7
Deviation (d)	3	6	6	3
$d^2 =$	9	36	36	9

$$\chi^2 = \sum \frac{d^2}{x}$$

$$\therefore \chi^2 = \frac{9}{90} + \frac{36}{30} + \frac{36}{30} + \frac{9}{10}$$

$$= \frac{1}{10} + \frac{12}{10} + \frac{12}{10} + \frac{9}{10}$$

$$= \frac{34}{10}$$

$$= 3.4$$

In this instance there are four classes of results (walnut, rose, pea and single) and this is equivalent to three degrees of freedom. We must therefore use this row to determine whether the deviations are significant. The values lies between 2.37 and 4.11 which is equivalent to a probability of 0.5 (50%) to 0.25 (25%). This deviation is not significant and is simply the result of statistical chance.

10.4   Origins of variation

Variation may be due to the effect of the environment on an organism. For example, the action of sunlight on a light-coloured skin may result in its becoming darker. Such changes have little evolutionary significance as they are not passed from one generation to the next. Much more important to evolution are the inherited forms of variation which result from genetic changes. These genetic changes may be the result of the normal and frequent reshuffling of genes which occurs during sexual reproduction, or as a consequence of mutations.

10.4.1 Environmental effects

We saw in the previous chapter that the final appearance of an organism (phenotype) is the result of its genotype and the effect of the environment upon it. If organisms of identical genotype are subject to different environmental influences, they show considerable variety. If one of a pair of genetically identical plants is grown in a soil deficient in nitrogen, it will not attain the height of the other grown in a soil with sufficient nitrogen. Because environmental influences are themselves very various, and because they often form gradations, e.g. temperature, light intensity, they are largely responsible for continuous variation within a population.

10.4.2 Reshuffling of genes

The sexual process in organisms has three inbuilt methods of creating variety:

1. The mixing of two different parental genotypes where cross-fertilization occurs.

2. The random distribution of chromosomes during metaphase I of meiosis.

3. The crossing over between homologous chromosomes during prophase I of meiosis.

These changes, which were dealt with in more detail in Section 8.4.2, do not bring about major changes in features but rather create new combinations of existing features.

Mutations

Any change in the structure or the amount of DNA of an organism is called a **mutation**. Most mutations occur in somatic (body) cells and are not passed from one generation to the next. Only those mutations which occur in the formation of gametes can be inherited. These mutations produce sudden and distinct differences between individuals. They are therefore the basis of discontinuous variation.

10.4.3 Changes in gene structure (point mutations)

A change in the structure of DNA which occurs at a single locus on a chromosome is called a **gene mutation** or **point mutation**. In Section 7.5 we saw that the genetic code, which ultimately determines an organism's characteristics, is made up of a specific sequence of nucleotides on the DNA molecule. Any change to one or more of these nucleotides, or any rearrangement of the sequence, will produce the wrong sequence of amino acids in the protein it makes. As this protein is often an enzyme, it may result in it having a different molecular shape and hence prevent it catalysing its reaction. The result will be that the end product of that reaction cannot be formed. This may have a profound effect on the organism. For example, a gene mutation may result

TABLE 10.3

Message on telegram	Equivalent form of gene mutation	Likely result of receiving the message
Meet station 2100 hours today	Normal	Individuals meet as arranged
Meet Met station 2100 hours today	Duplication	Individuals arrive at correct time, one at the prearranged station, the other at the nearest station on the Metropolitan line (or London police station)
Met station 2100 hours today	Deletion	
Meet bus station 2100 hours today	Addition	Individuals arrive on time but one at the bus station, the other at the prearranged station
Meet station 1200 hours today	Inversion	Individuals arrive at correct place – but 9 hours apart
Meet station 1100 hours today	Substitution	Individuals arrive at correct place but 12 hours apart
Meet sat on it 2100 hours today	Inversion (including translocation	Message incomprehensible

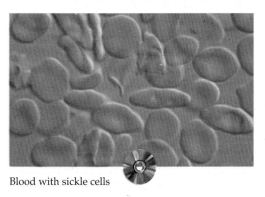

Blood with sickle cells

in the absence of pigments such as melanin. The organism will be unpigmented, i.e. an albino. There are many forms of gene mutation.

1. **Duplication** – a portion of a nucleotide chain becomes repeated.

2. **Addition (insertion)** – an extra nucleotide sequence becomes inserted in the chain.

3. **Deletion** – a portion of the nucleotide chain is removed from the sequence.

4. **Inversion** – a nucleotide sequence becomes separated from the chain. It rejoins in its original position, only inverted. The nucleotide sequence of this portion is therefore reversed.

5. **Substitution** – one of the nucleotides is replaced by another which has a different organic base.

To illustrate these different types, let us imagine each nucleotide is equivalent to a letter of the alphabet. The sequence of nucleotides therefore makes up a sentence or groups of sentences which can be understood by the cell's chemical machinery as the instructions for making specific proteins. If a mutation results in the instructions being incomprehensible, the cell will be unable to make the appropriate protein. In most cases this will result in the death of the cell or organism at an early stage. Sometimes, however, the mutation will result in inaccurate, and yet comprehensible, instructions being given. A protein may well be produced, but it is the wrong one. The defect may create some phenotypic change, but not of sufficient importance to cause the death of the organism.

Imagine a telegram to confirm the details of an earlier arrangement to meet at a pre-arranged station. If we alter just one or two letters each time, either the message may be totally incomprehensible or it may be understood by the receiver but not in the way intended by the sender. Table 10.3, gives some examples.

A gene mutation in the gene producing haemoglobin results in a defect called **sickle-cell anaemia**. The replacement of just one base in the DNA molecule results in the wrong amino acid being incorporated into two of the polypeptide chains which make up the haemoglobin molecule. The abnormal haemoglobin causes red blood cells to become sickle-shaped, resulting in anaemia and possible death. The detailed events are illustrated in Fig. 10.9.

The mutant gene causing sickle-cell anaemia is codominant. In the homozygous state, the individual suffers the disease and frequently dies. In the heterozygous state, the individual has 30–40% sickle cells, the rest being normal. This is called the sickle-cell trait. These individuals suffer less severe anaemia and rarely die from the condition. As they still suffer some disability, it might be expected that the disease would be very rare, if not completely eliminated, by natural selection. In parts of Africa, however, it is very common. The reason is that the malarial parasite, *Plasmodium*, cannot easily invade sickle cells. Individuals with either sickle-cell condition are therefore more resistant to severe attacks of malaria. In the homozygous condition, this resistance is insufficient to offset the

The DNA molecule which codes for the beta amino acid [ch]ain in haemoglobin has a mutation whereby the base [th]enine replaces thymine.

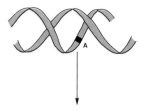

The mRNA produced has the triplet codon GUA (for [am]ino acid valine) rather than GAA (for amino acid glutamic [aci]d).

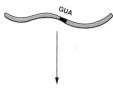

The beta amino acid chain produced has one glutamic [aci]d molecule replaced by a valine molecule.

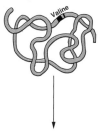

The haemoglobin molecule containing the abnormal beta [ch]ains forms abnormal long fibres when the oxygen level of [the] blood is low. This haemoglobin is called haemoglobin-S.

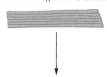

Haemoglobin-S causes the shape of the red blood cell to [be]come crescent (sickle) shaped.

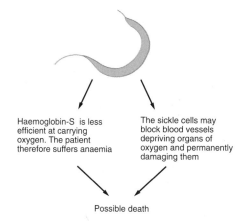

Haemoglobin-S is less efficient at carrying oxygen. The patient therefore suffers anaemia

The sickle cells may block blood vessels depriving organs of oxygen and permanently damaging them

Possible death

[F]ig. 10.9 Sequence of events whereby a gene [m]utation causes sickle-cell anaemia

considerable disadvantage of having sickle-cell anaemia. In the heterozygous condition (sickle-cell trait), however, the advantage of being resistant to severe malarial attacks outweighs the disadvantage of the mild anaemia the individual suffers. In malarial regions of the world the mutant gene is selected in favour of the one producing normal haemoglobin. Outside malarial regions, there is no advantage in being resistant to malaria, and the disadvantage of suffering anaemia results in selection *against* the mutant gene.

One relatively common gene mutation in European countries causes **cystic fibrosis**, which is the result of a recessive gene. Further details of this disorder are given on pages 124–5.

Dominant gene mutations are rarer but include **Huntington's disease**. This is characterized by involuntary muscular movement and progressive mental deterioration. The mutant gene is so rare (around 1 in 100 000 people carry it) that it occurs almost exclusively in the heterozygous state.

10.4.4 Changes in whole sets of chromosomes

Sometimes organisms occur that have additional whole sets of chromosomes. Instead of having a haploid set in the sex cells and a diploid set in the body cells, they have several complete sets. This is known as **polyploidy**. Where three sets of chromosomes are present, the organism is said to be **triploid**. With four sets, it is said to be **tetraploid**.

Polyploidy can arise in several different ways. If gametes are produced which are diploid and these self-fertilize, a tetraploid is produced. If instead the diploid gamete fuses with a normal haploid gamete, a triploid results. Polyploidy can also occur when whole sets of chromosomes double after fertilization.

Tetraploid organisms have two complete sets of homologous chromosomes and can therefore form homologous pairings during gamete production by meiosis. Triploids, however, cannot form complete homologous pairings and are usually sterile. They can only be propagated by asexual means. The type of polyploidy whereby the increase in sets of chromosomes occurs within the same species is called **autopolyploidy**. The actual number of chromosomes in an autopolyploid is always an exact multiple of its haploid number. Autopolyploidy can be induced by a chemical called **colchicine** which is extracted from certain crocus corms. Colchicine inhibits spindle formation and so prevents chromosomes separating during anaphase.

Sometimes hybrids can be formed by combining sets of chromosomes from species with different chromosome numbers. These hybrids are ordinarily sterile because the total number of chromosomes does not allow full homologous pairing to take place. If, however, the hybrid has a chromosome number which is a multiple of the original chromosome number, a new fertile species is formed. The species of wheat used today to make bread was formed in this way. The basic haploid number of wild grasses is seven. A tetraploid with 28 chromosomes called emmer wheat was accidentally cross-fertilized with a wild grass with 14 chromosomes. The resultant wheat with 42 chromosomes is today the main cultivated variety. Having a chromosome number which is a multiple of the original haploid number of 7, it is fertile. This form of polyploidy is called **allopolyploidy**.

Polyploidy is rare in animals, but relatively common in plants. Almost half of all flowering plants (angiosperms) are polyploids, including many important food plants. Wheat, coffee, bananas, sugar cane, apples and tomatoes all have polyploid forms. The polyploid varieties often have some advantage. Tetraploid apples, for example, form larger fruits and tetraploid tomatoes produce more vitamin C.

10.4.5 Changes in chromosome number

Sometimes it is an individual chromosome, rather than a whole set, which fails to separate during anaphase. If, for example, in humans one of the 23 pairs of homologous chromosomes fails to segregate during meiosis, one of the gametes produced will contain 22 chromosomes and the other 24, rather than 23 each. This is known as **non-disjunction** and is often lethal. The condition where an organism possesses an additional chromosome is represented as $2n + 1$; where one is missing, $2n - 1$. Where two additional chromosomes are present it is represented as $2n + 2$ etc.

One frequent consequence of non-disjunction in humans is **Down's syndrome** (mongolism). In this case the 21st chromosome fails to segregate and the gamete produced possesses 24 chromosomes. The fusion of this gamete with a normal one with 23 chromosomes results in the offspring having 47 ($2n + 1$) chromosomes. Non-disjunction does occur with other chromosomes but these normally result in the fetus aborting or the child dying soon after birth. The 21st chromosome is relatively small, and the offspring is therefore able to survive. Down's syndrome children have disabilities of varying magnitude. Typically they have a flat, broad face, squint eyes with a skin fold in the inner corner and a furrowed and protruding tongue. They have a low IQ and a short life expectancy.

Non-disjunction in the case of Down's syndrome appears to occur in the production of ova rather than sperm. Its incidence is related to the age of the mother. The chance of a teenage mother having a Down's syndrome child is only one in many thousand. A forty-year-old mother has a one in a hundred chance and by forty-five the risk is three times greater. The risk is unaffected by the age of the father.

Non-disjunction of the sex chromosomes can occur. One example is **Klinefelter's syndrome**. This may result in individuals who have the genetic constitution XXY, XXXY or XXXXY. These individuals are phenotypically male but have small testes and no sperm in the ejaculate. There may be abnormal breast development and the body proportions are generally female. The greater the number of Xs the more marked is the condition. As individuals are phenotypically male, this indicates that the presence of a Y chromosome is the cause of maleness. This is borne out by a second abnormality of the sex chromosomes. Individuals with **Turner's syndrome** have one missing X chromosome. Their genetic constitution is therefore

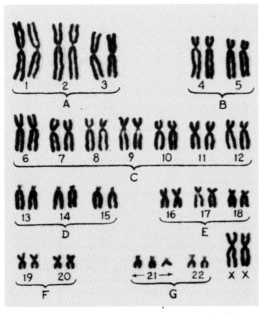

Karyotype of Down's syndrome

XO and they have only 45 (2n − 1) chromosomes. Individuals with this condition often do not survive pregnancy and are aborted. Those that do are phenotypically female, but small in stature and sexually immature. Despite having a single X chromosome, like males, they are female, indicating again that the Y chromosome is the cause of maleness.

10.4.6 Changes in chromosome structure

During meiosis it is normal for homologous pairs of chromosomes to form chiasmata. The chromatids break at these points and rejoin with the corresponding portion of chromatid on its homologous partner. It is not surprising that from time to time mistakes arise during this process. Indeed, it is remarkable that these chromosome mutations do not occur more frequently. There are four types:

1. **Deletion** – a portion of a chromosome is lost (Fig. 10.10a). As this involves the loss of genes, it can have a significant effect on an organism's development, often proving lethal.

2. **Inversion** – a portion of chromosome becomes deleted, but becomes reattached in an inverted position. The sequence of genes on this portion are therefore reversed (Fig. 10.10b). The overall genotype is unchanged, but the phenotype may be altered. This indicates that the sequence of genes on the chromosome is important.

3. **Translocation** – a portion of chromosome becomes deleted and rejoins at a different point on the same chromosome or with a different chromosome (Fig. 10.10c). The latter is equivalent to crossing over except that it occurs between non-homologous chromosomes.

4. **Duplication** – a portion of chromosome is doubled, resulting in repetition of a gene sequence (Fig. 10.10d).

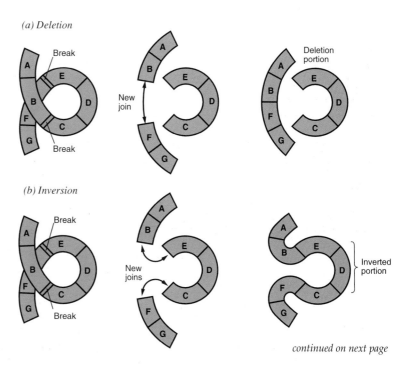

(a) Deletion

(b) Inversion

Fig. 10.10 *Diagrams illustrating the four types of chromosome mutation*

continued on next page

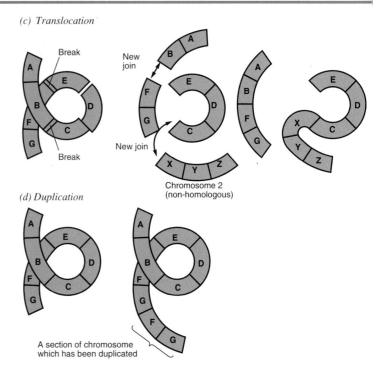

(c) Translocation

Break

New join

New join

Chromosome 2 (non-homologous)

(d) Duplication

A section of chromosome which has been duplicated

Fig. 10.10 (cont.) Diagrams illustrating the four types of chromosome mutation

10.5 Causes of mutations

Mutations occur continually. There is a natural mutation rate which varies from one species to another. In general, organisms with shorter life cycles, and therefore more frequent meiosis, show a greater rate of mutation. A typical rate of mutation is 1 c 2 new mutations per 100 000 genes per generation.

This natural mutation rate can be increased artificially by certain chemicals or energy sources. Any agent which induces mutations is called a **mutagen**. Most forms of high energy radiation are capable of altering the structure of DNA and thereby causing mutations. These include ultra-violet light, X-rays and gamma rays. High energy particles such as α and β particles and neutrons are even more dangerous mutagens.

A number of chemicals also cause mutations. We saw in Section 10.4.4 that colchicine inhibits spindle formation and so causes polyploidy. Other chemical mutagens include formaldehyde, nitrous acid and mustard gas.

10.6 Genetic screening and counselling

As our knowledge of inheritance has increased, more and more disabilities have been found to have genetic origins. Some of these disabilities cannot be predicted with complete accuracy. Ir Down's syndrome, for example, it is impossible to give a precise prediction of its occurrence for any individual. The risk for a mother of a particular age can, however, be calculated. Other

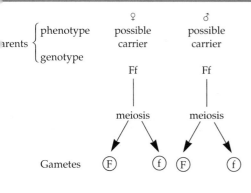

parents
{
phenotype — ♀ possible carrier — ♂ possible carrier

genotype — Ff — Ff
}

meiosis — meiosis

Gametes — (F) (f) (F) (f)

generation	♂ gametes	
	(F)	(f)
gametes (F)	FF	Ff
gametes (f)	Ff	ff

5% normal children
0% normal children, but carriers of the gene
5% children with cystic fibrosis

disabilities like haemophilia, cystic fibrosis, some forms of muscular dystrophy and Huntington's disease can be predicted fairly accurately, provided enough information on the history of the disease in the family is known. Genetic counselling has developed in order to research the family history of inherited disease and to advise parents on the likelihood of it arising in their children. Imagine a mother, whose family has a history of cystic fibrosis. If she herself is unaffected but possesses the gene, she can only be heterozygous for the condition. Suppose she wishes to produce children by a man with no history of the disease in his family. It must be assumed that he does not carry the gene for the disease and therefore none of the children will suffer from it, although they may be carriers. If, on the other hand, the potential father's family has a history of the disease, it is possible he too carries the gene. As we see from the genetic diagram opposite, it is possible to advise the parents that there is a one in four chance of their children being affected. The gene for cystic fibrosis is recessive and autosomal (i.e. not sex-linked).

On the basis of this advice the parents can choose whether or not to have children. With a very detailed knowledge of the disease in each of the parent's families it may even be possible to establish for certain whether they are carriers or not. It is now possible to carry out tests to establish with some accuracy whether an individual is heterozygous for the gene, and so to make precise predictions about the likelihood of having a child with cystic fibrosis (see Application on p. 124).

What then of the parents who have children knowing that there is a greater than usual possibility of them inheriting a genetic defect? Is there any means of establishing whether a child is affected, before it is born? The answer is yes, for some defects at least. Doctors can now diagnose certain genetic defects in a fetus, by studying samples of cells taken from the amniotic fluid which surrounds the fetus. In a process called **amniocentesis**. Certain genetic defects such as Down's syndrome can be detected directly as the additional chromosomes are easily seen. Biochemical tests on the cells, or even the amniotic fluid (which contains much fetal urine), may reveal other genetic defects. On the basis of these tests the parents can decide whether or not to have the pregnancy terminated. Details of the processes involved are given in Section 12.7.

10.6.1 Gene tracking

Gene tracking has proved a useful aid in genetic counselling. It is first necessary to find out on which chromosome a defective gene is located, something achieved through mapping chromosomes. The technique of making selected crosses of pure-breeding parents and collecting thousands of F_1 offspring in order to determine cross-over values and hence map chromosomes, may be appropriate with *Drosophila*, but can hardly be used with humans. Instead the inheritance of other, easily distinguished features such as blood groups, are traced in families to act as **genetic markers**. Study of the correlation between certain blood group alleles and the occurrence of a genetic disease can determine whether or not the gene for the disease is on the same chromosome as that for blood groups. If one genetic marker is not linked to the disease in question another must be tried and so on until the one which shows linkage with the disease is found. Linked markers are then used to work out whether or not someone carries a disease – this is the process of gene tracking. The technique is used with Huntington's disease (also called Huntington's chorea), a neuropsychiatric disorder leading to the loss of control of movements. It is caused by an autosomal dominant allele and sufferers are usually heterozygous. As the disease often fails to manifest itself until those affected are in their 40's or 50's, it is usually passed on to their children even before they realize they carry the gene. Any technique which allows the presence of the gene to be detected early helps individuals to decide whether or not to have children, and also shows the size of the risk of those children carrying the gene. Gene tracking has not only achieved this, but has also established that some sufferers of Huntington's disease are homozygous for the defect.

10.7 Questions

Read through the following account about DNA and genes and then write on the dotted lines the most appropriate word or words to complete the account.

The DNA molecule is composed of a large number of ...., each consisting of a nitrogenous base, a phosphate group and a 5-carbon sugar. A gene is a length of DNA that contains coded information leading to the synthesis of one ....

Changes in genes are known as ...., and can arise spontaneously or by exposure of the organism to certain chemicals or to .... The replacement of a base in the DNA strand by a different base is known as a .... This type of change to a gene can cause a condition known as ....

(*Total 6 marks*)

ULEAC January 1993, Paper I, No. 8

(a) Explain what is meant by a genetic disorder.
(5 marks)
(b) Discuss genetic disorders with reference to specific examples you have studied.
(18 marks)
(*Total 23 marks*)

UCLES (Modular) March 1992,
(Genetics & its applications), No. 2

Read the following passage and then answer the questions.

Evolutionary change develops when a mutation occurs and survives the selective process, that is, when it is found to be either neutral or advantageous. For example, a GCT codon might mutate to GAT and we would obtain leucine instead of arginine in the protein.

In about 20% of all mutations, because of the redundancy (degeneracy) of the code, a mutation might have no effect on protein structure. Thus a mutation from GCT to GCA would affect only the DNA and might well have no functional effects. No matter what the third base in the GC codon, we always obtain arginine in the protein.

The evolutionary process, then, involves a change (mutation) in the DNA which is incorporated into the ongoing gene pool of the evolving species, and which can be reflected by a corresponding change in the amino acid sequence of the particular protein coded for by that gene.

We might state as a basic rule that such a process will have to produce divergence when any two populations become isolated from one another. This is because the relative rarity of mutations and the finite size of populations make

it statistically improbable that identical changes will be available for natural selection to incorporate into the gene pools.
Adapted from 'A Molecular Approach to the Problem of Human Origins' by V. Sarich (1971).

(a) Explain what is meant by each of the following.
(i) 'A GCT codon' (line 4) (2 marks)
(ii) 'gene pool' (line 16) (2 marks)
(iii) 'divergence' (line 21) (2 marks)
(iv) 'natural selection' (line 26) (2 marks)
(b) (i) Suggest *two* ways in which 'a mutation from GCT to GCA' (line 10) might arise. (2 marks)
(ii) Explain why such a mutation might well have 'no functional effects' (line 11). (2 marks)
(c) (i) State three ways in which 'populations become isolated from one another' (lines 22–23). (3 marks)
(ii) Outline the possible consequences of the 'divergence' (line 21) that may result from such isolation. (3 marks)
(d) Give two examples of conditions which might make mutations more than a 'relative rarity' (line 23). (2 marks)
(*Total 20 marks*)

ULEAC January 1990, Paper I, No. 13

4. Wild type individuals of the fruit fly *Drosophila* have red eyes and straw-coloured bodies. A recessive allele of a single gene in *Drosophila* causes *glass eye* (g), and a recessive allele of a different gene causes *ebony body* (e).

A student carried out an investigation to test the hypothesis that the genes causing glass eye and ebony body show autosomal linkage. When she crossed pure-breeding wild type flies with pure-breeding flies having glass eye and ebony body, the F₁ flies all showed the wild type phenotype for both features. On crossing the F₁ flies among themselves, the student obtained the following results for the F₂ generation.

Eye	Body	Number of flies observed in F_2 generation (O)
Wild	Wild	312
Wild	Ebony	64
Glass	Wild	52
Glass	Ebony	107
Total		535

(a) Using appropriate symbols, write down the genotypes of the F_1 flies and the glass-eyed, ebony-bodied F_2 flies. *(2 marks)*

(b) In order to generate expected numbers of F_2 flies for use in a χ^2 test, the student used the *null hypothesis* that the genes concerned were *not* linked.

   (i) State the ratio of F_2 flies expected using the null hypothesis. *(1 mark)*

   (ii) Complete the table below to give the numbers of F_2 flies expected (E) using the null hypothesis, and the differences between observed and expected numbers (O–E).

Eye	Body	Number of flies observed (O)	Number of flies expected (E)	O–E
Wild	Wild	312		
Wild	Ebony	64		
Glass	Wild	52		
Glass	Ebony	107		

(c) (i) Use the formula *(2 marks)*

$$\chi^2 = \sum \frac{(O-E)^2}{E}$$

to calculate the value of χ^2. Show your working. *(2 marks)*

   (ii) How many degrees of freedom does this test involve? Explain your answer. *(2 marks)*

   (iii) For this number of degrees of freedom, χ^2 values corresponding to important values of P are as follows.

Value of P 0.99 0.95 0.05 0.01 0.001

Value of χ^2 0.115 0.352 7.815 11.34 16.27

What conclusions can be drawn concerning linkage of the **g** and **e** alleles? Explain your answer. *(3 marks)*
(Total 12 marks)

ULEAC January 1994, Paper III, No. 8

5. (a) Explain what is meant by each of the following terms, giving an appropriate example in each case.

   (i) Continuous variation *(2 marks)*

   (ii) Discontinuous variation *(2 marks)*

(b) Explain why populations of species on volcanic islands may differ from the mainland populations from which they originated. *(3 marks)*
(Total 7 marks)

ULEAC June 1993, Paper III, No. 5

6. The diagram below shows the sequence of amino acids in part of a molecule of haemoglobin. This sequence of amino acids was determined by the sequence of codons shown on the adjacent messenger RNA during the process of translation. The codon sequence on the messenger RNA was determined by the base sequence on the adjacent DNA strand during the process of transcription. The bases on this DNA strand have not been specified.

Chain of haemoglobin	Val	His	Leu	Thr	Pro	Glu	Glu

m-RNA   GUA CAU UUA ACU CCU GÅA GAG

DNA single strand

Key

Val = valine   Pro = proline
Thr = threonine   Leu = leucine
His = histidine   Glu = glutamic acid

(a) Write in the complementary base sequence on the DNA strand in the diagram. *(2 marks)*

(b) Comment on the fact that glutamic acid has two different codons. *(2 marks)*

(c) (i) If the base U was substituted for the base marked *, how would the haemoglobin chain be different? *(1 mark)*

   (ii) This substitution produces an abnormal form of haemoglobin called haemoglobin S. What condition is associated with this abnormal haemoglobin? *(1 mark)*

(d) Substitution of this type is an example of a 'point mutation of a gene'. Name *two* other ways in which a point mutation may occur in genes. *(2 marks)*
(Total 8 marks)

ULEAC June 1993, Paper I, No.

7. Read the passage, and then answer the questions that follow.

Haemoglobin, the red pigment in blood which transports oxygen, is a globular protein made up of four polypeptide chains. In an adult human, these consist of two alpha chains and two beta
5 chains.

    In most people, both copies of the gene coding for the beta polypeptide chain contain the base sequence -CCT-GAG-GAG-. This allele codes for the normal beta polypeptide chain. There is
10 another allele of this gene in which the base sequence is -CCT-GTG-GAG-. This allele codes for an abnormal beta polypeptide chain.

Haemoglobin containing this polypeptide chain does not function correctly. In people with two copies of the abnormal allele, the disease sickle cell anaemia results, in which the haemoglobin becomes fibrous at low oxygen concentrations, distorting the red blood cells and preventing oxygen transport. In people with one normal and one abnormal allele, only half of their haemoglobin is abnormal. They suffer from a mild form of the disease, called sickle cell trait.

Sickle cell anaemia and sickle cell trait are most common in populations who live in areas where malaria is prevalent, or in people who are descended from such populations. Although a strong selection pressure acts against people with sickle cell anaemia, sickle cell trait gives some protection against malaria. Natural selection therefore has not eliminated the sickle cell allele from these populations.

(a) (i) Write down the base sequence on mRNA transcribed from:

the part of the normal allele shown in line 8;

the part of the abnormal allele shown in line 11.

(ii) Explain how a small difference in base sequence can result in a large difference in the structure of a protein such as haemoglobin. *(6 marks)*

(b) If two people with sickle cell trait marry, what are the chances that their first child will have sickle cell anaemia? Explain your answer. *(4 marks)*

(c) Explain why, despite strong selection pressure against people with sickle cell anaemia, the sickle cell allele has not been eliminated in some populations. *(4 marks)*

(Total 14 marks)

UCLES (Modular) June 1992, (Foundation), No. 3

8. The diagram shows variation in butterfat content in the milk of two breeds of cattle, Red Dane and Jersey, and in a cross between them. Red Dane and Jersey cattle are homozygous for genes affecting butterfat content.

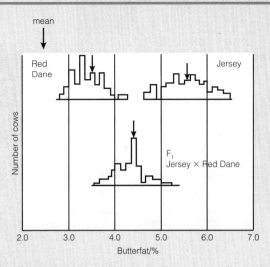

(a) What is the evidence that the variation in butterfat content in the parent breeds may be due to:
(i) genetic differences; *(2 marks)*
(ii) environmental differences? *(2 marks)*

(b) (i) What is meant by polygenic inheritance? *(1 mark)*
(ii) How do the results for the F_1 generation suggest that butterfat content of milk shows polygenic inheritance? *(2 marks)*

(Total 7 marks)

AEB June 1992, Paper I, No. 16

9. (a) With the aid of a suitable example in each case, explain what is meant by:
(i) continuous variation, and
(ii) discontinuous variation. *(4 marks)*

(b) Outline the sources and causes of genetic variation. *(10 marks)*

(c) (i) Distinguish between *natural selection* and *artificial selection*. *(3 marks)*
(ii) Briefly describe an example of artifical selection that has proved to be of importance to human society. *(3 marks)*
(iii) Describe one example of natural selection in action and explain how this could contribute to evolution. *(10 marks)*

(Total 30 marks)

Oxford June 1994, Paper 2, No. 7

Evolution

Evolution is the process by which new species are formed from pre-existing ones over a period of time. It is not the only explanation of the origins of the many species which exist on earth, but it is the one generally accepted by the scientific world at the present time.

11.1 Population genetics

To a geneticist, a population is an interbreeding group of organisms. In theory, any individual in the population is capable of breeding with any other. In other words, the genes of any individual organism are capable of being combined with the genes of any other. The genes of a population are therefore freely interchangeable. The total of all the alleles of all the genes in a population is called the **gene pool**. Within the gene pool the number of times any one allele occurs is referred to as its frequency.

11.1.1 Heterozygotes as reservoirs of genetic variation (the Hardy–Weinberg principle)

If one looks at a particular characteristic in a population, it is apparent that the dominant form expresses itself more often than the recessive one. In almost all human populations, for example, brown eyes occur more frequently than blue. It might be thought, therefore, that in time the dominant form would predominate to the point where the recessive type disappeared from the population completely. The proportion of dominant and recessive alleles of a particular gene remains the same, however. It is not altered by interbreeding. This phenomenon is known as the **Hardy–Weinberg principle**. It is a mathematical law which depends on four conditions being met:

1. No mutations arise.

2. The population is isolated, i.e. there is no flow of genes into, or out of, the population.

3. There is no natural selection.

4. The population is large and mating is random.

While these conditions are probably never met in a natural population, the Hardy–Weinberg principle nonetheless forms a basis for the study of gene frequencies.

1st allele	2nd allele	Frequency
A	A	$p \times p = p^2$
A	a	$p \times q$
a	A	$q \times p$
a	a	$q \times q = q^2$

The braces joining $p \times q$ and $q \times p$ give $2pq$.

To help understand the principle, consider a gene which has a dominant allele A and a recessive one a.

Let p = the frequency of allele A and

q = the frequency of allele a

In diploid individuals these alleles occur in the combinations given opposite.

As the homozygous dominant (AA) combination is 1/4 of the total possible genotypes, there is a 1/4 (25%) chance of a single individual being of this type. Similarly, the chance of it being homozygous recessive (aa) is 1/4 (25%) whereas there is a 1/2 (50%) chance of it being heterozygous. There is a 1/1 (100%) chance of it being any one of these three types. In other words:

homozygous dominant (1/4) + heterozygous 1/2 + homozygous recessive (1/4) = 1.0 (100%)

thus   AA + 2Aa + aa = 1.0 (100%)

and p^2 + $2pq$ + q^2 = 1.0 (100%)

The Hardy–Weinberg principle is expressed as:

$$p^2 + 2pq + q^2 = 1.0$$

(where p and q represent the respective frequencies of the dominant and recessive alleles of any particular gene).

The formula can be used to calculate the frequency of any allele in the population. For example, imagine that a particular mental defect is the result of a recessive allele. If the number of babies born with the defect is one in 25 000, the frequency of the allele can be calculated as follows:

The defect will only express itself in individuals who are homozygous recessive. Therefore the frequency of these individuals $(q^2) = 1/25\,000$ or $0.000\,04$.

The frequency of the allele (q) is therefore $\sqrt{0.000\,04}$

$= 0.0063$ approx.

As the frequency of both alleles must be 1.0, i.e. $p + q = 1.0$, then the frequency of the dominant allele (p) can be calculated.

$$p + q = 1.0$$
$$\therefore p = 1.0 - q$$
$$\therefore p = 1.0 - 0.0063$$
$$\therefore p = 0.9937$$

The frequency of heterozygotes can now be calculated.

From the Hardy–Weinberg formula, the frequency of heterozygotes is $2pq$, i.e. $2 \times 0.9937 \times 0.0063 = 0.0125$.

In other words, 125 in 10 000 (or 313 in 25 000) are carriers (heterozygotes) of the allele.

This means that in a population of 25 000 individuals, just one individual will suffer the defect but around 313 will carry the allele. The heterozygotes are acting as a reservoir of the allele, maintaining it in the gene pool. As these heterozygotes are normal, they are not specifically selected against, and so the allele remains. Even if the defective individuals are selectively removed, the frequency of the allele will hardly be affected. In our population of 25 000, there is one individual who has two recessive alleles and 313 with one recessive allele – a total of 315. The removal of the defective individual will reduce the number of alleles in the population by just 2, to 313. Even with the removal of all defective individuals it would take thousands of years just to halve the allele's frequency.

Occasionally, as in sickle cell anaemia (Section 10.4.3), the heterozygote individuals have a selective advantage. This is known as **heterozygote superiority**.

11.1.2 Genetic drift

The Hardy–Weinberg principle is only applicable to large populations. In small populations a situation called **genetic drift** arises. Consider an allele which occurs in 1% of the members of a species. In a population of one million, 10 000 individuals may be expected to possess this allele. Even if some of these fail to pass it on to their offspring, the vast majority are likely to do so. The proportion of individuals with the allele will not be significantly altered in the next generation. If, however, the population is much smaller, say 1000 individuals, only 10 will carry the allele. The effect of some of these failing to pass it on will have a marked effect on its frequency in the next generation. This drift in the frequency of the allele is greater the smaller the population. Indeed, in an extreme example, a population of just 100 will have only a single individual with the allele. If this individual fails to breed, the allele will be lost from the population altogether.

11.2 Evolution through natural selection (Darwin/Wallace)

Charles Darwin

Charles Darwin (1809–1882) became the naturalist on HMS *Beagle* which sailed in 1832 to South America and Australasia. On the voyage he had an excellent opportunity to examine a wide variety of living plants and animals and his knowledge of geology was invaluable for studying fossils he came across. He was struck by the remarkable likeness between the fossils he found and present-day organisms. At the same time he observed the differences in certain characters that occurred when otherwise similar animals lived in different environments. He noted such differences between the organisms of different continents and between the east and west coast of South America. What impressed him most were the distinct variations between the species which inhabited different islands in a small group 580 miles off the coast of Ecuador. These were the Galapagos Islands. In particular he studied the finches which inhabited each of the islands. While they all had a general resemblance to those on the mainland of Ecuador, they nevertheless differed in certain respects, such as the shape of their beaks. He considered that originally a few finches had strayed from the mainland to these volcanic islands, shortly after their formation. Encountering, as they did, a range of different foods, each type of finch developed a beak which was adapted to suit their diet. Following the five-year voyage, Darwin set about developing his views on the mechanism by which these changes occurred.

Quite independently of Darwin, Alfred Wallace had drawn his own conclusions on the mechanism of evolution. Wallace sent Darwin a copy of this theory and Darwin realized that they were in essence the same as his own. As a result, they jointly presented their findings to the Linnaean Society in 1858. A year later Darwin published his book *On the Origin of Species by Means of Natural Selection and the Preservation of Favoured Races in the Struggle for Life*. The essential features of the theory Darwin put forward are:

1. Overproduction of offspring

All organisms produce large numbers of offspring which, if they survived, would lead to a geometric increase in the size of any population.

2. Constancy of numbers

Despite the tendency to increase numbers due to overproduction of offspring, most populations actually maintain relatively constant numbers. The majority of offspring must therefore die, before they are able to reproduce.

3. Struggle for existence

Darwin deduced on the basis of **1** and **2** that members of the species were constantly competing with each other in an effort to survive. In this struggle for existence only a few would live long enough to breed.

4. Variation among offspring

The sexually produced offspring of any species show individual variations (Section 10.4) so that generally no two offspring are identical.

5. Survival of the fittest by natural selection

Among the variety of offspring there will be some better able to withstand the prevailing conditions than others. That is, some will be better adapted ('fitter') to survive in the struggle for existence. These types are more likely to survive long enough to breed.

6. Like produces like

Those which survive to breed are likely to produce offspring similar to themselves. The advantageous characteristics which gave them the edge in the struggle for existence are likely to be passed on to the next generation.

7. Formation of new species

Individuals lacking favourable characteristics are less likely to survive long enough to breed. Over many generations their numbers will decline. The individuals with favourable characteristics will breed, with consequent increase in their numbers. The inheritance of one small variation will not, by itself, produce a new species. However, the development of a number of variations in a particular direction over many generations will gradually lead to the evolution of a new species.

11.3 Natural selection

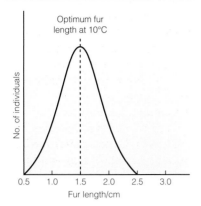

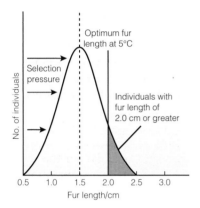

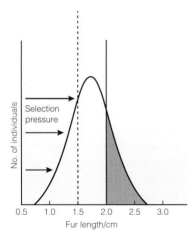

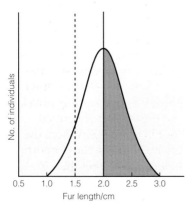

Fig. 11.1 Directional selection

The evolutionary theory of Darwin and Wallace is based on the mechanism of natural selection. Let us look more closely at exactly how this process operates.

Selection is the process by which organisms that are better adapted to their environment survive and breed, while those less well adapted fail to do so. The better adapted organisms are more likely to pass their characteristics to succeeding generations. Every organism is therefore subjected to a process of selection, based upon its suitability for survival given the conditions which exist at the time. The organism's environment exerts a **selection pressure**. The intensity and direction of this pressure varies in both time and space. Selection pressure determines the spread of any allele within the gene pool.

11.3.1 Types of selection

There are three types of selection which operate in a population of a given species.

Directional selection
When environmental conditions change, there is a selection pressure on a species causing it to adapt to the new conditions. Within a population there will be a range of individuals in respect of any one character. The continuous variation among individuals forms a normal distribution curve, with a mean which represents the optimum for the existing conditions. When these conditions change, so does the optimum necessary for survival. A few individuals will possess the new optimum and by selection these in time will predominate. The mean for this particular character will have shifted. An example is illustrated in Fig. 11.1.

Stabilizing selection
This occurs in all populations and tends to eliminate the extremes within a group. In this way it reduces the variability of a population and so reduces the opportunity for evolutionary change.

In our earlier example, we see that at 10 °C there was an optimum coat length of 1.5 cm. Individuals within the population, however, had a range of coat lengths from 0.5 cm to 3.0 cm. Under normal climatic circumstances the average

In a population of a particular mammal fur length shows continuous variation.

1. *When the average environmental temperature is 10 °C, the optimum fur length is 1.5 cm. This then represents the mean fur length of the population.*

2. *A few individuals in the population already have a fur length of 2.0 cm or greater. If the average environmental temperature falls to 5 °C these individuals are better insulated and so are more likely to survive to breed. There is a selection pressure favouring individuals with longer fur.*

3. *The selection pressure causes a shift in the mean fur length towards longer fur over a number of generations. The selection pressure continues.*

4. *Over further generations the shift in the mean fur length continues until it reaches 2.0 cm – the optimum length for the prevailing average environmental temperature of 5 °C. The selection pressure now ceases.*

temperature will vary from one year to the next. In a warm year with an average temperature of 15 °C the individuals with shorter fur may be at an advantage as they can lose heat more quickly. In these years the numbers of individuals with short fur increase at the expense of those with long fur. In cold years, the reverse is true and individuals with long fur increase at the expense of their companions with shorter coats. The periodic fluctuations in environmental temperature thus help to maintain individuals with very long and very short fur.

Imagine that the average environmental temperature was 10 °C every year and there were no fluctuations. Without the warmer years to give them an advantage in the competition with others in the population, the individuals with short hair would decline in numbers. Likewise the absence of colder years would reduce the number of long-haired individuals. The mean fur length would remain at 1.5 cm but the distribution curve would show a much narrower range of lengths (Fig. 11.2).

Disruptive selection
Although much less common, this form of selection is important in achieving evolutionary change. Disruptive selection may occur when an environmental factor takes a number of distinct forms. To take our hypothetical example, suppose the environmental temperature alternated between 5 °C in the winter and 15 °C in the summer, with no intermediate temperatures occurring. These conditions would favour the development of two distinct phenotypes within the population: one with a fur length of 2.0 cm (the optimum of an environmental temperature of 5 °C); the other with a fur length of 1.0 cm (optimum length at 15 °C). (Fig. 11.3)

11.3.2 Polymorphism

Polymorphism (*poly* – 'many', *morph* – 'form') is the word used to describe the presence of clear-cut, genetically determined differences between large groups in the same population. One well known example is the A B O blood grouping system found in human populations; another is the colour and banding patterns which arise in certain species of land snail.

Polymorphism in the land snail (Cepaea nemoralis)
The shells of the land snail *Cepaea nemoralis* have a variety of distinct colours including yellow, pink and brown. The snail shells may also be marked with dark bands, ranging in number from 1 to 5. Colour and banding are both genetically determined.

The snail is an important source of food for the song thrush. The thrush uses rocks as anvils upon which to smash the shells to enable them to eat the soft parts within. Examination of the smashed shells reveals that the proportions of each type differ according to the habitat around. Where the surrounding area is relatively uniform, e.g. grassland, the proportion of banded shells is greater among those smashed by thrushes. Where the surrounding area is less uniform, e.g. hedgerows, the proportion of unbanded shells is greater. How can we interpret these results?

It must be assumed that on uniform backgrounds, banded shells are more conspicuous and so more easily seen by

Initially there is a wide range of fur length about the mean of 1.5 cm. The fur lengths of less than 1.0 cm or greater than 2.0 cm in individuals are maintained by rapid breeding in years when the average temperature is much warmer or colder than normal.

When the average environmental temperature is consistently around 10 °C with little annual variation, individuals with very long or very short hair are eliminated from the population over a number of generations.

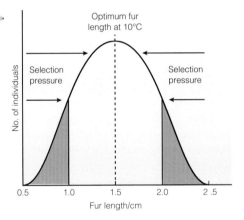

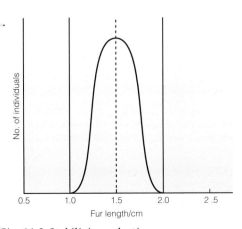

Fig. 11.2 Stabilizing selection

1. *When there is a wide range of temperatures throughout the year, there is continuous variation in fur length around a mean of 1.5 cm.*

2. *Where the summer temperature is static around 15°C and the winter temperature is static around 5°C, individuals with two distinct fur lengths predominate: 1.0 cm types which are active in summer and 2.0 cm types which are active in winter.*

3. *After many generations two distinct sub-populations are formed.*

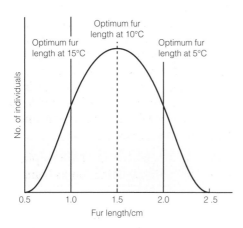

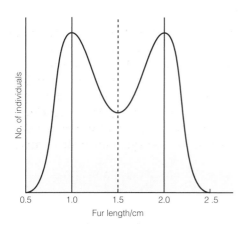

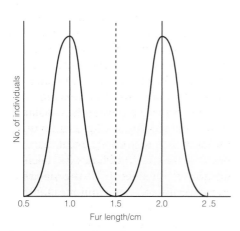

Fig. 11.3 Disruptive selection

predators such as thrushes. These shells form a disproportionately large number of smashed shells found around 'anvils' in this type of habitat. In less uniform backgrounds, bands on the shells tend to break up the shape of the animal, thus providing a form of disruptive camouflage. By contrast, the unbanded ones are not camouflaged. Being more conspicuous these forms are more frequently captured by thrushes and are consequently more common around nearby 'anvils'. As a result of this selection, the variety of snail shell commonly found around any anvil is correspondingly rare in the surrounding habitat.

Colour varieties (variously striped and unstriped) of white-lipped snail (*Cepaea hortensis*)

In a similar way, there is a correlation between shell colour and the predominant colour of the habitat background. Where the habitat is light in colour yellow-shelled snails are most common in the area, but remains of brown ones occur more frequently around 'anvils'. The reverse is true of habitats with dark backgrounds where brown snails predominate overall, but remains of yellows are more common around 'anvils'. Clearly the thrushes find brown snails easier to spot against a light background and yellow snails easier to spot against a dark one. In time one might expect this selection to eliminate the more conspicuous variety from any habitat. It appears, however, that the genes for colour and banding are closely linked with other genes which may confer advantages. In this way any evolutionary disadvantage due to colour or banding, is offset by these advantages, in a way comparable with the sickle cell gene (Section 10.4.3). When particular gene loci are so close together that crossing over almost never separates them, then these genes effectively act as a single unit. This unit is called a **super-gene** and is the possible explanation for the continued existence of certain polymorphic forms of *Cepaea* in particular habitats.

Polymorphism in the peppered moth (Biston betularia) *(Industrial melanism)*
Another example of polymorphism occurs in the peppered moth *Biston betularia*. It existed only in its natural light form until the middle of the last century. Around this time a melanic (black)

Peppered moth (*Biston betularia*) – normal and melanic forms on charred tree trunk (top) and on birch trunk (above)

variety arose as a result of a mutation. These mutants had doubtless occurred before (one existed in a collection made before 1819) but they were highly conspicuous against the light background of lichen-covered trees and rocks on which they normally rest. As a result, the black mutants were subject to greater predation from insect-eating birds, e.g. robins and hedge sparrows, than were the better camouflaged, normal light forms.

When in 1848 a melanic form of the peppered moth was captured in Manchester, most buildings, walls and trees were blackened by the soot of 50 years of industrial development. The sulphur dioxide in smoke emissions killed the lichens that formerly covered trees and walls. Against this black background the melanic form was less, not more conspicuous than the light natural form. As a result, the natural form was taken by birds more frequently than the melanic form and, by 1895, 98% of Manchester's population of the moth was of the melanic type.

Dr H.B.D. Kettlewell attempted to show that this change in gene frequency was the result of natural selection. He bred large stocks of both varieties of the moth. He then marked them and released them in equal numbers in two areas:

1. Birmingham (an area where soot pollution and high sulphur dioxide levels had resulted in 90% of the existing moths being of the melanic form).

2. Rural Dorset (where the absence of high levels of soot and sulphur dioxide meant lichen-covered trees and no record of the melanic form of the moth).

He recaptured samples of the moths using light traps and found that in polluted Birmingham over twice as many marked melanic moths were recaptured as normal light ones. The reverse was true for unpolluted Dorset where the normal light form was recaptured with twice the frequency of the melanic type.

Further studies have confirmed the findings that the melanic form has a selection advantage over the lighter form in industrial areas. In non-polluted areas the selection advantage lies with the lighter form.

11.3.3 Drug and pesticide resistance

Following the production of antibiotics in the 1940s, it was noticed that certain bacterial cells developed resistance to these drugs, i.e. the antibiotics failed to kill them in the normal way. Experiments showed that this was not a cumulative tolerance to the drug, but the result of chance mutation. This mutation in some way allowed the bacteria to survive in the presence of drugs like penicillin, e.g. by producing an enzyme to break it down. In the presence of penicillin non-resistant forms are destroyed. There is a selection pressure favouring the resistant types. The greater the quantity and frequency of penicillin use, the greater the selection pressure. The medical implications are obvious. Already the usefulness of many antibiotics has been destroyed by bacterial resistance to them. By 1950, the majority of staphylococcal infections were already penicillin-resistant.

The problem has been made more acute by the recent discovery that resistance can be transmitted between species. This means that disease-causing bacteria can become resistant to

PROJECT

The land snail *Cepaea nemoralis* possesses a shell with or without black bands

1. Collect data on land snails from two different localities, for example from a beech-wood and from under hedges, or from an oak-wood and from grassland.

2. Score the numbers of banded and unbanded snails from the two localities.

3. Analyze the data and suggest a hypothesis to explain the results.

a given antibiotic even before the antibiotic is used against them. As a result, certain staphylococci are resistant to all major antibiotics.

Resistance to insecticides has come about in a similar way. Within two years of using DDT, many insects had developed resistance to it, often independently in different parts of the world. Most common insect pests are now resistant to most insecticides. In many cases the presence of the insecticide switches on the gene present in the mutant varieties. This gene initiates the synthesis of enzymes which break down the insecticide. Apart from directly harmful insects, insect vectors have also acquired resistance. Examples include mosquitoes of the genus *Aedes*, which carry yellow fever, and of the genus *Anopheles* which carry malaria.

Resistance to myxomatosis in rabbits takes two forms. In one type a mutant gene renders the myxomatosis virus ineffective in some way. In the second form a mutant gene alters the rabbits' behaviour, in that they spend more time above ground and less in their burrows. The disease is spread by a vector, the rabbit flea. Normal rabbits live in crowded warrens underground where the flea can easily be transferred between individuals. The mutant variety, spending less time underground, has a reduced chance of being affected by fleas, and hence catching the myxomatosis virus. This variety is favourably selected whereas previously it was selected against because of the increased chance of predation due to its vulnerability when above ground.

11.3.4 Heavy metal tolerance in plants

Another example of natural selection occurs on spoil heaps which contain the waste material from mining activities. These heaps contain high concentrations of certain heavy metals, e.g. tin, lead, copper and nickel. In the concentrations found, these metals are toxic to most plants. Some varieties of grasses, e.g. *Festuca ovina* and *Agrostis tenuis*, have become genetically adapted to survive high levels of these metals. These plants are less competitive where the concentration of these metals is low and so do not always survive.

11.4 Artificial selection

Humans have cultivated plants and kept animals for about 10 000 years. Over much of this time they have bred them selectively. There have been two basic methods, each with a particular aim:

1. Inbreeding – When, by chance, a variety of plant or animal arose which possessed some useful character, it was bred with its close relatives in the hope of retaining the character for future generations. Inbreeding is still widely practised today, especially with dogs and cats.

One problem with inbreeding is that it increases the danger of a harmful recessive gene expressing itself, because there is greater risk of a double recessive individual arising (Section 9.1.3).

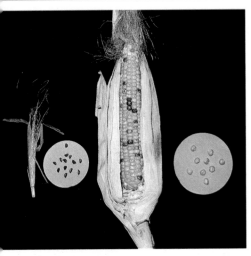

Selective breeding of corn. Note the small cob and seeds of the wild type (left) compared to the modern variety (right)

As a result, inbreeding is not usually carried out indefinitely but new genes are introduced by outbreeding with other stock. While this makes consistent qualities harder to achieve, it can lead to stronger, healthier offspring.

2. **Outbreeding** – This is carried out in order to improve existing varieties. Where two individuals of a species each have their own beneficial feature they are often bred together in order to combine the two. A racehorse breeder, for example, might cross a fast mare with a strong stallion in the hope of attaining a strong, fast foal. Outbreeding frequently produces tougher individuals with a better chance of survival, especially where many generations of inbreeding have taken place. This is known as **hybrid vigour**.

Extreme examples of outbreeding occur when individuals of different species are mated. Only rarely is this successful. When it is, the resulting offspring are normally sterile. These sterile hybrids may still be useful. Mules, produced from a cross between a horse and a donkey, have strength and endurance which make them useful beasts of burden.

The improvement of the human race by the selection or elimination of specific characters is called **eugenics**. To some, the idea of such selection is offensive but, as we saw in Section 10.6, genetic counselling is now fairly commonplace. Provided the individuals involved remain free to make their own choice about whether to have children, many see no harm in providing them with statistical information which might help them reach a decision.

11.5 Isolation mechanisms

Within a population of one species there are groups of individuals which breed with one another. Each of these breeding sub-units is called a **deme**. Although individuals within the deme breed with each other most of the time, it is still possible for them to breed with individuals of separate demes. There therefore remains a single gene pool. If demes become separated in some way, the flow of genes between them may cease. Each deme may then evolve along separate lines. The two demes may become so different that, even if reunited, they would be incapable of successfully breeding with each other. They would thus become separate species each with its own gene pool. The process by which species are formed is called **speciation** and depends on groups within a population becoming isolated in some way. There are two main forms of speciation:

11.5.1 Allopatric speciation

Allopatric speciation occurs as the result of two populations becoming geographically isolated. Any physical barrier which prevents two groups of the same species from meeting must prevent them interbreeding. Such barriers include mountain ranges, deserts, oceans, rivers, etc. The effectiveness of any

Did you know?

The Arctic tern migrates about 35 000 kilometres each year.

barrier varies from species to species. A small stream may separate two groups of woodlice, whereas the whole of the Pacific Ocean may fail to isolate some species of birds. A region of water may separate groups of terrestrial organisms, whereas land may isolate aquatic ones. The environmental conditions on either side of a barrier frequently differ. This leads to the group on each side adapting to suit its own environment – a process called **adaptive radiation**.

Imagine, for example, that climatic changes resulted in two areas becoming separated from one another by an area of arid grassland. A possible sequence of events which could lead to a new species being formed under these conditions is illustrated in Fig. 11.4.

Fig. 11.4 Speciation due to geographical isolation

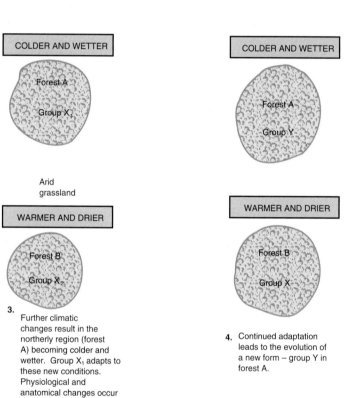

1. Species X occupies a forest area. Individuals within the forest form a single gene pool and freely interbreed.

2. Climatic changes to drier conditions reduce the size of the forest to two isolated regions. The distance between the two regions is too great for the two groups of species X to cross to each other.

3. Further climatic changes result in the northerly region (forest A) becoming colder and wetter. Group X₁ adapts to these new conditions. Physiological and anatomical changes occur in this group.

4. Continued adaptation leads to the evolution of a new form – group Y in forest A.

5. A return to the original climatic conditions results in regrowth of forest. Forests A and B are merged and groups X and Y are reunited. The two groups are no longer capable of interbreeding. They are now two species X and Y each with its own gene pool.

11.5.2 Sympatric speciation

Sympatric speciation occurs when organisms inhabiting the same area become reproductively isolated into two groups for reasons other than geographical barriers. Such reasons might include:

1. The genitalia of the two groups may be incompatible (mechanical isolation) – It may be physically impossible for the penis of a male mammal to enter the female's vagina.

2. The gametes may be prevented from meeting – In animals, the sperm may not survive in the female's reproductive tract or, in plants, the pollen tube may fail to grow.

3. Fusion of the gametes may not take place – Despite the sperm reaching the ovum, or the pollen tube entering the micropyle, the gametes may be incompatible and so will not fuse.

4. **Development of the embryo may not occur (hybrid inviability)** – Despite fertilization taking place, further development may not occur, or fatal abnormalities may arise during early growth.

5. **Polyploidy (hybrid sterility)** – When individuals of different species breed, the sets of chromosomes from each parent are obviously different. These sets are unable to pair up during meiosis and so the offspring cannot produce gametes. For example, the cross between a horse ($2n = 60$) and an ass ($2n = 66$) results in a mule ($2n = 63$). It is impossible for 63 chromosomes to pair up during meiosis. More details of polyploidy are given in Section 10.4.4.

6. **Behavioural isolation** – Before copulation can take place, many animals undergo elaborate courtship behaviour. This behaviour is often stimulated by the colour and markings on members of the opposite sex, the call of a mate or particular actions of a partner. Small differences in any of these may prevent mating. If a female stickleback does not make an appropriate response to actions of the male, he ceases to court her. The beak shape in many of Darwin's finches in the Galapagos Islands is the only feature which distinguishes the species. Individuals will only mate with partners having a similar beak to themselves. The song of a bird or the call of a frog must be exact if it is to elicit the appropriate breeding response from the opposite sex. The timing of courtship behaviour and gamete production is also important. If the breeding season of two groups (demes) does not coincide, they cannot breed. Different flowering times in plants may mean that cross-pollination is impossible. These are both examples of **seasonal isolation**.

Did you know?

The tambalacoque tree in Mauritius is threatened with extinction. It seems that its seeds only germinate if they have passed through the digestive system of a Dodo and the Dodo has been extinct for nearly 300 years. The seeds of the tree are being fed to turkeys in the hope that they will act as substitutes for the Dodo.

NOTEBOOK

Radiocarbon dating

Radiocarbon dating is a technique used to determine the age of an organism which died many, possibly thousands of, years ago. In the upper atmosphere unstable, radioactive ^{14}C atoms are formed when carbon atoms are bombarded by cosmic rays. These are oxidized and taken up as $^{14}CO_2$ by living organisms. The ^{14}C atoms emit β particles and decay to stable, nonradioactive ^{12}C. When an organism is living the ^{14}C is constantly replenished by take up from the environment but when it dies the ^{14}C gradually decays and is not replaced. By assessing the emission of β particles, and hence the activity of ^{14}C, it is possible to determine how long ago the organism died. The activity of ^{14}C decays by half every 5730 years; in other words its half-life is 5730 years. This means that after about 40 000 years its activity is so low that the measurement may not give a reliable date.

Many elements exist as several isotopes and release α and β particles as they change from an unstable to a stable form. By assessing the activity of several such isotopes dating can be cross-checked and thus made more reliable.

11.6   Questions

1. Explain the process of natural selection in relation to each of the following:
  (i)   melanic moths
  (ii) heavy metal tolerance
  (iii) bacterial resistance to antibiotics.
        (Total 15 marks)

SEB (revised) Specimen 1992, Higher Grade
Paper II, No. 13B

2. (a) Discuss **two** lines of evidence used to support the theory of evolution.   *(6 marks)*
   (b) Using appropriate examples, distinguish between continuous and discontinuous variation and briefly explain the inheritance of **each** type of variation.   *(4 marks)*
   (c) Describe how variation, selection and isolation may contribute to evolutionary change in living organisms.   *(10 marks)*
        (Total 20 marks)

NEAB June 1992, Paper IB, No. 1

3. The graphs show three basic types of natural selection. The shaded areas marked with arrows show the individuals in the populations which are being selected against.

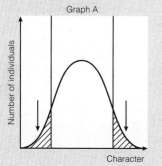

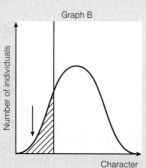

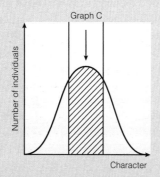

(a) What name is given to the type of selection shown in graph **A**?   *(1 mark)*
(b) Describe **one** specific example of the type of selection shown in graph **B**.
    In your answer:

(i)   name the organism;
(ii) describe the character selected.
        (2 marks)
(c) What will happen to the modal class in subsequent generations as a result of the type of selection shown in: graph **B**; graph **C**?   *(2 marks)*
        (Total 5 marks)

AEB June 1991, Paper I, No. 1

4. (a) The diagrams below show the original sequence of gene loci on a chromosome and the result of one form of mutation.

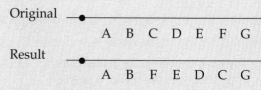

What term is used to describe the above mutation?   *(1 mark)*
(b) In Man, a single base pair substitution has produced a mutant allele which codes for the amino acid valine instead of glutamic acid; this results in the synthesis of abnormal haemoglobin.
    What effect does this mutation have on the structure of red blood cells?   *(1 mark)*
(c) The following information relates to three species of the buttercup genus, *Ranunculus*, which live close together on ridges and in furrows in grassland habitats.

Species	Chromosome number	Vegetative feature
R. bulbosus	2n = 16	corm present
R. acris	2n = 14	rhizome present
R. repens	2n = 32	rhizome present

The pattern of distribution in pasture fields is shown in the diagrams at the top of the next page.
(i)   In what way does the information given above provide evidence for 'adaptive radiation' in buttercup species?   *(2 marks)*
(ii) The three species do not hybridize.
      State *one* mechanism which can prevent hybridization between plant species that are closely related.   *(1 mark)*
(iii) Which tissue of a buttercup plant should be examined to determine the diploid chromosome number?   *(1 mark)*

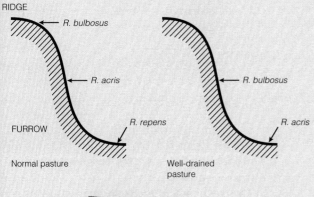

RIDGE

R. bulbosus

R. acris

R. repens

FURROW

Normal pasture

R. bulbosus

R. acris

Well-drained pasture

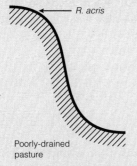

R. acris

Poorly-drained pasture

(Total 6 marks)

SEB (revised) Specimen 1992, Higher Grade Paper II, No. 7

5. The diagram below shows how bread wheat (*Triticum aestivum*) has arisen by polyploidy.

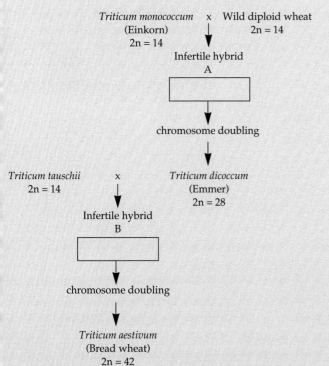

Triticum monococcum × Wild diploid wheat
(Einkorn)   2n = 14
2n = 14

Infertile hybrid
A

chromosome doubling

Triticum tauschii × Triticum dicoccum
2n = 14   (Emmer)
                    2n = 28

Infertile hybrid
B

chromosome doubling

Triticum aestivum
(Bread wheat)
2n = 42

(a) On the diagram, in the boxes provided, write in the chromosome numbers of the infertile hybrid A and the infertile hybrid B. (*2 marks*)

(b) (i) Suggest why hybrids A and B are infertile. (*1 mark*)

(ii) Suggest how doubling of the chromosomes might have occurred. (*2 marks*)

(ii) Explain how doubling of the chromosomes results in infertile hybrids becoming fertile. (*2 marks*)

(c) *Triticum durum* is another species of modern wheat. Suggest why it is important that there should be more than one species of wheat cultivated nowadays. (*1 mark*)

(Total 8 marks)

ULEAC June 92, Paper III, No. 3

6. (a) The following are types of selection:

directional   stabilising   disruptive

Identify the type of selection that would best describe each of the examples below.

(i) Early farmers selected cattle to produce some breeds that were horned and some that were hornless. (*1 mark*)

(ii) There is a higher death rate among very light and very heavy human babies than among those of average mass at birth. (*1 mark*)

(b) The two-spot ladybird exists in the two different colour forms shown in the diagram.

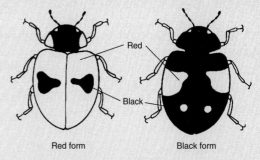

Red

Black

Red form   Black form

These forms are controlled by a single gene with the allele for black, B, dominant to that for red, b.

(i) Use the Hardy–Weinberg formula to predict the frequency of red and the frequency of black ladybirds in a population if the frequency of alleles B and b are the same. (*3 marks*)

(ii) Within Britain there is a difference in the frequency of the black form. In Northern regions the frequency of the black form is as high as 97% while in Southern regions it is below 10%. Suggest **two** reasons why it is **not** appropriate to use the Hardy–Weinberg formula to find the frequency of heterozygous black ladybirds in Britain. (*2 marks*)

(Total 7 marks)

AEB June 1993, Paper I, No. 17

209

7. (a) A particular species of insect may occur in either light or dark forms. The dark trait is dominant. In a certain population of 5000 such insects there are 950 which are dark. using the Hardy–Weinberg equation $(p^2 + 2pq + q^2 = 1)$ calculate the following and show your working:

  (i) the frequency of the 'light' allele.
  (2 marks)

  (ii) the frequency of the 'dark' allele.
  (1 mark)

  (iii) the number in the population which are heterozygous. *(2 marks)*

(b) State THREE conditions required for the use of the Hardy–Weinberg equation to be valid.
  (3 marks)

(c) Some years ago it was considered that the use of insecticides would result in the total eradication of mosquitoes. However mosquitoes are now more widespread than ever, and in a form resistant to the insecticides used. Describe how the insecticide resistance might have developed in mosquitoes. *(4 marks)*

  (Total 12 marks)

  NISEAC June 1993, Paper II, No. 7

8. In humans, the condition of cystic fibrosis is caused by an allele *a* which is recessive to the normal allele, *A*.

(a) Complete the spaces below and show the probability of two people, both heterozygous for this condition, having a child with cystic fibrosis.

	mother	father
Parental phenotypes	....	....
Parental genotypes	....	....
Gamete genotypes	....	....

Offspring genotypes ....

Probability of offspring having cystic fibrosis ....
  (2 marks)

(b) In Britain, approximately 1 in 25 people are heterozygous for this condition. What is the probability that any two parents are both heterozygotes? *(1 mark)*

(c) The frequency of the cystic fibrosis allele *a* in Britain is 0.02. Calculate the expected frequency of people who are:

  (i) homozygous *aa*;

  (ii) homozygous *AA*.

In each case, show your working. *(3 marks)*
  (Total 6 marks)

  AEB June 1992, Paper I, No. 17

9. (a) Give **two** of the assumptions that must be made about a population before applying the Hardy-Weinberg equation. *(2 marks)*

(b) In Europe, one person in 10 000 is lacking in body pigmentation (albino) and is homozygous for the recessive allele which controls this character. Using this information and the Hardy-Weinberg equation, calculate the frequency of people who are heterozygous for this allele in Europe. Show your working. *(4 marks)*
  (Total 6 marks)

  AEB June 1994, Common Paper I, No. 17

Reproduction, development and growth

No individual can live indefinitely. Some of its cells may be worn or damaged beyond repair or it may be killed by predators, disease or other environmental factors. If a species is to survive it must therefore produce new individuals. This is achieved in two ways:

1. Asexual reproduction – Rapidly produces large numbers of individuals, usually having an identical genetic composition to each other and to the single parent from which they are derived; gametes are never involved.

2. Sexual reproduction – Often less rapid, frequently involves two parents and produces offspring which are genetically different. The fusion of haploid nuclei is always involved. These nuclei are often contained in special cells called gametes.

Apart from purely increasing numbers, reproduction may involve one or more of the following:

(a) a means of increasing genetic variety and therefore helping a species adapt to changing environmental circumstances;

(b) the development of resistant stages in a life cycle which are capable of withstanding periods of drought, cold or other adverse conditions;

(c) the formation of spores, seeds or larvae which may be used to disperse offspring and so reduce intraspecific competition as well as capitalizing on any genetic variety among the offspring.

12.1 Comparison of asexual and sexual reproduction

Sexual reproduction always involves the fusion of nuclei which are often contained in special sex cells called gametes; asexual reproduction never does. If these gametes are produced by meiosis they will show considerable genetic variety (Section 8.4.2). The offspring resulting from the fusion of gametes will likewise show genetic variability and therefore be better able to adapt to environmental change. In other words, they have the capacity to evolve to suit new conditions. As asexual reproduction rarely involves meiosis the offspring are usually identical to each other and to their parents. While this lack of variety is a disadvantage in adapting to environmental change, it has one main advantage. If an individual has a genetic make-up

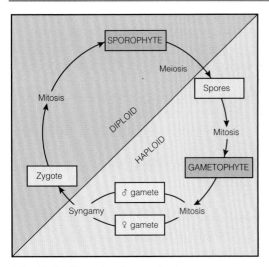

Fig. 12.1 *Life cycle of most plants*

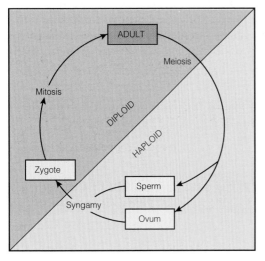

Fig. 12.2 *Life cycle of most animals*

which is suited to a particular set of conditions, asexual reproduction is a means by which large numbers of this successful type may be built up. A localized area can be rapidly colonized, something which is of particular advantage to plants.

It must be said that the differences in the variety of offspring resulting from sexual and asexual reproduction are not the same in all organisms. In mosses and ferns, for instance, the gametes are produced by a haploid gametophyte generation. Being haploid, its gametes can only be produced mitotically. Mitosis does not introduce genetic variety (Section 8.4.1) and these gametes are therefore identical. During asexual reproduction, mosses and ferns produce spores by meiosis. These spores show genetic variability. The usual differences between asexual and sexual reproduction are therefore reversed in mosses and ferns.

In all cases of asexual reproduction only one parent is involved but in sexual reproduction it may be one or two parents. If a species has male and female sex organs on separate individuals, it is said to be **dioecious**. In this case the male produces one gamete, the **sperm**, and the female another, the **ovum**. Only fusion between a sperm and an ovum can give rise to a new individual. Two parents are therefore always necessary for reproduction in these species. In some other species, one individual is capable of producing both male and female gametes. Such species are said to be **monoecious** or **hermaphrodite**. In most of these species the sperm from one individual fuses with the ovum from a separate individual. The fusion of gametes from two separate parents is called **cross-fertilization**. Provided the two parents are genetically different, greater variety of offspring results. In some species, however, sperm and ova from the same individual fuse. This is called **self fertilization** and involves a single parent; the degree of variety is therefore less. The process is nonetheless sexual as fusion of gametes is involved and as these are produced by meiosis some variety is still achieved.

It would be inaccurate to say that asexual reproduction always produces identical offspring. Mutations, although rare, nevertheless occur and so help create a little variety. Mutations also arise during sexual reproduction, indeed the greater complexity of the process means they arise more frequently.

To summarize, the main methods by which variety among offspring is achieved are:

1. The recombination of two different parental genotypes (genetic recombination).

2. During meiosis by:
   (a) random segregation of chromosomes on the metaphase plate (independent assortment);
   (b) crossing over during prophase I.

3. Mutations.

4. The effect of the environment on the genotype.

1 and **2** apply only to sexual reproduction; **3** applies to both, although the frequency is usually greater in sexual reproduction and **4** applies equally to asexual and sexual reproduction. Overall the processes of sexual reproduction are more complex than those of asexual reproduction. The events of meiosis are more complex than those of mitosis, and the processes of producing and transferring gametes are often complicated. Elaborate **courtship** and **mating rituals** are frequently part of sexual reproduction and serve to increase the likelihood of gametes successfully fusing (Section 12.7.1). These processes may take many months in some species and it is not therefore surprising that the offspring are often cared for by the parents. Where this is the case, the number of offspring is small to permit this parental care to be effective. By contrast, the process of asexual reproduction is normally more simple and straightforward. It is rapid and involves no parental care; the number of offspring is normally large.

On account of the variety of offspring produced and the consequent evolutionary potential, almost all organisms have a sexual phase at some stage in their life cycle. While simpler animals have retained the asexual process, most complex ones have abandoned it. A major disadvantage of being totally reliant on the sexual process is that it is difficult to maintain a favourable genotype. Once an organism has adapted to a particular set of conditions, sexual reproduction will tend to produce different offspring. These may not be as well adapted as identical copies of the parents would be. At least animals, with their ability to move from place to place, can search out conditions that suit any new variety. Plants do not exhibit locomotion and individuals must remain where they are. For this reason most have retained the asexual process as part of their life cycle. Hence, once a plant has successfully established itself in a suitable environment, it uses asexual means to rapidly establish a colony of identical, and therefore equally well-suited, individuals. Such a group has the advantage of reducing competition from other plant species, although with its identical genotypes it may be vulnerable to disease.

12.2 Asexual reproduction

As we have seen, asexual reproduction requires only a single parent and haploid gametes are not involved. There are seven major forms of asexual reproduction in organisms.

12.2.1 Binary or multiple fission

This occurs in single-celled organisms like bacteria and protozoans. The organism divides into two or more parts each of which leads a separate existence. Where the cell divides into two parts it is called binary fission and this typically occurs in bacteria.

The bacterial DNA replicates first and the nucleoplasm then divides into two, followed by the cell as a whole. Under favourable conditions (temperature around 20°C and an abundant supply of food) the daughter cells grow rapidly and may themselves divide within twenty minutes. Under unfavourable conditions some species develop a thick resistant wall around each daughter cell. The **endospore** thus formed is resistant to desiccation, extremes of temperature and toxic

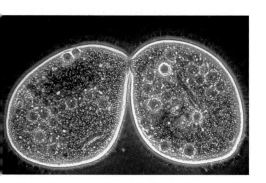

Binary fission in *Paramecium*

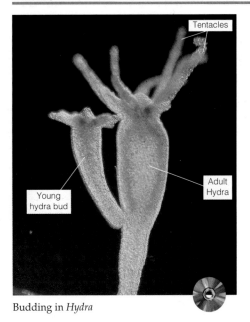

Budding in *Hydra*

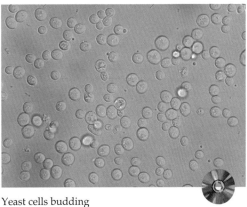

Yeast cells budding

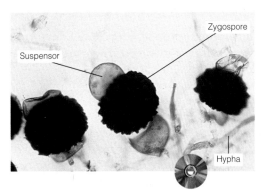

Zygospores of *Rhizopus nigra* germinate to form sporangia from which spores are released

chemicals. Only when favourable conditions return does a vegetative cell emerge to continue binary fission as before. It is these endospores which can survive certain forms of sterilization and so cause infection or disease. Multiple fission, or **schizogamy**, occurs when a cell divides into many parts rather than just two. In *Plasmodium*, the malarial parasite, for instance, the process occurs at a number of stages in the life cycle. These are described in Section 15.6.1.

12.2.2 Budding

An outgrowth develops on the parent and this later becomes detached and is then an independent organism. The process occurs in the Platyhelminthes (flatworms), some annelids (segmented worms) and cnidarians like *Hydra* and *Obelia*. In its simplest form it is little more than a type of binary fission, except that the two resultant cells are not of equal size but comprise a smaller bud cell, becoming detached from the larger parent cell. This occurs in *Saccharomyces* (yeast). In *Hydra* the buds arise near the centre of the parental column. They become highly differentiated, multicellular structures, and may develop their own buds before becoming detached.

12.2.3 Fragmentation

In one sense fragmentation is no more than a form of regeneration. If certain organisms are divided into sections, each portion will regenerate the missing parts thus giving rise to new individuals. If the division occurs as a result of injury then the process is regeneration. If however an organism regularly and spontaneously divides itself up in this way the process is fragmentation. Organisms exhibiting fragmentation must have relatively undifferentiated tissues and it is therefore limited to certain algae, sponges, cnidarians and flatworms. In *Spirogyra*, for example, portions of the filamentous alga break off when the filament reaches a certain length. These drift away, attach themselves elsewhere and begin vegetative growth again.

12.2.4 Sporulation

Sporulation is the formation of small unicellular bodies called **spores**, which detach from the parent and, given favourable conditions, grow into new organisms. Spores are usually small, light and easily dispersed. They are produced in vast numbers – a single mushroom for example may produce 500 000 spores a minute at the peak of its production. Sporulation occurs in bacteria, protozoans, algae, fungi, mosses and ferns (in one sense at least, all plants produce spores).

12.2.5 Vegetative propagation

In general, vegetative propagation involves the separation of a part of the parent plant which then develops into a new individual. Almost any part – root, stem, leaf or bud – may serve the purpose. They are often highly specialized for the task and bear little resemblance to the original plant organ from which they evolved; the potato, for instance, is actually a modified stem. No plant is likely to survive long in a changing world if it relies exclusively on asexual reproduction. Plants exhibiting vegetative propagation have not therefore abandoned the sexual process but continue to produce flowers in the usual way. In some plants, to ensure the continued survival of favourable genotypes, organs of vegetative propagation also act as **perennating organs** which lie in the soil over the winter. They are frequently swollen with excess food from the previous summer which is used to produce the new offspring the following year. One advantage is that growth can begin early in the spring using the stored food and the plant is therefore able to start photosynthesizing when there is little competition for light from other species. This is particularly important for small plants which live in deciduous woodland where they are subject to shading by their larger neighbours. Some organs of vegetative propagation are described in Table 12.1.

TABLE 12.1 **Organs of vegetative propagation and perennation**

Name	Example	Description of organ	Mechanism of action	Perennating organ
Bulb	Onion Garlic Daffodil Tulip	Underground, swollen, fleshy leaf-bases closely packed on a short stem, i.e. a bud	Apical and axillary buds among the leaves each give rise to more new plants	Yes
Corm	Crocus Montbretia Gladiolus Cyclamen	Underground, vertical swollen base of main stem	Buds develop in the axils of the scale leaves surrounding the corm. Each may develop into a new plant	Yes
Rhizome	Solomon's seal Iris Couch grass Canna	Underground horizontal branching stem	Stem grows and branches. At the tip of each branch a bud produces new vertical growth which gives rise to a new plant	Yes
Stem tuber	Potato Artichoke	Swollen tip of slender rhizome	Many slender rhizomes arise from axil of scale leaf. The tips swell to form a stem tuber, each giving rise to a new plant	Yes
Suckers	Mint Pear	Underground horizontal branches	A number of underground branches radiate laterally from the parent plant. The tips ultimately turn upwards out of the soil and develop into new plants	Yes
Runner	Creeping buttercup Strawberry	Thin lateral stems on the soil surface	A number of stems radiate from the parent plant. Adventitious roots arise at points along the stem and new plants arise from these	No
Offset	Leek	A short, stout lateral stem on the soil surface	Stem grows laterally along the soil and a single plant arises from a bud at the top of each stem	No
Stolon	Blackberry	A long vertical stem with little structural support	The stem grows vertically at first but then bends over until the tip touches the soil. Adventitious roots develop and at this point a new plant arises from a nearby lateral bud	No
Root tuber	Dahlia	Swollen fibrous root	The tuber stores food but the new plant arises from an axillary bud at the base of the old stem	Yes

12.2.6 Cloning

A group of genetically identical offspring produced by asexual reproduction is called a **clone**. The nucleus of every cell of an individual contains all the genetic information needed to develop the entire organism. It is therefore possible under suitable conditions to produce a whole organism from a single cell. If a cell divides mitotically it will produce a clone. If each cell of the clone is separated and allowed to develop into the complete organism, a group of genetically identical offspring is formed. This is known as cloning. (See Application on micropropagation on page 609.)

12.2.7 Parthenogenesis

Parthenogenesis is the further development of a female gamete in the absence of fertilization. As a gamete is produced the process is a modified form of sexual reproduction, even though only a single parent is involved. The parent is always diploid and the gametes are produced by mitosis or meiosis. In **diploid parthenogenesis** the gamete is produced by mitosis and therefore the offspring are likewise diploid. In aphids, diploid parthenogenesis is used as a means of rapidly building up numbers when conditions are favourable. During the summer, colonies of wingless females are formed by this means. In **haploid parthenogenesis** the gametes are produced by meiosis giving rise to haploid eggs which may develop directly into haploid offspring. In the honey bee, unfertilized haploid eggs develop into males or drones, while fertilized eggs develop into female bees. If these females are fed honey and pollen they develop into workers. If they are fed a special food called royal jelly, produced by the workers, a queen results.

Did you know?

A pair of aphids could give rise to 800 million tonnes of aphids in a single year – assuming all the offspring of succeeding generations survived.

12.3 Reproduction in flowering plants

As plants have evolved they have reduced the water dependent phase of their life cycle – the gametophyte – and made the sporophyte generation totally dominant. This has allowed them to colonize land much more efficiently. They have also developed a unique reproductive structure – the flower – which has contributed to the undoubted success of this group.

12.3.1 Floral structure

Sexual reproduction confers on organisms the advantages of increased variety. This variety is achieved during meiosis by:

(i) the independent assortment of chromosomes on the metaphase plate;
(ii) exchange of genetic material due to crossing over between homologous chromosomes at prophase I.

Both processes achieve variety even when the offspring result from self-fertilization. Cross-fertilization confers an additional source of variety, that of mixing two parental genotypes. As terrestrial plants are incapable of moving from place to place in

TABLE 12.2 **Some basic plant terminology**

Floral part	Collective names of parts	Name given to flower if floral part is absent
Carpel	Gynoecium	Staminate, e.g. *Zea*
Stamen	Androecium	Carpellate, e.g. *Zea*
Petal	Corolla	Apetalous, e.g. *Clematis*
Sepal	Calyx	Asepalous, e.g. most of the Umbelliferae
Corolla + calyx	Perianth	e.g. wild arum

order to transfer genetic material between individuals, this third source of variety would be denied them but for the assistance of some external agent like insects or the wind. This, however, exposes the vulnerable gamete to dangers such as drying out during its transfer. To overcome this, the Angiospermophyta have enclosed their male gamete within a spore, the **microspore** or **pollen grain**, a structure resistant to desiccation.

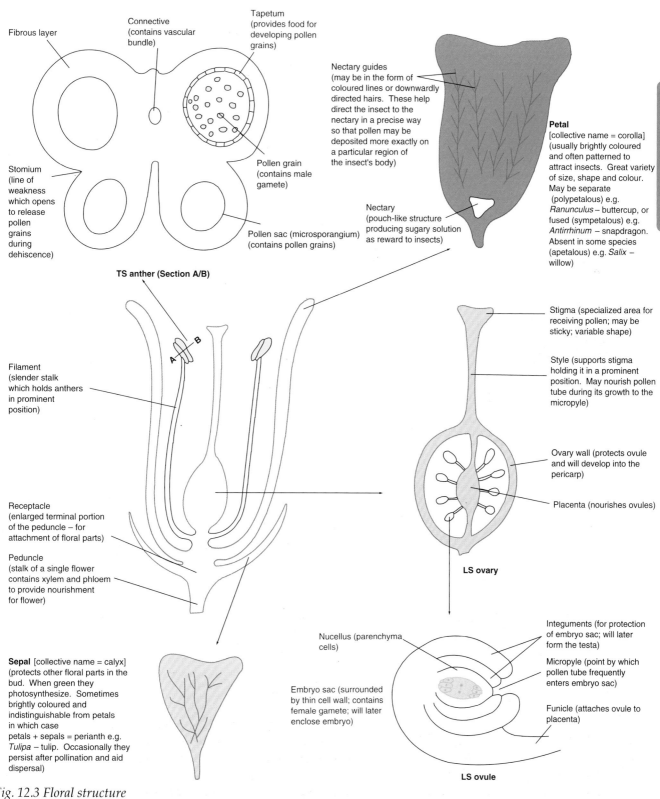

Fibrous layer

Connective (contains vascular bundle)

Tapetum (provides food for developing pollen grains)

Nectary guides (may be in the form of coloured lines or downwardly directed hairs. These help direct the insect to the nectary in a precise way so that pollen may be deposited more exactly on a particular region of the insect's body)

Petal [collective name = corolla] (usually brightly coloured and often patterned to attract insects. Great variety of size, shape and colour. May be separate (polypetalous) e.g. *Ranunculus* – buttercup, or fused (sympetalous) e.g. *Antirrhinum* – snapdragon. Absent in some species (apetalous) e.g. *Salix* – willow)

Stomium (line of weakness which opens to release pollen grains during dehiscence)

Pollen grain (contains male gamete)

Nectary (pouch-like structure producing sugary solution as reward to insects)

Pollen sac (microsporangium) (contains pollen grains)

TS anther (Section A/B)

Stigma (specialized area for receiving pollen; may be sticky; variable shape)

Style (supports stigma holding it in a prominent position. May nourish pollen tube during its growth to the micropyle)

Filament (slender stalk which holds anthers in prominent position)

Ovary wall (protects ovule and will develop into the pericarp)

Placenta (nourishes ovules)

Receptacle (enlarged terminal portion of the peduncle – for attachment of floral parts)

Peduncle (stalk of a single flower contains xylem and phloem to provide nourishment for flower)

LS ovary

Integuments (for protection of embryo sac; will later form the testa)

Micropyle (point by which pollen tube frequently enters embryo sac)

Nucellus (parenchyma cells)

Funicle (attaches ovule to placenta)

Sepal [collective name = calyx] (protects other floral parts in the bud. When green they photosynthesize. Sometimes brightly coloured and indistinguishable from petals in which case petals + sepals = perianth e.g. *Tulipa* – tulip. Occasionally they persist after pollination and aid dispersal)

Embryo sac (surrounded by thin cell wall; contains female gamete; will later enclose embryo)

LS ovule

ig. 12.3 *Floral structure*

The anthers comprise pollen sacs (usually 4) which contain a mass of diploid pollen mother cells.

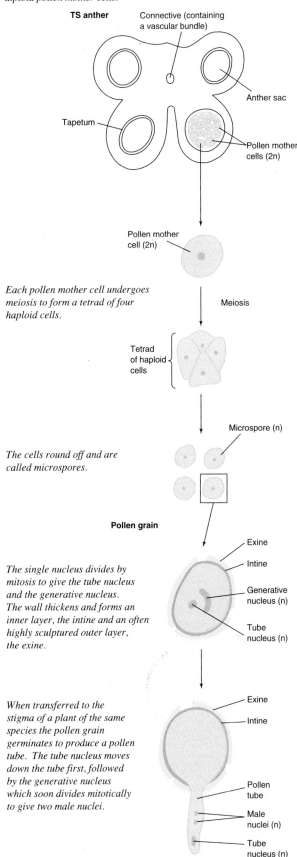

Each pollen mother cell undergoes meiosis to form a tetrad of four haploid cells.

Meiosis

The cells round off and are called microspores.

The single nucleus divides by mitosis to give the tube nucleus and the generative nucleus. The wall thickens and forms an inner layer, the intine and an often highly sculptured outer layer, the exine.

When transferred to the stigma of a plant of the same species the pollen grain germinates to produce a pollen tube. The tube nucleus moves down the tube first, followed by the generative nucleus which soon divides mitotically to give two male nuclei.

Fig. 12.4 Structure and development of the pollen grain

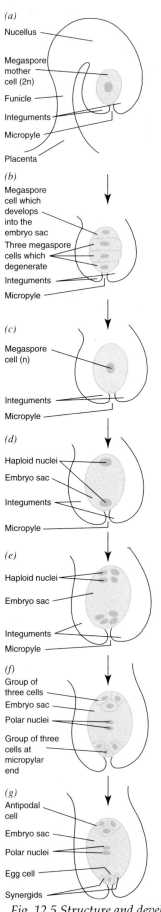

(a) The ovule consists of a mass of cells called the **nucellus** which is carried on a short stalk called the **funicle.** The nucellus is completely surrounded by two protective **integuments** except for a narrow channel at the tip called the **micropyle.** One cell of the nucellus becomes larger and more conspicuous than the rest. This is the **embryo sac mother cell.**

(b) The embryo sac mother cell divides meiotically to give four haploid **megaspore cells.**

(c) The three cells nearest the micropyle degenerate while the remaining one enlarges to form the **embryo sac.**

(d) The embryo sac nucleus divides by mitosis and the resultant nuclei migrate to opposite poles.

(e) Each nucleus undergoes two mitotic divisions to give a group of four haploid nuclei at each pole.

(f) One nucleus from each polar group moves to the centre of the embryo sac. These are the polar nuclei. The remaining nuclei develop cytoplasm around them and become separated by cell walls leaving two groups of three cells at each pole.

(g) The three cells at the opposite end to the micropyle are called **antipodal cells** and play no further rôle in the process. Of the three cells at the micropyle end, one, the **egg cell** remains, the other two, the **synergids,** degenerate.

Fig. 12.5 Structure and development of the ovule

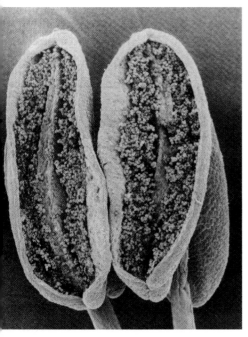

...nther showing dehiscence (scanning EM)
×60 approx)

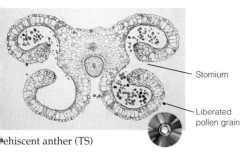

Stomium

Liberated
pollen grain

●ehiscent anther (TS)

A typical flower is made up of four sets of modified leaves: **carpels**, **stamens**, **petals** and **sepals**, all attached to a modified stem, the **receptacle**.

In the centre are the carpels which comprise a sticky **stigma** at the end of a slender stalk called the **style**. At the base of the carpel is the **ovary**, a hollow structure containing one or more **ovules**, each of which encloses the female gamete, the **egg nucleus**. The carpels may be separate as in *Ranunculus* (buttercup) or fused in groups as in *Tulipa* (tulip) where three carpels combine to give a single structure. Around the carpels are the **stamens**, each comprising a long stalk, the **filament**, at the end of which are the **anthers**. The anthers produce pollen grains which contain the male gametes.

The stamens are surrounded by the petals, brightly coloured leaf-like structures which attract insects. They may produce **nectar** and be scented. The outer ring of structures is the sepals. These are usually green and may photosynthesize, but their main function is to protect the other floral parts when the flower is a bud.

Occasionally the sepals are brightly coloured, e.g. in lilies, and help in insect attraction or may later assist in dispersal, e.g. in mulberry, where they become juicy and attract animals.

The four sets of floral parts are given collective names as shown in Table 12.2.

Some flowers like those of the buttercup exhibit **radial symmetry** whereas flowers of others such as broad bean are **bilaterally symmetrical**.

12.3.2 Pollination

Pollination is the transfer of pollen from anthers to stigmas. If the transfer occurs between two plants of different genetic make-up the process is **cross-pollination**. If the transfer takes place between flowers of identical genetic constitution, the process is **self-pollination**. It is common to think of self-pollination occurring within a single flower on a plant, such as garden peas, where the petals so enclose the stamens that the pollen has little chance of escaping. However, it also occurs if pollen is transferred between different flowers on the same plant. This may occur as an insect moves from flower to flower collecting nectar.

A third type occurs when pollen is transferred between flowers on two separate plants which are genetically identical. This is most common where groups of plants have arisen as a result of asexual reproduction, e.g. groups of daffodils or irises. As members of a single clone, these groups have individuals with identical genotypes.

The design of any individual flower is related to the precise agent used to transfer pollen. If the plant is insect-pollinated its bright colour, patterns and scent attract potential pollinating insects. They receive nectar or excess pollen which encourages them to seek out a similar flower and thereby transfer more pollen.

As colour and scent have no bearing on wind direction, wind-pollinated flowers are dull, unattractive and without scent. Indeed as petals may shelter the reproductive structures from the wind they are frequently dispensed with altogether, leaving the anthers and stigmas exposed. (See Figs. 12.6 and 12.8.)

PROJECT

Pollen grains found in peat can tell us what types of vegetation were growing in the past

1. Examine peat samples from different depths under the microscope.

2. Make drawings of the different kinds of pollen grains that you find.

3. Do the percentages of each type of pollen grain differ in the various samples?

4. Try to find out from the pollen grains what types of plants were growing when peat was formed.

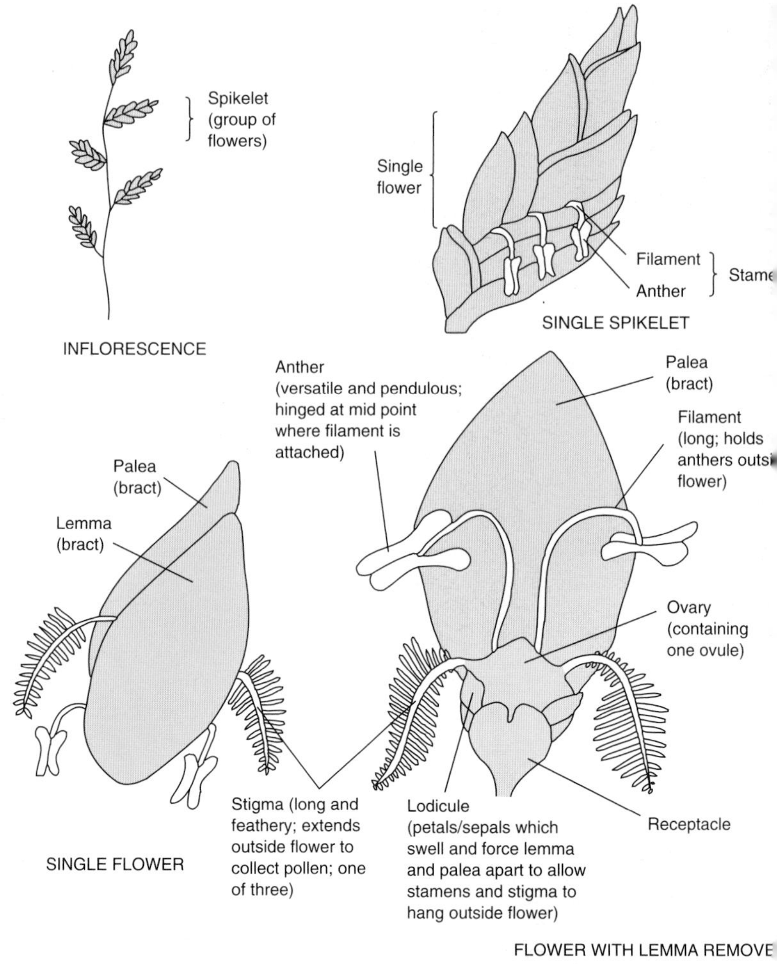

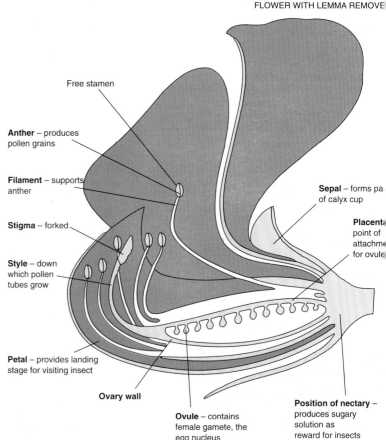

INFLORESCENCE

SINGLE SPIKELET

Spikelet (group of flowers)

Single flower

Filament
Anther } Stamen

Anther (versatile and pendulous; hinged at mid point where filament is attached)

Palea (bract)

Filament (long; holds anthers outside flower)

Palea (bract)

Lemma (bract)

Ovary (containing one ovule)

Stigma (long and feathery; extends outside flower to collect pollen; one of three)

Lodicule (petals/sepals which swell and force lemma and palea apart to allow stamens and stigma to hang outside flower)

Receptacle

SINGLE FLOWER

Fig. 12.6 Rye grass (Lolium perenne)

FLOWER WITH LEMMA REMOVED

Free stamen

Anther – produces pollen grains

Filament – supports anther

Stigma – forked

Style – down which pollen tubes grow

Petal – provides landing stage for visiting insect

Ovary wall

Sepal – forms part of calyx cup

Placenta – point of attachment for ovule

Ovule – contains female gamete, the egg nucleus

Position of nectary – produces sugary solution as reward for insects

Fig. 12.7 Sweet pea (Lathyrus odoratus)

Did you know?

In Australia when it is very hot the nectar in some flowers ferments and turns into alcohol. The bees get 'drunk' and are not allowed back into the hive.

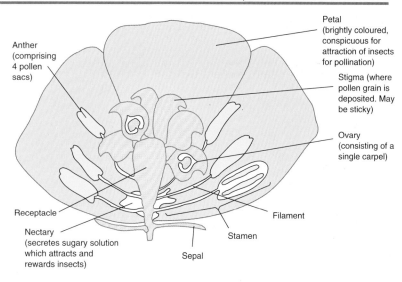

Fig. 12.8 Half-flower of buttercup (Ranunculus), *a regular (actinomorphic) flower*

Labels on Fig. 12.8:
- Anther (comprising 4 pollen sacs)
- Petal (brightly coloured, conspicuous for attraction of insects for pollination)
- Stigma (where pollen grain is deposited. May be sticky)
- Ovary (consisting of a single carpel)
- Receptacle
- Filament
- Nectary (secretes sugary solution which attracts and rewards insects)
- Stamen
- Sepal

ind dispersal of pollen from male tkins of alder tree

ild sage about to be pollinated by a bee

owering shoot of grass releasing its ollen

TABLE 12.3　**Comparison of wind- and insect-pollinated flowers**

Wind-pollinated flowers (anemophilous)	Insect-pollinated flowers (entomophilous)
e.g. Rye grass (*Lolium perenne*)	e.g. Buttercup (*Ranunculus repens*)
Plants often occur in dense groups covering large areas	Plants often solitary or in small groups
Flowers occur in groups (inflorescences) on the plant (e.g. Graminae)	Flowers may occur on the plant as inflorescences (e.g. apple) but may also be solitary (e.g. tulip)
Flowers are often unisexual with an excess of male flowers	Mostly bisexual (hermaphrodite) flowers
Petals are dull and much reduced in size	Petals are large and brightly coloured to make them conspicuous to insects
No scent or nectar is produced	Flowers produce scent and/or nectar to attract insects
Stigmas often protrude outside the flower on long styles	Stigmas lie deep within the corolla
Stigmas are often feathery, giving them a large surface area to filter pollen from the air	Stigmas are relatively small as the pollen is deposited accurately by the pollinating insects
Anthers dangle outside the flower on long filaments so the pollen is easily released into the air	Anthers lie inside the corolla so the pollinating insect brushes against them when collecting the nectar
Enormous amounts of pollen are produced to offset the high degree of wastage during dispersal	Less pollen is produced as pollen transfer is more precise and so entails less wastage
Pollen is smooth, light and small and sometimes has 'wing-like' extensions to aid wind transport	Pollen is larger and often bears projections which help it adhere to the insect

12.4 Fertilization and development in flowering plants

12.4.1 Fertilization

In angiosperms the female gamete is protected within the carpel and the male gamete can only reach it via the **pollen tube**. On landing upon the stigma the pollen grains absorb water and germinate to give the pollen tube. The tube pushes between the loosely packed cells of the style, the **tube nucleus** preceding the **male nuclei**. The rôle of the tube nucleus is to control the growth of the pollen tube and it plays no part in fertilization. While the initial growth into the style may be the result of a negative aerotropic response, it is thought that the tube then shows a positive chemotropic response to some substance produced in the embryo sac. The secretion of pectases by the pollen tube may soften the middle lamellae of the cells in the style and so assist its growth towards the micropyle – the usual point of entry to the embryo sac. Many tubes grow down the style simultaneously and where an ovary has many ovules a separate one penetrates each. On entering the embryo sac the tube nucleus, its work done, degenerates and the two male nuclei enter. One male nucleus fuses with the egg cell to give a diploid zygote; the other fuses with the two polar nuclei to form the primary endosperm nucleus, which is triploid. This double fertilization is peculiar to flowering plants.

Before fertilization	After fertilization
Ovary and contents	Fruit
Ovary wall	Pericarp (wall of fruit)
Ovule	Seed
Integuments	Testa (seed coat)

Germinating pollen grain
Stigma (may be sticky to help pollen adhere)
Pollen tube
Style
Male nuclei
Ovary wall (will form f...
Tube nucleus (controls growth of pollen tube)
Locule (ca... of ovary in which ovu... is suspend...
Antipodal cell
Nucellus (parenchy... cells)
Polar nuclei
Egg cell (female gamete)
Embryo sa...
Synergids (non-functional egg cells)
Integumen... (will later form testa of the see...
Funicle (attaches ovule to ovary wall)
Micropyle (pollen tube frequently enters embryo sac through h...
Placenta (attaches ovary to plant)

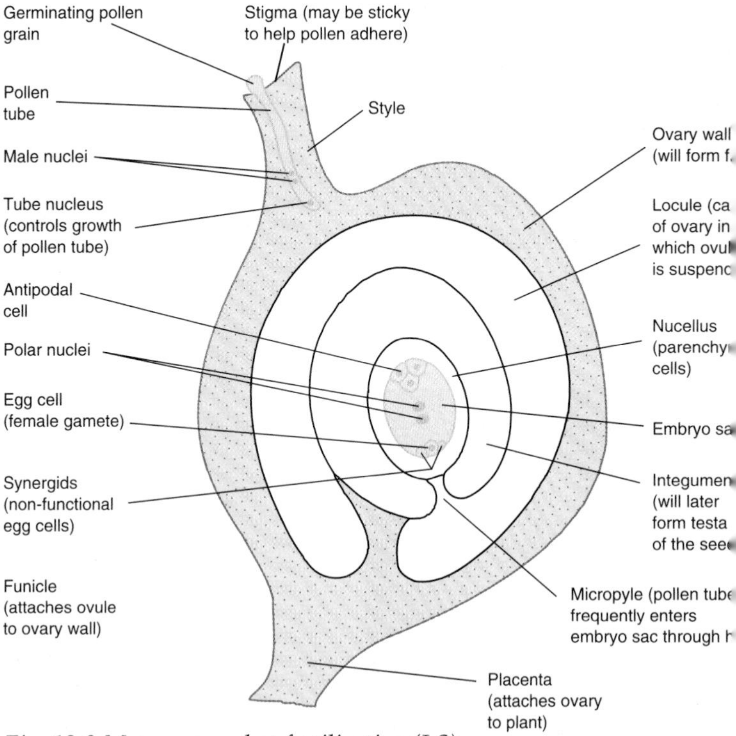

Fig. 12.9 Mature carpel at fertilization (LS)

12.4.2 Methods of preventing inbreeding

There can be no doubt that in some plants self-pollination occurs more or less regularly. However, there appears to be a general tendency to avoid self-pollination since this is a form of

Did you know?

Rafflesia flowers are more than 1 metre across and weigh more than 7 kg. They smell of rotting meat which attracts pollinating flies.

Did you know?

The flower of the bee orchid resembles a female bee. This attracts male bees who collect pollen while trying to mate with the flower.

PROJECT

1. Collect wind-dispersed seeds/fruits from, for example, dandelion, thistle, sycamore, lime, ash, elm, etc. You can keep them for some time in a closed dry jar.

2. Set up experiments to compare times from release to landing under different wind conditions.

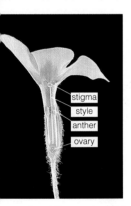
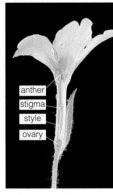

n-eyed and thrum-eyed *Primula* flowers (LS)

Did you know?

The coco de mer palm of the Seychelles has the biggest seed in the plant kingdom. It weighs up to 27 kg.

inbreeding and would very quickly reduce the variability of a population and hence its potential for evolutionary change. There are four main methods of avoiding self-pollination and thereby promoting outbreeding.

1. The stamens and stigma of a flower mature at different times. If the stamens ripen before the stigma is in a condition to receive pollen, the flower is **protandrous**, e.g. white dead nettle (*Lamium*). If the stigma and ovule ripen before the stamens, the flower is **protogynous**, e.g. plantain (*Plantago*). Protandry is more common than protogyny.

2. If a plant has separate male and female flowers it is said to be **monoecious**, e.g. maize (*Zea*). This condition clearly limits the possibility of self-pollination and in *Zea* the number of seeds produced by self-pollination is reduced to less than 1%.

3. The structure of the flower itself makes self-pollination unlikely. Some flowers, e.g. *Iris*, have a stigmatic flap which is exposed to the pollen on the back of a visiting insect. The insect collects pollen from the stamens and closes the flap as it withdraws from the flower, thus protecting the stigmatic surface from its own pollen.

4. A **dioecious** species is one in which some individual plants have either all male or all female flowers. Completely dioecious plants are rare. It is more usual for a plant to be predominantly, although not completely, of one sex, e.g. plantain (*Plantago*) and ash (*Fraxinus*).

Preventing self-pollination is not the only means of preventing inbreeding. In many plants self-pollination occurs but there is a mechanism to prevent this leading to successful fertilization of the ovule and production of a seed. This is known as **incompatibility**. For example, in pears the pollen only becomes functional if the stigmatic surface on which it lands has a different genetic composition. *Primula vulgaris* (primrose) is a dimorphic plant in which there are two types of flower, pin-eyed and thrum-eyed. These differ in the length of the style (**heterostyly**) as well as in the size and chemical composition of their pollen.

These differences do not prevent self-pollination but the yield of seed produced by self-pollination is very poor.

12.4.3 Development of fruits and seeds

Following fertilization, the zygote divides rapidly by mitosis and develops into the embryo, which then differentiates into a young shoot, called the **plumule**, a young root, the **radicle** and seed leaves known as **cotyledons**. The primary endosperm nucleus also divides mitotically to give a mass of cells, the **endosperm**. This forms the food source for the growing embryo. In some species, e.g. maize (*Zea mays*), the endosperm remains while in others, e.g. peas (*Pisum*), it is quickly absorbed by and stored in the cotyledons. Other parts develop as shown opposite and in Fig. 12.10, on the next page.

The most common food stored in seeds is carbohydrate. This is usually in the form of starch but some seeds, e.g. maize and peas, store quantities of sugar. Many young seeds store sugar but this changes to starch as they mature. Lipids are often stored

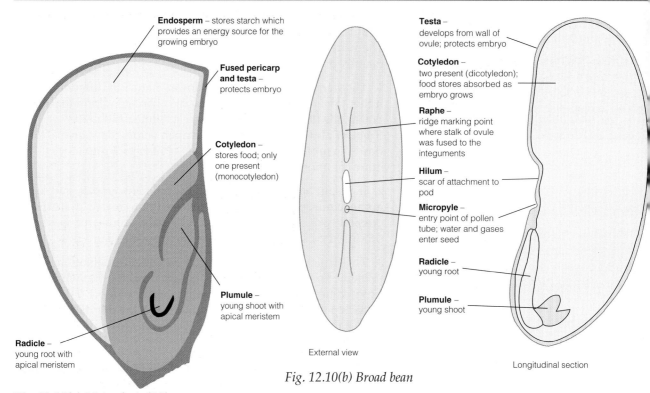

Fig. 12.10(a) Maize fruit (LS)

Fig. 12.10(b) Broad bean

PROJECT

Hawthorn seeds are dispersed by birds and have to pass through their intestines before they germinate

Test this statement by proposing a hypothesis and designing appropriate experiments.

(Hint: compare percentage germination of seeds soaked in different concentrations of hydrochloric acid, enzymes, etc.)

in the cotyledons and may form a high percentage of the dry weight, e.g. 60% in walnuts and coconuts; 40% in sunflowers. Other economically important examples are peanuts, soya bean and castor oil seeds. Proteins are found to a lesser extent in seed but wheat has an aleurone layer and protein is stored in the cotyledons of legumes and nuts.

12.4.4 Dormancy

The water content of seeds at between 5–10% is very low and is the major factor in preventing them germinating. As a rule, the addition of water in the presence of oxygen and a favourable temperature is enough to break this dormancy (Section 12.4.5). Some seeds, however, still fail to germinate for one reason or another:

1. Light is necessary for the germination of certain seeds, e.g. lettuce.

2. A sustained period of cold is needed to make some seeds of temperate climates germinate (Section 29.3). This helps to ensur that seeds do not germinate in late summer or during mild winter spells, thus making the young plant vulnerable to frosts at a later date.

3. Conversely a few seeds will not germinate unless subject to the heat of a flash-fire.

4. A period of time is necessary to permit internal chemical changes to take place before other seeds germinate.

5. The seed coat may be impermeable to water and/or gases an time may be needed for it to decay and break. In many seeds physical abrasion or partial digestion in the intestines of an animal help break dormancy by weakening the testa.

6. Another type of dormancy is brought about by the presence of natural chemical inhibitors.

12.4.5 Germination and early growth in flowering plants

Germination is the onset of growth of the embryo and requires water, oxygen and a temperature within a certain range (normally 5–40°C). In some seeds light is also required. Under these conditions the seed takes up water rapidly, initially by imbibition and later by osmosis. This water causes the seed contents to swell and so ruptures the **testa** (seed coat). At the same time the water activates enzymes in the seed which hydrolyse insoluble storage material into soluble substances which can be easily transported. In this way proteins are converted into amino acids, carbohydrates such as starch are converted into glucose, and fats are converted into fatty acids and glycerol. The soluble products of these conversions are transported to the growing point of the embryo. The glucose, fatty acids and glycerol provide respiratory substrates from which energy for growth is released. Glucose is also used in the formation of cellulose cell walls. The amino acids are used to form new enzymes and structural proteins within new cells.

Early growth results in the **plumule** and the **radicle** growing rapidly. The radicle grows downwards and the plumule upwards. In sunflowers the **cotyledons** may be carried up and out of the soil by this growth (**epigeal germination**), in which case they form the first photosynthetic structure. In some other plants, such as broad beans and wheat the cotyledons remain below the soil surface (**hypogeal germination**).

12.5   Reproduction in mammals

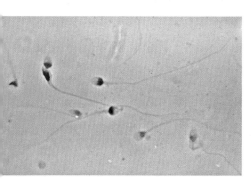

perm (× 600 approx.)

We saw earlier in this chapter that one major advantage of sexual reproduction is the genetic variety it creates, and that the extent of this variety is greater the more diverse the parental genotypes are. Animals, with their capacity for locomotion, are able to move far afield in their search for mates and so reproduce with individuals outside their family groups. This produces a greater degree of outbreeding, greater mixing of genes within the gene pool and hence greater variety. In mammals the gametes are differentiated into a small motile male gamete or **sperm** which is produced in large numbers, and a larger, non-motile food-storing female gamete or **ovum** which is produced in much smaller numbers.

12.5.1 Gametogenesis

Gametogenesis is the formation of gametes. In the case of sperm production it is called **spermatogenesis** and where eggs are formed it is called **oogenesis**. Both types involve a multiplication phase, a growth phase and a maturation phase as depicted in Fig. 12.11 on the next page.

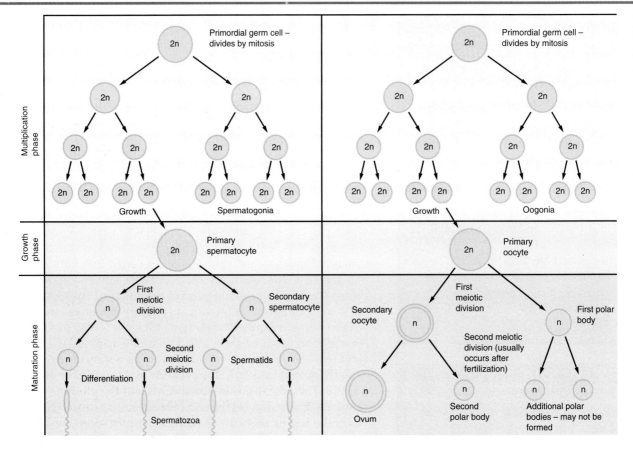

Fig. 12.11(a) Spermatogenesis – formation of sperm

Fig. 12.11(b) Oogenesis – formation of ova

The organs which produce gametes are called **gonads** and are of two types: the ovaries which produce ova and the testes which produce sperm. In some animals, e.g. mammals, the reproductive and excretory systems are closely associated with one another; in such cases they are often represented together as the urinogenital system.

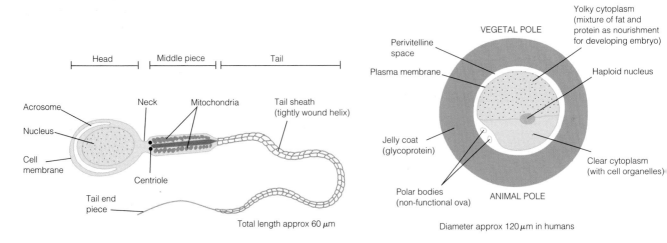

Fig. 12.12(a) Human spermatozoan based on electron micrograph

Fig. 12.12(b) A generalized egg cell

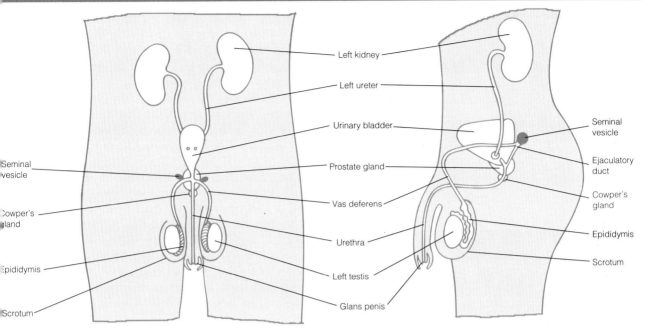

Fig. 12.13 Male urinogenital system (simplified) – front view

Male urinogenital system – side view

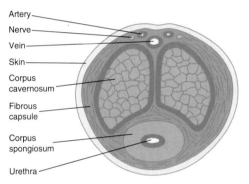

Fig. 12.14 Penis (TS)

12.5.2 Human male reproductive system

The male gonads, the **testes**, develop in the abdominal cavity and descend into an external sac, the **scrotum**, prior to birth. The optimum temperature for sperm development is around 35°C, about 2°C below normal human body temperature. The testes can be kept at this temperature by the contraction and relaxation of muscle in the scrotal wall. When the temperature of the testes exceeds 35°C the muscles relax, holding them away from the body to assist cooling. In colder conditions the muscles contract to bring the testes as close to the abdominal cavity as is necessary to maintain them at the optimum temperature. Each testis is suspended by a spermatic cord composed of the sperm duct or vas deferens, spermatic artery and vein, lymph vessels and nerves, bound together by connective tissue. A single testis is surrounded by a fibrous coat and is separated internally by septa into a series of lobules (Fig. 12.15).

Within each lobule are convoluted **seminiferous tubules**, the total length of which is over 1 km. Between the tubules are the **interstitial cells** which secrete the hormone **testosterone**. Each seminiferous tubule is lined by **germinal epithelium** which, by a series of divisions, gives rise to sperm, a process taking 8–9 weeks (Fig. 12.16).

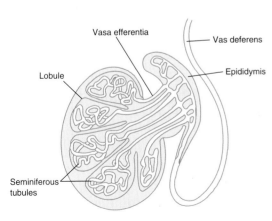

Fig. 12.15 Testis (LS)

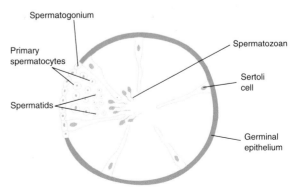

Fig. 12.16 Seminiferous tubule ×500 (TS)

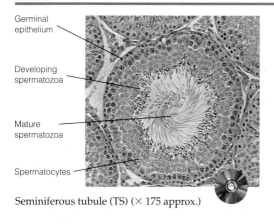

Germinal epithelium

Developing spermatozoa

Mature spermatozoa

Spermatocytes

Seminiferous tubule (TS) (× 175 approx.)

Did you know?

During his lifetime one male human may produce as many as 12 000 000 000 000 (1.2×10^{13}) sperm.

The seminiferous tubules merge to form small ducts called the **vasa efferentia**, which in turn join up to form a six-metre-long coiled tube called the **epididymis**. The sperm are stored here, gaining motility over a period of 18 hours. From the epididymis leads another muscular tube, the **vas deferens**, which carries the sperm towards the urethra. Before it joins the urethra it combines with the duct leading from the **seminal vesicle**, forming the **ejaculatory duct**. The seminal vesicles produce a mucus secretion which aids sperm mobility. The ejaculatory duct then passes through the **prostate gland** which produces an alkaline secretion that neutralizes the acidity of any urine in the urethra as well as aiding sperm mobility. Below the prostate gland is a pair of **Cowper's glands** which secrete a sticky fluid into the urethra. The resultant combination of sperm and secretions is called **semen**. The semen passes along the **urethra**, a muscular tube running through the **penis**. The penis comprises three cylindrical masses of spongy tissue covered by an elastic skin (Fig. 12.14).

The end of the penis is expanded to form the **glans penis**, a sensitive region covered by loose retractable skin, the **prepuce** or **foreskin**. The foreskin is sometimes removed surgically, for medical or religious reasons, in a small operation called circumcision.

12.5.3 Human female reproductive system

The female gonads, the **ovaries**, lie suspended in the abdominal cavity by the ovarian ligaments. The external coat is made up of **germinal epithelium** which begins to divide to form ova while the female is still a fetus. At birth around 400 000 cells have reached prophase of the first meiotic division and are called **primary oocytes**. Each month after puberty, one of these cells completes its development into an ovum. As each cell takes some time to complete this change, the ovary consists of a number of oocytes at various stages of development. These

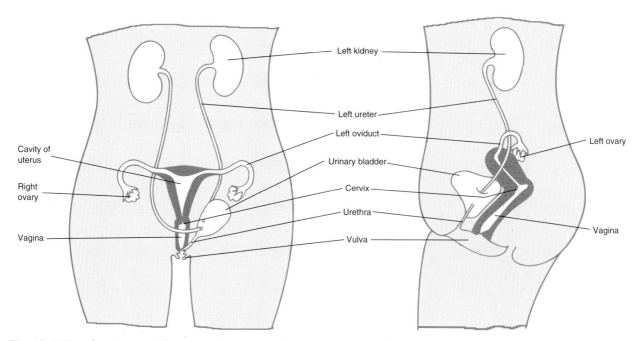

Left kidney

Left ureter

Left oviduct

Urinary bladder

Cervix

Urethra

Vulva

Left ovary

Vagina

Cavity of uterus

Right ovary

Vagina

Fig. 12.17 Female urinogenital system – front view

Female urinogenital system – side view

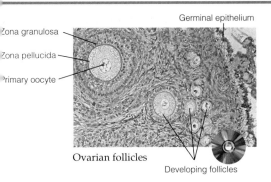

Germinal epithelium

Zona granulosa

Zona pellucida

Primary oocyte

Ovarian follicles

Developing follicles

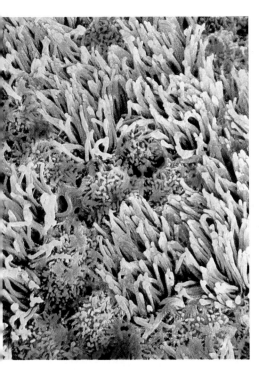

False colour SEM of epithelium of Fallopian tube

oocytes lie in a region of fibrous tissue, the stroma, which fills the rest of the ovary. The largest and most mature of the cells are called **Graafian follicles** which are fluid-filled sacs each containing a secondary oocyte. A mature Graafian follicle can reach a diameter greater than 1 cm before it releases its ovum. Once the ovum is released the empty follicle develops into a yellow body called the **corpus luteum** (Fig. 12.18).

Close to the ovary is the funnel-shaped opening of the **oviduct** or **Fallopian tube**. The opening has fringe-like edges called **fimbriae**. The oviducts are about 10 cm long and have a muscular wall lined with a mucus-secreting layer of ciliated epithelium. They open into the **uterus**, or womb, which is a pear-shaped body about 5 cm wide and 8 cm in length, held in position by ligaments joined to the pelvic girdle. It has walls of unstriated muscle and is lined internally by a mucus membrane called the **endometrium**. The uterus opens into the **vagina** through a ring of muscle, the **cervix**. The vagina has a wall of unstriated muscle with an inner mucus membrane lined by stratified epithelium. The vagina opens to the outside through the **vulva**, a collective name for the external genital organs. These consist of two outer folds of skin, the **labia majora**, covering two inner, more delicate folds, the **labia minora**. Anterior to the vaginal opening is a small body of erectile tissue, the **clitoris**, which is homologous to the penis of the male. Between the vaginal opening and the clitoris is the opening of the urethra.

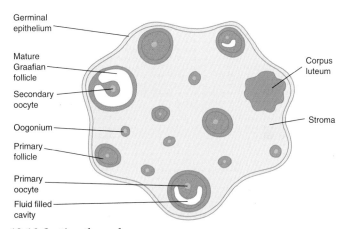

Germinal epithelium

Mature Graafian follicle

Secondary oocyte

Oogonium

Primary follicle

Primary oocyte

Fluid filled cavity

Corpus luteum

Stroma

Fig. 12.18 Section through ovary

12.6 Sexual cycles

Many animals have cycles of sexual activity in both males and females. These cycles often occur so that fertilization takes place at a time which gives the offspring the best chance of survival, e.g. the offspring are produced at a time when the climate and food availability are most favourable.

In mammals these cycles are of three main types:

1. The female undergoes a single period of sexual activity during the year, e.g. deer (**monoestrus**).

2. The female undergoes a number of periods of sexual activity during the year, each separated by a period of sexual inactivity, e.g. horses (**polyoestrus**).

3. The female has a more or less continuous cycle of activity where the end of one cycle is followed immediately by the start of the next, e.g. humans.

12.6.1 The menstrual cycle

In human females the onset of the first menstrual cycle is called **menarche** and represents the start of puberty. This takes place around the age of 12 years although the age varies widely between individuals. The menstrual cycle, which lasts about 28 days, continues until the **menopause** at the age of 45–50 years. The events of the cycle are controlled by hormones to ensure that the production of an ovum is synchronized with the readiness of the uterus to receive it, should it be fertilized. The start of the cycle is taken to be the initial discharge of blood known as **menstruation**, as this event can be easily identified. This flow of blood, which lasts about five days, is due to the lining of the uterus being shed, along with a little blood.

During the following days the lining regenerates in readiness for a fertilized ovum. By day 14 it has thickened considerably and the Graafian follicle releases its ovum into the oviduct, the process being called **ovulation**. The ovum is moved down the oviduct mostly by muscular contractions of the oviduct wall, although the beating of the cilia may also assist. The journey to the uterus takes about three days during which time the ovum may be fertilized. If it is not, the ovum quickly dies and passes out via the vagina. The uterine lining is maintained for some time but finally breaks down again about 28 days after the start of the cycle.

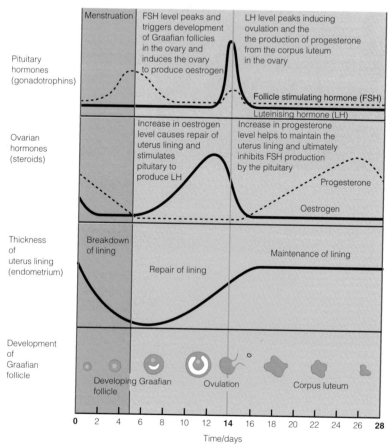

Fig. 12.19 Diagram summarizing the events of the menstrual cycle

12.6.2 Hormonal control of the menstrual cycle

The control of the menstrual cycle is an excellent example of hormone interaction. The action of one hormone is used to stimulate or inhibit the production of another. There are four hormones involved, two produced by the anterior lobe of the pituitary gland at the base of the brain, and two produced by the ovaries. The production of the hormones from the ovaries is stimulated by the pituitary hormones. These hormones are thus referred to as the gonadotrophic hormones. The two gonadotrophic hormones are **follicle stimulating hormone (FSH)** and **luteinizing hormone (LH)**. These stimulate the ovaries to produce **oestrogen** and **progesterone** respectively.

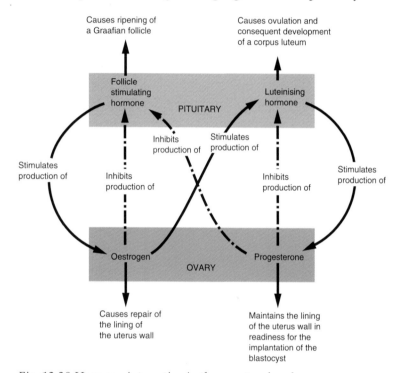

Fig. 12.20 Hormone interaction in the menstrual cycle

The functions of these hormones are as follows:

Follicle stimulating hormone
1. Causes Graafian follicles to develop in the ovary.

2. Stimulates the ovary to produce oestrogen.

Oestrogen
1. Causes repair of the uterus lining following menstruation.

2. Stimulates the pituitary to produce luteinizing hormone.

Luteinizing hormone
1. Causes ovulation to take place.

2. Stimulates the ovary to produce progesterone from the corpus luteum.

Progesterone
1. Causes the uterus lining to be maintained in readiness for the blastocyst (young embryo).

2. Inhibits production of FSH by the pituitary.

The hormones are produced in the following sequence: FSH, oestrogen, LH, progesterone. Progesterone at the end of the sequence inhibits the production of FSH. In turn, the production of the other hormones stops, including progesterone itself. The absence of progesterone now means that the inhibition of FSH ceases and so oestrogen production commences again. In turn, all the other hormones are produced. This alternate switching on and off of the hormones produces a cycle of events – the menstrual cycle. At the end of their fertile period, **the menopause**, women often experience a number of symptoms which can sometimes be relieved by the use of **hormone replacement therapy (HRT)** details of which are given in Section 28.3.

12.6.3 Artificial control of the menstrual cycle

The artificial control of the menstrual cycle has two main purposes: firstly as a contraceptive device by preventing ovulation and secondly as a fertility device by stimulating ovulation.

The contraceptive Pill
The Pill contains both oestrogen and progesterone and when taken daily it maintains high levels of these hormones in the blood. These high levels inhibit the production of the gonadotrophic hormones from the pituitary, and the absence of LH in particular prevents ovulation. The Pill is normally taken for 21 consecutive days followed by a period of 7 days without it, during which the uterus lining breaks down and a menstrual period occurs. The 'morning after' Pill contains the synthetic oestrogen, diethylstibestrol which is thought to prevent implantation of the fertilized ovum if it is present. Both types of Pill are very effective forms of contraception. These and other methods of contraception are reviewed in Table 12.4 on page 234.

Fertility drugs
A fertility drug may induce ovulation in one of two ways:

1. It may provide gonadotrophins such as FSH which stimulate the development of Graafian follicles.

2. It may provide some chemical which inhibits the natural production of oestrogen. As oestrogen normally inhibits FSH production, the level of FSH increases and Graafian follicles develop.

Fertility drugs frequently result in multiple births.

12.6.4 Male sex hormones

Although male humans do not have a sexual cycle similar to that of females, they nonetheless produce some gonadotrophic hormones from the anterior lobe of the pituitary gland. Follicle stimulating hormone (FSH) stimulates sperm development. Luteinizing hormone (LH) stimulates the interstitial cells between the seminiferous tubules of the testis to produce **testosterone**. For this reason, in the male, LH is often called **interstitial cell stimulating hormone (ICSH)**. Testosterone is the most important of a group of male hormones or **androgens**. It is first produced in the fetus, where it controls the development of

APPLICATION

Premenstrual syndrome

The natural fluctuation in sex hormone levels during the menstrual cycle produces a wide range of side effects especially in the days leading up to the menstrual period. Over a hundred such effects have been identified, ranging from breast tenderness, bloating due to water retention, headache and pelvic pain through to irritability, depression, clumsiness and a craving for sweet foods. Such is the range of symptoms that the term pre-menstrual tension (PMT) is misleading and PMS is now preferred. It is estimated that 70% of fertile women suffer from PMS to a varying degree. A variety of drugs may be used to alleviate the symptoms but longer term treatment often involves providing oestrogen in patch, pill or implant form.

the male reproductive organs. An increase in production takes place at puberty and causes an enlargement of the reproductive organs and the development of the secondary sex characteristics. Removal of the testes (castration) prevents these changes taking place. It was used in the past as a rather drastic means of preventing choirboys' voices from breaking. Castration of animals is still practised in order to help fatten them and make the meat less tough. (See Fig. 12.21 for details of the control of male sex hormones.)

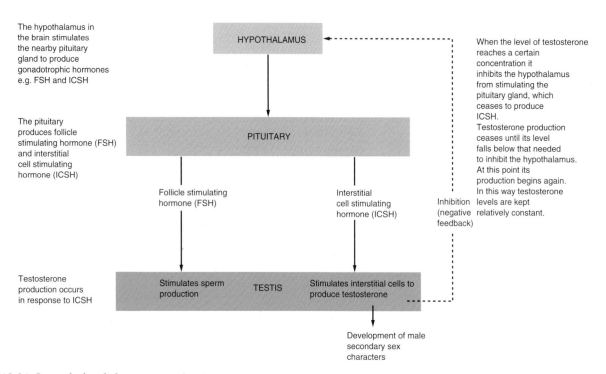

Fig. 12.21 Control of male hormone production

TABLE 12.4 **Birth control**

Method	How it works	Effectiveness	Advantages	Disadvantages
Sterilization	**Male (vasectomy)** – The vasa deferentia (the ducts carrying sperm from the testes to the urethra) are cut and tied off	100%	No artificial appliance is involved. Once the operation has been performed there is no further cost	Irreversible in normal circumstances
	Female (tubal ligation) – The oviducts are cut and tied off			
Prevention of ovulation	**Oral contraceptive** – Contains artificial oestrogen and progesterone	99%	Very reliable if taken regularly	A slightly higher than normal risk of thrombosis. Occasional side effects e.g. nausea, breast tenderness and water retention
	Injection contraceptive (e.g. Depo-provera) – Injection given by doctor about every 3 months	100%	Almost totally reliable	Hormonal surge on injection may produce side-effects e.g. irregular menstrual bleeding
	Implant contraceptive (e.g. Norplant) – Implant placed under the skin which releases artificial oestrogen and progesterone	100%	Almost totally reliable. Each implant lasts 5 years	Can cause irregular menstrual bleeding
Prevention of implantation	**Morning-after Pill** – Contains high level of oestrogen. Taken orally	Not widely used but probably 99–100%	Can be used after rather than before intercourse	High dose of oestrogen can produce side-effects. Not suitable for regular use
	Intra-Uterine Device (loop, coil) – A device usually made of plastic and/or copper which is inserted into the womb by a doctor and which prevents implantation	99–100%	Once fitted, no further action is required except for annual check-ups	Possible menstrual discomfort. The device may be displaced or rejected. Must be inserted by a trained practitioner. Only really suitable for women who have had children
	Intra-vaginal ring – Ring shaped device which releases artificial progesterone. Placed in vagina	Very reliable	Very reliable	Long term health effects yet to be assessed
Barriers which prevent sperm reaching an egg	**Female (diaphragm, cap)** – A dome-shaped sheet of thin rubber with a thicker spring rim which is inserted into the vagina, over the cervix. Best used with spermicide	Very reliable	Reliable. Available for use by all women	Must be inserted prior to intercourse and should be removed 8–24 hours after intercourse. Initial fitting must be by a trained practitioner
	Female (condom) – Sheath of thin rubber with two springy rings. Smaller inserted into vagina, larger remains outside	Very reliable	Readily available, quite easy to fit. Gives some protection against sexually transmitted diseases	May reduce enjoyment of intercourse
	Male (condom, sheath) – a sheath of thin rubber unrolled onto the erect penis prior to intercourse. Semen is collected in teat at the tip. Best used with a spermicide.	Very reliable	Easily available, no fitting or instruction by others needed. Available for use by all men. Gives some protection against sexually transmitted disease including AIDS	May reduce the sensitivity of the penis and so interfere with enjoyment
	Spermicide – Cream, jelly or foam inserted into vagina. Only effective with a mechanical barrier. Kills sperm	Not reliable alone	Easy to obtain and simple to use	Not effective on its own. May occasionally cause irritation
Natural method	**Rhythm method** – Refraining from intercourse during those times in the menstrual cycle when conception is most likely	Variable – not very reliable	No appliance required. Only acceptable method to some religious groups	Not reliable. Restricts times when intercourse can take place. Unsuitable for women with irregular cycles

12.6.5 Factors affecting breeding cycles

In many animals environmental factors affect breeding cycles. In birds the gonads are very small outside the breeding season. This reduction in mass assists flight. The seasonal growth of the gonads prior to mating occurs in response to increasing day length, i.e. in the spring. The same stimulus promotes testosterone production in the stickleback (*Gasterosteus aculeatus*). Temperature and availability of food are other factors which affect sexual activity, e.g. sexual activity in the minnow (*Couesius plumbius*) is stimulated by a rise in temperature. In many species the production of testosterone increases in response to these factors. In male deer, for example, the increase in testosterone leads to growth of the antlers, changes in the voice and aggressive behaviour towards other males. In domestic cattle, the mere sight of a cow is sufficient to cause a massive rise in ICSH, and hence testosterone level, in the bull.

12.7 Fertilization and development in mammals

12.7.1 Courtship

In many species it is necessary for both partners to follow a specific pattern of behaviour before mating can occur. Courtship behaviour as it is called is developed in sexually mature individuals. In this way matings between sexually immature individuals, which cannot produce offspring, are avoided. This ensures that the often scarce sites for raising young are only occupied by pairs which have a good chance of producing offspring. On reaching sexual maturity many species develop easily recognizable features which are sexually attractive to a potential partner. These are referred to as the secondary sex characteristics. They take a variety of forms, including bright plumage in many birds, the mane of a lion, the comb and spurs of a cockerel and territory marking in dogs. In humans, secondary sex characteristics include the growth of pubic hair in both sexes, increased musculature, growth of facial hair and deepening of the voice in males and development of the breasts and broadening of the hips in females. Apart from preparing the female for child-bearing the changes in many animals help to distinguish males and females. In this way time and energy are not wasted on the fruitless courting of members of the same sex or sexually immature individuals of the opposite sex.

The females of many species undergo a cycle of sexual activity during which they are only capable of conceiving for a very brief period. Courtship behaviour is used by the male to determine whether the female is receptive or not. If she responds with the correct behavioural actions, courtship continues and is likely to result in fertilization. If she is not receptive, she exhibits a different pattern of behaviour and the male ceases to court her, turning his attentions elsewhere.

235

12.7.2 Mating

Under a variety of erotic conditions the blood supply to the genital regions increases. In females the process is slower than in males and results in the clitoris and labia becoming swollen with blood. At the same time the walls of the vagina secrete a lubricating fluid which assists the penetration of the penis. The fluid also neutralizes the acidity of the vagina which would otherwise kill the sperm. In males the increased blood supply results in the spongy tissues of the penis becoming swollen with blood, making it hard and erect. In this condition it more easily enters the vagina. By repeated thrusting of the penis within the vagina the sensory cells in the glans penis are stimulated. This leads to reflex contractions of muscles in the epididymis and vas deferens. The sperm are thus moved by peristalsis along the vas deferens and into the urethra. Here they mix with the secretions from the seminal vesicles, prostate and Cowper's glands. The resultant semen is forced out of the penis by powerful contractions of the urethra, a process called **ejaculation**. This is accompanied by **orgasm**, a sensation of extreme pleasure as a result of physiological and emotional release. The female orgasm is similarly intense resulting from the contraction of the muscles of the vagina and uterus although there is no associated expulsion of fluid. The process of mating, also known as **copulation** or **coitus**, results in internal fertilization and is an adaptation to life on land. The sperm, which require a liquid environment in which to swim, are never exposed to the drying effect of air.

Did you know?

Wild boar produce 0.5 l of semen in a single ejaculation.

12.7.3 Semen

In humans, each ejaculation consists of approximately $5\,cm^3$ of semen. While it contains around 500 million sperm they comprise only a tiny percentage of the total volume, the majority being made up of the fluids secreted by the seminal vesicles, prostate and Cowper's glands. The semen therefore contains:

1. **Sperm**.

2. **Sugars** which nourish the sperm and help to make them mobile.

3. **Mucus** which forms a semi-viscous fluid in which the sperm swim.

4. **Alkaline chemicals** which neutralize the acid conditions encountered in the urethra and vagina, which could otherwise kill the sperm.

5. **Prostaglandins**, hormones which help sperm reach the ovum by causing muscular contractions of the uterus and oviducts.

12.7.4 Fertilization

The force of ejaculation of the semen from the penis is sufficient to propel some sperm through the cervix into the uterus, with the remainder being deposited at the top of the vagina. The sperm swim up through the uterus and into the oviducts by the lashing movements of their tails. The speed with which they reach the top of the oviducts indicates that muscular contractions of the uterus and oviduct are also involved. The egg or ovum released from the Graafian follicle of the ovary is metabolically inactive and dies within 24 hours unless fertilized. The ovum is surrounded by up to 2000 **cumulus cells** which aid its movement towards the uterus by giving the cilia which line the oviduct a large mass to 'grip'. The cumulus cells may also provide nutrients to the ovum. As the journey to the uterus takes three days in humans, it follows that fertilization must take place in the top third of the oviduct if the ovum is still to be alive when the sperm reaches it. In mammals there is no evidence that the ovum attracts the sperm in any way; their meeting would appear to be largely a matter of chance. Of the 500 million sperm in the ejaculate only a few hundred reach the ovum, and only one actually fertilizes it.

The fertilized ovum is called a **zygote**. The fertilizing sperm firstly releases **acrosin**, a trypsin-like enzyme, from the acrosome. This softens the plasma membrane which covers the ovum. Inversion of the acrosome results in a fine needle-like filament developing at the tip of the sperm and this pierces the already softened portion of the plasma membrane. An immediate set of changes occurs which thickens the membrane and so ensures that no other sperm can penetrate the egg. This is essential to prevent a 'multinucleate' fertilized egg; such cells normally degenerate after a few divisions. The thickened membrane is now called the **fertilization membrane**. The sperm discards its tail, and the head and middle piece enter the cytoplasm. The second meiotic division of the ovum nucleus normally occurs immediately following the penetration of the sperm. The sperm and ovum nuclei fuse, restoring the diploid state. A spindle forms, the two sets of chromosomes line up and the cell undergoes mitotic division at once.

If the ovum is not fertilized it quickly dies and in humans the lining of the uterus is later shed to give the menstrual flow. In the female horseshoe bat, to achieve the earliest possible fertilization in spring, mating takes place in the autumn but the female stores the sperm in a thick plug of mucus. In spring the plug dissolves releasing the sperm for fertilization. This is called **delayed fertilization** and depends on the sperm surviving considerably longer than the 2–3 days which is normal in a human.

12.7.5 Causes of infertility and its cures

There are a number of reasons why a couple may have difficulty conceiving a baby:

1. Blocked oviducts – These may prevent ova and sperm meeting, in which case an operation may be undertaken to unblock the tubes or in vitro fertilization can be attempted.

2. An irregular menstrual cycle – This may make the chance of fertilization remote and hormone treatment necessary to regularize the cycle.

3. Incorrect frequency and/or timing of intercourse may make conception unlikely and couples may need to be counselled on

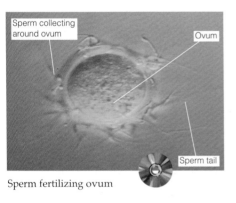

Sperm collecting around ovum

Ovum

Sperm tail

Sperm fertilizing ovum

APPLICATION

In vitro fertilization and test tube babies

After first being achieved in rabbits in 1959, in vitro fertilization or IVF, was successfully performed between human sperm and ova by Drs Edwards, Bavister and Steptoe ten years later. The development of these zygotes and their successful transfer into the uterus of the mother, called **embryo transfer** or **ET**, took a number of years but finally, in 1978, the first test tube baby was born.

The success of this technique owes as much to the development of a suitable medium in which the sperm, ova and embryo can survive and grow, as to the clinical techniques of obtaining ova and implanting the embryo. Such a medium must have not only a pH, osmotic potential and ionic concentration similar to that of blood, but also contain the patient's serum as a source of protein and other macromolecules. Glucose, lactate and pyruvate are other essential components.

The process begins with a fertility drug being administered to the potential mother to increase her ova production. Around six of these are collected using a fine needle, via the vagina. Some 100 000 sperm, collected from the potential father's semen sample by centrifugation, are added to the ova in a Petri dish. When the embryos are two days old a few are transferred into the mother's uterus where, if all goes well, one will develop normally. More than one is used to guard against some not implanting successfully.

A major cause of infertility is blocked oviducts which therefore prevent ova and sperm meeting in natural circumstances. IVF has solved this problem in some cases, allowing both parents to contribute genetically to the offspring and almost all embryo development to take place inside the natural mother. IVF clinics are now common throughout the UK and despite their low success rate, at 10%, make a major contribution to providing otherwise childless couples with much wanted children. For those with other forms of infertility the technique is unsuitable.

the most appropriate time (the middle of the menstrual cycle) to have sexual intercourse in order to increase the possibility of fertilization.

4. Non-production of ova – This affects a few females making it impossible for them to contribute genetically to their offspring. Adoption or the use of a donated ovum from another female for in vitro fertilization are the possible alternatives. Artificial insemination of a surrogate mother with the potential father's sperm is another option.

5. Non-production of sperm – Some men produce no sperm, or so few that there is little realistic prospect of conception. Donated semen from another male can be used to artificially inseminate the woman.

6. Impotence – Some men are unable to erect the penis and/or ejaculate semen. The cause is often psychological, or the result of prolonged drug or alcohol abuse. In these cases counselling and guidance can sometimes remedy the problem. In other cases an implant may be used which can be pumped up as required. Alternatively an injection of a drug at the base of the penis will raise blood pressure within it and so create an erection.

Some of the above causes of infertility may be the result of certain diseases or infections. Sexually transmitted diseases such as gonorrhoea can cause sterility, especially in females; mumps, if contracted in adult life, sometimes makes males infertile. Even when conception occurs, a few women are not able to sustain the pregnancy because either the embryo does not implant in the uterus lining, or having implanted, it is later miscarried. For some, the solution is to use in vitro fertilization (see Application page 238) but rather than implant the embryo into the natural mother, it is transferred to a different female. The process whereby one woman carries a fertilized egg for another through to birth, is known as **surrogacy**.

Surrogate motherhood, in vitro fertilization and artificial insemination all raise complex legal and moral issues. Should the surrogate mother or sperm donor have any legal rights over the offspring they helped produce? To what extent should the natural mother be able to influence the behaviour of the surrogate mother during pregnancy – should she be able to insist on abstinence from smoking or drinking, both of which could damage the fetus? What details, if any, should a potential mother be entitled to know about the donor of the sperm to be used in artificial insemination? Should the excess embryos which result from in vitro fertilization be used for the purposes of medical research? These are just a few of the issues which have been raised by recent scientific research into the causes of, and cures for, infertility.

12.7.6 Implantation

Following fertilization, the zygote divides (cleavage) mitotically until a hollow ball of cells, the **blastocyst**, is produced. It takes three days to reach the uterus and a further three or four days to become implanted in the lining of the uterus. The outer layer of cells of the blastocyst, called the **trophoblast**, develops into the embryonic membranes, the **chorion** and the **amnion**. The chorion develops villi which grow into the surrounding uterine tissue from which they absorb nutrients. These villi form part of the **placenta** which is connected to the fetus by the **umbilical cord**. The amnion develops as a membrane around the fetus and encloses the amniotic fluid, a watery liquid which protects the fetus by cushioning it from physical damage. In badgers, mating and fertilization occur in midsummer and development takes place up to the blastocyst stage. This however does not become implanted until the late winter or early spring, after which development proceeds normally. A similar process occurs in polar bears and allows offspring to be weaned when food is plentiful, regardless of the time of fertilization. This is called **delayed implantation**.

Screening in early pregnancy

Screening is usually offered when a family has a history of an inherited disorder or when the mother is comparatively old and is therefore more likely to give birth to a child with Down's syndrome.

Amniocentesis is carried out from 10 weeks' gestation. A small quantity of amniotic fluid is taken through a hypodermic needle and the fetal cells contained in it are separated from the liquid by centrifugation. The cells are then cultured and their chromosomes examined.

Chorionic villus sampling is also carried out after 10 weeks' gestation. A sample of cells is taken from the chorion, the developing placenta, using a plastic catheter inserted through the vagina under ultrasound guidance. Preliminary results can be obtained within a day but the procedure is thought to increase the risk of miscarriage by about 4% (compared to a 1% increased risk with amniocentesis).

Coelocentesis is a new, relatively untried technique which involves the removal of cells from the coelomic cavity surrounding the amniotic sac. It can be carried out before 10 weeks' gestation and is thought to present less risk to the unborn child.

12.7.7 The placenta

The chorionic villi will develop about 14 days after fertilization and represent the beginning of the placenta. It rapidly develops into a disc of tissue covering 20% of the uterus. The capillaries of the mother and fetus come into close contact without actually combining.

12.7.8 Functions of the placenta

1. It allows exchange of materials between the mother and fetus without the two bloods mixing. This is necessary as the fetal blood may be different from that of the mother due to the influence of the father's genes. If incompatible bloods mix they agglutinate (clot), causing blockage in vital organs such as the kidney, possibly resulting in death.

2. Oxygen, water, amino acids, glucose, essential minerals, etc. are transferred from maternal to fetal blood to nourish the developing fetus.

3. Carbon dioxide, urea and other wastes are transferred from fetal to maternal blood to allow their excretion by the mother and prevent harmful accumulation in the fetus.

4. It allows certain maternal antibodies to pass into the fetus, providing it with some immunity against disease. Such immunity is termed **passive natural immunity** as, while the antibodies are naturally produced, they are not formed by the fetus itself. The immunity only lasts for a few months after birth although this period may be extended by antibodies provided in the mother's milk.

Did you know?

The longest pregnancy of any mammal is that of the Asiatic elephant with an average of 609 days and maximum of 760 days.

The chorionic villi present a large surface area for the exchange of materials by diffusion across the chorionic membrane. In some mammals the maternal and fetal bloods flow in opposite directions. This counter-current flow leads to more efficient exchange as described later, in Section 20.2.4

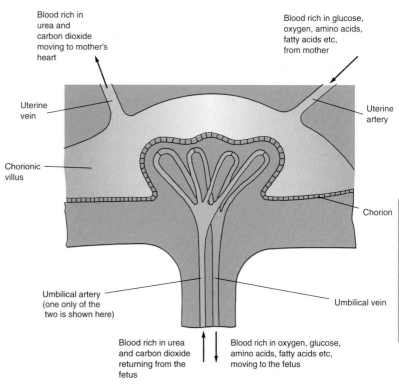

Blood rich in urea and carbon dioxide moving to mother's heart

Blood rich in glucose, oxygen, amino acids, fatty acids etc, from mother

Uterine vein

Uterine artery

Chorionic villus

Chorion

Umbilical artery (one only of the two is shown here)

Umbilical vein

Blood rich in urea and carbon dioxide returning from the fetus

Blood rich in oxygen, glucose, amino acids, fatty acids etc, moving to the fetus

Fig. 12.22 The mammalian placenta

5. It protects the fetus by preventing certain pathogens (disease-causing organisms) and their toxins from crossing the placenta. This protection is by no means complete. Notable exceptions include toxins of the Rubella (German measles) virus which can cross the placenta causing physical and mental damage to the fetus, and the HIV virus which can also pass into the fetus.

6. In a similar way it acts as a barrier to those maternal hormones and other chemicals in the mother's blood which could adversely affect fetal development. Again the protection is not complete and substances like nicotine, alcohol and heroin can all enter the fetus causing lasting damage.

7. As the two blood systems are not directly connected, the placenta permits them to operate at different pressures without harm to mother or fetus.

8. As the pregnancy progresses the placenta increasingly takes over the rôle of hormone production. In particular it produces progesterone which prevents ovulation and menstruation. It also secretes **human choriogonadotrophin** (HCG), a hormone whose presence in the urine of pregnant women is the basis of most pregnancy tests (see Application page 416).

Did you know?

The human uterus increases 500 times in size during pregnancy, from 10 cm³ to 5000 cm³

12.7.9 Birth (parturition)

During pregnancy the placenta continues to produce progesterone and small amounts of oestrogen. The amount of progesterone decreases during pregnancy while oestrogen increases. These changes help to trigger the onset of birth. As the end of the gestation period nears, the posterior lobe of the

241

APPLICATION

Drugs across the placenta

Drugs taken by a pregnant woman can cross the placenta if their molecules are small enough. The effect they have on the fetus depends on the nature of the drug, the dose taken and the stage of pregnancy. Obviously the best way to avoid damaging the unborn child is to avoid all drugs but this may not be possible if a serious condition needs treatment. The effects of a few legal and illegal drugs on the development of the fetus will be considered.

Tetracyclines: When given in early pregnancy these antibiotics may cause cataracts and bone abnormalities. Later on they are stored in the bones and also in the teeth where they cause yellow staining.

Cytotoxic drugs: These drugs are used in the treatment of cancers but they must be avoided during pregnancy because of their effect on dividing cells.

Aspirin: This rarely causes any harm but large doses towards the end of pregnancy interfere with the production of prostaglandins and thus delay the onset and progress of labour.

Cannabis and LSD: Both these drugs may cause growth retardation and LSD causes chromosome damage.

Opiates: Babies whose mothers are addicted to heroin become addicted in the uterus. They suffer withdrawal symptoms and growth retardation and may die of their addiction.

Alcohol: Alcohol abuse may lead to mid-term abortion and premature labour. Babies may be retarded both physically and mentally.

Nicotine: Smoke contains many substances which cross the placenta and damage the fetus. Even passive smoking causes an alteration in the heart rate and breathing pattern of the fetus and the constriction of blood vessels in the placenta leads to slower growth. Smoking leads to an increased risk of spontaneous abortion, congenital abnormalities, still birth and mental and physical retardation in later childhood.

pituitary produces the hormone **oxytocin** which causes the uterus to contract. These contractions increase in force and frequency during labour.

The process of birth can be divided into three stages:

1. The dilation of the cervix, resulting in loss of the cervical plug ('the show') and the rupture of the embryonic membranes ('breaking of the waters').

2. The expulsion of the fetus.

3. The expulsion of the placenta ('afterbirth') which is eaten by most mammals.

APPLICATION

Infections across the placenta

The most significant infections which cross the placenta to affect the fetus are rubella and syphilis.

Rubella: If the rubella virus is contracted during the first 12 weeks of pregnancy the risk of having a child with a congenital abnormality is between 5 and 12 times greater. Growth of the early fetal organs may be disorganized leading to possible damage of the eyes, ears, heart and other organs.

Syphilis: The bacterium *Treponema pallida* which causes syphilis can only cross the placenta after the twentieth week of pregnancy but it causes either death in the uterus or the birth of a child with congenital syphilis.

Babies are unable to manufacture **antibodies** in the uterus or for about 6 weeks after birth. Immunity to many diseases is transferred to them by the passage of the mother's antibodies across the placenta. However, the antibodies to tuberculosis and whooping cough cannot cross the placenta and young babies require protection against these two diseases.

12.7.10 Lactation

During pregnancy the hormones progesterone and oestrogen cause the development of lactiferous (milk) glands within the mammary glands. Following birth, the anterior lobe of the pituitary gland produces the hormone **prolactin** which causes the lactiferous glands to begin milk production. Suckling by the offspring causes the reflex expulsion of this milk from the nipple of the mammary glands. The first formed milk, called **colostrum**, is mildly laxative and helps the baby expel the bile which has accumulated in the intestines during fetal life. As well as essential nutrients, the milk contains antibodies which give some passive immunity to the newly born.

12.7.11 Parental care

Some organisms produce vast numbers of offspring, the cod (*Gadus gadus*), for example, may produce over one million eggs at a time. In such organisms there is little or no parental care and the majority of offspring fail to reach maturity, most being consumed by predators. In birds and mammals the tendency is to reduce the number of offspring but to expend much time and energy in caring for them in order to ensure a high survival rate. This is especially important in groups such as primates where learning plays an important rôle in their development. Only through extended care of the young and close association with adults is there sufficient opportunity and time to acquire through learning the necessary skills for adult life. The onset of sexual maturity is often delayed in these species to allow time for learning to take place.

In mammals, the provision of milk is the most obvious example of parental care. As the offspring develop they are

Did you know?

The sex of hatching sea turtles depends on their incubation temperature:

28°C – all male
30°C – equal numbers of male and female
32°C – all female

gradually introduced to other, more solid types of food, a process called weaning. In many animals the parents singly, or in pairs, collect the appropriate food for the offspring. This food may be partly digested, e.g. regurgitated from the crop in birds.

Many animals provide a nest in which to raise their young. Here the offspring may be raised in the relative safety of a warm, dry environment remote from predators.

12.8 Growth

The growth in size of an individual cell is limited by the distance over which the nucleus can exert its control. For this reason, when single-celled organisms reach a maximum size they divide to give two separate individuals. In order to attain greater size, organisms become multicellular. While being large and multicellular can present some problems, these are easily outweighed by the advantages conferred:

1. Cells may become differentiated in order to perform a particular function.

2. Specialized cells performing one particular function lead to greater efficiency.

3. It is possible to store more materials and so be better able to withstand periods when these are scarce.

4. If some cells are damaged, enough may still remain to carry out the repair.

5. Some processes require a range of conditions, e.g. digestion often has an acid and an alkaline phase. It is easier to separate regions of opposing conditions in a multicellular organism than it is in a single cell.

6. Larger organisms may have a competitive advantage, e.g. large plants compete better for light than small ones.

7. Large size may provide some protection from predators because the organisms are simply too large to ingest.

12.8.1 Measurement of growth

Growth is estimated by measuring some parameter (variable) over a period of time. The parameter chosen depends upon the organism whose growth is to be measured. It may be appropriate to measure the weight of a mouse; but this method would be impractical for an oak tree. Mass and length are most often used, but these may be misleading. A bush for example, while not increasing in height, may continue to grow in size by spreading sideways. Area or volume give a more accurate indication of growth but are often impractical to measure. The measurement of mass has its problems. If an organism takes in a large amount of water its mass may increase markedly, and yet such a temporary increase could not be considered as growth. For this reason two types of mass are recognized:

1. Fresh mass – This is the mass of the organism under normal conditions. It is easy to measure and doing so involves no damage to the organism. It may, however, be inaccurate due to temporary fluctuations in water content.

2. Dry weight – This involves removing all water by drying, before weighing. It is difficult to carry out and permanently destroys the organisms involved, but does give an accurate measure of growth.

It is sometimes possible to measure one part of an organism, e.g. the girth of a tree; the length of the tail of a rat. Provided this part grows in proportion to the complete organism, increases in its size will reflect those of the individual as a whole. Groups of organisms are sometimes used rather than an individual. For instance, if the growth of peas was measured using dry mass, it would be necessary to grow a large population of the plants. Growth could be estimated by removing say ten plants every day, drying and weighing them. Provided each sample is large enough to average out individual differences in growth, a good estimate of the growth rate can be found. The growth of a population of yeast can be measured by counting the number of cells in a known, and very small, volume of the medium in which the yeast is growing.

12.8.2 Growth patterns

When any parameter of growth is measured against set intervals of time, **a growth curve** is produced. For many populations, organisms or organs, this curve is S-shaped and is called a **sigmoid curve**. It represents slow growth at first, because there are so few cells initially that even when they are dividing rapidly the actual increase in size is small.

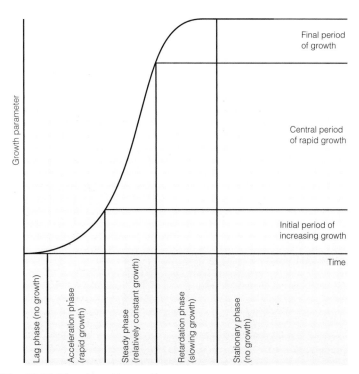

Fig. 12.23 The sigmoid growth curve

As the number of cells becomes greater the size increases more quickly because there are more cells carrying out division. There is a limit to this rapid phase of growth. This limit may be imposed by the genotype of the individual, which specifies a certain maximum size, or any external factors, such as shortage of food. Whatever the cause, the growth rate decreases until it ceases altogether. At this point cells are still dividing, but only at a rate which replaces those which have died. The size of the organism therefore remains constant.

While the sigmoid curve forms the basis of most growth curves, it may be modified in certain circumstances. In humans, for example, there are two phases of rapid growth; one during the early years of life, the other during adolescence. Between these two phases there is a period of relatively slow growth. The growth curve therefore resembles two sigmoid curves, one on top of the other.

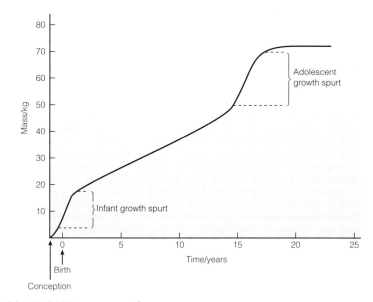

Fig. 12.24 Human growth curve

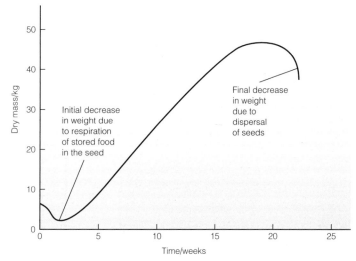

Fig. 12.25 Annual plant growth curve, e.g. Pea plant (Pisum sativum

In annual plants, the growth curve is typically sigmoid, except there may be an initial decrease in mass during the early stages of germination. This occurs as the food reserves in the seed are respired in order to produce the roots and leaves. Once the leaves begin to photosynthesize growth proceeds in a sigmoid fashion. However, with the liberation of fruits and seeds at the end of the growing period, the mass of the plant may decrease prior to its death. When there is a natural limit on growth, as in annual plants, they are said to show **limited growth**. In these cases the growth curve flattens out, or even decreases prior to the organism's death.

In perennial plants, the growth pattern is an annnual series of sigmoid curves. During spring when the temperature and light intensity are relatively low, there is less photosynthesis, and growth is slow. In summer, with higher temperatures and more light, the rate of photosynthesis increases and growth is rapid. The falling temperatures and lower light intensities of autumn again reduce the rate of photosynthesis and hence growth. During winter in temperate regions there is no growth in deciduous plants and so the curve flattens out. The following spring the process is repeated. The overall shape of these annual sigmoid curves is itself sigmoid, except that many perennial plants show **unlimited growth**, i.e. they grow continuously throughout their lives, and the curve therefore never flattens out.

A very different growth curve is exhibited by many arthropods. As their exoskeleton is incapable of expansion, they have to moult periodically during growth. Before a new exoskeleton has fully hardened it is capable of some expansion. During this time the insect may take up water in order to expand the exoskeleton as much as possible. This means that once it has hardened there is still some room for growth. Measuring the fresh mass as the growth parameter therefore gives the unusual growth pattern shown in Fig. 12.27. This type of growth is called **intermittent growth**. If dry mass is measured, a normal sigmoid curve is obtained.

e annual growth follows a normal sigmoid curve.
riations occur from one year to the next according to
ironmental conditions. In a cold, dry year for example
re will be less growth than in a mild, wet one.

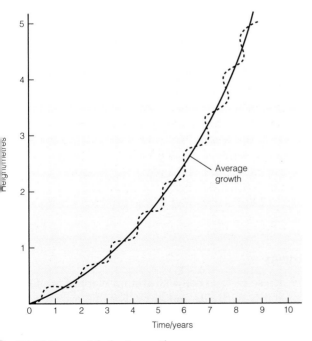

g. 12.26 Perennial plant growth curve

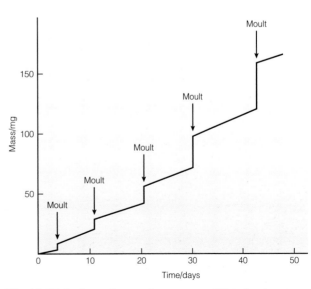

Fig. 12.27 Arthropod growth curve, e.g. Waterboatman (Notonecta glauca)

247

Certain organs of an individual grow at the same rate as the organism as a whole. This is called **isometric growth**. Other organs grow at a different rate from the entire organism. This is called **allometric growth**. The leaves of most plants exhibit isometric growth and their growth curve is typically sigmoid (Fig. 12.28).

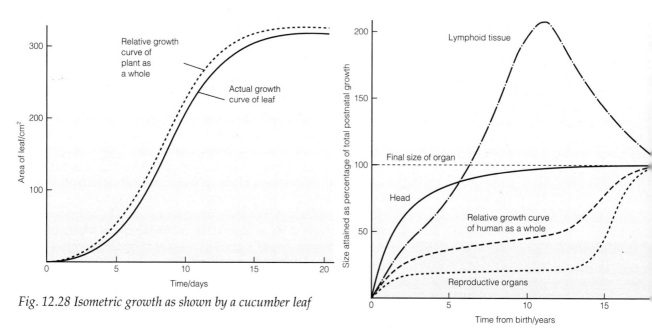

Fig. 12.28 *Isometric growth as shown by a cucumber leaf*

Fig. 12.29 *Allometric growth as shown by human organs and tissues*

In animals, organs often exhibit allometric growth. Lymph tissue, which produces white blood cells to fight infection, grows rapidly in early life when the risk of disease is greater as immunity has not yet been acquired. By adult life the mass of lymph tissues is less than half of what it was in early adolescence. The reproductive organs grow very little in early life but develop rapidly with the onset of sexual maturity at puberty. Fig. 12.29 illustrates allometric growth in some human organs.

12.8.3 Rate of growth

The actual growth of an organism is the cumulative increase in size over a period of time. A small annual plant, for example, might grow as shown in Fig. 12.30(a), in which case a typical sigmoid growth curve results. The rate of growth is a measure of size increase over a series of equal time intervals. If instead of measuring the actual height of the plant we measure the increase in height over each three-day period, a set of results like that shown in Fig. 12.30(b) is obtained. These produce a bell-shaped graph as shown.

12.8.4 Meristems

The presence of a semi-rigid cell wall around plant cells effectively restricts their ability to divide and grow. For this reason, unlike animals, plants retain groups of immature cells which form the only actively growing tissues. These tissues are called **meristems**. Three types of meristems are generally recognized:

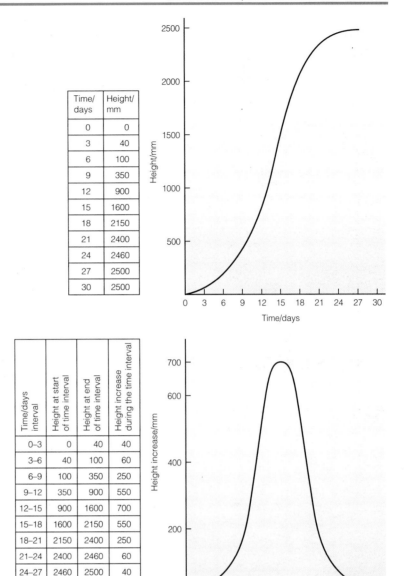

Time/ days	Height/ mm
0	0
3	40
6	100
9	350
12	900
15	1600
18	2150
21	2400
24	2460
27	2500
30	2500

Time/days interval	Height at start of time interval	Height at end of time interval	Height increase during the time interval
0–3	0	40	40
3–6	40	100	60
6–9	100	350	250
9–12	350	900	550
12–15	900	1600	700
15–18	1600	2150	550
18–21	2150	2400	250
21–24	2400	2460	60
24–27	2460	2500	40
27–30	2500	2500	0

Fig. 12.30 Comparison of (a) actual growth curve and (b) rate of growth curve

1. **Apical meristems** – These are found at the tips of roots and shoots and are responsible for primary growth of the plant. They increase its length.

2. **Lateral meristems** – These are found in a cylinder towards the outside of stems and roots. They are responsible for secondary growth and cause an increase in girth.

3. **Intercalary meristems** – These are found at the nodes in monocotyledonous plants. They allow an increase in length in positions other than the tip.

12.8.5 Growth and metamorphosis in insects

During their life cycles insects undergo **metamorphosis**. This is the series of changes which take place between larval and adult forms.

Eggs

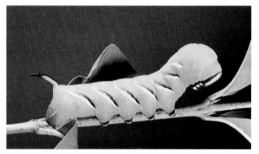

Larva (caterpillar)

Pupa in soil

Adult

Stages of development of privet hawk moth

1. Hemimetabolous (incomplete metamorphosis) – The eggs hatch into **nymphs** which clearly resemble the adults except that they are smaller, lack wings and are sexually immature. There are a number of nymphal stages between which moulting occurs. Examples of hemimetabolous insects include locusts and cockroaches.

2. Holometabolous (complete metamorphosis) – The eggs hatch into **larvae** which differ considerably from the adults. Each larva undergoes a series of moults until it changes its appearance and becomes a dormant stage known as a **pupa**. After much reorganization of the tissues within the pupa, the adult **(imago)** emerges. Examples of holometabolous insects include moths, butterflies and flies.

The control of insect metamorphosis involves two main hormones: **moulting hormone (ecdysone)** and **juvenile hormone (neotonin)**. Moulting hormone is produced by a gland in the first thoracic segment called the **prothoracic gland**. Juvenile hormone is produced by a region behind the brain known as the **corpus allatum**. The production of both hormones is controlled by neurosecretory cells in the brain. All moults require moulting hormone. If juvenile hormone is present in high concentrations larval moults occur, which mean the insect remains as a larva. If only low concentrations of juvenile hormone are present a pupal moult occurs and the larva metamorphoses into a pupa. In the complete absence of juvenile hormone, the pupa metamorphoses into the imago (adult). The events are summarized in Fig. 12.31.

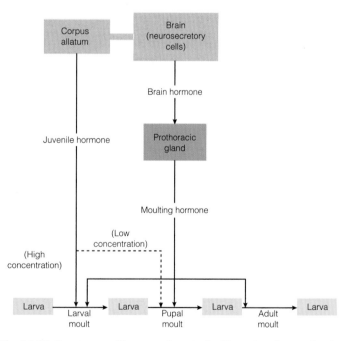

Fig. 12.31 Summary of hormonal control of insect metamorphosis

12.9   Questions

Write an essay on the roles of hormones in human reproduction.   *(Total 30 marks)*

ULEAC June 1993, Paper III, No. 9(b)

Describe the structure and functions of the human placenta.   *(Total 25 marks)*

O & C SEB June 1992, Paper 2/6, No. 1

Describe the process of reproduction in a flowering plant and comment on the ways in which reproduction in humans differs from that in flowering plants.   *(Total 30 marks)*

ULEAC January 1993, Paper III, No. 9(b)

Write an essay on gametes and their formation.
(Total 24 marks)

AEB June 1992, Paper II, No. 5B

The diagram shows the structure of part of a human placenta.

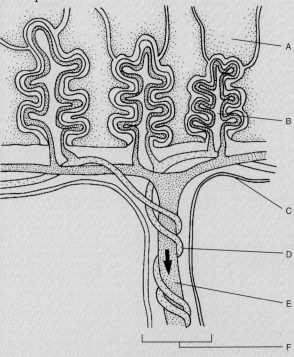

(a)   (i)   Name the structures labelled **A** to **F**.
    (ii)   Which of these structures is composed of maternal tissue?   *(7 marks)*
(b) Explain how gas exchange occurs across the placenta.   *(4 marks)*

If a pregnant woman develops high blood pressure, particularly in the later stages of pregnancy, doctors may consider inducing the birth.

(c) Suggest why maternal high blood pressure is dangerous for the fetus.   *(2 marks)*

(d) The placenta functions as an endocrine organ. Comment on this statement.

(4 marks)
(Total 17 marks)

UCLES (Modular) December 1992,
(Growth, development and reproduction), No. 2

6.   *(a)*   (i)   State *one* function of luteinizing hormone (LH).   *(1 mark)*
    (ii)   On what day of a typical 28-day menstrual cycle is the level of circulating LH at its highest?   *(1 mark)*

(b) The secretion of LH is controlled by *LH releasing factor* (LRF) a small peptide produced in the hypothalamus. LRF has been synthesized in the laboratory.

The graphs below show the effect of synthetic LRF on LH secretion in a healthy woman. Synthetic LRF was injected on day 3 of the menstrual cycle and on day 11 of the menstrual cycle. The same dose of LRF was injected on each occasion.

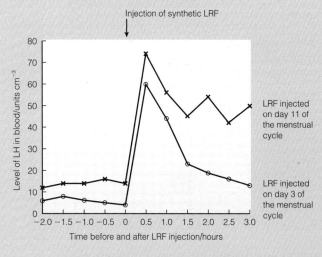

    (i)   Comment on the differences in response to synthetic LRF on days 3 and 11 of the menstrual cycle.

(4 marks)

    (ii)   Suggest *one* possible medical application of synthetic LRF.   *(1 mark)*

(c) LRF triggers LH secretion by binding to receptor molecules on the surface of LH-secreting cells. It has been suggested that synthesis of substances with a molecular structure similar, but not identical, to LRF may have applications in contraception.

Explain the physiological basis of this suggestion.   *(3 marks)*

(d) In the hypothalamus, LRF is secreted directly into capillaries which join to form a short *portal vein* leading to the anterior pituitary. In the pituitary, this portal vein branches into another capillary network, supplying the cells which secrete LH. This system is illustrated in the simplified diagram below.

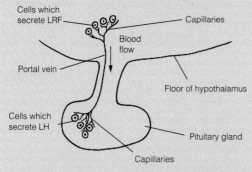

(i) How does the transport of LRF to its target organ differ from the transport of a typical hormone such as LH to its target organ? *(2 marks)*

(ii) What advantages may be gained from the system described for the transport of LRF? *(2 marks)*

(Total 14 marks)

ULEAC June 1992, Paper III, No. 8

7. In some women, infertility may be treated with the drug clomiphene. The graph shows the blood-oestrogen levels in a woman during and after treatment with clomiphene.

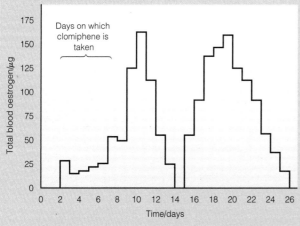

(a) Clomiphene stimulates the production of gonadotrophic hormones from the anterior pituitary gland.

(i) From the timing of the treatment shown in the graph, name the gonadotrophic hormone whose secretion is stimulated by the clomiphene. *(1 mark)*

(ii) Explain how stimulation of this hormone brings about the change in oestrogen secretion shown in the graph. *(2 marks)*

(b) (i) On what day would ovulation be most likely to occur? *(1 mar)*

(ii) Give a reason for your answer. *(1 mar)*

(Total 5 mark)

AEB November 1992, Paper I, No.

8. The diagram below shows the structure of the flower of a lupin, a member of the Papilionaceae.

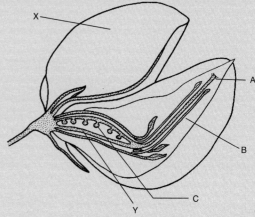

(a) Name the structures labelled A, B and C. *(3 mark)*

(b) Describe the process of pollination in flower of the Papilionaceae. *(3 mark)*

(c) What happens to the structures labelled X and Y after pollination and fertilization? *(2 mark)*

(Total 8 mark)

ULEAC January 1994, Paper III, No.

9. *Primula* plants produce flowers which are typically pollinated by insects such as bees. Two different type of flower, pin-eyed and thrum-eyed, are produced. The diagrams show the structure of the two types of flower, and their pollen grains and stigmatic surfaces

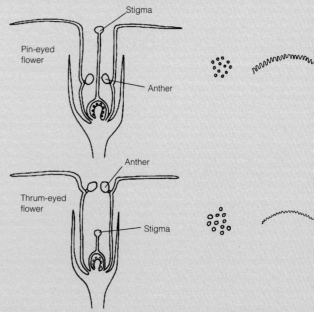

(a) Using the information in the diagrams, suggest **two** ways in which cross-pollination between the two types of flower is favoured.
(*2 marks*)

n experiment was carried out to find out the ercentage of successful fertilization in flowers which ere artificially pollinated. The results are shown in e table.

Type of pollination	Percentage of pollinated flowers which produced seed
Thrum pollen on pin stigma	67
Pin pollen on thrum stigma	61
Thrum pollen on thrum stigma	7
Pin pollen on pin stigma	35

(b) Describe, briefly, what you would need to do to ensure that pollination occurred only as intended in the experiment. (*3 marks*)

(c) (i) Suggest a *non-structural* mechanism to account for the lower success of fertilization between flowers of the same type.

(ii) Refer to the data in the table. Suggest a hypothesis which might account for the greater success of fertilisation from the artificial self-pollination of pin flowers than from that of thrum flowers.
(*4 marks*)
(*Total 9 marks*)

NEAB June 1991, Paper IIA, No. 1

0. The diagram shows a section through the ovary and ollen tube of a flowering plant just before fertilization.

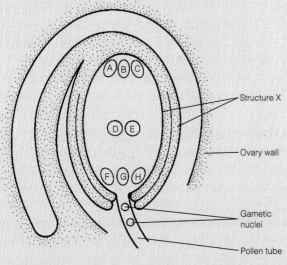

(a) Which of the nuclei labelled **A** to **H** fuse with a gametic nucleus to form:
(i) the endosperm; (ii) the zygote?
(*2 marks*)

(b) After fertilization, into what structure does each of the following develop:
(i) the ovary wall; (ii) the structure labelled **X**? (*2 marks*)

(c) In this plant, the diploid number of chromosomes is 14. How many chromosomes would you expect to find in: (i) a nucleus in one of the cells of structure **X**; (ii) the nucleus labelled C; (iii) a nucleus in the endosperm? (*3 marks*)
(*Total 7 marks*)

AEB June 1992, Paper I, No. 6

11. The diagram shows the various stages of spermatogenesis in a mammal.
(a) Name in the spaces provided adjacent to the diagram the types of cell shown in **B**, **C**, **D** and **E** and the type of division occurring between stages **D** and **E**.

(b) State **two** ways in which *spermatogenesis* differs from *oogenesis*.

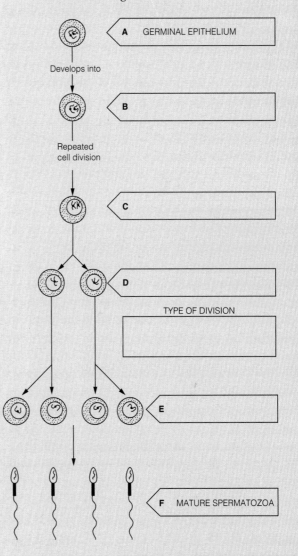

UCLES June 1990, Paper III, No. 2

12. The diagram shows a mature mammalian sperm cell.

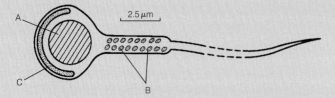

(a) Describe how features **A**, **B** and **C** enable the sperm cell to carry out its function. *(6 marks)*

(b) Sexual reproduction in animals requires an aquatic environment for the male gametes.

   (i) How is an aquatic environment achieved in the reproduction of mammals?
(2 marks)

   (ii) Give **one** explanation why it is necessary to have an aquatic environment for the male gametes. *(1 mark)*

The graph shows changes in the mean diameters of follicles and corpora lutea in the ovary of a pig over a period of 40 days.

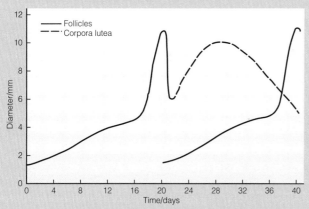

(c)   (i) Between which days is fertilization most likely to occur? Explain your answer. *(3 marks)*

   (ii) In the cycle represented in the graph, fertilization did not occur. Give **two** pieces of evidence from the graph to support this. *(2 marks)*

   (iii) Describe the part played by hormones in controlling the changes in size of the follicles. *(4 marks)*

(d)   (i) Explain why it is wrong to say that pollen is the male gamete of a flowering plant. *(1 mark)*

   (ii) Give **three** similarities between the pollen grain of a flowering plant and a mammalian sperm cell. *(3 marks)*

   (iii) Describe how the pollen from a wind-pollinated plant differs from that from an insect-pollinated plant. *(2 marks)*

(Total 24 marks)

AEB June 1990, Paper II, No. 1

13. (a) Using **named** examples, explain what is meant by the terms *complete metamorphosis* and *incomplete metamorphosis*. *(6 mark*

(b)   (i) Sketch a growth curve (time/body length) for an insect showing incomplete metamorphosis.

   (ii) Explain why an increase in body lengt of an insect is not an accurate measure of its daily growth.

   (iii) Suggest how daily growth of an insect could be measured. *(8 mark*

(Total 14 mark

UCLES (Modular) March 199
(Growth, development and reproduction), No.

14. (a) Explain the terms *absolute growth rate* and *specific growth rate*. *(4 mark*

The table shows the height of an annual flowering plant measured over a period of 90 days from germination.

Time/days	Height/mm	Absolute growth rate
0	0	
10	20	
20	80	
30	280	
40	760	
50	1240	
60	1660	
70	1960	
80	2000	
90	2000	

(b)   (i) Use the data from the table to calculate the absolute growth rates. Write these in the column provided, adding suitable units to the heading of this column. *(3 mark*

   (ii) Use your results to plot an absolute growth rate curve. *(3 mark*

(c) From your graph determine:

   (i) the period during which the greatest increase in height occurred;

   (ii) the period during which least growth occurred. *(2 mark*

(d) Suggest what might happen to the height of the plant if the readings were continued for further 90 days. Explain your answer.
(2 mark

(Total 14 mark

UCLES (Modular) December 199.
(Growth, development and reproduction), No.

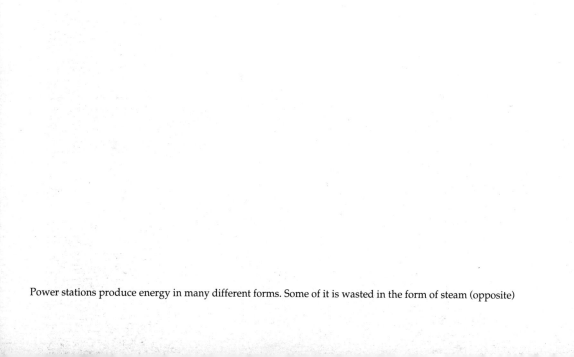

Power stations produce energy in many different forms. Some of it is wasted in the form of steam (opposite)

Part III

ENERGETICS

13 Energy and organisms

Energy is defined as the 'capacity to do work'. It exists in a number of different forms: heat, light, electrical, magnetic, chemical, atomic, mechanical and sound. The laws which apply to energy conversions are the **laws of thermodynamics**.

13.1 First law of thermodynamics

This states: **energy cannot be created or destroyed but may be converted from one form into another**. Energy may also be stored. Water in a lake high up on a mountain is an example of stored energy. The energy it possesses is called **potential energy**. If the water is released from the lake it begins to flow downhill, and the energy of its motion is called **kinetic energy**. The stream of moving water may be used to drive a turbine which produces electricity (hydroelectric power). The kinetic energy is thus converted to electricity. This electricity may in turn be converted into light (light bulb), heat (electric fire/cooker), sound (CD or cassette player) etc. During these changes not all the energy is converted into its intended form; some is lost as heat. By 'lost' we mean the energy is no longer available to do useful work because it is distributed evenly. Energy which is available to do work under conditions of constant temperature and pressure is called **free energy**. Reactions which liberate energy are termed **exothermic**, those which absorb free energy are termed **endothermic**.

13.2 Second law of thermodynamics

This states: **all natural processes tend to proceed in a direction which increases the randomness or disorder of a system**. The degree of randomness is called **entropy**. A highly ordered system has low entropy whereas a disordered one, with its high degree of randomness, has high entropy. We saw in the Introduction that entropy and free energy are inversely related. Systems with high entropy have little free energy, those with low entropy have more free energy. We also saw in the Introduction that the ability of living systems to maintain low entropy is what distinguishes them from non-living systems. The fact that living systems can decrease their entropy does not mean that they fail to obey the second law of thermodynamics.

The reason that they are able to reduce their entropy is that they take in useful energy from their surroundings and release it in a less useful form. While the organism's entropy decreases, that of its surroundings increases to an even greater extent. The organism and its environment represent one system, the total entropy of which increases. The second law of thermodynamics is therefore not violated.

13.3 Energy and life

There are three stages to the flow of energy through living systems:

1. The conversion of the sun's light energy to chemical energy by plants during photosynthesis.

2. The conversion of the chemical energy from photosynthesis into ATP – the form in which cells can utilize it.

3. The utilization of ATP by cells in order to perform useful work.

The chemical reactions which occur within organisms are collectively known as **metabolism**. They are of two types:

1. The build-up of complex compounds from simple ones. These synthetic reactions are collectively known as **anabolism**.

2. The breakdown of complex compounds into simple ones. Such reactions are collectively known as **catabolism**.

A typical chemical reaction may be represented as:

$$A \rightarrow B + C$$

In this case A represents the **substrate** and B and C are the **products**. If the entropy of C and B is greater than A then the reaction will proceed naturally in the direction shown. A reaction which involves an increase in entropy is said to be **spontaneous**. The free energy of the products is less than that of the substrate. The word spontaneous could be misleading because the reaction is not instantaneous. Before any chemical reaction can proceed it must initially be activated, i.e. its energy must be increased. The energy required is called the **activation energy**. Once provided, the activation energy allows the products to be formed with a consequent loss of free energy and increase in entropy (Fig. 13.1). Chemical reactions are reversible and therefore C and B can be synthesized into A. Such a reaction is not, however, spontaneous and requires an external source of energy if it is to proceed. Most biological processes are in fact a cycle of reversible reactions. Photosynthesis and respiration, for example, are basically the same reaction going in opposite directions.

$$\text{Energy} + 6CO_2 + 6H_2O \underset{\text{respiration}}{\overset{\text{photosynthesis}}{\rightleftarrows}} C_6H_{12}O_6 + 6O_2$$

As there is inevitably some loss of free energy in the form of heat each time the reaction is reversed, the process cannot

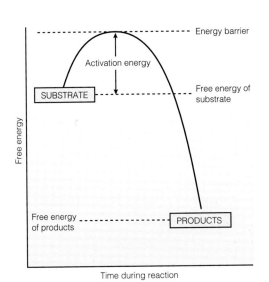

Fig. 13.1 Concept of activation energy

continue without a substantial input of energy from outside the organisms. The ultimate source of this energy is the light radiation of the sun. The way in which organisms obtain their energy for metabolic and other processes is probably more important in determining their design than any other single factor. The fundamental differences between plants and animals are a result of their modes of nutrition.

Plants obtain their energy from the sun and use it to combine carbon dioxide and water in the synthesis of organic molecules. As the raw materials are readily available almost everywhere, there is no necessity for plants to move to obtain their nutrients. Indeed, in order to obtain sufficient light plants need to have a large surface area. They therefore need to be as large as possible in order to compete with other plants for light. For this reason many plants are large. Locomotion for these plants would not only be difficult and slow, it would also be very energy-consuming. Plants therefore do not exhibit locomotion.

Animals obtain their energy from complex organic compounds. These occur in other organisms which must be sought. Most animals therefore exhibit locomotion in order to obtain their food. To help animals move from place to place they have developed a wide range of locomotory mechanisms. They are therefore more complex, and variable, in their design than plants. In carrying out locomotion, animals require a complex nervous system to coordinate their actions and a range of sense organs to help them to manoeuvre and to locate food.

Being sessile, plants do not require nervous systems and sense organs. Having a large surface area to obtain light energy means that plants have no need for separate specialized surfaces for obtaining respiratory gas. Any additional oxygen needed, over and above that produced in photosynthesis, can be obtained by diffusion through the leaves and roots which already provide a substantial surface area. Animals, by contrast, being compact to assist locomotion, require specialized respiratory surfaces with a large surface area to compensate for their small external area.

Even in reproduction, the differences between plants and animals can be related to the means by which they obtain energy. As their method of nutrition favours being stationary, plants have to use an external agent to transfer the male gametes from one individual to another during sexual reproduction. Insects and wind are the main agents of pollination. Animals, being capable of locomotion, utilize it in finding a mate and therefore male gametes are either introduced directly into the females, as in terrestrial organisms, or released in the vicinity of the female, as in some aquatic ones. In either case, male and female are in close proximity.

In this section of the book, we shall look at how energy is obtained and utilized by organisms.

14 Autotrophic nutrition (photosynthesis)

In Chapter 13 we saw that living systems differ from non-living ones in their ability to replace lost energy from the environment and so maintain themselves in an ordered condition (low entropy). Photosynthesis is the means by which this energy is initially obtained by living systems. All life is directly or indirectly dependent on this most fundamental process in living organisms. It provides part of the air we breathe, the food we eat and the fossil fuels we burn.

Autotrophic (*auto* – 'self'; *trophic* – 'feeding') organisms use an inorganic form of carbon, such as carbon dioxide, to make up complex organic compounds. These complex compounds are more ordered and so possess more energy. In autotrophs this energy is provided from two sources: light and chemicals. The processes involved are termed photosynthesis and chemosynthesis respectively.

Photosynthesis is much the more common of the two processes. It is principally important because:

1. It is the means by which the sun's energy is captured by plants for use by all organisms.

2. It provides a source of complex organic molecules for heterotrophic organisms.

3. It releases oxygen for use by aerobic organisms.

AUTOTROPHIC NUTRITION

light energy   chemical energy

PHOTOSYNTHESIS   CHEMOSYNTHESIS
All green plants   Certain bacteria

14.1   Leaf structure

The leaf is the main photosynthetic structure of a plant, although stems, sepals and other parts may also photosynthesize. It is adapted to bring together the three raw materials, water, carbon dioxide and light, and to remove the products oxygen and glucose. The structure of the leaf is shown in Fig. 14.1, on page 263.

Considering that all leaves carry out the same process, it is perhaps surprising that they show such a wide range of form. This range of form is often the consequence of different environmental conditions which have nothing directly to do with photosynthesis. In dry areas, for example, leaves may be small in size with thick cuticles and sunken stomata to help reduce water loss. The presence of spines to deter grazing by herbivores is not uncommon. Other differences in leaf shape are a result of the plant living in a sunny or shady situation.

Did you know?

Leaves of the Raphia palm found in tropical forests can be 22 metres long.

The equation for photosynthesis may be summarized as:

$$6CO_2 + 6H_2O + sunlight \xrightarrow{chlorophyll} C_6H_{12}O_6 + 6O_2$$

$$carbon\ dioxide + water + sunlight \xrightarrow{chlorophyll} glucose + oxygen$$

$$gas + liquid + energy \xrightarrow{chlorophyll} liquid + gas\ (solution\ in\ water)$$

The adaptations of the leaf to photosynthesis are therefore:

1. To obtain energy (sunlight).

2. To obtain and remove gases (carbon dioxide and oxygen).

3. To obtain and remove liquids (water and sugar solution).

14.1.1 Adaptations for obtaining energy (sunlight)

As sunlight is the energy source which drives the photosynthetic process, it is often the factor which determines the rate of photosynthesis. To ensure its efficient absorption the leaf shows many adaptations:

1. **Phototropism** causes shoots to grow towards the light in order to allow the attached leaves to receive maximum illumination.

2. **Etiolation** causes rapid elongation of shoots which are in the dark, to ensure that the leaves are brought up into the light as soon as possible.

3. **Leaves arrange themselves into a mosaic**, i.e. they are arranged on the plant in a way that minimizes overlapping and so reduces the degree of shading of one leaf by another.

4. **Leaves have a large surface area** to capture as much sunlight as possible. They are held at an angle perpendicular to the sun during the day to expose the maximum area to the light. Some plants, e.g. the compass plant, actually 'track' the sun by moving their leaves so they constantly face it during the day.

5. **Leaves are thin** – If they were thicker, the upper layers would filter out all the light and the lower layers would not then photosynthesize.

6. **The cuticle and epidermis are transparent** to allow light through to the photosynthetic mesophyll beneath.

7. **The palisade mesophyll cells are packed with chloroplasts** and arranged with their long axes perpendicular to the surface. Although there are some air spaces between them, they still form a continuous layer which traps most of the incoming light. In some plants this layer is more than one cell thick.

8. **The chloroplasts within the mesophyll cells can move** – This allows them to arrange themselves into the best positions within a cell for the efficient absorption of light.

9. **The chloroplasts hold chlorophyll in a structured way** – The chlorophyll within a chloroplast is contained within the grana, where it is arranged on the sides of a series of unit membranes. The ordered arrangement not only presents the maximum

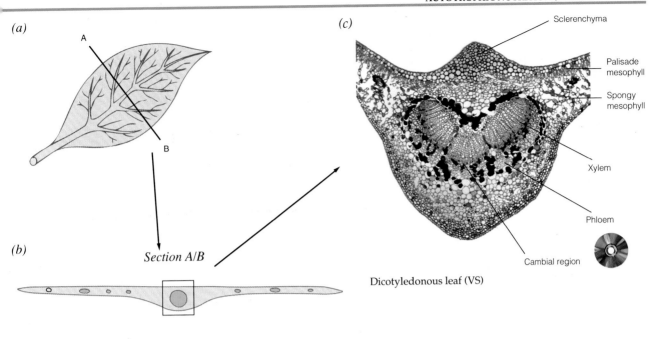

(a)

(b)

Section A/B

(c)

Sclerenchyma

Palisade mesophyll

Spongy mesophyll

Xylem

Phloem

Cambial region

Dicotyledonous leaf (VS)

(d) Dicotyledonous leaf (VS) (×40 approx.)

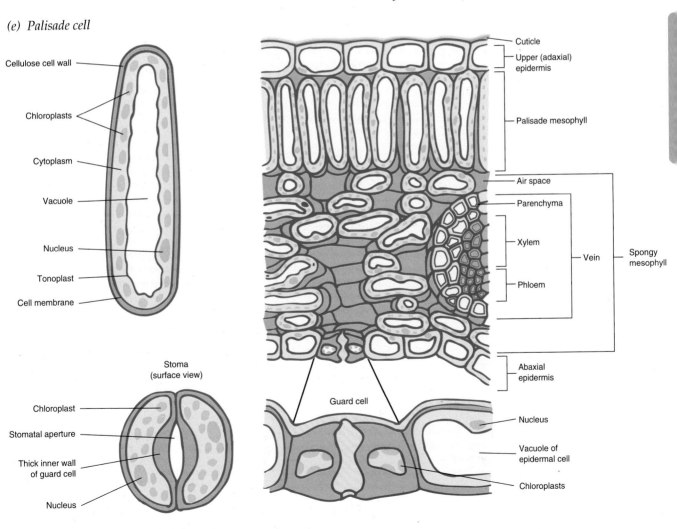

(e) Palisade cell

Cellulose cell wall

Chloroplasts

Cytoplasm

Vacuole

Nucleus

Tonoplast

Cell membrane

Stoma (surface view)

Chloroplast

Stomatal aperture

Thick inner wall of guard cell

Nucleus

Cuticle

Upper (adaxial) epidermis

Palisade mesophyll

Air space

Parenchyma

Xylem

Phloem

Vein

Spongy mesophyll

Abaxial epidermis

Nucleus

Vacuole of epidermal cell

Chloroplasts

Guard cell

Fig. 14.1 The structure of the leaf (continued on next page)

263

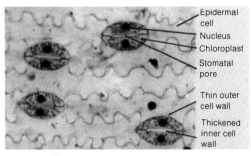

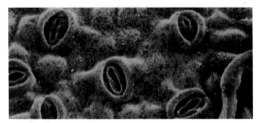

(g) Surface view of stomata

Stomata in surface view (scanning EM) (× 600 approx.)

(h)

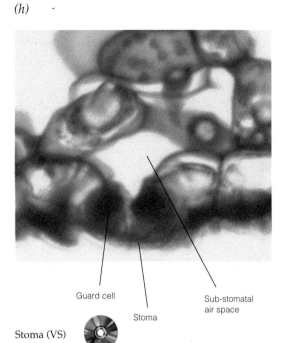

Guard cell   Stoma   Sub-stomatal air space

Stoma (VS)

Fig. 14.1 The structure of the leaf (continued)

amount of chlorophyll to the light but also brings it in close proximity to other pigments and substances which are necessary for its functioning. The structure of a chloroplast is shown in Fig. 4.5, Section 4.2.4.

14.1.2 Adaptations for obtaining and removing gases

As gases diffuse relatively rapidly, the leaf has no special transport mechanism for carbon dioxide and oxygen. The leaf does, however, show a number of adaptations which ensure rapid diffusion of these gases:

1. **Numerous stomata** are present in the epidermis of leaves. There may be tens of thousands per cm^2 of leaf surface, which itself represents a very considerable area. Stomata are minute pores in the epidermis which, when open, permit unrestricted diffusion into and out of the leaf.

2. **Stomata can be opened and closed** – Plants need to be relatively impermeable to gases in order to prevent water loss, and yet they need the free entry of carbon dioxide for photosynthesis. To overcome this problem, they have stomatal pores which are bounded by two **guard cells**. Alterations in the turgidity of these cells open and close the stomatal pore, thus controlling the uptake of carbon dioxide and the loss of water. The detail of stomatal control is given in Section 22.4.1. Stomata open in conditions which favour photosynthesis and at this time some water loss is unavoidable. When photosynthesis cannot take place, e.g. at night, they close, thus reducing considerably the loss of water. At times of considerable water loss, the stomata may close anyway, regardless of the demands for carbon dioxide.

3. Spongy mesophyll possesses many airspaces – The mesophyll layer on the underside of the leaf has many air spaces. These communicate with the palisade layer and the stomatal pores. There is hence an uninterrupted diffusion of gases between the atmosphere and the palisade mesophyll. During photosynthesis carbon dioxide diffuses in and oxygen out of this layer. The air spaces avoid the need for these gases to diffuse through the cells themselves, a process which would be much slower. The palisade mesophyll also possesses air spaces to permit rapid diffusion around the cells of which it is made.

14.1.3 Adaptations for obtaining and removing liquids

As water is a liquid raw material for photosynthesis and as the sugar produced is carried away in solution, the leaf has to be adapted for the efficient transport of liquids.

1. A large central midrib is possessed by most dicotyledonous leaves. This contains a large vascular bundle comprising xylem and phloem tissue. The xylem permits water and mineral salts to enter the leaf and the phloem carries away sugar solution, usually in the form of sucrose.

2. A network of small veins is found throughout the leaf. These ensure that no cell is ever far from a xylem vessel or phloem sieve tube, and hence all cells have a constant supply of water for photosynthesis and a means of removing the sugars they produce. The xylem, and any sclerenchyma associated with the vascular bundle, also provide a framework of support for the leaf, helping it to present maximum surface area to the light.

14.2 Mechanism of light absorption

14.2.1 The nature of light

There are three features of light which make it biologically important:

1. Spectral quality (colour).

2. Intensity (brightness).

3. Duration (time).

To be of use as an energy source for organisms, light must first be converted to chemical energy. Radiant energy comes in discrete packets called **quanta** (Planck's quantum theory). A single quantum of light is called a **photon**. Light also has a **wave** nature and so forms a part of the electromagnetic spectrum. Visible light represents that part of this spectrum which has a wavelength between 400 nm (violet) and 700 nm (red).

The wavelength of light is the distance between successive peaks along a wave (Fig. 14.2) and this is inversely proportional to its frequency, i.e. the smaller the wavelength, the greater the frequency. The amount of energy is inversely proportional to the wavelength, i.e. light with a short wavelength has more energy than light with longer wavelengths. In other words, a photon of blue light (wavelength 400 nm) has more energy than a photon of red light (wavelength 700 nm).

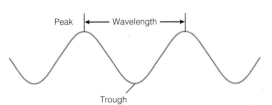

Fig. 14.2 Wave function of light

265

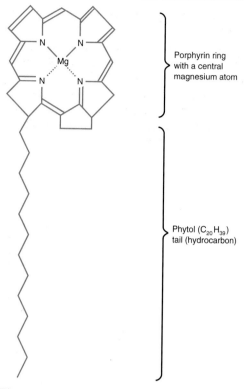

Fig. 14.4 *The general shape of the chlorophyll molecule*

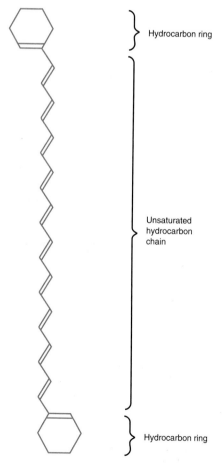

Fig. 14.5 *The general shape of a carotenoid molecule*

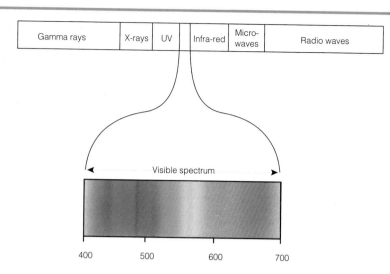

Fig. 14.3 *The visible section of the electromagnetic spectrum*

14.2.2 The photosynthetic pigments

A variety of pigments are involved in photosynthesis of which **chlorophyll** is by far the most important. There are a number of different chlorophylls, with chlorophylls *a* and *b* being the most common. Chlorophylls absorb light in the blue-violet and the red regions of the visible spectrum. The remaining light, in the green region of the spectrum, is reflected and gives chlorophyll its characteristic colour. All chlorophylls comprise a complex ring system called a **porphyrin ring** and a long hydrocarbon 'tail' (Fig. 14.4). This 'tail' is lipid soluble (hydrophobic) and is therefore embedded in the thylakoid membrane. The porphyrin ring is hydrophilic and lies on the membrane surface.

There is a second group of pigments involved in photosynthesis – the **carotenoids**. There are many types but their basic structure comprises two small rings linked by a long hydrocarbon chain (Fig. 14.5). The colour, which ranges from pale yellow through orange to red, depends upon the number of double bonds in the chain. The greater the number of double bonds, the deeper the colour.

The colour of carotenoids is normally masked in photosynthetic tissues by chlorophyll. Their colour does, however, become apparent when chlorophyll breaks down prior to leaf fall. The characteristic red, orange and yellow colours of leaves in the autumn are attributable to carotenoids as are many flower and fruit colours. They absorb light in the blue-violet range of the spectrum. There are two main types of carotenoids: the **carotenes** and the **xanthophylls**. A common example of a carotene is β-carotene which gives carrots their familiar orange colour. It is easily formed into two molecules of vitamin A, making carrots a useful source of this vitamin to animals.

14.2.3 Absorption and action spectra

If a pigment such as chlorophyll is subjected to different wavelengths of light, it absorbs some more than others. If the degree of absorption at each wavelength is plotted, an

absorption spectrum of that pigment is obtained. The absorption spectra for chlorophylls *a* and *b* are given in Fig. 14.6.

An **action spectrum** plots the biological effect of different wavelengths of light – in this case the effectiveness of different wavelengths of light in bringing about photosynthesis. As Fig. 14.6 shows, the action spectrum for photosynthesis is closely correlated to the absorption spectra for chlorophylls *a* and *b* and carotenoids. This suggests these pigments are those responsible for absorbing the light used in photosynthesis.

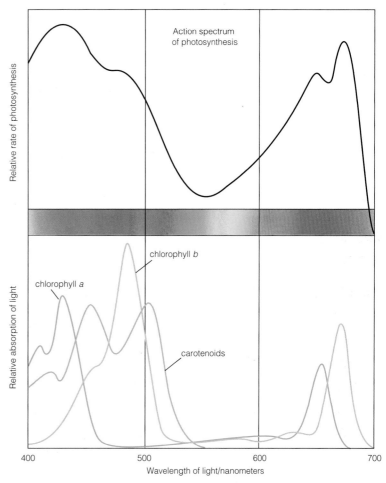

Fig. 14.6 Action spectrum for photosynthesis and absorption spectra for common plant pigments

14.3   Mechanism of photosynthesis

The overall equation for photosynthesis is:

$$6CO_2 + 6H_2O \xrightarrow[\text{chlorophyll}]{\text{sunlight}} C_6H_{12}O_6 + 6O_2$$

carbon dioxide + water ⟶ glucose + oxygen

Photosynthesis is essentially a process of energy transduction. Light energy is firstly converted into electrical energy and finally into chemical energy. It has three main phases:

1. **Light harvesting**. Light energy is captured by the plant using a mixture of pigments including chlorophyll.

2. **The light dependent stage (photolysis)** in which a flow of electrons results from the effect of light on chlorophyll and so causes the splitting of water into hydrogen ions and oxygen.

3. **The light independent stage** during which these hydrogen ions are used in the **reduction of carbon dioxide** and hence the manufacture of sugars.

14.3.1 Light harvesting

Within the thylakoid membranes of the chloroplast, chlorophyll molecules are arranged along with their accessory pigments into groups of several hundred molecules. Each group is called an **antenna complex**. Special proteins associated with these pigments help to funnel photons of light entering the chloroplast

NOTEBOOK

Chromatography

Chromatography is a means of separating one type of molecule from another. It involves moving the mixture, normally as a liquid or a gas, over a stationary phase embedded in cellulose or silica. The separation may depend on a range of chemical and physical properties of the molecules such as solubility and molecular mass.

Essentially there are two basic ways of carrying out the separation. **Paper chromatography** is often used in schools and colleges to separate photosynthetic pigments, sugars or amino acids. The mixture is 'spotted' near one end of a paper strip and then dipped into a solvent which moves up the paper by capillarity, carrying the molecules with it.

Instead of using paper a thin layer of silica may be formed on an inert solid support. This is called **thin layer chromatography**.

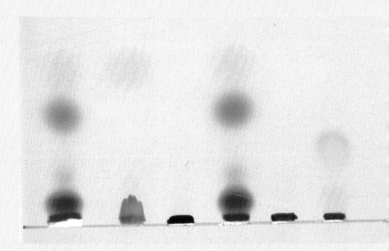

Chromatography plate at end of run

on to special molecules of chlorophyll *a*, known as the **reaction centre chlorophyll molecule**. On striking this molecule an electron in its orbit is raised to a higher energy level, thus initiating a flow of electrons.

There are two types of reaction centre which differ in both their chlorophylls and their functions. These are known as **photosystem I (PSI)** and **photosystem II (PSII)**.

14.3.2 Light dependent stage (photolysis)

The light dependent stage of photosynthesis occurs in the thylakoids of the chloroplasts and involves the splitting of water by light – **photolysis of water**. In the process, ADP is converted to ATP. This addition of phosphate is termed **phosphorylation** and as light is involved it is called **photophosphorylation**. These processes are brought about by two photochemical systems which are summarized in Figs. 14.7 and 14.8.

In the process summarized in Fig. 14.8, electrons from chlorophyll are passed into the light independent reaction via $NADPH + H^+$. They are replaced by electrons from another source – the water molecule. The same electrons are *not* recycled

Column chromatography

The second, more commonly used method involves the mobile phase flowing over a supporting matrix held in a glass or metal tube. This is known as **column chromatography**. There have been many recent advances in the development of new matrices so that the liquid can now be pumped through under high pressure and very small fractions can be separated in miniature columns.

If chromatography is to be a really useful biochemical tool it is necessary to link separation to detection. This may be simply on the basis of colour, as with photosynthetic pigments.

Proteins, peptides and nucleic acids can be detected by their ability to absorb light in the ultra-violet region of the spectrum. Other molecules, such as amino acids, are colourless but can be made to form coloured derivatives if treated with particular chemicals, e.g. ninhydrin causes amino acids to form purple derivatives.

In paper chromatography the identification of a particular molecule is usually made on the basis of the distance travelled by the substance in relation to the distance moved by the solvent. Each molecule can then be referred to by its R_f value (retardation factor) expressed as:

$$\frac{\text{Distance travelled by a compound}}{\text{Distance travelled by solvent front}}$$

If a mixture of radioactive compounds is separated the molecules can be detected by their emission of radioactivity (see *Using isotopes as tracers* on page 4). This was the basis of Calvin's work on the light independent stages of photosynthesis (page 271).

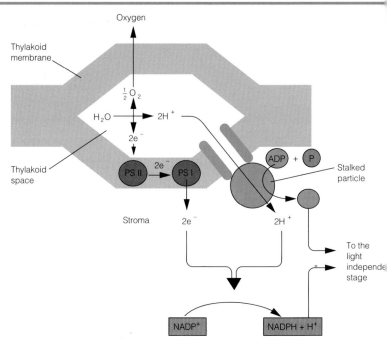

Fig. 14.7 Summary of the events of the light dependent stage and their locations within the chloroplast

back into the chlorophyll. This method of ATP production is thus called **non-cyclic photophosphorylation**. It is alternatively called the **Z-scheme** because the zig-zag route of the electrons in the diagram resembles a Z on its side.

There is a second method by which ATP can be generated. The electrons from the pigment system may return to the chlorophyll directly, via the electron carrier system, forming ATP in the process. Such electrons are recycled, harnessing energy from light and generating ATP. This is called **cyclic photophosphorylation** and involves only photosystem I. No reduced NADP is produced during cyclic photophosphorylation.

1. *Light energy is trapped in photosystem II and boosts electrons to a higher energy level.*
2. *The electrons are received by an electron acceptor.*
3. *The electrons are passed from the electron acceptor along a series of electron carriers to photosystem I. The energy lost by the electrons is captured by converting ADP to ATP. Light energy has thereby been converted to chemical energy.*
4. *Light energy absorbed by photosystem I boosts the electrons to an even higher energy level.*
5. *The electrons are received by another electron acceptor.*
6. *The electrons which have been removed from the chlorophyll are replaced by pulling in other electrons from a water molecule.*
7. *The loss of electrons from the water molecule causes it to dissociate into protons and oxygen gas.*
8. *The protons from the water molecule combine with the electrons from the second electron acceptor and these reduce **nicotinamide adenine dinucleotide phosphate**.*
9. *Some electrons from the second acceptor may pass back to the chlorophyll molecule by the electron carrier system, yielding ATP as they do so. This process is called **cyclic photophosphorylation**.*

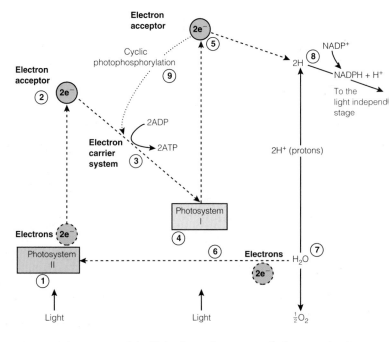

Fig. 14.8 Summary of the light dependent stage of photosynthesis

14.3.3 The light independent stage

The light independent stage of photosynthesis occurs in the stroma of the chloroplasts and it takes place whether or not light is present. The details of this stage were analysed by Melvin Calvin and his co-workers and the process is often called the **Calvin cycle** (Figs. 14.9 and 14.10). It is basically the reduction of carbon dioxide using the reduced nicotinamide adenine dinucleotide phosphate (NADPH + H$^+$) and ATP from the light reaction. The carbon dioxide is initially fixed by combining it with a 5-carbon compound – **ribulose bisphosphate** with the aid of an enzyme called **ribulose bisphosphate carboxylase oxygenase** – thankfully abbreviated to **RUBISCO**.

1. Carbon dioxide diffuses into the leaf through the stomata and dissolves in the moisture on the walls of the palisade cells. It diffuses through the cell membrane, cytoplasm and chloroplast membrane into the stroma of the chloroplast.
2. The carbon dioxide combines with a 5-carbon compound called **ribulose bisphosphate** to form an unstable 6-carbon intermediate.
3. The 6-carbon intermediate breaks down into two molecules of the 3-carbon **glycerate 3-phosphate (GP)**.
4. Some of the ATP produced during the light dependent stage is used to help convert GP into **triose phosphate** (glyceraldehyde 3-phosphate – GALP).
5. The reduced nicotinamide adenine dinucleotide phosphate (NADPH + H$^+$) from the light dependent reaction is necessary for the reduction of the GP to triose phosphate. NADP$^+$ is regenerated and this returns to the light dependent stage to accept more hydrogen.
6. Pairs of triose phosphate molecules are combined to produce an intermediate hexose sugar.
7. The hexose sugar is polymerized to form starch which is stored by the plant.
8. Not all triose phosphate is combined to form starch. A portion of it is used to regenerate the original carbon dioxide acceptor, ribulose bisphosphate. Five molecules of the 3-carbon triose phosphate can regenerate three molecules of the 5-carbon ribulose bisphosphate. More of the ATP from the light dependent reaction is needed to provide the energy for this conversion.

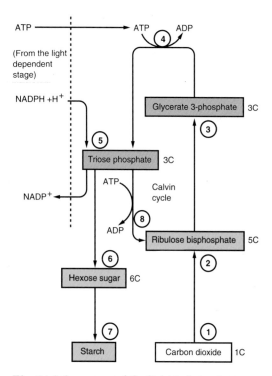

Fig. 14.9 Summary of the light independent stage of photosynthesis

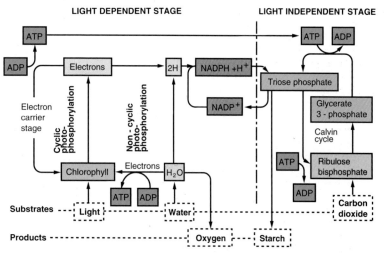

Fig. 14.10 Summary of photosynthesis

14.3.4 Fate of photosynthetic products

From the products of photosynthesis a totally autotrophic plant must synthesize all organic materials necessary for its survival. Many of the raw materials for the synthesis of these organic molecules come from intermediates of the Calvin cycle.

Synthesis of other carbohydrates
The triose phosphate of the Calvin cycle can be synthesized into hexose sugars such as glucose and fructose by a reversal of the stages of glycolysis which occur during respiration (Section 16.2). The monosaccharides glucose and fructose may be combined to give the disaccharide **sucrose** which is the main form in which carbohydrate is transported throughout the plant in the phloem. The glucose may, on the other hand, be polymerized into **starch** for storage. In some plants, e.g. *Dahlia*, it is the fructose which is polymerized, in which case the storage polysaccharide formed is **inulin**. The glucose may alternatively be polymerized into another polysaccharide, **cellulose**. This makes up over 50% of plant cell walls. The chemical structure of polysaccharides is given in Section 2.5.

Synthesis of lipids
As we saw in Section 2.6, all lipids are esters of fatty acids, with glycerol as the most common alcohol found. Glycerate 3-phosphate (GP) may be converted to acetyl coenzyme A which in turn is used to synthesize a variety of fatty acids in chloroplasts as well as cytoplasm. Triose phosphate is easily converted into glycerol. The lipids are formed by combination of appropriate fatty acids with glycerol. As well as being important storage substances, especially in seeds, lipids are a major constituent of cell membranes and their waxy derivatives make up the waterproofing cuticle. Fatty acids provide some flower scents which are used to attract insects for pollination.

Synthesis of proteins
Conversion of the glycerate 3-phosphate of the Calvin cycle into acetyl coenzyme A is the starting point for amino acid synthesis. The acetyl CoA enters the Krebs cycle and from its intermediates a wide variety of amino acids can be made by transamination reactions. The nitrogen necessary for these reactions is derived from the nitrates absorbed by plant roots and subsequently reduced. The amino acids are polymerized into proteins. Details of all these processes are given in Section 7.6.

Proteins are essential for growth and development and make up a major structural component of the cell, especially the cell membrane. All enzymes are proteins and they may also be used as storage material.

14.4 Factors affecting photosynthesis

The rate of photosynthesis is affected by a number of factors, the level of which determine the yield of material by a plant. Before reviewing these factors it is necessary to understand the principle of limiting factors.

14.4.1 Concept of limiting factors

In 1905, F. F. Blackman, a British plant physiologist, measured the rate of photosynthesis under varying conditions of light and carbon dioxide supply. As a result of his work he formulated the **principle of limiting factors**. It states: **At any given moment, the rate of a physiological process is limited by the one factor which is in shortest supply, and by that factor alone**.

In other words, it is the factor which is nearest its minimum value which determines the rate of a reaction. Any change in the level of this factor, called the **limiting factor**, will affect the rate of the reaction. Changes in the level of other factors have no effect. To take an extreme example, photosynthesis cannot proceed in the dark because the absence of light limits the process. The supply of light will alter the rate of photosynthesis – more light, more photosynthesis. If, however, more carbon dioxide or a higher temperature is supplied to a plant in the dark, there will be no change in the rate of photosynthesis. Light is the limiting factor, therefore only a change in its level can affect the rate.

PROJECT

All of these factors can be investigated, using hydrogencarbonate indicator solution and leaves or aquatic plants. You can also use a colorimeter and appropriate filters to distinguish between fine differences in colour of the indicator solution.

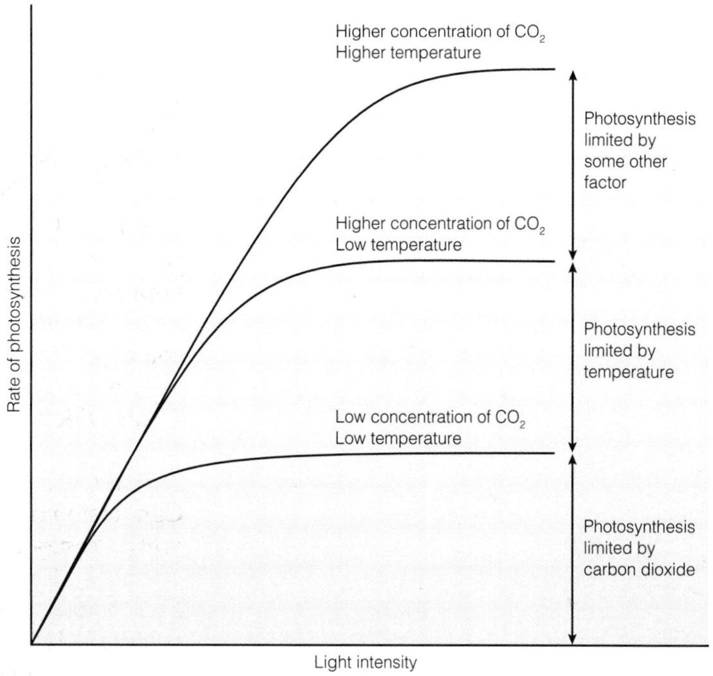

Fig. 14.11 The concept of limiting factors as illustrated by the levels of different conditions on the rate of photosynthesis

If the amount of light given to a plant is increased, the rate of photosynthesis increases up to a point and then tails off. At this point some other factor, such as the concentration of carbon dioxide, is in short supply and so limits the rate. An increase in carbon dioxide concentration again increases the amount of photosynthesis until some further factor, e.g. temperature, limits the process. These changes are illustrated in Fig. 14.11.

14.4.2 Effect of light intensity on the rate of photosynthesis

The rate of photosynthesis is often measured by the amount of carbon dioxide absorbed or oxygen evolved by a plant. These forms of measurement do not, however, give an absolute measure of photosynthesis because oxygen is absorbed and carbon dioxide is evolved as a result of cellular respiration. As light intensity is increased, photosynthesis begins, and some carbon dioxide from respiration is utilized in photosynthesis and so less is evolved. With a continuing increase in light intensity a point is reached where carbon dioxide is neither evolved nor absorbed. At this point the carbon dioxide produced in respiration exactly balances that being used in photosynthesis. This is the **compensation point**. Further increases in light intensity result in a proportional increase in the rate of photosynthesis until **light saturation** is reached. Beyond this point further increases in light intensity have no effect on the rate of photosynthesis. If, however, more carbon dioxide is made available to the plant further increases in light intensity do increase the rate of photosynthesis until light saturation is again reached, only this time at a higher light intensity. At this point the carbon dioxide concentration, or possibly some new factor such as temperature, is limiting the process. These relationships are represented graphically in Fig. 14.12.

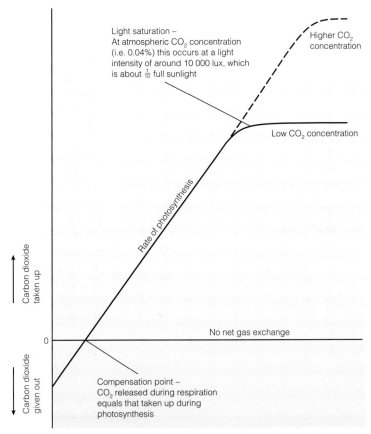

Fig. 14.12 Graph showing the effect of light intensity on the rate of photosynthesis, as measured by the amount of CO_2 exchanged

14.4.3 Effect of carbon dioxide concentration on the rate of photosynthesis

Carbon dioxide is one of the less common gases in the atmosphere. At its normal concentration of 0.04% it is present at only one fifteenth of the concentration of argon, a so-called 'rare' gas. Despite occasional local fluctuations in carbon dioxide concentration, there is remarkable consistency in its concentration throughout the atmosphere at any one time. This consistency is the result of a fine balance between carbon dioxide taken up by green plants during photosynthesis and that released by all organisms during respiration. There has been a slight increase in the carbon dioxide concentration in the atmosphere in recent times and this has been attributed to the burning of fossil fuels. Coal and oil represent stores of carbon resulting from the fossilization of organisms. Burning these fuels oxidizes the carbon to carbon dioxide and so increases its concentration in the atmosphere.

The normally low atmospheric carbon dioxide concentration of 0.04% is a major limiting factor to photosynthesis. The optimum concentration for a sustained high rate of photosynthesis is 0.1% and some greenhouse crops like tomatoes are grown in carbon dioxide enriched environments to provide greater yields. The effect on the rate of photosynthesis of increasing the carbon dioxide concentration is illustrated in Figs. 14.11 and 14.12.

14.4.4 Effect of temperature on the rate of photosynthesis

We have seen that the photochemical reaction or light stage of photosynthesis is unaffected by temperature, but that the light independent (Calvin cycle) is temperature dependent. Provided the light intensity and concentration of carbon dioxide are not limiting, the rate of photosynthesis is found to increase proportionately with an increase in temperature. The minimum temperature at which photosynthesis can take place is 0 °C for most plants, although some arctic and alpine varieties continue to do so below this level. The rate of photosynthesis at these temperatures is very low. The rate approximately doubles for each rise of 10 °C up to an optimum temperature, which varies from species to species. Above the optimum temperature, the rate of increase is reduced until a point is reached above which there is no increase in photosynthesis. The optimum photosynthetic rate for most plants is around 25 °C. Above these levels further temperature increases lead to a levelling off and then a fall in the rate of photosynthesis. The fall occurs at temperatures too low for it to be entirely accounted for by the denaturation of enzymes.

14.4.5 Effect of inorganic ions on the rate of photosynthesis

In the absence of certain inorganic ions, such as iron, chlorophyll cannot be synthesized. Other ions, like nitrogen and magnesium, are an integral part of the chlorophyll molecule and their absence likewise prevents its formation. Where plants are grown on soils deficient in any one of these minerals, the chlorophyll concentration is reduced and the leaves become yellow, a condition called **chlorosis**. Under these circumstances the rate of photosynthesis is substantially reduced.

14.4.6 Other factors affecting the rate of photosynthesis

Of the other factors which may affect the rate of photosynthesis, water is by far the most important. As a substrate in the process its deficiency will clearly reduce the rate of photosynthesis. The problem is that water has so many functions in a plant that it is impossible to directly relate its availability to the rate of photosynthesis. There are many specific chemical compounds which prevent photosynthesis, often by inhibiting enzyme action. Examples include cyanide and dichlorophenyl dimethyl urea (DCMU). Even certain pollutants such as sulphur dioxide are known to reduce photosynthetic rate.

14.5   Chemosynthesis

TABLE 14.1   **Examples of chemosynthetic bacteria**

Substrate	Main product	Example
Ammonium (NH_4^+)	Nitrite (NO_2^-)	*Nitrosomonas*
Nitrite (NO_2^-)	Nitrate (NO_3^-)	*Nitrobacter*
Sulphur (S)	Sulphate (SO_4^{2-})	*Thiobacillus*
Ferrous (Fe^{2+})	Ferric (Fe^{3+})	*Ferrobacillus*
Hydrogen (H_2)	Water (H_2O)	*Hydrogenomonas*

Some autotrophic bacteria carry out a similar process to photosynthesis but the energy is derived from the oxidation of inorganic chemicals and therefore takes place in the absence of light. The process is called **chemosynthesis**. Some examples are given in Table 14.1. Organisms using the oxidation of inorganic chemicals as a source of energy are known as **chemoautotrophs** in contrast to those using light which are called **photoautotrophs**.

Chemosynthetic bacteria perform a vitally important function in helping to recycle valuable minerals. Their rôles in the nitrogen and other cycles, details of which are given in Section 17.2.2, help to maintain soil fertility.

14.6   Questions

1. Investigations of photosynthesis have involved the use of isotopes, centrifugation and two-dimensional chromatography.

   (a) Describe, in outline only, the principles involved in the use of these techniques in such investigations.   (12 marks)
   (b) Explain how the use of each one of these techniques has increased our knowledge of photosynthesis.   (8 marks)
            (Total 20 marks)

   NEAB June 1990, Paper IB, No. 4

2. (a) Describe fully the way in which energy is trapped by green plants and stored by means of the light stage of photosynthesis.
            (16 marks)
   (b) Outline the ways in which the products of the light stage are directly and indirectly important for animals.   (4 marks)
            (Total 20 marks)

   WJEC June 1993, Paper A1, No. 1

3. Algae were supplied with a radioactive isotope of carbon, ^{14}C, and allowed to photosynthesize. After a period of time, the light was switched off and the algae left in the dark. The graph shows the relative amounts of some radioactively labelled compounds over the period of the experiment.

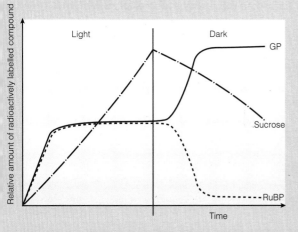

Explain the changes in relative amounts of each of the following substances after the light was switched off:

   (a) glycerate 3-phosphate, GP (phosphoglycerate, PGA);   (2 marks)
   (b) ribulose bisphosphate (RuBP);   (2 marks)
   (c) sucrose.   (1 mark)
            (Total 5 marks)

   AEB June 1991, Paper I, No. 5.

4. The important pigments in most chloroplasts are the yellow-green chlorophyll *a*, the blue-green chlorophyll *b* and the orange carotenoids (mainly carotene). The diagram below shows (left *y* axis) the *absorption spectra* of these pigments.

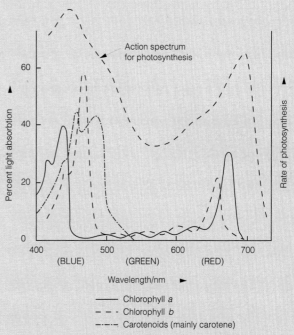

Absorption spectra of chlorophyll and carotenoid pigments and action spectrum for photosynthesis.

   (a) Using the information shown, state concisely one piece of evidence which suggests that absorption of white light by chloroplasts is not uniform over the whole spectrum.
            (1 mark)
   (b) Which colour of light is best absorbed by the carotenoids?   (1 mark)
   (c) Why do most plants characteristically have a green colour?   (1 mark)
   (d) Chlorophyll *a* and chlorophyll *b* are almost identical molecules, both functioning in photosynthesis. How do the absorption spectra of the two differ?   (2 marks)
   (e) Variegated leaves have non-green, non-photosynthetic parts and such areas may appear pale yellow to dark yellow-orange. Suggest a reason for these facts with regard to energy absorption and utilization.
            (2 marks)

The diagram also shows (right *y* axis) the *action spectrum* for photosynthesis which is the amount of photosynthesis occurring in a green plant when illuminated by lights of different wavelengths (but of equal intensity).

(f) Suggest one suitable method of measuring and thus obtaining the experimental results necessary to produce the action spectrum graph. *(1 mark)*

(g) Describe the relationship between the absorption spectrum and the action spectrum. *(2 marks)*

(h) Briefly indicate what role carotenoid pigments in plastids are thought to play in photosynthesis. *(2 marks)*

(i) If, in the laboratory, you wish to extract chlorophyll pigments you will need to use an organic solvent such as ether (ethoxyethane) or acetone (propanone). Such substances also dissolve lipids. Relate these two pieces of information to explain briefly why such a solvent is necessary in the extraction of the pigments. *(2 marks)*

(Total 14 marks)

Oxford June 1990, Paper I, No. 9

5. (a) The technique of paper chromatography can be used to separate and identify different photosynthetic pigments.

    Describe a method which could be used to prepare a solution of photosynthetic pigments from fresh geranium leaves. *(2 marks)*

    (b) The diagram below shows the result of a chromatography experiment using the solution of photosynthetic pigments.

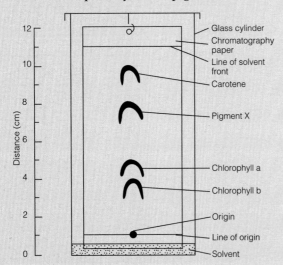

(i) Describe the technique used to apply the pigment solution to the point of origin. *(2 marks)*

(ii) Why has carotene a position closer to the solvent front than the other pigments? *(1 mark)*

(iii) The R_f value can be used to identify the different pigments.

$$R_f = \frac{\text{distance travelled by pigment front from origin}}{\text{distance travelled by solvent front from origin}}$$

Which pigment has an R_f value of 0.4? *(1 mark)*

(iv) Name pigment X. *(1 mark)*

(c) It has been suggested that leaves of the copper beech have different photosynthetic pigments from those occurring in oak leaves.

Design an investigation to test this hypothesis. *(4 marks)*

(Total 11 marks)

SEB (revised) May 1991 Higher Grade, Paper I, No. 14

6. The apparatus illustrated below may be used to measure the rate of photosynthesis of the aquatic plant *Elodea*.

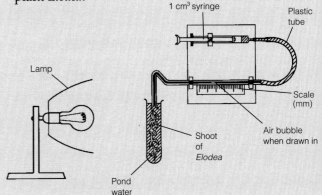

(a) Suggest **one** item of equipment which could be added to the apparatus shown to improve the reliability of the experiment, and give its purpose. *(2 marks)*

(b) Why is a small volume of sodium hydrogencarbonate solution usually added to the pond water before photosynthetic rates are measured? *(1 mark)*

(c) State **two** precautions that would be necessary to ensure that the results obtained by different students with the same piece of apparatus, were comparable. *(2 marks)*

Distance of lamp from plant/cm	Length of bubble after 5 minutes/mm	Mean length of bubble after 5 minutes/mm
5	60, 55, 28, 62, 63, 68.	
10	3, 36, 39, 33, 24, 28.	
20	7, 10, 5, 6, 9, 8.	
40	1, 3, 1, 2, 1, 4.	

he results shown in the table were collected from a umber of students carrying out an experiment to etermine the effect of altering the light intensity on he rate of photosynthesis.

(d) (i) Identify any anomalous results in the table and draw a box around each.

(ii) Excluding these anomalous results from your calculations, find the mean bubble length for **each** distance and enter this value in the table.   (*3 marks*)

(e) (i) What is the relationship between the distance of the lamp from the plant and the mean bubble length over the range of 10 to 40 cm distance?

(ii) Do the 5 cm results fit this relationship? Explain your answer.   (*3 marks*)

(f) (i) Explain how you would calculate the **volume** of gas produced per hour from these values.

(ii) Suggest **one** possible reason why the volume of gas collected in this experiment may not truly represent the rate of photosynthesis.   (*2 marks*)

(*Total 13 marks*)

NEAB June 1992, Paper IIA, No. 3

An investigation was carried out into the effect of arbon dioxide concentration and light intensity on he productivity of lettuces in a glasshouse. The roductivity was determined by measuring the rate of arbon dioxide fixation in milligrams per dm^2 leaf rea per hour.

xperiments were conducted at three different light ntensities, 0.05, 0.25 and 0.45 (arbitrary units), the ighest approximating to full sunlight. A constant emperature of 22 °C was maintained throughout.

he results are given in the table below.

Carbon dioxide concentration/ ppm	Productivity at different light intensities/mg $dm^{-2}h^{-1}$		
	At 0.05 units light intensity	At 0.25 units light intensity	At 0.45 units light intensity
300	12	25	27
500	14	30	36
700	15	35	42
900	15	37	46
1100	15	37	47
1300	12	31	46

(a) For the experiment at 0.25 units light intensity, describe and comment on the effect on the productivity of the lettuces of increasing carbon dioxide concentration in the range

(i) 300 to 900 ppm   (*2 marks*)

(ii) 900 to 1300 ppm   (*2 marks*)

(b) (i) A carbon dioxide concentration of 300 ppm is approximately equivalent to that in atmospheric air.

For each of the three light intensities, work out the maximum increase in productivity that was obtained compared with that at 300 ppm and use it to calculate the percentage increase in productivity at each light intensity.

1 At 0.05 units light intensity
2 At 0.25 units light intensity
3 At 0.45 units light intensity   (*3 marks*)

(ii) Comment on the effect on productivity of changing light intensity.   (*2 marks*)

(c) Explain why the carbon dioxide concentration affects the productivity of plants.   (*3 marks*)

(d) State why the temperature should be kept constant during this experiment.   (*1 mark*)

(e) Suggest why, even with artificial lighting, glasshouse crops generally need to have more carbon dioxide added when temperatures are low, than when temperatures are high.   (*2 marks*)

(*Total 15 marks*)

ULEAC June 1992, Paper I, No. 12

8. The diagram below shows part of a leaf.

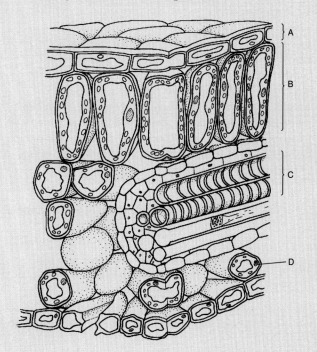

(a) (i) Name the cells A, B and C.   (*3 marks*)

(ii) Name a structure present in cell B but absent from cell A.   (*1 mark*)

(b) Describe how water moves from C into D and then into an air space.   (*3 marks*)

(c) Name *two* substances which diffuse out through stomata in the light. (*2 marks*)

(d) State *two* ways in which the shape of cell B is adapted for its functions in a leaf. (*2 marks*)

(*Total 11 marks*)

London January 1990, Paper I, No. 1

9. Read through the following account of photosynthesis, then write on the dotted lines the most appropriate word or words to complete the account.

Photosynthesis is a type of .... nutrition, involving the synthesis of organic molecules from inorganic materials. The process involves two types of reactions, light-dependent and light-independent.

In the light-dependent reactions, light energy is absorbed by chlorophyll molecules located on the .... of the chloroplasts; .... and .... are produced and oxygen gas is given off a by-product.

In the light-independent reactions, ... accepts molecules of carbon dioxide, which together with the products of the light-dependent reactions, results in the formation of .... This compound can be converted to .... or used to regenerate the carbon dioxide acceptor molecule.

(*Total 7 marks*)

ULEAC June 1993, Paper I, No. 8

10. Summary
This question is about photosynthesis. You are asked to recall information about both the 'light' and the 'dark' reactions of photosynthesis. There is a short passage comparing the 'dark' reactions of cacti with plants with which you should be familiar. There are questions concerning the differences in the photosynthetic reactions found between these plants, and concerning the adaptations of cacti to dry conditions. Finally you are asked to comment on the biological problems associated with the destruction of tropical rain forests.

SECTION A

(a) (i) Draw a labelled diagram of a longitudinal section through an angiosperm chloroplast, indicating typical overall dimensions.

(ii) State the location in the chloroplast of the following processes:
1. photoactivation of chlorophyll,
2. regeneration of ribulose bisphosphate.
(*5 marks*)

(b) (i) Explain what is meant by the photoactivation of chlorophyll.

(ii) Describe how, as a consequence of the photoactivation of chlorophyll, both ATP and $NADPH_2$ may be formed in the chloroplast.

(iii) What is the third major chemical product of the light dependent reactions of photosynthesis? (*5 mark*)

SECTION B
In most plants of temperate zones the first stable intermediate of CO_2 fixation is a 3-carbon compound for this reason such plants are termed C3 plants. The fixation of CO_2 in these plants, although sometimes described as a 'dark' reaction, rapidly comes to a halt in dark conditions.

In many cactus plants, whose stomata are open only at night, the mechanism of CO_2 fixation is different. Incoming CO_2 is attached to a 3-carbon compound by a cytoplasmic enzyme and a 4-carbon compound, malate, is formed. Malate is stored in the cell vacuole until daylight. It is then decomposed, releasing CO_2 which enters the chloroplast; here it is fixed by the mechanism normally used by C3 plants.

(a) Outline the formation of the first stable intermediate of CO_2 fixation in the 'dark' reactions of C3 plants. (*2 mark*)

(b) Explain why the fixation of atmospheric CO cannot proceed
(i) during the day in a cactus plant and
(ii) at night in a C3 plant. (*2 mark*)

(c) Using information in the passage above, sta**FOUR** differences, apart from timing, between the processes of CO_2 fixation in C3 plants and cacti. (*2 mark*)

(d) The pattern of stomatal opening and closure is an adaptation to life in a hot, dry environment. Suggest **FOUR** additional adaptations of cacti. (*2 mark*)

(e) The large scale destruction of rain forests in South America is giving biologists cause for concern. Suggest **FOUR** adverse biological consequences of this destruction. (*2 mark*)

(*Total 20 mark*)

NISEAC June 1992, Paper III, No. 1

11. The following passage contains **eight errors**, the first of which has been circled and the **correct** term written in the margin opposite. Identify the remaining **seven** errors and write down a term which would correct the accuracy of the passage. (*7 mark*)

The third simultaneous phase of

(respiration) occurring in the — photosynthes

light is the oxidation of carbon

dioxide to the phosphorylated

3-carbon sugar. This conversion is

brought about in part of a cyclic

series of enzyme reactions known

as the Hill cycle. ATP and reduced

NAD produced during light harvesting and energy transduction are essential components. The primary carboxylation reaction is between a two-carbon compound and carbon dioxide. The reaction is catalysed by a hydrolase enzyme. The first stable product is a six carbon compound which is phosphorylated and reduced using ADP and reduced NADP to form a different three-carbon compound. For every three molecules of carbon dioxide and three molecules of the five-carbon compound, six molecules of this three-carbon compound are synthesized. One of these six is the net production of photosynthesis, the other five molecules being used in the regeneration of three molecules of the five carbon acceptor which re-enters the cycle.

WJEC June 1992, Paper A2, No. 12

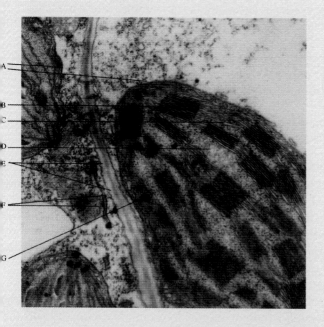

12. An electronmicrograph of parts of three chloroplasts in neighbouring cells of a maize leaf is shown.

(a) (i) Identify features **A**, **B**, **C**, **D**, **E** and **F**.
            (3 marks)
     (ii) State briefly the functions of features
           E and **F**.   *(2 marks)*
    (iii) Which feature is the site of the light-
           independent stage of photosynthesis?
                                                (1 mark)
(b) The appearance of the large chloroplast
    shows that the leaf has been kept in the dark.
    How can you tell this?   *(1 mark)*
(c) Feature G contains triglycerides. Give a
    simple diagram showing the structure of a
    triglyceride molecule.   *(1 mark)*

The blue dye DCPIP can be converted to colourless reduced DCPIP by gaining electrons. This is summarised below.

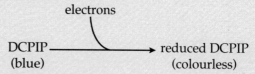

A suspension of chloroplasts was made by grinding fresh leaves in buffer solution and centrifuging the mixture. Tubes were then prepared and treated in the following way.

Tube	Contents	Treatment	Colour at start	Colour after 20 minutes
A	1 cm^3 chloroplast suspension 5 cm^3 DCPIP	illuminated strongly	blue/green	green
B	5 cm^3 DCPIP 1 cm^3 buffer solution	illuminated strongly	blue	blue
C	1 cm^3 chloroplast suspension 5 cm^3 DCPIP	left in the dark	blue/green	blue/green

(d) (i) In tube A, from where do the electrons
          come that reduce the DCPIP?   *(1 mark)*
     (ii) What normally happens to these electrons
           in a photosynthesizing leaf?   *(1 mark)*
(e) The chloroplast suspension may be
    contaminated with mitochondria. Explain
    the evidence from this investigation that the
    presence of mitochondria was not responsible
    for the reduction of the DCPIP.   *(2 marks)*
(f) Name the carbon dioxide acceptor in the
    light-independent stage of photosynthesis.
                                          (1 mark)
                                   (Total 13 marks)

AEB Summer 1994 (AS), Paper 1, No. 13

See *Further questions* pp. 614–20, questions 1, 8, 10, 12.

Heterotrophic nutrition

Heterotrophic organisms consume complex organic food material. This food originates with autotrophic organisms which synthesize it from simple inorganic raw materials. There are a number of forms of heterotrophic nutrition:

1. Holozoic nutrition – Involves the consumption of complex food which is broken down inside the organism into simple molecules which are then absorbed. Most animals feed in this way, utilizing a specialized digestive system. Insectivorous plants are partly holozoic.

2. Saprobiontic nutrition (sometimes called saprotrophic or saprophytic nutrition) – Involves the consumption of complex organic food from the bodies of decaying organisms. The food is either already in a soluble form or it is digested externally into simple molecules which then diffuse into the saprobiont. There is no digestive system. Some bacteria and fungi feed in this way.

3. Parasitism – Involves feeding on complex organic food derived from other living organisms. There is a close association between the parasite, which benefits, and the host which is harmed. Food is usually obtained in soluble form and so if a digestive system is present it is very simple. A few parasites ingest solid food and therefore possess digestive systems.

4. Mutualism – Here again there is a close association between members of two species, but in this case both derive some benefit from the relationship.

5. Commensalism – In this close association, one member benefits while the other neither benefits nor is it harmed.

The term **symbiosis** is often used to describe two species which live together in an intimate relationship which entails one living in or on the body of another. As such the term is a general one covering the more specific associations of parasitism, mutualism and commensalism.

Associations between organisms are never static nor do they fall into clearly defined groups. It is often difficult to say to which category a relationship should be assigned – indeed a number of associations change in the course of time. The parasite *Armillaria mellea* (honey fungus) for example eventually kills the tree on which it lives but then continues to live as a saprobiont on its remains.

15.1 Holozoic nutrition

Holozoic organisms obtain their energy from the consumption of complex organic food which is digested within their bodies. Their nutrition involves most, if not all, of the following stages:

1. **Obtaining the food** – May involve movement of the organism to the food source.

2. **Ingestion** – Organisms use a variety of feeding mechanisms which depend upon the size and nature of the food.

3. **Physical (mechanical) digestion** – By means of a variety of structures including teeth, radula and gizzard.

4. **Chemical digestion** – A process largely carried out by enzymes.

5. **Absorption** – Useful soluble materials must be absorbed from the digestive system into the body tissues.

6. **Assimilation** – The materials absorbed must enter individual cells and be incorporated into them.

7. **Elimination (egestion)** – Unwanted material which has been ingested must be removed from the body.

   Holozoic organisms can be classified according to the type of food ingested. Organisms which feed on living or recently dead plant material are called **herbivores**, while those feeding on living or recently dead animals are called **carnivores**. Organisms that feed on a diet combining plant and animal material are called **omnivores**. Some holozoic organisms consume liquid material (**liquid feeders**) but the majority take in particles of solid food and are known as **phagotrophs**. Phagotrophs may take in relatively large particles in which case the organisms are called **macrophagous feeders**, or very small particles in which case they are called **microphagous feeders**. Examples of microphagous feeders are mussels which use their sheet-like gills as a fine meshwork which strains tiny particles from the water. Cilia on these gills draw water across them and the particles become trapped on the mucus which covers the gills. This food-laden mucus is drawn to the mouth where it is ingested.

15.2 Diet

All organisms require a constant supply of essential nutrients. What these nutrients are and the amounts of each required by an organism varies from species to species. In mammals **carbohydrates** and **fats** are needed in relatively large quantities as sources of energy, and **proteins** are needed in large amounts for growth and repair. **Vitamins** and **minerals** are required in much smaller quantities for a variety of specific functions. **Water** is a vital constituent of the diet and **dietary fibre (roughage)** is necessary for efficient digestion.

15.2.1 Carbohydrates and fats (energy requirements)

Details of the chemistry of carbohydrates and fats is given in Chapter 2. The main function of both is to provide energy. The amount of energy in food is expressed in **joules**. (It was previously measured in calories. In books and magazines concerned with diet the term calorie is still commonly used. One calorie is equal to 4.18 joules.) To measure the amount of energy in different foods, a given mass is burned in oxygen in a piece of apparatus called a **bomb calorimeter**. The total heat generated gives a measure of the food's energy content (also called its **calorific value**).

TABLE 15.1 **Recommended daily intake of energy according to age, activity and sex**

Age/years	Average body weight/kg	Degree of activity/ circumstances	Energy requirement/kJ	
			Male	**Female**
1	7	Average	3200	3200
5	20	Average	7500	7500
10	30	Average	9500	9500
15	45	Average Sedentary	11500 11300	11500 9000
25	65 (male) 55 (female)	Moderately active Very active Sedentary	12500 15000 11000	9500 10500 9000
50	65 (male) 55 (female)	Moderately active Very active	12000 15000	9500 10500
75	63 (male) 53 (female)	Sedentary	9000	8000
Any	–	During pregnancy	–	10000
Any	–	Breast feeding	–	11500

The energy required by an organism varies with sex, size, age and activity. Table 15.1 provides examples of the recommended daily energy intake for humans of various ages. Ideally two thirds of this should be derived from carbohydrates and the remainder from fats.

Much attention has been focused recently on the correlation between a high fat intake in the diet and heart disease. It is always difficult to draw relationships directly between one type of food and the incidence of a specific disease because foods contain a wide variety of substances. In addition, factors such as exercise, stress and smoking affect an individual's health and the way food is utilized. It does, however, seem that a high intake of fats, especially saturated fats (see Section 2.6.1), is a contributory factor in causing heart disease.

15.2.2 Proteins

The chemistry of proteins is given in Section 2.7. As a last resort, the body may respire proteins to provide energy, but their main function is as a source of amino acids which are used to synthesize new proteins. These proteins are used in metabolism, growth and repair. Plants are able to synthesize all their own amino acids but animals are more limited. Humans, for example require nine amino acids, called **essential amino acids**, in the diet. Although plant food contains proportionately fewer proteins, a properly balanced vegetable diet can nevertheless provide all the essential amino acids. It is only where there is a dependence on just one or two plant foods as sources of proteins that malnutrition results.

15.2.3 Vitamins

Vitamins are a group of essential organic compounds which are needed in small amounts for normal growth and metabolism. If the diet lacks a particular vitamin, a disorder called a **deficiency disease** results. The vitamins required vary from species to species. Table 15.2 lists those needed in a human diet and the rôles they play. Vitamins are normally classified as **water soluble** (vitamins C and the B complex) or **fat soluble** (vitamins

TABLE 15.2 **Vitamins required in the human diet**

Vitamin/name	Fat/water soluble	Major food sources	Function	Deficiency symptoms
A_1 Retinol	Fat soluble	Liver, vegetables, fruits, dairy foods	Maintains normal epithelial structure. Needed to form visual pigments	Dry skin. Poor night vision
B_1 Thiamin	Water soluble	Liver, legumes, yeast, wheat and rice germ	Coenzyme in cellular respiration	Nervous disorder called beri-beri. Neuritis and mental disturbances. Heart failure
B_2 Riboflavin	Water soluble	Liver, yeast, dairy produce	Coenzymes (flavo-proteins) in cellular respiration	Soreness of the tongue and corners of the mouth
B_3 (pp factor) Niacin	Water soluble	Liver, yeast, wholemeal bread	Coenzyme (NAD, NADP) in cellular metabolism	Skin lesions known as pellagra. Diarrhoea
B_5 Pantothenic acid	Water soluble	Liver, yeast, eggs	Forms part of acetyl coenzyme A in cellular respiration	Neuromotor disorders, fatigue and muscle cramps
B_6 Pyridoxine	Water soluble	Liver, kidney, fish	Coenzymes in amino acid metabolism	Dermatitis. Nervous disorders
B_{12} Cyanocobalamine	Water soluble	Meat, eggs, dairy food	Nucleoprotein (RNA) synthesis. Needed in red blood cell formation	Pernicious anaemia. Malformation of red blood cells
Biotin	Water soluble	Liver, yeast. Synthesized by intestinal bacteria	Coenzymes in carboxylation reactions	Dermatitis and muscle pains
Folic acid	Water soluble	Liver, vegetables, fish	Nucleoprotein synthesis. Red blood cell synthesis	Anaemia
C Ascorbic acid	Water soluble	Citrus fruits, tomatoes, potatoes	Formation of connective tissues, especially collagen fibres	Non-formation of connective tissues. Bleeding gums – scurvy
D Calciferol	Fat soluble	Liver, fish oils, dairy produce. Action of sunlight on skin	Absorption and metabolism of calcium and phosphorus, therefore important in formation of teeth and bones	Defective bone formation known as rickets
E Tocopherol	Fat soluble	Liver, green vegetables	Function unclear in humans. In rats it prevents haemolysis of red blood cells	Anaemia
K Phylloquinone	Fat soluble	Green vegetables. Synthesized by intestinal bacteria	Blood clotting	Failure of blood to clot

A, D, E and K). Whereas excess water-soluble vitamins are simply excreted in urine, fat-soluble vitamins tend to accumulate in fatty tissues of the body, and may even build up to lethal concentrations if taken in excess.

15.2.4 Minerals

The principal mineral ions required by plants and animals and their functions are listed in Chapter 2 in Table 2.1. The principal minerals required in the human diet, and their sources, are further summarized in Table 15.3.

15.2.5 Water

Water makes up about 70% of the total body weight of mammals and serves a wide variety of important functions which are discussed more fully in Section 22.1.8. Table 15.4 gives the daily water balance in a human not engaged in active work, i.e. there is no excessive sweating.

15.2.6 Dietary fibre (roughage)

Fibre is indigestible material which passes through the alimentary canal almost unchanged. As it does not cross an epithelial lining of the gut it never actually enters the body. It is

TABLE 15.3 **Some essential minerals required in the human diet**

Mineral	Major food source	Function
Macronutrients Calcium (Ca^{2+})	Dairy foods, eggs, green vegetables	Constituent of bones and teeth, needed in blood clotting and muscle contraction. Enzyme activator
Chlorine (Cl^-)	Table salt	Maintenance of anion/cation balance. Formation of hydrochloric acid
Magnesium (Mg^{2+})	Meat, green vegetables	Component of bones and teeth. Enzyme activator
Phosphate (PO_4^{3-})	Dairy foods, eggs, meat, vegetables	Constituent of nucleic acids, ATP, phospholipids (in cell membranes), bones and teeth
Potassium (K^+)	Meat, fruit and vegetables	Needed for nerve and muscle action and in protein synthesis
Sodium (Na^+)	Table salt, dairy foods, meat, eggs, vegetables	Needed for nerve and muscle action. Maintenance of anion/cation balance
Sulphate (SO_4^{2-})	Meat, eggs, dairy foods	Component of proteins and coenzymes
Micronutrients (trace elements) Cobalt (Co^{2+})	Meat	Component for vitamin B_{12} and needed for the formation of red blood cells
Copper (Cu^{2+})	Liver, meat, fish	Constituent of many enzymes. Needed for bone and haemoglobin formation
Fluorine (F^-)	Many water supplies	Improves resistance to tooth decay
Iodine (I^-)	Fish, shellfish, iodized salt	Component of the growth hormone, thyroxine
Iron (Fe^{2+} or Fe^{3+})	Liver, meat, green vegetables	Constituent of many enzymes, electron carriers, haemoglobin and myoglobin
Manganese (Mn^{2+})	Liver, kidney, tea and coffee	Enzyme activator and growth factor in bone development
Molybdenum (Mo^{4+})	Liver, kidney, green vegetables	Required by some enzymes
Zinc (Zn^{2+})	Liver, fish, shellfish	Enzyme activator, involved in the physiology of insulin

TABLE 15.4 **Human daily water balance**

Process	Water uptake /cm³	Water output /cm³
Drinking	1450	–
In food	800	–
From respiration	350	–
In urine	–	1500
In sweat	–	600
Evaporation from lungs	–	400
In faeces	–	100
TOTAL	2600	2600

not a metabolic product and its removal from the body is therefore called **egestion** or **elimination** and not excretion. Although it does not have a metabolic function, fibre is essential to the efficient working of the alimentary canal. It gives bulk to the material within the intestines, absorbing water and making the contents much more solid. In this form it stimulates peristalsis and is easier to move along the intestines. It thus helps prevent constipation and other intestinal disorders. Fibre consists mostly of the cellulose cell walls of plants. In humans, the removal of much fibre from processed food has led to an increase in intestinal disorders. This has shown the value of fibre and led to an emphasis on high-fibre diets as an aid to healthy living.

15.2.7 Milk

As milk is the only food received by mammals in the period after birth, it follows that it must provide all essential materials for growth and development. In this sense it is a balanced diet in itself. It cannot, however, sustain healthy development indefinitely for these reasons:

1. **It contains little if any iron** – This is no problem to a new-born baby as it accumulates iron from its mother before birth. This store cannot last indefinitely and alternative sources of iron are necessary in later life.

2. **It contains no fibre** – We saw in Section 15.2.6 the necessity of fibre and the problems associated with its long-term absence from the diet.

3. **It contains a high proportion of fat** – For a young, actively growing organism this is ideal, but as it grows the energy demand is reduced. This could lead to an increase in weight due to storage of the excess fat and a consequent increased risk of heart disease. For the early years, and as a supplement to the human diet in later life, milk nevertheless plays an invaluable rôle.

15.3 Principles of digestion

If it is not already, food must be made small enough to be ingested by holozoic organisms. This may involve the use of teeth or other organs designed to break up food into small pieces. This **mechanical breakdown** also has the effect of giving the food a larger surface area which aids later digestion.

Food comprises relatively few building blocks, largely monosaccharides, amino acids, fatty acids and glycerol which are arranged into an almost infinite variety of macromolecules which meet the needs of the organism they make up. What suits one organism however does not necessarily suit another. The food ingested must therefore be broken down further into its component parts so that they can be rebuilt into the macromolecules and structures of the organism ingesting them. The food must, in any case, be made small enough to pass across cell membranes. This breakdown is mainly achieved through hydrolysis reactions speeded up by enzymes and is termed **chemical digestion**.

APPLICATION

Food additives

PRAWN COCKTAIL

INGREDIENTS
Maizemeal, Vegetable Oil
and Hydrogenated
Vegetable Oil, Starch, Prawn
Cocktail Flavour (Acidity
Regulator – E262), Flavour
Enhancer (621), Citric Acid,
Flavouring, Artificial
Sweetener (Saccharin),
Sugar, Salt, Colours
(E110, E160b).

Ingredients on a packet of crisps

We are all exposed to food additives which have been used in increasing numbers and volumes during the past few decades. Most of the permitted additives are regulated by the European Community (EC) and labelled with the prefix E and a number. The EC coding groups classes of additives with similar functions together.

During recent years people have grown wary of the proliferation of 'E numbers' appearing on ingredients' lists of prepared foods. Some manufacturers have countered opposition by listing the full names of additives rather than using their 'E numbers', but many have begun to reduce the number of artificial additives used.

It must be realized that not all additives are harmful. In many cases foods would be more susceptible to bacterial infections if they were not used and the health of the general population might suffer. Many of us would certainly have to change our life styles if all additives were eliminated. There would be fewer convenience foods and low-fat spreads, more foods would have to be sold close to their point of production or manufacture and shopping trips would need to be more frequent. However, a balance should be reached and there is no substitute for hygiene and adequate cooking. Preservatives may be needed but does it really matter what colour our food is? Tartrazine (E102) is one of the most common food colourings, used in such things as orange juice, sauces and fish fingers, but it is now known that it can trigger hyperactivity and people with asthma suffer adversely.

A brief summary of the major groups of food additives is given in the boxes:

Flavour enhancers and sweeteners (E620–E637)
Flavour enhancers have no flavour of their own but they make the flavours of other foods stronger. The best known is **monosodium glutamate**, which has been used by the Chinese for centuries. Sweeteners are used in nearly all processed foods, both sweet and savoury, and about 60% of the sugar consumed in the UK is in processed foods.

Texture enhancers (E322–E495)
These include thickeners as well as emulsifiers and stabilizers. Emulsifiers are used to bind together fat and water, and stabilizers prevent them from separating out again. Together, they are useful in making low-fat spreads.

Antioxidants (E300–E321)
Antioxidants prevent oils and fats becoming rancid on contact with air. The natural antioxidant, vitamin E, is destroyed in processing and is therefore replaced by synthetic antioxidants.

Colourings (E100–E180)
These are purely cosmetic and rarely add nutritional value; in fact they may disguise poor quality foods. They are banned from baby foods.

Preservatives (E200–E297)
These, perhaps the most easily justified additives, only make up 1% of all additives used. They do not spoil the texture, appearance or flavour as many old-fashioned methods of preservation do. However, they do mean that food can be kept bacteriologically safe for longer, even though its nutritional value may decline.

Synthetic flavourings
These make up a high proportion of additives used and they are vital to modern food processing. There are between 3000 and 6000 in use today. They are not subject to any regulation and do not have to be identified in detail on labels. They are used to replace the natural flavour lost during processing, and their use means that flavour is no longer an infallible guide to the food's quality.

As we have seen in Section 3.2.1 enzymes are specific in the reactions they accelerate and therefore many are needed to completely break down a large macromolecule. Typically one enzyme breaks up a molecule into smaller sections and then others reduce these parts to their basic components. The carbohydrate starch, for example, has alternate glycosidic bonds hydrolysed by **amylase** to yield the disaccharide maltose, which is then further hydrolysed to its glucose monosaccharides by the enzyme **maltase**. Protein molecules, being larger and varied, require groups of enzymes to digest them. These are called **peptidases**. One group hydrolyses the peptide bonds between amino acids in the central region of molecules; these are called **endopeptidases**. Another group then hydrolyses the peptide bonds on the terminal amino acids of these portions, progressively reducing them to their individual amino acids. These are called **exopeptidases** and are of two types. The **aminopeptidases** work at the end of the chain that has an amino acid with a free amino ($-NH_2$) group whereas the **carboxypeptidases** work at the opposite end where the amino acid has a free carboxyl ($-COOH$) group (see Fig. 15.1). Fats, being smaller molecules can be broken down into fatty acids and monoglycerides (a single fatty acid linked to a glycerol) by the one enzyme **lipase**.

As we also saw in Section 3.2.6, enzymes differ in the conditions under which they operate most efficiently, not least in their optimum pH. Clearly it is impossible for one region of the digestive system to be both acid and alkaline at the same time, and so an alimentary canal evolved, a tube which was long enough to possess different regions each with its own set of conditions. To pass through each region food has to be moved along the alimentary canal by **peristalsis**. Digestion complete, **absorption** of the useful products must take place and the undigested food, dead cells and bacteria which aid digestion, must be eliminated – a process called **egestion**.

PROJECT

Right-handed people find it easier to brush the teeth on the left so that more teeth on the right suffer from dental caries. The opposite is true for left-handed people.

1. Tidy up this statement into a testable hypothesis.

2. Test your hypothesis using fellow students as 'subjects'.

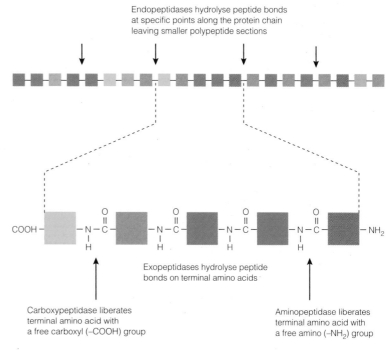

Fig. 15.1 Action of endo- and exo-peptidases

15.4 Digestion in humans

Digestion in humans takes place along a muscular tube which measures around 10 m in length in adults. Associated with it are a variety of secretory glands, some of which are embedded in the wall of the alimentary canal, others are separate from it but connected to it by a duct.

15.4.1 Digestion in the mouth

Mechanical breakdown of food begins in the mouth or **buccal cavity**. Humans are omnivores and hence have an unspecialized diet of mixed animal and plant origin. Their teeth reflect this lack of specialization, all types being present and developed to a similar extent. Apart from assisting speech, the tongue also manipulates the food during chewing and so ensures it is well mixed with **saliva** produced from three pairs of **salivary glands** (Fig. 15.2). Around 1.0–1.5 dm^3 of saliva are produced daily. Saliva contains:

1. **Water** – Over 99% of saliva is water.

2. **Salivary amylase** – A digestive enzyme which hydrolyses starch to maltose.

3. **Mineral salts** (e.g. sodium hydrogencarbonate) – This helps to maintain a pH of around 6.5–7.5 which is the optimum for the action of salivary amylase.

4. **Mucin** – A sticky material which helps to bind food particles together and lubricate them to assist swallowing.

Taste buds on the tongue allow food to be selected – unpleasant tasting food being rejected. The thoroughly chewed food is rolled into a **bolus** and passed to the back of the mouth for swallowing.

15.4.2 Swallowing and peristalsis

The bolus is pushed by the tongue to the back of the mouth and then into the **pharynx** where the **oesophagus** (leading to the stomach) meets with the trachea (which leads to the lungs). A variety of reflexes ensure that food when swallowed passes down the oesophagus and not the trachea. One such reflex is the closure of the opening into the larynx (which leads to the trachea). This opening, called the **glottis**, is covered by a structure known as the **epiglottis** when food is passed to the back of the mouth. The opening to the nasal cavity is closed by the **soft palate**. In this way, which is illustrated in Fig. 15.2, the bolus enters the oesophagus, a muscular tube lined with stratified epithelium and mucus glands. Lubricated by the mucus secreted by these glands, the bolus passes to the stomach by means of a wave of muscular contraction which causes constriction of the oesophagus behind the bolus. As this constriction passes along the oesophagus it pushes the bolus before it, down to the stomach. This process, which continues throughout the alimentary canal, is called **peristalsis**.

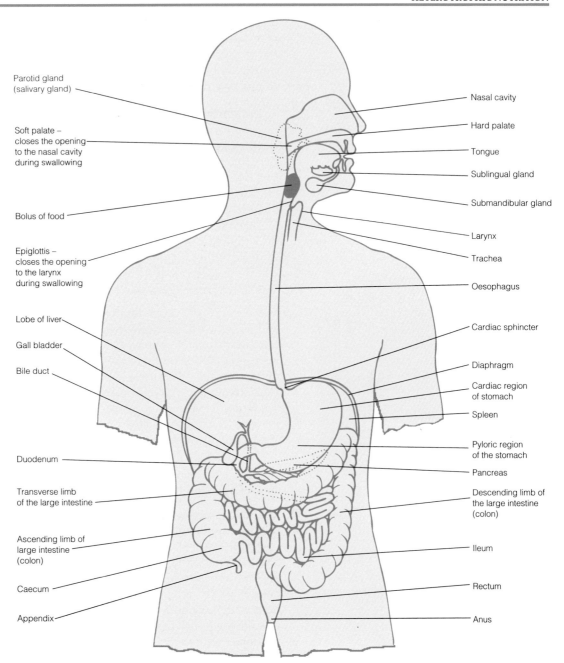

Fig. 15.2 Human digestive system

15.4.3 Digestion in the stomach

The stomach is roughly J-shaped, situated below the diaphragm. It is a muscular sac with a folded inner layer called the **gastric mucosa**. Embedded in this is a series of **gastric pits** which are lined with secretory cells (Fig. 15.3). These produce **gastric juice** which contains:

1. **Water** – The bulk of the secretion is water in which are dissolved the other constituents.

2. **Hydrochloric acid** – This is produced by **oxyntic cells** and with the water forms a dilute solution giving gastric juice its pH of around 2.0. It helps to kill bacteria brought in with the food and activates the enzymes pepsinogen and prorennin. It also initiates the hydrolysis of sucrose and nucleoproteins.

(a) Entire stomach

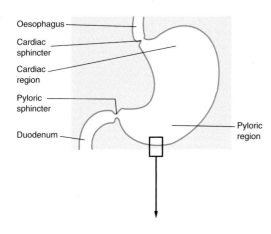

(b) Part of the stomach wall

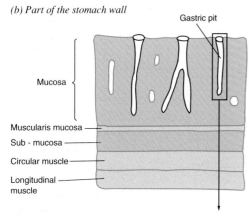

(c) Detail of gastric gland

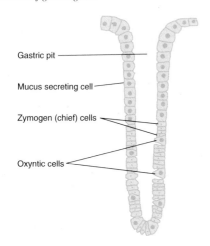

Fig. 15.3 Structure of the human stomach

3. Pepsinogen – This is produced by the **zymogen** or **chief cells** in an inactive form to prevent it from hydrolysing the proteins of the cells producing it. Once in the stomach it is activated to **pepsin** by hydrochloric acid. Pepsin is an endopeptidase which hydrolyses protein into polypeptides.

4. Prorennin – This too is produced by zymogen cells and is an inactive form of **rennin**, an enzyme which coagulates milk by converting the soluble **caseinogen** into the insoluble **casein**. It is therefore especially important in young mammals. Prorennin, too, is activated by hydrochloric acid.

5. Mucus – This is produced by **goblet cells** and forms a protective layer on the stomach wall, thus preventing pepsin and hydrochloric acid from breaking down the gastric mucosa (i.e. prevents autolysis). If the protection is not effective and the gastric juice attacks the mucosa, an ulcer results. Mucus also helps lubricate movement of food within the stomach.

During its stay in the stomach, food is thoroughly churned and mixed with gastric juice by periodic contractions of the muscular stomach wall. In this way a creamy fluid called **chyme** is produced. Relaxation of the pyloric sphincter and contraction of the stomach allow the chyme to enter the duodenum. The chyme from any one meal is released gradually over a period of 3–4 hours. This enables the small intestine to work on a little material at a time and provides a continuous supply of food for absorption throughout the period between meals.

15.4.4 Digestion in the small intestine

In humans the small intestine is over 6 m in length and its coils fill much of the lower abdominal cavity. It consists of two main parts: the much shorter **duodenum** where most digestion occurs and the longer **ileum** which is largely concerned with absorption. The walls of the small intestine are folded and possess finger-like projections called **villi**. The villi contain fibres of smooth muscle and regularly contract and relax. This helps to mix the food with the enzyme secretions and keep fresh supplies in contact with the villi, for absorption. The digestive juices which operate in the small intestine come from three sources: the liver, the pancreas and the intestinal wall.

Bile juice
Bile juice is a complex green fluid produced by the liver. It contains no enzymes but possesses two other substances important to digestion.

1. Mineral salts (e.g. sodium hydrogencarbonate) – These help to neutralize the acid chyme from the stomach and so create a more neutral pH for the enzymes of the small intestine to work in.

2. Bile salts – sodium and potassium glycocholate and taurocholate – They **emulsify** lipids, breaking them down into minute droplets. This is a physical, not a chemical change, which provides a greater surface area for pancreatic lipase to work on.

The liver performs other functions, some associated with digestion, and these are detailed in Section 25.7.2.

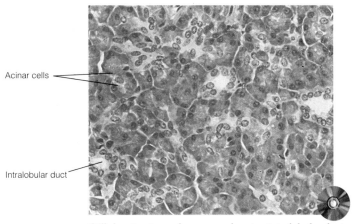

Acinar cells

Intralobular duct

Cellular structure of pancreas showing intralobular ducts

Pancreatic juice

The pancreas is situated below the stomach and is unusual in that it produces both an exocrine secretion, the pancreatic juice, and an endocrine secretion, the hormone insulin. The endocrine function is not directly concerned with digestion and is described in Section 25.3. Pancreatic juice, in addition to water, contains:

1. Mineral salts (e.g. sodium hydrogencarbonate) – Help to neutralize acid chyme from the stomach and so provide a more neutral pH in which the intestinal enzymes can operate.

2. Proteases – These include **trypsinogen** which, when activated by enterokinase from the intestinal wall, forms the endopeptidase called **trypsin** which hydrolyses proteins into peptides. Trypsin also activates another protease in the secretion, chymotrypsinogen into **chymotrypsin**; this too converts proteins into peptides. Also present is the exopeptidase called **carboxypeptidase** which converts peptides into smaller peptides and some amino acids.

3. Pancreatic amylase – Completes the hydrolysis of starch to maltose which began in the mouth.

4. Lipase – Breaks down fats into fatty acids and monoglycerides (glycerol + one fatty acid) by hydrolysis.

5. Nuclease – Converts nucleic acids into their constituent nucleotides.

Intestinal juice (succus entericus)

The mucus and sodium hydrogencarbonate in intestinal juice are made by coiled **Brunner's glands** whereas the enzymes are produced by the breakdown (lysis) of cells at the tips of the villi.

1. Mucus – Helps to lubricate the intestinal walls and prevent autolysis.

2. Mineral salts (e.g. sodium hydrogencarbonate) – Produced by the Brunner's glands in order to neutralize the acid chyme from the stomach and so provide a more suitable pH for the action of enzymes in the intestine.

3. Proteases (erepsin) – These include the exopeptidase called **aminopeptidase**, which converts peptides into smaller peptides and amino acids, and **dipeptidase**, which hydrolyses dipeptides into amino acids.

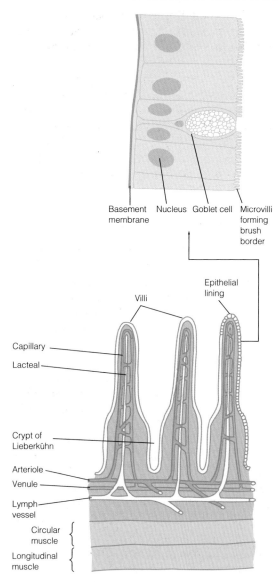

Basement membrane Nucleus Goblet cell Microvilli forming brush border

Epithelial lining

Villi

Capillary

Lacteal

Crypt of Lieberkühn

Arteriole

Venule

Lymph vessel

Circular muscle

Longitudinal muscle

Fig. 15.4 Intestinal wall showing villi (LS)

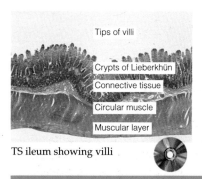

Tips of villi

Crypts of Lieberkühn

Connective tissue

Circular muscle

Muscular layer

TS ileum showing villi

Did you know?

Folds, villi and microvilli increase the internal surface area of the intestine from about 4 m² to 250 m² – greater than the area of a tennis court.

4. Enterokinase – A non-digestive enzyme which activates the trypsinogen produced by the pancreas.

5. Nucleotidase – Converts nucleotides into pentose sugars, phosphoric acid and organic bases.

6. Carbohydrases – These include **amylase**, which helps complete the hydrolysis of starch to maltose; **maltase**, which hydrolyses maltose to glucose; **lactase**, which hydrolyses the milk sugar lactose into glucose and galactose; and **sucrase**, which hydrolyses sucrose into glucose and fructose.

15.4.5 Absorption and assimilation

Digestion results in the formation of relatively small, soluble molecules which, provided there is a concentration gradient, could be absorbed into the body through the intestinal wall by diffusion. This, however, would be slow and wasteful and in any case, if the epithelial lining were permeable to molecules such as glucose, it could just as easily result in it diffusing out of the body when the concentration in the intestines was too low. For these reasons most substances are absorbed by **active transport** (Section 4.3.4) which only allows inward movement. Efficient uptake is often dependent on the presence of other factors. For example, glucose and amino acid absorption appear to be linked to the movement of sodium ions across the membranes of epithelial cells; calcium ion absorption requires the presence of vitamin D.

Efficient absorption is also dependent on a large surface area being available. The wall of the ileum achieves this in four ways:

1. It is very long – almost 6 m in humans and up to 45 m in cattle.

2. Its walls are folded (**folds of Kerkring**) to provide large internal projections.

3. The folds themselves have numerous tiny finger-like projections called **villi** (Fig. 15.4).

4. The epithelial cells lining the villi are covered with minute projections about 0.6 μm in length, called **microvilli** (not to be confused with cilia). These collectively form a **brush border**.

Sugars, amino acids and other water-soluble materials such as minerals enter the blood capillaries of the villi. From here they enter arterioles which later merge to form the hepatic portal vein which carries blood to the liver. In general, the level of different absorbed foods in the hepatic portal vein varies, depending on the type of food eaten and the interval since ingestion. It is the main rôle of the liver to regulate these variations by storing excess where the level of a substance is above normal and by releasing its store when its level in the hepatic portal vein is low. For this reason, blood from the intestines is sent to the liver for homeostatic regulation before it passes to other organs where fluctuations in blood composition could be damaging. The liver is also able to break down any harmful substances absorbed, a process called **detoxification**. The fatty acids and glycerol from lipid digestion enter the epithelial cells lining the villi where they recombine into lipids. These then enter the l[...] than the blood capillaries. From here they are transpo[...] lymph vessels before later joining the venous system of the [...] blood near the heart.

The passage of food from the ileum into the large intestine or colon is controlled by the **ileo-caecal valve**. The caecum in humans is little more than a slight expansion between the small and large intestine and the appendix is a small blind-ending sac leading from the caecum. In humans neither structure performs any important digestive function but, as we shall see in Section 15.8, they are of considerable importance to herbivorous mammals.

15.4.6 Water reabsorption in the large intestine

Most of the water drunk by humans is absorbed by the stomach. The large intestine or **colon** is partly responsible for reabsorbing the water from digestive secretions. With the gastric and intestinal juices each producing up to $3\,dm^3$ (litre) of secretion every day and the saliva, pancreatic and bile juices each adding a further $1.5\,dm^3$ the total volume of digestive secretions may exceed $10\,dm^3$. As most of this volume is water, it follows that the body cannot afford to allow it simply to pass out with the faeces. While most water is absorbed in the ileum, the large intestine plays an important rôle in reabsorbing the remainder. In doing so it changes the consistency of the faeces from liquid to semi-solid.

Within the large intestine live a huge population of bacteria, such as *Escherichia coli*, which in humans synthesize a number of vitamins including biotin and vitamin K. Deficiency of these vitamins is therefore rare, although orally administered antibiotics may destroy most of the bacteria and so create a temporary shortage. The vitamins produced are absorbed by the wall of the large intestine with water and some mineral salts. This wall is folded to increase the surface area available for absorption. Excess calcium and iron salts are actively transported from the blood into the large intestine for removal with the faeces.

15.4.7 Elimination (egestion)

The semi-solid faeces consist of a small quantity of indigestible food (fibre) but mostly comprise the residual material from the bile juice and other secretions, cells sloughed off the intestinal wall, a little water and immense numbers of bacteria. The wall of the large intestine produces mucus which, in addition to lubricating the movement of the faeces, helps to bind them together. After 24–36 hours in the large intestine the faeces pass to the rectum for temporary storage before they are removed through the anus, a process known as **defaecation**. Control of this removal is by two sphincters around the **anus**, the opening of the rectum to the outside.

As much of the material making up the faeces is not the result of metabolic reactions within the body, it is said to be eliminated or egested rather than excreted. However, cholesterol and bile pigments from the breakdown of haemoglobin are metabolic products and are therefore excretory.

Obesity

In Britain 20–30% of adults are overweight. An obese person has a larger number of fat cells than a thin person who may use the energy from food more efficiently so that there is little surplus to be converted to fat. Overeating often has a behavioural component, the habit being learnt in childhood or being used as a 'comfort'. Obese people have an increased risk of strokes, gall stones, diabetes and coronary heart disease.

Anorexia nervosa

People suffering from anorexia, most commonly white teenage girls, have an abnormal fear of becoming obese and so do not eat in spite of feeling hungry. As they lose weight their periods stop and their metabolic rate falls very low. It is a difficult condition to treat needing medical help and careful counselling.

Bulimia

People suffering from bulimia have not lost their appetites but tend to 'binge' on food, eating huge amounts at a time. This is then followed by forced vomiting or the use of laxatives as the fear of becoming overweight overwhelms them. The condition can lead to a serious loss of water and electrolytes and needs both medical and psychological treatment.

15.5   Nervous and hormonal control of secretions

The production of a digestive secretion must be timed to coincide with the presence of food in the appropriate region of the gut. In mammals the production of digestive secretions is under both nervous and hormonal control.

Nervous stimulation occurs even before the food reaches the mouth. The sight, smell or even the mere thought of food is sufficient to cause the salivary glands to produce saliva. This response is a conditioned reflex and is explained more fully in Section 27.7.3. Once in the mouth, contact of food with the tongue causes it to transmit nervous impulses to the brain. The brain in turn sends impulses which stimulate the salivary glands to secrete saliva. This is an unconditioned reflex response. At the same time the brain stimulates the stomach wall to secrete gastric juice, a response reinforced by nervous impulses transmitted as the food is swallowed. The stomach is thus prepared to digest the food even before it reaches it. Once initiated, the response will continue for up to an hour. The stretching of the stomach due to the presence of food within it stimulates production of gastric juice after this time.

TABLE 15.5 **Summary of digestion**

Organ/ secretion	Production induced by	Site of action	pH of secretion	Contents	Effect
Salivary glands produce saliva	Visual or olfactory expectation and reflex stimulation	Mouth	About neutral	Salivary amylase	Amylose(starch) → maltose
				Mineral salts	Produce optimum pH for amylase action
				Mucin	Binds food particles into a bolus
Gastric glands in stomach wall produce gastric juice	Presence of food in mouth and swallowing. Presence of food in stomach. Hormones – gastrin and enterogasterone from stomach wall	Stomach	Very acid	Pepsin(ogen)	Proteins → peptides
				(Pro)rennin	Caseinogen → casein
				Hydrochloric acid	Activates pepsinogen and prorennin. Produces optimum pH for action of these enzymes
				Mucus	Lubrication and prevention of autolysis
Liver produces bile juice	Secretion stimulates production of bile and cholecystokinin causes it to be released	Duodenum	Neutral	Bile salts	Emulsify fats
				Mineral salts	Neutralize acid chyme
				Bile pigments	Excretory products from breakdown of haemoglobin
				Cholesterol	Excretory product
Pancreas produces pancreatic juice	Secretion stimulates production of mineral salts and pancreozymin production of enzymes	Duodenum	Neutral	Trypsin (ogen)	Protein → peptides + amino acids activates chymotrypsinogen
				Chymotrypsin (ogen)	Peptides → smaller peptides + amino acids
				Carboxypeptidase	Peptides → smaller peptides + amino acids
				Amylase	Amylose (starch) → maltose
				Lipase	Fats → fatty acids + glycerol
				Nuclease	Nucleic acids → nucleotides
				Mineral salts	Neutralize acid chyme
Wall of small intestine produces intestinal juice (succus entericus)	Presence of food stimulates the intestinal lining	Duodenum and ileum	Alkaline	Aminopeptidase	Peptides → amino acids
				Dipeptidase	Dipeptides → amino acids
				Enterokinase	Activates trypsinogen
				Nucleotidase	Nucleotides → organic base + pentose sugar + phosphate
				Maltase	Maltose → glucose
				Lactase	Lactose → glucose + galactose
				Sucrase	Sucrose → glucose + fructose
				Mineral salts	Neutralize acid chyme

Hormonal control of secretions begins with the presence of food in the stomach. This stimulates the stomach wall to produce a hormone called **gastrin** which passes into the bloodstream. Gastrin continues to stimulate the production of gastric juice for up to four hours. Because fat digestion takes longer and requires less acidic conditions, its presence in the stomach initiates the production of **enterogasterone** from the stomach wall. This hormone reduces the churning motions of the stomach and

APPLICATION

Food poisoning

Food poisoning may be caused by a number of different bacteria including salmonellae, *Campylobacter*, *Listeria* and *Clostridium*. Salmonellae, particularly *Salmonella enteritidis*, are major causes of human food-borne disease. Although the bacteria are easily destroyed by heating, cross-contamination from raw to cooked foods, recontamination after heating and undercooking of contaminated foods can all lead to food poisoning.

Campylobacter is extremely sensitive to heating and drying and is able to grow only under a relatively restricted range of conditions. However *C. jejuni* and *C. coli*, which are the most common, are responsible for over 40 000 reported cases of human food-borne disease annually in the UK.

Listeria monocytogenes is a general environmental contaminant and traces may be found on virtually all raw agricultural products and many processed ones. It is generally regarded to be of particular concern in chilled foods because of its abilities to grow at refrigeration temperatures, and to survive and grow over a wider range of pH values than most food-poisoning bacteria. *Listeria monocytogenes* can be destroyed by heating at 70 °C for two minutes. Immuno-compromised individuals are particularly susceptible to listeriosis, and meningitis, septicaemia and infection of the fetus may occur.

Clostridium botulinum produces one of the most powerful toxins known to humans. If consumed, the toxin can cause serious illness or even death. Some varieties of *Clostridium botulinum* are of particular concern to the chilled food industry because of their ability to grow slowly at refrigeration temperatures. There is a very widespread occurrence of *Clostridium botulinum* in the environment and it has been found in many foods. However, food poisoning outbreaks are rare and are generally the consequence of faulty food processing or inadequate refrigeration.

The main symptoms of most food poisoning include diarrhoea, abdominal pain, sickness and fever; the fluid loss caused may even be fatal to vulnerable groups. Prevention is the main weapon used against the disease and includes strict hygiene precautions on farms, at abattoirs and in the handling of meat, both in shops and at home. Thorough cooking of meat and eggs is especially important.

decreases the flow of the acid gastric juice. As stomach ulcers are irritated by gastric juice, sufferers are often urged to drink milk. Being rich in fat, it reduces the production of gastric juice.

When food leaves the stomach and enters the duodenum, it stimulates the production of two hormones from the duodenal wall. **Secretin**, via the bloodstream, travels to the liver where it causes the production of bile and to the pancreas, where it stimulates the secretion of mineral salts. **Cholecystokinin-pancreozymin** causes the gall bladder to contract (releasing the bile juice into the duodenum) and stimulates the pancreas to secrete its enzymes.

15.6 Parasitism

PROJECT

Students who are involved in sport are more likely to suffer from athlete's foot

Test this hypothesis.

Parasitism is an association between two organisms in which one, the parasite, is metabolically dependent on the other, the host. This is invariably a nutritional dependence, the parasite absorbing either host tissues and fluids, or the contents of the host's intestine. In this relationship the host is harmed in some way. It is often difficult to distinguish between parasitism and predation or scavenging. However, most parasitologists feel that a parasitic relationship is one in which the parasite spends a significant length of time feeding on the host. A biting fly would not be considered a parasite by most biologists, but a leech would.

Another difference between a parasite and a predator is that the host can produce an immune response to a parasite but not to a predator. A parasite's success may be measured by its ability to resist this immune reaction.

There are two main categories of parasites: **endoparasites** which live inside the body of the host and **ectoparasites** which live on the outside. In both cases the parasite needs to be able to maintain its position in or on the host. Any organism must be able to reproduce and the offspring must be able to find a suitable habitat in which to develop. This is a particular problem for parasites which often need to produce large numbers of eggs or spores in order to ensure the success of a few. Parasitic life cycles may include elaborate mechanisms for successful transmission, often with several larval stages.

Plant parasites are sometimes divided into two groups according to the way they obtain their energy from cells. **Biotrophs** (living feeders) obtain their energy only from living host cells, whereas **necrotrophs** (dead feeders) can obtain their energy from cells which they have killed.

Parasites display most, but not necessarily all, of the following features:

1. They have agents for penetration of the host.

2. They have a means of attachment to the host.

3. They have protection against the host's immune responses.

4. They show degeneration of unnecessary organ systems.

5. They produce many eggs, seeds or spores.

6. They have a vector or intermediate host.

7. They produce resistant stages to overcome the period spent away from the host.

Let us now look at some examples of parasites to illustrate these features.

15.6.1 *Plasmodium* (malarial parasite)

Plasmodium is a protoctistan of the phylum Apicomplexa. Four species are parasitic in humans, all causing forms of malaria. Malaria is a debilitating fever and *Plasmodium falciparum* probably causes more human deaths in the tropics than any other organism. The malarial parasite has a very complex life

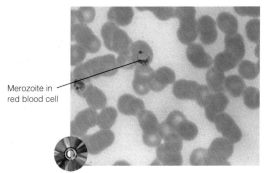
Merozoite in red blood cell

Plasmodium merozoites at signet ring stage

cycle involving an asexual stage in the liver and red blood cells of humans and a sexual stage which begins in humans and continues in mosquitoes of the genus *Anopheles*.

Plasmodium, in common with almost all parasites, has very rapid phases of multiplication to produce vast numbers of offspring. Three stages, the sporozoite, merozoite and zygote, must also be able to penetrate the cells of their host. It also shows an additional parasitic feature, the use of a **vector**. A vector is a secondary host which, because it feeds on the primary host, ensures the transmission of the parasite. The life cycle of *Plasmodium vivax* will be described briefly in order to illustrate these parasitic features but a detailed consideration is beyond the scope of this book.

Following the bite of an infected mosquito, sickle-shaped forms of the parasite, called **sporozoites**, enter the human blood. Within half an hour of infection these enter the liver cells where they undergo a prolonged period of rapid division. Numerous **merozoites** are produced which infect other liver cells and which may also enter red blood corpuscles. Within the erythrocytes, the multiplication continues and further merozoites are released into the blood. Eventually asexual reproduction is replaced by sexual reproduction and some of the merozoites develop into gametes. These remain dormant in the blood unless they are taken up by a mosquito. Female *Anopheles* mosquitoes need a meal of blood before ovulation and if they bite an infected human the gametes which are ingested become active. Fertilization takes place in the stomach of the mosquito and the zygote burrows into the stomach wall, encysts and meiosis takes place. The zygote then divides asexually to produce large numbers of sporozoites. These are released into the body cavity and many eventually reach the salivary glands. The life cycle continues when the mosquito bites a human and releases these sporozoites into the human blood.

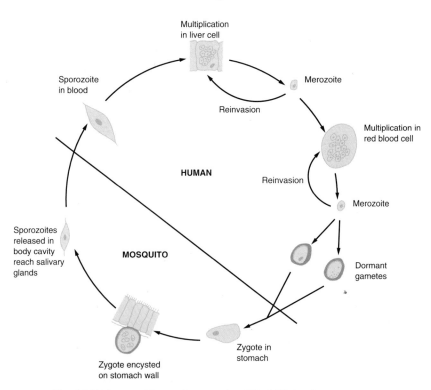

Fig. 15.5 Plasmodium vivax – *simplified life cycle*

Although large-scale eradication programmes have been in action for many years, malaria is still the most important transmissable human disease. Control of the secondary host, the mosquito, is vital if malaria is to be contained. Within the human host the parasite is most susceptible to drugs during the sporozoite and gamete stages but merozoites in the liver form a reservoir of the disease, causing relapses when resistance is low. People who suffer from sickle-cell anaemia have a higher resistance to malaria than those with normal erythrocytes. This is discussed fully in Section 10.4.3.

15.6.2 Parasitic flatworms

Two groups of Platyhelminthes (flatworm) are parasitic: the Cestoda, or tapeworms, and the Trematoda, or flukes. The parasitic features of *Fasciola* (liver fluke) and *Taenia* (tapeworm) will be considered in this section.

Both have complex life cycles, that of *Fasciola* involving a vertebrate primary host and an invertebrate secondary host. The many eggs which are produced inside the primary host give rise to a series of larval stages, one of which enters the secondary host where its development continues. *Fasciola hepatica* is found in sheep and cattle world-wide and may cause epidemics of 'liver rot'. The life cycle of *Taenia* involves two vertebrate hosts.

The liver fluke (*Fasciola hepatica*)

Fasciola hepatica adults are leaf-shaped and capable of limited movement. They live in the liver and bile ducts to which they may attach by means of two suckers. They feed on blood and liver cells. A considerable part of the body is occupied by the hermaphrodite reproductive organs. Numerous eggs are produced and these pass down the bile duct to the intestine and leave the primary host in the faeces. From the egg hatches a free-swimming **miracidium** larva which produces enzymes capable of penetrating the foot of a snail, the secondary host. The parasite undergoes further larval stages within the secondary host before leaving it as a **cercaria** larva. This larva encysts on vegetation and only develops further when eaten by sheep or

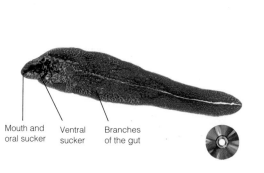

Mouth and
oral sucker   Ventral
sucker   Branches
of the gut

Adult *Fasciola*

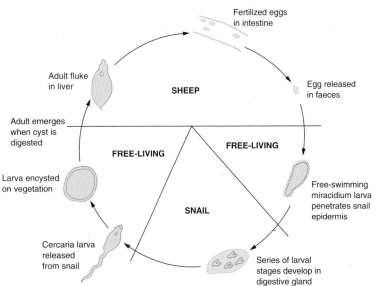

Fig. 15.6 Fasciola hepatica – *simplified life cycle*

301

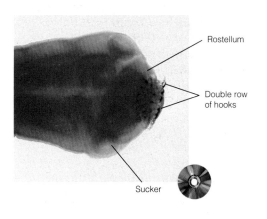

Scolex of *Taenia*

cattle. In the gut of the primary host, the cyst bursts and larvae bore through to the liver and gain their adult form. *Fasciola hepatica* is especially prevalent in regions of high rainfall and poor drainage.

The tapeworm (*Taenia*)

The adults of *Taenia solium*, the pork tapeworm, live attached to the intestinal mucosa of humans. The intermediate host is a pig. Tapeworms consist of a **scolex** followed by a series of segments known as **proglottids**. The scolex bears four suckers at the sides and a crown of hooks at the top. Although it is small it provides a firm attachment to the wall of the intestine and prevents the worm from being dislodged by the host's peristaltic movements.

Behind the scolex the narrow neck region gives rise to proglottids by a continuous process of budding. As the individual segments grow and mature they are pushed back from the scolex. The adult worm may be over three metres long.

Tapeworms are highly specialized endoparasites and neither scolex nor proglottids contain a mouth or alimentary canal. Predigested food can be absorbed over the entire body surface, facilitated by the worm's large surface-area to volume ratio; each proglottid is no more than 1 mm thick. The worms have a thick cuticle and produce inhibitory substances to prevent their digestion by the host's enzymes. Simple nerve fibres and a pair of excretory canals run the length of the worm but most of its anatomy is concerned with reproduction. Each mature proglottid contains both male and female reproductive organs and, following fertilization, eggs receive shells and yolk before passing to the uterus where they accumulate. The sex organs degenerate leaving an egg-packed uterus which fills the proglottid. At intervals these segments, known as **gravid proglottids**, break off the chain and are expelled with the host's faeces. Each segment may contain up to 40 000 eggs which survive until eaten by the secondary host because they have resistant shells. Further development only takes place when the eggs are consumed by pigs. Embryos then emerge and move into the animal's muscles where they remain dormant until the 'meat' is eaten by humans and the worm once again becomes active. Tapeworms are contracted by eating undercooked, infected meat.

TABLE 15.6 **Parasitic diseases**

Parasite	Major group	Primary host	Disease caused	Secondary host, if any
Phytophthora infestans	Fungi	Potato	Potato blight	–
Puccinia graminis	Fungi	Wheat	Black stem rust	–
Eimeria	Apicomplexa	Poultry	Coccidiosis	–
Plasmodium	Apicomplexa	Humans	Malaria	*Anopheles* mosquito
Schistosoma	Platyhelminthes	Humans	Bilharzia (schistosomiasis)	Fresh-water snail
Fasciola	Platyhelminthes	Sheep	Liver rot	Snail
Taenia solium	Platyhelminthes	Humans	Occasionally cysticercosis	Pig
Wucheria bancrofti	Nematoda	Humans	Elephantiasis	Mosquito
Oncocerca volvulus	Nematoda	Humans	River blindness	Blackfly

The adult worms cause little discomfort in a human, who is usually aware of the parasite only by the presence of creamy-white segments in the faeces. However, pork tapeworm has a unique feature in its life history which makes its eradication essential. If the eggs are eaten by a human they develop in just the same way as if they were eaten by a pig. This causes human **cysticercosis** with the dormant embryos encysting in various organs and damaging the surrounding tissue. The adults can be eradicated from humans by the use of appropriate drugs and this should reduce the frequency of cysticercosis. Treatment must also be accompanied by thorough meat inspection and public health measures. There must be adequate sewage-treatment plants and the prohibition of the discharge of raw sewage into inland waters or the sea.

Beef tapeworm (*T. saginata*) has a similar life cycle to *T. solium* but cattle provide the intermediate host. *T. saginata* is more widespread world-wide but it is less serious because human cysticercosis does not arise in infections with this species.

Both *Fasciola* and *Taenia* have structural modifications for their parasitic mode of life. They have suckers for attachment and a body covering which helps protect them from the host's immune responses or digestive enzymes. *Fasciola* has a relatively well developed gut, and feeds on cells predominantly by extracellular digestion. *Taenia* having no gut, absorbs predigested food. Most parasitic Platyhelminthes are hermaphrodite and all have systems capable of producing large numbers of eggs. A combination of sexual and asexual stages in order to multiply rapidly is ideal for a parasite.

15.7 Saprobiontism

Saprobionts are heterotrophic organisms which obtain carbon by absorption from dead organisms or organic wastes. True saprobionts do not invade living tissues but some parasitic fungi, e.g. *Ceratocystis ulmi* which causes Dutch elm disease, are able to feed saprobiontically when they have killed their host.

15.7.1 Extracellular digestion by saprobionts

The majority of saprobionts are bacteria or fungi and they carry out their digestion by secreting enzymes to the outside of their cells. These enzymes, which include amylases, cellulases, lignases and proteases, break down the material on which the saprobiont grows. The organism is thus surrounded by a solution of monosaccharides and amino acids which are then absorbed. An example of a saprobiont feeding in this way is the fungus *Rhizopus* which secretes enzymes from the tips of its hyphae.

15.7.2 Economic importance of saprobionts

If nutrients are not to become exhausted, carbon, nitrogen and other elements contained in dead organisms must be made available to living plants and animals. Saprobiontic bacteria and fungi play an essential rôle in recycling these chemicals as they break down dead organic material (Section 17.2).

Similar organisms cause spoilage of food as they utilize the organic compounds in such things as stored fruit and grain. Both recycling and food spoilage involve a large number of different saprobionts.

As saprobionts break down organic materials to obtain carbon, they form a number of by-products many of which can be used by humans as the basis for various industrial processes, such as brewing, baking and cheese-making (Section 30.3).

15.8 Mutualism

Lichen (*Xanthoria*)

Hydra with *Chlorella* algae

Mutualism is an association between two different organisms in which neither is harmed and both may benefit in some way. Examples of mutualism include:

1. Lichen – This is an association between an alga, usually *Trebouxia*, and a fungus, usually an ascomycete. The benefit to the alga is not clear but the fungus receives and uses the products of algal photosynthesis.

2. Hydra-Chlorella symbiosis – *Chlorella* are unicellular green algae found in the endoderm of the cnidarian *Hydra*. They are able to photosynthesize and supply *Hydra* with maltose.

3. Mycorrhizas – These are structures formed by the association of roots of plants with fungi. They are common in many higher plants, including pine, oak, beech and birch. The fungus is dependent on symbiosis for carbon nutrition and the higher plant for inorganic nutrients.

15.8.1 Digestion of cellulose by microorganisms

Lacking the ability to produce their own cellulase, herbivorous mammals are reliant upon bacteria and protoctista to carry out cellulose breakdown for them. The herbivore must provide a region of the alimentary canal for these microorganisms to inhabit. This region must be separate from the main canal in order that food can be kept there long enough for the microorganisms to carry out the breakdown. Being separate, this compartment can also be kept free of the mammal's own digestive enzymes, which might otherwise destroy the microorganisms, and at a suitable pH for their activity. The accommodation for these cellulose-digesting bacteria and protoctists takes two main forms in mammals.

In **ruminants**, e.g. cattle, sheep and deer, a complex four-chambered stomach is present. When swallowed, the food enters the first two chambers, the **rumen** and **reticulum**. It is here that the microorganisms carry out extracellular digestion of the cellulose by secreting cellulase. The products of this digestion are either absorbed by the walls of the rumen and reticulum which have villi or honey-combed ridges for this purpose, or are absorbed by the microorganisms which are later digested. The waste gases, largely carbon dioxide and methane, are expelled via the mouth.

Root of pine tree

Hyphae of fungus

Mycorrhiza ectotrophic on pine root

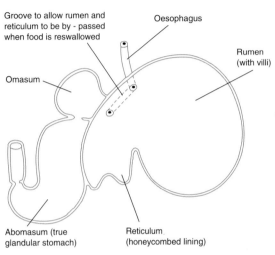

Fig. 15.7 *The ruminant stomach*

The relationship between the mammal and microorganisms is **mutualistic** as both gain benefit. The mammal acquires the products of cellulose breakdown which it could not obtain alone, and the microorganisms receive a constant supply of food and a warm, sheltered environment in which to live. After some hours, and usually in the relative safety of some sheltered position, the herbivore regurgitates the food into the mouth, where it thoroughly chews it – 'chewing the cud'. On being re-swallowed, the food enters the final two chambers of the stomach, the **omasum** and the **abomasum** (true stomach) where the usual process of protein digestion in acid conditions takes place.

Also present in the rumen are protein synthesizing bacteria which use ammonia as their source of nitrogen. These protein-rich bacteria are engulfed by the protoctists which in turn are digested by the ruminant's enzymes further along the alimentary canal. In this way ruminants are able to obtain a valuable source of protein allowing them to survive on a diet which might otherwise prove too low in protein for a healthy existence.

In rabbits and horses, the **caecum** and **appendix** are much enlarged and accommodate the microorganisms. Some absorption of the products of this digestion takes place through the walls of the caecum. In rabbits the yield is improved by the re-swallowing of the material from the caecum after it has left the anus – a process known as **coprophagy**.

15.8.2 Mutualism and the nitrogen cycle

Many angiosperms and a few conifers form swellings called **nodules** on their stems, roots or leaves. These nodules contain microorganisms which are capable of **nitrogen fixation**. The best known examples are the root nodules formed in plants belonging to the Papilionaceae (Leguminosae) such as peas, beans and clover. In this case the nitrogen-fixing microorganism is the bacterium *Rhizobium*. This association provides the bacteria with a carbon source and the plant with a source of nitrates, independent of their abundance in the soil. The development of a nodule is similar to the development of a lateral root except that at an early stage the central cells are filled with bacterial cells enclosed in a membrane. *Rhizobium* thus remains extracellular.

It has been estimated that mutualistic organisms fix about 100 million tonnes of nitrogen per year. Without this recycling of atmospheric nitrogen the level of soil nitrates would be far too low to support the present vegetation cover. Leguminous crops improve soil fertility and, in terms of efficiency of fixation, the biological process compares favourably with the commercial manufacture of nitrogenous fertilizer.

The exact site of nitrogen fixation has been the subject of much research. Neither the plant nor *Rhizobium* alone are capable of nitrogen fixation. Fixation depends on the mutualistic relationship, and a possible site for it is the membrane which separates the two organisms. The nitrogen cycle is considered in more detail in Section 17.2.2.

15.9   Questions

1. Discuss adaptations for digestion and absorption of food, illustrating your answer by reference to humans, a *named* saprophyte (saprobiont) and a *named* parasite.

(*Total 30 marks*)

ULEAC June 1993, Paper I, No. 15(a)

2. A person eats a piece of cheese consisting mainly of fats and protein. Describe the processes which enable these fats and proteins to be digested and absorbed.

(*Total 20 marks*)

NEAB June 1991, Paper IB, No. 2

3. Write an essay on the adaptations of parasites to their way of life.

(*Total 24 marks*)

AEB November 1993, Paper II, No. 5A

4. Complete the table which refers to enzyme activity in the mammalian gut.

Name of enzyme	Site of production	Substrate	Product(s)
entero kinase	Duodenal mucosa	Trypsinogen	Trypsin
Lipase	Pancreas	Fat	Fatty acids Glycerol
Lactase	Ileum	Lactose	Galactose Glucose

(*Total 5 marks*)

AEB June 1991, Paper I, No. 7

5. Different concentrations of maltose were placed in the small intestine of a mammal. The amounts of glucose appearing in the blood and in the small intestine were measured. The results are shown in the graph.

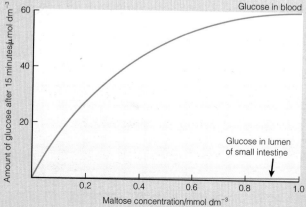

(a)   (i) Name the blood vessel most likely to have been sampled for glucose. (*1 mark*)

(ii) Describe the chemical process by which a molecule of maltose is converted into two molecules of glucose. (*2 marks*)

(b) As a result of this experiment it was suggested that the enzyme involved in the breakdown of maltose is located in the cell-surface membrane of the epithelial cells of the small intestine.

(i) Explain the evidence from the graph that supports the view that maltose breakdown does *not* occur in the lumen of the small intestine. (*2 marks*)

(ii) Suggest an explanation for the shape of the curve showing the change in the amount of glucose in the blood. (*3 marks*)

The drawing has been made from an electron micrograph of part of an epithelial cell from the small intestine. The magnification of the drawing is ×20 000.

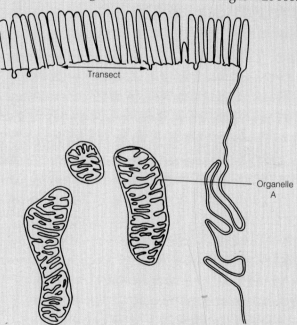

(c)   (i) Name the organelle labelled A. (*1 mark*)

(ii) Describe the part played by organelle A in the uptake of substances from the small intestine. (*3 marks*)

(d)   (i) Calculate the actual length of one microvillus in micrometres. Show your working. (*2 marks*)

(ii) By how many times does the possession of microvilli increase the surface of the cells along the transect shown in the drawing? Explain how you arrived at your answer. (*3 marks*)

The photograph shows the same tissue as the drawing but it is taken from an intestine that has been infected with the bacterium, *Escherichia coli*.

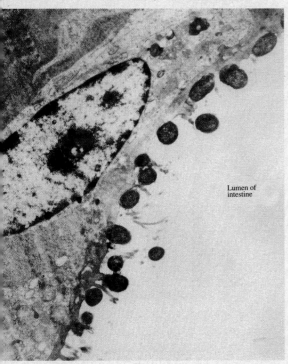

Lumen of intestine

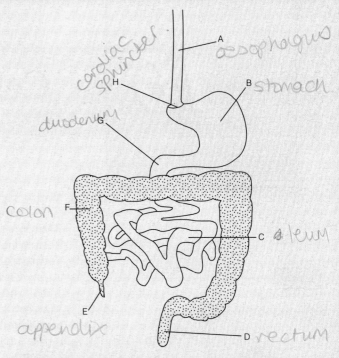

(e) (i) This photograph was taken using a transmission electron microscope. Describe the main difference in the way in which transmission and scanning electron microscopes function. *(2 marks)*

(ii) Explain why it is not possible to see the detail shown in this photograph when using a light microscope at the same magnification. *(2 marks)*

(f) (i) Describe how the bacteria have damaged the tissue. *(1 mark)*

(ii) Suggest **one** way in which this damage might lead to diarrhoea in the animal concerned. *(2 marks)*

(Total 24 marks)

AEB November 1992, Paper II, No. 5

6. The diagram shows the human gastrointestinal tract.
(a) Name the parts labelled **A** to **H**. *(4 marks)*
(b) Different regions of the tract have different pH values.
   (i) Suggest why it is important to have a particular pH in a particular region.
   (ii) Name a region which has: a low pH; a high pH. *(5 marks)*
(c) Use **one** of the letters **A** to **H**, to indicate the main region where absorption of the following compounds occurs
   (i) digested protein; (ii) glucose;
   (iii) water. *(5 marks)*

Vitamins differ from most other nutrients in that, with the exception of vitamin D, they cannot be synthesized by humans and so must be obtained from the diet.

(d) (i) State **three** consequences of vitamin A deficiency.

(ii) Suggest why deficiency of vitamin A is more likely to occur in vegetarians.
(4 marks)
(Total 18 marks)

UCLES (Modular) December 1992,
(Human Health and Disease), No. 1

7. Read through the following account of the human ileum, and then write on the dotted lines the most appropriate word or words to complete the account.

The inner lining of the ileum is composed of ....

epithelium, the cells of which have .... on their free

surfaces, thus increasing the surface area for

absorption. Further increase in surface area is brought

about by the folding of the mucosa into numerous

small projections called .... Between these

projections are the ...., which secrete intestinal

juice. Enzymes in intestinal juice include ....,

which catalyses the hydrolysis of .... into ....
(Total 7 marks)

ULEAC January 1994, Paper I, No. 2

See *Further questions* pp. 614–20, questions 4, 9, 19.

16 Cellular respiration

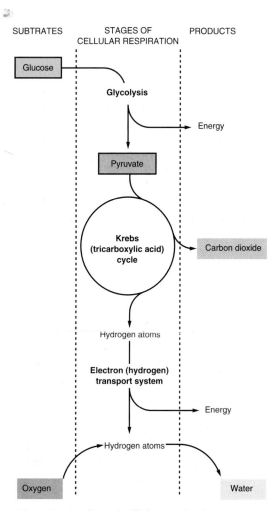

SUBTRATES	STAGES OF CELLULAR RESPIRATION	PRODUCTS

Glucose

Glycolysis

→ Energy

Pyruvate

Krebs (tricarboxylic acid) cycle

→ Carbon dioxide

Hydrogen atoms

Electron (hydrogen) transport system

→ Energy

Hydrogen atoms

Oxygen

Water

Fig. 16.1 Outline of cellular respiration

In Chapter 13 we saw that living systems require a constant supply of energy to maintain low entropy and so ensure their survival. This energy initially comes from the sun and is captured in chemical form by autotrophic organisms during the process of photosynthesis (Chapter 14). While the carbohydrates, fats and proteins so produced are useful for storage and other purposes, they cannot be directly used by cells to provide the required energy. The conversion of these chemicals into forms like adenosine triphosphate, which can be utilized by cells, occurs during respiration.

Whatever form the food of an organism initially takes, it is converted into carbohydrate, usually the hexose sugar glucose, before being respired. Most respiration is the oxidation of this glucose to carbon dioxide and water with the release of energy, and the process can be conveniently divided into two parts:

1. **Cellular (internal or tissue) respiration** – the metabolic processes within cells which release the energy from glucose.

2. **Gaseous exchange (external respiration)** – the processes involved in obtaining the oxygen for respiration and the removal of gaseous wastes.

Gaseous exchange is dealt with in Chapter 20. Cellular respiration is the subject of this chapter and can be divided into three stages:

1. Glycolysis

2. Krebs (tricarboxylic acid) cycle

3. Electron (hydrogen) transport system.

The relationship of these stages in cellular respiration is outlined in Fig. 16.1.

16.1 Adenosine triphosphate (ATP)

Adenosine triphosphate (ATP) is the short-term energy store of all cells. It is easily transported and is therefore the universal energy carrier.

16.1.1 Structure of ATP

ATP is formed from the nucleotide adenosine monophosphate (Fig. 16.2) by the addition of two further phosphate molecules. Its structure is shown in Fig. 16.2.

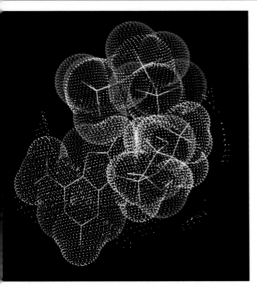

Computer graphics representation of ATP

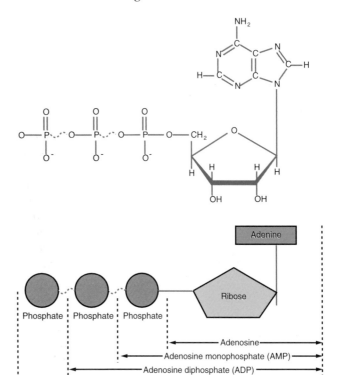

Fig. 16.2 Structure of adenosine triphosphate

16.1.2 Importance of ATP

The hydrolysis of ATP to ADP is catalyzed by the enzyme ATP-ase and the removal of the terminal phosphate yields 30.6 kJ mol^{-1} of free energy. Further hydrolysis of ADP to AMP yields a similar quantity of energy, but the removal of the last phosphate produces less than half this quantity. For this reason the last two phosphate bonds are often termed high energy bonds on account of the relatively large quantity of energy they yield on hydrolysis. This is misleading in that it implies that all the energy is stored in these bonds. The energy is in fact stored in the molecule as a whole, although the breaking of the bonds initiates its release.

AMP and ADP may be reconverted to ATP by the addition of phosphate molecules in a process called **phosphorylation**, of which there are two main forms:

1. Photosynthetic phosphorylation – occurs during photosynthesis in chlorophyll-containing cells (Chapter 14).

2. Oxidative phosphorylation – occurs during cellular respiration in all aerobic cells.

The addition of each phosphate molecule requires 30.6 kJ of energy. If the energy released from any reaction is less than this, it cannot be stored as ATP and is lost as heat. The importance of ATP is therefore as a means of transferring free energy from energy-rich compounds to cellular reactions requiring it. While not the only substance to transfer energy in this way, it is by far the most abundant and hence the most important.

16.1.3 Uses of ATP

A metabolically active cell may require up to two million ATP molecules every second. ATP is the source of energy for:

1. **Anabolic processes** – It provides the energy needed to build up macromolecules from their component units, e.g.
 – polysaccharide synthesis from monosaccharides
 – protein synthesis from amino acids
 – DNA replication.

2. **Movement** – It provides the energy for many forms of cellular movement including:
 – muscle contraction
 – ciliary action
 – spindle action in cell division.

3. **Active transport** – It provides the energy necessary to move materials against a concentration gradient, e.g. ion pumps.

4. **Secretion** – It is needed to form the vesicles necessary in the secretion of cell products.

5. **Activation of chemicals** – It makes chemicals more reactive, enabling them to react more readily, e.g. the phosphorylation of glucose at the start of glycolysis.

16.2   Glycolysis

Glycolysis (*glyco* – 'sugar'; *lyso* – 'breakdown') is the breakdown of a hexose sugar, usually glucose, into two molecules of the three-carbon compound **pyruvate (pyruvic acid)**. It occurs in all cells; in anaerobic organisms it is the only stage of respiration. Initially the glucose is insufficiently reactive and so it is phosphorylated prior to being split into two triose sugar molecules. These molecules yield some hydrogen atoms which may be used to give energy (ATP) before being converted into pyruvate. During its formation, the ATP used in phosphorylating the glucose is regenerated. Glycolysis takes place in the cytoplasm of the cell and its main stages are outlined opposite.

   Each glucose molecule produces two molecules of glycerate 3-phosphate and there is therefore a pair of every subsequent molecule for each glucose molecule. The energy yield is a net gain of two molecules of ATP (Stage 7). The two pairs of hydrogen atoms produced (Stage 5) may yield a further six ATPs (see Section 16.4), giving an overall total of eight ATPs.

Stages of glycolysis

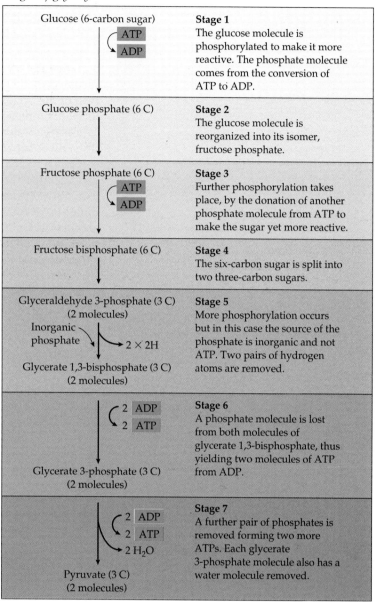

	Stage 1
Glucose (6-carbon sugar) ATP → ADP	The glucose molecule is phosphorylated to make it more reactive. The phosphate molecule comes from the conversion of ATP to ADP.
Glucose phosphate (6 C)	**Stage 2** The glucose molecule is reorganized into its isomer, fructose phosphate.
Fructose phosphate (6 C) ATP → ADP	**Stage 3** Further phosphorylation takes place, by the donation of another phosphate molecule from ATP to make the sugar yet more reactive.
Fructose bisphosphate (6 C)	**Stage 4** The six-carbon sugar is split into two three-carbon sugars.
Glyceraldehyde 3-phosphate (3 C) (2 molecules) Inorganic phosphate → 2 × 2H Glycerate 1,3-bisphosphate (3 C) (2 molecules)	**Stage 5** More phosphorylation occurs but in this case the source of the phosphate is inorganic and not ATP. Two pairs of hydrogen atoms are removed.
2 ADP → 2 ATP Glycerate 3-phosphate (3 C) (2 molecules)	**Stage 6** A phosphate molecule is lost from both molecules of glycerate 1,3-bisphosphate, thus yielding two molecules of ATP from ADP.
2 ADP → 2 ATP → 2 H_2O Pyruvate (3 C) (2 molecules)	**Stage 7** A further pair of phosphates is removed forming two more ATPs. Each glycerate 3-phosphate molecule also has a water molecule removed.

16.3 Krebs (tricarboxylic acid) cycle

Although glycolysis releases a little of the energy from the glucose molecule, the majority still remains 'locked-up' in the pyruvate. These molecules enter the mitochondria and, in the presence of oxygen, are broken down to carbon dioxide and hydrogen atoms. The process is called the **Krebs cycle**, after its discoverer Hans Krebs. There are a number of alternative names, notably the **tricarboxylic acid cycle (TCA cycle)** and **citric acid cycle**. While the carbon dioxide produced is removed as a waste product, the hydrogen atoms are oxidized to water in order to yield a substantial amount of free energy. Before pyruvate enters the Krebs cycle it combines with a compound called coenzyme A to form **acetyl coenzyme A**. In the process, a molecule of carbon dioxide and a pair of hydrogen atoms are removed. The 2-carbon acetyl coenzyme A now enters the Krebs cycle by combining with the 4-carbon **oxaloacetate (oxaloacetic acid)** to give the 6-carbon **citrate (citric acid)**. Coenzyme A is reformed and may be used to combine with a further pyruvate molecule. The citrate is degraded to a 5-carbon α-**ketoglutarate (α-ketoglutaric acid)** and then the 4-carbon oxaloacetate by the progressive loss of two carbon dioxide molecules, thus completing the cycle. For each turn of the cycle, a total of four pairs of hydrogen atoms are also formed. Of these, three pairs are combined with the hydrogen carrier **nicotinamide adenine dinucleotide (NAD)** and yield three ATPs for each pair of hydrogen atoms. The remaining pair combines with a different hydrogen carrier, **flavine adenine dinucleotide (FAD)** and yields only two ATPs. In addition, each turn of the cycle produces sufficient energy to form a single molecule of ATP. It must be remembered that all these products are formed from a single pyruvate molecule of which two are produced from each glucose molecule. The total yields from a single glucose molecule are thus double those stated. The significance of this will become apparent when considering the total quantity of energy released (Section 16.6).

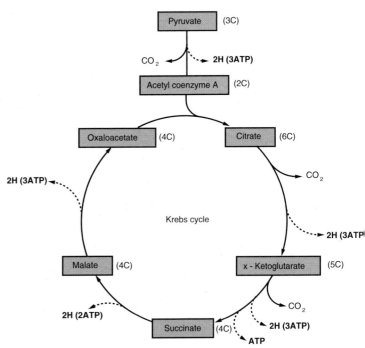

Fig. 16.3 Summary of Krebs cycle

16.3.1 Importance of Krebs cycle

The Krebs cycle plays an important rôle in the biochemistry of a cell for three main reasons:

1. **It brings about the degradation of macromolecules** – The 3-carbon pyruvate is broken down to carbon dioxide.

2. **It provides the reducing power for the electron (hydrogen) transport system** – It produces pairs of hydrogen atoms which are ultimately the source of metabolic energy for the cell.

3. **It is an interconversion centre** – It is a valuable source of intermediate compounds used in the manufacture of other substances, e.g. fatty acids, amino acids, chlorophyll (Section 16.7

16.4   Electron transport system

The electron transport system is the means by which the energy, in the form of hydrogen atoms, from the Krebs cycle, is converted to ATP. The hydrogen atoms attached to the hydrogen carriers NAD and FAD are transferred to a chain of other carriers at progressively lower energy levels. As the hydrogens pass from one carrier to the next, the energy released is harnessed to produce ATP. The series of carriers is termed the **respiratory chain**. The carriers in the chain include **NAD**, **Flavo Protein**, **coenzyme Q** and iron-containing proteins called **cytochromes**. Initially hydrogen atoms are passed along the chain, but these later split into their protons and electrons, and only the electrons pass from carrier to carrier. For this reason, the pathway can be called the electron, or hydrogen, transport system. At the end of the chain the protons and electrons recombine, and the hydrogen atoms created link with oxygen to form water. This formation of ATP through the oxidation of the hydrogen atoms is called **oxidative phosphorylation**. It occurs in the mitochondria.

The rôle of oxygen is to act as the final acceptor of the hydrogen atoms. While it only performs this function at the end of the many stages in respiration, it is nevertheless vital as it drives the whole process. In its absence, only the anaerobic glycolysis stage can continue. The transfer of hydrogen atoms to oxygen is catalyzed by the enzyme **cytochrome oxidase**. This enzyme is inhibited by cyanide, so preventing the removal of hydrogen atoms at the end of the respiratory chain. In these circumstances the hydrogen atoms accumulate and aerobic respiration ceases, making cyanide a most effective respiratory inhibitor.

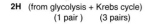

2H   (from glycolysis + Krebs cycle)
     (1 pair)   (3 pairs)

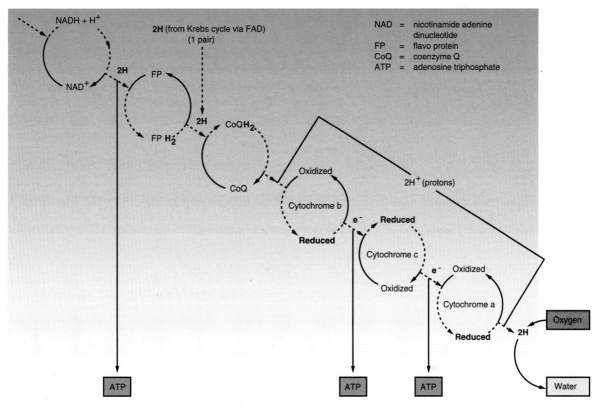

ig. *16.4 Summary of the electron transport system*

16.4.1 Mitochondria and oxidative phosphorylation

Mitochondria are present in all eukaryotic cells where they are the main sites of respiratory activity. Highly active cells, requiring much energy, characteristically have numerous large mitochondria packed with cristae. Such cells include:

Liver cells – Energy is required to drive the large and varied number of biochemical reactions taking place there.

Striated muscle cells – Energy is needed for muscle contraction, especially where this is rapid, e.g. flight muscle of insects.

Sperm tails – These provide energy to propel the sperm.

Nerve cells – Mitochondria are especially numerous adjacent to synapses where they provide the energy needed for the production and release of transmitter substances.

NOTEBOOK

Oxidation, reduction and energy

Many everyday processes such as burning, rusting and respiration are the result of substances combining with oxygen. These reactions also release energy. Fuel in a car engine for example, petrol (made almost entirely of hydrogen and carbon) is mixed with air in the carburettor, the oxygen of which combines with the petrol to form oxides of both hydrogen and carbon when ignited by a spark. The reaction is **exothermic**, i.e. it releases much energy, which is used to propel the car.

$$\text{Hydrocarbon} + \text{Oxygen} \longrightarrow \text{Carbon dioxide} + \text{Water} + \text{Energy}$$
$$\text{(Petrol)} \qquad \text{(From air)} \qquad \text{(Oxide of carbon)} \quad \text{(Oxide of hydrogen)}$$

Respiration is essentially the same process with the carbon and hydrogen in our food being substituted for the petrol.

The process by which substances combine with oxygen is called **oxidation** and the substances to which oxygen is added are said to be **oxidized**. However, as one substance gains oxygen another must lose it. We call the process by which oxygen is lost **reduction** and say that the substance losing oxygen has been **reduced**. Just as oxidation involves energy being given out, so reduction involves being taken in.

Oxidation and reduction therefore always take place together; as one substance is oxidized so another must be reduced. We call these chemical reactions **redox** reactions (**red**uction + **ox**idation).

In many redox reactions oxygen is reduced by the addition of hydrogen to make water, e.g. in respiration. For this reason reduction is sometimes described as the **gain** of hydrogen and oxidation as the **loss** of hydrogen.

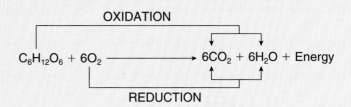

$$\text{OXIDATION}$$
$$C_6H_{12}O_6 + 6O_2 \longrightarrow 6CO_2 + 6H_2O + \text{Energy}$$
$$\text{REDUCTION}$$

Intestinal epithelial cells – Many mitochondria occur beneath the microvilli on these cells to provide energy for the absorption of digested food by active transport.

We saw in Section 4.2.5 that mitochondria have an inner membrane which is folded to form cristae in order to increase its surface-area. The cristae are lined with stalked particles. Within the inner mitochondrial membrane there appears to be a mechanism which actively transports protons (H^+) from the matrix into the space between the inner and outer membranes of the organelle. This creates an electrochemical gradient of hydrogen ions across the inner membrane. According to the **chemi-osmotic theory** put forward by the British biochemist Peter Mitchell, in 1961, it is the energy of this 'charged' membrane which is used to synthesize ATP. Basically the

A closer investigation of redox reactions shows that when a substance is oxidized it **loses electrons** and when it is reduced it **gains electrons**. This is the modern definition of the two processes.

	Reduction	Oxidation
Oxygen	lost	gained
Hydrogen	gained	lost
Electrons	gained	lost
Energy	absorbed	released

We can now see a pattern emerging which is helpful when considering the biochemical reactions of metabolism. The build up or synthesis of substances (**anabolism**) involves the reduction of molecules and hence an intake of energy, whereas the breakdown or degradation of substances (**catabolism**) involves the oxidation of molecules and a consequent release of energy. How then does this help our understanding of biological molecules? Clearly substances rich in hydrogen or electrons have more to lose, i.e. they can more easily become oxidized and since oxidation involves the release of energy these substances are more 'energy rich'. Conversely substances rich in oxygen are more likely to be reduced – a process involving the absorption of energy. From the point of view of our food, molecules with much hydrogen and little oxygen have the greatest potential to provide energy. So which foods are these? Let us consider two types: fats and carbohydrates. Typical examples are given in the table, left.

While the ratio of hydrogen to carbon is about 2:1 in both cases, there is proportionally more oxygen in the carbohydrates. This is because they comprise many H—C—OH groups whereas fats have H—C—H groups. In other words the carbohydrates are already partially oxidized and therefore can undergo less further oxidation than fats. Since oxidation releases energy, carbohydrates have less to release – typically 17 kJ per gram compared to 38 kJ for one gram of fat. In terms of energy then, it is the relative amount of hydrogen and oxygen a food molecule contains which is important; the carbon simply acts as a 'skeleton' to which these atoms are attached.

Typical fat	Typical carbohydrate
Stearic acid	Glucose
Formula: $C_{17}H_{35}COOH$	Formula: $C_6H_{12}O_6$

315

hydrogen atoms are picked up by NAD in the matrix and later split into protons and electrons. The protons enter the space between the inner and outer membrane of the mitochondrion while the electrons pass along the cytochromes located within the inner membrane. The protons flow back to the matrix via the stalked granules due to their high concentration in the intermembrane space. This flow acts as the driving force to combine ADP with inorganic phosphate and so synthesize ATP. ATP-ase associated with the stalked granules catalyzes this reaction. These protons then recombine in the matrix with the electrons and the hydrogen atoms so formed then combine with oxygen to form water.

In addition to carrying out oxidative phosphorylation, the mitochondria perform the reactions of the Krebs cycle. The enzymes for these reactions are mostly found within the matrix, with a few, like succinic dehydrogenase, attached to the inner mitochondrial membrane.

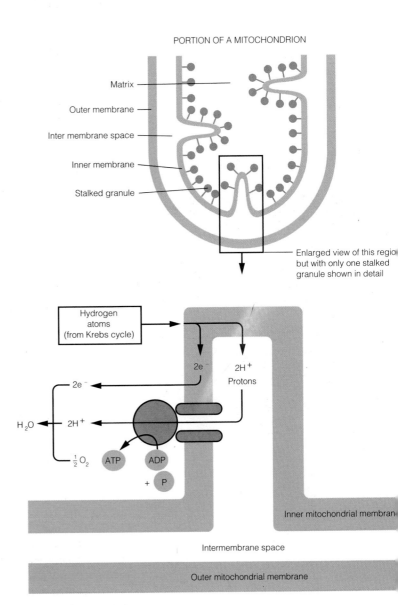

Fig. 16.5 The synthesis of ATP according to the chemi-osmotic theory of Mitchell

16.5   Anaerobic respiration (anaerobiosis)

Present thinking suggests that life originated in an atmosphere without oxygen and the first forms of life were therefore anaerobic. Many organisms today are also anaerobic; indeed, some find oxygen toxic. These forms are termed **obligate anaerobes**. Most anaerobic organisms will, however, respire aerobically in the presence of oxygen, only resorting to anaerobiosis in its absence. These forms are **facultative anaerobes**. The cells of almost all organisms are capable of carrying out anaerobic respiration, for a short time at least. From what we have so far learnt it is clear that, in the absence of oxygen, the Krebs cycle and electron transport system cannot operate. Only glycolysis can take place. This yields a little ATP directly (two molecules for each glucose molecule) and a total of two pairs of hydrogen ions. In the previous section we saw that these hydrogen ions possess much free energy. In the absence of oxygen, however, this energy cannot be released. Nevertheless, these hydrogen ions must be removed if glycolysis is to continue. They are accepted by the pyruvate formed at the end of glycolysis, to give either ethanol (alcohol) or lactate, in a process called **fermentation**. Neither process yields any additional energy; both are merely mechanisms for 'mopping-up' the hydrogen ions.

16.5.1 Alcoholic fermentation

In alcoholic fermentation the pyruvate from glycolysis is first converted to ethanal (acetaldehyde) through the removal of a carbon dioxide molecule.

$$CH_3COCOOH \longrightarrow CH_3CHO + CO_2$$
pyruvate   ethanal   carbon dioxide

The ethanal then combines with the hydrogen ions, which are transported by the hydrogen carrier NAD, to form the alcohol, ethanol.

$$NADH + H^+ \qquad NAD^+$$
$$CH_3CHO \longrightarrow CH_3CH_2OH$$
ethanal   ethanol

This form of fermentation occurs in yeast, where the alcohol produced may accumulate in the medium around the cells until its concentration rises to a level which prevents further fermentation, and so kills the yeast. The ethanol cannot be further broken down to yield additional energy.

The overall equation is:

$$C_6H_{12}O_6 \longrightarrow 2CH_3CH_2OH + 2CO_2$$
glucose   ethanol   carbon dioxide

Under anaerobic conditions, e.g. waterlogging of plant roots, the cells of higher plants may temporarily undergo this form of fermentation. Alcoholic fermentation is of considerable economic importance to humans. It is the basis of the brewing industry, where the ethanol is the important product, and of the baking industry, where the carbon dioxide is of greater value (Section 30.3).

An industrial fermenter

16.5.2 Lactate fermentation

In lactate fermentation the pyruvate from glycolysis accepts the hydrogen atoms from NADH + H$^+$ directly.

$$CH_3COCOOH \xrightarrow{\hspace{1cm} NADH + H^+ \hspace{0.5cm} NAD^+ \hspace{1cm}} CH_3CHOHCOOH$$

pyruvate   lactate
(2-hydroxypropanoic acid)

Unlike alcoholic fermentation, the lactate can be further broken down, should oxygen be made available again, thus releasing its remaining energy. Alternatively it may be resynthesized into carbohydrate, or excreted.

This form of fermentation is common in animals. Clearly any mechanism which allows an animal to withstand short periods without oxygen (anoxia) has great survival value. Animals living in environments of fluctuating oxygen levels, such as a pond or river, may benefit from the temporary use of it, as might a baby in the period during and immediately following birth. A more common occurrence of lactate fermentation is in a muscle during strenuous exercise. During this period, the circulatory system may be incapable of supplying the muscle with its oxygen requirements. Lactate fermentation not only yields a little energy, but removes the pyruvate which would otherwise accumulate. Instead lactate accumulates, and while this in time will cause cramp and so prevent the muscle operating, tissues have a relatively high tolerance to it.

In the process of lactate fermentation, the organism accumulates an **oxygen debt**. This is repaid as soon as possible after the activity, by continued deep and rapid breathing following the exertion. The oxygen absorbed is used to oxidize the lactate to carbon dioxide and water, thereby removing it, and at the same time replenishing the depleted stores of ATP and oxygen in the tissue. In some organisms such as parasitic worms where the food supply is abundant, the lactate is simply excreted, obviating the need to repay an oxygen debt.

16.6   Comparison of energy yields

Let us now compare the total quantity of ATP produced by the aerobic and anaerobic pathways.

Aerobic respiration
The ATP is derived from two sources: directly by phosphorylation of ADP and indirectly by oxidative phosphorylation using the hydrogen ions generated during glucose breakdown.

The figures given represent the yield for each pyruvate molecule which subsequently enters the Krebs (TCA) cycle. As there are two pyruvate molecules formed for each glucose molecule (Section 16.3), all these figures must be doubled ($\times 2$) to give the quantities formed per glucose molecule.

The energy yield for each molecule of NADH + H$^+$ is three ATPs whereas for FADH$_2$ it is only two ATPs (Section 16.4).

TABLE 16.1   ATP yield during aerobic respiration of one molecule of glucose

Respiratory process	Number of reduced hydrogen carrier molecules formed	Number of ATP molecules formed from reduced hydrogen carriers	Number of ATP molecules formed directly	Total number of ATP molecules
Glycolysis (glucose → pyruvate)	$2 \times (NADH + H^+)$	$2 \times 3 = 6$	2	8
pyruvate → acetyl CoA	$1 \times (NADH + H^+)(\times 2)$	$2 \times 3 = 6$	0	6
Krebs (TCA) cycle	$3 \times (NADH + H^+)(\times 2)$ $1 \times FADH_2 (\times 2)$	$6 \times 3 = 18$ $2 \times 2 = 4$	$1(\times 2)$	24
		Total ATP =		38

The total of thirty-eight ATPs produced represents the maximum possible yield; the actual yield may be different depending upon the conditions in any one cell at the time. For example, the two $NADH + H^+$ may enter the mitochondria in two different, indirect ways. Depending on the route taken, they may yield only four ATPs, rather than six ATPs as shown in Table 16.1.

Each ATP molecule will yield 30.6 kJ of energy. The total energy available from aerobic respiration is $38 \times 30.6 = 1162.8$ kJ. Compared to the total energy available from the complete oxidation of glucose of 2880 kJ, this represents an efficiency of slightly over 40%. This may not appear very remarkable, but it compares very favourably with machines – the efficiency of a car engine is around 25%.

Anaerobic respiration
We have seen in the previous section that only glycolysis occurs during anaerobiosis and that the $NADH + H^+$ it yields is not available for oxidative phosphorylation. The total energy released is therefore restricted to the two ATPs formed directly. With each providing 30.6 kJ of energy, the total yield is a mere 61.2 kJ. Compared to the 2880 kJ potentially available from a molecule of glucose, the process is a little over 2% efficient. It must, however, be borne in mind that in lactate fermentation all is not lost, and the lactate may be reconverted to pyruvate by the liver, and so enter the Krebs cycle, thus releasing its remaining energy.

16.7   Alternative respiratory substrates

Sugars are not the only material which can be oxidized by cells to release energy. Both fats and protein may, in certain circumstances, be used as respiratory substrates, without first being converted to carbohydrate. The alternative pathways are shown on Fig. 16.6 on the next page.

16.7.1 Respiration of fat

The oxidation of fat is preceded by its hydrolysis to glycerol and fatty acids. The glycerol may then be phosphorylated and converted into the triose phosphate glyceraldehyde 3-phosphate. This can then be incorporated into the glycolysis pathway and subsequently the Krebs cycle. The fatty acid component is progressively broken down in the matrix of the mitochondria into 2-carbon fragments which are converted to acetyl coenzyme A. This then enters the Krebs cycle with consequent release of its energy. The oxidation of fats has the advantage of producing a large quantity of hydrogen ions. These can be transported by hydrogen carriers and used to produce ATP in the electron (hydrogen) transport system. For this reason, fats liberate more than double the energy of the same quantity of carbohydrate.

16.7.2 Respiration of protein

Protein is another potential source of energy but is only used in cases of starvation. It must first be hydrolysed to its constituent amino acids which then have their amino (NH_2) group(s) removed in the liver – a process called **deamination**. The remaining portions of the amino acids then enter the respiratory pathway at a number of points depending on their carbon content: 5-carbon amino acids (e.g. glutamate) and 4-carbon amino acids (e.g. aspartate) are converted into the Krebs cycle intermediates, α-ketoglutarate and oxaloacetate respectively; 3-carbon amino acids like alanine are converted to pyruvate ready for conversion to acetyl coenzyme A. Other amino acids with larger quantities of carbon undergo transamination reactions to convert them into 3, 4 or 5-carbon amino acids.

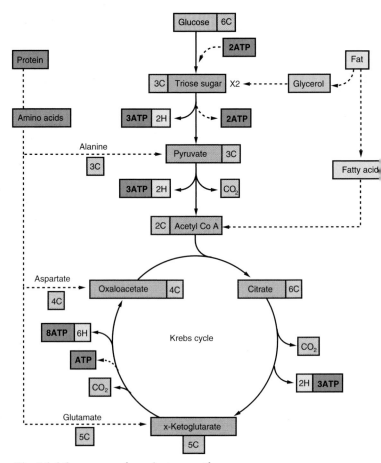

Fig. 16.6 Summary of respiratory pathways

16.8 Respiratory quotients

The **respiratory quotient (RQ)** is a measure of the ratio of carbon dioxide evolved by an organism to the oxygen consumed, over a certain period.

$$RQ = \frac{CO_2 \text{ evolved}}{O_2 \text{ consumed}}$$

For a hexose sugar like glucose, the equation for its complete oxidation is:

$$C_6H_{12}O_6 + 6O_2 \longrightarrow 6CO_2 + 6H_2O$$

The RQ is hence: $\dfrac{6CO_2}{6O_2} = 1.0$

In fats, the ratio of oxygen to carbon is far smaller than in a carbohydrate. A fat therefore requires a greater quantity of oxygen for its complete oxidation and thus has a RQ less than one.

$$C_{18}H_{36}O_2 + 26O_2 \longrightarrow 18CO_2 + 18H_2O$$
stearic acid

$$RQ = \frac{18CO_2}{26O_2} = 0.7$$

The composition of proteins is too varied for them to give the same RQ, but most have values around 0.9.

Organisms rarely, if ever, respire a single food substance, nor are substances always completely oxidized. Experimental RQ values therefore do not give the exact nature of the material being respired. Most resting animals have RQs between 0.8 and 0.9. With protein only respired during starvation, this must be taken to indicate a mixture of fat and carbohydrate as the respiratory substrates.

16.9   Questions

1. (a) Outline the process of anaerobic respiration in cells. (*13 marks*)

 (b) (i) Discuss the possible advantages and disadvantages of anaerobic respiration in yeasts and humans. (*6 marks*)

   (ii) Explain the economic importance of anaerobic respiration in yeasts. (*4 marks*)

   (*Total 23 marks*)

   UCLES (Modular) June 1993,
   (Energy in living organisms), No. 1

2. (a) Make a labelled diagram to show the structure of the organelle in which the Krebs cycle and electron transfer occur. Indicate on the diagram where **each** of these processes takes place. (*4 marks*)

 (b) (i) Describe the rôles in the operation of the Krebs cycle and the electron transport system of
   (A) coenzyme A,
   (B) NAD,
   (C) cytochromes,
   (D) oxygen,
   (E) ADP.

   (ii) Name **one** of the substances, (A)–(E), that performs a similar role in photosynthesis. (*12 marks*)

 (c) Explain the inhibition of the Krebs cycle and the electron transfer system which occurs when there is a limited supply of
   (i) oxygen,
   (ii) ADP or inorganic phosphate. (*4 marks*)
   (*Total 20 marks*)

   NEAB June 1993, Paper IB, No. 6

3. In the process of glycolysis, the formation of pyruvate involves the chemical reaction:

$$NAD^+ + 2H \longrightarrow NADH + H^+$$

What happens to the NADH in
 (a) an animal cell which is respiring aerobically, (*2 marks*)
 (b) a yeast cell which is respiring anaerobically? (*2 marks*)
   (*Total 4 marks*)

   AEB June 1991, Paper I, No. 8

4. Read through the following account of cellular respiration, and then write on the dotted lines the most appropriate word or words to complete the account.

The initial phase in the breakdown of glucose, a process known as ...., takes place in the .... of the cell and eventually results in the production of two molecules of .... from each molecule of glucose. In most organisms, this product then enters the second phase of cellular respiration, known as the .... cycle. This cycle occurs under .... conditions in specific organelles, called the .... During both phases, hydrogen atoms are removed from the substrate and passed to coenzymes such as .... These reactions are catalyzed by enzymes called .... In the respiratory process, energy is released and is used to synthesize energy rich molecules of .... from .... and ...., thereby storing energy for future use.

(*Total 11 marks*)

ULEAC June 1991, Paper I, No. 4

5. Succinate dehydrogenase is an enzyme in the Krebs (TCA) cycle which converts succinate into fumarate by dehydrogenation.

Two experiments were carried out investigating the changes in oxygen concentration in a suspension of isolated liver mitochondria, by placing the suspension in a reaction chamber containing an electrode which measured changes in oxygen concentration. In both experiments, A and B, the mitochondria were kept in a buffer solution containing sucrose and inorganic salts. In experiment B, succinate, a Krebs cycle intermediate, was included, and malonate was added 6 minutes after the experiment had begun. Malonate is a competitive inhibitor of succinate dehydrogenase.

The results are shown in the table below.

Time/ minute	Oxygen concentration/percentage saturation	
	Experiment A	**Experiment B**
0	100	100
2	97	89
4	94	78
6	92	66 (malonate added)
8	89	61
10	86	57
12	83	53

 (a) Plot these data in a suitable form on graph paper. (*5 marks*)
 (b) Explain why the oxygen concentration decreased in both experiments. (*3 marks*)
 (c) Compare the rates of oxygen utilization in experiments A and B during the first six minutes of the experiment and suggest an explanation for any differences you observe. (*3 marks*)

(d) Explain why, in experiment B, the rate of oxygen consumption changed after the addition of malonate. *(3 marks)*

(Total 14 marks)

ULEAC January 1994, Paper I, No. 11

. The diagram below shows stages in cellular respiration.

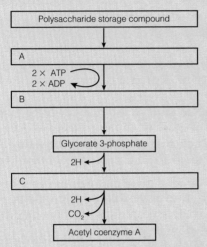

Polysaccharide storage compound

A

$2 \times$ ATP
$2 \times$ ADP

B

Glycerate 3-phosphate

2H

C

2H
CO_2

Acetyl coenzyme A

(a) Write the name of *one* missing compound into each of the boxes A, B and C. *(3 marks)*

(b) (i) What is the name given to the series of reactions from A to C? *(1 mark)*

(ii) Where do these reactions occur in a eukaryotic cell? *(1 mark)*

(c) What happens to the hydrogen atoms removed during this process? *(1 mark)*

(d) Name *one* polysaccharide storage compound found in each of the following.

(i) Mammals *(1 mark)*

(ii) Flowering plants *(1 mark)*

(Total 8 marks)

ULEAC January 1994, Paper I, No. 6

7. *(a)* (i) What is meant by R.Q. (respiratory quotient)? *(2 marks)*

(ii) If the R.Q. of tissue X = 1, of tissue Y = 0.7, and of tissue Z = 2.8, what can you conclude about respiration occurring in tissues X, Y and Z? *(3 marks)*

(b) (i) Describe the method and results of a practical experiment that you have performed in the laboratory to measure quantitatively the respiration rate of some plant or animal material. *(8 marks)*

(ii) State clearly **two** important precautions or problems it is necessary to consider in order to achieve valid results in the experiment you have described in *(b)* (i). *(2 marks)*

(c) Explain the significance of the following in the process of aerobic respiration:

(i) the **use** of ATP (adenosine triphosphate) in glycolysis,

(ii) acetyl CoA,

(iii) NAD (nicotinamide adenine dinucleotide),

(iv) cytochromes,

(v) oxygen. *(15 marks)*

(Total 30 marks)

Oxford, June 1992, Paper II, No. 7

8. The apparatus shown was used in an experiment to investigate the respiratory metabolism of yeast.

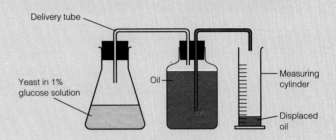

Delivery tube

Yeast in 1% glucose solution

Oil

Measuring cylinder

Displaced oil

(a) Suggest why a smaller amount would be displaced if water were used instead of oil. *(1 mark)*

Using this apparatus the results shown in the graph were obtained. Yeast was added at time 0.

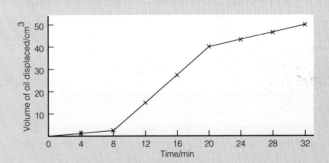

(b) Suggest *one* explanation, in each case, for the rate of carbon dioxide production (volume of oil displaced)

(i) during the first 8 minutes;

(ii) between 8 and 20 minutes;

(iii) between 20 and 32 minutes. *(3 marks)*

(c) Predict what would happen to the shape of the graph if the experiment had been continued beyond 32 minutes. Give a reason for your prediction. *(2 marks)*

In a separate experiment, the ability of the yeast to metabolize a range of different carbohydrates, all at the same concentration, was investigated. The results obtained after 20 minutes are given in the table overleaf.

Carbohydrate	Volume of oil displaced/cm³
Glucose	8.7
Sucrose	5.9
Lactose	0.9
Maltose	6.1
Starch	0.8
Fructose	8.9
None	0.9

(d) Suggest **two** factors, other than carbohydrate concentration and pH, which would need to be standardized in this investigation.   (2 marks)

(e) What was the purpose of the experiment without any carbohydrate added?   (1 mark)

(f) Glucose and fructose give comparable and relatively high values for the volume of carbon dioxide produced. Give explanations for the values obtained for
  (i)   sucrose and maltose,
  (ii) starch and lactose.   (2 marks)
     (Total 11 marks)

NEAB June 1990, Paper IIA, No. 2

9. Read the following passage.

All animals can withstand short periods of total oxygen lack but some species, such as the goldfish, can survive several days under anaerobic conditions. These may occur when the pond in which the
5 goldfish lives is frozen over in winter.

In the complete absence of oxygen, the rate of glucose consumption would be 19 times greater than the aerobic rate if the demand for ATP were the same and the normal glucose-lactate pathway were used.
10 This raised glucose requirement could be satisfied by mobilizing the glycogen in the liver but this would soon deplete the reserves. An alternative is to use glycogen from peripheral tissues. Goldfish store large amounts of glycogen in their skeletal muscles.
15 The goldfish may also use alternative energy-yielding pathways. As an example, conversion of the amino acid aspartate to succinate yields one molecule of ATP for each molecule of aspartate. The succinate produced can later be fed into the Krebs cycle.
20 Lactate is toxic if it accumulates and can create metabolic disruption as well as pH and osmotic problems. If animals respire anaerobically, the lactate must be dealt with, for example, pH changes can be reduced by the buffering capacity of the blood and
25 tissues.

Goldfish have two mechanisms for controlling lactate accumulation under anaerobic conditions. In one of these, the lactate is converted in the pathway:

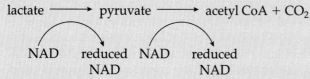

The acetyl coA is then converted to ethanol which is
30 largely excreted or diffuses out into the surrounding water. This is a wasteful process but it seems to be reasonable as a means of survival in the anaerobic conditions of a winter freeze.

The second mechanism involves minimizing the
35 production of waste products by depressing metabolism. The reduction in metabolic rate seen in goldfish at 4 °C may be linked to an increase in the tolerance of anaerobic conditions to eleven days.

When the period of oxygen shortage ends,
40 facultative anaerobes such as goldfish need to re-establish metabolic homeostasis rapidly. The lactate must be cleared and the storage depots of glycogen must be replenished. The lactate is reconverted to glucose and then to glycogen in the
45 liver and the muscles. In the presence of oxygen most tissues are able to oxidize lactate completely.

(Adapted from: *A Companion to Animal Physiology*)

Using information in the passage and your own knowledge, answer the following questions.

(a) Give the meaning of the following words as used in the passage:
  (i)   peripheral (line 13);   (1 mark)
  (ii) facultative (line 40);   (1 mark)

(b) Explain how a winter freeze could lead to anaerobic conditions for a goldfish.   (2 marks)

(c) Explain why 'the rate of glucose consumption would be 19 times greater' in the complete absence of oxygen (line 7).   (2 marks)

(d) Explain why large stores of glycogen enable tissues to tolerate anaerobic conditions for longer periods of time (lines 13 and 14). (2 marks)

(e) How does lactate create osmotic problems (lines 21 and 22)?   (2 marks)

(f) Anaerobic respiration can cause pH problems (lines 21 and 22). Describe **three** ways in which the blood buffers pH.   (4 marks)

(g) What type of enzyme (hydrolase, oxido-reductase or transferase) controls the reaction

    lactate → pyruvate

(line 28)? Give a reason for your answer.
     (2 marks)

(h) Give **two** reasons why depression of metabolism (line 35) reduces the problem of living in anaerobic conditions.   (2 marks)

(i) Why is it necessary to specify the temperature when determining the increase in tolerance to anaerobic conditions (line 37)?   (2 marks)

(j) Draw a simple flow chart to summarize the possible fates of lactate described in the passage. For each reaction path, say whether aerobic or anaerobic conditions are required.
     (4 marks)
     (Total 24 marks)

AEB November 1992, Paper II, No. 2

17

Energy and the ecosystem

Organisms live within a relatively narrow sphere over the earth's surface; it is less than 20 km thick, extending about 8 km above sea level and 10 km below it. The total volume of this thin film of land, water and air around the earth's surface is called the **biosphere**. It consists of two major divisions, the aquatic and terrestrial environments, with the aquatic environment being subdivided into freshwater, marine and estuarine. The terrestrial portion of the biosphere is subdivided into **biomes** which are determined by the dominant plants found there. It is, of course, largely climatic conditions which determine the dominant plant type of a region, and hence the biome. Tropical rain forests, for example, occur where the climate is hot and wet all through the year and are characterized by dense, lush vegetation and an immense variety of species. By contrast tundra occurs where the ground is frozen for much of the year and hence the vegetation is sparse with little variety of species. Deserts are the result of a lack of usable water, either because rain falls all too rarely and soon evaporates in the heat (hot deserts) or because the water is permanently frozen (cold deserts). Temperate deciduous forests occur where the rain is intermittent, the winters are cold and the summers are warm. It is equally possible to subdivide the terrestrial biosphere into **geographical zones**, e.g. Africa, Australia, North America, South America, Antarctica, etc. In this case the divisions are made by barriers like oceans or mountain ranges.

A biome can be further divided into **zones** which consist of a series of small areas called **habitats**. Examples of habitats include a rocky shore, a freshwater pond and a beech wood.

Within each habitat there are **populations** of individuals which collectively form a **community** (see Section 17.6). An individual member of the community is usually confined to a particular region of the habitat, called the **microhabitat**. The position any species occupies within its habitat is referred to as its **ecological niche**. It represents more than a physical area within the habitat as it includes an organism's behaviour and interactions with its living and non-living environment. As Section 17.6.8 shows, no two species can occupy the same ecological niche.

The inter-relationship of the living (**biotic**) and non-living (**abiotic**) elements in any biological system is called the **ecosystem**. There are two major factors within an ecosystem:

1. The flow of energy through the system.

2. The cycling of matter within the system.

APPLICATION

Tropical rain forests

Erosion after logging

Tropical rain forests have been estimated to contain 50% of the world's standing timber. They represent a huge store for carbon and sink for carbon dioxide and their destruction may increase atmospheric concentrations of carbon dioxide by 50%. They are important in conserving soil nutrients and preventing large-scale erosion in regions of high rainfall. They contain a large gene pool of plant resources and their potential for the production of food, fibre and pharmaceutical products is not known. At present rates of destruction, all tropical rain forest, except that in reserves, will have disappeared by the middle of the twenty-first century.

The traditional small-scale slash-and-burn agriculture of the hunter–gatherer cultures caused few environmental problems but attempts to exploit the forest on a large scale have been disastrous. The problems have been most apparent in the rain forests of Brazil and other countries of the Amazon basin. The forest has been cleared for rubber plantations, timber extraction, planting of cash crops like cocoa, coffee and oil palm, extraction of minerals, especially bauxite, and for pulp and paper manufacture. The consequences of deforestation have been soil infertility, floods, soil erosion and increased sediment in rivers. The scale of the problem is huge and there seems to be no decrease in the rate of deforestation as yet. However, some areas are being set aside as reserves and some attempts have been made at replanting.

It is feasible to consider the biosphere as a single ecosystem because, in theory at least, energy flows through it and nutrients may be recycled within it. However, in practice there are much smaller units which are more or less self-contained in terms of energy and matter. A freshwater pond, for example, has its own community of plants to capture the solar energy necessary to supply all organisms within the habitat, and matter such as nitrogen and phosphorus is recycled within the pond with little or no loss or gain between it and other habitats. It is often easier to consider these smaller units as single ecosystems.

17.1 Energy flow through the ecosystem

The study of the flow of energy through the ecosystem is known as **ecological energetics**. All the energy utilized by living organisms is ultimately derived from the sun but as little as 1% of its total radiant energy is actually captured by green plants for distribution throughout the ecosystem. This relatively small amount is nonetheless sufficient to support all life on earth. It may at first glance appear inefficient, but it must be remembered that factors other than light intensity often limit photosynthesis – carbon dioxide, temperature, water and mineral availability, to name a few.

Did you know?

A large swarm of locusts can consume 20 000 tonnes of grain and vegetables in one day.

17.1.1 Food chains

Because green plants manufacture sugars from simple raw materials utilizing solar energy, they are called **primary producers**. All are autotrophic and they include some bacteria as well as green plants.

Organisms that are unable to utilize light energy for the synthesis of food must obtain it by consuming other organisms. These are heterotrophs and include all animals as well as fungi and some bacteria. If they feed off the primary producers they are called **primary consumers**. These are typically herbivores but also include plant parasites.

Some heterotrophs, the carnivores, feed on other heterotrophs. These are called **secondary consumers** if they feed on a herbivore, and **tertiary consumers** if they feed on other carnivores. There is hence a type of feeding hierarchy with the primary producers at the bottom and the consumers at the top. The energy is therefore passed along a chain of organisms, known as a **food chain**. Each feeding level in the chain is called a **trophic level**. Only a small proportion of the available energy is transferred from one trophic level to the next. Much energy is lost as heat during the respiratory processes of each organism in the chain. It is this loss of energy at each stage which limits the length of food chains. It is rare for this reason to find chains with more than six different trophic levels.

On the death of the producers and consumers, some energy remains locked up in the complex organic compounds of which they are made. This is utilized by further groups of organisms which break down these complex materials into simple components again and in doing so contribute to the recycling of nutrients. The majority of this work is achieved by the saprobiontic fungi and bacteria, called **decomposers**, and to a lesser extent by certain animals called **detritivores**. The trophic relationships of these groups are shown on Fig. 17.1 on the next page.

Trophic efficiency
This is the percentage of the energy at one trophic level which is incorporated into the next trophic level. The values differ from one ecosystem to another with some of the highest values, around 40%, occurring in oceanic food chains. At the other

TABLE 17.1   Trophic levels of a food chain in each of 5 different habitats

Trophic level	Habitat				
	Grassland	Woodland	Freshwater pond	Rocky marine shore	Ocean
Quaternary consumers (3° carnivores)	Mammal e.g. stoat	Bird e.g. thrush	Large fish e.g. pike	Bird e.g. gull	Marine mammal e.g. seal
Tertiary consumers (2° carnivores)	Reptile e.g. grass snake	Arachnid e.g. spider	Small fish e.g. stickleback	Crustacean e.g. crab	Large fish e.g. herring
Secondary consumers (1° carnivores)	Amphibian e.g. toad	Carnivorous insect e.g. ladybird	Annelid e.g. leech	Carnivorous mollusc e.g. whelk	Small fish e.g. sand eel larvae
Primary consumers (herbivores)	Insect larva e.g. caterpillar	Herbivorous insect e.g. aphid	Mollusc e.g. freshwater snail	Herbivorous mollusc e.g. limpet	Zooplankton e.g. copepods
Primary producers (e.g. photosynthetic organisms)	Grass e.g. *Festuca*	Tree e.g. oak leaves	Aquatic plant e.g. *Elodea*	Seaweed e.g. sea lettuce	Phytoplankton e.g. diatom

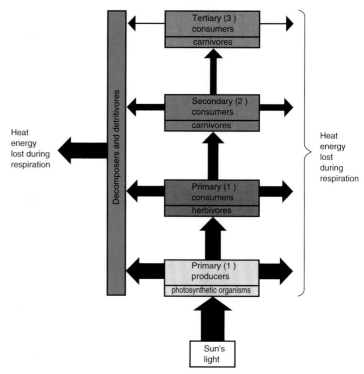

Fig. 17.1 Energy flow through different trophic levels of a food chain

extreme values under 1% have been calculated for small
mammals at the top of food chains, e.g. the shrew.

A classic piece of research carried out by Odum at Silver Springs
in Florida provided the following figures for trophic efficiency:

Photosynthesis – 1.2%
Primary consumers (herbivores) – 15.9%
Secondary consumers (1st level carnivores) – 4.5%
Tertiary consumers (2nd level carnivores) – 6.7%.

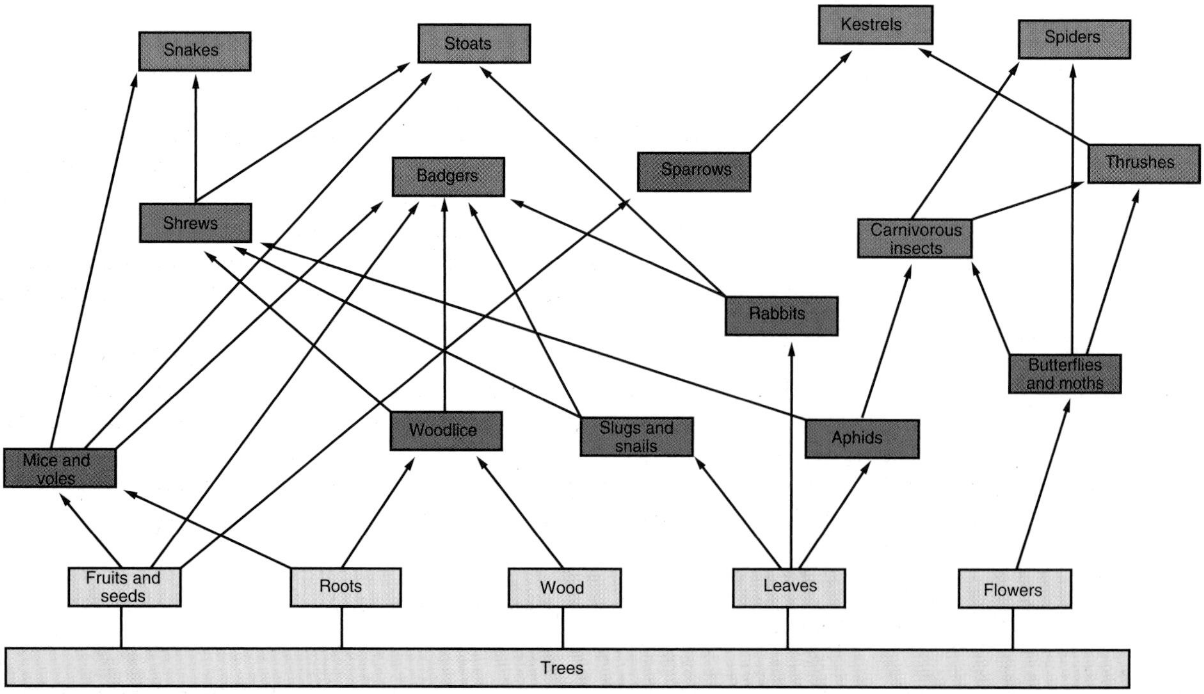

Fig. 17.2 Simplified food web based on a woodland habitat

17.1.2 Food webs

With rare exceptions, the diet of an individual is not restricted to a single food. Most animals feed on many different types. In the same way, an individual is normally a potential meal for many different species. The idea of a food chain as a sequence of species which feed exclusively off the individuals below them in the series is clearly oversimplified. Individual food chains interconnect in an intricate and complex way. A single species may form part of many different chains, not always occupying the same trophic level in each chain. Fig. 17.2 gives an example of a simplified food web for a woodland habitat.

17.1.3 Primary producers

It is the rôle of the photosynthetic organisms which make up the primary producers to manufacture organic substances using light, water and carbon dioxide. The rate at which they produce this organic food per unit area, per unit time, is called **gross primary productivity**. Not all this food is stored; around 20% is utilized by the plant, mainly during respiration. The remainder is called **net primary productivity**. It is this food which is available to the next link in the food chain, namely the primary consumers (herbivores). The net primary productivity depends upon climatic and other factors which affect photosynthesis. It is reduced in certain conditions such as cold, drought, absence of essential minerals, low light intensity, etc. The type of primary producer varies from habitat to habitat and some examples are given in Table 17.1. Chemosynthetic bacteria (Section 14.5) must be considered as primary producers as they do provide energy for ecosystems, although from inorganic chemicals rather than light. Their overall contribution to the provision of the energy in a given ecosystem is very small compared to that provided by photosynthetic organisms.

17.1.4 Consumers

Those consumers which feed on the primary producers are called **primary consumers** or **herbivores**. Of the energy absorbed by a herbivore, only around 30% is actually used by the organism, the remainder being lost as urine and faeces. Some of this 30% is lost as heat, leaving even less of the net productivity of the primary producer to be incorporated into the herbivore and so made available to the next animal in the food chain – the **secondary consumer**. The secondary consumers are the **carnivores** and they often show greater efficiency than herbivores in incorporating available energy into themselves. This is largely because their protein-rich diet is much more easily digested. Not all secondary and tertiary consumers are predators; parasites and scavengers may also fall into these categories depending on the nature of their food.

17.1.5 Decomposers and detritivores

A dead organism contains not only a potential source of energy but also many valuable minerals. Decomposers **(lysotrophs)** are saprobiontic microorganisms which exploit this energy source

Saprobiontic fungi are important decomposers of wood

by breaking down the organic compounds of which the organism is made. In so doing they release valuable nutrients like carbon, nitrogen and phosphorus which may then be recycled (Section 17.2). Apart from dead organisms they also decompose the organic chemicals in urine, faeces and other wastes.

Detritus is the organic debris from decomposing plants and animals and is normally in the form of small fragments. It forms the diet of a group of animals called **detritivores**. They usually differ from decomposers in being larger and in digesting food internally rather than externally. Examples of detritivores include earthworms, woodlice, maggots, dog whelks and sea cucumbers.

17.1.6 Ecological pyramids

Pyramids of numbers

If a bar diagram is drawn to indicate the relative numbers of individuals at each trophic level in a food chain, a diagram similar to that shown in Fig. 17.3(a) is produced. The length of each bar gives a measure of the relative numbers of each organism. The overall shape is roughly that of a pyramid, with primary producers outnumbering the primary consumers which in turn outnumber secondary consumers. Accepting that there is inevitably some loss when energy is transferred from one trophic level to the next in a food chain, it follows that to support an individual at one level requires more energy from the individual at the level below to compensate for this loss. In most instances this can only be achieved by having more individuals at the lower level (Fig. 17.3(a)).

The use of pyramids of numbers has drawbacks, however:

1. All organisms are equated, regardless of their size. An oak tree is counted as one individual in the same way as an aphid.

2. No account is made for juveniles and other immature forms of a species whose diet and energy requirements may differ from those of the adult.

3. The numbers of some individuals are so great that it is impossible to represent them accurately on the same scale as other species in the food chain. For example, millions of blackfly may feed on a single rose-bush and this relationship cannot be effectively drawn to scale on a pyramid of numbers.

These problems may create some different shaped 'pyramids'. Take for example the food chain:

oak tree → aphid → ladybird

The pyramid of numbers produced is illustrated in Fig. 17.3(b). It clearly bulges in the middle.

The food chain:

sycamore → caterpillar → protozoan parasites (of the caterpillar)

produces a complete inversion of the pyramid as illustrated in Fig. 17.3(c).

These difficulties are partly overcome by the use of a pyramid of biomass, instead of one of numbers.

Did you know?

284 different species of insect can survive on a single oak tree.

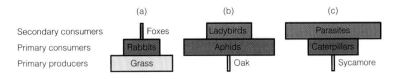

Fig 17.3 Pyramids of numbers

Pyramids of biomass

To overcome some of the problems of using pyramids of numbers, the dry mass of all the organisms at each trophic level may instead be estimated. The relative biomass is represented by bars of proportionate length. While this method is an improvement, it too has its drawbacks:

1. It is impossible to measure exactly the biomass of all individuals in a population. A small sample is normally taken and measured. This sample may not be representative.

2. The time at which a sample is taken may affect the result. Figures for a deciduous tree in summer may be very different from those in winter. What the sample measures is only the amount of material present at a particular instant. This is called the **standing crop** and gives no indication of total productivity. A young tree, for example, is the result of the accumulation of many years' growth, but it may not yet have seeded and produced offspring. A diatom, itself much smaller than the tree, may however have produced many times the tree's biomass in the same period of time. Such anomalies can occasionally lead to inverted pyramids of biomass, e.g. in oceans at certain times of the year zooplankton biomass exceeds phytoplankton biomass, although over the year as a whole the reverse is true.

Pyramids of energy

An energy pyramid overcomes the main drawbacks of the other forms of ecological pyramid. Here the bar is drawn in proportion to the total energy utilized at each trophic level. The total productivity of the primary producers of a given area (e.g. one square metre) can be measured for a given period (e.g. one year). From this, the proportion of it utilized by the primary consumer can be calculated, and so on up the food chain. The pyramids produced do not show any anomalies, but obtaining the necessary data can be a complex and difficult affair. Once again the unifying nature of energy in ecology is apparent.

17.2 The cycling of nutrients

Energy exists in a number of forms, only some of which can be utilized by living organisms. In an ecosystem, energy is obtained almost entirely as light and this is converted to chemical energy which then passes along the food chain. During chemical reactions in living organisms some of this energy is lost as heat – 'lost', because heat is a form of energy which is dissipated to the environment and cannot be re-used by organisms. It is exactly because this heat cannot be recycled that energy flows through ecosystems in one direction only. Like energy, minerals such as carbon, nitrogen and phosphorus exist in different forms. Unlike energy in ecosystems, these forms can be continuously recycled and so used repeatedly by organisms. Most nutrient or mineral cycles have two components:

1. A geological component – This includes rock and other deposits in the oceans and the atmosphere. These form the major reservoirs of the mineral.

2. A biological component – This includes those organisms which in some way help to convert one form of the mineral into another and so recycle it. It therefore includes the producers, the consumers and especially the decomposers.

17.2.1 The carbon cycle

Despite containing less than 0.04% carbon dioxide, the atmosphere acts as the major pool of carbon. The turnover of this carbon dioxide is considerable. It is removed from the air by the photosynthetic activities of green plants and returned as a result of the respiration of all organisms. This may lead to short-term fluctuations in the proportions of oxygen and carbon dioxide in the atmosphere. For example the concentration of CO_2 varies daily, being up to 40 parts per million (ppm) higher at night, and seasonally – it is around 16 ppm lower in summer than in winter. These differences are accounted for by changes in the rate of photosynthesis. Overall however there is a long-term global balance between the two gases. Heterotrophic organisms (animals and fungi) obtain their carbon by eating plants, directly or indirectly. At times in the past, large quantities of dead organisms accumulated in anaerobic conditions and so were prevented from decaying. In time they formed coal, oil and other fossil fuels. The combustion of these fuels returns more carbon dioxide to the atmosphere and has resulted in a rise in its level from an estimated 265 parts per million (ppm) in 1600 to 315 ppm

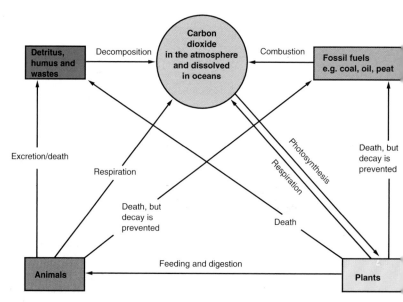

Fig. 17.4 The carbon cycle

in 1958 and to 350 ppm today. Another factor contributing to a rise in atmospheric carbon dioxide has been the clearance of forests. As net users of carbon dioxide trees make a major contribution to the reduction of atmospheric carbon dioxide, especially in tropical rain forests where wet, hot conditions throughout the year favour rapid photosynthesis. It has been estimated that removal of these trees has produced 30% of the increase in atmospheric carbon dioxide which, as a 'greenhouse gas', contributes to global warming (see Section 18.4.3).

17.2.2 The nitrogen cycle

In Chapter 2 (Table 2.1) we saw the range of biologically important chemicals which contain nitrogen, and it is clear from these just how essential a mineral it is to all organisms. Although the atmosphere contains 78% nitrogen, very few organisms can use this gaseous nitrogen directly. Instead they depend upon soil minerals, especially nitrates, as their source of nitrogen. The supply of these nitrates is variable, not least because they are soluble and therefore easily leached from the soil. For this reason a deficiency of nitrates is often the limiting factor to plant growth, and hence the size of the ecosystem it supports. Surrounded by an atmosphere abundant in nitrogen, the growth of many plants is stunted by a lack of it. This is the main reason that it is commonly applied in the form of artificial fertilizers.

Deforestation, especially of tropical rain forests, has also led to nitrate deficiency. With the trees gone there has been large scale erosion due to the high rainfall in these areas washing away the soil. Previously the trees not only acted as a canopy preventing the rain beating directly on the soil, but they also absorbed much of the water which otherwise washes off the surface taking the soil with it. Their roots, in addition, helped to bind the soil particles together. The combination of soil erosion and leaching of nitrates in these areas has impoverished the land making it unfit for vegetation.

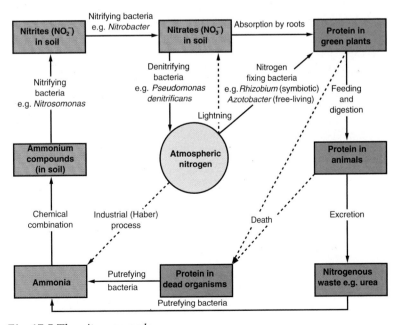

Fig. 17.5 The nitrogen cycle

Nitrification

This is the name given to the series of reactions involved in the oxidation of ammonia to nitrates. The process is carried out in two stages:

1. The oxidation of ammonia or ammonium compounds to nitrites by free-living bacteria e.g. *Nitrosomonas*.

$$\underset{\text{ammonia}}{2NH_3} + \underset{\text{oxygen}}{3O_2} \longrightarrow \underset{\text{nitrite}}{2NO_2^-} + \underset{\text{hydrogen ions}}{2H^+} + \underset{\text{water}}{2H_2O}$$

2. The oxidation of the nitrite by other free-living bacteria e.g. *Nitrobacter* and *Nitrococcus*.

$$\underset{\text{nitrite}}{2NO_2^-} + \underset{\text{oxygen}}{O_2} \longrightarrow \underset{\text{nitrate}}{2NO_3^-}$$

In both cases these chemosynthetic bacteria carry out these processes as a means of obtaining their respiratory energy.

Denitrification

This is the process by which nitrate in the soil is converted into gaseous nitrogen, and thereby made unavailable to the majority of plants. It is carried out by anaerobic bacteria like *Pseudomonas denitrificans* and *Thiobacillus denitrificans*. The necessary anaerobic conditions are more likely in waterlogged soil. Where this occurs, denitrifying bacteria thrive and by converting nitrates to atmospheric nitrogen they reduce soil fertility. It is to avoid this that farmers and gardeners plough or dig up their land in order to improve drainage and aeration, so avoiding anaerobic conditions.

Nitrogen fixation

Nitrogen fixation is the opposite to denitrification in that it converts gaseous nitrogen into a form in which it can be utilized by plants, usually organic nitrogen-containing chemicals. It is carried out by both free-living organisms and organisms living in symbiotic association with leguminous plants.

Free-living **nitrogen-fixing bacteria** include *Azotobacter* and *Clostridium* and **nitrogen-fixing blue-green bacteria** include *Nostoc*. Both types reduce gaseous nitrogen to ammonia which they then use to manufacture their amino acids. Around 90% of the total nitrogen fixation is carried out by free-living microorganisms such as these and they make a worthwhile contribution to soil fertility.

Symbiotic nitrogen-fixing bacteria like *Rhizobium* live mostly in association with leguminous plants such as beans, peas and clover although a few types associate with non-leguminous species. They live in special swollen areas on roots, known as **root nodules**. Details of this relationship are given in Section 15.8.1. The raising of an appropriate leguminous crop as part of a crop rotation scheme has long been used as a means of improving soil fertility. The crop fixes nitrogen during its growth and may later be ploughed into the soil so that its decay can slowly release much needed nitrogen for use by later crops.

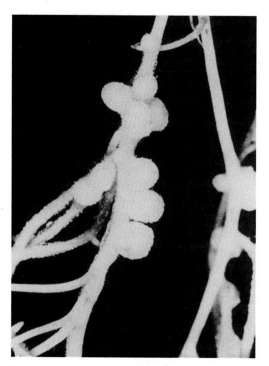

Legume with root nodules

17.3 Ecological factors and their effects on distribution

An organism's environment may be divided into two main parts: the non-living or **abiotic** component and the living or **biotic** component. They work both separately and jointly to influence the distribution and behaviour of organisms in the ecosystem. To help consider these influences we shall examine both components although, for convenience, the abiotic portion will be divided into **climatic** factors and **edaphic** (soil) factors.

17.3.1 Edaphic factors

Soil possesses both living and non-living components. The living portion comprises plant roots and an immense population of microorganisms and small animals. The non-living portion includes particles ranging in size from boulders to fine clay. In addition, there are minerals, water, organic matter and gases. Soil is the result of the weathering of rock which takes two forms:

1. **Physical weathering** – the mechanical breakdown of rock as a result of the action of water, frost, ice, wind and other rocks.

2. **Chemical weathering** – the chemical breakdown as a result of water, acids, alkalis and minerals attacking certain rock types.

The general appearance of a soil seen in vertical section and called a **soil profile** is given in Fig. 17.6.

We have seen that the nature of any ecosystem is dependent upon the type of primary producer and its productivity. Both these factors are, in turn, largely determined by the properties of the soil on which the producer grows. The factors which determine a soil's properties are briefly described below.

Particle size and nature

The size of the constituent mineral particles of soil probably affects its properties, and hence the type of plant which grows on it, more than any other single factor. Soils are classified according to the size of their particles as shown in Table 17.2.

The **texture** of a soil is determined by the relative proportions of sand, silt and clay particles and this affects the agricultural potential of a soil. A clay soil, with its many tiny particles, has the advantage over a sandy soil, with its coarse particles, in holding water more readily and being less likely to have its minerals leached. On the other hand, it may easily become compacted, reducing its air content; it is slower to drain, colder and more difficult to work, especially when wet. A fuller comparison of clay and sandy soils is given in Table 17.3.

The nature of the particles as well as their size affects soil properties. Sand and silt are mainly silica (SiO_2) which is inert. Clay particles, however, have negative charges and these react with minerals in the soil, especially cations. This helps to prevent these nutrient minerals from being leached.

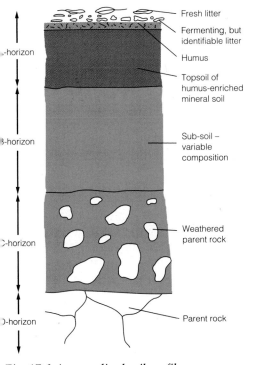

Fresh litter

Fermenting, but identifiable litter

Humus

Topsoil of humus-enriched mineral soil

Sub-soil – variable composition

Weathered parent rock

Parent rock

A-horizon

B-horizon

C-horizon

D-horizon

Fig. 17.6 A generalized soil profile

TABLE 17.2 **Classification of soil particles according to size**

Particle size (diameter in mm)	Particle type
2.00–0.200	Coarse sand
0.20–0.020	Fine sand
0.02–0.002	Silt
<0.002	Clay

TABLE 17.3 **A comparison of clay and sandy soils**

Clay soil	Sandy soil
Particle size is less than 0.002 mm (2 μm)	Particle size from 0.02 mm to 2.0 mm
Small air spaces between particles giving poor aeration	Large air spaces between particles giving good aeration
Poor drainage; soil easily compacted	Good drainage; soil not compacted
Good water retention leading to possible waterlogging	Poor water retention and no waterlogging
Being a wet soil, evaporation of water causes it to be cold	Less water evaporation and therefore warmer
Particles attract many mineral ions and so nutrient content is high	Minerals are easily leached and so mineral content is low
Particles aggregate together to form clods, making the soil heavy and difficult to work	Particles remain separate, making the soil light and easy to work

Did you know?

The number of dormant seeds found in 2.4 ha of soil at Rothampstead Horticultural Research Centre was 300 million.

Organic (humus) content

This includes all dead plant and animal material as well as some animal waste products. Dead animals, leaves, twigs, roots and faeces are broken down by the decomposers and detritivores into a black, amorphous material called **humus**. It has a complicated and variable chemical make-up and is often acidic. It acts rather like a sponge in retaining water and in this way improves the structure of sandy soils. It is equally beneficial to a clay soil where it helps to lighten it by breaking up the clods and thereby improving aeration and drainage. Its slow breakdown releases valuable minerals in both types of soil. This breakdown is carried out by aerobic decomposers and thereby ceases in waterlogged conditions due to the lack of oxygen. In these circumstances the partly decomposed detritus accumulates as **peat**.

Water content

The water content of any well-drained soil varies markedly. Any freely drained soil which holds as much water as possible is said to be at **field capacity**. The addition of more water which cannot drain away leads to waterlogging and anaerobic conditions. Plants able to tolerate these conditions include the rushes (*Juncus* sp), sedges (*Carex* sp) and rice. They have air spaces among the root tissues which allow some diffusion of oxygen from the aerial parts to help supply the roots.

Air content

The space between soil particles is filled with air, from which the roots obtain their respiratory oxygen by direct diffusion. It is equally essential to the aerobic microorganisms in the soil which decompose the humus. They make heavy demands upon the available oxygen and may create anaerobic conditions.

Mineral content

As shown in Chapter 2 (Table 2.1), a wide variety of minerals is necessary to support healthy plant growth. Different species make different mineral demands and therefore the distribution of plants depends to some extent on the mineral balance of a particular soil. Some plants have particular nutrient requirements; the desert shrub, *Atriplex*, for example, requires sodium, a mineral not essential to most species.

Biotic content

Soils contain vast numbers of living organisms. They include bacteria, fungi and algae as well as animals like protozoans, nematodes, earthworms, insects and burrowing mammals. Bacteria and fungi carry out decomposition, while burrowing animals such as earthworms improve drainage and aeration by forming air passages in the soil. Earthworms also improve fertility by their thorough mixing of the soil which helps to bring leached minerals from lower layers within reach of plant roots. They may improve the humus content through their practice of pulling leaves into their burrows. By passing soil through their bodies they may make its texture finer.

pH

The pH of a soil influences its physical properties and the availability of certain minerals to plants. Plants such as heathers, azaleas and camelias grow best in acid soils, while dog's mercury and stonewort prefer alkaline ones. Species which are tolerant to extremes of pH can become dominant in certain areas because competing species find it hard to survive in these extreme conditions. The dominance of heathers on upland moors is, in part, due to their ability to withstand very low soil pH. Most plants, however, grow best in an optimum pH close to neutral.

Temperature

All chemical and biological activities of a soil are influenced by temperature. The temperature of a soil may be different from that of the air above it. Evaporation of water from a soil may cool it to below that of the air whereas solar radiation may raise it above air temperature. Germination and growth depend on suitable temperatures and the optimum varies from species to species. The activity of soil organisms is likewise affected by changes in temperature, earthworms becoming dormant at low temperatures, for example.

Topography

Three features of topography may influence the distribution of organisms:

1. **Aspect** (slope) – South-facing slopes receive more sunlight, and are therefore warmer than north-facing ones (in the northern hemisphere).

2. **Inclination** (steepness) – Water drains more readily from steep slopes and these therefore dry more quickly than ones with a shallower gradient.

3. **Altitude** (height) – At higher altitudes the temperature is lower, the wind speed is greater and there is more rainfall.

17.3.2 Climatic factors

The world's major biomes are largely differentiated on the basis of climate. From the warm, humid tropical rain forest to the cold arctic tundra it is the prevailing weather conditions which determine the predominant flora and hence its attendant fauna. Each climatic zone has its own community of plants and animals which are suited to the conditions. The adaptations of plants and animals to these conditions are dealt with elsewhere in this book and what follows is merely a general review of the major climatic variables within ecosystems.

Light

As the ultimate source of energy for ecosystems, light is a fundamental necessity. Light is not only needed for photosynthesis, however. It plays a role in such photoperiodic behaviour as flowering in plants, and reproduction, hibernation and migration in animals. Phototaxis and phototropism in plants as well as visual perception in animals require light. There are three aspects to light – its wavelength, its intensity and its duration. The influence of these on photosynthesis is dealt with in Chapter 14, and on photoperiodism in Chapter 29.

Tropical rain forest

Tundra

Temperature

Just as the sun is the only source of light for an ecosystem, so it is the main source of heat. The temperature range within which life exists is relatively small. At low temperatures ice crystals may form within cells, causing physical disruption, and at high temperatures enzymes are denatured. Fluctuation in environmental temperature is more extreme in terrestrial habitats than aquatic ones because the high heat capacity of water effectively buffers the temperature changes in aquatic habitats. The actual temperature of any habitat may differ in time according to the season and time of day, and in space according to latitude, slope, degree of shading or exposure, etc.

Water

Water is essential to all life and its availability determines the distribution of terrestrial organisms. The adaptations of terrestrial organisms to conserve water are discussed in Chapters 22 and 23. Even aquatic organisms do not escape problems of water shortage. In saline conditions water may be withdrawn from organisms osmotically, thus necessitating adaptations to conserve it. The salinity of water is a major factor in determining the distribution of aquatic organisms. Some fish, e.g. roach and perch, live exclusively in fresh water; others, like cod and herring, are entirely marine. A few fish, like salmon and eels, are capable of tolerating both extremes during their life.

Air and water currents

Air movements may affect organisms indirectly, for example by evaporative cooling or by a change in humidity. They may also affect them directly by determining their shape; the development of branches and roots of trees in exposed situations is an example. Wind is an important mechanism for dispersing seeds and spores. In the same way that the air currents determine the distribution of certain species in terrestrial habitats, so too do water currents in aquatic ones.

Humidity

Humidity has a major bearing on the rate of transpiration in plants and so affects their distribution. Although to a lesser degree, it affects the distribution of some animals by affecting the rate of evaporation from their bodies.

17.3.3 Biotic factors

Relationships between organisms are obviously varied and complex and are detailed throughout the book, but a brief outline of a few major biotic factors which affect organisms' distribution is given below.

Competition

Organisms compete with each other for food, water, light, minerals, shelter and a mate. They compete not only with members of other species – **interspecific competition** – but also with members of their own species – **intraspecific competition**. Where two species occupy the same ecological niche, the interspecific competition leads to the extinction of one or the other – the **competitive exclusion principle** (Section 17.6.8).

Predation

The distribution of a species is determined by the presence or absence of its prey and/or predators. The predator–prey relationship is an important aspect in determining population size (Section 17.6.7).

Antibiosis

Organisms sometimes produce chemicals which repel other organisms. These may be directed against members of their own species. Many mammals, for example, use chemicals to mark their territories, with the intention of deterring other members of the species from entering. Some ants produce a type of external hormone called a **pheromone** when they are in danger and, in sufficient concentrations, this warns off other members of the species. The chemicals may also be directed against different species. Many fungi, e.g. *Penicillium*, produce **antibiotics** to prevent bacterial growth in their vicinity.

Dispersal

Many organisms depend upon another species to disperse them. Plants in particular use a wide variety of animal species to disperse their seeds.

Pollination

Angiosperms utilize insects to transfer their pollen from one member of a species to another, and a highly complex form of interdependence between these two groups has developed. This is discussed in Section 12.3.2.

Hoverfly mimicking a wasp

Mimicry

Many organisms, for a variety of reasons, seek to resemble other living organisms. Warning mimicry is used by certain flies which resemble wasps. Potential predators are warned off the harmless flies, fearing they may be stung.

Human influence

Humans influence the distribution of other organisms more than any other single species. As hunters, fishers, farmers, developers and polluters, to name a few activities, they dictate which organisms grow where. Some aspects of these influences are considered in Chapter 18.

17.3.4 Species diversity index

The number and range of different species found in an ecosystem is called its **species diversity**. A measure of species diversity is helpful when considering the interaction of the edaphic, climatic and biotic factors which influence an ecosystem. In general, a stable ecosystem has a wide range of different species each with a similar population size. A less stable ecosystem, i.e. one which is under stress due to pollution or extreme climatic conditions, has just a few species with very large populations.

One method of measuring species diversity is the **Simpson index**. It is most often used to estimate plant diversity and involves counting the numbers of each type found in a given area. The diversity is then calculated using the formula:

$$D = \frac{N(N-1)}{\Sigma n(n-1)}$$

where
D = diversity index
N = total number of plants
n = total number of species
Σ = sum of.

17.4 Ecological techniques

17.4.1 Measurement of environmental parameters

It is essential in any ecological study to be able to measure variations in environmental conditions both in time and space. The main parameters which the ecologist measures are:

Temperature

In ecological studies the precise temperature at any one moment is of little value. Of much greater significance are the **diurnal (daily)** and **seasonal** temperature variations. These variations can be measured using a mercury thermometer by taking readings at regular intervals. The best method of determining the highest and lowest temperature over a period of time is to use a maximum–minimum thermometer which leaves a marker at the highest temperature recorded and another at the lowest. For relatively inaccessible positions, e.g. under small stones, at different depths in a lake, a miniaturized **thermistor** may be used. This electrical instrument measures resistance, which changes with temperature. Using calibration scales it is possible to determine the actual temperature which corresponds to a given resistance.

pH

pH is a measure of the acidity or alkalinity of a solution. It is most easily determined by use of **universal indicator**, either as a liquid or impregnated on test paper. More accurate is the use of a **pH meter**. The probe of the instrument is rinsed in distilled water (pH 7 – neutral) before being placed in the test solution. The reading is noted. A calibration graph or scale must be made by taking readings of a series of buffer solutions of known pH. From this calibration graph the pH corresponding to any reading can be determined.

Light

Two aspects of light, its duration and intensity, are generally important to ecological studies. The duration of daylight hours can be determined astronomically and is predictable for any location. The values are available for any given day, from most diaries. The intensity of light is most easily measured by use of an ordinary photographic light meter.

Humidity

Atmospheric humidity is normally expressed as **relative humidity**, i.e. the water content of a given volume of air relative to the same volume of fully saturated air. It is measured using a

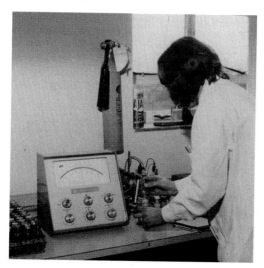
pH meter in use

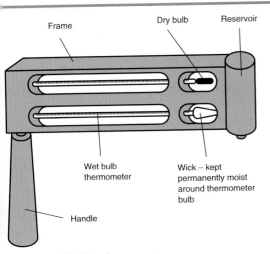

Fig. 17.7 Whirling hygrometer

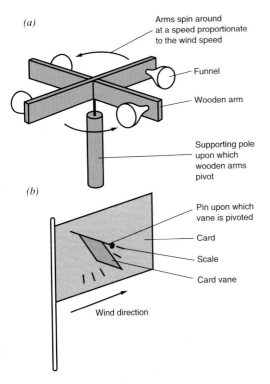

Fig. 17.8 Simple wind gauges

whirling hygrometer. This has two thermometers, the bulb of one being kept dry while the other is permanently wet. Both are mounted in a frame (Fig. 17.7). The hygrometer is rotated in the air until both thermometers give a constant reading. The wet bulb thermometer will always give a lower reading than the dry bulb one, due to the cooling effect of the evaporating water – the less humid the air, the more evaporation and the greater the temperature difference between the two thermometers. The actual humidity is determined by reading off the temperatures on a special scale.

Wind and water speed

Wind speed is measured by an instrument called an **anemometer**. Simple, but effective, wind gauges can be improvised and two examples are given in Fig. 17.8. These may not provide a direct measure of wind speed but will give comparative readings, e.g. inside and outside a wood, or at various vertical heights.

The easiest method of determining water speed is to time the movement of a floating object over a measured distance. To avoid inaccuracies caused by the wind blowing the portion floating above the water, it should be weighted so that it hardly breaks the surface. A specimen tube partly filled with water works admirably.

Salinity

The salinity of a water sample is best measured using a **conductivity meter**. This instrument measures the conductivity between two probes; the greater the salinity the greater the conductivity. It is also possible to determine the concentration of a particular ion. For example, chloride concentration can be determined by titration against silver nitrate solution, using potassium chromate as an indicator.

Oxygen level

An instrument known as an **oxygen meter** can be used to give a measurement of the oxygen concentration in a water sample. This has now largely replaced the chemical technique involving titration, known as the Winkler method.

17.4.2 Sampling methods

It is virtually impossible to identify and count every organism in a habitat. For this reason only small sections of the habitat are usually studied in detail. Provided these are representative of an area as a whole, any conclusions drawn from the findings will be valid. There are four basic sampling techniques.

Quadrats

A quadrat (Fig. 17.9) is a sturdily built wooden frame, often designed so it can be folded to make it more compact for storage and transport. It is placed on the ground and the species present within the frame are identified and their abundance recorded. Where the species are small and/or densely packed, one or more of the smaller squares within the frame may be used rather than the quadrat as a whole.

PROJECT

Use quadrats to investigate the distribution of plant or animal species in two different localities, for example:

(a) sheltered and exposed rocky shores
(b) different types of wood
(c) grazed and ungrazed grassland, etc.

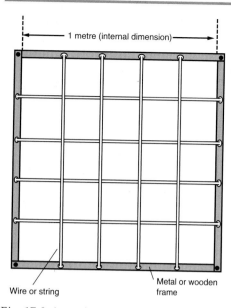

Fig. 17.9 A quadrat frame

Sampling with a quadrat may be random or systematic. **Random sampling** can be as simple as throwing a quadrat over one's shoulder and counting the species within it wherever it falls. Even with the best of intentions it is difficult not to introduce an element of personal bias using this method. A better form of random sampling is to lay out two long tape measures at right angles to each other, along two sides of the study area. Using random numbers generated on a computer or certain calculators, a series of coordinates can be obtained. The quadrat is placed at the intersection of each pair of coordinates and the species within it recorded. **Systematic sampling** involves placing the quadrat at regular intervals, for example, along a transect. It is sometimes necessary to sample the same area over many years in order to investigate seasonal changes or monitor ecological succession. In these circumstances a rectangular area of ground may be marked out by boundary stakes which are connected by rope. This is known as a **permanent quadrat**.

Point frames

A point frame, or point quadrat (Fig. 17.10), consists of vertical legs across which is fixed a horizontal bar with small holes along it. A long metal pin, resembling a knitting needle, is placed in each of the holes in turn. Each time the pin touches a species, it is recorded. The point frame is especially useful where there is dense vegetation as it can sample at many different levels.

PROJECT

Use transects to investigate the distribution of plant or animal species, for example:

(a) down a seashore
(b) from the middle to the edge of a small wood
(c) across sand dunes
(d) across a peat bog
(e) across a path in grassland, etc.

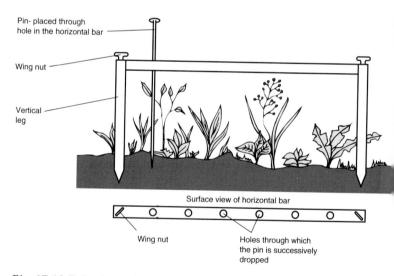

Fig. 17.10 Point frame (point quadrat)

Line transect

A line transect is used so that systematic sampling of an area can be carried out. A string or tape is stretched out along the ground in a straight line. A record is made of the organisms touching or covering the line all along its length, or at regular intervals. This technique is particularly useful where there is a transition of flora and/or fauna across an area, down a sea shore for example. If there is any appreciable height change along the transect, it is advisable to construct a profile of the transect to indicate the changes in level. This is especially important where vertical height is a major factor in determining the distribution of

Using a quadrat along a belt transect

TABLE 17.4 Collecting apparatus of general use

Apparatus	Purpose
Specimen tube (small)	With tight-fitting caps and of varying sizes, they can be used to contain small organisms while they are identified, or used to transport them
Screw-topped jars (large)	To collect and transport smaller organisms. Also useful for aquatic organisms
Polythene bags (various sizes)	To collect plant material, soil samples and other non-living material
Forceps	For transferring plants and hard-bodied animals as well as non-living materials, e.g. stones
Paint brush	For transferring small, delicate or soft-bodied organisms, e.g. aphids
Bulb pipette	For transferring small aquatic organisms
Pooter (aspirator) (Fig. 17.11)	For collecting and transferring small animals, e.g. insects and spiders
Widger	A useful all-purpose tool, similar to a spatula, used for digging, levering and transferring material
Sieve	For sifting sand, soil, mud and pond-water for small organisms
Hand lens	To magnify features of organisms to help in their identification
Enamel dish	For sorting specimens

species. On a sea shore, for example, the height above the sea affects the duration of time any point is submerged by the tide. This has a considerable bearing on the species that can survive at that level. So, the distribution of species is related to the vertical height on the shore rather than the horizontal distance along it. This form of transect is called a **profile transect**.

Belt transect

A belt transect is a strip, usually a metre wide, marked by putting a second line transect parallel to the other. The species between the lines are carefully recorded, working a metre at a time. Another method is to use a frame quadrat in conjunction with a single line transect. In this case the quadrat is laid down alongside the line transect and the species within it recorded. It is then moved its own length along the line and the process repeated. This gives a record of species in a continuous belt, but the quadrat may also be used at regular intervals, e.g. every 5 m, along the line.

17.4.3 Collecting methods

Reliable methods of collecting organisms are an essential part of ecology. Because they photosynthesize, plants always occur in the light and are hence visible and usually large and stationary. All these factors make them easy to find and collect. Animals by contrast may live underground, in crevices, or simply be camouflaged; this makes them less conspicuous. Even when seen, the animal's ability to move from place to place presents problems of capture. Collecting all organisms within a habitat is normally impractical and therefore small areas are selected. Organisms should be identified on site, but if removing them is unavoidable, as few as possible should be taken and for as short a time as possible and always returned to the same location. When collecting specimens, as much information as possible should be recorded at the time. This should include details of the time, date, location, substrate, climate and any other relevant data.

In addition to collecting specimens by hand or with the use of a pooter, there are other pieces of apparatus of varying degrees of complexity which may be used to lure and trap animals.

Beating tray

This is a fabric sheet on a collapsible frame. It is held under a part of a bush or a tree which is then shaken or disturbed with a stick. The organisms which are dislodged are collected by hand or with the aid of a pooter (Fig. 17.11). It is used to collect small non-flying terrestrial organisms, e.g. beetles, spiders, caterpillars.

Light traps

Any light source will attract certain nocturnal flying insects. A very simple light trap may be made by placing a vertical sheet at the side of a light source, and a horizontal one beneath it. This is necessary as some insects prefer to rest vertically and others horizontally. More effective traps involve **mercury vapour lamps**. These emit much ultraviolet light which is particularly attractive to nocturnal insects such as moths.

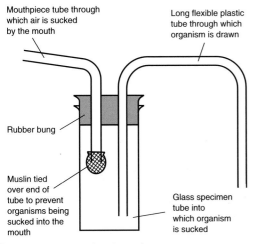

Mouthpiece tube through which air is sucked by the mouth

Long flexible plastic tube through which organism is drawn

Rubber bung

Muslin tied over end of tube to prevent organisms being sucked into the mouth

Glass specimen tube into which organism is sucked

Fig. 17.11 Pooter (aspirator)

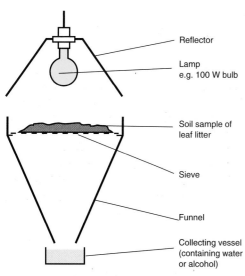

Fig. 17.12(a) Tullgren funnel

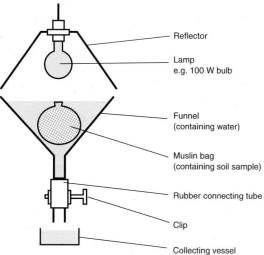

Fig. 17.12(b) Baermann funnel

Tullgren funnel

This is used to extract small animals from a sample of soil or leaf litter. The soil sample is placed on a coarse sieve and light and moderate heat are used to drive the animals downwards through the sieve. (Fig. 17.12(a)). They fall into a funnel which directs them into a collecting vessel.

Baermann funnel

This is again used to extract soil animals and is particularly effective for worms, especially nematodes. The soil sample is contained within a muslin bag which is then submerged in water in a funnel (Fig. 17.12(b)). A tungsten bulb may be used as a source of heat which, along with the water, induces the organisms to leave the sample. They collect in the neck of the funnel from where they can be periodically removed.

Mammal traps

The best live trap is the **Longworth trap**. It is placed in situations which small mammals such as mice and voles frequent, like a runway. It comprises a metal box with a single entrance which closes firmly behind the mammal when it enters. The box is baited with bedding and the appropriate food to entice the animal to enter. The behaviour of many small mammals creates problems when this trap is used to assess population sizes, because some individuals, called 'trap-shy', never enter the trap while others, called 'trap-happy', actively seek them out for the meal and bed they provide.

Pitfall traps

A jam-jar or similar vessel is sunk into the ground with its rim level with the soil. It is baited with the appropriate food, e.g. decaying meat to attract scavenging insects such as beetles, or honey to attract ants. Having fallen in, the insects are unable to climb the smooth walls of the jar to escape.

Netting

This is a popular method of capture and takes many forms. Hand-held nets with short handles give greater precision for catching insects in flight. With some insects it is better to stalk them until they settle before netting them. These nets are called **kite nets**. A more robust form, called the **sweep net**, is used to collect insects from foliage. It is swept along grass or through bushes, dislodging insects which fall into the net. This net may also be used to collect aquatic animals by sweeping it through streams or ponds. A **plankton net** is made of bolting silk because its fine mesh, while allowing water through, traps even microscopic organisms. It has a wide mouth held open by a circular metal frame and narrows down to a small collecting jar at the other end in which the plankton accumulate. The net is towed slowly through the water, usually behind a small boat.

17.5 Estimating population size

To count accurately every individual of any species within a habitat is clearly impractical, and yet much applied ecology requires information on the size of animal and plant populations. It is necessary therefore to use sampling techniques in particular ways in order to make estimates of the size of any population. The exact methods used depend not only on the nature of the habitat but also on the organism involved. Whereas, for example, it may be useful to know the number of individuals in an animal population, this may be misleading for a plant species, where the percentage cover may be more relevant.

17.5.1 Using quadrats

By sampling an area using quadrats and counting the number of individuals within each quadrat, it is possible to estimate the total number of individuals within the area. If, for example, an area of 1000 m² is studied and 100 quadrats, each 1 m², are sampled, it follows that a total of 100 m² of the area has been sampled. This represents one tenth of the total. The total number of individuals of a species in all 100 quadrats must therefore be multiplied by ten to give an estimate of the total population of that species in the area. The use of quadrats in estimating population size is largely confined to plants and sessile, or very slow-moving animals. Faster-moving animals would simply disperse upon being disturbed.

17.5.2 Capture–recapture techniques

The capture–recapture method of estimating the size of a population is useful for mobile animals which can be tagged or in some other way marked. A known number of animals are caught, clearly marked and then released into the population again. Some time later, a given number of individuals is collected randomly and the number of marked individuals recorded. The size of the population is calculated on the assumption that the proportion of marked to unmarked individuals in this second sample is the same as the proportion of marked to unmarked individuals in the population as a whole. This, of course, assumes that the marked individuals released from the first sample distribute themselves evenly among the remainder of the population and have sufficient time to do so. This may not be the case, due to deaths, migrations and other factors. Another problem is that while the tag or label may not itself be toxic, it often renders the individual more conspicuous and so more liable to predation. In this case the number of marked individuals surviving long enough to be recaptured is reduced and the size of the population will consequently be over-estimated. 'Trap-shy' and 'trap-happy' individuals (see Section 17.4.3) will also adversely influence the estimate. The population size can be estimated using the calculation below:

quadrat in use

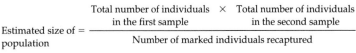

$$\text{Estimated size of population} = \frac{\substack{\text{Total number of individuals} \\ \text{in the first sample}} \times \substack{\text{Total number of individuals} \\ \text{in the second sample}}}{\text{Number of marked individuals recaptured}}$$

The estimate calculated is called the **Lincoln index**.

The method can be used on a variety of animals; arthropods may be marked on their backs with non-toxic dabs of paint, fish can have tags attached to their opercula, mammals may have tags clipped to their ears and birds can have their legs ringed.

17.5.3 Abundance scales

The population size may be fairly accurately determined by making some form of **frequency assessment**. These methods are subjective and involve an experimenter making some estimate of the number of individuals in a given area, or the percentage cover of a particular species. This is especially useful where individuals are very numerous, e.g. barnacles on a rocky shore, or where it is difficult to distinguish individuals, e.g. grass plants in a meadow. The assessments are usually made on an abundance scale of five categories. These are given, with two examples of how they are used for different species, in Table 17.5.

TABLE 17.5 **Abundance scales for two rocky shore species as devised by Crisp and Southward (1958)**

Abundance group	Symbol	Scale used for the limpet (*Patella*)	Scale used for barnacle e.g. *Chthamalus stellatus*
Abundant	A	Over $50\,m^{-2}$	Over $1\,cm^{-2}$ (rocks well covered)
Common	C	$10–50\,m^{-2}$	$10–100\,dm^{-2}$ (up to one third of rock covered)
Frequent	F	$1–10\,m^{-2}$	$1–10\,dm^{-2}$ (never less than 10 cm apart)
Occasional	O	Less than $1\,m^{-2}$	$10–100\,m^{-2}$ (few less than 10 cm apart)
Rare	R	Only a few found in 30 minutes' searching	Less than $1\,m^{-2}$ (only a few found in 30 minutes searching)

17.6 Populations and communities

In previous chapters we reviewed the variety of organisms. These organisms live, not in isolation, but as part of populations and communities.

A **population** is a group of individuals of the same species, all occupying a particular area at the same time.

A **community** comprises all the plants and animals which occupy a particular area. Communities therefore consist of a number of populations.

17.6.1 Population growth

Provided the birth rate exceeds the death rate, a population will grow in size. If only a few individuals are present initially, the rate of growth will be very slow. This is called the **lag phase**. As numbers increase, more individuals become available for reproduction and the population grows at an ever increasing rate, provided no factor limits growth. This is called the

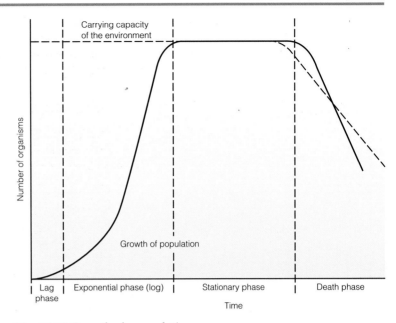

Fig. 17.13 Growth of a population

exponential phase. Growth cannot continue indefinitely because there is a limit to the number of individuals any area can support. This limit is called the **carrying capacity** of the area. Beyond this point certain factors limit further population growth. The size of the population may then stabilize at a particular level. This is called the **stationary phase**. The high population level may, however, cause the carrying capacity of the environment to decline. In these circumstances the population level falls. This is called the **death phase**.

The factors which limit the growth of a particular population are collectively called the **environmental resistance**. Such factors include predation, disease, the availability of light, food, water, oxygen and shelter, the accumulation of toxic waste and even the size of the population itself.

17.6.2 Density-dependent growth

In this type of growth a population reaches a certain size and then remains stable. It is referred to as density-dependent because the size (or density) of the population affects its growth rate. Typical density-dependent factors are food availability and toxic waste accumulation. In a small population, little food is used up and only small amounts of waste are produced. The population can continue to grow. At high population densities the availability of food is reduced and toxic wastes build up. These cause the growth of the population to slow, and eventually stabilize at a particular level.

17.6.3 Density-independent growth

In this type of growth a population increases until some factor causes sudden reduction in its size. Its effect is the same regardless of the size of the population, i.e. it is independent of the population density. A typical density-independent factor is temperature. A sudden fall in temperature may kill large numbers of organisms regardless of whether the population is large or small at the time. Environmental catastrophes such as fires, floods or storms are other density-independent factors.

17.6.4 Regulation of population size

The maximum possible number of offspring varies considerably from species to species. It may be as little as one offspring in two years in some mammals, or as great as one million eggs in a single laying in certain molluscs like the oyster. The term **fecundity** is used to describe the reproductive capacity of individual females of a species. In mammals the **birth rate** or **natality** is used to measure the fecundity. On the other hand the number of individuals of a species which die from whatever cause, is called the **death rate** or **mortality**. Clearly the size of a population is regulated by the balance between its fecundity and its mortality. However, there are other influences on the size of a population. Two such influences are immigration and emigration.

Immigration occurs when individuals join a population from neighbouring ones. **Emigration** occurs when individuals depart from a population. The emigrants may either enter an existing neighbouring population or, as in the swarming of locusts and bees, they may form a new population. Factors such as overcrowding often act as a stimulus for emigration. Unlike the periodic seasonal movements which occur in migration, emigration is a non-reversible, one-way process.

The size of a population may fluctuate on a regular basis, called a **cycle**. These fluctuations are normally the consequence of regular seasonal changes, such as temperature or rainfall. At other times the population may be subject to sudden and unexpected fluctuations. While both types of fluctuation are usually due to a number of factors, there is often one, called the **key factor**, which is paramount in bringing about the change.

17.6.5 Control of human populations

Most animal populations are kept in check by food availability, climate, disease or predators. Populations frequently increase in size rapidly and then undergo a sudden 'crash' during which there is a dramatic reduction in numbers.

In increasingly more regions of the world human knowledge, expertise and technology are succeeding in reducing the impact of the climate and disease. As the top organism in many food chains, humans have little to fear from predators. Even in food production humans have made considerable advances, although in parts of South America, Asia and Africa famine remains a major check to population growth. As a result, the human population as a whole has grown virtually unchecked in recent times. Figures for the past rate of increase in human populations can only be estimated, but it seems probable that prior to 1600 it had taken around 2000 years to double the world's population. By 1850, it had doubled again. It took just 80 years to complete the next doubling in 1939 and a mere 50 years to double again. At present the world population increases by about 1 500 000 every week, equivalent to about 150 people every minute.

Did you know?

The world's population is growing each year by more than the total population of Britain.

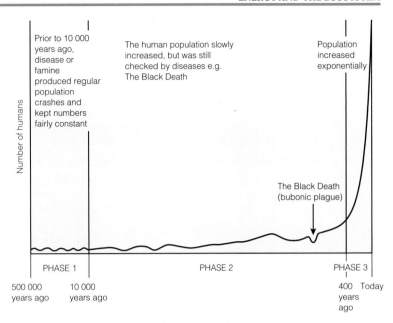

Fig. 17.14 The growth of the human population

In 1798, Thomas Robert Malthus, an English economist, published an essay on population in which he suggested that while the world's food supply would increase arithmetically, the human population would do so geometrically. The so-called **Malthusian principle** suggested famine as the inevitable consequence of this state of affairs. Despite many important agricultural advances, much of the world's population is still undernourished. It is inconceivable that the present rate of population growth can be sustained for much longer. War, famine or disease will inevitably curb further increases unless humans reduce their birth rate by appropriate forms of birth control. The variety of birth control methods available are given in Chapter 12 (Table 12.4). The solution may seem simple enough, but opposition to birth control is often deeply rooted in personal, social, religious or traditional belief.

17.6.6 Competition

Individuals of species in a population are continually competing with each other, not only for nutrients but also for mates and breeding sites. This competition between individuals of the same species is called **intraspecific competition**. Individuals are also in continual competition with members of different species for such factors as nutrients, space and shelter. Competition between individuals of different species is called **interspecific competition**.

17.6.7 Predation

Predator–prey relationships are important in producing cyclic change in the size of a population. By eating their prey, predators remove certain members of a population and so reduce their numbers. As the size of the prey population diminishes, the predators experience greater competition with each other for the remaining prey. The predator population therefore diminishes as some individuals are unable to obtain

349

enough food to sustain them. The reduction in the predator population results in fewer prey being taken and so allows their numbers to increase again. This increase in its turn leads to an increase in the predator population.

The number of predators is usually less than the number of prey.
The shape of the two curves is similar, but there is a time lag between the two; the curve for the predators lags behind that of the prey.

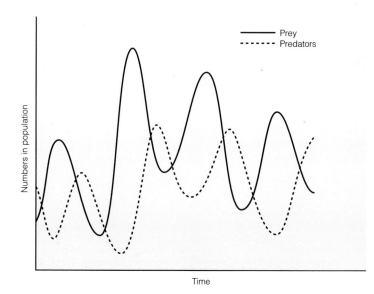

Fig. 17.15 Relationship between prey and predator populations

Typical examples of the relationship shown in Fig. 17.15, include the lynx preying upon the Canadian snowshoe hare and *Hydra* preying upon the water flea *Daphnia*. It must be said that predator–prey relationships alone are not responsible for fluctuations in the numbers in a population; disease and climatic factors also play a rôle. Nevertheless, predation is significant in the regulation of natural populations. The type of cyclic fluctuation shown in Fig. 17.15 plays an important role in evolution. The periodic population crashes create selection pressure whereby only those individuals who are able to escape predation, or withstand disease or adverse climatic conditions, will survive to reproduce. The population thereby evolves to be better adapted to the prevailing conditions.

17.6.8 Competitive exclusion principle

In 1934, a Russian biologist, C. F. Gause, experimented on two species of *Paramecium*. He grew *P. caudatum* and *P. aurelia* both separately and together. When grown together, the two species competed for the available food. After a few days the population of *P. caudatum* began to decline, and after three weeks all its members had died. It seemed that the two species were in such close competition that only one could survive. This became known as the **competitive exclusion principle** or **Gause's principle**. It states that only one species (population) in a given community can occupy a given ecological niche at any one time. Fig. 17.16 summarizes the results of Gause's experiments with *Paramecium*. Although *P. aurelia* survives, its population is reduced in the presence of *P. caudatum*, compared to its population when grown alone.

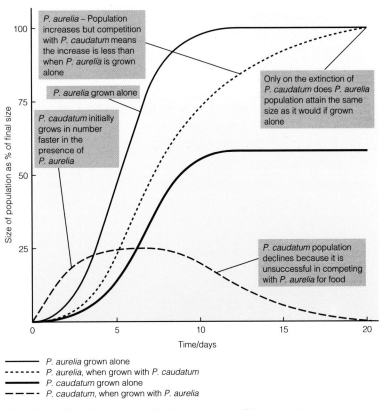

Fig. 17.16 Population growth of two species of Paramecium grown
separately and together

Why *P. aurelia* is more successful in the long term may be
because of its smaller size. Being smaller, it requires less food
and is better able to survive when food is scarce. Its success may,
however, be due to a faster reproductive rate or greater
efficiency in obtaining its food. The reasons for success are hard
enough to isolate in the laboratory; it is considerably more
difficult to do so for a wild population.

17.6.9 Biological control

The effect of the predator–prey relationship in regulating
populations has been exploited by humans as a method of
controlling various pests. Biological control is a means of
managing populations of organisms which compete for human
food or damage the health of humans or livestock. The aim is to
bring the population of a pest down to a tolerable level by use of
its natural enemies. A beneficial organism (the **agent**) is
deployed against an undesirable one (the **target**). A typical
situation is where a natural predator of a harmful organism is
introduced in order to reduce its numbers to a level where they
are no longer harmful. The aim is not to eradicate the pest;
indeed, this could be counter-productive. If the pest was reduced
to such an extent that it no longer provided an adequate food
source for the predator, then the predator in its turn would be
eradicated. The few remaining pests could then increase their
population rapidly, in the absence of the controlling agent. The
ideal situation is where the controlling agent and the pest exist in
balance with one another, but at a level where the pest has no
major detrimental effect.

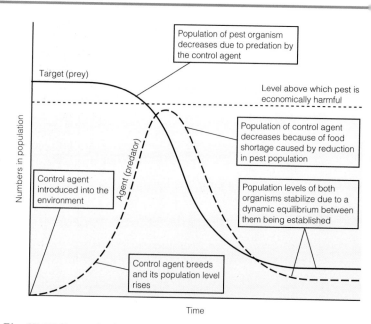

Fig. 17.17 *General relationships between pest and control agent populations in biological control*

TABLE 17.6 **Some examples of biological control**

Target (pest)	Harmful effects of pest	Control agent	Method of action
Scale insect (*Icerya*)	Kills citrus fruit trees	Ladybird (*Rodolia*)	Ladybird uses scale insect as a food source
Codling moth (*Crypto-phlebia*)	Ruins orange crop	African wasp (*Tricho-gamatoidea*)	Wasp parasitizes moth eggs
Mosquito (*Anopheles*)	Vector of malarial parasite (*Plasmo-dium*)	Hydra (*Chloro-hydra*)	Hydra is a predator of mosquito larvae
Snail (*Biomphal-aria*)	Vector of *Schistosoma* which causes bilharzia	Snail (*Marisa*)	Control agent snail is predatory on the snail vector
Prickly pear (*Opuntia*)	Makes land difficult to farm by restricting access	Cochineal insect (*Dactylo-pius*)	*Opuntia* is a food source for the insect
Larvae of many butterflies and moths	Consume the foliage of many economic-ally important plants	Bacterium (HD-1 strain of *Bacillus thuringien-sis*)	Bacterium parasitizes the larvae of moths and butterflies

Biological control was originally used against insect and weed pests of economically important crops. In more recent times its use has broadened to include medically important pests such as snails and even vertebrate pests. In the same way the type of controlling agent has become more diverse and the following are now employed: bacteria, viruses, fungi, protozoans, nematodes, insects and even amphibians and birds. These agents are sometimes used in combination. Certain nematodes carry bacteria which are deadly to many insects and their larvae. These nematodes have been nicknamed 'biological exocets' because of their ability to seek out their insect larvae hosts. The nematodes enter the insects' bodies releasing a fatal cargo of bacteria. Insects controlled in this way include the black vine weevil. Table 17.6 lists some examples of biological control.

One interesting and unusual form of biological control takes place in Australia. Cattle dung there presents a problem because two major pests, the bush fly and buffalo fly, lay their eggs in it. In addition, the dung carries the eggs of worms which parasitize the cattle. The indigenous dung beetles which are adapted to coping with the fibrous wastes of marsupials are ineffective in burying the soft dung of cattle. The introduction of an African species of dung beetle, which bury the dung within forty-eight hours, has been effective in controlling the flies. By burying the dung before the flies can mature, or before the parasitic worm can develop and reinfect cattle, they have controlled the populations of these pests.

17.6.10 Communities and succession

A community is all the plants and animals which occupy a particular area. The individual populations within the community interact with one another. The community is a constantly changing dynamic unit, which passes through a number of stages from its origin to its climax. The transition from one stage to the next is called **succession**.

Imagine an area of bare rock. One of the few kinds of organisms capable of surviving on such an inhospitable area is lichen. The symbiotic relationship between an alga and a fungus which makes up a lichen, allows it to survive considerable drying out. As the first organisms to bring about **colonization** of a new area, the lichens are called **pioneers** or the **pioneer community**.

The weathering of any rock produces a sand or soil, but in itself this is inadequate to support other plants. With the decomposing remains of any dead lichen, however, sufficient nutrients are made available to support a community of small plants. Mosses are typically the next stages in the succession, followed by ferns. With the continuing erosion of the rock and the increasing amounts of organic material available from these plants, a thicker layer of soil is built up. This will then support smaller flowering plants such as grasses and, by turn, shrubs and trees. In Britain the ultimate community is most likely to be deciduous oak woodland. The stable state thus formed comprises a balanced equilibrium of species with few, if any, new varieties replacing those established. This is called the **climax community**. This community consists of animals as well as plants. The animals have undergone a similar series of successional stages, largely dictated by the plant types available. Within the climax community there is normally a **dominant** plant and animal species, or sometimes two or three **co-dominant** species. The dominant species is normally very prominent and has the greatest biomass.

The succession described above, where bare rock or some other barren terrain is first colonized, is called **primary succession**. If, however, an area previously supporting life is made barren, the subsequent recolonization is called **secondary succession**. Secondary succession occurs after a forest fire or the clearing of agricultural land. Spores, seeds and organs of vegetative propagation may remain viable in the soil, and there will be an influx of animals and plants through dispersal and migration from the surrounding area. In these circumstances the succession will not begin with pioneer species but with organisms from subsequent successional stages.

Around 4000 years ago much of lowland Britain was a climax community of oak woodland, but most of this forest was cleared to allow grazing and cultivation. The many heaths and grasslands which we now refer to as 'natural' are the result of this clearance and subsequent grazing by animals. These are not true climax communities but sub-climax ones resulting from human activities. Because the normal succession has been artificially changed it is often referred to as a **deflected succession** and the resultant sub-climax is called a **plagioclimax**.

A series of successional stages is called a **sere**. There are a number of different seres according to the environment being colonized. A **hydrosere** refers to a series of successions in an aquatic environment and a **halosere** to one in a saltmarsh.

An oak wood is a climax community

17.7 Questions

1. *Either* (a) Discuss the productivity of ecosystems with reference to pyramids of numbers, biomass and energy. (*30 marks*)

   Or (b) Discuss the roles of bacteria and fungi in the recycling of materials within ecosystems. (*30 marks*)
   (*Total 30 marks*)

   ULEAC January 1994, Paper I, No. 15

2. (a) Describe the way in which you have estimated the size of a population. Explain fully the reasons for using the technique you have described. (*8 marks*)

   (b) Indicate the factors which might influence population density in an area, distinguishing between density-dependent and density-independent factors. (*8 marks*)

   (c) Suggest, with **one** example, the practical value of studying changes in the population densities of organisms. (*4 marks*)
   (*Total 20 marks*)

   WJEC June 1993, Paper A1, No. 7

3. (a) Explain the meaning of the following ecological terms.
   (i) ecosystem
   (ii) community
   (iii) population
   (iv) habitat (*4 marks*)

   (b) Select **three** abiotic factors and describe how **each** factor affects the distribution of a named organism.

   You may choose the same organism or different organisms in relation to each abiotic factor. The distribution of the organism or organisms chosen should be affected significantly by the selected abiotic factor. (*6 marks*)

   (c) Pyramids of number, biomass and energy often are used in ecological studies.

   Explain the basis of **each** type of pyramid and give a range of examples to show the various forms of pyramids that can be obtained. Comment on the usefulness of these pyramids in ecological studies. (*10 marks*)
   (*Total 20 marks*)

   NEAB June 1992, Paper IB, No. 8

4. Any habitat may be said to have a *carrying capacity* and this is defined as the maximum population of a given species that can be sustained there.

   The graph below shows changes in the numbers of wild sheep on a large island during the 100 years following their introduction to the island.

   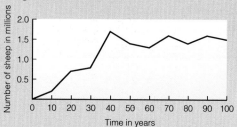

   (a) From the graph, estimate the carrying capacity of this island for wild sheep. (*1 mark*)

   (b) Suggest *two* factors that may determine the carrying capacity of this island for wild sheep. (*2 marks*)

   (c) Comment on the pattern of population change after carrying capacity had been reached. (*3 marks*)

   (d) Suggest *three* factors that may have influenced the length of time taken for the sheep population to reach carrying capacity. (*3 marks*)
   (*Total 9 marks*)

   ULEAC June 1990, Paper I, No. 5

5. In an experiment investigating the rate of disappearance of leaf litter, oak leaf discs were placed in nylon mesh bags and buried in newly cultivated pasture.

   The table shows the disappearance of oak leaf discs from bags made from 7 mm and 0.5 mm mesh over a period of months.

Month	Percentage oak leaf area remaining in bags of mesh size:	
	7 mm	0.5 mm
June	100	100
August	81	94
October	30	91
December	13	66
February	9	62
April	6	60

Ref: Ecology of Woodland Processes by J.R. Packham and D.J.L. Harding (1982)

(a) Plot the data on graph paper. *(5 marks)*

(b) (i) Describe the effect of mesh size on the rate of disappearance of leaf litter between June and October. *(2 marks)*

   (ii) Suggest an explanation for this. *(2 marks)*

   (iii) Explain the variation in the rate of disappearance of litter from the 0.5 mm mesh bags during the period of the experiment. *(2 marks)*

(c) (i) One estimate of the nitrogen content of leaf litter is $11 \, g \, m^{-2}$ of leaf. Calculate the total amount of nitrogen $(g \, m^{-2})$ in 50 leaf discs each 1.25 cm in radius contained in the 7 mm mesh bag. Assume that the area of each disc is πr^2 and that $\pi = 3.14$. Show your working. *(4 marks)*

(Total 15 marks)

WJEC June 1993, Paper A2, No. 5

. The table summarizes some results of an ecological investigation into the effects of keeping sheep at different population densities. All figures have been converted to units of energy and are expressed in $MJ \times 1000 \, hectare^{-1} \, year^{-1}$.

Amount of energy/$MJ \times 1000 \, hectare^{-1} \, year^{-1}$			
	10 sheep per hectare	20 sheep per hectare	30 sheep per hectare
used for growth by sheep	43	83	97
respired by sheep	39	74	66
lost in faeces	25	50	60
in dead plant material	51	25	9

(a) From the table, complete the word equation to show the efficiency with which sheep convert the food they eat into body material.

Efficiency of conversion = _____ $\times 100\%$
(1 mark)

(b) Calculate the mean amount of food energy eaten by one sheep at each population density. Show your working. *(2 marks)*

(c) What is the effect of increasing population density on the amount of energy passing to decomposers? *(1 mark)*

(d) Explain how your answers to questions *(b)* and *(c)* would help a farmer to determine the population density at which to keep sheep. *(2 marks)*

(Total 6 marks)

AEB November 1993, Paper I, No. 11

7. The diagram shows some aspects of the relation between a leguminous plant, clover, and the bacterium, *Rhizobium.*

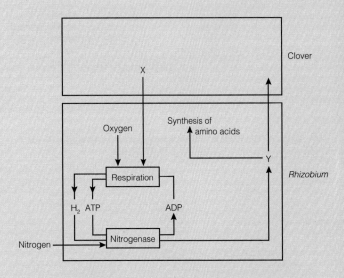

(a) Name:
   (i) Substance **X**; (ii) Substance **Y**. *(2 marks)*
(b) What name is given to the relation between these two organisms? *(1 mark)*

The graphs show an effect of applying nitrogen-containing fertilizer on two plots of land. No fertilizer was applied to plot **A** while plot **B** had 247 kg of fertilizer added per hectare each year since 1864. The vertical axis shows the proportion of different species growing on the plot that were grasses or legumes.

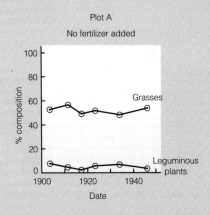

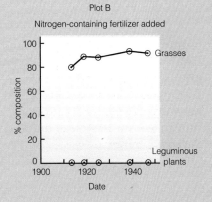

(c) Explain why there were:
  (i) more species of grass in plot **B**;
  (1 mark)
  (ii) more species of leguminous plants in plot **A**.
  (2 marks)
  (Total 6 marks)

  AEB June 1990, Paper I, No. 15

8. Study the energy flow model of an ecosystem and then answer the questions that follow.

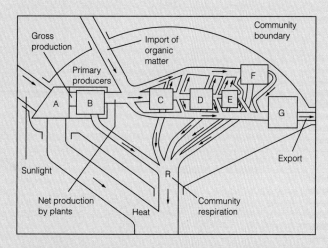

A Photosynthesis
B Respiratory process
C Herbivores
D Carnivores
E Top carnivores
F Decomposers
G Storage of dead organic matter

(a) What is the prime source of energy for the ecosystem?
  (1 mark)
(b) Account for the difference between gross production and net production by plants.
  (2 marks)
(c) How is energy transferred between each stage/trophic level?
  (2 marks)
(d) Explain why the width of the energy flow bands becomes progressively narrower as energy flows through the ecosystem.
  (2 marks)
(e) Suggest an explanation for the limit on the total number of trophic levels to four or five at most in a community.
  (2 marks)
(f) Distinguish between autotrophic and heterotrophic modes of nutrition.
  (3 marks)
(g) Briefly describe the rôle of decomposers in the community.
  (3 marks)
  (Total 15 marks)

  UCLES June 1992, Paper II, No. 3

9. The statements A, B, C and D in the list below represent food chains in different habitats.
A Grass → Rabbits → Fox
B Tree → Insects on tree → Birds
C Rose bush → Aphids → Parasites of aphids
D Unicellular → Marine invertebrate → Fish
    algae   larvae
  (Phytoplankton) (Zooplankton)
The diagrams 1, 2, 3 and 4 represent pyramids of numbers.

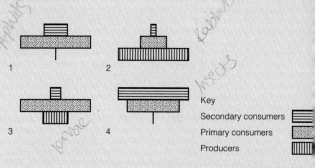

(a) For each pyramid of numbers (1, 2, 3 and 4) write the letter of the appropriate food chain (A, B, C or D).
  (3 marks)
(b) In pyramid 3 the producers are small in size and relatively few in number. Suggest why, nevertheless, they can support the large population of primary consumers.
  (2 marks)
(c) In deep sea or deep lake ecosystems there are no producers present.
  (i) What would be the food source for the primary consumers in a food chain in this ecosystem?
  (1 mark)
  (ii) Suggest *one* reason why there are no producers present.
  (1 mark)
  (Total 7 marks)

  ULEAC January 1993, Paper I, No. 1

10. The diagram opposite represents some of the more important inter-relationships between living organisms and the occurrence of compounds of nitrogen, carbon and oxygen in the environment.

Study the diagram carefully and then answer the questions that follow.

(a) What is the name of the process indicated by arrow 1?
  (1 mark)
(b) State **two** environmental conditions under which you would expect the process represented by arrow 2 to be at a **maximum**.
  (1 mark)
(c) State **two** conditions within animal B that could increase the rate of the process represented by arrow 3.
  (1 mark)

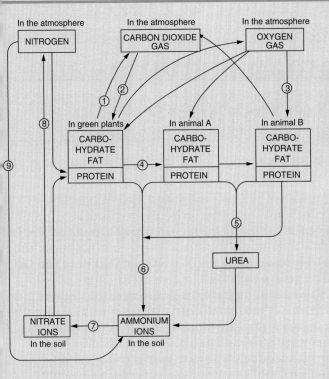

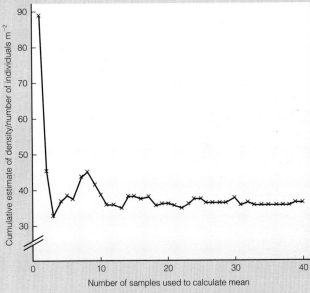

(d) What type of nutrition is represented by arrow 4? *(1 mark)*

(e) Where would the process represented by arrow 5 occur? *(1 mark)*

(f) What general name is given to the organisms responsible for the process represented by arrow 6? *(1 mark)*

(g) What single biochemical term summarizes process ⑤ (the conversion of protein to urea)? *(1 mark)*

(h) Which one of the following terms best describes the biochemical process represented by arrow 7? *(1 mark)*
hydrolysis; nitrogen fixation; symbiosis; oxidation; condensation.

(i) What general name is given to the organisms carrying out process ⑦? *(1 mark)*

(j) What process is indicated by arrow 8? *(1 mark)*

(k) What is the purpose of process ⑧ to the living organisms involved? *(1 mark)*

(l) Give the name of a genus of bacterium that could carry out the process represented by arrow 9. *(1 mark)*

(Total 12 marks)

Oxford June 1992, Paper I, No. 1

11. One method of sampling populations in an ecological study is to use a quadrat. A quadrat is an area of known size. Individuals of a population are not distributed uniformly, so if an estimate of density is to be made, a system of *random sampling* must be adopted.

In order to determine how many samples need to be taken, it is helpful to plot the *cumulative estimate* of density against the number of samples taken. Cumulative density is found by dividing the total number of individuals per unit area by the number of samples taken. An example of such a graph is shown below.

(a) (i) From the graph, find the minimum number of samples required to give a reliable estimate of the mean density. *(1 mark)*

(ii) Explain how you arrived at your answer. *(2 marks)*

(b) (i) Describe how random sampling can be carried out. *(2 marks)*

(ii) Suggest *one* limitation of random sampling methods. *(1 mark)*

(c) (i) Describe how you could use a quadrat to estimate the percentage frequency of plants in a habitat. *(3 marks)*

(ii) Suggest *one* way in which quadrats could be used to study changes in plant populations. *(3 marks)*

(d) Describe *two* methods you could use for sampling the animals present in a *named* habitat. *(4 marks)*

(Total 16 marks)

ULEAC June 1993, Paper I, No. 13

See *Further questions*, pages 614–20, question 14.

18 Human activities and the ecosystem

The effect of human activities on the environment is proportional to the size of the human population. As we saw in Section 17.6.5, the size of the human population has been rising exponentially and is presently increasing at the rate of one and a half million people each week, equivalent to 150 people per minute. The reasons for this increase are many but include more intensive forms of food production and better medical care. The latter has given many individuals a greater life expectancy and, more importantly, reduced child mortality. Reducing the mortality of individuals who have passed child-bearing age has little significance on the population, but reducing mortality among children means more people reach sexual maturity and so are able to produce offspring. The effect of this on the population is significant.

18.1 The impact of pre-industrial humans on the environment

18.1.1 Humans as hunters

As their population was small, pre-industrial humans did not have a great impact on the environment. Early humans hunted, fished and removed trees to make fires and shelters but, being nomadic, they did not remain in one place long enough to have a significant effect. When they moved on, the natural environment rapidly recovered. Early humans did use fire and this may have accidentally got out of hand and burned down large areas of forest. Equally it may have been used deliberately to flush out prey to enable it to be captured. Humans may thereby have created some grasslands at the expense of the forest, but their total impact was small.

18.1.2 Humans as shepherds

In time humans domesticated animals such as sheep, cattle, goats, llamas and alpacas. These herbivorous species required large areas of grassland on which to graze. To extend the grasslands humans deliberately burnt large areas of trees causing **deforestation**. Domesticated species like cattle became symbols of wealth and power, as well as suppliers of milk and meat. This led to the build-up of large herds and subsequent overgrazing with resultant loss of soil fertility leading to erosion.

APPLICATION

Desertification

Grazing in a desertified area

Desertification is the term used to describe severe land degradation which turns semi-arid areas into deserts. There are many possible causes but they normally involve climatic changes and/or an increase in the human population above the carrying capacity (Section 17.6.1) of the land. Extended periods of drought make plant regeneration difficult and bare soils are exposed to erosion. Increased human populations put additional pressures on the vegetation as shrubs and trees are cut for fuel and the land is overgrazed. The further loss of ground vegetation allows wind erosion to remove the soil and makes replanting difficult. Little can be done to combat these problems in the short term if there is not adequate water, although there is evidence of recovery in the longer term. In the more favourable semi-arid areas with sufficient water, skilful engineering schemes and a thorough knowledge of plant nutrition and soil structure have enabled some tree planting programmes to be successful in halting the spread of desertification.

Effect of irrigation from the Nile

These activities, which began in the Mediterranean and Near East, may have contributed to the development of many of the desert regions of these areas. Animal domestication also led to the extinction of their wild ancestors, probably through competition. Aurochs and European bison may have suffered this fate.

18.1.3 Humans as farmers

The advent of agriculture marked the most significant event of pre-industrial human impact on the environment. It probably originated in the Near East and involved the deliberate sowing of seeds to produce a crop. If humans were to enjoy the fruits of this labour, they needed to harvest the crop at some later date. They therefore had to remain in one place or risk losing the crop to other animals. For the first time humans formed permanent settlements. They built shelters for themselves and their animals, and barns to store their crops. This required much wood and led to further deforestation. More importantly, humans cleared much forest to provide a greater area for sowing crops. With no knowledge of minerals, they continued to grow the same crop on the same piece of land for many years, thus depleting it of essential nutrients. The soil could no longer support life, and this was a major factor in the formation of desert areas.

The extent of deforestation caused by humans is clearly illustrated in the United States of America. When first settled by Europeans in the early seventeenth century, there were an estimated 170 million hectares of forest. Now there are eight million. Most of this clearance was carried out to permit the cultivation of corn and wheat in the north, and tobacco and cotton in the south.

18.2   Exploitation of natural resources

Prior to the industrial revolution, the energy expended in the production of food by humans and their beasts of burden came from the food itself. Much of the crop therefore went into producing the next one. With the industrial revolution came machines which carried out ploughing, sowing, harvesting, etc. Instead of food, these machines used fossil fuels like coal, and more recently oil, as energy sources. A very much smaller proportion of the crop harvested was therefore needed to produce the next one. More food was left and a larger population could be supported. This partly explains the exponential rise in the size of the human population since the industrial revolution. The use of fertilizers, pesticides and better crops are other significant factors.

Humans are dependent for their survival on the earth's resources, and these take two forms: renewable and non-renewable

18.2.1 Renewable resources

Renewable resources, as the name suggests, can be replaced. They are things which grow, and are materials based on plants or animals, e.g. trees and fish. They are not, however, produced in limitless quantities and their supply is ultimately exhausted if the rate at which they are removed exceeds that at which they have been produced. Renewable resources have a **sustainable yield**. This means that the amount removed (yield) is equal to, or less than, the rate of production. If the trees in a forest take 100 years to mature, then one hundredth of the forest may be felled each year without the forest becoming smaller. A sustainable yield can be taken indefinitely.

Whilst wood is a renewable resource, its production is not without ecological problems. Trees grow relatively slowly and so give a small yield for a given area of land. For this reason, it is not economic to use fertile farmland for their cultivation. Instead, poorer quality land typical of upland areas is often used. As conifers grow more rapidly, these softwood species are more often cultivated than indigenous hardwoods such as elm, oak, ash and beech. Large areas in Scotland, Wales and the Lake

Rows of larch planted on a hillside

Thirty-tonne catch of fish by deep-sea trawler

District have become **afforested**. The trees are often grown in rows and many square miles are covered by the same species. Not only does this arrangement have an unnatural appearance, but the density of the trees permits little, if anything, to grow beneath them and the forest floor is a barren place. There is little diversity of animal life within these forests. The demand for wood, not only for construction but also for paper, necessitates this intensive form of wood production.

Another renewable resource is fish. Unlike many renewable resources used by humans, fish are not generally farmed. For the most part humans remove them from the seas with no attempt to replace stocks by breeding. The replacement is left to nature. As the seas are considered a common resource for all, no previous attempt has been made to control the amount of fish removed by each country. While fishing was carried out by small boats, working locally, its impact on stocks was negligible because a sustainable yield was removed. Modern fishing methods involve large factory ships, capable of travelling thousands of miles and catching huge hauls of fish, which can be processed and frozen on board. Sonar equipment, echo sounders and even helicopters may be used in locating shoals. These methods have led to **over-fishing**, because sustainable yields have been exceeded and stocks depleted. It takes many years for such stocks to recover. Some controls now exist and international agreement has been reached on **quotas** of fish which each country can take. These quotas are often bitterly disputed and the difficulty of enforcement has led to many being ignored. There are regulations concerning the **mesh size of nets**. If the mesh is sufficiently large, younger, and therefore smaller, fish escape capture. These survive to grow larger and, more importantly, are able to reach sexual maturity. These fish can then spawn thus ensuring some replenishment of the stock. Other methods of control include **close seasons** for fishing (usually during a particular species' breeding season) and **exclusion zones** where fishing is banned completely. One species to have suffered from over-fishing is the North Sea herring. Depletion of its stocks have made fishing it practically uneconomic in recent years.

18.2.2 Non-renewable resources

These are resources which, for all practical purposes, are not replaced as they are used. Minerals such as iron and fuels like coal and oil are non-renewable. There is a fixed quantity of these resources on the planet and in time they will be exhausted. Oil and natural gas supplies are unlikely to last more than 50 years, although much depends upon the rate at which they are burned.

Mineral and ore extraction have been carried out for a considerable time with important metals such as iron, copper, lead, tin and aluminium being mined. In theory these metals can be recycled, but in practice this is often difficult or impossible for various reasons:

1. The metal may be oxidized or otherwise converted into a form unsuitable for recycling. Iron for example rusts.

2. The quantities of the metal within a material may be so small that it is not worthwhile recovering it. The thin layer of tin on most metal cans is not economically worth recovering.

APPLICATION

Fuel from oilseed rape

Oilseed rape in flower

Over a million tonnes of rapeseed are grown in Britain each year, almost all of it for processing into consumer goods. Rapeseed oil, from *Brassica napus* has, in the past, been turned into soap, lubricants and synthetic rubber. It can be used as a heating oil and, as rape methyl ester (RME), is now a renewable replacement for diesel. Most RME is manufactured by the Italian company Novamont and it has been used to run taxis, lorries, ferries and public transport in Italy and other parts of Europe. The production of biodiesel from rapeseed is expected to rise as it provides an environmentally friendly alternative to diesel. It emits fewer sooty particles and no sulphur dioxide. Recently an English farmer claimed to get 60 miles to the gallon using refined rapeseed oil in his unadapted Opel Ascona car.

Part of the Chernobyl nuclear reactor after the explosion in April 1986

3. The metal is often combined with many other materials, including other metals, which make it difficult to separate. A motor car may contain small amounts of many metals including zinc, lead, tin, copper and aluminium.

The supply of many ores is becoming more scarce and the belief that supply would satisfy demand for the foreseeable future is being questioned. The lead, tin, copper, gold and silver mines of Wales, Cornwall and the Lake District have almost entirely ceased their activities. As the supply is reduced, the price increases and it could be that sources previously considered uneconomic may prove worthwhile exploiting again.

Fossil fuels are continually being formed, but the process is so slow compared to their rate of consumption that for all practical purposes they may be considered as a non-renewable resource. Over 80% of the world's consumption of fossil fuels occurs in developed countries, where only 25% of its population lives. The burning of fossil fuels produces a range of pollutants and even their extraction is not without its hazards. As the supply of these fuels is becoming rapidly depleted, humans have sought alternative energy sources. Nuclear power is a potentially long-term supplier of energy, but it has inherent dangers as the accident at Chernobyl in Russia in April 1986 illustrated. It is therefore treated by the public with some suspicion. Attempts continue to be made to harness wind, wave and solar energy effectively. In the end it could be **biological fuels** that humans may have to look to to supply their growing energy needs. The energy content of the organic matter produced annually by photosynthesis exceeds annual human energy consumption by 200 times. The main end-product of this photosynthesis is cellulose, most of which is unused by humans. Some of it can be burnt as wood or straw to provide heat or electricity. Much can be converted to other fuels like methane (CH_4), methanol (CH_3OH), ethanol (C_2H_5OH) and other gases. These processes need not use valuable food resources; the energy may be obtained from plants with no food value or from the discarded parts of food plants, estimated to total 20 million tonnes dry mass year^{-1} in the UK alone. The gasohol programme in Brazil, where sugar cane wastes are used to produce a motor vehicle fuel, is an example of this (Section 30.5.1). Wastes such as animal manure (45 million tonnes dry mass year^{-1} in the UK), human sewage (6 million tonnes dry mass year^{-1} in the UK) and other domestic and industrial wastes (30 million tonnes dry mass year^{-1} in the UK) could be converted to useful fuels like methane by biogas digesters (Section 30.5.2). These conversions can be carried out by bacteria, often as part of fermentation reactions. The day may not be far away when large industrial plants convert these wastes into useful fuels and energy forms and where crops cultivated entirely for conversion to fuels are commonplace.

18.3 Pollution

Pollution is a difficult term to define. It has its origins in the Latin word *polluere* which means 'contamination of any feature of the environment'. Any definition of pollution should take account of the fact that:

1. It is not merely the addition of a substance to the environment but its addition at a rate faster than the environment can accommodate it. There are natural levels of chemicals such as arsenic and mercury in the environment, but only if these levels exceed certain critical values can they be considered pollutants.

2. Pollutants are not only chemicals; forms of energy like heat, sound, α-particles, β-particles and X-rays may also be pollutants.

3. To be a pollutant, a material has to be potentially harmful to life. In other words, some harmful effect must be recognized.

Using the above criteria, it is arguable that there is such a thing as natural pollution. We know for example that sulphur dioxide, one product of the combustion of fossil fuels, is a pollutant, and yet 70% of the world's sulphur dioxide is the result of volcanic activity. To avoid 'natural pollution' some scientists like to add a fourth criterion, namely that pollution is only the result of human activities.

18.4 Air pollution

Did you know?

In 1307 a resident of London was executed for causing air pollution.

The layer of air which supports life extends about 8 km above the earth's surface and is known as the **troposphere**. While there may be small localized variations in the levels of gases in air, its composition overall remains remarkably constant. Almost all air pollutants are gases added to this mixture. Air pollution has existed since humans first used fire but it is only since the industrial revolution in the nineteenth century that its effects have become significant. Almost all air pollution is the result of burning fossil fuels, either in the home, by industry or in the internal combustion engine.

Air pollution from a coking plant

TABLE 18.1 **Tolerance of moss and lichen species to sulphur dioxide**

Annual average sulphur dioxide concentration in $\mu g\, m^{-3}$	Species tolerant and therefore able to survive
Greater than 60	Lecanora conizaeoides (lichen) Lecanora dispersa (lichen) Ceratodon purpureus (moss) Funaria hygrometrica (moss)
Less than 60	Parmelia saxatilis (lichen) Parmelia fulginosa (lichen)
Less than 45	Grimma pulvinata (lichen) Hypnum cupressiforme (moss)

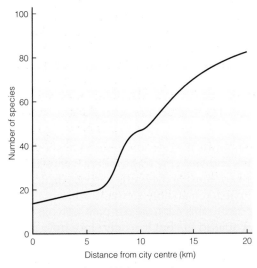

Fig. 18.1 Number of lichen species as one moves from the centre of Newcastle upon Tyne

18.4.1 Smoke

Smoke is tiny particles of soot (carbon) suspended in the air, which are produced as a result of burning fossil fuels, particularly coal and oil. It has a number of harmful effects:

1. When breathed in, smoke may blacken the alveoli, causing damage to their delicate epithelial linings. It also aggravates respiratory ailments, e.g. bronchitis.

2. While it remains suspended in the air, it can reduce the light intensity at ground level. This may lower the overall rate of photosynthesis.

3. Deposits of smoke, or more particularly soot and ash, may coat plant leaves, reducing photosynthesis by preventing the light penetrating or by blocking stomata.

4. Smoke, soot and ash become deposited on clothes, cars and buildings. These are costly to clean.

18.4.2 Sulphur dioxide

Fossil fuels contain between 1 and 4% sulphur and as a result around 30 million tonnes of sulphur dioxide is emitted from the chimneys of Europe each year. Much of this combines readily with other chemicals like water and ammonia and is quickly deposited. It may increase soil fertility in areas where sulphates are deficient, or even help to control diseases such as blackspot of roses by acting as a fungicide. Nevertheless its effects, especially in high concentrations, are largely harmful:

1. It causes irritation of the respiratory system and damage to the epithelial lining of the alveoli. It can also irritate the conjunctiva of the eye.

2. It reduces the growth of many plants, e.g. barley, wheat, lettuce, while others such as lichens may be killed.

The tolerance of lichen and moss species to sulphur dioxide is very variable and makes them useful **indicator species** for measuring sulphur dioxide pollution. Table 18.1 shows the tolerance of some mosses and lichen species to sulphur dioxide.

As one moves from the centre of a major industrial city like Newcastle upon Tyne, the concentration of sulphur dioxide falls rapidly. At the same time the number of species of lichen and moss increases. In the centre only the most tolerant species are found, whereas on the outskirts less tolerant ones also occur (Fig. 18.1). Using Table 18.1, we can see that if an area of a city possesses *Lecanora dispersa* and *Funaria hygrometrica* but none of the other species, then the levels of sulphur dioxide must exceed $60\,\mu g\, m^{-3}$.

If all species in the table are present, the sulphur dioxide level must be less than $45\,\mu g\, m^{-3}$.

Much of the sulphur dioxide released into the atmosphere returns to earth as gas or minute particles (dry deposition) but about one third dissolves in rain water. The sulphur dioxide and water combine to form sulphurous and sulphuric acids. The rain therefore has a low pH and is known as **acid rain**. The oxides of nitrogen are other pollutants which contribute to acid rain. Indeed, while the contribution of sulphur dioxide has

TABLE 18.2   Sources of acidifying gases

Source	Percentage contribution	
	Nitrogen oxides	Sulphur dioxide
Motor vehicles	45	1
Power stations	37	71
Industry	12	19
Domestic	3	5
Other sources	3	4

diminished due to the industrial recession, that from nitrogen oxides has increased due to the increase in motor vehicle use. Table 18.2 shows the relative amounts of acidifying gases from different sources. Due to the prevailing winds, much of the sulphur dioxide from Europe, including that from Britain, is carried over Scandinavia. It is here that acid rain causes the greatest problems. Coniferous trees are particularly vulnerable and considerable damage has been caused to some forests. Lakes in the region are extremely acid and many species within them have been killed, largely as a result of the accumulation of aluminium leached from soils as a result of acid rain. This affects aquatic organisms' gills and their osmoregulatory mechanisms. Many countries have committed themselves to reducing the level of sulphur dioxide emissions, largely through changing to 'cleaner' fuels such as natural gas or by fitting desulphurization units to remove sulphur dioxide from the flue gases at power stations.

18.4.3 Carbon dioxide

Carbon dioxide is formed during the respiration of organisms, and by the burning of fossil fuels. That produced as a result of respiration is taken up by plants during photosynthesis, ensuring it does not accumulate. The additional carbon dioxide produced in the burning of fossil fuels has caused a rise in atmospheric carbon dioxide concentration. Scientists believe that this change in air composition prevents more of the sun's heat escaping from the earth, much in the way the glass in a greenhouse does. They argue that the rise in temperature that this so-called **greenhouse effect** produces will cause expansion of the oceans and the gradual melting of the polar ice caps with a consequent rise in sea level. This would in turn cause flooding of low-lying land, upon which, as it happens, many of the world's capital cities lie. The greenhouse effect is neither new, nor all bad. Indeed it is its influence which maintains the earth's surface at an average of $15\,°C$ rather than $-18\,°C$ which would be the case in the absence of greenhouse gases. The problem lies in the additional greenhouse gases which have been released over the past 200 years. While water vapour, methane and nitrogen oxides are all greenhouse gases, it is the influence of carbon dioxide that has been most significant in contributing to global warming. Estimates of the warming which is attributable to carbon dioxides vary from 50–70%. While the other greenhouse gases are present in much lower concentrations than carbon dioxide they are much more efficient at absorbing infra-red radiation and hence have a potentially greater influence on the greenhouse effect. Carbon dioxide however remains the greatest influence, not just because of its higher concentration but also the fact that it remains in the atmosphere longer – on average each molecule remains for 100 years, compared to 10 years for methane and a few months for carbon monoxide.

18.4.4 Carbon monoxide

Carbon monoxide occurs in exhaust emissions from cars and other vehicles. It is poisonous on account of having an affinity for haemoglobin some 250 times greater than that of oxygen.

Upon combining with haemoglobin, it forms a stable compound which is not released and prevents oxygen combining with it. Continued inhalation leads to death as all haemoglobin becomes combined with carbon monoxide, leaving none to transport oxygen. In small concentrations it may cause dizziness and headache. Even on busy roads levels of carbon monoxide rarely exceed 4%, and it does not accumulate due to the action of certain bacteria and algae which break it down, according to the equation:

$$4CO + 4H_2O \rightarrow 4CO_2 + 8H^+ + 8e^-$$

carbon monoxide | water | carbon dioxide | protons | electrons

hydrogen

Cigarette smoking is known to increase the carbon monoxide concentration of the blood; up to 10% of a smoker's haemoglobin may be combined with carbon monoxide at any one time.

18.4.5 Nitrogen oxides

Nitrogen oxides, like nitrogen dioxide, are produced by the burning of fuel in car engines and emitted as exhaust. In themselves they are poisonous, but more importantly they contribute to the formation of **photochemical smog**. Under certain climatic conditions pollutants become trapped close to the ground. The action of sunlight on the nitrogen oxides in these pollutants causes them to be converted to **peroxyacyl nitrates (PAN)**. These compounds are much more dangerous, causing damage to vegetation, and eye and lung irritation in humans.

Photochemical smog in Rio de Janeiro

18.4.6 Lead

The toxicity of lead has been known for some time. It has long been used in making water pipes and water obtained through these may be contaminated with it. As lead is not easily absorbed from the intestines this does not present a major health hazard. Much more dangerous is the lead absorbed from the air by the lungs. Most lead in the air is emitted from car exhausts. **Tetraethyl lead (TEL)** is added to petrol as an **anti-knock** agent to help it burn more evenly in car engines. Each year in Britain alone, around 50 000 tonnes of lead are added to the atmosphere in this way. While much of this is deposited close to roads, that which remains in the atmosphere and is absorbed by the lungs could have the following adverse effects:

1. Digestive problems, e.g. intestinal colic.
2. Impairing the functioning of the kidney.
3. Nervous problems, including convulsions.
4. Brain damage and mental retardation in children.

Anti-knock agents which do not contain lead exist and in some countries legislation permits only this type. The British Government has made price incentives on unleaded fuel, but latest research shows unleaded fuel may contain higher levels of benzene – a pollutant more harmful than lead.

18.4.7 Control of air pollution

On 9 December 1952, foggy conditions developed over London. Being very cold, most houses kept fires burning, with coal as the major fuel. The smoke from these fires mixed with the fog and was unable to disperse, resulting in a smog which persisted for four days. During this period some 4000 more people died than would be expected at this time of the year. Most of these additional deaths were due to respiratory disorders. These alarming consequences of smog prompted the government to seek ways of controlling smoke emissions from chimneys. This led ultimately to the **Clean Air Act** of 1956. Among other things this created smokeless zones, in which only smoke-free fuels could be burned. Grants were made available to assist with the cost of having fires converted to take these smokeless fuels. For many years now most cities have been smokeless and the smogs, once a common feature of winter, no longer occur.

Other methods of controlling air pollution include the use of non-lead anti-knock agents and the removal of pollutants such as sulphur dioxide before smoke is emitted from chimneys. The latter is achieved by passing the smoke through a spray of water in which much of the sulphur dioxide dissolves. The use of electric cars is a further means of limiting air pollution.

18.4.8 Ozone depletion

Between 15 and 40 kilometres above the earth is a layer of ozone which is formed by the effect of ultra-violet radiation on oxygen molecules. In this way, a large amount of the potentially harmful ultra-violet radiation is absorbed and so prevented from reaching the earth's surface. There is evidence that this beneficial ozone layer is being damaged by atmospheric pollution to the point where a hole in it has appeared over the Antarctic and possibly the Arctic too.

A number of pollutants can affect the ozone layer, the **chlorofluorocarbons (CFCs)** being the best known. CFCs are used in refrigerators, as propellants in aerosol sprays, and make up the bubbles in many plastic foams, e.g. expanded polystyrene. They are remarkably inert and therefore reach the upper stratosphere unchanged. Along with other ozone depleting gases such as **nitrous oxide** (NO), CFCs are contributing to global warming – the so-called 'greenhouse effect' (Section 18.4.3). In addition, the ultra-violet radiation causes skin cancer: an increase in the incidence of this disease is already evident.

TABLE 18.3 **Estimated relative importance of various gases to the greenhouse effect**

Gas	% contribution
Carbon dioxide	71
CFCs	10
Methane	9
Carbon monoxide	7
Oxides of nitrogen	3

18.5 Water pollution

Pure water rarely, if ever, exists naturally. Rain water picks up additives as it passes through the air, not least sulphur dioxide (Section 18.4.2). Even where there is little air pollution, chlorides and other substances are found in rain water. As water flows from tributaries into rivers it increasingly picks up minerals,

organic matter and silt. If not the result of human activities, these
may be considered as natural additives and therefore not
pollutants. For domestic use alone, each individual in Britain
uses an average of 150 litres of water each day.

18.5.1 Sewage and its disposal

Sewage is quite simply anything which passes down sewers. It
has two main origins: from industry and from the home
(domestic). Domestic effluent is 95–99% water, the remainder
being organic matter. In itself, the organic material is harmless,
but it acts as a food source for many saprobiontic organisms,
especially bacteria. Where oxygen is available, aerobic
saprobionts decompose the organic material – a process called
putrefaction – and in so doing use up oxygen. This creates a
biochemical oxygen demand (BOD).

Where sewage is deposited untreated into relatively small
volumes of water, i.e. rivers and lakes rather than the oceans, the
BOD may be great enough to remove entirely the dissolved
oxygen. This causes the death of all aerobic species, including
fish, leaving only anaerobic ones. The BOD is offset by new
oxygen being dissolved, and in fast-moving, shallow, turbulent
streams this is sufficient to prevent anaerobic conditions.

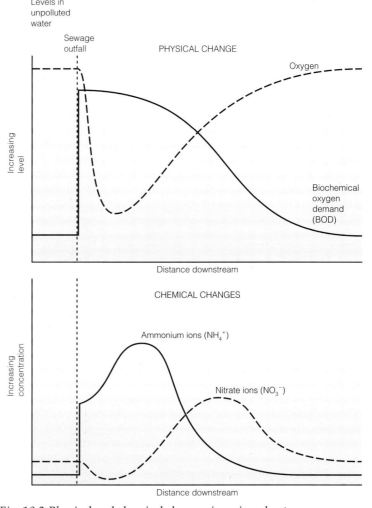

Fig. 18.2 *Physical and chemical changes in a river due to sewage
effluent*

Outlet pipe discharging sewage on to a beach

Unfortunately many centres of population are situated near river estuaries where the waters are slower moving, deeper and less turbulent. The amount of oxygen dissolving is much less and so any untreated sewage added to these waters quickly results in them becoming anaerobic. With only around $5 \, cm^3$ of dissolved oxygen in each dm^3 (litre) of fresh water, every individual human produces enough organic matter each day to remove the oxygen from $9000 \, dm^3$ of water. Where untreated sewage enters a river it creates a BOD which gradually decreases further downstream as organic material is decomposed. Part of this sewage is combined nitrogen; each human produces $8 \, g$ of this daily, mostly in the form of urea and uric acid. This combined nitrogen is converted to ammonia by bacteria. While the ammonia may be toxic, its effects are temporary, as nitrifying bacteria rapidly oxidize it to nitrates. These relationships are illustrated in Fig. 18.2.

The chemical and physical changes brought about by sewage are accompanied by changes in the fauna and flora of the water. Where the level of organic material is high, saprobiontic bacteria concentrations, including filamentous bacteria known as **sewage fungus**, increase as they feed on the sewage. The algal levels initially fall, possibly due to the sewage reducing the amount of light which penetrates the water. Further down stream the algal levels rise above normal because the bacterial breakdown of the sewage releases many minerals, including nitrates. These minerals, which previously limited algal growth, now allow it to flourish. As the minerals are used up, algal population levels return to normal.

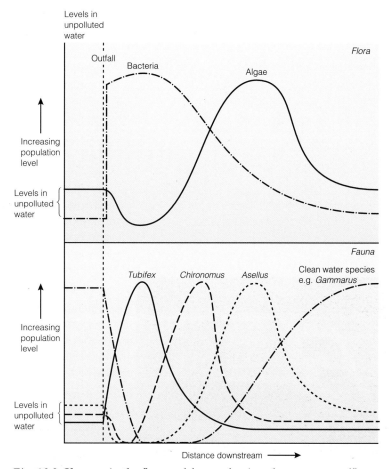

Fig. 18.3 Changes in the flora and fauna of a river due to sewage effluent

The population levels of animal species vary according to the level of oxygen in the water. Most tolerant of low oxygen levels are worms of the genus *Tubifex* whose haemoglobin has a particularly high affinity for oxygen which it obtains even at very low concentrations. These worms can therefore survive close to a sewage outfall; indeed, as other species cannot survive there, *Tubifex* are free from competitors and predators and so their numbers increase greatly.

Further down stream, as oxygen levels rise, other species such as the larvae of the midge *Chironomus* are also able to tolerate low oxygen levels. These compete with *Tubifex* for the small amount of available oxygen, and the worm population is reduced as a consequence. A continuing rise in oxygen level further from the outfall results in the appearance of species like the water louse, *Asellus*. Its presence adds to the competition, causing reduction in the populations of *Tubifex* and *Chironomus*. Finally, as the sewage is completely decomposed, oxygen levels in the water return to normal and clean-water species, like the freshwater shrimp, *Gammarus*, are present again. The ecological equilibrium is restored and population levels return to those found above the outfall. These changes in fauna and flora are illustrated in Fig. 18.3.

These organisms act as indicator species for polluted water. Where repeated additions of sewage occur at different points along the river, the water may be anaerobic for much of its length. In addition to the death of aerobic species, these conditions can result in the build-up of ammonia and hydrogen sulphide from anaerobic decomposition of sewage. These chemicals are toxic and result in an almost lifeless river. This was the situation with most large British rivers until the introduction of **sewage treatment works**. These works not only remove organic material but also potentially dangerous pathogenic organisms such as those causing cholera and typhoid.

The process of sewage treatment is outlined in Fig. 18.4 on page 372. It consists of a series of stages:

1. Screening – Large pieces of debris are filtered off to prevent them blocking pipes and equipment in the treatment works. This filtering is performed by a screen of metal rods, about 2 cm apart. The debris which is trapped on the screen is periodically scraped off and either buried or broken up into smaller pieces ready to undergo normal sewage treatment. Alternatively, the sewage enters a machine called a comminuter which reduces all the sewage into pieces small enough to enter the treatment works without risk of blockage.

2. Detritus removal – The sewage enters a tank or channel in which the rate of flow is reduced sufficiently to allow heavy inorganic material such as grit to deposit out. The lighter organic matter is, however, carried along in the water flow. The material that settles out is called **detritus** and can be dumped without further treatment.

3. Primary sedimentation – The sewage flows into large tanks which have a conical shaped base with a central exit pipe. The flow across these tanks is very slow, and may take several days. Fine silt and sand along with any organic material settle out and become deposited at the bottom of these tanks. The addition of ferric chloride, which causes flocculation, assists sedimentation

Sewage treatment works – primary sedimentation

in these tanks. The material which settles out is called **sludge** and is periodically pumped from the bottom of the tank to sludge digestion tanks. The sewage which has had most solid material extracted is now known as **effluent**. It is removed from the top of the sedimentation tanks and either enters **activated sludge tanks** or passes through **percolating filters**.

4a. Activated sludge method – The effluent is inoculated with aerobic microorganisms which break down dissolved organic material. It then flows into long channels through which air is blown in a fine stream of bubbles from the bottom. This provides oxygen for aerobic microorganisms rapidly to decompose the organic matter into carbon dioxide and some nitrogen oxides. One problem with this method is that detergents in the sewage can cause foaming.

4b. Percolating filter method – The alternative to the activated sludge method is to spray the effluent on to beds of sand, clinker and stones in which live a large variety of aerobic organisms, especially bacteria.

In both the above processes microorganisms oxidize the various dissolved substances. Urea for example may be decomposed as follows:

(i) $$CO(NH_2)_2 + H_2O \xrightarrow{\text{bacteria producing urease}} 2NH_3 + CO_2$$
urea   water   ammonia   carbon dioxide

(ii) $$2NH_3 + 3O_2 \xrightarrow{\textit{Nitrosomonas}} 2NO_2^- + 2H^+ + H_2O$$
ammonia   oxygen   nitrite   hydrogen   water
                                                                            ions

(iii) $$2NO_2^- + O_2 \xrightarrow{\textit{Nitrobacter}} 2NO_3^-$$
nitrite   oxygen   nitrate

5. Final (humus) sedimentation – The effluent from the sludge tanks or percolating filters contains a large number of microorganisms. It therefore passes into further sedimentation tanks to allow these organisms to settle out. The sediment, known as **humus**, is then passed into the sludge treatment tanks.

6. Fine filters – These remove any suspended particles in the effluent which may then be safely discharged into rivers.

7. Sludge digestion – The sludge and humus are pumped into large covered tanks where they are hydrolysed into simpler compounds, leaving gases such as methane, and a digested sludge.

8. Use of methane for generating power – The methane produced during sludge digestion is usually collected and used as a fuel to drive turbines in a power-house. The electricity generated can be used to power the equipment and lights at the sewage works, making them, in some cases, self-sufficient in energy.

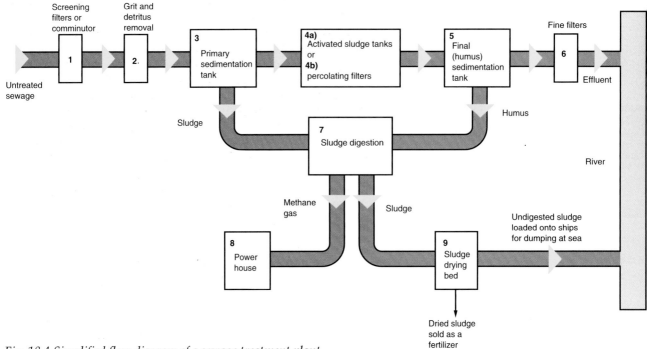

Fig. 18.4 Simplified flow diagram of a sewage treatment plant

9. Sludge drying beds – The sludge is led off into large tanks where its water content is reduced by air-drying. The resultant semi-solid material may either be loaded on to ships for dumping at sea or sold as fertilizer.

The removal of solid material during sewage treatment is highly efficient, being reduced from $400 \, mg \, dm^{-3}$ in untreated sewage to $10 \, mg \, dm^{-3}$ once treated. Similarly, the amount of organic carbon is reduced from $250 \, mg \, dm^{-3}$ to $20 \, mg \, dm^{-3}$. Ninety-nine per cent of complex chemicals like the pesticide DDT are also removed. Potential pollutants such as zinc, copper and phosphorus may only be 50% removed. The phosphorus is a particular problem as it is widely used as a water softener in detergents and is therefore present in high concentrations in sewage. Most pathogenic organisms are removed by sewage treatment although *Salmonella paratyphi* (causes paratyphoid) and *Enteramoeba histolytica* (causes dysentery) may survive in small numbers. The eggs of worm parasites like *Ascaris* and *Taenia* have also been found in sewage works' effluent.

18.5.2 Toxic chemicals

There is a large variety of toxic chemicals released into rivers and seas around the world. These include copper, zinc, lead, mercury and cyanide. Fish are killed by fairly low concentrations of copper but algae are even more susceptible and die at concentrations as low as one part in two million.

Mercury is a particularly hazardous chemical as it forms strong complexes with the —SH groups in proteins, and so causes disruption of cell membranes and denaturation of enzymes. The kidney, liver and brain are most affected, resulting in loss of sensation, paralysis and death. An estimated 250 tonnes of mercury enters the world's oceans each year from the natural weathering of rocks. A further 3000 tonnes is added as

Did you know?

Lead poisoning in Lake Kariba in Zimbabwe resulted in floppy-trunk syndrome in elephants due to degeneration of peripheral nerves.

the result of environmental pollution. One graphic illustration of the effect of mercury occurred during the 1950s in the Japanese fishing village of Minamata where mercury, used as a catalyst in a nearby plastics factory, was discharged into the sea. In itself not especially toxic, the mercury accumulated in mud on the sea-bed. Here anaerobic methane-producing bacteria converted it to dimethyl mercury, a neurotoxin. The bacteria were consumed by filter-feeding shellfish which formed a major food source for the inhabitants of the village. Numbness, locomotory disorders, convulsions and blindness were experienced by the villagers, forty-six of whom died as a result of mercury poisoning. One hundred and twenty people died in similar circumstances at the nearby village of Niigata. Even babies born to mothers from the village some three years later showed mental retardation as a result of the poisoning.

18.5.3 Eutrophication by sewage and fertilizer

Eutrophication is a natural process during which the concentration of salts builds up in bodies of water. It occurs largely in lakes and the lower reaches of rivers, and the salts normally accumulate until an equilibrium is reached where they are exactly counterbalanced by the rate at which they are removed. Lakes and rivers with low salt concentrations are termed **oligotrophic** and the salts are frequently the factor limiting plant growth. Waters with high concentrations are termed **eutrophic** and here there is much less limitation on growth. **Algal blooms** occur where the waters become densely populated with species of blue-green bacteria in particular. The density of these blooms increases to a point where light is unable to penetrate to any depth. The algae in the deeper regions of the lake are therefore unable to photosynthesize, and die. Decomposition of these dead organisms by saprobiontic bacteria creates a considerable biochemical oxygen demand (BOD) resulting in deoxygenation of all but the very upper layers of the water. As a consequence all aerobic life in the lower regions dies.

The salts necessary for eutrophication of lakes and rivers are largely nitrates and phosphates and come from three sources:

1. **Leaching from the surrounding land** – This natural process is slow and is offset by the removal of salts as water drains from lakes or rivers.

2. **Sewage** – Even when treated, sewage effluent contains much phosphate as a result of the decomposition of detergents and washing powders (Section 18.5.1).

3. **Fertilizers** – An increasing quantity of inorganic fertilizer is now applied to farmland to increase crop yield. A major constituent of these fertilizers is nitrate. As this is highly soluble it is readily leached and quickly runs off into lakes and rivers.

18.5.4 Oil

The effects of oil pollution are localized, but nonetheless serious. Oil is readily broken down by bacteria, especially when thoroughly dispersed. Most oil pollution is either the result of illegal washing at sea of storage tanks of oil tankers or accidental spillage. The first major oil pollution incident in Great Britain

Cormorant killed by the *Braer* oil spill

occurred in 1967 when the Torrey Canyon went aground off Land's End. It released 120 000 tonnes of crude oil which was washed up on many Cornish beaches. Sea birds are particularly at risk because the oil coats their feathers, preventing them from flying; it also reduces their insulatory properties, causing death by hypothermia. The Torrey Canyon incident alone is estimated to have killed 100 000 birds. On shores, the oil coats seaweed, preventing photosynthesis, and covers the gills of shellfish, interfering with feeding and respiration. The effects are, however, temporary and shores commonly recover within two years. Detergents, used to disperse oil, can increase the ecological damage as they are toxic. With larger 'super-tankers', the potential danger from oil pollution is increased. The wrecking of the Amoco Cadiz off the Brittany coast in 1978 with the release of 200 000 tonnes of crude oil made the Torrey Canyon incident appear small by comparison. In 1989 Exxon Valdez spilt 38 000 tonnes in Prince William Sound, Alaska and in 1993 the Braer spilt 84 000 tonnes in the Shetlands. The long-term effects of these spills in such environmentally sensitive areas are yet to be seen.

18.5.5 Thermal pollution

All organisms live within a relatively narrow range of temperature. Wide fluctuations in temperature occur more often in terrestrial environments as the high specific heat of water buffers temperature changes. For this reason aquatic organisms are less tolerant of temperature fluctuations. Most thermal pollution of water is the result of electricity generation in power-stations. The steam used to drive the turbines in these stations is condensed back to water in large cooling towers. The water used in the cooling process is consequently warmed, being discharged at a temperature some 10–15 °C higher than when removed from the river. Although warmer water normally contains less dissolved oxygen, the spraying of water in cooling towers increases its surface area and thereby actually increases its oxygen content. The main effect of thermal pollution is to alter the ecological balance of a river by favouring warm-water species at the expense of cold-water ones. Coarse fish such as roach and perch may, for example, replace salmon and trout.

18.6 Terrestrial pollution

Pollution of land may be separated into two parts:

1. The dumping of wastes and deposits.
2. The use of pesticides.

18.6.1 The dumping of wastes and deposits

Many commercial activities result in the production of large volumes of solid waste material. For the most part this waste is dumped in pits or heaps. **Spoil heaps** consist of waste material from various mining activities, like gravel digging. While they

...urial site for low-level radioactive nuclear waste

may be unsightly they rarely present a direct hazard. **Slag heaps** are wastes from ore-digging and metal refining activities, and particularly from the mining of coal. The heaps themselves, especially when they result from mining metal ores, are toxic owing to the high concentration of heavy metal ions. Vegetation is difficult to grow on them and only varieties carefully selected for their tolerance of a particular metal will survive. Some 400 million tonnes of these wastes are piled up each year. They are unsightly and, in the absence of vegetation, may be unstable. This instability brought tragedy to the Welsh village of Aberfan when, on Friday 21 October 1966, a slag heap above the town, destabilized by heavy rain, began to slip. It engulfed the primary school and a number of houses, killing 116 children and 28 adults.

Domestic rubbish once contained a high proportion of ash, cinders and other solid, non-combustible material. It was frequently dumped in old quarries and pits and on low-lying land where it formed a relatively stable base upon which top soil could be placed and the land used for recreational or agricultural purposes. Present-day rubbish contains little ash or cinder and comprises mostly combustible material like paper and plastics. Incineration is often the best way to dispose of this rubbish, with the heat produced possibly being used to generate electricity. An incinerator at Edmonton in North London supplies electricity to the National Grid and so makes a sizeable profit from the disposal of its waste. Even when rubbish is dumped, the high content of organic material leads to decomposition by bacteria. These produce methane and other gases and there are plans at one large tip in the Midlands to collect the gas and burn it to produce electricity. It is hoped that enough will be generated to supply up to 10 000 homes. More details about the production of methane and alcohol from waste are given in Section 30.6.

18.6.2 Pesticides

It is difficult to define what exactly is a 'pest', but it is generally accepted to be an organism which is in competition with humans for food or soil space, or is potentially hazardous to health. It may even be an organism which is simply a nuisance and so causes annoyance. Pesticides are poisonous chemicals which kill pests, and they are named after the pests they destroy; hence insecticides kill insects, fungicides kill moulds and other fungi, rodenticides kill rodents such as rats and mice, and herbicides kill weeds. Unlike other pollutants, where their poisonous nature is an unfortunate and unwanted property, pesticides are quite deliberately produced and dispersed in order to exploit their toxicity.

An ideal pesticide should have the following properties:

1. It should be **specific**, in that it is toxic only to the organisms at which it is directed and harmless to all others.

2. It should **not persist** but be unstable enough to break down into harmless substances. It is therefore temporary and has no long-term effect.

3. It should **not accumulate** either in specific parts of an organism or as it passes along food chains.

375

TABLE 18.4 **Some major pesticides**

Name of pesticide	Type of pesticide	Additional information
Inorganic pesticides Calomel (mercuric chloride)	Fungicide	Used for dusting seeds to control transmission of fungal diseases
Copper compounds (e.g. copper sulphate)	Fungicide and algicide	One of the first pesticides ever used was Bordeaux mixture (copper sulphate + lime)
Sodium chlorate	Herbicide	Used to clear paths of weeds. Persistent, although not very poisonous
Organic pesticides Organo-phosphorus compounds (e.g. malathion and parathion)	Insecticides	Although very toxic they are not persistent and therefore not harmful to other animals if used responsibly. May kill useful insects such as bees, however
Organo-chlorine compounds (e.g. DDT, BHC, dieldrin, aldrin)	Insecticides	DDT is fairly persistent and accumulates in fatty tissue as well as along food chains. Aldrin may persist for more than 10 years. Resistance to them is now common. Most kill by inhibiting the action of cholinesterase
Hormones (e.g. 2,4-D, 2,4,5-T)	Herbicides	Selective weedkillers which kill broad leaved species. Stimulate auxin production and so disrupt plant growth. May contain a dangerous impurity – dioxin

Pesticides have been used for some time. A mixture of copper sulphate and lime, called Bordeaux mixture, was used over 100 years ago to control fungal diseases of vines. The problem is that in an attempt to produce food more economically and control human disease, pesticides have been used in large amounts in most regions of the world. A summary of some major pesticides is given in Table 18.4.

Most pesticides are not persistent. Warfarin, for example, readily kills any rodent which eats it, but as it is quickly broken down inside the rodent's body, it is harmless to anything which eats the corpse, e.g. maggots. Some pesticides, dichlorodiphenyl-trichlorethane (DDT), for example, are unfortunately persistent. First synthesized in 1874, its insecticidal properties were not appreciated until 1939. It was used extensively during the Second World War, in which it played a vital rôle in controlling lice, fleas and other carriers of disease. It was subsequently used in killing mosquitoes and so helped control malaria. Not only is DDT persistent, it also accumulates along food chains. If, for example, garden plants are sprayed with it in order to control greenfly, some of the flies will survive despite absorbing the DDT. These may then be eaten by tits who further concentrate the chemical in their bodies, especially in the fat tissues where it accumulates. If a number of tits, each containing DDT, are consumed by a predator, e.g. a sparrowhawk, the DDT builds up in high enough concentrations to kill the bird. Even where the concentrations are not sufficient to kill, they may still cause harm. It is known that DDT can alter the behaviour of birds, sometimes preventing them building proper nests. It may cause them to become infertile and can result in the egg shells being so thin that they break when the parent bird sits on them during

Weed control. Sugar beet crop in which weeds on the right have been controlled by chemicals and those on the left are untreated

incubation. In Britain these effects led to a marked decline in the 1950s and 60s of populations of peregrine falcons, sparrowhawks, golden eagles and other predatory birds. As a consequence, Britain, along with many other countries, restricted the use of DDT with the result that populations of these birds have now recovered.

Owing to the persistence of DDT, it remains in the environment despite the death of the organism containing it. With over one million tonnes of the chemical having already been used it now occurs in all parts of the globe and is found in almost all animals. Indeed, many humans contain more DDT than is permitted by many countries in food for human consumption.

With such widespread use of DDT, it is not surprising that selection pressure has resulted in insect varieties which are able to break it down and so render it useless. The development of **resistance** is now common among insect disease vectors like mosquitoes, and has set back prospects of eradicating malaria.

Herbicides make up 40% of the world's total pesticide production, and in developed countries the figure exceeds 60%. Some herbicides like paraquat kill all vegetation. While paraquat is highly poisonous it is rapidly broken down by bacteria and rendered harmless. Other weedkillers are selective, destroying broad-leaved plants (mostly dicotyledons) but not narrow-leaved ones (mostly monocotyledons). As most cereal crops are narrow-leaved and the weeds that compete with them are broad-leaved, such selective weedkillers are extensively used. They are similar to the plant's natural hormones, auxins, and as such are quickly broken down and rendered harmless. The two best known examples are 2,4-dichlorophenoxyacetic acid (2,4-D) and 2,4,5-trichlorophenoxyacetic acid (2,4,5-T). In the production of 2,4,5-T an impurity called **dioxin** is formed. Dioxin is one of the most toxic compounds known, a single gram being sufficient to kill in excess of 5000 humans. Even in minute quantities it may cause cancer, a skin disorder called chloracne and abnormalities in unborn babies. The chemical gained notoriety when used as a defoliant by the US army during the Vietnam war in the 1970s. It was a constituent of 'Agent Orange', 50 million dm^3 (litres) of which were sprayed over jungle areas to cause the leaves to drop so that enemy camps could be revealed. The dioxin produced physical and mental defects in children born in the area, as well as in those born to American servicemen working in the region. In 1976, an accident at a factory in Seveso, Italy, resulted in the release of dioxin into the atmosphere. Despite evacuation of the area, thousands of people suffered with chloracne, miscarriages, cancer and fetal abnormalities.

18.7 The impact of agriculture on the environment

We all need to eat to live and with an ever increasing population the need to produce sufficient food to meet the growing needs of the world's population has led to intensification of agricultural practices. Land is artificially prevented from reaching its climax vegetation through regular grazing, ploughing and the use of fertilizers and pesticides. In the United Kingdom agricultural food production has been doubled over the last 40 years. This has been achieved in a number of ways:

1. **Improved strains of plant and animal species** – Through artificial selection and genetic engineering the productivity of most crop plants and livestock animals has been increased.

2. **Greater use of fertilizers and pesticides** – There has been almost a ten fold increase in the use of artificial fertilizers over the past 50 years.

3. **Increased mechanization and use of biotechnology** – There have been major technological advances in machines used to sow, fertilize, harvest and transport crops as well as advances in the use of technology in controlling the harvesting and the conditions under which crops are stored. Animals are often reared under the optimum conditions for growth which are carefully controlled. There has been a consequent reduction in the number of farm labourers employed.

4. **Changes in farm practices and consequent increase in farm size** – There has been a trend to arable, rather than pastoral, farming. Sugar beet and oilseeds are increasingly grown instead of turnips and rye. Fields have become larger to accommodate modern machinery and so hedgerows have increasingly been removed. Wetland areas and ponds have been drained to increase the area of productive land.

Such has been the success of agricultural production in Europe that there are now surpluses of foods such as beef, dairy products and cereals. To reduce these surpluses farmers may, if they choose, be paid to **set-aside** up to 20% of their land for purposes other than food production, e.g. for planting woodland.

The demands of agriculture often conflict with the need for conservation. One example is **hedgerow removal**. It has been estimated that each year in the UK some 8000 km of hedgerows are removed. On the one hand the farmer may seek to remove hedges because:

1. They harbour pests, diseases and weeds, especially over winter.
2. They take up space which could otherwise be used to cultivate a crop.
3. They impede use of and accessibility for large machinery.
4. They reduce crop yields by absorbing moisture and nutrients

On the other hand the hedges have conservation value:

1. They are a habitat for a rich and diverse variety of plant and animal species.
2. They produce food for many birds and other animals which do not actually live in the hedgerows.
3. They act as corridors along which many species move and disperse themselves.
4. They act as wind-breaks, often preventing soil erosion by the wind.
5. They add diversity and interest to the landscape.

Did you know?

Satellites show that 10 000 km of hedgerow disappeared between 1990 and 1993 in Britain.

Farmland with small fields and many hedgerows

Large area of arable land without hedgerows

18.8 Conservation

There has been a growing interest in conservation as a result of increasing pressures placed upon the natural environment, the widespread loss of natural habitats and the growing numbers of extinct and endangered species. As early as 1872, the Yellowstone National Park in the USA was established in order to protect a particularly valuable natural environment. Australia (1886) and New Zealand (1894) established national parks soon after. It was not until 1949 that the first national park in Britain was established, but prior to that many societies such as the Royal Society for the Protection of Birds (1889) and the National Trust (1895) had been set up to promote conservation. There are now a large number of agencies responsible for conservation in one form or another. These include international groups like the World Wide Fund for Nature, large national bodies such as the Department of the Environment (DoE), Nature Conservancy Council (NCC), and the National Trust (NT); commercial organizations like the water authorities and the Forestry Commission; charitable groups like the Royal Society for the Protection of Birds (RSPB) as well as County Trusts for Nature Conservation and Farming and Wildlife Advisory Groups. The main impetus for conservation has come as a result of the pressures created by an ever-increasing human population – likely to be 6000 million before the end of the century.

18.8.1 Endangered species

Many species have become **extinct**, i.e. they have not been definitely located in the wild during the past 50 years. Others are **endangered**, i.e. they are likely to become extinct if the factors causing their numbers to decline continue to operate. At least 25 000 plant species are considered to be endangered.

There are a number of reasons why organisms become endangered:

1. **Natural selection** – It is, and always has been, part of the normal process of evolution that organisms which are genetically better adapted replace ones less well adapted.

2. **Habitat destruction** – Humans exploit many natural habitats, destroying them in the process. Timber cutting destroys forests and endangers species like the orang-utan. Industrial and agricultural development threaten many plant species of the Amazon forest. Clearing of river banks destroys the natural habitat of the otter, and modern farming methods remove hedgerows and drain wetlands, endangering the species which live and breed there.

3. **Competition from humans and their animals** – Where a species is restricted to a small area, e.g. the giant tortoises in the Galapagos Islands, they are often unable to compete with the influx of humans and their animals. Because their habitat is restricted, in this case by water, they cannot escape.

4. **Hunting and collecting** – Humans hunt tigers for sport, crocodiles for their skins, oryx as trophies, elephants for ivory, whales for oil and rhinoceros for their horn. Other organisms are

Heather moorland in Dartmoor National Park, Devon

collected for the pet trade, e.g. tamarins and parrots; and for research purposes, e.g. frogs. These are in addition to the numerous species hunted purely as food.

5. Destroyed by humans as being a health risk – Many species are persecuted because they carry diseases of domesticated species, e.g. badgers (tuberculosis of cattle) and eland (various cattle diseases).

6. Pollution – Oil pollution threatens some rare species of sea birds. The build-up of certain insecticides along food chains endangers predatory birds like the peregrine falcon and the golden eagle (Section 18.6.2).

18.8.2 Conservation methods

To combat the pressures listed above a number of conservation techniques are used:

1. Development of national parks and nature reserves – These are habitats legally safeguarded and patrolled by wardens. They may preserve a vulnerable food source, e.g. in China areas of bamboo forest are protected to help conserve the giant panda. In Africa game parks help to conserve endangered species such as the African elephant. Efforts are being made to conserve the dwindling areas of tropical rainforest. Planning authorities have greater powers to control developments and activities within these areas.

2. Planned land use – On a smaller scale, specific areas of land may be set aside for a designated use. The types of activities permitted on the land are carefully controlled by legislation. Such areas include Green Belts, Areas of Outstanding Natural Beauty, Sites of Special Scientific Interest, and country parks. Some places are designated as Environmentally Sensitive Areas (ESAs) and farmers or other landowners may be compensated for restricting activities which might conflict with conserving the natural habitats in the region.

3. Legal protection for endangered species – It is illegal to collect or kill certain species, e.g. the koala in Australia. In Britain, the Wildlife and Countryside Act gives legal protection to many plants and animals. Even legislation such as the Clean Air Act, may indirectly protect some species from extinction. Despite stiff penalties, such laws are violated because of the difficulty of enforcing them.

4. Commercial farming – The development of farms which produce sought-after goods, e.g. mink farming, deer farming, may produce enough material to satisfy the market and so remove the necessity to kill these animals in the wild.

5. Breeding in zoos and botanical gardens – Endangered species may be bred in the protected environment of a zoo and when numbers have been sufficiently increased they may be reintroduced into the wild. In the same way plant species may be protected in botanical gardens.

6. Removal of animals from threatened areas – Organisms in habitats threatened by humans, or by natural disasters such as floods, may be removed and resettled in more secure habitats.

Conservation of fenlands

Wicken Fen

Surrounding the Wash in eastern England there was once a huge area of waterlogged marsh and peatland supporting a rich and unique flora. This fenland has been systematically drained to be used for farming, resulting in a shrinkage of the soil so that the level has fallen about 4.5 metres. For some years now attempts have been made to conserve the remaining patches of undrained land but it is difficult to maintain, or reintroduce if necessary, the original varieties of plants. The fen is now higher than the surrounding farmland and is becoming acidic as the topsoil is leached. This leads to invasion by untypical acid-loving species like *Sphagnum* and the bog myrtle. As water drains off the fen more has to be pumped on to it along special channels to maintain the correct conditions for the vegetation. Typical fenland plants are the sedge, *Cladium mariscus*, and the reed, *Phragmites communis* and these must be cut back every four years in the spring or summer to prevent invasion by scrub such as buckthorn and willows.

ane toad eating pygmy possum. Introduced as a eans of biological control the cane toad is now a edator of native species.

ns for waste recycling

7. Control of introduced species – Organisms introduced into a country by humans often require strict control if they are not to out-compete the indigenous species. Feral animals (domesticated individuals which escape into the wild) must be similarly controlled.

8. Ecological study of threatened habitats – Careful analysis of all natural habitats is essential if they are to be managed in a way that permits conservation of a maximum number of species.

9. Pollution control – Measures to control pollution such as smoke emissions, oil spillage, over use of pesticides, fertilizer run-off, etc, all help to prevent habitat and species destruction. This is especially important in sensitive and vulnerable areas such as river estuaries and salt marshes.

10. Recycling – The more material which is recycled, the less need there is to obtain that material from natural sources e.g. through mining. These activities often destroy sensitive habitats either directly, or indirectly through the dumping of waste which is toxic or the development of roads to transport the products. This can be especially true of metal ores which are often found in mountainous regions, many of which are home to rare species.

11. Education – It is of paramount importance to educate people in ways of preventing habitat destruction and encouraging the conservation of organisms.

Did you know?

It takes 15 000 recycled sheets of A4 paper to save a tree.

18.9   Questions

1. (a) Explain what is meant by the *greenhouse effect*. (5 marks)
   (b) Describe the causes of the recent increase in this effect. (9 marks)
   (c) Discuss the possible global consequences of this increase. (9 marks)
   (Total 23 marks)

   UCLES (Modular) June 1992,
   (Energy in Living Organisms), No. 2

2. Write an essay on the advantages and disadvantages of chemicals in agriculture.

   AEB June 1990, Paper II, No. 5A

3. Summary

   This question concerns both atmospheric and terrestrial pollution. You are asked to recall information concerning atmospheric pollutants and their effects on ecosystems and human health. In addition you have to present tabulated information drawn from a field experiment in the form of a suitable graph, and draw conclusions from these data. Finally you are asked to suggest ways of improving the usefulness of the field observations.

   ## PART A

   (a) Give THREE examples of chemicals which seriously pollute the atmosphere and name ONE major source of EACH pollutant. (2 marks)
   (b) Describe the effects that your chosen pollutants are likely to have on ecosystems and/or human health. (6 marks)

   ## PART B

   The marine mussel *Mytilus edulis* is a mollusc which lives at high densities on the sea shore. In an experiment designed to test the effects of oil pollution, three different areas of a mussel bed were treated as follows.

   AREA 1: covered in crude oil followed by washing with water.

   AREA 2: washing with dispersant (a chemical used to disperse oil slicks) followed by washing with water.

   AREA 3: covered in crude oil followed by washing with dispersant followed by washing with water.

   The density of mussels in each of these areas was estimated in the following months as the percentage cover in a single 1 m² quadrat. These results are presented in the table.

   (c) Present these data in a suitable graphical form. (6 marks)
   (d) What conclusions can you draw from your graph? (3 marks)

Percentage cover of mussels in experiment plots

| Months after treatment | AREA | | |
	1 (oil + wash)	2 (dispersant + wash)	3 (oil + dispersant + wash)
0	68	81	70
6	65	83	3
10	52	33	8
14	46	35	12
23	49	43	17

(e) What changes would you make in the procedure and/or measurements to increase the usefulness of this survey? (3 marks)
(Total 20 marks)

NISEAC June 1990, Paper III, No. 2.

4. In most countries, the microbiological process which most affects the quality of life is the disposal of industrial and domestic waste.
   (a) Briefly define the objectives of a sewage treatment process. (3 marks)
   (b) Describe in detail the various stages by which raw sewage may be treated. (12 marks)
   (c) What is done with the materials produced at the end of the treatment process? (2 marks)
   (d) Some synthetic chemicals are difficult to remove in a sewage treatment works. Suggest why this is and what problems can be caused if they are not eliminated. (3 marks)
   (Total 20 marks)

O&C SEB June 1993, Paper 2/8, No.

5. Read the information below and answer the questions which follow.

   Ideally insecticides should kill harmful insects without causing harm to beneficial insects or other animals, including humans. Chemicals used as insecticides frequently act on the nervous system. The nervous system of insects functions in essentially the same way as that of mammals, so many chemicals targeted against insects are also likely to affect humans.

   The relative toxicity of various insecticides to mammals and to insects is usually expressed as the *average lethal dose* or LD50. These figures indicate the dose required, either by the mouth (oral) or through the body surface (dermal), to kill half of the animals in the test group.

   The table below shows the LD50 values for various insecticides tested on adult houseflies and on rats.

   DDT and gamma HCH are described as persistent insecticides and can persist for decades in an active orm. Malathion and parathion will remain active for only a few months after use.

arathion and malathion are applied in aerosol form ut gamma HCH is applied in a powder form. DDT nay be applied in aerosol or powder form but is now anned from use.

Insecticide	LD50/mg insecticide kg⁻¹ body mass		
	Rat (oral)	Rat (dermal)	Houseflies (dermal)
DDT	120–180	18	18
Malathion	2800	50	50
Gamma HCH	90	1.5	1.5
Parathion	8	3	3

nswer the following questions referring to the nformation given above where relevant.

(a) (i) Which insecticide would be the safest to use in the home to control houseflies? Explain your answer. (3 marks)

(ii) Which insecticide would be the most dangerous if used in the kitchen? Explain your answer. (2 marks)

(b) Many insecticides, such as malathion, are applied as aerosols but gamma HCH is only applied as a powder. Suggest why this is so. (2 marks)

(c) (i) Suggest one advantage and one disadvantage of using a persistent insecticide such as gamma HCH. Explain your answers. (4 marks)

(ii) DDT is no longer used as an insecticide. Suggest three reasons why this is so. (3 marks)

(d) Biological control of pests is frequently preferable to chemical control. Briefly describe one example of the biological control of a pest. (2 marks)

(Total 16 marks)

ULEAC June 1993, Paper I, No. 14

. The data in the table summarize the distribution of ecanora muralis in and around an industrial city in the orth of England. Lecanora muralis is a lichen which haracteristically grows on walls and can be used as n indicator species. It is particularly susceptible to ulphur dioxide (SO₂) pollution in the atmosphere.

(a) Explain the term *indicator species*. (1 mark)

(b) How does the pH of the rain water correlate with the mean annual sulphur dioxide concentration at different distances from the city centre? (2 marks)

(c) (i) Comment on the results shown in the table.

(ii) Near the city centre, lichens are found only on asbestos-cement roof tiles. Suggest a reason for this. (5 marks)

Distance from city centre /miles	Distribution of *Lecanora muralis*	Mean annual SO₂ concentration /μg m⁻³	Rain water pH
0–1.5	Absent	Over 240	4.4–4.7
1.5–2.5	On asbestos-cement tile roofs	200–240	4.7–4.9
2.5–3.5	On asbestos-cement tile and sheet roofs	170–200	4.9–5.1
3.5–5.5	On asbestos-cement roofs, cement, concrete and mortar	125–170	5.1–5.5
Over 5.5	On asbestos-cement roofs, cement and concrete	Under 125	Over 5.5

(from Seaward, M.R.D. In: Brown et al (1976))

(d) (i) State the major source of atmospheric sulphur dioxide pollution in the United Kingdom at the present time.

(ii) Explain your answer. (2 marks)

(e) Suggest **two** possible ways in which atmospheric sulphur dioxide pollution may be reduced. (2 marks)

(f) State **four other** effects which increased levels of atmospheric sulphur dioxide are having on the environment. (4 marks)

(g) Motor cars are another major source of pollution. Comment on this statement as fully as you can. (4 marks)

(Total 20 marks)

UCLES (Modular) November 1992, (Conservation and principles of ecology), No. 1

7. Samples of water were taken from a river at a factory outflow and at a number of points down stream. A clean freshwater stream entered the river some distance below the factory outflow.

The table below shows the degree of pollution, determined by the oxygen content of the water, at each sample point.

Distance from outflow/m	Oxygen content/arbitrary units
0	0.12
100	0.13
200	0.11
300	0.14
400	0.20
500	0.24
600	0.28
700	0.34
800	0.40
900	0.62
1000	1.02
1100	1.16
1200	1.18

(a) Plot the data on graph paper. (*4 marks*)

(b) (i) Use the graph to suggest the distance from the outflow at which the fresh water entered the river. (*1 mark*)

   (ii) Explain the reasons for your choice. (*2 marks*)

(c) There is an inverse relationship between the reading on the scale of oxygen content and the level of pollution. A reading of 1.2 on the scale for the oxygen content indicates zero pollution and a reading of 0.0 indicates 100% pollution.
Calculate the percentage pollution at 550 m down stream from the factory outflow. Show your working. (*3 marks*)

(d) Dead fish were seen in the water just below the outflow. Give *two* possible explanations for this observation. (*2 marks*)

(e) Suggest *two* ways in which the river pollution caused by the factory outflow could be reduced. (*2 marks*)
(*Total 14 marks*)

ULEAC June 1992, Paper I, No. 11

8. Nitrate fertilizer is applied to many crops. If too little is applied, yields are low. If excess is applied, yields may also be low and nitrate may accumulate in crops, rivers and ponds.
A simple test kit now enables farmers to obtain instant measurements of nitrate concentrations in plant cells and in water samples. Using such a kit, nitrate concentrations were measured in lettuce crops, river water and pond water in an intensively farmed area. The results are summarized in the table below.

Sample	Nitrate level in mg dm^{-3}
Lettuce cells: inner leaves	500
Lettuce cells: outer leaves	750
River water	150
Pond water	250

(a) Suggest why application of (i) too little and (ii) excess nitrate fertilizer may lead to lower crop yields.
   (i) Too little. (*3 marks*)
   (ii) Excess. (*2 marks*)

(b) Suggest how farmers may benefit by readily obtaining measurements of nitrate levels in growing crops. (*2 marks*)

(c) (i) Suggest reasons for the difference in cell nitrate concentration between inner and outer lettuce leaves as shown in the table. (*3 marks*)

   (ii) Suggest why nitrate concentration in the pond water is higher than that in the river water. (*2 marks*)

(d) Explain how nitrate added to agricultural land may reach rivers and ponds. (*2 marks*)

(e) Describe possible effects of increased nitrate concentrations in ponds and rivers on the following organisms.
   (i) Algae (*2 marks*)
   (ii) Saprobiontic bacteria (*2 marks*)
   (iii) Fish (*2 marks*)
(*Total 20 marks*)

ULEAC June 1990, Paper I, No. 1

9. The graph shows the effect of a single application of a pesticide on the numbers of predatory soil mites and their principal prey, springtails.

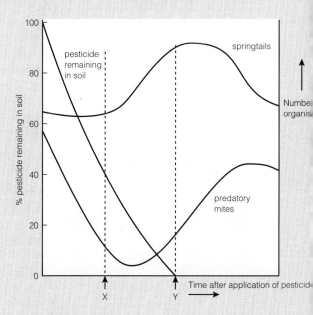

(a) Briefly describe the pattern of changes that might be expected in the numbers of mites and springtails if the soil were left untreated. (*2 marks*)

(b) Suggest an explanation for the changes in the numbers of mites and springtails over the following periods:
   (i) from application of the pesticide to time **X**; (*1 mark*)
   (ii) between time **X** and time **Y**. (*3 marks*)
(*Total 6 marks*)

AEB November 1993, Paper I, No.

False colour SEM of ciliated epithelium lining the trachea of a rat (opposite)

Part IV

TRANSPORT AND EXCHANGE MECHANISMS

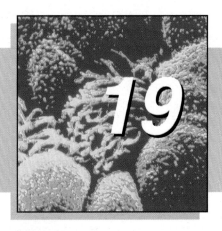

19

Why organisms need transport and exchange mechanisms

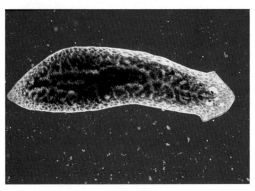

Dugesia – free-living cold water flatworm

All organisms need to exchange materials between themselves and their environment. Respiratory gases and the raw materials for growth must pass into an organism and waste products must be removed. This exchange is carried out passively by diffusion and osmosis and actively by active transport, pinocytosis and phagocytosis (Section 4.3). To be efficient, exchange mechanisms require the surface-area over which transfer occurs to be large when compared to the volume of the organism. Where diffusion is involved the exchange surface needs to be moist and the distance across which diffusion occurs must be as small as possible. In small organisms such as protozoans and unicellular algae their surface-area is sufficiently large compared with their volume to allow efficient exchange of most materials over the whole surface of their bodies.

When organisms became multicellular and so grew in size, they could only meet their exchange demands by simple diffusion if their requirements were very modest, for example if they had a very low metabolic rate. An increase in size inevitably meant an increased distance from the surface to the centre of the organism. Even if sufficient exchange occurred at the surface, the centre could still be starved of raw materials, because the rate of delivery was inadequate to supply the demand, if it was dependent on diffusion alone. One means of overcoming this problem is to become flattened in shape, so ensuring that no part of the body is far from the surface which supplies its nutrients. This explains the shape of the flatworms (Platyhelminthes). A further solution is to leave the central region of the organism hollow, or fill it with non-metabolizing material.

Further increases in size and/or metabolic rate necessitated the development of specialized exchange surfaces to compensate for a smaller surface area to volume ratio and/or an increased oxygen demand. In insects, flight necessitated a high metabolic rate and hence a more efficient delivery of oxygen to the tissues and subsequent removal of carbon dioxide. To achieve this they developed tubular ingrowths, the tracheae, which carried the air directly to the respiring tissues. These have the advantage of allowing oxygen and carbon dioxide to diffuse through a gaseous medium, rather than through the aqueous medium of cells – a much slower process. In addition, mass flow of the air is possible and this too speeds the movement of gases. With tracheae there is no need for a circulatory system to carry respiratory gases. While a blood system is present in insects, it possesses no respiratory pigments and its purpose is to carry nutrients, wastes and phagocytic cells.

Did you know?

The largest mammal is 70 million times bigger than the smallest (mass of the smallest shrew is 2 g; the largest whale is 140 tonnes)

Albino axolotl (*Ambystoma mexicanum*)

Where large size is combined with a high metabolic rate, both specialized exchange surfaces and an efficient means of transport become essential. In water the exchange surfaces for respiratory gases take the form of gills. In their simplest forms these are branched external outpushings of the body wall, as in amphibians like the axolotl. When covered, gills require a means of ventilation to supply them with fresh respiratory medium. This ventilation may be carried out by muscular action or by cilia. In fish gills form highly branched, blood-filled extensions around gill slits which lead from the pharynx. A regular current of water is pumped over these internal gills. While gills are in principle adequate exchange surfaces for terrestrial organisms, they suffer the disadvantage of lacking support in the less dense medium of air. They therefore collapse, reducing their surface-area and making them inefficient. An additional problem is that respiratory surfaces need to be kept moist to allow efficient diffusion. Gills could be kept moist, but due to their positions they would lose intolerably large quantities of water through evaporation. The solution for large, highly metabolizing, terrestrial organisms was to develop lungs. These comprise tiny elastic sacs, the alveoli, which are supported by connective tissue. The tubes, called bronchi, leading to these sacs are supported by cartilagenous rings to prevent collapse and the lungs as a whole are supported and protected by a bony cage of ribs. Being located deep within the body and communicating to the outside only by means of a narrow tube, the trachea, evaporative losses are kept to a minimum. The linings of the alveoli are thin and well supplied with blood. Muscular action ensures constant ventilation of the lungs.

With increasing size and specialization of organisms, tissues and organs became increasingly dependent upon one another. Materials needed to be exchanged not only between organs and the environment, but also between different organs. To this end, animals developed circulatory systems. These comprise a fluid which either flows freely over all cells (open system) or is confined to special vessels which communicate within diffusing distance of cells (closed system). The fluid is circulated by cilia, body muscle or a specialized pump (the heart) or some combination of these mechanisms. Closed blood systems are used by the larger, more highly evolved animals as they allow greater control of the distribution of the blood, making them more efficient at meeting changes in the demands of different tissues. Control of the heart beat assists this process.

The blood itself must be adapted to transport a wide variety of substances. Many are simply dissolved in a watery solution (the plasma), but others like oxygen are carried by special chemicals (respiratory pigments); these may be contained in specialized cells (red blood cells). Being distributed to all parts of the body, the blood is ideally situated to convey the body's defence and immune system (the white blood cells). The liquid nature of blood, so necessary for rapid transport around the body, suffers the disadvantage that it leaks away when damage is caused to the cavities or vessels containing it. Consequently a mechanism has evolved to ensure rapid clotting in these circumstances.

In plants their method of nutrition, necessitating as it does the need to capture light, means that they have an exceedingly large surface-area for this purpose. This same surface therefore serves

Giant redwoods (*Sequoiadendron giganteum*)

for the exchange of gases. As plants do not carry out locomotion, their metabolic rate is relatively low compared to most animals and therefore diffusion suffices. In addition, all respiring tissues are near to the surface of the plant. Large trees, for example, possess dead xylem tissue at the centre of their trunks and large branches. Respiratory gases therefore need only be conveyed very short distances and there is no specialized system for their transport – diffusion suffices. The same is not true of water and photosynthetic products. These often need to be transported the total length of the plant, 100 m or more in some cases.

In an attempt to gain a competitive advantage in the struggle for light, many plants have evolved to be very tall. The water required for photosynthesis is, however, obtained from the roots, which are firmly anchored in the soil. A transport system is therefore necessary to convey water from the roots to the leaves. At the same time, the sugars manufactured in the leaves must be transported in the opposite direction to sustain their respiration.

Plants, unlike animals, do not possess contractile cells like muscles. They therefore depend largely upon passive rather than active mechanisms for transporting materials. The evaporation of water from stomata creates an osmotic gradient across the leaf which draws in water from the xylem. Xylem forms a continuous unimpeded column of narrow tubes from the roots to the leaves. Owing to its cohesive properties, removal of water at the top of this column pulls up water from the bottom in a continuous stream. An osmotic gradient is responsible for the movement of water from the soil into the roots and across the cortex to the xylem. Only in the process of getting water into the xylem in the root is energy expended by the plant. The flow of sugars from the leaves to the roots is less clearly understood. Theories like 'mass flow' involve a passive movement; others, such as the 'transcellular strand theory', suggest an active mechanism.

20 Gaseous exchange

20.1 Respiratory surfaces

All aerobic organisms must obtain regular supplies of oxygen from their environment and return to it the waste gas carbon dioxide. The movement of these gases between the organism and its environment is called **gaseous exchange**. Gaseous exchange always occurs by **diffusion** over part or all of the body surface. This is called a **respiratory surface** and in order to maintain the maximum possible rate of diffusion respiratory surfaces have a number of characteristics.

1. **Large surface-area to volume ratio** – This may be the body surface in small organisms or infoldings of the surface such as lungs and gills in larger organisms.

2. **Permeable**

3. **Thin** – Diffusion is only efficient over distances up to 1 mm since the rate of diffusion is inversely proportional to the square of the distance between the concentrations on the two sides of the respiratory surface.

4. **Moist** – since oxygen and carbon dioxide diffuse in solution.

5. **Efficient transport system** – This is necessary to maintain a diffusion gradient and may involve a vascular system.

The relationship between some of these factors is expressed as **Fick's law** which states:

Diffusion is proportional to

$$\frac{\text{surface area} \times \text{difference in concentration}}{\text{thickness of membrane}}$$

Organisms can obtain their gases from the air or from water. The oxygen content of a given volume of water is lower than that of air, therefore an aquatic organism must pass a greater volume of the medium over its respiratory surface in order to obtain enough oxygen.

TABLE 20.1 **Water and air as respiratory media**

Property	Water	Air
Oxygen content	Less than 1%	21%
Oxygen diffusion rate	Low	High
Density	Relative density of water about 1000 times greater than that of air at the same temperature	
Viscosity	Water much greater, about 1000 times that of air	

20.2 Mechanisms of gaseous exchange

As animals increase in size most of their cells are some distance from the surface and cannot receive adequate oxygen. Many larger animals also have an increased metabolic rate which increases their oxygen demand. These organisms need to develop specialized respiratory surfaces such as gills or lungs. These surfaces allow the gases to enter and leave the body more rapidly. There remains the problem of transporting the gases between the respiring cells and the respiratory surface. Generally the gases are carried by the blood vascular system. The presence of respiratory pigments like haemoglobin increase the oxygen-carrying capacity of the blood (Section 21.2.1). The diffusion gradients may be further maintained by ventilation movements, e.g. breathing.

Gaseous exchange will be considered in detail for a number o different organisms. They will serve to demonstrate the differences between aquatic and terrestrial organisms and also show the problems associated with increased size.

20.2.1 Small organisms

Small organisms have a large surface-area to volume ratio and do not require specialized structures for gaseous exchange. *Amoeba* are less than 1 mm in diameter and gases diffuse over their whole surface. Cnidarians, like the sea anemone *Actinia*, are hollow and have all their cells in contact with the water which surrounds them. Platyhelminthes (flatworms) also rely on diffusion over the whole body surface and this is facilitated by their flattened shape which considerably increases their surface-area to volume ratio. All these organisms must live in water from which they obtain dissolved oxygen; they would rapidly desiccate in a terrestrial environment.

20.2.2 Flowering plants

Plants have a low metabolic rate, requiring less energy per unit volume than animals. Unicellular algae employ the whole surface of the cell for gaseous exchange. In the larger flowering plants this is not possible because their outer surfaces are waterproof to prevent the desiccation which results from living on land. Gases pass through small pores in the leaves and green stems. These pores are the **stomata** whose structure is illustrated in Section 14.1 and the mechanisms of which are described in Chapter 22. Woody plants still have stomata in the leaves but the stems have small areas of loosely packed bark cells called **lenticels**. Gases diffuse through the stomata and lenticels. Within the plant, oxygen moves through the intercellular air spaces to the respiring cells. Carbon dioxide moves in the reverse direction. Both gases move by diffusion through the spaces and then through the moist cell walls into the respiring cells themselves. Cells which contain chloroplasts have a further source of oxygen because it is released as a waste product of photosynthesis and immediately taken up by the mitochondria. Similarly, the carbon dioxide released from the mitochondria car be used by the chloroplasts for photosynthesis. The rate of photosynthesis is affected by the intensity of light and so the carbon dioxide used and oxygen released by this process vary considerably during the day. The balance of respiratory and photosynthetic gases therefore changes. This is referred to in Section 14.4.2 and also in Fig. 14.12.

PROJECT

Use your knowledge of colour changes in hydrogencarbonate indicator solution to investigate the effect of temperature variations on the rate of respiration in pond snails.

Elongated lenticel

Bark of silver birch showing lenticels

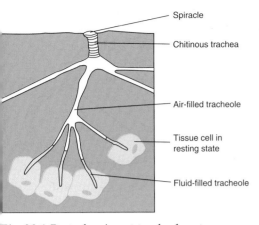

Spiracle

Chitinous trachea

Air-filled tracheole

Tissue cell in resting state

Fluid-filled tracheole

Fig. 20.1 Part of an insect tracheal system

20.2.3 Insects

For insects, diffusion of gases over the whole body surface is no longer possible. This is because they are predominantly a terrestrial group and therefore need to conserve water. This they do by having an exoskeleton and waterproof cuticle which are largely impermeable to gases. There is also a conflict between the requirement of a large surface-area for gaseous exchange and the need to conserve water. Gases enter and leave through pores called **spiracles**. Spiracles are usually found in pairs in ten of the body segments and are not just simple holes. Each is surrounded by hairs which help to retain water vapour and may be closed by a system of valves operated by tiny muscles. Respiring cells inside the insect release carbon dioxide and as this accumulates it is detected by chemoreceptors and the spiracles open.

The spiracles open into a complex series of tubes running throughout the body. These tubes are called **tracheae**. They are supported by rings of chitin which prevent their collapse when the pressure inside them falls. The tracheae divide to form smaller **tracheoles** extending right into the tissues. Respiratory gases are carried in the **tracheal system** between the environment and the respiring cells; they are not transported by the blood. The tracheal system carries oxygen rapidly to the cells and allows the insects to develop high metabolic rates. The ends of the fine tracheoles are fluid-filled. At rest the tissue cells have a lower solute concentration than the fluid in the tracheoles. As activity increases the muscles respire anaerobically and lactic acid accumulates. This raises the solute concentration of these cells above that of the fluid in the tracheoles. Water therefore moves out of the tracheoles into the muscle cells by osmosis. As water is lost, air is drawn further into the tracheoles, making more oxygen available for cellular respiration. When activity ceases the metabolites are oxidized, the solute concentration of the muscle cells is lowered and the water re-enters the tracheoles. The system is ventilated by contraction of the abdominal muscles of the insect flattening the body. This reduces the volume of the tracheal system. The volume increases again as the elastic nature of the body returns the insect and tracheal system to their original shape. Larger insects, such as locusts, have some of the tracheae expanded to form air-sacs which act as bellows.

The tracheal system provides an extremely efficient means of gaseous exchange, but it does have its limitations. Since it relies entirely on diffusion for the gases to move from the environment to the respiring cells, insects are not able to attain a large size. In addition, the chitinous linings of the tracheae must be moulted with the rest of the exoskeleton.

20.2.4 Bony fish

Bony fish have four pairs of bony branchial arches supporting gill lamellae. These lamellae form a double row arranged in a V-shape and bear gill plates at right angles to their surface. There are no branchial valves but the gill slits are covered by a bony flap called an **operculum**. This helps to protect the delicate gills but also plays a part in their ventilation.

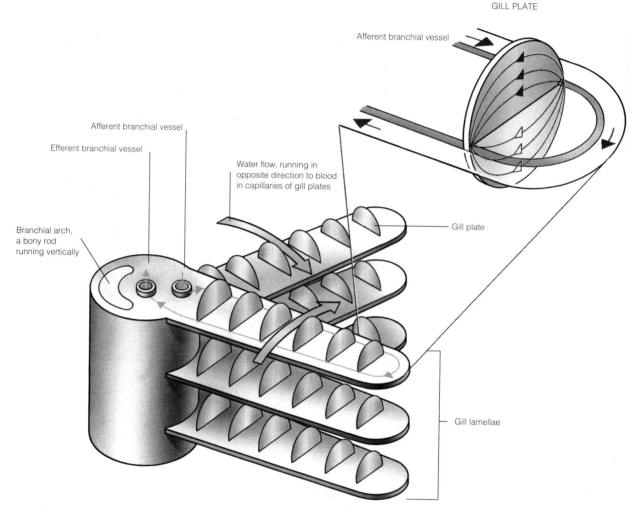

GILL PLATE

Afferent branchial vessel

Afferent branchial vessel

Efferent branchial vessel

Water flow, running in opposite direction to blood in capillaries of gill plates

Gill plate

Branchial arch, a bony rod running vertically

Gill lamellae

Fig. 20.2 Water flow over gill lamellae in a bony fish

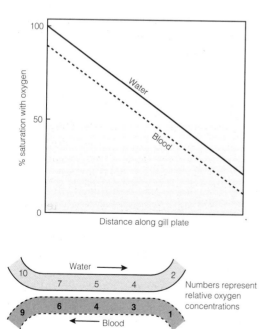

Fig. 20.3 Counter-flow in the gills of a bony fish

Deoxygenated blood enters the gill capillaries via the afferent branchial vessels. Oxygenated blood leaves in the efferent branchial artery to join the dorsal aorta along which blood passes to the rest of the body.

The gills of bony fish demonstrate extremely well the **counter current principle**. The essential feature of this is that the blood and water flow over the gill plates in opposite directions. This allows a fairly constant diffusion gradient to be maintained between the blood and the water, right across the gill. It ensures that blood which is already partly loaded with oxygen meets water which has had very little oxygen removed from it. Similarly, blood with very low oxygen saturation meets water which has already had much of its oxygen removed. (See Fig. 20.3.)

This mechanism allows bony fish to achieve 80% absorption of oxygen, compared to about 50% in the parallel flow system of a dogfish. The overlapping ends of the gill lamellae also slow down the passage of the water so that there is a greater time for diffusion to occur. Alternation of a buccal pressure pump and an opercular suction pump allows water to be drawn over and between the gills more or less continuously. To take in water, the floor of the buccal cavity is lowered, this increases its volume

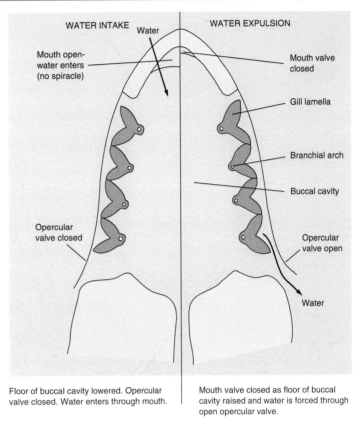

WATER INTAKE   Water   WATER EXPULSION

Mouth open-
water enters
(no spiracle)

Mouth valve
closed

Gill lamella

Branchial arch

Buccal cavity

Opercular
valve closed

Opercular
valve open

Water

| Floor of buccal cavity lowered. Opercular valve closed. Water enters through mouth. | Mouth valve closed as floor of buccal cavity raised and water is forced through open opercular valve. |

Fig. 20.4 Ventilation of gills in a bony fish

and the pressure within it decreases. Water enters through the mouth. At the same time the operculum is pressed close to the body and tiny opercular muscles increase the volume of the opercular cavity slightly. This causes some water to move out of the buccal cavity and into the opercular region, bathing the gills. Water is expelled by raising the floor of the buccal cavity. As pressure inside the buccal cavity increases, flaps of skin close the mouth and the water is forced over the gills and out of the body under the free edge of the operculum.

20.2.5 Mammals

Lungs are the site of gaseous exchange in mammals. They are found deep inside the thorax of the body and so their efficient ventilation is essential. The lungs are delicate structures and, together with the heart, are enclosed in a protective bony case, the **rib cage**. There are twelve pairs of ribs in humans, all attached dorsally to the thoracic vertebrae. The anterior ten pairs are attached ventrally to the sternum. The remaining ribs are said to be 'floating'. The ribs may be moved by a series of intercostal muscles. The thorax is separated from the abdomen by a muscular sheet, the **diaphragm**.

Air flow in mammals is **tidal**, air entering and leaving along the same route. It enters the nostrils and mouth and passes down the **trachea**. It enters the lungs via two **bronchi** which divide into smaller **bronchioles** and end in air-sacs called **alveoli**. These regions are illustrated in Fig. 20.5.

Trachea

Pulmonary
artery

Resin cast of pulmonary arteries (red) and trachea
and bronchi (clear)

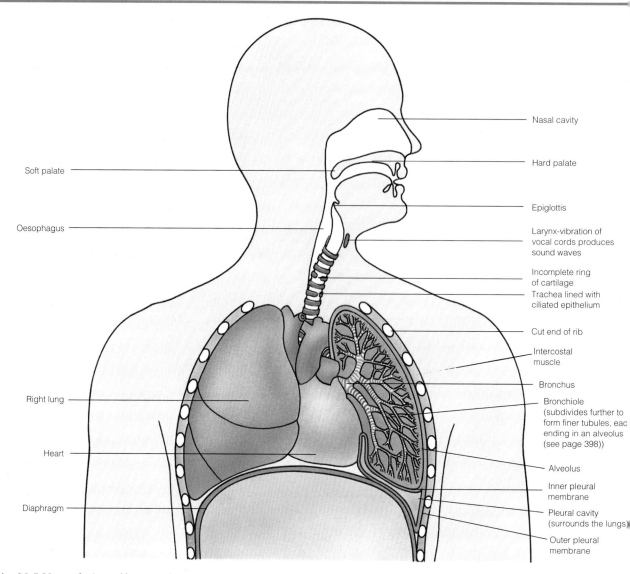

Fig. 20.5 Ventral view of human thorax

Labels (clockwise from top right):
- Nasal cavity
- Hard palate
- Epiglottis
- Larynx-vibration of vocal cords produces sound waves
- Incomplete ring of cartilage
- Trachea lined with ciliated epithelium
- Cut end of rib
- Intercostal muscle
- Bronchus
- Bronchiole (subdivides further to form finer tubules, eac ending in an alveolus (see page 398))
- Alveolus
- Inner pleural membrane
- Pleural cavity (surrounds the lungs)
- Outer pleural membrane
- Diaphragm
- Heart
- Right lung
- Oesophagus
- Soft palate

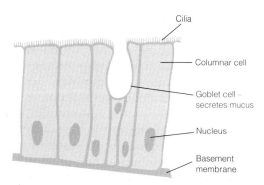

Fig. 20.6 Ciliated epithelium (LS)

Labels:
- Cilia
- Columnar cell
- Goblet cell – secretes mucus
- Nucleus
- Basement membrane

Regions of the respiratory system
Within the nasal channels mucus is secreted by goblet cells in th ciliated epithelium (Fig. 20.6). This mucus traps particles and th cilia move them to the back of the buccal cavity where they are swallowed. The mucus also serves to moisten the incoming air and it is warmed by superficial blood vessels. Within this region there are also olfactory cells which detect odours.

Air then passes through the pharynx and past the **epiglottis**, flap of cartilage which prevents food entering the trachea. The **larynx**, or voice box, at the anterior end of the trachea is a box-like, cartilagenous structure with a number of ligaments, the **vocal cords**, stretched across it. Vibration of these cords when ai is expired produces sound waves. The trachea is lined with ciliated epithelium and goblet cells. The mucus traps particles and the cilia move them to the back of the pharynx to be swallowed. The trachea is supported by incomplete rings of cartilage which prevent the tube collapsing when the pressure inside it falls.

The trachea divides into two **bronchi**, one entering each lung. The bronchi are also supported by cartilage. The **bronchioles** branch throughout the lung; as the tubes get finer cartilagenous support gradually ceases. These eventually end in alveoli. Each

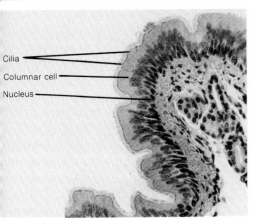

Section of human trachea showing ciliated epithelium (× 150 approx.)

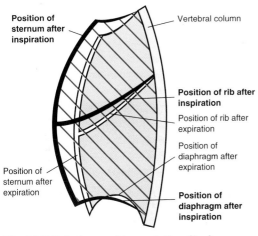

Fig. 20.7 Relative positions of ribs, diaphragm and sternum after breathing in and out

lung is surrounded by an air-tight cavity called the **pleural cavity**. This is bounded by two membranes, or **pleura**, which secrete **pleural fluid** into the cavity. The fluid is a lubricant, preventing friction when the lungs expand at inspiration. Pressure in the pleural cavity is always about 500 Pa lower than in the lungs and this allows them to expand and fill the thorax.

Breathing in (inspiration) in humans

In order for air to enter the lungs from the exterior the pressure inside the lungs must be lower than atmospheric pressure. This lowering of pressure is brought about as follows.

When the external intercostal muscles contract and the internal intercostal muscles relax, the ribs move upwards and outwards (anteriorly and ventrally). The diaphragm muscle contracts and flattens. These two movements cause the volume of the thorax to increase and therefore the pressure inside it falls. The elastic lungs expand to fill the available space and so their volume increases and the pressure within them falls. This causes air to rush into the lungs from the exterior.

Breathing out (expiration) in humans

Breathing in is an active process but breathing out is largely passive. The volume of the thorax is decreased as the diaphragm muscle relaxes and it resumes its dome-shape. The external intercostal muscles also relax, allowing the ribs to move downwards (posteriorly) and inwards (dorsally). They may be assisted by contraction of the internal intercostal muscles. As the volume of the thorax decreases, the pressure inside it increases and air is forced out of the lungs as their elastic walls recoil.

Exchange at the alveoli

Each minute alveolus (diameter 100 μm) comprises **squamous epithelium** and some elastic and collagen fibres. Squamous epithelial cells form a single layer attached to a basement membrane. In surface view the cell outlines are irregular and closely packed. The cells are shallow, the central nucleus often forming a bump in the surface. Adjacent cells may be joined by strands of cytoplasm. Such epithelia form ideal surfaces over which diffusion can occur and so are important not only in the alveoli of the lungs, but also in the Bowman's capsule and in capillary walls (Fig. 20.8). Each alveolus is surrounded by a network of blood capillaries which come from the pulmonary artery and unite to form the pulmonary vein. These capillaries are extremely narrow and the red corpuscles (erythrocytes) are squeezed as they pass through. This not only slows down the passage of the blood, allowing more time for diffusion, but also results in a larger surface-area of the red blood cell touching the endothelium and thus facilitates the diffusion of oxygen. The oxygen in the inspired air dissolves in the moisture of the alveolar epithelium and diffuses across this and the endothelium of the capillary into the erythrocyte. Inside the red blood cell, the oxygen combines with the respiratory pigment **haemoglobin** to form **oxyhaemoglobin** (Section 21.2.1). Carbon dioxide diffuses from the blood into the alveolus to leave the lungs in the expired air.

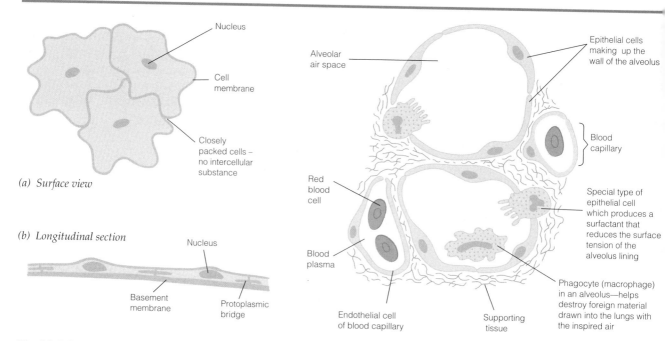

Nucleus

Cell
membrane

Closely
packed cells –
no intercellular
substance

(a) Surface view

(b) Longitudinal section

Nucleus

Basement
membrane

Protoplasmic
bridge

Fig. 20.8 *Squamous epithelium*

Alveolar
air space

Epithelial cells
making up the
wall of the alveolus

Blood
capillary

Red
blood
cell

Special type of
epithelial cell
which produces a
surfactant that
reduces the surface
tension of the
alveolus lining

Blood
plasma

Endothelial cell
of blood capillary

Supporting
tissue

Phagocyte (macrophage)
in an alveolus—helps
destroy foreign material
drawn into the lungs with
the inspired air

Fig. 20.10 *Arrangement of cells and tissues in mammalian lung alveoli*

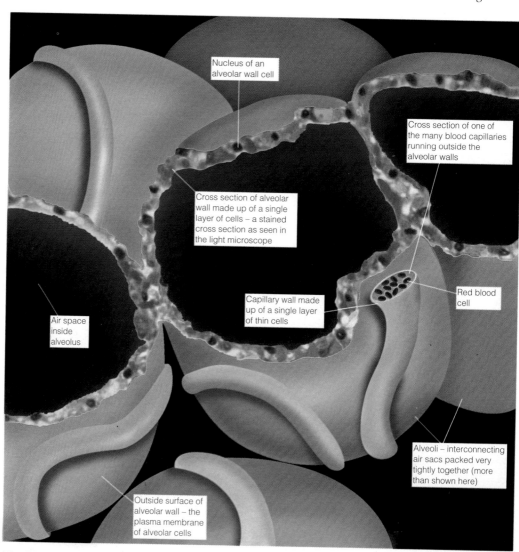

Nucleus of an
alveolar wall cell

Cross section of one of
the many blood capillaries
running outside the
alveolar walls

Cross section of alveolar
wall made up of a single
layer of cells – a stained
cross section as seen in
the light microscope

Red blood
cell

Capillary wall made
up of a single layer
of thin cells

Air space
inside
alveolus

Alveoli – interconnecting
air sacs packed very
tightly together (more
than shown here)

Outside surface of
alveolar wall – the
plasma membrane
of alveolar cells

Fig. 20.9 *External appearance of a group of alveoli*

Magnification ×300 (approx.)

Emphysema

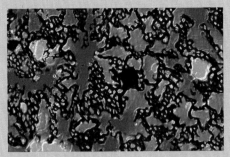

Normal healthy lung tissue

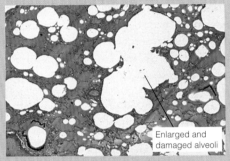

Enlarged and damaged alveoli

Lung tissue damaged by emphysema

One in every five smokers will develop the crippling lung disease called emphysema which, together with other related obstructive lung disorders, kills 20 000 people a year in Britain alone. The disease develops over a period of 20 or so years and it is impossible to diagnose until the lungs have been irreversibly damaged. In its early stages the only symptom is a slight breathlessness but as this gets progressively worse many people are so disabled that they cannot even get out of bed. People with emphysema usually die of respiratory failure, often accompanied by infection. A small number die of heart failure as the heart becomes enlarged and overworked trying to pump blood through arteries which have become constricted as a result of lack of oxygen.

Healthy lungs contain large quantities of elastic connective tissue comprising predominantly the protein elastin. This tissue expands when we breathe in and returns to its former size when we breathe out. In emphysematous lungs the elastin has become permanently stretched and the lungs are no longer able to force out all the air from the alveoli. Little if any exchange of gases can take place across the stretched and damaged air sacs.

The damage is brought about by abnormally high levels of elastase, an enzyme formed in some of the white blood cells, which breaks down elastin. Elastase also degrades other proteins so that, in the latter stages of the disease, breakdown of lung tissue results in large, non-functional holes in the lung.

In healthy lungs elastin is not broken down because a protein inhibitor (PI) inhibits the action of the enzyme elastase. However, in smokers it has been suggested that the oxidants in cigarette smoke inactivate PI, resulting in greater elastase activity and hence a breakdown of elastin.

Elastase is produced by phagocytes which need it so they can migrate through tissue to reach sites of infection. This is part of the body's normal inflammatory response. In smokers, where a large number of phagocytic cells are attracted to the lungs by the particulate materials in smoke, a combination of the release of elastase and a low level of its natural inhibitor leads to a lot of tissue degradation.

Smoking obviously causes much of the damage associated with emphysema but why is it that not all smokers suffer to the same extent? It is possible that the one in five who develop emphysema do so because they have defective repair mechanisms and are unable to counteract the considerable cell damage which occurs during the development of the disease. This may be the result of a genetic defect which limits cell division or production of abnormal connective tissue proteins during smoke-induced stress.

Emphysema cannot be cured and the disease cannot be reversed. The only way to minimize the chance of getting it is not to smoke at all, or to give up – the function cannot be restored to smoke-damaged lungs but giving up can significantly reduce the rate of further deterioration.

20.3   Control of ventilation in humans

Ventilation of the respiratory system in humans is primarily controlled by the **breathing centre** in a region of the hindbrain called the medulla oblongata. The ventral portion of this centre controls inspiratory movements and is called the **inspiratory centre**; the remainder controls breathing out and is called the **expiratory centre**. Control also relies on **chemoreceptors** in the **carotid and aortic bodies** of the blood system. These are sensitive to minute changes in the concentration of carbon dioxide in the blood. When this level rises, increased ventilation of the respiratory surfaces is required. Nerve impulses from these chemoreceptors stimulate the inspiratory centre in the medulla. Nerve impulses pass along the phrenic and thoracic nerves to the diaphragm and intercostal muscles. Their increased rate of contraction causes faster inspiration. As the lungs expand, **stretch receptors** in their walls are stimulated and impulses pass along the vagus nerve to the expiratory centre in the medulla. This automatically 'switches off' the inspiratory centre, the muscles relax and expiration takes place. The stretch receptors are no longer stimulated, the expiratory centre is 'switched off' and the inspiratory centre 'switched on'. Inspiration takes place again. This complex example of a feedback mechanism is illustrated in Fig. 20.11 and further examples of homeostatic control are considered in Chapter 25.

The breathing centre may also be stimulated by impulses from the forebrain resulting in a conscious increase or decrease in breathing rate.

The main stimulus for ventilation is therefore the change in carbon dioxide concentration and stimulation of stretch receptors in the lungs; changes in oxygen concentration have relatively little effect. At high altitudes the reduced atmospheric pressure makes it more difficult to load the haemoglobin with oxygen. In an attempt to obtain sufficient oxygen a mountaineer

Did you know?

Snakes have only one lung.

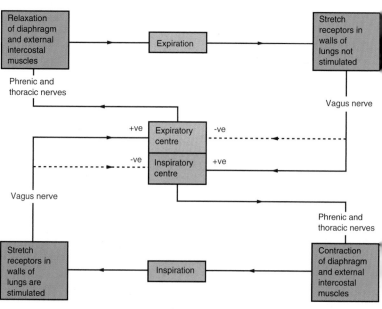

Fig. 20.11 Control of ventilation

takes very deep breaths. This forces more carbon dioxide out of the body and the level of carbon dioxide in the blood therefore falls. The inspiratory centre is no longer stimulated and breathing becomes increasingly laboured, causing great fatigue. Given time, humans can adapt to these conditions by excreting more alkaline urine. This causes the pH of the blood to fall, the chemoreceptors are stimulated and so is the inspiratory centre.

20.4 Measurements of lung capacity

Human lungs have a volume of about $5\,dm^3$ but in a normal breath only about $0.45\,dm^3$ of this will be exchanged **(tidal volume)**. During forced breathing the total exchanged may rise to $3.5\,dm^3$ **(vital capacity)** which leaves a **residual volume** of about $1.5\,dm^3$. These terms, and others associated with lung capacity, are illustrated in Fig. 20.12.

Air that reaches the lungs on inspiration mixes with the residual air so that it does not 'stagnate' but is gradually changed. This mixing of relatively small volumes of fresh air with a much larger volume of residual air keeps the level of gases in the alveoli more or less constant.

Measurements of respiratory activity may be made using a spirometer attached to a kymograph which records all its movements.

The ventilation rate is calculated as the number of breaths per minute × tidal volume.

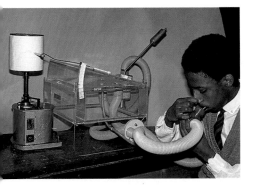

spirometer attached to a kymograph

TABLE 20.2 **Comparison of inspired, alveolar and expired air**

Gas	% Composition by volume		
	Inspired air	Alveolar air	Expired air
Oxygen	20.95	13.8	16.4
Carbon dioxide	0.04	5.5	4.0
Nitrogen	79.01	80.7	79.6

PROJECT

Use the spirometer to find the effect of exercise on the rate and depth of breathing.

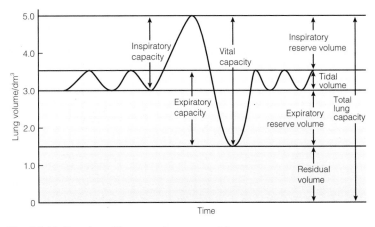

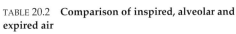

Fig. 20.12 Graph to illustrate lung capacities

APPLICATION

Smoking

Tobacco is responsible for 15–20% of all deaths in Britain, amounting to about 100 000 people every year. The three main diseases closely linked with smoking are:

- coronary heart disease
- lung cancer
- chronic bronchitis

Other smoking-related conditions include:

- emphysema (see page 399)
- cancer of mouth, throat, bladder and pancreas
- other cardiovascular diseases
- peptic ulcers
- narrowing of blood vessels in limbs
- damage to the unborn child

Tobacco smoke is a mixture of chemicals, a number of which interact with each other, multiplying their effects. The three constituents which do most harm are nicotine, carbon monoxide and tars.

Nicotine:

- is quickly absorbed into blood, reaching the brain in 30 seconds
- causes platelets to become sticky leading to clotting
- stimulates production of adrenalin leading to increased heart rate and raised blood pressure which puts an extra strain on the heart

Carbon monoxide:

- combines with haemoglobin to form carboxyhaemoglobin, therefore lowering oxygen-carrying capacity of the blood
- may aggravate angina
- seems to slow growth of the fetus

Tars:

- form an aerosol of minute droplets which enter the respiratory system causing thickening of the epithelium leading to chronic bronchitis
- paralyse cilia so dust, germs and mucus accumulate in lungs leading to infection and damage
- contain carcinogens – heavy smokers have a 25% greater risk of cancer than non-smokers.

20.5   Questions

. *(a)* Describe in detail the process of gaseous
  exchange in a healthy person.   *(13 marks)*
  (b) Explain how gaseous exchange in the lungs
  is impaired by cigarette smoking.   *(10 marks)*
                              (Total 23 marks)

> *UCLES (Modular) March 1992,*
> *(Human health and disease), No. 1*

. Discuss adaptations for gas exchange illustrating
your answer by reference to a mammal, a flowering
plant and an insect.

> *(Total 30 marks)*

> *ULEAC June 1993, Paper I, No. 15(b)*

. Describe the pathways and mechanisms involved
n the movement of oxygen in the mammal from
        (i)   atmosphere to alveolus;
        (ii) alveolus to lung capillary;
        (iii) lung capillary to placenta;
        (iv) placenta to fetus.

> *NEAB June 1990, Paper IB, No. 1*

. *(a)* Three cubical blocks of gelatine were
  prepared as shown in the diagram.

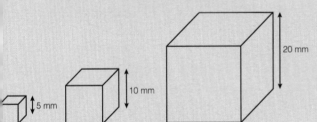

20 mm

10 mm

5 mm

   (i)   Construct a table to record side length,
         surface-area, volume, and surface-
         area/volume ratio for the blocks.
         Calculate all values for **each** block and
         enter them in your table.
   (ii) What is the relationship between side
         length and surface-area/volume ratio?
   (iii) The blocks were immersed in a solution
         of methylene blue (which stains
         gelatine) for five minutes, removed,
         dried and cut in half.

   Draw diagrams to show the probable
   appearance of the cut surfaces of the
   smallest and the largest blocks. *(5 marks)*

(b) The drawings illustrate the size and shape of
Amoeba (a protozoan) and *Planaria* (a
platyhelminth).

Amoeba   *Planaria*

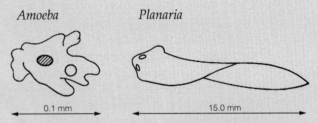

0.1 mm   15.0 mm

For each animal, briefly explain why simple
diffusion provides an adequate gaseous
exchange between the organism and its
environment.   *(4 marks)*
(c) Outline how a continuous supply of oxygen
reaches all tissues of an earthworm. *(3 marks)*
(d) The diagram shows a gill plate of a bony fish.

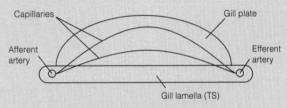

Capillaries   Gill plate

Afferent   Efferent
artery   artery

Gill lamella (TS)

   (i)   Draw arrows on the capillaries to
         indicate the direction of blood flow.
   (ii) Draw and label an arrow to indicate the
         direction of water flow over the gill
         plate.
   (iii) Explain the advantage of this
         arrangement.   *(3 marks)*
                              (Total 15 marks)

> *WJEC June 1992, Paper A2, No. 6*

5. Consider the spirometer trace below.

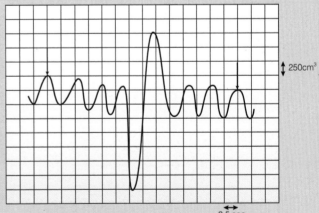

250cm³

2.5 sec

(a) What is the tidal volume?
(b)   (i) How much oxygen has been consumed
            between the arrows on the trace?
        (ii) What is the mean rate of oxygen
             consumption over this period?
                              (Total 3 marks)

> *NISEAC June 1992, Paper I, No. 17*

6. The respiratory surfaces of teleost fish consist of numerous gill lamellae.

(a) Explain how the following features of gill lamellae make gas exchange more efficient.
  (i) The lamellae being folded into many gill plates *(2 marks)*
  (ii) Blood flowing through the plates in the opposite direction to water flowing over the plates *(2 marks)*

(b) The following table summarizes some features of the gill lamellae of three species of teleost fish.

Species of fish	Thickness of lamellae in μm	Distance between lamellae in μm	Distance between blood and surrounding water in μm
Eel	26	30	6
Trout	12	35	3
Herring	7	20	1

Comment on these figures, given that herring are very active swimmers, trout are moderately active, and eels are relatively inactive. *(4 marks)*

(c) Suggest *two* reasons why gill lamellae would not provide an efficient respiratory surface on land. *(2 marks)*
(Total 10 marks)

ULEAC June 1991, Paper I, No. 7

7. (a) Describe the mechanism by which air is made to enter the human lungs. *(5 marks)*

(b) On its way to the lungs air passes over a ciliated mucous epithelium.
State *two* ways in which the air is changed as it passes over the epithelium. *(2 marks)*

(c) Photograph A below shows the surfaces of these epithelial cells in a young adult. Photograph B shows the surfaces of these epithelial cells in a young adult who is a heavy smoker.

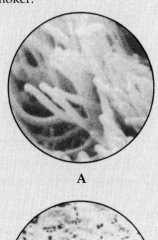

A

B

(i) Describe how smoking appears to have altered the structure of these cells. *(2 marks)*

(ii) Suggest how the functioning of the cells lining the respiratory tract might be altered by this damage. *(3 marks)*
(Total 12 marks)

ULEAC June 1990, Paper I, No. 9

See *Further questions*, pages 614–20, questions 6, 12, 17.

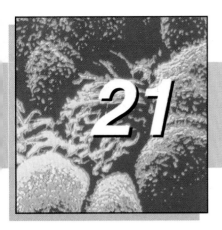

Blood and circulation (transport in animals)

As all cells are bathed in an aqueous medium, the delivery of materials to and from these cells is carried out largely in solution. The fluid in which the materials are dissolved or suspended is blood. The cellular components of mammalian blood are described in Section 21.1. While a number of ideas on blood were put forward by Greek and Roman scientists, it was the English physician William Harvey (1578–1657) who first showed that it was pumped into arteries by the heart, circulated around the body and returned via veins.

21.1   Structure of blood

Blood comprises a watery **plasma** in which are a variety of different cells. The majority of cells present are **erythrocytes** or red blood cells which are biconcave discs about 7 μm in diameter. They have no nucleus and are formed in the bone marrow. The remaining cells are the larger, nucleated white cells or **leucocytes**. Most of these are also made in the bone marrow. There are two basic types of leucocyte. **Granulocytes** have granular cytoplasm and a lobed nucleus; they can engulf bacteria by phagocytosis. Some of them are also thought to have antihistamine properties. **Agranulocytes** have a non-granular cytoplasm and a compact nucleus. Some of these also ingest bacteria but the **lymphocytes**, made mainly in the thymus gland and lymphoid tissues, produce **antibodies**. More sparsely distributed in the plasma are tiny cell fragments called **platelets**. These are important in the process of blood clotting.

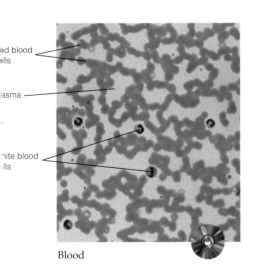

Blood

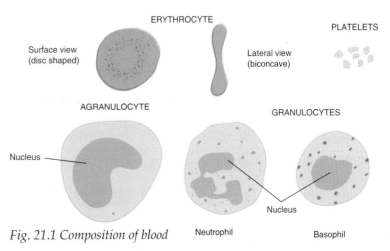

Fig. 21.1 Composition of blood

405

Formation of blood

In the fetus red blood cells are formed in the liver, but in adults production moves to bones, such as the cranium, sternum, vertebrae and ribs, which have red bone marrow. White cells lik[e] lymphocytes are formed in the thymus gland and lymph nodes whereas other types are formed in bones, e.g. the long bones of the limbs, which have white bone marrow.

21.2 Functions of blood

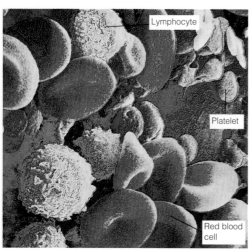

False-colour SEM of human blood cells

Blood performs two distinct functions: the transport of material (summarized in Table 21.1) and defence against disease.

21.2.1 Respiratory pigment – haemoglobin

The solubility of oxygen is low, with only $0.58\,\text{cm}^3$ dissolving in $100\,\text{cm}^3$ of water at $25\,°\text{C}$. At human body temperature ($37\,°\text{C}$) the quantity is even less, just $0.46\,\text{cm}^3$, because the solubility decreases as the temperature increases. Vertebrates, and many invertebrates, have evolved a group of coloured proteins capabl[e] of loosely combining with oxygen, in order to increase the oxygen-carrying capacity of the blood. These are known as **respiratory pigments**. With a few exceptions, the pigments with large relative molecular mass (RMM) are found in the plasma while those of smaller RMM occur within cells to prevent them being lost by ultrafiltration in the kidneys.

The important property of respiratory pigments is their abilit[y] to combine readily with oxygen where its concentration is high, i.e. at the respiratory surface, and to release it as readily where its concentration is low, i.e. in the tissues.

TABLE 21.1 **Summary of the transport functions of blood**

Materials transported	Examples	Transported from	Transported to	Transported in
Respiratory gases	Oxygen	Lungs	Respiring tissues	Haemoglobin in red blood cells
	Carbon dioxide	Respiring tissues	Lungs	Haemoglobin in red blood cells. Hydrogen carbonate ions in plasma
Organic digestive products	Glucose	Intestines	Respiring tissues/liver	Plasma
	Amino acids	Intestines	Liver/body tissues	Plasma
	Vitamins	Intestines	Liver/body tissues	Plasma
Mineral salts	Calcium	Intestines	Bones/teeth	Plasma
	Iodine	Intestines	Thyroid gland	Plasma
	Iron	Intestines/liver	Bone marrow	Plasma
Excretory products	Urea	Liver	Kidney	Plasma
Hormones	Insulin	Pancreas	Liver	Plasma
	Anti-diuretic hormone	Pituitary gland	Kidney	Plasma
Heat	Metabolic heat	Liver and muscle	All parts of the body	All parts of the blood

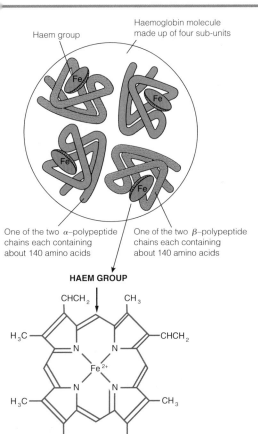

Haem group

Haemoglobin molecule made up of four sub-units

One of the two α–polypeptide chains each containing about 140 amino acids

One of the two β–polypeptide chains each containing about 140 amino acids

HAEM GROUP

ig. 21.2 The structure of haemoglobin

The best known and most efficient respiratory pigment is **haemoglobin**. It occurs in most animal phyla, protozoans and even a few plants. The haemoglobin molecule is made up of an iron porphyrin compound – the **haem** group – and a protein – **globin**. The haem group contains a ferrous iron atom, which is capable of carrying a single oxygen molecule. Different haemoglobins have a different number of haem groups and so vary in their ability to carry oxygen. A single molecule of human haemoglobin, for example, has a RMM of 68 000 and possesses four haem groups. It therefore is capable of carrying four molecules of oxygen. The arrangement of the haemoglobin molecule is given in Fig. 21.2.

21.2.2 Transport of oxygen

An efficient respiratory pigment readily picks up oxygen at the respiratory surface and releases it on arrival at tissues. This may appear contradictory as a substance with a high affinity for oxygen is unlikely to release it easily. Respiratory pigments overcome the problem by having a high affinity for oxygen when its concentration is high, but this is reduced when the oxygen concentration is low. Oxygen concentration is measured by partial pressure, otherwise called the **oxygen tension**. Normal atmospheric pressure is approximately 100 kiloPascals. As oxygen makes up around 21% of the atmosphere, the oxygen tension (partial pressure) of the atmosphere is around 21 kPa.

When a respiratory pigment such as haemoglobin is exposed to a gradual increase in oxygen tension it absorbs oxygen rapidly at first, but more slowly as the tension continues to rise. This relationship between the oxygen tension and the saturation of haemoglobin is called the **oxygen dissociation curve** and is illustrated below.

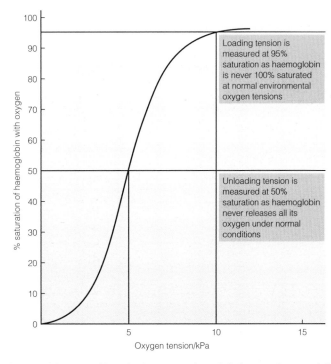

Loading tension is measured at 95% saturation as haemoglobin is never 100% saturated at normal environmental oxygen tensions

Unloading tension is measured at 50% saturation as haemoglobin never releases all its oxygen under normal conditions

Fig. 21.3 Oxygen dissociation curve for adult human haemoglobin

The different haemoglobins found in animals vary in their affinity for oxygen. The oxygen dissociation curves for a number of animals are given in Figs. 21.4 and 21.5. Before attempting to explain the significance of these differences, it should be noted that:

1. The more the dissociation curve of a particular pigment is displaced to the right, the less readily it picks up oxygen, but the more easily it releases it.

2. The more the dissociation curve of a particular pigment is displaced to the left, the more readily it picks up oxygen, but the less readily it releases it.

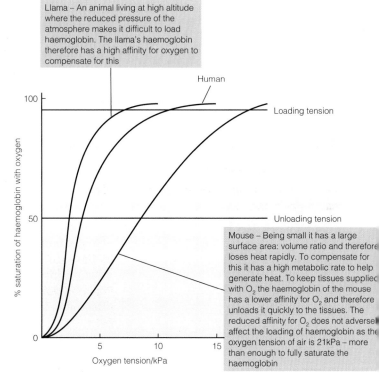

Fig. 21.4 Oxygen dissociation curves for the haemoglobin of three mammals

The release of oxygen from haemoglobin is facilitated by the presence of carbon dioxide – a phenomenon known as the **Bohr effect**. Where carbon dioxide concentration is high, i.e. in respiring tissues, oxygen is released readily; where carbon dioxide concentration is low, i.e. at the respiratory surface, oxygen is taken up readily. These effects are shown in Fig. 21.6.

The Bohr effect is not shown by the haemoglobin of animals like *Arenicola* which live in environments with low oxygen tensions. As the habitat of these animals frequently has a high level of carbon dioxide, the Bohr effect would cause the dissociation curve to shift to the right, reducing the haemoglobin's affinity for oxygen. As the environment contains little oxygen, this reduced affinity for it would prevent them taking it up in sufficient quantities for their survival.

Not only do respiratory pigments vary between species, there are often different types in the same species. In humans, for example, the fetus has a haemoglobin which differs in two of the four polypeptide chains from the haemoglobin of an adult.

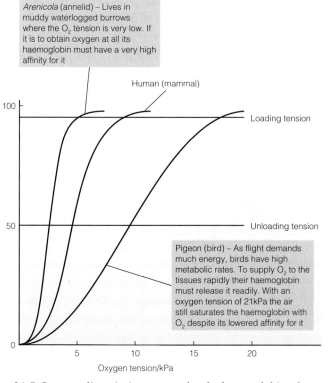

Arenicola (annelid) – Lives in muddy waterlogged burrows where the O_2 tension is very low. If it is to obtain oxygen at all its haemoglobin must have a very high affinity for it

Human (mammal)

Loading tension

Unloading tension

Pigeon (bird) – As flight demands much energy, birds have high metabolic rates. To supply O_2 to the tissues rapidly their haemoglobin must release it readily. With an oxygen tension of 21kPa the air still saturates the haemoglobin with O_2 despite its lowered affinity for it

Oxygen tension/kPa

Fig. 21.5 Oxygen dissociation curves for the haemoglobin of three animals from different groups

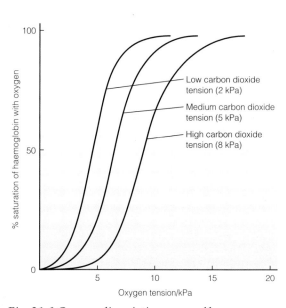

% saturation of haemoglobin with oxygen

Low carbon dioxide tension (2 kPa)

Medium carbon dioxide tension (5 kPa)

High carbon dioxide tension (8 kPa)

Oxygen tension/kPa

Fig. 21.6 Oxygen dissociation curve of human haemoglobin, illustrating the Bohr effect

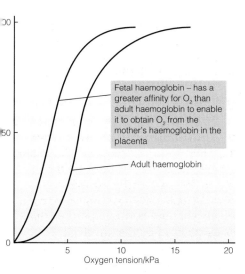

Fetal haemoglobin – has a greater affinity for O_2 than adult haemoglobin to enable it to obtain O_2 from the mother's haemoglobin in the placenta

Adult haemoglobin

Oxygen tension/kPa

Fig. 21.7 Comparison of the oxygen dissociation curves of adult and fetal haemoglobin

This gives the fetal haemoglobin a dissociation curve to the left of that of the adult and therefore a greater affinity for oxygen (Fig. 21.7). Only in this way can the fetal haemoglobin absorb oxygen from the maternal haemoglobin in the placenta. At birth the production of fetal haemoglobin gives way to that of the adult type.

Another respiratory pigment in vertebrates is **myoglobin**. It consists of a single polypeptide chain and a single haem group, rather than the four found in haemoglobin. Like fetal haemoglobin, myoglobin has a dissociation curve displaced to the left of that of the adult and therefore has a greater affinity for oxygen (Fig. 21.8). Myoglobin occurs in the muscles of all vertebrates, where it acts as a store of oxygen. In periods of extreme exertion, when the supply of oxygen by the blood is insufficient to keep pace with demand, the oxygen tension of muscle falls to a very low level. At these very low oxygen tensions, myoglobin releases its oxygen to keep the muscles working efficiently. Once exercise has ceased the myoglobin store is replenished from the haemoglobin in the blood. Being red, myoglobin is largely responsible for the red colour of meat. The breast meat of much poultry is white as it is made up of the muscles which operate the wings. As most poultry is non-flying these muscles are relatively inactive and so have no need of an oxygen-storing pigment like myoglobin. By contrast, the breast meat of flighted birds is very dark.

Haemoglobin has a greater affinity for carbon monoxide than it does for oxygen. When carbon monoxide is inhaled, even in small quantities, it combines with haemoglobin in preference to oxygen to form a stable compound, **carboxyhaemoglobin**. The carbon monoxide is not released at normal atmospheric oxygen

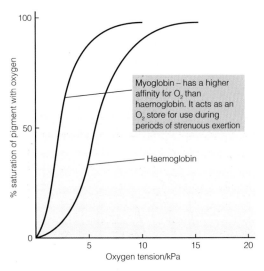

Fig. 21.8 Comparison of the oxygen dissociation curves of human haemoglobin and myoglobin

tensions and the haemoglobin is therefore permanently prevented from transporting oxygen. Obviously if sufficient carbon monoxide is inhaled vital tissues become deprived of oxygen, resulting in death from **asphyxia**.

21.2.3 Transport of carbon dioxide

Carbon dioxide is more soluble than oxygen in water , but its transport in solution is still inadequate to meet the needs of mo organisms. There are three methods of carrying carbon dioxide from the tissues to the respiratory surface.

1. **In aqueous solution** – A small amount, around 5% of carbor dioxide is transported in physical solution in blood plasma.

2. **In combination with haemoglobin** – A little carbon dioxide around 10%, will combine with the amino groups (–NH_2) in the four polypeptide chains which make up each haemoglobin molecule (Hb).

$$Hb\text{-}N\begin{array}{c}H\\ \\H\end{array} + CO_2 \rightleftharpoons Hb\text{-}N\begin{array}{c}H\\ \\COO^-\end{array} + H^+$$

haemoglobin	+	carbon dioxide	carbamino haemoglobin	+	hydrogen ions

3. **In the form of hydrogen carbonate** – The majority of the carbon dioxide (85%) produced by the tissues combines with water to form carbonic acid. This reaction is catalysed by the zinc-containing enzyme **carbonic anhydrase**. The carbonic acid dissociates into hydrogen and hydrogen carbonate ions.

$$H_2O + CO_2 \underset{\text{anhydrase}}{\overset{\text{carbonic}}{\rightleftharpoons}} H_2CO_3 \rightleftharpoons H^+ + HCO^-_3$$

water	carbon dioxide	carbonic acid	hydrogen ion	hydroge carbonat ion

The above reactions take place in red blood cells. The hydrogen ions produced combine with haemoglobin which loses its oxygen. The oxygen so released diffuses out of the red blood ce through the capillary wall and tissue fluid into a respiring tissu cell. The hydrogen carbonate ions diffuse out of the red blood cell into the plasma where they combine with sodium ions from the dissociation of sodium chloride to form sodium hydrogen carbonate. It is largely in this form that the carbon dioxide is carried to the respiratory surface where the processes are reversed, releasing carbon dioxide which diffuses out of the body. The loss of negatively charged hydrogen carbonate ions from the red blood cells is balanced by the inward diffusion of negative chloride ions from the dissociation of the sodium chloride. In this way the electrochemical neutrality of the red blood cell is restored. This is known as the **chloride shift** and is illustrated in Fig. 21.9.

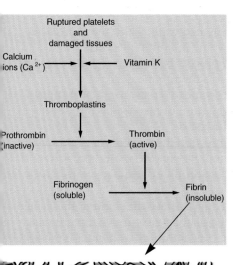

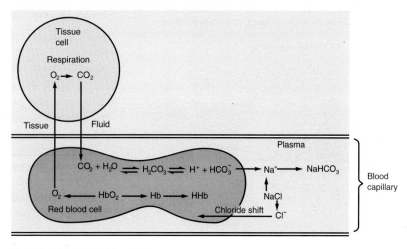

Fig. 21.9 The chloride shift

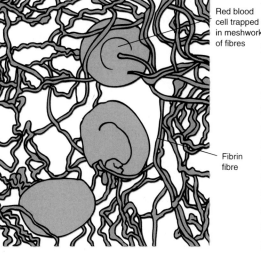

Fig. 21.10 The clotting process – a summary of the main stages

21.2.4 Clotting of the blood

If a blood vessel is ruptured it is important that the resultant loss of blood is quickly arrested. If not, the pressure of the blood in the circulatory system could fall dangerously low. At the same time it is important that clotting does not occur during the normal circulation of blood. If it does, the clot might lodge in some blood vessel, cutting off the blood supply to a vital organ and possibly resulting in death from **thrombosis**. For this reason the clotting process is very complex, involving a large number of stages. Only under the very specific conditions of injury are all stages completed and clotting occurs. In this way the chances of clotting taking place in other circumstances is reduced. The following account includes only the major stages, of what is a more complex process.

Cellular fragments in the blood called **platelets (thrombocytes)** are involved in the clotting or **coagulation** of the blood. At the site of a wound the damaged cells and ruptured platelets release **thromboplastins**. The platelets attract **clotting factors** which create a cascade effect whereby each activates the next in the chain. Amongst these is factor VIII, the absence of which, due to a sex-linked genetic defect, is the cause of haemophilia (Section 9.4.2). At the end of these chain reactions factor X is produced which in the presence of calcium ions and vitamin K causes the inactive plasma protein, **prothrombin**, to become converted to its active form, **thrombin**. This in turn converts another plasma protein, the soluble **fibrinogen**, to **fibrin**, its insoluble form. The fibrin forms a meshwork of threads in which red blood cells become trapped. These dry to form a clot beneath which repair of the wound takes place. The clot not only prevents further blood loss, it also prevents entry of bacteria which might otherwise cause infection. The clotting process is summarized in Fig. 21.10.

Clotting of blood is prevented by substances such as oxalic acid, which precipitates out the calcium ions as calcium oxalate, and heparin, which inhibits the conversion of prothrombin to thrombin. These substances are known as **anticoagulants**.

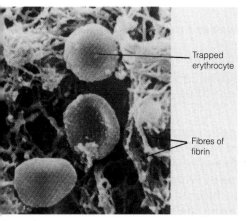

The clotting process (scanning EM) (× 2500 approx.)

411

1. *The neutrophil is attracted to the bacterium by chemoattractants. It moves towards the bacterium along a concentration gradient.*

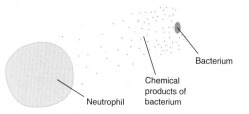

2. *The neutrophil binds to the bacterium.*

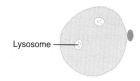

3. *Lysosomes within the neutrophil migrate towards the phagosome formed by pseudopodia engulfing the bacterium.*

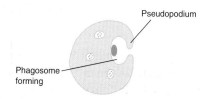

4. *The lysosomes release their lytic enzymes into the phagosome where they break down the bacterium.*

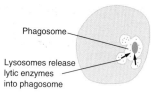

5. *The breakdown products of the bacterium are absorbed by the neutrophil.*

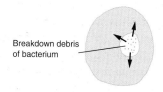

Fig. 21.11 Summary of phagocytosis of a bacterium by a neutrophil

21.2.5 Defence against infection – phagocytosis

Two types of white cell, the neutrophils and monocytes, are capable of amoeboid movement. Both types carry out **phagocytosis**. This is the process by which large particles are taken up by cells via plasma membrane-derived vesicles. White cells carry out phagocytosis for two reasons: to protect the organism against pathogens and to dispose of dead, dying or damaged cells and cellular debris.

In protecting against infection the phagocyte is attracted to chemicals produced naturally by bacteria. The recognition is aided by the presence of **opsonins** – plasma proteins which attach themselves to the surface of the bacteria. The phagocytes have specific proteins on their surface that bind to these chemo-attractants. This causes the phagocyte to move towards the bacteria, possibly along a concentration gradient. The phagocyte strongly adheres to a bacterium on reaching it. This stimulates the formation of pseudopodia which envelop the bacterium, forming a vacuole called a **phagosome**. **Lysosomes** within the phagocyte migrate towards the phagosome into which they release lytic enzymes that break down the bacterium. The breakdown products are finally absorbed by the phagocyte. Fig. 21.11 summarizes the process.

Some phagocytic cells called **macrophages** are found throughout body tissues. They are part of the **reticulo-endothelial system** and are mostly concentrated in lymph nodes and in the liver.

Phagocytosis causes **inflammation** at the site of infection. The hot and swollen area contains many dead bacteria and phagocytes which are known as **pus**. Inflammation results when **histamine** is released as a result of injury or infection. This causes dilation of blood capillaries from which plasma, containing antibodies, escapes into the tissues. Neutrophils also pass through the capillary walls in a process called **diapedesis**.

21.3   The immune system

Immunity is the ability of an organism to resist disease. It involves the recognition of foreign material and the production of chemicals which help to destroy it. These chemicals, called antibodies, are produced by lymphocytes of which there are two types: **T-lymphocytes**, which are formed in bone marrow but mature in the thymus gland, and **B-lymphocytes**, which are formed and mature in the bone marrow.

21.3.1 Self and non-self antigens

Effective defence of the body against infection lies in the ability of the lymphocyte to recognize its own cells and chemicals (self) and to distinguish these from cells and chemicals which are foreign to it (non-self). Therefore it must be able to recognize everything which exists in nature as any chemical or cell can potentially invade the body. Cell surfaces are complicated three-dimensional structures. Each lymphocyte has, somewhere on its surface, receptors which fit exactly into one small part (perhaps only a few amino acids) of every cell. Clearly there are many different types of lymphocyte.

The different types of lymphocyte are derived from **stem cells** in the bone marrow, special cells in the embryo which also make red blood cells and platelets. The stem cells, on dividing, actually lose most of their DNA to the lymphocytes, i.e. they donate their genes to the lymphocytes. This they do randomly – dealing out genes in a way similar to dealing a hand of cards. Consider the vast number of combinations of cards in a typical hand. Clearly with hundreds of genes, rather than 52 cards, being randomly distributed there are at least 100 million different types of receptors that can be generated – each one able to fit a different chemical shape.

How then do lymphocytes distinguish between their own cells and those that are foreign? In the embryo the lymphocytes are constantly colliding with their own cells' shapes. Since infection in the uterus is rare, any lymphocytes whose receptors exactly fit cells must be the ones that recognize their own cells. These lymphocytes then either die or are suppressed ensuring that the body's own cells will not be attacked. The body has thus become **self-tolerant**. The remaining lymphocytes have receptors which fit chemical shapes of non-self material. Any material with one of these shapes to which lymphocyte receptors adhere is called an **antigen**.

Immune responses

Once a lymphocyte has become attached to its complementary antigen it multiplies rapidly by mitosis to give a clone of identical lymphocytes. This is known as the **primary immune response** and typically takes a few days during which time the invading pathogen often multiplies and so gives rise to the symptoms of the disease it causes. Some of the lymphocytes in this clone change into cells known as **memory cells**. Unlike the other lymphocytes from the clone which die within a few days these memory cells survive much longer – often for years. Further infections by the same pathogen cause these memory cells to divide immediately. Their numbers therefore build up as fast, if not faster, than those of the invader and so the pathogen is repelled before it can induce the symptoms of the disease. This is called the **secondary immune response** and explains why we only suffer some diseases once in a lifetime, despite frequent exposure to them. Each time a pathogen enters the body more and more memory cells are built up, making future infection even less likely. This progressive increase in the level of immunity to a disease is known as **adaptive** or **acquired immunity**.

413

The variable region differs with each antibody. It has a shape which exactly fits an antigen. Each antibody therefore can bind to two antigens.

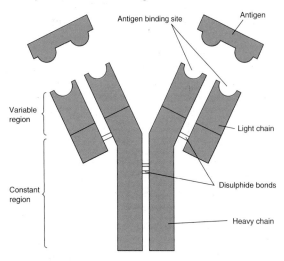

Fig. 21.12 Structure of an antibody

B-lymphocytes and humoral immunity

B-lymphocytes are so called because they both originate and mature in the bone marrow. When a B-lymphocyte recognizes antigen it divides to form a clone. As most microorganisms possess more than one antigen then many lymphocytes are activated to produce clones in a process known as **polyclonal activation**. Most of the cells in these clones are plasma cells which begin to secrete **antibodies** at the rate of thousands each second.

Antibodies are large protein molecules which comprise four polypeptide chains. One pair of chains is long – **heavy chains**; the other pair is shorter – **light chains**. They are arranged in a Y-shape and have two sites called **binding sites** which fit exactly into the antigen.

There are a number of different antibodies each performing different function:

1. Agglutination – Because each antibody has two different binding sites it can join to two antigens on two different pathogens. In this way the pathogens can be joined together in clumps making them more vulnerable to attack from other types of antibody.

2. Precipitation – Some antibodies bind together soluble antigens into large units which are thus precipitated out of solution. As such they are more easily ingested by phagocytes.

3. Neutralization – Certain antibodies bind to toxic molecules produced by a pathogen and in doing so neutralize their harmful effects.

4. Lysis – Antibodies which are attached to a pathogen act as binding sites for a number of blood proteins which are collectively known as the **complement system**. Some of these proteins are enzymes which cause the breakdown of the pathogen.

T-lymphocytes and cell-mediated immunity

T-lymphocytes are so called because while produced in bone marrow they mature in the thymus gland. Once attached to the antigen T-lymphocytes divide to form a clone, the cells of which then differentiate into different cell types:

1. T-helper cells – These produce chemicals which activate other white cells such as phagocytes to engulf harmful material. The chemicals attach themselves to the foreign material and so label them as requiring phagocytosis. These labelled chemicals are called **opsonins** (from the Greek meaning 'ready for the table'). The T-helper cells also activate B-lymphocytes to divide to produce plasma cells as well as assisting the T-killer cells to destroy pathogens. Clearly they are essential to a successful immune response – a fact borne out by the consequences of them being rendered inoperable by the Human Immunodeficiency Virus (HIV) leading to AIDS.

2. T-cytotoxic cells (killer cells) – These kill body cells which have become invaded by viruses. They force cylindrical proteins through the cell membrane causing the cell to burst. Since viruses require host cells to reproduce, this sacrifice effectively prevents multiplication of the virus.

3. T-suppressor cells – Once an infection has been eliminated these cells suppress the activities of the lymphocytes and so maintain control of the immune system.

21.3.2 Monoclonal antibodies and their applications

We have seen in Section 21.3.1 that foreign material entering the body will possess more than one antigen and so induces a number of different B-lymphocytes to multiply, producing clones of themselves. The many clones then produce a range of antibodies known as **polyclonal antibodies**.

It is obviously of considerable therapeutic value to be able to produce antibodies outside the body, but until recently the inability to sustain the growth of B-lymphocytes prevented this. However, a cancer of B-lymphocytes produces myeloma cells which continue to divide indefinitely. These can be fused in the laboratory, using polyethylene glycol, with each specific B-lymphocyte to produce cells, called **hybridoma cells**, which produce antibodies of one type only – **monoclonal antibodies**.

The large scale production of antibodies using hybridoma cells is used medically to treat a range of infections. This is not their only application, however. Because they are specific to a single chemical (antigen), to which they become attached, they can be used to separate a particular chemical from a complex mixture. To do this, the monoclonal antibody for the required chemical is immobilized on resin beads which are then packed in a column. The mixture is passed over the beads and only the required chemical becomes attached to the antibodies. The chemical may then be obtained in a pure state by washing the beads with a solution which causes the antibodies to release it.

Monoclonal antibodies are also used in **immunoassays**. Here, the antibody is labelled in some way, e.g. radioactively or by a fluorescent dye, so that it can easily be detected. When added to a test sample they will attach to their specific antigen. Washing in solutions which remove only unattached antibodies leaves only those attached to the antigen. The amount of these in the sample is then apparent from the degree of radioactivity or fluorescence. For example, the presence of a particular pathogen in a blood sample can be detected by use of the appropriate monoclonal antibody tagged with a fluorescent dye.

Another technique is to immobilize the antibodies and pass the solution under test over them. Suppose we are testing for chemical X. If it is present in the solution it will attach to the antibody. A second type of antibody which has an enzyme attached is then added. It combines only with those original antibodies which are linked to chemical X. By adding a substrate which the enzyme causes to change colour, the amount of chemical X will be apparent by the extent of any colour change. This technique, called **Enzyme Linked Immunosorbant Assay (ELISA)**, has many uses including detecting drugs in athletes' urine, pregnancy testing kits and detecting the Human Immuno-deficiency Virus (the AIDS test). It is also possible to link anti-cancer drugs to monoclonal antibodies which are attracted to cancer cells – the so-called '**magic bullets**'. An even more sophisticated technique is to tag monoclonal antibodies with an enzyme which converts an inactive form of the cytotoxic drug (**the prodrug**) into an active form. Once injected these antibodies link to the cancer cells. The prodrug is then administered in a relatively high dose as it is harmless in its inactive state. In the vicinity of normal cells the drug remains ineffective but in the presence of the cancer cells the enzyme on the attached antibody activates the drug which acts upon the cells, killing them. The technique is called **ADEPT (Antibody Direct Enzyme Prodrug Therapy)**.

Pregnancy testing

Home pregnancy testing kits make use of immobilized antibodies on a urine dipstick to detect traces of human chorionic gonadotrophin (hCG), a hormone released from the placenta. Antibodies 'tagged' with blue latex combine with the hormone to produce a readily visible result.

5. Further row of antibodies which combine with latex-tagged antibodies *without* hCG attached

Negative result (blue line in this position)

4. Complex held by immobilized antibodies to form a blue line

Positive result (blue line in this position)

3. hCG bound to antibody to form a complex

2. Antibody tagged with blue latex particle

1. Urine sample containing hCG

Home pregnancy testing kit

21.3.3 Types of immunity and immunization

There are two basic types of immunity, passive and active.

Passive immunity is the result of antibodies being passed in an individual in some way, rather than being produced by the individual itself. This passive immunity may occur naturally in mammals when, for example, antibodies pass across the placenta from a mother to her fetus or are passed to the newborn baby in the mother's milk. In both cases the young developing mammal is afforded some protection from disease until its own immune system is fully functional.

Alternatively, passive immunity may be acquired artificially by the injection of antibodies from another individual. This occurs in the treatment of tetanus and diphtheria in humans, although the antibodies are acquired from other mammals, e.g. horses. In all cases, passive immunity is only temporary.

Active immunity occurs when an organism manufactures its own antibodies. Active immunity may be the natural result of an infection. Once the body has started to manufacture antibodies in response to a disease-causing agent, it may continue to do so for a long time after, sometimes permanently. It is for this reason that most people suffer diseases such as mumps and measles only once. It is possible to induce an individual to produce antibodies even without them suffering disease. To achieve this the appropriate antigen must be injected in some way. This is the basis of **immunization (vaccination)** of which there are a number of different types depending on the form the antigen takes.

APPLICATION

Allergies

An excessive reaction of the body's immune system to certain substances is known as an allergy. When a foreign substance is detected for the first time lymphocytes produce antibodies, some of which bind to the surface of other white cells called mast cells. If the same substance is encountered again it binds to the antibodies on the mast cells causing the release of chemicals called mediators. The most important mediator is **histamine**. This chemical can produce rash, swelling, narrowing of the airways and a drop in blood pressure. These effects are important in protecting against infection but they may also be triggered inappropriately in allergy. One of the most common allergic disorders, hayfever, is caused by an allergic reaction to inhaled grass pollen leading to allergic rhinitis – swelling and irritation of the nasal passages and watering of the nose and eyes.

Antihistamines are the most widely used drugs in the treatment of allergic reactions of all kinds. Their main action is to counter the effects of histamine by blocking its action on H_1 receptors. These receptors are found in various body tissues, particularly the small blood vessels in the skin, nose and eyes. This helps prevent the dilation of the vessels, thus reducing redness and swelling. Antihistamines pass from the blood into the brain where their blocking action on histamine activity produces general sedation, and repression of various brain functions, including the vomiting and coughing mechanisms.

1. **Living attenuated microorganisms** – Living pathogens which have been treated, e.g. by heating, so that they multiply but are unable to cause the symptoms of the disease. They are therefore harmless but nonetheless induce the body to produce appropriate antibodies. Living attenuated microorganisms are used to immunize against measles, tuberculosis, poliomyelitis and rubella (German measles).

2. **Dead microorganisms** – Pathogens are killed by some means and then injected. Although harmless they again induce the body to produce antibodies in the same way it would had they been living. Typhoid, cholera and whooping cough are controlled by this means.

3. **Toxoids** – The toxins produced by some diseases, e.g. diphtheria and tetanus, are sufficient to induce antibody production by an individual. To avoid these toxins causing the symptoms of the disease they are first detoxified in some way, e.g. by treatment with formaldehyde, and then injected.

4. **Extracted antigens** – The chemicals with antigenic properties may be extracted from the pathogenic organisms and injected. Influenza vaccine is produced in this way.

5. **Artificial antigens** – Through genetic engineering it is now possible to transfer the genes producing antigens from a pathogenic organism to a harmless one which can easily be

grown in a laboratory. Mass production of the antigen is then possible in a fermenter ready for separation and purification before use. Vaccines used in the treatment of hepatitis B can be produced in this way.

21.3.4 Acquired Immune Deficiency Syndrome (AIDS)

Acquired Immune Deficiency Syndrome or AIDS is caused by the **Human Immunodeficiency Virus (HIV)** – a retrovirus, more details of which are given in Section 5.2.2. HIV infects T-helper cells (Section 21.3.1) which are essential to cell mediated immunity and so the body's immune system is rendered ineffective, not only against HIV but other infections. Hence AIDS victims frequently die, usually within two years of developing the disease, due to opportunist pathogens which take advantage of impaired resistance.

Once infected with HIV an individual is said to be **HIV positive**, a condition which persists throughout their life. As the virus remains dormant for about eight years, on average, an HIV positive person does not suffer any symptoms during this period but can act as a carrier, often unwittingly spreading the disease. The virus can be detected in virtually all body fluids of an HIV positive individual. However, since it is only in blood, semen or vaginal fluid that the concentration is high enough to infect others, it is spread through sexual intercourse or transfer of infected blood from one person to another – such as, when drug users share a hypodermic needle, or from mother to baby during childbirth. Faeces, urine, sweat, saliva and tears have such a low incidence of HIV in an infected person that contact with these presents practically no risk of contracting AIDS. In any case, the virus quickly dies outside the human body, and therefore even blood, semen and vaginal secretions must be transferred directly. Thus, the risk from contaminated clothing, etc. is negligible.

Preventative precautions such as restricting the number of sexual partners, using a condom during sexual intercourse, and intravenous drug users not sharing a needle, are at present the best means of containing the disease. While antibiotics may be used to help combat the opportunistic infections there is, as yet, no cure for AIDS.

21.3.5 Blood groups

Blood groups are an example of an antigen–antibody system. The membrane of red blood cells contains polysaccharides which act as antigens. They may induce the production of antibodies when introduced into another individual. While there are over twenty different blood grouping methods, the ABO system, first discovered by Landsteiner in 1900, is the best known. In this system there are just two antigens, A and B, which determine the blood group (see left). For each of these antigens there is an antibody, which is given the corresponding lower case letter. The presence of an antigen and its corresponding antibody together causes an immune response resulting in the clumping together of red cells (**agglutination**) and their ultimate breakdown (**haemolysis**). For this reason, an individual does not produce antibodies corresponding to the antigens present but

Blood group	Antigens present
A	A
B	B
AB	A and B
O	None

Blood group	Antigen	Antibodies
A	A	b
B	B	a
AB	A and B	None
O	None	a and b

produces all others as a matter of course. These antibodies are present in the plasma. The composition of each blood group is therefore as given opposite.

In transfusing blood from one person, the **donor**, to another, the **recipient**, it is necessary to avoid bringing together corresponding antigens and antibodies. However, if only a small quantity of blood is to be transfused, then it is possible to add the antibody to the antigen, because the donor's antibodies become so diluted in the recipient's plasma that they are ineffective. It is not, however, feasible to add small quantities of antigen to the corresponding antibody as even a tiny amount of antigen will cause an immune response. This, after all, is why small numbers of invading bacteria are immediately destroyed as part of the body's defence mechanism. It is therefore possible to safely add antibody a to antigen A and antibody b to antigen B, in small quantities, but not the reverse. For this reason, blood group O, with no antigens present, may be given in small amounts to individuals of all other blood groups. Group O is therefore referred to as the **universal donor**. Individuals of this group are, however, restricted to receiving blood from their own group. In the same way group AB, with no antibodies, may receive blood with either antigen. In other words, group AB can receive blood from all groups, and is therefore termed the **universal recipient**. They can, however, only donate to their own group. When agglutination occurs between two groups, they are said to be **incompatible**. Table 21.2 shows the compatibility of blood groups in the ABO system.

TABLE 21.2   **Compatibility of blood groups in the ABO system**

				Recipient's blood group			
	Group			A	B	AB	O
	Group	Antigens		A	B	A and B	None
		Antigens	Antibodies	b	a	None	a and b
Donor's blood group	A	A	b	✓	✗	✓	✗
	B	B	a	✗	✓	✓	✗
	AB	A and B	None	✗	✗	✓	✗
	O	None	a and b	✓	✓	✓	✓

✓ Compatible – bloods do not clot   ✗ Incompatible – bloods clot

Despite this knowledge and careful matching of blood groups there continued to be inexplicable failures of transfusions up to 1940. It was then that Landsteiner discovered a new antigen (actually a system of antigens) in rhesus monkeys, which was also present in humans. This became known as the **rhesus system** and the antigen as **antigen D**. Where an individual possesses the antigen he or she is said to be **rhesus positive**; where it is absent he or she is **rhesus negative**. There is no naturally occurring antibody to antigen D, but if blood with the antigen is transfused into a person without it (rhesus negative), antibody d production is induced in line with the usual immune response. For this reason, before transfusion, blood is matched with respect to the rhesus factor as well as the ABO system.

One problem associated with the rhesus system arises in pregnancy. As blood groups are genetically determined (Section

9.5.3), it is possible for the fetus to inherit from the father a blood group different from that of the mother. The fetus may, for example, be rhesus positive while the mother is rhesus negative. Towards the end of pregnancy, and especially around birth, fragments of blood cells may cross from the fetus to the mother. The mother responds by producing the rhesus antibody (d) in response to the rhesus antigen (D) on the fetal red blood cells. These antibodies are able to cross the placenta. As the build-up and transfer of rhesus antibodies takes some time, and as the problem only arises during the latter stages of pregnancy, their concentration is rarely sufficient to have any effect on the first child. The production of rhesus antibodies by the mother continues for only a few months, but subsequent fetuses may again induce production and are therefore subject to a greater influx of rhesus antibodies. These break down the fetal red blood cells – a condition known as **haemolytic disease of the newborn**. It requires a number of fetal blood transfusions throughout the pregnancy if it is not to prove fatal. Knowledge that the mother is rhesus negative can, however, avert the danger. If this is the case, and the father is known to be rhesus positive, a potential problem exists. In this event rhesus antibodies (d) from blood donors, are injected into the mother immediately after the first birth. These destroy any fetal cell fragments with antigen D, which may have entered her blood, before they induce the mother to manufacture her own antibodies. The injected antibodies are soon broken down by the mother, and in the absence of new ones being produced subsequent fetuses are not at risk.

The proportion of different blood groups varies throughout the world. In the British population the proportions are O–46%, A–42%, B–9%, AB–3%, although there are variations between different areas. In England, for example, numbers with group A slightly exceed group O. Some South American tribes are exclusively group O whereas some North American Indian tribes are three quarters group A. Over a third of European gypsies have group B. The proportion of rhesus positive individuals in most groups is between 75% and 85%.

21.3.6 Chemotherapy and immunity

The use of chemicals to prevent and cure diseases and disorders is known as **chemotherapy**. The earliest examples used natural chemicals extracted from plants, but nowadays the chemicals are largely synthetically manufactured. Among the most widely used synthetic drugs are the **sulphonamides**. These were found to be effective against certain bacterial infections during the mid 1930s and were used extensively against urinary and bowel infections as well as pneumonia. The development of strains of bacteria resistant to sulphonamides and the wider use of antibiotics have reduced the importance of these drugs.

Antibiotics are chemicals produced by various fungi and bacteria, which suppress the growth of other microorganisms. Since the discovery by Sir Alexander Fleming, in 1928, of **penicillin**, many other types of antibiotics have been marketed, e.g. ampicillin, streptomycin and chloramphenicol. All interfere with some stage of bacterial metabolism and so suppress their growth. Penicillin for example prevents the synthesis of certain

APPLICATION

Tissue compatibility and rejection

We have seen that blood, if adequately matched, can be transfused from one person to another. It should therefore be equally feasible to transplant organs in the same way. The problem lies in the complexity of organs; they possess a far greater number of antigens and so perfect cross-matching can rarely be achieved. In the absence of perfect cross-matching the recipient treats the donated organ as foreign material and so an immune response is initiated. The organ is therefore rejected. Despite these difficulties there have been major advances in the grafting and transplanting of tissues. Clearly if a tissue is grafted from one part of an organism to another, there are no problems of rejection as all material is genetically identical and so compatible. Skin is frequently grafted by this means. Equally transplants between genetically identical individuals like identical twins do not present problems of rejection. Unfortunately, most humans do not have genetically identical brothers or sisters and so depend upon organs from others when the need for a transplant arises. To minimize the chances of rejection, careful matching takes place, to find tissues which are as nearly compatible as possible. This minimizes the extent of the immune response, reducing the risk of rejection. Such compatible tissues are often, but not always, found in close relatives. In addition the recipient is treated with **immunosuppressant drugs** which lower the activity of their natural immune response, so delaying rejection long enough for the transplanted tissue to be accepted. The problem with these drugs is that the recipient is vulnerable to other infections and even minor ones can prove fatal.

More recently two techniques have been developed which suppress the T-lymphocytes responsible for the rejection response while having little effect on the B-cells which produce antibodies. This helps the patient to maintain a resistance to infection. The first technique employs Orthoclone OKT-3, a monoclonal antibody (see Section 21.3.2). Although this has some side effects they are reversible and the antibody has been used successfully to combat kidney rejection. In the second technique human T-lymphocytes are injected into a horse which then produces antibodies to them, known as anti-lymphocyte immunoglobulin (ALG). This is purified and, when injected into a transplant patient, is effective at combating rejection.

components of the bacterial cell wall, while streptomycin, chloramphenicol and tetracycline inhibit mRNA at the ribosomes thus preventing protein synthesis. Polymixin and amphotericin interfere with the proper functioning of the bacterial membrane. Because bacterial cells are prokaryotic, these effects are not experienced by the eukaryotic cells of the host and so antibiotics may be safely used throughout the body. Once again, the development of resistant species has reduced the effectiveness of some antibiotics (see Section 11.3.3).

21.4   The circulatory system

Only very small animals, where cells are never far from the outside, exist without a specialized transport system. The larger and more active an animal, the more extensive and efficient is its transport system. These systems frequently incorporate a pump, valves and an elaborate means of controlling distribution of the blood. Very simple animals have an **open blood system**. Here the blood moves freely over the tissues, through a series of spaces known collectively as the **haemocoel**. The blood is not confined to vessels. In its simplest form, e.g. in nematodes, the blood is transported haphazardly by the muscular movements of the body as it moves around. In most arthropods, including insects, there is some circulation of the blood by a tubular heart, which pumps it into the haemocoel. In all open blood systems the blood is moved at very low pressure and there is little control over its distribution.

To allow more rapid transport and greater control of distribution, larger and more active organisms have evolved **closed blood systems**. Here blood is confined to vessels. The pumping action of the heart sustains high pressure within these vessels and a combination of vasodilation, vasoconstriction and valves ensures a much more controlled distribution of blood. Closed systems occur in cephalopod molluscs, echinoderms, annelids and vertebrates.

Within vertebrates, there are two different systems of circulating the blood. In fish, the blood passes from the heart, over the gills and then to the rest of the body before returning to the heart. As the blood passes only once through the heart during a complete circulation of the body it is called a **single circulation**. The disadvantage of this system is that the resistance created by the fine network of blood capillaries in the gills causes the blood pressure to drop from around 11 kPa as it leaves the heart to 7 kPa as it leaves the gills. This blood then meets more resistance as it passes through the capillaries of the body tissues and its pressure is even further reduced. The flow is therefore rather sluggish. To overcome this problem, other vertebrates have developed a **double circulation**. Here the blood is returned to the heart after passing over the respiratory surface and before it is pumped over the body tissues. This helps to sustain a high blood pressure and so allows more rapid circulation. In mammals and birds the complete separation of the heart into two halves allows oxygenated and deoxygenated blood to be kept separate. This improves the efficiency of oxygen distribution, something which is essential to sustain the higher metabolic rate of these endothermic animals.

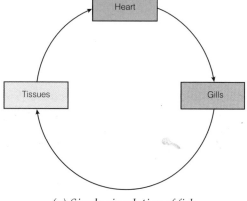

(a) Single circulation of fish

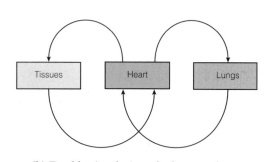

(b) Double circulation of other vertebrates

Fig. 21.13 Single and double circulations

21.4.1 Blood vessels

In a closed circulation there are three types of vessel. **Arteries** carry blood away from the heart ('a' for 'artery' = 'a' for 'away' from the heart), **veins** carry blood to the heart whereas the much smaller **capillaries** link arteries to veins. A comparison of the structure of these three vessels is given in Table 21.3.

The diameter of arteries and veins gradually diminishes as they get further from the heart. The smaller arteries are called **arterioles** and the smaller veins are called **venules**.

TABLE 21.3   A comparison of arteries, veins and capillaries

Artery	Vein	Capillary
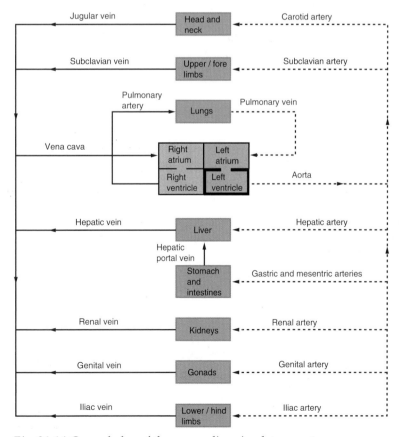		
Thick muscular wall Much elastic tissue Small lumen relative to diameter Capable of constriction Not permeable Valves in aorta and pulmonary artery only Transports blood from the heart Oxygenated blood except in pulmonary artery Blood under high pressure (10–16 kPa) Blood moves in pulses Blood flows rapidly	Thin muscular wall Little elastic tissue Large lumen relative to diameter Not capable of constriction Not permeable Valves throughout all veins Transports blood to heart Deoxygenated blood except in pulmonary vein Blood under low pressure (1 kPa) No pulses Blood flows slowly	No muscle No elastic tissue Large lumen relative to diameter Not capable of constriction Permeable No valves Links arteries to veins Blood changes from oxygenated to deoxygenated Blood pressure reducing (4–1 kPa) No pulses Blood flow slowing

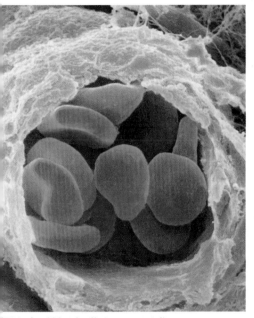

SEM of red blood cells in an arteriole

21.4.2 Mammalian circulatory system

The purpose of the mammalian circulatory system is to carry blood between various parts of the body. To this end, each organ has a major artery supplying it with blood from the heart and a major vein which returns it. These arteries and veins are usually

Fig. 21.14 General plan of the mammalian circulatory system

named by preceding them with the adjective appropriate to that organ, e.g. each kidney has a renal artery and renal vein. A general plan of the mammalian circulation is given in Fig. 21.14.

The flow of blood is maintained in three ways:

1. The pumping action of the heart – This forces blood through the arteries into the capillaries.

2. Contraction of skeletal muscle – The contraction of muscles during the normal movements of a mammal squeeze the thin-walled veins, increasing the pressure of blood within them. Pocket valves in the veins ensure that this pressure directs the blood back to the heart.

3. Inspiratory movements – When breathing in, the pressure in the thorax is reduced. This helps to draw blood towards the heart, which is within the thorax.

21.5   Heart structure and action

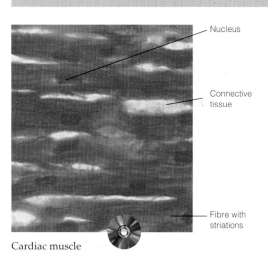

Cardiac muscle

Nucleus

Connective tissue

Fibre with striations

A pump to circulate the blood is an essential feature of most circulatory systems. These pumps or hearts generally consist of a thin-walled collection chamber – the **atrium** or **auricle** – and a thick-walled pumping chamber – the **ventricle**. Between the two are valves to ensure the blood flows in one direction, namely, from the atrium to the ventricle. In fish, with their single circulation, the heart has two chambers only. With the evolution of a double circulation came the development of two atria, one to receive blood from the systemic (body) circulation and the other to receive it from the pulmonary (lungs) circulation. In amphibia and most reptiles there is a three-chambered heart – two atria and a single ventricle. This has the disadvantage of allowing oxygenated blood from the pulmonary system to mix with deoxygenated blood from the systemic system. Ridges within the ventricle minimize this mixing so that in frogs, for example, three quarters of the oxygenated blood from the pulmonary system is pumped into the systemic one. Being ectothermic, and therefore usually having a lower metabolic rate, allows this inefficiency to be tolerated. In the endothermic mammals and birds, however, the ventricle is completely partitioned into two, allowing complete separation of oxygenated and deoxygenated blood. This four-chambered heart is really two two-chambered hearts side by side.

21.5.1 Structure of the mammalian heart

The mammalian heart consists largely of **cardiac muscle**, a specialized tissue which is capable of rhythmical contraction and relaxation over a long period without fatigue. Its structure is shown in Fig. 21.15. The muscle is richly supplied with blood vessels and also contains connective tissue which gives strength and helps to prevent the muscle tearing. The mammalian heart is made up of two thin-walled atria which are elastic and distend as blood enters them. The left atrium receives oxygenated blood from the pulmonary vein while the right atrium receives deoxygenated blood from the vena cava. When full, the atria

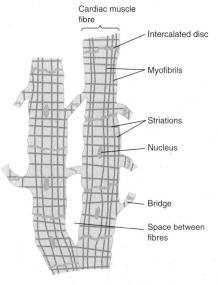

Cardiac muscle fibre

Intercalated disc

Myofibrils

Striations

Nucleus

Bridge

Space between fibres

Fig. 21.15 Cardiac muscle (LS)

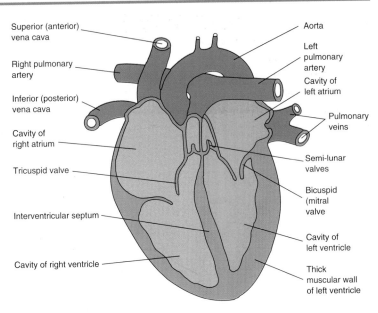

Fig. 21.16 *The structure of the mammalian heart as seen in vertical section from the ventral side*

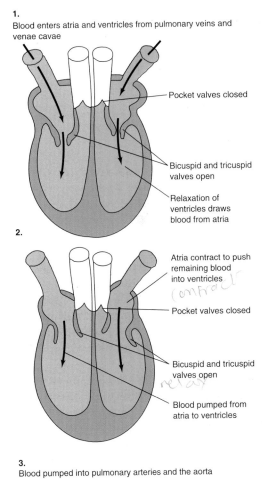

1.
Blood enters atria and ventricles from pulmonary veins and venae cavae

Pocket valves closed

Bicuspid and tricuspid valves open

Relaxation of ventricles draws blood from atria

2.

Atria contract to push remaining blood into ventricles

Pocket valves closed

Bicuspid and tricuspid valves open

Blood pumped from atria to ventricles

3.
Blood pumped into pulmonary arteries and the aorta

Pocket valves open

Bicuspid and tricuspid valves closed

Ventricles contract

1. *Diastole*
   Atria are relaxed and fill with blood. Ventricles are also relaxed.

2. *Atrial systole*
   Atria contract pushing blood into the ventricles. Ventricles remain relaxed.

3. *Ventricular systole*
   Atria relax. Ventricles contract pushing blood away from heart through pulmonary arteries and the aorta.

Fig. 21.17 *The cardiac cycle*

contract together, forcing the remaining blood into their respective ventricles. The right ventricle then pumps blood to the lungs. Owing to the close proximity of the lungs to the heart, the right ventricle does not need to force blood far and is much less muscular than the left ventricle which has to pump blood to the extremities of the body. To prevent backflow of blood into the atria when the ventricles contract, there are valves between the atria and ventricles. On the right side of the heart these comprise three cup-shaped flaps, the **tricuspid valves**. On the left side of the heart only two cup-shaped flaps are present; these are the **bicuspid** or **mitral valves**. To prevent these valves inverting under the pressure of blood, they are attached to papillary muscles of the ventricular wall by fibres known as the **chordae tendinae**. Fig. 21.16 illustrates the structure of the heart.

Blood leaving the ventricle is prevented from returning by pocket valves in the aorta and pulmonary artery. These close when the ventricles relax.

21.5.2 Control of heart beat (cardiac cycle)

All vertebrate hearts are **myogenic**, that is, the heart beat is initiated from within the heart muscle itself rather than by a nervous impulse from outside it. Where it is initiated by nerves, as in insects, the heart is said to be **neurogenic**.

The initial stimulus for a heart beat originates in a group of histologically different cardiac muscle cells known as the **sino-atrial node (SA node)**. This is located in the wall of the right atrium near where the vena cavae enter it. The SA node determines the basic rate of heart beat and is therefore known as the **pacemaker**. In humans, this basic rate is 70 beats/minute but can be adjusted according to demand by stimulation from the autonomic nervous system. A wave of excitation spreads out from the SA node across both atria, causing them to contract more or less at the same time.

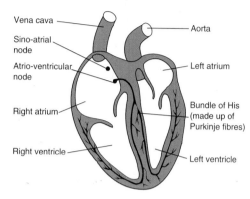

Fig. 21.18 VS through mammalian heart to show position of sino-atrial node, atrio-ventricular node and bundle of His

Labels: Vena cava, Sino-atrial node, Atrio-ventricular node, Right atrium, Right ventricle, Aorta, Left atrium, Bundle of His (made up of Purkinje fibres), Left ventricle

The wave of excitation reaches a similar group of cells known as the **atrio-ventricular node (AV node)** which lies between the two atria. To allow blood to be forced upwards into the arteries, the ventricles need to contract from the apex upwards. To achieve this the new wave of excitation from the AV node is conducted along **Purkinje fibres** which collectively make up the **bundle of His**. These fibres lead along the interventricular septum to the apex of the ventricles from where they radiate upwards. The wave of excitation travels along these fibres, only being released to effect muscle contraction at the apex. The ventricles contract simultaneously from the apex upwards. These events are known as the **cardiac cycle** and are summarized in Fig. 21.17.

21.5.3 Factors modifying heart beat

The human heart normally contracts 70 times a minute, but this can be varied from 50 to 200 times a minute. In the same way the volume of blood pumped at each beat can be varied. The volume pumped multiplied by the number of beats in a given time is called the **cardiac output**. Changes to the cardiac output are effected through the autonomic nervous system. Within the medulla oblongata of the brain are two centres. The **cardio-acceleratory centre** is linked by the sympathetic nervous system to the SA node. When stimulated these nerves cause an increase in cardiac output. The **cardio-inhibitory centre** is linked by parasympathetic fibres within the vagus nerve, to the SA node, AV node and bundle of His. Stimulation from these nerves decreases the cardiac output.

Which of these centres stimulates the heart depends on factors like the pH of the blood. This in turn depends upon its carbon dioxide concentration. Under conditions of strenuous exercise, the carbon dioxide concentration of the blood increases as a consequence of the greater respiratory rate. The pH of the blood is therefore lowered. Receptors in a swelling of the carotid artery called the **carotid body**, detect this change and send nervous impulses to the cardio-acceleratory centre which increases the heart beat, thereby increasing the rate at which carbon dioxide is delivered to the lungs for removal. A fall in carbon dioxide level (rise in pH) of blood causes the carotid receptors to stimulate the cardio-inhibitory centre, thus reducing the heart beat.

Another means of control is by stretch receptors in the aorta, carotid artery and vena cava. When the receptors in the aorta and carotid artery are stimulated, it indicates that there is distention of these vessels as a result of increased blood flow in them. This causes the cardio-inhibitory centre to stimulate the heart to reduce cardiac output. Stimulation of receptors in the vena cava indicates increased blood in this vessel, probably as a result of muscular activity increasing the rate at which blood is returned from the tissues. Under these conditions the cardiac centres in the brain increase the cardiac output.

21.5.4 Maintenance and control of blood pressure

Changes in cardiac output will alter blood pressure, which must always be maintained at a sufficiently high level to permit blood to reach all tissues requiring it. Another important factor in

APPLICATION

Artificial pacemakers

In Britain about 10 000 people a year receive an artificial pacemaker. The operation takes less than an hour and is performed under local anaesthetic.

Pacemakers are made up of a pulse generator and two electrodes. The pulse generator is about the size of a thin matchbox and weighs 20 – 60 g. It is powered by a lithium battery and is implanted under the patient's skin. The electrodes are placed intravenously into the right atrium and the right ventricle. Disease or ageing can damage the heart's natural pacemaker and the conduction of impulses through the heart, causing an abnormally slow heart beat. The artificial pacemaker overcomes this by generating electrical impulses artificially and conducting them to the muscles of the heart on demand or when the heart misses a beat.

Pacemakers

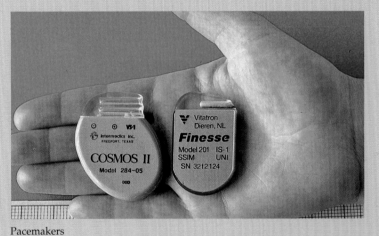

Fig. 21.19 Pressure changes in the atria, ventricles and aorta during one cardiac cycle

APPLICATION

Electrocardiogram

An ECG (electrocardiogram) trace follows the electrical activity of nerves and muscles in the heart. There are various characteristic patterns or waves. For diagnostic purposes doctors use three of the waves.

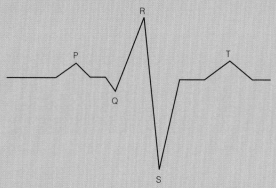

The P wave charts current flow through the atria from the SA node to the AV node. The QRS complex follows the spread of depolarization through the ventricles. Finally the T wave results from currents generated during ventricular repolarization.

Various waves can be picked up as electrical echoes from the heart by connecting the ECG machine's leads to the skin at different sites on the body. By fixing two or three electrodes to the wrists and ankles, doctors get a frontal view of the heart's activity by charting the direction of current flow in two dimensions. Together with horizontal-plane electrodes (six ECG leads strapped across the chest), the ECG can give a three-dimensional picture of the heart's electrical activity, and hence its health.

controlling blood pressure is the diameter of the blood vessels. When narrowed – **vasoconstriction** – blood pressure rises; when widened – **vasodilation** – it falls. Vasoconstriction and vasodilation are also controlled by the medulla oblongata, this time by the **vasomotor centre**. From this centre nerves run to the smooth muscles of arterioles throughout the body. Pressure receptors, known as **baroreceptors**, in the carotid artery detect blood pressure changes and relay impulses to the vasomotor centre. If blood pressure falls, the vasomotor centre sends impulses along sympathetic nerves to the arterioles. The muscles in the arterioles contract, causing vasoconstriction and a consequent rise in blood pressure. A rise in blood pressure causes the vasomotor centre to send messages via the parasympathetic system to the arterioles, causing them to dilate and so reduce blood pressure.

A rise in blood carbon dioxide concentration also causes a rise in blood pressure. This increases the speed with which blood is delivered to the lungs and so helps remove the carbon dioxide more quickly. Hormones like adrenaline similarly raise blood pressure.

21.5.5 Heart disease

As the organ pumping blood around the body, any interruption to the heart's ceaseless beating can have serious, often fatal, consequences. There are many defects and disorders, some acquired, like atherosclerosis, others congenital, such as a hole-in-the-heart. Some affect the pacemaker leading to an irregular heart rhythm, others the valves allowing blood to 'leak' back into the atria when the ventricles contract. By far the most common is **coronary heart disease** which affects the pair of blood vessels – the coronary arteries – which serve the heart muscle itself. There are three ways in which blood flow in these arteries may be impeded:

Coronary thrombosis – a blood clot which becomes lodged in a coronary vessel.

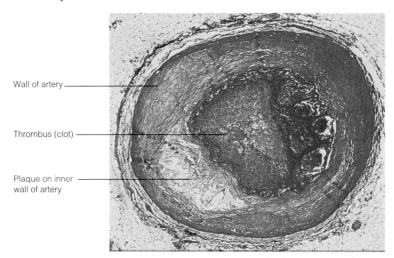

Wall of artery

Thrombus (clot)

Plaque on inner wall of artery

Human coronary artery containing a thrombus

Atherosclerosis – narrowing of the arteries due to thickening of the arterial wall caused by fat, fibrous tissue and salts being deposited on it. The condition is sometimes referred to as hardening of the arteries.

Spasm – repeated contractions of the muscle in the coronary artery wall.

It is often a combination of these factors, rather than one in isolation, which results in a **heart attack**. If the main coronary artery is blocked, the whole of the heart muscle or **myocardium** may be deprived of blood, resulting in death. If only a branch vessel is affected the loss of blood supply affects only a portion of the myocardium and, after a period of severe chest pain and temporary incapacitation followed by a number of days of complete rest, recovery normally follows. **Angina** is the result of reduced blood flow in the coronary arteries due to atherosclerosis, but sometimes the result of thrombosis or spasm. Chest pain, and breathlessness often occur when an angina sufferer is undertaking strenuous physical effort.

Many factors are known to increase the risk of coronary heart disease. Smoking is a major contributor increasing the likelihood of both thrombosis and atherosclerosis. A raised level of fat, especially cholesterol, in the blood is a major cause of atherosclerosis. Saturated fat of the type found in most meat and animal products, such as milk, is particularly dangerous. High blood pressure or hypertensive disease, a high level of salt in the diet and diabetes are other factors which contribute to atherosclerosis and hence coronary heart disease. Stress is suspected of increasing the risk of heart disease, but scientific evidence is hard to come by as it is difficult to measure stress levels accurately. There appears to be an inherited factor with individuals with a family history of heart disease being more susceptible to heart attacks. Older people are more at risk than younger ones, males more at risk than females. One thing generally accepted is that exercise can reduce the risk of coronary heart disease.

21.6   Lymphatic system

The lymphatic system consists of widely distributed **lymph capillaries** which are found in all tissues of the body. These capillaries merge to form **lymph vessels** which possess valves and whose structure is similar to that of veins. The fluid within these vessels, the **lymph**, is therefore carried in one direction only, namely, away from the tissues. The lymph vessels from the right side of the head and thorax and the right arm combine to form the **right lymphatic duct** which drains into the right subclavian vein near the heart. The lymph vessels from the rest of the body form the **thoracic duct** which drains into the left subclavian vein.

Along the lymph vessels are series of **lymph nodes**. These contain a population of phagocytic cells, e.g. lymphocytes which remove bacteria and other foreign material from lymph. During infection these nodes frequently swell. Lymph nodes are the major sites of lymphocyte production.

The movement of lymph through the lymphatic system is achieved in three ways:

1. Hydrostatic pressure – The pressure of tissue fluid leaving the arterioles helps push lymph along the lymph system.

2. Muscle contraction – The contraction of skeletal muscle compresses lymph vessels, exerting a pressure on the lymph within them. The valves in the vessels ensure that this pressure pushes the lymph in the direction of the heart.

3. Inspiratory movements – On breathing in, pressure in the thorax is decreased. This helps to draw lymph towards the vessels in the thorax.

Lymph is a milky liquid derived from tissue fluid. It contains lymphocytes and is rich in fats obtained from the lacteals of the small intestines. These fats might damage red blood cells and so are carried separately, until they are later added to the general circulation in safe quantities.

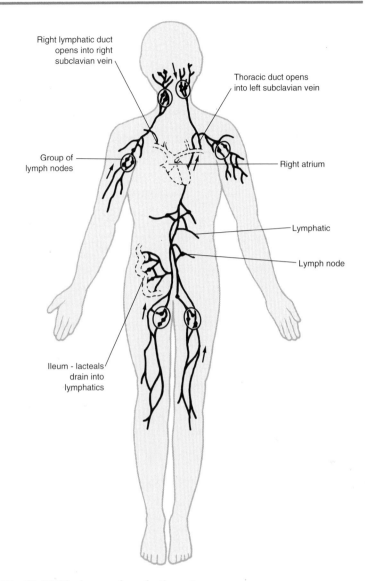

Right lymphatic duct opens into right subclavian vein

Thoracic duct opens into left subclavian vein

Group of lymph nodes

Right atrium

Lymphatic

Lymph node

Ileum - lacteals drain into lymphatics

Fig. 21.20 The human lymphatic system

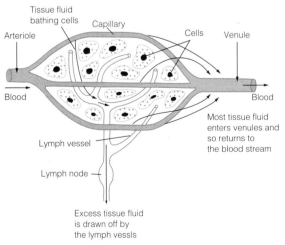

Tissue fluid bathing cells

Capillary

Cells

Venule

Arteriole

Blood

Blood

Most tissue fluid enters venules and so returns to the blood stream

Lymph vessel

Lymph node

Excess tissue fluid is drawn off by the lymph vessls

Fig. 21.21 Formation and destination of tissue fluid

21.6.1 Tissue fluid and its formation

As blood passes from arterioles into the narrow capillaries, a hydrostatic pressure is created which helps fluid escape through the capillary walls. This fluid is called **tissue (intercellular) fluid** and it bathes all cells of the body. It contains glucose, amino acids, fatty acids, salts and oxygen which it supplies to the tissues, from which it obtains carbon dioxide and other excretory material. Tissue fluid is thus the means by which materials are exchanged between blood and tissues.

The majority of this tissue fluid passes back into the venules by osmosis. The plasma proteins, which did not leave the blood, exert an osmotic pressure which draws much of the tissue fluid back into the blood. The fluid which does not return by this means passes into the open-ended lymph capillaries, from which point it becomes known as lymph.

21.7   Questions

1. Describe how **each** of the following function in the transport of vital materials to and from mammalian tissues:

(i)   the capillaries;   *(6 marks)*
(ii)   the red blood cells;   *(8 marks)*
(iii)   the lymphatic system.   *(6 marks)*
(Total 20 marks)

WJEC June 1990, Paper A1, No. 6

2.   *(a)* Describe the events of the cardiac cycle.
(12 marks)
(b) Explain how the heart action is controlled.
(11 marks)
(Total 23 marks)

UCLES (Modular) June 1992,
(Transport, regulation and control), No. 2

3.   *(a)* Distinguish between antibodies and antigens.
(3 marks)
(b) Outline the defence mechanisms that operate following the entry of a pathogenic bacterium into the body of a mammal.
(8 marks)
(c) Give reasons why some grafts in humans are successful whilst others are not.   *(6 marks)*
(d) Some organs for human transplants are in short supply. Do you think it feasible that organs from other animals could be transplanted into humans and function normally? Give your reasons.   *(3 marks)*
(Total 20 marks)

NEAB June 1993, Paper IB, No. 8

4. Read through the following account of blood clotting in a mammal, and then write on the dotted lines the most appropriate word or words to complete the account.

For blood to clot, the soluble plasma protein called .... has to be converted to its insoluble form, ... Cell fragments called .... release thromboplastin on exposure to .... This causes an enzyme, ..., to be produced from an inactive precursor. Numerous other factors, including .... ions, are needed to activate the enzyme. The cell fragments involved are produced in the ...., and the plasma proteins are produced in the ....   *(Total 8 marks)*

ULEAC January 1990, Paper I, No. 2

5. The graph below shows changes in pressure in the heart and aorta during the course of a heart beat.

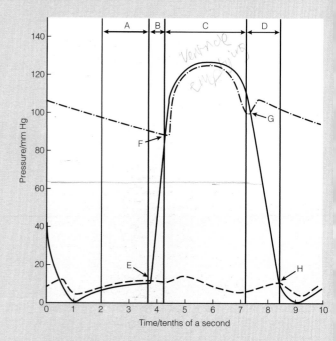

(a) Pressure was recorded in three regions: left atrium, left ventricle and aorta. In the table below, indicate which line you think represents **each** region:

Line on graph	Region
— · — · — · — · —	
——————	
— — — — — — —	

(b) On the graph, which one of the periods (labelled A–D) or points (labelled E–H), identifies **each** of the following?
The time when
(i)   the atrio-ventricular valve closes
(ii)   the semi-lunar valve closes
(iii)   the ventricle is filling with blood
(iv)   the ventricle is emptying
(v)   the volume of the ventricle is not changing.   *(5 marks)*

(c)   (i)   From the graph, estimate the **diastolic** arterial blood pressure that would be measured in this subject in an artery where it branches from the aorta.
(ii)   The total period displayed in the graph is actually slightly more than the duration of a single heart beat. Calculate the pulse rate in beats per minute for this subject.   *(2 marks)*

(d) In an experiment to discover whether the drug caffeine caused an increase in heart rate, subjects had their pulse rates taken before and after drinking cups of coffee.
  (i) It was decided to adjust the amount of coffee to be drunk by each subject in proportion to body mass. Explain why this procedure is necessary.
  (ii) Suggest **two** reasons why different subjects might respond in different ways even after this adjustment had been made.
  (iii) Describe a suitable control for this experiment. (5 marks)
  (Total 13 marks)

NEAB June 1992, Paper IIA, No. 4

Diagrams A, B and C below show cross sections of three different types of blood vessel. They are not drawn to the same scale.

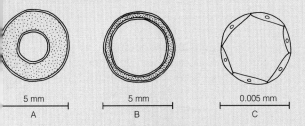

(a) Identify blood vessels A, B and C. (3 marks)
(b) State *two* ways in which vessel A is adapted for its functions. (2 marks)
(Total 5 marks)

ULEAC June 1992, Paper III, No. 1

The diagram below shows a vertical section of the human heart. The labels A, B and C show the positions of regions concerned with coordinating the contraction of heart muscle during the cardiac cycle.

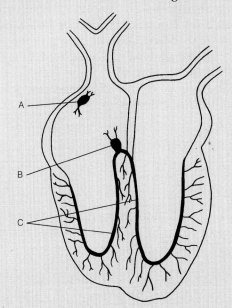

(a) Name the coordinating regions found at A, B and C. (3 marks)
(b) State the roles of regions A, B and C in coordinating the cardiac cycle. (3 marks)
(Total 6 marks)

ULEAC June 1993, Paper III, No. 4

8. (a) The diagram below shows the relationship between the circulatory system, the tissues and the lymphatic system in a mammal. The arrows indicate the direction of movement of the fluids involved.

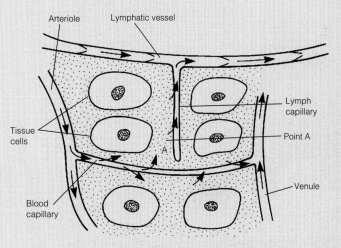

  (i) Name **one** factor responsible for the movement of fluid from the blood capillary into the intercellular spaces at point A. (1 mark)
  (ii) Explain how the circulation of fluid in the lymphatic system is maintained. (2 marks)
  (iii) Name the specialized lymph vessels found in the villi of the small intestine. (1 mark)
(b) The following diagram represents the heart and major blood vessels in the incomplete double circulatory system of an amphibian.

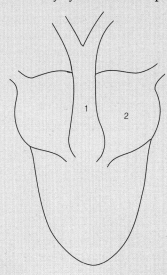

(i) By inserting arrows on the diagram, show the direction of blood flow at points 1 and 2. *(1 mark)*

(ii) State **one** major structural difference between the heart of an amphibian and that of a mammal. *(1 mark)*

(iii) State **two** ways in which the complete double circulatory system of a mammal is considered to be more efficient than the incomplete double circulation of an amphibian. *(2 marks)*

(Total 8 marks)

SEB May 1991 Higher Grade, Paper II, No. 4

9. The diagram below represents cells as seen in a stained blood film.

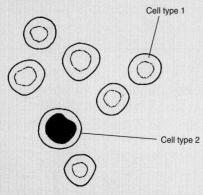

Cell type 1

Cell type 2

(a) Describe one feature of cell type 1 which enables it to carry out its function. Explain your answer. *(2 marks)*

(b) (i) State the main site of production of cell type 1 in an adult. *(1 mark)*

(ii) Name an important mineral required for adequate production of cell type 1 and state why this mineral is essential. *(2 marks)*

(iii) What is the fate of both the mineral and the pigment released as a result of the destruction of cell type 1? *(2 marks)*

(c) Cells of type 2 are involved in the immune response to a non-self antigen and may be present in B or T cell form.

(i) What is meant by the term *non-self antigen*? *(1 mark)*

(ii) Describe the role of B and T cell forms in the immune response. *(2 marks)*

SEB Human Biology Specimen, 1992,
Higher Grade II, No. 5

10. The percentage saturation of human haemoglobin at different partial pressures of oxygen, measured in kPa, was compared with the percentage saturation of mouse haemoglobin at the same temperature and pressure.
The results are shown in the table below.

Partial pressure oxygen/kPa	Percentage saturation of haemoglobin with oxygen	
	Human haemoglobin	Mouse haemoglobin
1	8	3
3	40	10
5	71	25
7	85	50
9	92	75
11	96	90
13	98	97
15	98	98

(a) (i) Using the data in the table, plot oxygen haemoglobin dissociation curves for the human and the mouse haemoglobin on graph paper. *(5 marks)*

(ii) In the human, assume the partial pressure of oxygen in the lung is 13.3 kPa and in the blood in the pulmonary artery it is 5.3 kPa. Calculate the increase in percentage saturation of the blood as it passes through the lung capillaries. Show your working. *(2 marks)*

(b) (i) Describe how the oxygen haemoglobin dissociation curve for the mouse differs from that for the human. *(2 marks)*

(ii) Suggest how these differences might be related to the difference in size of the mouse and the human. *(3 marks)*

(c) Suggest what effect a decrease in temperature might have on the dissociation of oxyhaemoglobin. Give a reason to support your answer. *(2 marks)*

(d) (i) How would an increase in the acidity of the blood affect the ability of the haemoglobin to associate with oxygen in both animals? *(1 mark)*

(ii) How would this affect the shape of the dissociation curves? *(1 mark)*

(Total 16 marks)

ULEAC January 1994, Paper III, No. 7

See *Further questions*, pages 614–20, questions 6, 7.

22 Uptake and transport in plants

For plants there are certain advantages in large size. They can, for example, compete more readily for light. As a result, many trees are tall, some exceeding 100 m. The leaves, as the sites of photosynthesis, must be in these aerial parts to obtain light. The water so essential to photosynthesis is, however, collected by the roots which may be some considerable distance beneath the soil surface. An efficient means of transporting this water, and certain minerals, to the leaves is necessary. The sugars formed as a result of photosynthesis in the leaves must be transported in the opposite direction to sustain respiration in the roots. In the absence of muscle or other contractile cells, plants depend to a large extent on passive rather than active means of transport.

22.1 The water molecule

Water is the most abundant liquid on earth and is essential to all living organisms. It is, however, no ordinary molecule. It possesses some unusual properties as a result of the hydrogen bonds which readily form between its molecules. These properties make water an ideal constituent of living things.

22.1.1 Structure of the water molecule

The water molecule is made up of two atoms of hydrogen and one of oxygen. Its basic atomic structure is given in Chapter 1, Fig. 1.4. This is a simplified representation for in practice the two hydrogen atoms are closer together, as shown in Fig. 22.1. By weight, 99.76% of water molecules consist of $^1H_2^{16}O$; the remainder is made up of various isotopes such as ^2H and ^{18}O. The commonest isotope is deuterium (^2H) and when this is incorporated into a water molecule it is known as **heavy water**, as a result of its greater molecular mass. Heavy water may be harmful to living organisms.

22.1.2 Polarity and hydrogen bonding

The distribution of the charges on a water molecule is unequal. The charge on the hydrogen atoms is slightly positive while that on the oxygen atom is negative (Fig. 22.1). Molecules with unevenly distributed charges are said to be **polar**. Attraction of oppositely charged poles of water molecules causes them to group together. The attractive forces form **hydrogen bonds**.

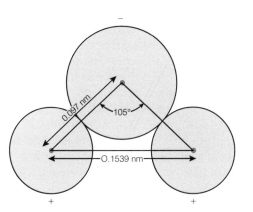

Fig. 22.1 Structure of water molecule

435

Although individual bonds are weak, they collectively form important forces which hold water molecules together. This makes water a much more stable substance than would otherwise be the case.

22.1.3 Thermal properties

Due to the hydrogen bonds, more energy is required to separate water molecules. It therefore requires more heat than expected to convert liquid water into vapour; thus evaporation of sweat is an effective means of cooling in mammals. For this reason, the boiling point of water is much higher than expected. In the same way, its **specific heat** is abnormally high. More heat is needed to raise the temperature of a given mass of water. Water is heated and cooled more slowly than expected – it in effect buffers sharp temperature changes.

At temperatures of $0\,°C$ and below, water forms the crystalline substance ice. The arrangement of the water molecules in ice makes it less dense than liquid water. As a result, ice will form at the surface of a body of water such as a pond or lake. This insulates the water beneath and thus prevents the whole of the body of water freezing solid. Living organisms can therefore survive beneath the ice providing there is sufficient food and oxygen available. Water has its maximum density at $4°C$; above this temperature the additional heat breaks some hydrogen bonds and so the water molecules are less densely compacted.

22.1.4 Dissociation, pH and buffers

There is a natural tendency for water molecules to dissociate into ions.

$$H_2O \rightleftharpoons H^+ + OH^-$$
water   hydrogen ion   hydroxyl ion

The positively charged hydrogen ions become attached to the slightly negatively charged oxygen atom of another water molecule.

$$H^+ + H_2O \rightleftharpoons H_3O^+$$
hydrogen ion   water molecule   oxonium ion

The complete reaction may be summarized thus:

$$2H_2O \rightleftharpoons H_3O^+ + OH^-$$
water   oxonium ion   hydroxyl ion

At $25\,°C$ the concentration of oxonium ions (hydrogen ions) in pure water is $10^{-7}\,mol\,dm^{-3}$ and this is given the value of 7 on the **pH scale**. When the concentration of hydrogen ions in a solution is $10^{-3}\,mol\,dm^{-3}$ this represents a pH of 3. The pH scale ranges from 1 (very acid) to 14 (very alkaline). The pH scale is logarithmic and so a pH of 6 is ten times more acid than one of pH 7, pH 5 is one hundred times more acid than pH 7 and so on. The pH within most cells is in the range 6.5–8.0. It remains fairly constant because substances within the cells act as **buffers**. Buffer solutions do not appreciably change their pH, despite the addition of small amounts of acids or bases. This is important in cells, because fluctuations in pH could affect the efficiency with which their enzymes work. Apart from dissociating itself, water readily causes the dissociation of other substances. This makes it an excellent solvent.

22.1.5 Colloids

A solid placed in a liquid may dissolve. If not, it may float or sink. In some cases an intermediate situation may arise whereby the solid becomes finely dispersed as particles throughout the liquid. These particles, normally 1–100 nm in diameter, are known as the **disperse phase** while the liquid around them is called the **dispersion medium**. Collectively they form a **colloid**. Many protein and polysaccharide molecules form colloids. Their most important feature is the large surface area of contact between the particles and the liquid. Cytoplasm is a colloid, made up largely of protein molecules dispersed in water.

22.1.6 Cohesion and surface tension

Cohesion is the tendency of molecules of a substance to attract one another. The magnitude of this attractive force depends upon the mass of the particles and their distance apart. Gases, with their smaller molecular masses, have small cohesive forces. In liquids, the cohesive forces are much greater. Unlike gases, liquids cannot be expanded or compressed to any degree. Hydrogen bonding increases the cohesive forces between water molecules. One effect of these large cohesive forces in water is that the molecules are pulled inwards towards each other, so forming spherical drops rather than spreading out in a layer. The inward pull of the water molecules creates a skin-like layer at the surface. This force is called **surface tension**.

The cohesive forces between molecules accounts for the upward pull of water in xylem when evaporation occurs at the leaves. It is surface tension which allows insects like pond-skaters to walk on the surface of the water.

22.1.7 Adhesion and capillarity

Adhesion is the tendency of molecules to be attracted to ones of a different type. Considerable adhesive forces exist between the walls of xylem vessels and the water within them. The magnitude of these forces has been estimated at up to 100 000 kPa. **Capillarity** is also the result of intermolecular forces between various molecules. If one end of an open glass tube is held vertically beneath the surface of water, the liquid can be seen to rise up the tube. The smaller the diameter of the tube the higher it rises. Xylem vessels, with their diameters around 0.02 mm, have considerable capillarity forces which contribute to the movement of water up a plant. Capillarity also plays an important rôle in the upward movement of water in soil.

22.1.8 The importance of water to living organisms

Life arose in water and many organisms still live surrounded by it. Those that left water to colonize land nonetheless keep their cells bathed in it. Water is therefore the main constituent of all organisms – in jellyfish up to 98% and in most herbaceous plants 90%. Even mammals consist of around 65% water. The importances of water to organisms are many and there is only room here to list a few.

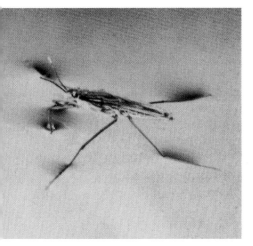

pond-skater walking on water

Did you know?

All living things contain water. A lettuce leaf is 94% water, a human being 60–70% and a pine tree 55%.

Metabolic rôle of water

1. Hydrolysis – Water is used to hydrolyse many substances, e.g. proteins to amino acids, fats to fatty acids and glycerol and polysaccharides to monosaccharides.

2. Medium for chemical reactions – All chemical reactions take place in an aqueous medium.

3. Diffusion and osmosis – Water is essential to the diffusion of materials across surfaces such as the lungs or alimentary canal.

4. Photosynthetic substrate – Water is a major raw material in photosynthesis.

Water as a solvent
Water readily dissolves other substances and therefore is used for:

1. Transport – Blood plasma, tissue fluid and lymph are all predominantly water and are used to dissolve a wide range of substances which can then be easily transported.

2. Removal of wastes – Metabolic wastes like ammonia and urea are removed from the body in solution in water.

3. Secretions – Most secretions comprise substances in aqueous solution. Most digestive juices have salts and enzymes in solution; tears consist largely of water and snake venoms have toxins in suspension in water.

Water as a lubricant
Water's properties, especially its viscosity, make it a useful lubricant. Lubricating fluids which are mostly water include:

1. Mucus – This is used externally to aid movement in animals, e.g. snail and earthworm; or internally in the vagina and gut wall.

2. Synovial fluid – This lubricates movement in many vertebrate joints.

3. Pleural fluid – This lubricates movement of the lungs during breathing.

4. Pericardial fluid – This lubricates movement of the heart.

5. Perivisceral fluid – This lubricates movement of internal organs, e.g. the peristaltic motions of the alimentary canal.

Supporting rôle of water
With its large cohesive forces water molecules lie close together. Water is therefore not easily compressed, making it a useful means of supporting organisms. Examples include:

1. Hydrostatic skeleton – Animals like the earthworm are supported by the pressure of the aqueous medium within them.

2. Turgor pressure – Herbaceous plants and the herbaceous parts of woody ones are supported by the osmotic influx of water into their cells.

3. Humours of the eye – The shape of the eye in vertebrates is maintained by the aqueous and vitreous humours within them. Both are largely water.

4. **Amniotic fluid** – This supports and protects the mammalian fetus during development.

5. **Erection of the penis** – The pressure of blood, a largely aqueous fluid, makes the penis erect so that it can be introduced into the vagina during copulation.

6. **Medium in which to live** – Water provides support to the organisms which live within it. Very large organisms, e.g. whales, returned to water as their sheer size made movement on land difficult.

Miscellaneous functions of water

1. **Temperature control** – Evaporation of water during sweating and panting is used to cool the body.

2. **Medium for dispersal** – Water may be used to disperse the larval stages of some terrestrial organisms. In mosses and ferns it is the medium in which sperm are transferred. The build-up of osmotic pressure helps to disperse the seeds of the squirting cucumber.

3. **Hearing and balance** – In the mammalian ear the watery endolymph and perilymph play a rôle in hearing and balance.

22.2 Simple plant tissues

Simple plant tissues each consist of only one type of cell. They are normally grouped according to the degree of thickening present in the cell wall.

22.2.1 Parenchyma

Parenchyma cells are usually spherical although their shape may be distorted by pressure from adjacent cells. These unspecialized living cells form the bulk of packing tissue within the plant. Parenchyma cells are metabolically active and may also store food. When tightly packed and turgid they provide support for herbaceous plants. Air spaces around parenchyma cells allow exchange of gases to take place. (See Fig. 22.2.)

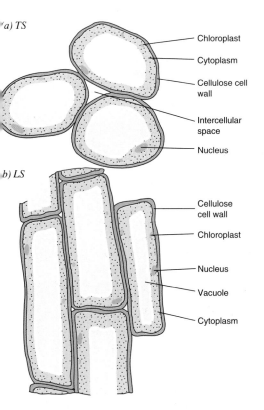

(a) TS

- Chloroplast
- Cytoplasm
- Cellulose cell wall
- Intercellular space
- Nucleus

(b) LS

- Cellulose cell wall
- Chloroplast
- Nucleus
- Vacuole
- Cytoplasm

Fig. 22.2 Parenchyma

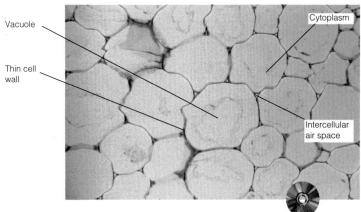

Parenchyma cells (TS) (×200 approx.)

Labels: Vacuole, Thin cell wall, Cytoplasm, Intercellular air space

(a) TS

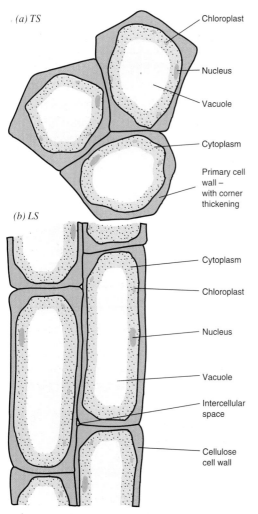

Chloroplast

Nucleus

Vacuole

Cytoplasm

Primary cell wall – with corner thickening

(b) LS

Cytoplasm

Chloroplast

Nucleus

Vacuole

Intercellular space

Cellulose cell wall

Fig. 22.3 Collenchyma

(a) TS

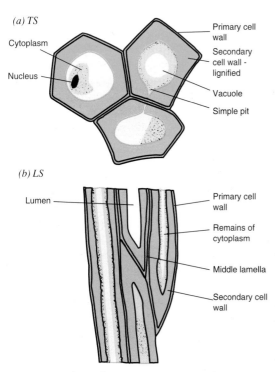

Cytoplasm

Nucleus

Primary cell wall

Secondary cell wall - lignified

Vacuole

Simple pit

(b) LS

Lumen

Primary cell wall

Remains of cytoplasm

Middle lamella

Secondary cell wall

Fig. 22.4 Sclerenchyma

22.2.2 Collenchyma

Collenchyma cells are living and have cell walls with additional cellulose deposited in the corners. This provides them with extra mechanical strength. They are elongated and important in growing stems since they are able to stretch. They are often found just under the epidermis of a stem or in the corners of angular stems. (See Fig. 22.3.)

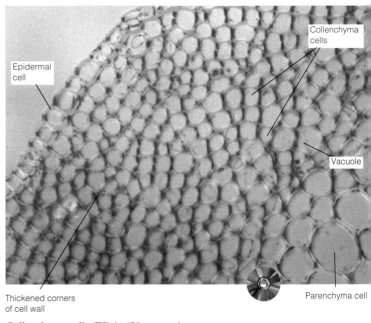

Collenchyma cells

Epidermal cell

Vacuole

Thickened corners of cell wall

Parenchyma cell

Collenchyma cells (TS) (×450 approx.)

22.2.3 Sclerenchyma

Mature sclerenchyma cells are dead and therefore incapable of growth. They develop fully when the growth of surrounding tissues is complete. Sclerenchyma cells have large deposits of lignin

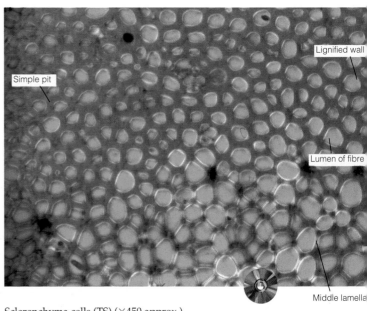

Lignified wall

Simple pit

Lumen of fibre

Middle lamella

Sclerenchyma cells (TS) (×450 approx.)

on the primary cell wall and the cell contents are lost. In places, lignin is not deposited due to the presence of **plasmodesmata** in the primary cell wall; such regions are called **pits**. (See Fig. 22.4.)

Some sclerenchyma cells are roughly spherical and are known as **sclereids**. These are usually found in small groups in fruits and seeds, cortex, pith and phloem.

Elongated sclerenchyma cells are called **fibres** and they provide the main supporting tissue of many mature stems. They may form a cylinder below the epidermis, are found in xylem and phloem and sometimes as masses associated with vascular bundles.

22.3 Water relations of a plant cell

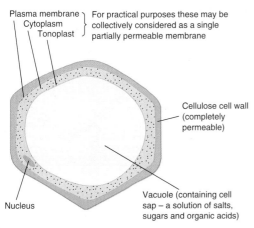

Plasma membrane ⎫ For practical purposes these may be
Cytoplasm ⎬ collectively considered as a single
Tonoplast ⎭ partially permeable membrane

Cellulose cell wall (completely permeable)

Nucleus

Vacuole (containing cell sap – a solution of salts, sugars and organic acids)

Fig. 22.5 Typical plant cell showing osmotically important structures

For practical purposes the plant cell can be divided into three parts (Fig. 22.5):

1. The vacuole – This contains an aqueous solution of salts and sugars and organic acids.

2. The cytoplasm surrounded by membranes – The inner membrane is called the **tonoplast** and the outer one, the **plasma membrane**; both are partially permeable.

3. The cell wall – This is made of cellulose fibres and is completely permeable to even large molecules. It may be impregnated with substances like lignin, in which case it is impermeable to molecules.

We saw in Section 4.3.3 that **water potential (Ψ)** is the capacity of a system to lose water. The water potential of pure water at standard temperature and pressure (25 °C and 100 kPa) is zero.

The presence of solute molecules in the vacuole of a plant cell makes its water potential more negative (lower). The greater the concentration of solutes, the more negative is the water potential. This change in water potential as a consequence of the presence of solute molecules is called the **solute potential (Ψ_s)**. (This was previously called 'osmotic potential'.) As the solute molecules always lower the water potential, the value of the solute potential is always negative.

If a plant cell is surrounded by pure water, the water potential of the vacuole, containing solute molecules as it does, will be more negative than the surrounding medium. Water therefore enters by osmosis, thus creating a hydrostatic pressure which pushes outwards on the cell wall. This is known as the **pressure potential (Ψ_p)**. In this instance, as in most plant cells, the pressure potential is positive. In xylem vessels, however, where transpiration is pulling water up the plant, the pressure potential is negative.

The relationship between water potential, solute potential and pressure potential is shown in the equation:

$$\Psi = \Psi_s + \Psi_p$$

water potential · solute potential · pressure potential

441

If the same cell is placed in a solution which has a more negative solute potential than that of its cell sap, it will lose water by osmosis. In this case the external solution has a more negative water potential than the internal solution of the cell sap. As a result, water is drawn out of the cell by osmosis and the protoplast (the living part of the cell) shrinks. As it does so the

Fig. 22.6 Chart to show differences between cells placed in external solutions of different water potential

Water potential (ψ) of external solution compared to internal solution	Less negative (higher)	Equal	More negative (lower)
Net movement of water	Enters cell	Neither enters nor leaves cell	Leaves cell
Protoplast	Swells	No change	Shrinks
Condition of cell	Turgid	Incipient plasmolysis	Plasmolysed (flaccid)
	Protoplast pushed up against the cell wall Vacuole containing cell sap Nucleus Cellulose cell wall Cytoplasm	Protoplast beginning to pull away from the cell wall	Protoplast completely pulled away from the cell wall External solution which enters through the permeable cell wall

NOTEBOOK

Negative values for water potential – designed to confuse?

Not really – consider this:

The highest possible value of water potential is that of pure water – namely zero. Add solute to the water and its water potential is lowered, i.e. it has a negative value. All solutions therefore have negative water potentials. A somewhat crazy situation you may think until you understand that water potential is actually the capacity of a system to lose water.

Think of it as a person gambling at a casino with no money (not that many casinos would tolerate the idea!). If the gambler loses £5 he is in debt to the casino, i.e. he has −£5. The more he loses the higher is his debt and the lower is the level of his money. In terms of the level of his money −£10 is lower than −£5 (although the number 10 is actually bigger than 5). The bigger the debt, the lower the level of his money. At the same time the bigger the gambler's debt, the greater need he has for money to flow into his account. So it is with water potential (except that water is the currency rather than pound coins!). The more negative the water potential the greater the water 'debt' and the more readily water flows into the system.

A water potential of −20 kPa is lower than one of −10 kPa and therefore water moves from the solution at −10 kPa to that at −20 kPa just as a temperature of −20 °C is lower than one of −10 °C and heat will move from a system at −10 °C to one at −20 °C.

pressure potential decreases. A point is reached where the protoplast no longer presses on the cell wall, and hence the pressure potential falls to zero. This point is termed **incipient plasmolysis**. Any further loss of water causes the protoplast to shrink more and so pull away from the cell wall. This condition is called **plasmolysis** and the cell is said to be **flaccid**. It is important to remember that water will always move from a less negative (higher) water potential to a more negative (lower) one. Pure water has a water potential of zero.

22.4   Transpiration

The evaporation of water from plants is called **transpiration**. This evaporation takes place at three sites:

1. Stomata – Most water loss, up to 90%, takes place through these minute pores which occur mostly, but not exclusively, on leaves. Some are found on herbaceous stems.

2. Cuticle – A little water, perhaps 10% is lost through the cuticle, which is not completely impermeable to gases. The thicker the cuticle the smaller the water loss.

3. Lenticels – Woody stems have a superficial layer of cork which considerably reduces gas exchange over their surface. At intervals the cells of this layer are loosely packed, appearing externally as raised dots. These are the lenticels through which gaseous exchange, and hence water loss, may occur. The amount of transpiration through these is relatively insignificant.

Transpiration would appear to be the inevitable, if undesirable, consequence of having leaves punctured with stomata. The necessity of stomata for photosynthesis is obvious. It is unfortunate that any opening in a leaf which allows gases in will inevitably allow water out. The plant has to balance carefully the conflicting needs of obtaining carbon dioxide for photosynthesis with the inevitable loss of water this creates. We shall see in the next section how this is achieved by controlling the opening and closing of stomata. Transpiration is not essential as a means of bringing water to the leaves. This could occur by purely osmotic means. As water was used for photosynthesis in a leaf mesophyll cell, its water potential would become more negative (lower). Water could then enter this cell from adjacent ones whose water potential would also become more negative (lower). They in turn would draw it from adjacent cells and so on down to the root hair cells of the plant, which would draw their water from the soil solution. While mineral salts are drawn up the plant in the transpiration stream, these could be diffused, or actively transported, in the absence of transpiration. Mineral uptake from the soil is largely independent of the rate of transpiration and could certainly occur in its absence. The cooling effect of transpiration would only be beneficial when environmental temperatures were high and even then its effects would be insignificant. The definition of transpiration as a 'necessary evil' is therefore not without some justification.

(a) Stoma closed – guard cells less turgid

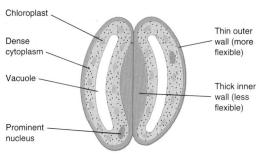

Chloroplast

Dense cytoplasm

Vacuole

Prominent nucleus

Thin outer wall (more flexible)

Thick inner wall (less flexible)

(b) Stoma open – guard cells more turgid

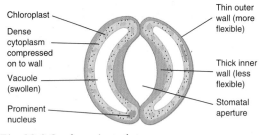

Chloroplast

Dense cytoplasm compressed on to wall

Vacuole (swollen)

Prominent nucleus

Thin outer wall (more flexible)

Thick inner wall (less flexible)

Stomatal aperture

Fig. 22.8 Surface view of stoma

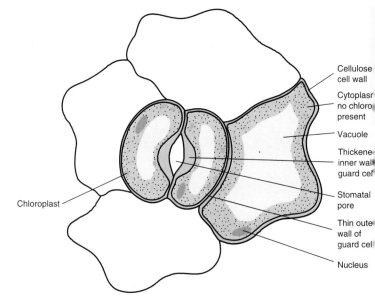

Cellulose cell wall

Cytoplasm no chloro present

Vacuole

Thickene inner wall guard cel

Stomatal pore

Thin oute wall of guard cel

Nucleus

Chloroplast

Fig. 22.7 Epidermis and stoma

Stoma is closed in the dark, but in the presence of light ATPase is stimulated to convert ATP to ADP and so provide the energy to pump out hydrogen ions (protons) from the guard cells. These protons return on a carrier which also brings chloride ions (Cl^-) with it. At the same time potassium ions (K^+) also enter guard cells.

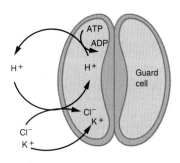

As a result of this influx of ions, the water potential of the guard cells becomes more negative (lower) causing water to pass in by osmosis. The resultant increase in pressure potential causes the stoma to open.

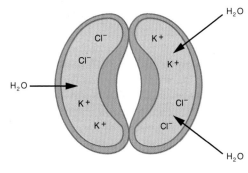

In the dark, the movement of ions and water is reversed.

Fig. 22.9 Mechanism of stomatal opening

22.4.1 Stomatal mechanism

Stomata occur primarily on leaves where they are normally mor abundant on the abaxial (lower) surface, where there are typically around 180 per mm^2. They may even be entirely absen on the adaxial (upper) surface. The structure of a stoma is given in Chapter 14, Fig. 14.1. It consists of a pair of specialized epidermal cells, the **guard cells** which surround a small pore a few microns wide known as the **stomatal aperture**. Guard cells are different from normal epidermal cells in being kidney-shaped. Unlike other epidermal cells, they possess chloroplasts and have denser cytoplasm with a more prominent nucleus. The inner walls of guard cells are thicker and less elastic than the outer ones. It is this which makes the inner wall less able to stretch and results in the typical kidney shape. Moreover, any increase in the volume of a guard cell, owing, for example, to th osmotic uptake of water, causes increased bowing of the cell owing to the greatest expansion occurring in the outer wall. When this occurs in the two guard cells of a stoma, the stomatal aperture enlarges (Fig. 22.8).

Even when all stomata on a leaf are fully open, the total area of the apertures rarely exceeds 2% of the leaf area.

It is possible to show that opening of a stoma is the result of pressure potential changes within the guard cells by puncturing turgid guard cells using a micro-needle. This releases the pressure potential and the cells collapse, closing the stoma.

What exactly causes changes in pressure potential within guard cells? While there are a number of factors which influence pressure potential, they all operate through changes in the wate potential. It has long been observed that stomata open in the light and close in the dark. In the light it is thought that potassium ions (K^+) and chloride ions (Cl^-) enter the guard cells lowering their water potential so causing water to enter by osmosis. As a result the stoma opens. The mechanism involves a proton pump and is described in more detail in Fig. 22.9.

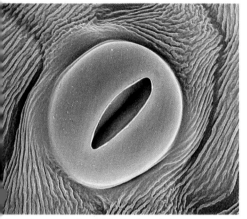

False colour scanning EM of a closed stoma

22.4.2 Movement of water across the leaf

Under normal circumstances the humidity of the atmosphere is less than that in the sub-stomatal air-space. Provided the stomata are open (see Section 22.4.1) the less negative (higher) water potential in the sub-stomatal air-space causes water vapour to diffuse out into the atmosphere through the stomatal aperture. Provided there is some air movement around the leaf, the water vapour is swept away once it leaves the stomata. The water lost from the sub-stomatal air-space is replaced by more evaporating from the spongy mesophyll cells surrounding the space. The water is brought to the spongy mesophyll cells from xylem in the leaf – in three ways:

1. The apoplast pathway – Most water travels from cell to cell via the cell wall, which is made up of cellulose fibres, between which are water-filled spaces. As the water evaporates into the sub-stomatal air-space from the wall of one cell, it creates a tension which pulls in water from the spaces in the walls of surrounding cells. The pull is transmitted through the plant by the cohesive forces between the water molecules which, due to hydrogen bonding, are particularly strong.

2. The symplast pathway – Some water is lost to the sub-stomatal air-space from the cytoplasm of cells surrounding it. The water potential of this cytoplasm is thereby made more negative (lower). Between adjacent cells are tiny strands of cytoplasm, known as **plasmodesmata**, which link the cytoplasm of one cell to that of the next. Water may pass along these plasmodesmata from adjacent cells with a less negative (higher) water potential. This loss of water makes the water potential of this second cell more negative (lower), which may in turn replace it with water from other cells with less negative (higher) water potentials. In this way a water potential gradient is established between the sub-stomatal space and the xylem vessels of the leaf. The symplast pathway carries less water than the apoplast pathway, but is of greater importance than the vacuolar pathway.

3. The vacuolar pathway – A little water passes by osmosis from the vacuole of one cell to the next, through the cell wall, membranes and cytoplasm of adjacent cells. In the same way as the symplast pathway, a water potential gradient between the xylem and the sub-stomatal air-space exists. It is along this gradient that the water passes.

In Fig. 22.10 water leaving cell C causes its water potential to become more negative. Assuming all three cells were originally of equal water potential, then, compared to cell C, cell B now has a less negative water potential. Water therefore flows from cell B to cell C. In the same way, loss of water from B to C causes water to enter B from A. Remember that this mechanism applies only to the symplast and vacuolar pathways. The apoplast pathway is due to cohesion tension and is independent of a water potential gradient.

22.4.3 Structure of xylem

Xylem consists of parenchyma cells and fibres together with two specialized types of cells: vessels and tracheids. These tissues are both dead and serve the dual rôle of support and water

Cell A

Cell B

Cell C

Cellulose cell wall (fibres with water filled spaces between)

Cytoplasm

Vacuole (with cell sap comprising a dilute solution of salts and sugars)

Plasmodesmata (fine strands of cytoplasm linking adjacent cells)

⟶ Apoplast pathway
⟶ Symplast pathway
⟶ Vacuolar pathway

Fig. 22.10 Alternative routes for water transport across cells

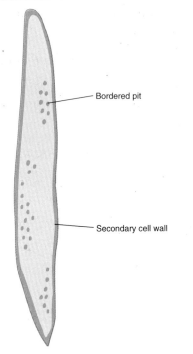

Fig. 22.11 Tracheid (LS)

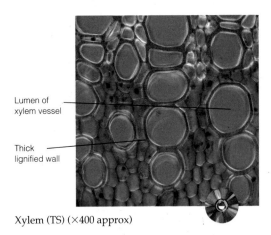

Xylem (TS) (×400 approx)

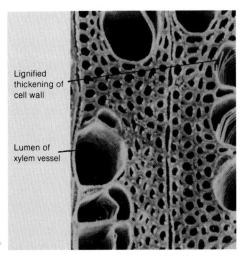

Wood of alder showing vessels
(scanning EM) (× 300 approx.)

transport. The types of **vessel** found depend upon the degree and nature of the cell wall thickening. In the **protoxylem** the lignin is deposited in rings or spirals so the cell is still capable of expansion.

In **metaxylem** there is more extensive lignification arranged in patterns known as **reticulate**, **scalariform** or **pitted**. All vessels are made up of cells whose cross walls have broken down, resulting in long tubes ideal for carrying water.

Tracheids are spindle-shaped cells arranged in rows with the ends of the cells overlapping. The cells have heavily lignified walls and so there are no cell contents. They provide mechanical strength and support to the plant. (See Fig. 22.11.)

22.4.4 Movement of water up the stem

Water moves up the stem and into the leaves through xylem vessels and tracheids. Xylem and phloem together form **vascular bundles** which are arranged mostly towards the outside of the stem. The reason for this is that the vascular bundles, along with associated sclerenchyma, give support to the stems of herbaceous plants. The main forces acting on stems are lateral ones, owing largely to the wind. These forces are best resisted by an outer cylinder of supporting tissue. Hence the vascular bundles are predominantly at the periphery of stems as shown in Fig. 22.12.

The evidence supporting the view that xylem carries water up the stem includes:

1. A leafy shoot is cut under water containing a dye, e.g. eosin. The cut end is kept in the solution of the dye and left for a few hours. The shoot is removed and cut at various levels up the stem. Only the xylem is found to be stained red, indicating that it alone transports water. If left long enough in the dye, the veins of the leaves also become stained.

2. A ring of tissue removed from the outside of a woody stem does not affect the flow of water up the stem, provided only the bark, including the phloem, are removed. If, however, the outer layers of xylem are also removed, upward transport ceases and the leaves wilt.

3. If the cut end of a shoot is placed in a solution of a metabolic poison, the uptake of the solution continues as normal. If the process were an active one, it would be expected that the poison would kill the cells in which it travelled and so prevent water transport. As the process continues, it must be assumed that it occurs passively. As xylem cells are dead, these would seem the most likely sites of this passive process.

4. Plants which are allowed to draw up fatty solutions soon wilt. It is found by staining and microscopic examination that these fats have blocked the lumina of the xylem cells. The wilting must therefore be the result of this blockage and hence xylem must be the route by which water rises up the stem.

The theory of the mechanism by which water moves up the xylem is known as the **cohesion–tension theory**. The transpiration of water from the leaves draws water across the leaf (Section 22.4.2). This water is replaced by that entering the mesophyll cells from the xylem by osmosis. As water molecules leave xylem cells in the leaf, they pull up other water molecules.

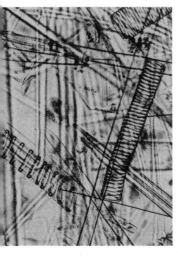

ylem macerate (× 150 approx.)

This pulling effect, known as the **transpiration pull**, is possible because of the large cohesive forces between water molecules. The pull creates a tension in the xylem cells which, if cut, draw in air rather than exude water. Such is the cohesive force of this column of liquid that it is sufficient to raise water to heights in excess of 100 m, i.e. large enough to supply water to the top of the tallest known trees, the Californian redwoods.

Other forces may contribute to the movement of water up the stems of plants. These make only a small contribution to transport in large trees but may be significant in smaller herbaceous plants. **Adhesion forces** between the water molecules and the walls of xylem vessels help water to rise in xylem – a phenomenon known as **capillarity**. This force can cause water to rise to a height of 3 m. Adhesion however also causes a frictional drag on the upward flow of water in the xylem. Its overall influence is therefore minimal and may even be detrimental in a large tree. **Root pressure**, which is discussed in Section 22.6.3, also contributes.

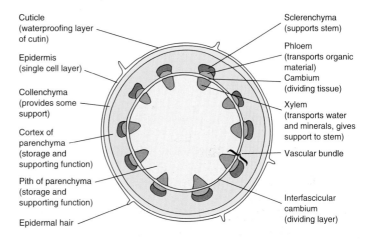

Fig. 22.12 (a) TS through a typical dicotyledonous stem

PROJECT

1. Make stained transverse sections of a variety of herbaceous plants, e.g. those with stems of round section, square section, etc.

2. Compare the distribution of simple plant tissues in the stems.

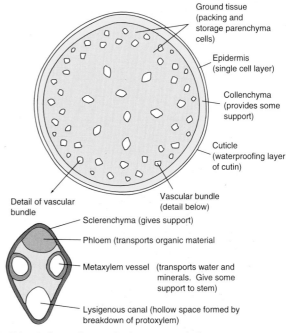

Fig. 22.12 (b) TS through a typical monocotyledonous stem

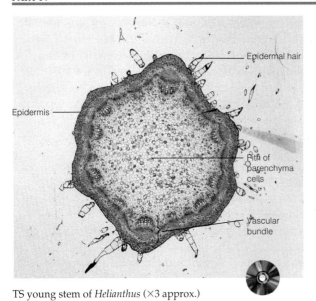

TS young stem of *Helianthus* (×3 approx.)

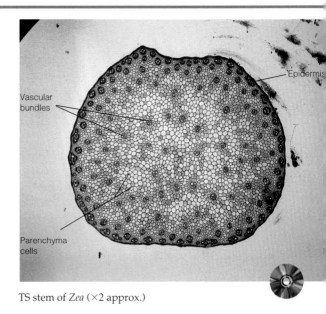

TS stem of *Zea* (×2 approx.)

22.4.5 Xylem structure related to its rôle of water transport

The relationship between structure and function is always a close one and this is especially true of xylem vessels and tracheids. The structure of xylem is given in Section 22.4.3. Correlations between xylem structure and its function of water transport include the six points below.

1. Both vessels and tracheids consist of long cells (up to 5 mm in length) joined end to end. This allows water to flow in a continuous column.

2. The end walls of xylem vessels have broken down to give an uninterrupted flow of water from the roots to the leaves. Even in tracheids where end walls are present, large bordered pits reduce the resistance to flow caused by the presence of these walls.

3. There are pits at particular points in the lignified wall which permit lateral flow of water where this is necessary.

4. The walls are impregnated with lignin, making them especially rigid to prevent them collapsing under the large tension forces set up by the transpiration pull.

5. The impregnation of the cellulose walls with lignin increases the adhesion of water molecules and helps the water to rise by capillarity.

6. The narrowness of the lumina of vessels and tracheids (0.01–0.2 mm in diameter) increases the capillarity forces.

22.4.6 Measurement of transpiration

Transpiration in a cut leafy shoot can be measured using a potometer, a diagram of which is given in Fig. 22.13. To be strictly accurate, the instrument measures the rate of water uptake of a shoot, but in practice this is almost exactly the same as the rate of transpiration. A little of the water taken up may be used in photosynthesis and other metabolic processes, but the vast majority is transpired. The experiment is carried out as follows:

1. Cut a leafy shoot off a plant. As it is under tension, cutting the shoot will cause air to enter the xylem so, if possible, firstly hold the part where the cut is to be made under water. If this is not feasible, it will be necessary to trim back the cut shoot a few centimetres to remove the xylem containing air.

2. Submerging the potometer, fill it with water, using the syringe to help pump out any air bubbles. Fit the leafy shoot to the rubber tube, ensuring a tight fit.

3. Remove the apparatus from the water and allow excess water to drain off. Gently shake the shoot to remove as much water as possible.

4. Seal joints around the rubber tube with vaseline to keep the apparatus watertight.

5. Introduce an air bubble into the water column by using the syringe to push the water almost to the end of the capillary tube. Leave a small air-space. Place the open end of the capillary tube in a vessel of water and draw up more water behind the air-space.

6. When the shoot is dry, the syringe may be depressed with the tap open until the air bubble in the capillary tube is pushed back to the zero mark. The tap should then be turned to close off the syringe.

7. Measure the distance moved by the air bubble in the calibrated capillary tube in a given time. Repeat the procedure a number of times, using the syringe to return the air bubble to zero each time.

8. Calculate the water uptake in mm^3 min^{-1} using the average of the results obtained.

9. The experiment can be repeated under differing conditions, e.g. in light and dark, at different air temperatures and humidities, in still and moving air.

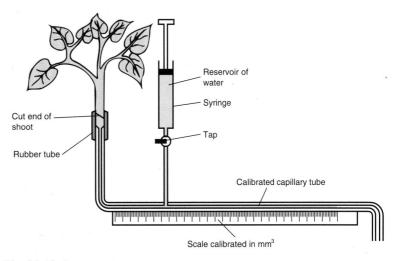

Fig. 22.13 A potometer

22.5 Factors affecting transpiration

A number of factors influence the rate of transpiration. These may be divided for convenience into external (environmental) factors and internal ones related to the structure of the plant itself.

22.5.1 External factors affecting transpiration

External factors include all aspects of the environment which alter the diffusion gradient between the transpiring surface and the atmosphere. Among these are:

1. Humidity – The humidity, or vapour pressure, of the air affects the water potential gradient between the atmosphere within the leaf and that outside. When the external air has a high humidity, the gradient is reduced and less water is transpired. Conversely, low humidities increase the transpiration rate.

2. Temperature – A change in temperature affects both the kinetic movement of water molecules and the relative humidity of air. A rise in temperature increases the kinetic energy of water molecules and so increases the rate of evaporation of water. At the same time it lowers the relative humidity of the air. Both changes increase the rate of transpiration. A fall in temperature has the reverse effect, namely, a reduction in the amount of water transpired.

3. Wind speed – In the absence of any air movement the water vapour which diffuses from stomata accumulates near the leaf surface. This reduces the water potential gradient between the moist atmosphere in the stomata and the drier air outside. The transpiration rate is thus reduced. Any movement of air tends to disperse the humid layer at the leaf surface, thus increasing the transpiration rate. The faster the wind speed, the more rapidly the moist air is removed and the greater the rate of transpiration.

4. Light – The stomata of most plants open in light and close in the dark. The suggested mechanism by which these changes are brought about is given in Section 22.4.1. It follows that, up to a point, an increase in light intensity increases the transpiration rate and vice versa.

5. Water availability – A reduction in the availability of water to the plant, for example as the result of a dry soil, means there is a reduced water potential gradient between the soil and the leaf. The transpiration rate is reduced as a result.

22.5.2 Internal factors affecting transpiration

There are a number of anatomical and morphological features of plants which also influence the transpiration rate. Many specific adaptations of plants designed to reduce water loss are dealt with in Section 22.5.3. Only general features applicable to all plants are dealt with here.

1. Leaf area – As a proportion of water loss occurs through the cuticle, the greater the total leaf area of a plant, the greater the rate of transpiration regardless of the number of stomata present. In addition, any reduction in leaf area inevitably involves a reduction in the total number of stomata.

2. Cuticle – The cuticle is a waxy covering over the leaf surface which reduces water loss. The thicker this cuticle the lower the rate of cuticular transpiration.

3. Density of stomata – The greater the number of stomata for a given area the higher the transpiration rate. Stomatal density on the abaxial (lower) epidermis of plants may vary from around $2000 \, cm^2$ in an oat (*Avena sativa*) leaf to $45\,000 \, cm^2$ on an oak *Quercus* spp.) leaf.

4. Distribution of stomata – In most dicotyledonous plants the leaves are positioned with their adaxial (upper) surfaces towards the light. The upper surfaces are subject to greater temperature rises than the lower ones owing to the warming effect of the sun. Transpiration is therefore potentially greater from the upper surface. Many plants, like the oak (*Quercus* spp.) and the apple (*Malus* spp.) limit their stomata entirely to the abaxial (lower) surface to reduce their overall water loss.

22.5.3 Xerophytic adaptations

Xerophytes are plants which have adapted to conditions of unfavourable water balance, i.e. conditions where the rate of loss is potentially greater than the availability of water. Most plants live in areas where ample water is available. These are known as **mesophytes**. Xerophytic plants have evolved a wide range of features designed to reduce the rate of transpiration. These are known as **xeromorphic** features. These adaptations are not confined to xerophytic plants but may also occur in mesophytes. By no means all xeromorphic adaptations occur in plants of hot, dry desert regions. Many species in cold regions cannot supply adequate water to the leaves as the soil water is frozen, and so unobtainable, for much of the year. Other species with a reasonable supply of water may suffer excessive loss from the leaves as a result of living in exposed, windy situations. These too exhibit xeromorphic features. Other plants live in salt marshes where the concentration of salts in the soil make the obtaining of water difficult. These **halophytes**, as they are called, also exhibit xeromorphic features.

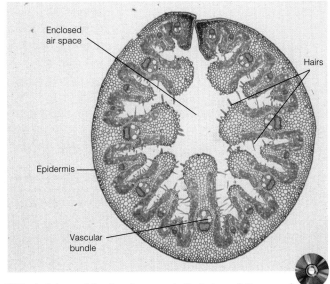

TS leaf of *Ammophila*, showing xerophytic features (×3 approx.)

451

Adaptations in xerophytes take a number of general forms:

1. Reduction in the transpiration rate – Clearly anything which lowers the rate at which the plant loses water helps to conserve it when in short supply,

2. Storage of water – In plants living where water supply is intermittent, e.g. in the desert, there is considerable advantage in rapidly absorbing the water when available and storing it for use during periods of drought. Plants which store water are termed **succulents**. The adaptations of these plants are not limited to specialized water storage tissue but include mechanisms for rapid water absorption, and reducing the rate at which it is lost.

3. Resistance to desiccation – Some species exhibit a remarkable tolerance to water loss and resistance to wilting.

The xeromorphic adaptations of plants are summarized in Table 22.1.

22.6 Uptake of water by roots

If a plant is to survive, the large quantities of water lost through transpiration must be replaced. Water absorption is largely carried out by the younger parts of roots which bear extensions of the epidermal cells, known as **root hairs**. These root hairs only remain functional for a few weeks, being replaced by others formed on the younger regions nearer the growing apex. As the root becomes older the epidermis is replaced by a layer known as the **exodermis** through which some water absorption takes place.

22.6.1 Root structure

Beneath the epidermis of a young root is a broad cortex of parenchyma cells. The vascular tissue occurs as a central column rather than a peripheral cylinder as in stems. The reason is that roots are subject to pulling forces in a vertical direction rather than the lateral forces experienced by stems. Vertical forces are better resisted by a central column of supporting tissue such as xylem. As roots, embedded as they are in soil, are subject to fewer stresses and strains than aerial parts of the plant, the other major supporting tissue, sclerenchyma, is reduced or absent.

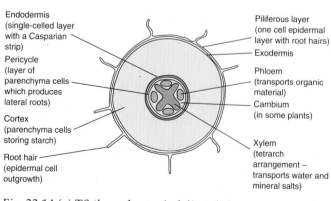

Fig. 22.14 (a) TS through a typical dicotyledonous root, as seen under a microscope (100×)

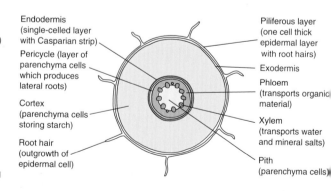

Fig. 22.14 (b) TS through a typical monocotyledonous root, as seen under a microscope (100×)

TABLE 22.1 Xeromorphic adaptations of plants

General form	Specific adaptation	Examples	Mechanism by which adaptation functions
Reduction of the transpiration rate	Thick cuticle	Evergreens, especially gymnosperms	Reduces cuticular transpiration by forming a waxy barrier preventing water loss
	Rolling of leaves	*Ammophila* (marram grass); *Calluna* (ling)	Moist air is trapped within the leaf, preventing water diffusing out through stomata which are confined to the inner surface
	Layer of protective hairs on leaf (pubescence)	*Ammophila* (marram grass); *Calluna* (ling)	Moist air is trapped in the hair layer, increasing the length of the diffusion path, so reducing transpiration
	Depression of stomata	*Pinus* (pine); *Ilex* (holly)	Lengthens the diffusion path by trapping still, moist air above the stomata, so reducing transpiration
	Reduction of surface-area/volume ratio of leaves	*Pinus* (pine)	The leaves are small and circular in cross-section to reduce the transpiration area. The shape also gives structural rigidity to help prevent wilting
	Absence of leaves	Most cacti, e.g. *Opuntia* (prickly pear)	Dispensing with leaves altogether limits water loss to the stems which have considerably fewer stomata. These stems are flattened to provide an adequate area for photosynthesis
	Orientation of leaves	*Lactuca* (compass plant)	The positions of the leaves are constantly changed so that the sun strikes them obliquely. This reduces their temperature and hence the transpiration rate
	More negative (lower) water potential of cell sap	Many xerophytes and most halophytes, e.g. *Salicornia* (glasswort)	The cells accumulate salts which make their water potential more negative (lower). This makes it more difficult for water to be drawn from them
Succulence	Succulent leaves	E.g. *Bryophyllum*	Stores water
	Succulent stems	Most cacti, e.g. *Opuntia* (prickly pear)	Stores water
	Closing of stomata during daylight	Most cacti and other C_4 plants	The more efficient use of carbon dioxide by C_4 plants allows them to keep stomata closed during much of the day, so reducing transpiration
	Shallow but extensive root systems	Most succulents	Allows efficient absorption of water over a wide area when the upper layers of the soil are moistened by rain
Resistance to desiccation	Reduction of transpiration surface through loss or adaptation of leaves	*Berberis* (barberry); many cacti	Leaves reduced to spines to protect plant from grazing. Flattened stems perform photosynthesis
		Ruscus (butcher's broom)	Leaves are lost and photosynthesis is performed by a flattened stem known as a cladode
		Acacia	The lamina of the leaf is lost and the petiole becomes flattened to carry out photosynthesis
	Lignification of leaves	*Hakea*	Lignified tissue supports the leaf, preventing it wilting in times of drought and thereby allowing it to continue photosynthesis
	Reduction in cell size	Many xerophytes	The proportion of cell wall material is greater with many small cells. This gives additional support, making the plant less liable to wilt

Around the central vascular tissue is a ring one cell thick known as the **endodermis**. These living cells are elongated vertically. Part of the wall is impregnated with **suberin**. This forms a distinctive band known as the **Casparian strip**. Inside the endodermis is the **pericycle**, a layer of parenchyma cells between the endodermis and the vascular tissue. It is from the pericycle that lateral roots originate. The internal structure of roots is shown in Fig. 22.14.

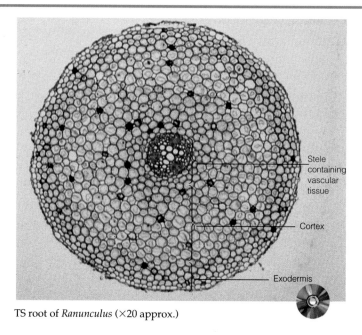

TS root of *Ranunculus* (×20 approx.)

Stele containing vascular tissue

Cortex

Exodermis

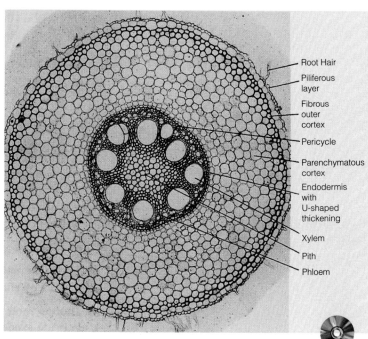

TS root of *Zea* (×35 approx.)

Root Hair

Piliferous layer

Fibrous outer cortex

Pericycle

Parenchymatous cortex

Endodermis with U-shaped thickening

Xylem

Pith

Phloem

22.6.2 Mechanisms of uptake

The same three pathways which are responsible for movement of water across the leaf also bring about its movement in the root:

1. The apoplast pathway – Compared to the leaf (Section 22.4.2) this pathway has one major difference in the root. Movement through the cell walls is prevented by the suberin of the Casparian strip in the endodermis. The water is thereby forced to enter the living protoplast of the endodermal cell, as the only available route to the xylem (Fig. 22.15).

The significance of this is that salts may then be actively secreted into the vascular tissue from the endodermal cells. This makes the water potential in the xylem more negative (lower),

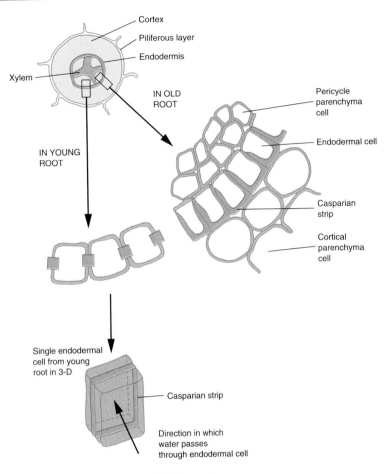

Cortex

Piliferous layer

Endodermis

Xylem

IN OLD
ROOT

IN YOUNG
ROOT

Pericycle
parenchyma
cell

Endodermal cell

Casparian
strip

Cortical
parenchyma
cell

Single endodermal
cell from young
root in 3-D

Casparian strip

Direction in which
water passes
through endodermal cell

Fig. 22.15 *Water transport in the root*

causing water to be drawn in from the endodermis. While the mechanism has yet to be proven scientifically, there is some circumstantial evidence supporting this view:

(a) There are numerous starch grains in endodermal cells which could act as an energy source for the process.

(b) Depriving roots of oxygen prevents water being exuded from cut stems. Lowering the temperature reduces the rate of exudation.

(c) Treating roots with metabolic poisons like cyanide also prevents water being exuded from cut stems.

Whatever the process, it is clearly an active one.

2. The symplast pathway — This operates in the same way as in the leaf (Section 22.4.2). Water leaving the pericycle cells to enter the xylem causes the water potential of the pericycle cells to become more negative (lower). These therefore draw in water from adjacent cells which in turn have a more negative (lower) water potential. In this way a water potential gradient is established across the root from the xylem to the root hair cells, which draws water across it.

455

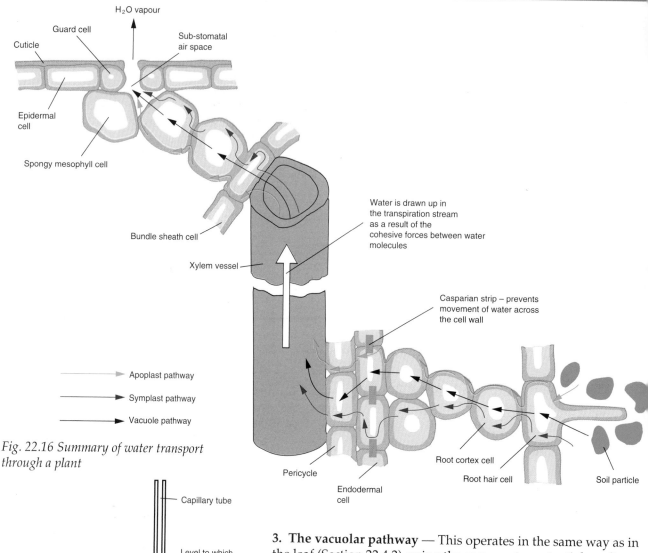

Fig. 22.16 Summary of water transport through a plant

Apoplast pathway

Symplast pathway

Vacuole pathway

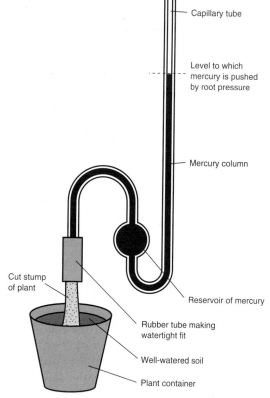

Fig. 22.17 Manometer for measuring root pressure

3. The vacuolar pathway — This operates in the same way as in the leaf (Section 22.4.2), using the same water potential gradient described above.

A summary diagram of the passage of water throughout the plant is given in Fig. 22.16.

22.6.3 Root pressure

The soil solution is normally very dilute and hence has a less negative (higher) water potential than the solution in the root hair cells. As a result, water enters the root hair cells by osmosis and their water potential becomes less negative (higher) as a consequence. Water therefore enters an adjacent cortical cell. This water potential gradient creates a force known as root pressure. If the stem of a plant is cut near to the roots, this root pressure causes water to be exuded from the cut stump. The process probably involves the pumping of salts into the xylem (see Section 22.6.2) as it is halted by the presence of metabolic poisons. In some plants, **root pressure** may be sufficiently large to force liquid out of pores on the leaves called **hydathodes**. The process is known as **guttation**. Root pressure may contribute to the movement of water up the stem, especially in herbaceous plants, but its contribution is far less than that of transpiration. An instrument for measuring root pressure is illustrated in Fig. 22.17.

22.7 Uptake and translocation of minerals

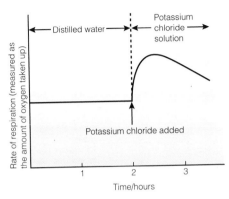

Fig. 22.18 Relationship between the rate of respiration and mineral uptake

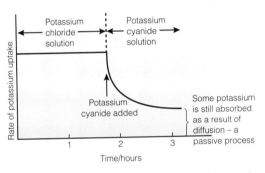

Fig. 22.19 Effect of the respiratory inhibitor cyanide on mineral uptake

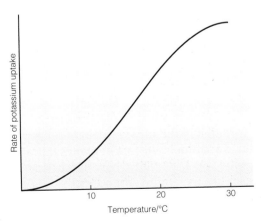

Fig. 22.20 Effect of temperature on the rate of potassium uptake in plants

All plants require a supply of minerals, the functions of which are summarized in Chapter 2, Table 2.1. These minerals are largely absorbed by roots through root hairs, although leaves can also absorb them if sprayed with a suitable solution. Such sprays are called **foliar feeds**.

22.7.1 Mechanisms of mineral uptake

Minerals may be absorbed either passively or actively:

1. Passive absorption – If the concentration of a mineral in the soil solution is greater than its concentration in a root hair cell, the mineral may enter the root hair cell by **diffusion**.

2. Active absorption – If the concentration of a mineral in the soil solution is less than that in a root hair cell it may be absorbed by active transport, details of which are given in Section 4.3.4. Most minerals are absorbed in this way. The process is selective. Requiring energy as it does, the rate of absorption is dependent upon respiration.

Fig. 22.18 shows that when salts are added to plants growing in water, their respiration rate increases, presumably in order to provide the energy for their absorption. The addition of the respiratory inhibitor potassium cyanide prevents active mineral uptake, leaving only absorption by passive means (Fig. 22.19). Increases in temperature increase the rate of respiration and hence the rate of mineral uptake (Fig. 22.20).

Once absorbed, the mineral ions may move along the cell walls (apoplast pathway) by either diffusion or mass flow. In the latter case they are carried along in solution by the water being pulled up the plant in the transpiration stream. When these minerals reach the endodermis the Casparian strip prevents further movement along the cell walls. Instead, the ions enter the cytoplasm of the cell from where they diffuse or are actively transported into the xylem. Minerals may alternatively pass through the cytoplasm of cortical cells (symplast pathway) to the xylem into which they diffuse or are actively pumped.

22.7.2 Transport of minerals in the xylem

Analysis of the contents of xylem vessels reveals the presence of mineral salts and water, although sugars and amino acids may also be present. The evidence supporting the rôle of xylem in transporting minerals includes:

1. The presence of mineral ions in xylem sap.

2. A similarity between the rate of mineral transport and the rate of transpiration.

3. Evidence that other solutes, e.g. the dye eosin, are carried in the xylem (Section 22.4.4).

4. Experiments using radioactive tracers (Fig. 22.21). The interpretation of the experiments is that where lateral transfer of minerals can take place, minerals pass from the xylem to the phloem. Where it is prevented, the transport of minerals takes place almost exclusively in the xylem.

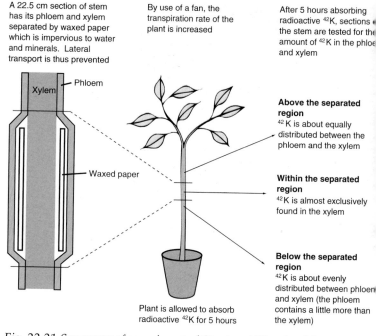

A 22.5 cm section of stem has its phloem and xylem separated by waxed paper which is impervious to water and minerals. Lateral transport is thus prevented

By use of a fan, the transpiration rate of the plant is increased

After 5 hours absorbing radioactive ^{42}K, sections of the stem are tested for the amount of ^{42}K in the phloem and xylem

Xylem — Phloem

— Waxed paper

Plant is allowed to absorb radioactive ^{42}K for 5 hours

Above the separated region
^{42}K is about equally distributed between the phloem and the xylem

Within the separated region
^{42}K is almost exclusively found in the xylem

Below the separated region
^{42}K is about evenly distributed between phloem and xylem (the phloem contains a little more than the xylem)

Fig. 22.21 Summary of experiment of Stout and Hoagland (1939) to demonstrate that minerals are translocated up the plant in the xylem

Once in the xylem, minerals are carried up the plant by the mass flow of the transpiration stream. Once they reach the places where they will be utilized, called **sinks**, they either diffuse or are actively transported into the cells requiring them.

22.8 Translocation of organic molecules

The organic materials produced as a result of photosynthesis need to be transported to other regions of the plant where they are used for growth or stored. This movement takes place in the phloem.

22.8.1 Evidence for transport of organic material in the phloem

The evidence supporting the view that organic material formed as a result of photosynthesis is carried in the phloem includes:

1. When phloem is cut, the sap which exudes is rich in organic materials such as carbohydrates. The fact that sap is exuded suggests the contents of the phloem are under pressure.

2. The sugar content of phloem varies in relation to environmental conditions. Where conditions favour photosynthesis, the concentration of sugar in phloem increases. There is also a diurnal variation in the sugar content of phloem which reflects the diurnal variation in the rate of photosynthesis in relation to light intensity. Fig. 22.22 shows how the sucrose content of leaves increases to a maximum around 1500 hours, as a result of the high light intensity and temperature favouring photosynthesis at this time. This peak of sugar concentration is reflected in the phloem of the stem a little time later. Little variation of sucrose concentration in the xylem takes place.

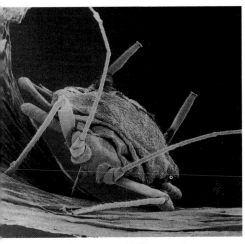

Aphid feeding showing needle-like mouthparts penetrating phloem

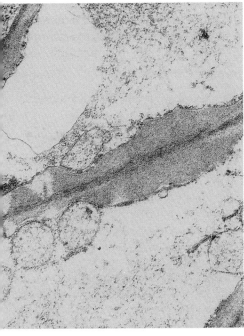

Sieve plate, at the junction between two sieve elements, in leaf of *Zinnia elegans* (EM) (× 21 000 approx.)

3. Removal of a complete ring of phloem from around a stem causes an accumulation of sugars above the ring, indicating that their downward progress has been interrupted.

4. If radioactive $^{14}CO_2$ is given to plants as a photosynthetic substrate, the sugars later found in the phloem contain ^{14}C. When phloem and xylem are separated by waxed paper, the ^{14}C is almost entirely found in the phloem.

5. Aphids have needle-like mouthparts with which they penetrate phloem in order to obtain sugars. If a feeding aphid is anaesthetized and then the mouthparts cut from the body, they remain as tiny tubes from which samples of the phloem contents exude. Analyses of these exudates confirm the presence of carbohydrates and amino acids, and a diurnal variation in their concentrations.

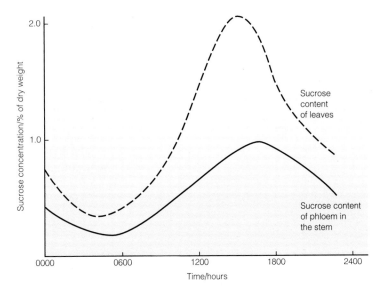

Fig. 22.22 Diurnal variation in the sugar content of leaves and the phloem in stems

22.8.2 Structure of phloem sieve tubes

Phloem tissue is living and comprises **sieve tubes** (or sieve elements), phloem parenchyma and phloem fibres. In angiosperms, specialized parenchyma cells known as **companion cells** are always found associated with sieve tube cells.

The sieve tube cells are the only components of phloem obviously adapted for the longitudinal flow of material. They are elongated with a characteristic series of pores 2–6 μm in diameter in the end walls. These are lined with the polysaccharide **callose** and form a **sieve plate**. The sieve tube cells have a well defined plasma membrane and their cytoplasm contains numerous plastids and mitochondria. Within the lumen of the cells are longitudinal strands of cytoplasm 1–7 μm wide. They are made up of **phloem protein**. The strands are continuous from cell to cell through the pores of the sieve plate and are known as **transcellular strands**. There is some question as to the existence of these strands as some researchers believe them to be no more than artefacts. Mature sieve tube cells lack a nucleus and are called **sieve tube elements**.

459

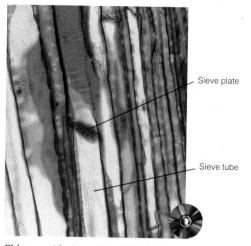

Phloem, with sieve plates (TS) (×400 approx.)

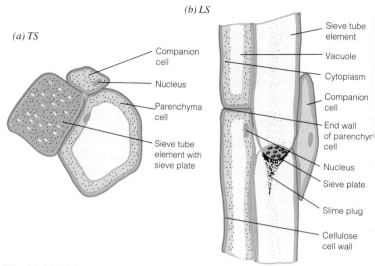

Fig. 22.23 Phloem

The companion cells have a thin cellulose cell wall and dense cytoplasm. Within the cytoplasm is a large nucleus, numerous mitochondria, plastids, small vacuoles and an extensive rough endoplasmic reticulum. Companion cells are metabolically active. They are closely associated with the sieve tube element with which they communicate by means of numerous plasmodesmata.

The structure of a sieve tube element and companion cell as seen under an electron microscope is given in Fig. 22.24.

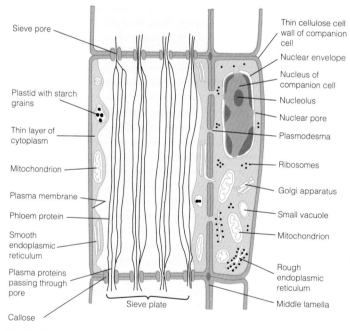

Fig. 22.24 Structure of a longitudinal section of a sieve tube element and companion cell, as seen by an electron microscope

22.8.3 Mechanisms of translocation in phloem

There is much controversy regarding the mechanism by which materials are translocated in phloem. One thing, however, is agreed: the observed rate of flow is much too fast for diffusion to be the cause. The theories put forward include the following hypotheses.

Mass flow (pressure flow) hypothesis

Photosynthesis forms soluble carbohydrates like sucrose. Photosynthesizing cells in the leaf therefore have their water potentials made more negative (lower) by the accumulation of this sucrose. As a result, water which has been transported up the stem in the xylem enters these cells. This causes an increase in their pressure potential. At the other end of the plant, in the roots, sucrose is either being utilized as a respiratory substrate or is being converted to starch for storage. The sucrose content of these cells is therefore low, giving them a less negative (higher) water potential and a consequently lower pressure potential. There is therefore a gradient of pressure potential between *the source* of sucrose (the leaves) and its point of utilization – *the sink* (the roots and other tissues). The two are linked by the phloem and as a result liquid flows from the leaves to other tissues along the sieve tube elements. A simple physical model to illustrate this mechanism is given in Fig. 22.25.

Evidence supporting the mass flow theory includes:

1. There is a flow of solution from phloem when it is cut or punctured by the stylet of an aphid.

2. There is some evidence of concentration gradients of sucrose and other materials, with high concentrations in the leaves and lower concentrations in roots.

3. Some researchers have observed mass flow in microscopic sections of living sieve elements.

4. Viruses or growth chemicals applied to leaves are only translocated downwards to the roots when the leaf to which they are applied is well illuminated and therefore photosynthesizing. When applied to shaded leaves, no downward translocation occurs.

It is likely that the sucrose produced in mesophyll cells as a result of photosynthesis needs to be actively transported against a concentration gradient into the sieve elements. The energy for this process is provided by ATP.

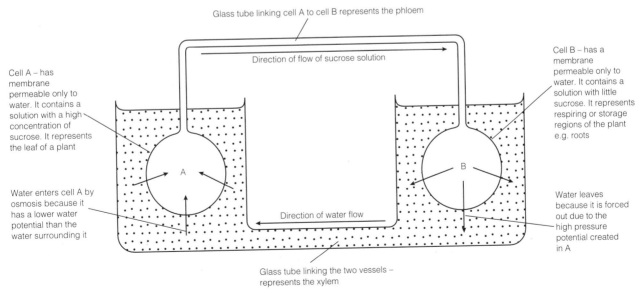

Fig. 22.25 *Physical model to illustrate the mass flow theory of translocation in phloem*

One criticism of the mass flow theory is that it offers no explanation for the existence of sieve plates which act as a series of barriers impeding flow. Indeed, as the process is passive after the initial stage, there is no necessity for the phloem to be a living tissue at all. One suggested function of the sieve plates is a means of sealing off damaged sieve tube elements. As the material within the elements is under pressure, any damage could lead to wasteful loss of sugar solution. It has been observed that once an element is damaged the sieve plate is quickly sealed by deposition of callose across the pores. Another criticism is that while the theory would *suggest* that all materials being transported in the phloem would travel at the same speed, in practice sugars and amino acids move at different rates.

Electro-osmosis hypothesis

Originally put forward by Spanner in 1958, and since modified on several occasions, the theory proposes that potassium ions are actively transported by companion cells, across the sieve plate. The movement of these ions draws polar water molecules across the plate. The movement is still one of mass flow, but the theory at least offers some function for the sieve plates and explains the high metabolic rate observed in companion cells. However, there is no consistent evidence of a potential difference existing across sieve plates.

Transcellular strand hypothesis – cytoplasmic streaming

Thaine in 1962 proposed that transcellular strands, which extend from cell to cell via pores in the sieve plate, carry out a form of cytoplasmic streaming. The solutes move in this cytoplasmic stream which can occur between the strands or through them as they are in fact tiny tubules about 20 nm in diameter. The process, being active, accounts for the many mitochondria in both sieve tube elements and companion cells. It will, however, require more positive proof of the existence of the actual process before it becomes widely accepted.

22.9   Questions

(a) Give details of the main cell types which make up the xylem tissue in the stem of flowering plants.   (*8 marks*)

(b) Indicate how the structure and distribution of xylem is related to its functions.   (*4 marks*)

(c) Explain how the movement of materials in xylem comes about.   (*8 marks*)

(*Total 20 marks*)

WJEC June 1993, Paper AI, No. 3

. Describe water transport in an Angiosperm (flowering plant) from the point where it enters the roots until it is released into the air from the leaves. What are the advantages to the plant of this movement of water?   (*Total 17 marks*)

SEB May 1991 Higher Grade, Paper II, No. 13A

(a) Give an illustrated account of the structure of the phloem tissue in a herbaceous flowering plant.   (*9 marks*)

(b) Discuss the theories that attempt to explain the transport of organic materials by the phloem.   (*11 marks*)

(*Total 20 marks*)

WJEC June 1991, Paper A1, No. 1

. The table below refers to xylem and phloem tissue n flowering plants. If the statement is correct, place a ick (✓) in the appropriate box and if the statement is ncorrect, place a cross (✗) in the appropriate box.

Statement	Xylem	Phloem
May contain tracheids		
Contains cells with living contents		
Contains lignified cells		
Transports organic products of photosynthesis		
Unidirectional transport		
Transport inhibited by metabolic poisons		

(*Total 6 marks*)

ULEAC January 1993, Paper III, No. 2

5. The diagram below shows a longitudinal section of some collenchyma cells in a young stem of a plant.

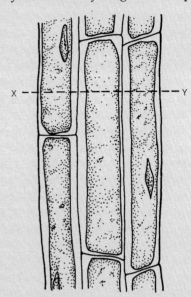

(a) Draw the cells as they would appear in the transverse section along the dotted line X–Y. Your drawing must be to the same scale.   (*3 marks*)

(b)   (i) State *one* region in a stem where collenchyma is found.   (*1 mark*)

(ii) Give *one* feature shown in the diagram that is characteristic of collenchyma.   (*1 mark*)

(iii) Explain how the structure of collenchyma cells enables them to carry out their function in the young stem of a plant.   (*3 marks*)

(*Total 8 marks*)

ULEAC June 1993, Paper III, No. 2

6. Diagrams A and B represent transverse sections from the stem and root of a flowering plant showing the distribution of the main tissue types (not to scale).

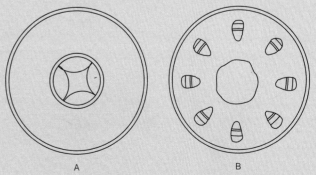

A   B

(a)   (i) Which diagram (A or B) is the root section? Explain your choice.

(ii) Using clear label lines and the letters given below, indicate on either diagram (A or B) the tissue **most** involved with each of the following activities.

upward transport of water   W
transport of sugars   S
cell division   D

(iii) The tissue that transports water upwards can readily be differentially stained. Which substance in the tissue is responsible for this effect?   (5 marks)

(b) In a study of water movement in plants, the cut end of a leafy shoot was placed in a dilute solution of a dye. Sections were later obtained from various parts of the shoot and examined for presence of the dye, which is carried up with the water.

(i) Give **three** environmental conditions that would increase the rate of movement of dye up the stem.

(ii) Over a period of several hours, deposits of the dye accumulate in the leaves of the plant. Explain why this occurs.

(iii) Suggest why a cut shoot is used in this experiment and not an intact plant complete with its root system.

(iv) It has been suggested that the upward movement of water observed is largely a passive process. How would you modify the above experiment to test this hypothesis?   (5 marks)

(Total 10 marks)

NEAB June 1992, Paper IIA, No. 1

7. The diagram shows changes in pressure potential and solute potential in a plant cell.

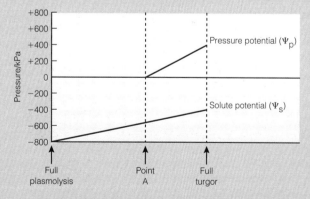

(a) What name is given to describe the state of the cell at Point **A**?   (1 mark)

(b) Write an equation to express the relation between the terms:
Water potential (Ψ); solute potential (Ψ_s); pressure potential (Ψ_p).   (1 mark)

(c) What is the value (in kPa) of the water potential of this cell at:
(i) full turgor; (ii) point A?   (2 marks)

(d) Complete the diagram by drawing a line showing the values of the pressure potential between full plasmolysis and point A.   (1 mark)

(Total 5 marks)

AEB June 1991, Paper I, No. 1

8. The diagram below shows two adjacent plant cells A and B. The values of their pressure potentials (Ψ_p) and solute potentials (Ψ_s) are given in kPa.

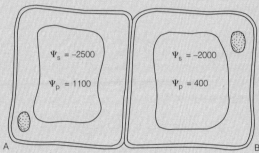

(a) In which direction would water flow between the two cells? Give a reason for your answer.   (2 marks)

(b) Complete the table below to show the values of the pressure potential Ψ_p and the water potential Ψ of each of the cells when equilibrium has been reached. Assume that changes in the value of Ψ_s are negligible.   (4 marks)

Potential	Cell A	Cell B
Pressure potential Ψ_p		
Water potential Ψ		

(c) The cells are then immersed in distilled water and again allowed to reach equilibrium. Assume that changes in the value of Ψ_s are negligible.

(i) State the new value of the water potential Ψ of the two cells.   (1 mark)

(ii) State the new value of the pressure potential Ψ_p of cell A.   (1 mark)

(Total 8 marks)

ULEAC January 1993, Paper III, No. 1

9. In an experiment to determine the water potential of a plant tissue, 1 cm strips of *Narcissus sp.* flower were totally immersed in a series of sucrose solutions of known molarity contained in covered petri dishes. After 24 hours at 20 °C the length of the tissue was measured. The results are shown in the table at the top of the next page.

Concentration of sucrose solution (mol dm⁻³)	0.2	0.3	0.4	0.5	0.6	0.7	0.8
Length of flower strip after 24 hours (cm)	1.9	1.7	1.2	0.8	0.75	0.7	0.7
Change (+ or −) in length of flower strip (cm)							

(a) (i) Complete the entries in the table.

(ii) Plot a graph to show the change in flower strip length brought about by different concentrations of sucrose.

(iii) Use your graph in order to state the molarity of a sucrose solution which would have the same water potential as the flower tissue. *(6 marks)*

(b) (i) An alternative method for determining the water potential of plant tissue is to weigh thin slices before and after immersion in sucrose solutions. Suggest **two** necessary precautions when using this method.

(ii) Explain why it is difficult to make an accurate determination of the water potential that exists in the tissues of an intact plant. *(4 marks)*

(c) The mean solute (osmotic) potential of cell sap is usually determined by the method of incipient plasmolysis.

(i) Label on a copy of the diagram: A – a plasmolysed cell; B – a turgid cell.

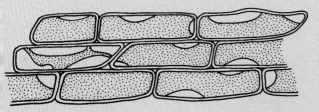

(ii) State the proportion of cells that would be plasmolysed at the state of incipient plasmolysis. *(2 marks)*

(d) (i) Consider two adjacent vacuolated cells X and Y. Cell X has a solute potential of −1800 kPa and a pressure potential of 800 kPa. Cell Y has a water potential of −800 kPa. Calculate the water potential of cell X.

(ii) Complete the following statements.

'Cell .... has the higher water potential.'

'Water will move from .... into cell ....' *(3 marks)*

(iii) What would be the water potential of both cells at equilibrium? *(1 mark)*
(Total 16 marks)

WJEC June 1992, Paper A2, No. 8

10. Water loss from the leaves of two different species of plants, *Species A* and *Species B*, was compared using two experimental methods.

Experiment 1
Leafy shoots were taken, one from each plant. They were weighed, hung in air and re-weighed at 15-minute intervals. The results are shown in the table (− means that a result is not available).

Time after start of experiment/minutes	Mass of leafy shoot/g	
	Species A	*Species B*
0	210.0	240.0
15	195.3	−
30	184.4	−
45	176.4	−
60	170.1	−
75	166.3	213.1

(a) (i) Plot a graph of mass against time for *Species A*.

(ii) Account for the shape of the curve for *Species A*. *(5 marks)*

(b) (i) Calculate the percentage change in mass for each species after 75 minutes.

(ii) Explain which of these two species you would expect to be more successful in an arid habitat. *(3 marks)*

Experiment 2
Different leafy shoots from each species were placed in a simple potometer, as shown below.

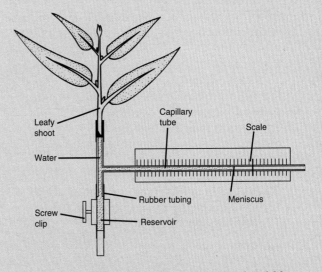

(c) (i) State **two** precautions that should be taken when using a potometer, to ensure that the results obtained are comparable for the two different species.

465

(ii) What **three** measurements are required to calculate the rate of water taken up in $cm^3 min^{-1} g^{-1}$ in *Experiment 2*?

(iii) Give **one** reason why the values obtained for water loss in *Experiment 2* were likely to have been higher than those from *Experiment 1* for the same species. *(6 marks)*

(Total 14 marks)

NEAB June 1990, Paper IIA, No. 6

11. In an investigation into mineral uptake in maize roots, maize plants were placed in solutions containing radioactively-labelled potassium ions. In one batch of experiments, (A), a small quantity of cyanide was added to the solution, but in the other batch, (B), it was not.

At regular intervals the plants were removed from the solution and the amount of radioactively-labelled potassium *remaining in the solution* was measured. The results are shown in the table below.

Time from start of the experiment/minutes	Units of radioactive potassium remaining in solution	
	(A)	**(B)**
0	2.50	2.50
15	2.20	1.10
30	2.12	0.99
60	2.11	0.80
120	2.11	0.40

(a) Outline briefly the way such an investigation should be carried out to ensure that the results of different experiments are comparable. *(3 marks)*

(b) On squared paper, plot a graph of the **uptake** of radioactive potassium from the solutions in A and B. *(4 marks)*

(c) Account for the shapes of the **two** curves and for any differences between them. *(4 marks)*

It is believed that in maize mineral ions move passively across the root cortex to the endodermis in one of two ways:

either through protoplasmic connections between adjacent cells;

or between adjacent cell walls.

(d) Explain in outline how, using radioactive ions, you might find out which of these methods operates in maize. *(2 marks)*

(Total 13 marks)

NEAB June 1993, Paper IIA, No. 3

12. Read the passage below and answer the questions which are based on it. (Adapted from 'Plants and Water' by James Sutcliffe.)

Without water, life as we know it could not exist. Living organisms originated in an aqueous environment and they are, despite millions of years of evolution, still absolutely dependent on
5 water in a variety of ways. In plants, water participates in a number of chemical reactions including photosynthesis. Much of the water in plants occurs in sap vacuoles, where it is largely responsible for maintaining the rigidity of cells
10 and hence the rigidity of the plant as a whole. Water provides a medium for the movement of the soluble products of photosynthesis and for the transport of minerals.

In addition, the structure of plants is
15 profoundly affected by the amount of water in the external environment. Aquatic plants (hydrophytes) show several adaptations to their watery environment. In some, the leaves and stems develop as thread-like structures. The
20 cuticle is often very thin and the stomata are usually absent or non-functional. Many hydrophytes show an enormous development of intercellular spaces. Water-conducting and mechanical supporting tissues are considerably
25 reduced and tend to be concentrated towards the centre of the stem.

The habit of xerophytic plants is very different from that of hydrophytes. Xerophytes are able to tolerate, at least temporarily, very dry conditions
30 and consequently show quite different structural adaptations. Their root systems develop in such a way as to ensure an adequate supply of water from soils with a low water availability. In addition, the aerial parts of the plants exhibit a
35 number of features which minimize water loss by evaporation. Such features include the possession of a thick cuticle and a reduced leaf area. In some cacti, in fact, the leaves are reduced to spines.

Although water is vitally important to plants,
40 the amount retained, except in submerged plants, is only a small fraction of the total absorbed by the roots. By far the greater proportion is transported to the aerial parts from which it
45 evaporates into the surrounding air. This evaporative loss of water from plants – transpiration – is a process which has often been described as a 'necessary evil'.

(a) Select, from the passage, three ways in which plants are 'still absolutely dependent on water' (lines 4 and 5). *(2 marks)*

(b) What term may be used to describe the state of 'rigidity of cells' (line 9)? *(1 mark)*

(c) Select a phrase from the passage which describes the process of translocation. *(1 mark)*

(d) State one advantage afforded to those aquatic plants which have 'thread-like structures' (line 19). *(1 mark)*

(e) Explain why 'the stomata are usually absent or non-functional' in aquatic plants (lines 20 and 21). *(1 mark)*

(f) What advantage is gained by aquatic plants having 'an enormous development of intercellular spaces' (lines 23 and 24)? *(1 mark)*

(g) What tissue in plants performs the function of water-conduction and mechanical support (lines 23 and 24)? *(1 mark)*

(h) State **two** ways, other than those mentioned in the passage, which enable xerophytic plants to minimize water loss. *(2 marks)*

(i) Describe **two** adaptations of the root systems of plants which grow in soils with a low water availability and explain how **each** assists uptake of available water. *(2 marks)*

(j) The author states that transpiration has 'often been described as a necessary evil' (lines 47 and 48).
  (i) In what way could transpiration be described as 'necessary'? *(1 mark)*
  (ii) For what reason could transpiration be described as 'evil'? *(1 mark)*

(k) Use the information in paragraph 2 to draw a labelled transverse section of a hydrophyte stem. *(3 marks)*
(Total 17 marks)

SEB May 1990 Higher Grade, Paper II, No. 13

2. You are advised to read the whole of this question before starting your answer. It is not assumed you will have previously studied the various hypotheses of stomatal movement.

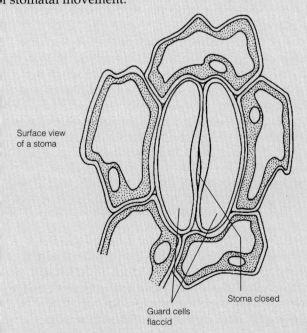

Surface view of a stoma

Guard cells flaccid

Stoma closed

(a) Stomata open when the guard cells take in water. With reference to the diagram and your knowledge of the forces controlling water movement in cells, explain how an increase in cell sap concentration of the guard cells leads to stomatal opening. *(3 marks)*

(b) Three hypotheses that have been put forward to explain the increase in cell sap concentration leading to stomatal opening are given below.

Hypothesis 1. In the light the acidic gas carbon dioxide is removed and starch is converted to glucose-1-phosphate.

$$\text{Starch + inorganic phosphate} \underset{\text{acid}}{\overset{\text{alkaline}}{\rightleftharpoons}} \text{glucose-1-phosphate}$$

  (i) As water potential depends upon the total number of molecules and ions in solution, explain why this reaction as presented above is unlikely to cause stomata to open. *(3 marks)*

Hypothesis 2. Guard cells have chloroplasts and may produce sugars in the light.
However, it is also suggested that these cells do not contain all of the enzymes required for the light independent (dark) stage of photosynthesis.

  (ii) What are the products of the light *dependent* stage of photosynthesis? *(3 marks)*

  (iii) What would be the implication of guard cells being able to carry out this stage but not the light independent stage? *(3 marks)*

Hypothesis 3. Potassium ions accumulate in the guard cells in the light and this causes the stomata to open.

  (iv) Using your knowledge of cell membrane structure, explain how potassium ions may enter the cell. *(3 marks)*

  (v) If movement of potassium ions causes stomata to open and close, explain why it may be advantageous for guard cells to possess chloroplasts even if the dark stage of photosynthesis is not completed. *(2 marks)*

(c) Define the term *hypothesis*. *(1 mark)*
(Total 21 marks)

O&CSEB June 1994, Paper I, No 3

See *Further questions*, pages 614–20, questions 1, 3, 10, 13.

23 Osmoregulation and excretion

23.1 Introduction

Animals living under terrestrial conditions tend to lose water by evaporation. Those living in surroundings more concentrated (hypertonic) than their tissue fluids lose water by osmosis. Animals in a dilute (hypotonic) environment have to face the problem of water flooding into the body by osmosis. Both problems have led to structural and physiological adaptations in order to maintain the balance of water and solutes. These homeostatic processes are termed **osmoregulation**. Homeostasis is the subject of Chapter 25.

The complex chemical reactions which occur in all living cells produce a range of waste products which must be eliminated from the body in a process known as **excretion**. Most nitrogenous waste comes from the breakdown of excess proteins which cannot be stored in the body. The form of these excretory products is influenced partly by the availability of water for their excretion. Animals living under conditions of water shortage cannot afford to lose large volumes of water in order to remove their nitrogenous waste. If water is plentiful it may be used to facilitate excretion.

It is important not to confuse the terms excretion, secretion and elimination. **Excretion** is the expulsion from the body of the waste products of metabolism. **Secretion** is the production by the cells of substances useful to the body, such as digestive juices or hormones. **Elimination**, or **egestion** (see Section 15.4.7), is the removal of undigested food and other substances which have never been involved in the metabolic activities of cells.

23.2 Excretory products

All animals produce carbon dioxide as a waste product of aerobic respiration and the elimination of this is dealt with in Chapters 20 and 21. Other excretory products include bile pigments, water and mineral salts. However, in this section we shall concentrate on the variety of nitrogenous excretory products released by animals. There are three main waste products of nitrogenous metabolism: ammonia, urea and uric acid. No animal excretes one of these to the exclusion of the others but the predominance of one over the others is determined by three factors:

1. The production of enzymes necessary to convert ammonia into either urea or uric acid.

2. The availability of water in the habitat for the removal of the nitrogenous excretory material.

3. The animal's ability to control water loss or uptake by the body.

Many aquatic animals excrete mainly ammonia and are called **ammoniotelic**. Other aquatic animals and some terrestrial forms excrete predominantly urea and are said to be **ureotelic**. The remaining terrestrial animals are **uricotelic**, excreting mainly uric acid.

23.2.1 Ammonia

Ammonia is derived from the breakdown of proteins and nucleic acids in the body. Ammonia is very toxic and is never allowed to accumulate within the body tissues or fluids. It is extremely soluble and diffuses readily across cell membranes. In spite of its toxicity it is the main excretory product of marine invertebrates and all freshwater animals.

23.2.2 Urea

Urea, $CO(NH_2)_2$, is formed by the combination of two molecules of ammonia with one of carbon dioxide.

However, its synthesis in living tissues is much more complex than this simple equation suggests. Urea is produced by a cyclic process known as the **urea** or **ornithine cycle**, details of which are given in Section 25.7.2. Urea is much less toxic than ammonia and, although it is less soluble, less water is needed for its elimination because the tissues can tolerate higher concentrations of it.

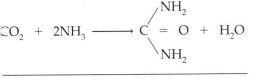

23.2.3 Uric acid

Uric acid is a more complex molecule than urea; it is a purine in the same group as adenine and guanine. Like urea, it involves the expenditure of quite considerable energy in its formation, but this is outweighed by the advantages it confers. Uric acid is virtually insoluble in water and is non-toxic. It requires very little water for its removal from the body and it is therefore a suitable product for animals living in arid conditions, e.g. terrestrial reptiles and insects. Containing little, if any, water, its storage within organisms does not greatly increase their mass. This is an advantage to flying organisms, e.g. birds and insects. It is removed as a solid pellet or thick paste.

23.3   Osmoregulation and excretion in animals

23.3.1 Protozoa

All protozoa are small unicells with a relatively large surface-area to volume ratio. They are separated from their surroundings only by a plasma membrane through which

APPLICATION

Gout

Gout is a disorder which arises when the blood contains increased levels of uric acid. An excess of uric acid can be caused either by increased production or by an impairment of kidney function. The disorder tends to run in families and is far more common in men. The risk of attack is increased by high alcohol intake, the consumption of red meat and some other foods, and by obesity.

When its concentration in the blood is excessive uric acid crystals may form in various parts of the body, especially in the joints of the foot, knee and hand, causing intense pain and inflammation known as gouty arthritis. Crystals may also form in the kidneys as kidney stones.

Drugs may be prescribed to treat an attack of gout or to prevent recurrent attacks that could lead to deformity of the affected joints and kidney damage, but changes in diet and a reduction in alcohol consumption may be an important part of treatment.

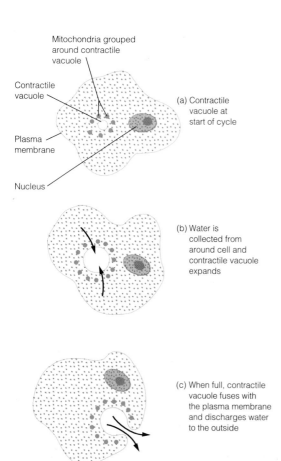

Fig. 23.1 Osmoregulation in a protozoan, e.g. Amoeba (arrows indicate flow of water)

carbon dioxide and the waste nitrogenous product, ammonia, can readily diffuse. Protozoa living in sea water are isotonic with their environment, have no net gain or loss of water and so do not have any osmoregulatory structures. Freshwater protozoa, on the other hand, are hypertonic to their surroundings and water constantly enters the cell by osmosis. If the protozoan cell is not to increase its volume this water must be removed by the constant filling and bursting at the surface of the **contractile vacuole**. These vacuoles are almost always found in freshwater protozoa, although their size, position and structure is extremely variable. The water which enters the cell by osmosis must be moved into the contractile vacuole across its bounding membrane. It is likely that small vesicles collect water from within the cell. These then pump ions back into the cytoplasm so that this is not lost. The vesicles then fuse with the contractile vacuole which expands as a consequence. This process requires the expenditure of energy and it is significant that there are always large numbers of mitochondria around the contractile vacuoles. These mitochondria are the site of many of the respiratory enzymes and it has been found that treatment with cyanide, which prevents oxidative phosphorylation by inhibiting the cytochrome system, stops the activity of the contractile vacuole and leads to swelling of the cell. Ammonia may be found in the water expelled by the contractile vacuole but any excretory rôle is probably secondary since ammonia diffuses so readily through the cell membrane.

23.3.2 Insects

Insects are among the most successful terrestrial animals. The problem of preventing water loss by evaporation is overcome by the waterproof waxy cuticle on the exoskeleton and a tracheal respiratory system (Chapter 20). The problem of water loss by excretion is overcome by the production of uric acid. Uric acid is

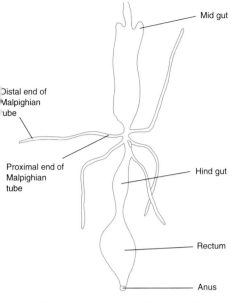

Fig. 23.2 Position of the Malpighian tubules

concentrated in the **Malpighian tubules**. These form a bunch of blind-ending tubes which project into the blood-filled body cavity (haemocoel) from the junction of the mid gut and hind gut (see Fig. 23.2).

Insect cells release uric acid into the blood where it combines with potassium and sodium hydrogen carbonates and water to form potassium and sodium urate, carbon dioxide and water. Potassium and sodium urate are actively taken up by the Malpighian tubules and the water then enters by osmosis. As the soluble potassium and sodium urate pass down the tubules, they combine with carbon dioxide and water (from respiration) to form hydrogen carbonates and uric acid. At the proximal end of the Malpighian tubule the walls have many microvilli and it is here that hydrogen carbonates are actively reabsorbed into the haemocoel, lowering the osmotic pressure within the Malpighian tubule, so that water passes out by osmosis. There is a subsequent lowering of pH and concentration of the uric acid which precipitates out as crystals. These crystals pass into the rectum where they mix with the waste materials from the digestive system. Further water reabsorption takes place through the rectal epithelium, so that a very concentrated excretory product is eliminated from the body. These processes are illustrated in Fig. 23.3.

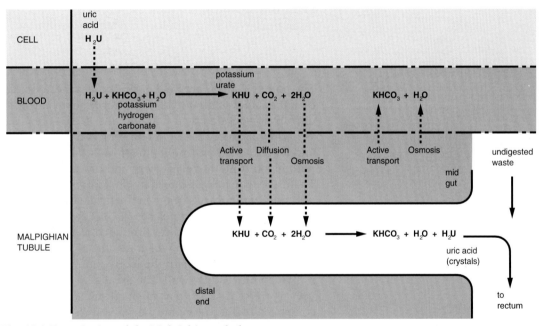

Fig. 23.3 Functioning of the Malpighian tubule

23.3.3 Fish

Freshwater fish have body fluids which are hypertonic to their surroundings and they are therefore subject to the osmotic uptake of water. Most of this water enters the fish through the highly permeable gills, the body itself being covered with impermeable scales and mucus. The excess water is removed from the body by the kidneys which have many large glomeruli. A large volume of glomerular filtrate is produced from which salts are selectively reabsorbed into the blood, resulting in the production of copious amounts of very dilute urine. Since water is plentiful, the nitrogenous excretory product is ammonia. Some of this ammonia is excreted by the kidneys, but as it is so soluble and diffuses readily, most of it is expelled by the gills.

Marine bony fish have body fluids which are hypotonic to sea water. There is little movement of water through the scale-covered body but the fish is liable to water loss by osmosis across the highly permeable gills. In order to maintain sufficient water inside the body these fish drink sea water and secretory cells in the gut actively absorb the salts and transfer them to the blood. The chloride secretory cells in the gills remove the sodium and chloride ions from the blood and other ions like sulphate and magnesium are removed by the kidney. The kidneys of marine bony fish are adapted to produce very small amounts of urine since the animal already has a problem of excessive water loss.

23.4 The mammalian kidney

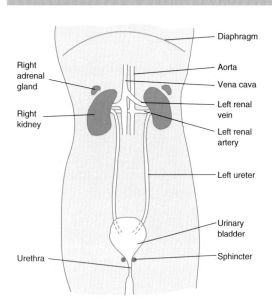

Fig. 23.4 Position of the kidneys in humans

In all vertebrates the main organ of nitrogenous excretion is the kidney. Kidneys are composed of a number of basic units called **nephrons**. In mammals, these nephrons are particularly numerous, with long tubules for water reabsorption.

23.4.1 Gross structure of the kidney

The paired kidneys are held in position in the abdominal cavity by a thin layer of tissue called the peritoneum and they are usually surrounded by fat. In humans, each kidney is about 7–10 cm long and 2.5–4.0 cm wide, packed with blood vessels and an estimated one million nephrons. Each kidney is supplied with blood from the renal artery and drained by a renal vein. The urine which is produced by the kidney is removed by a ureter for temporary storage in the urinary bladder. A ring of muscle called a sphincter closes the exit from the bladder. Sense cells in the bladder wall are stimulated as the bladder fills, triggering a reflex action which results in relaxation of the bladder sphincter and simultaneous contraction of the smooth muscle in the bladder wall. The expulsion of urine from the body via the urethra is known as **micturition**. Although micturition is controlled by the autonomic nervous system, humans learn to control it by voluntary nervous activity.

Within each kidney there are a number of clearly defined regions. The outer region, or **cortex**, mainly comprises Bowman's capsules and convoluted tubules with their associated blood supply. This gives the cortex a different appearance from the inner **medulla** with its loops of Henle, collecting ducts and blood vessels. These structures in the medulla are in groups known as **renal pyramids** and they project into the **pelvis** which is the expanded portion of the ureter.

23.4.2 Structure of the nephron

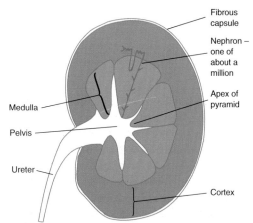

Fig. 23.5 Mammalian kidney to show position of a nephron (LS)

The main regions of the mammalian nephron are shown in Fig. 23.6, with details in Figs. 23.7 and 23.8. Basically it comprises a

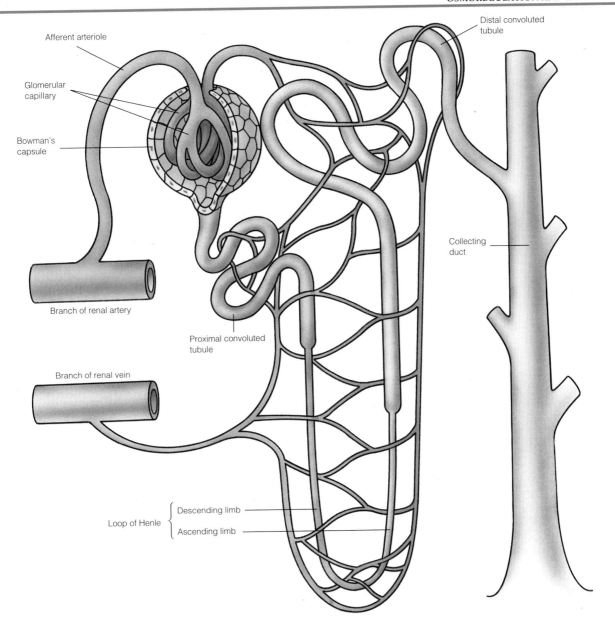

Fig. 23.6 Regions of the nephron

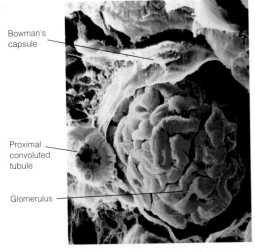

Scanning EM of nephron showing glomerulus

glomerulus and a long tubule with several clearly defined regions. The **glomerulus** is a mass of blood capillaries which are partially enclosed by the blind-ending region of the tubule called the **Bowman's capsule**. The blood supply to the glomerulus is from the afferent arteriole of the renal artery; blood leaves the glomerulus via the narrower efferent arteriole. The inner, or visceral, layer of the Bowman's capsule is made up of unusual cells called **podocytes** (see Fig. 23.7), while the outer layer is unspecialized squamous epithelial cells. The remaining regions of the nephron are the proximal convoluted tubule, whose surface-area is increased by the presence of microvilli, the descending and ascending limbs of the loop of Henle, which function as a counter-current multiplier, the distal convoluted tubule and the collecting duct.

473

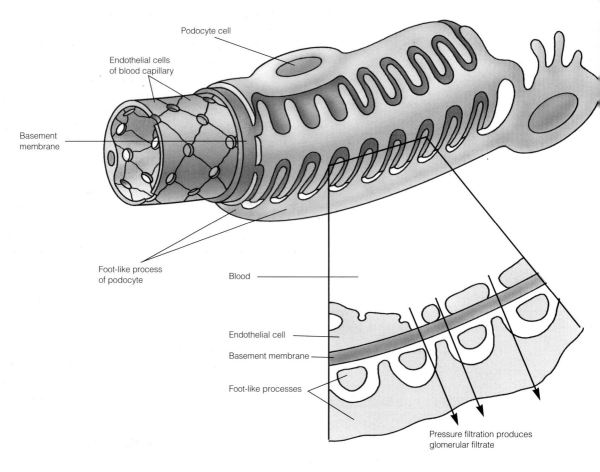

Fig. 23.7 Podocyte

23.4.3 Functions of the nephron

Apart from being organs of nitrogenous excretion, the kidneys also play a major rôle in maintaining the composition of the body fluids in a more or less steady state, in spite of wide fluctuations in water and salt uptake. This dual function of excretion and osmoregulation is best studied by a detailed consideration of the functioning of one nephron, the main regions of which are shown in Fig. 23.6.

Ultrafiltration in the Bowman's capsule
The cup-shaped Bowman's capsule encloses a mass of capillaries, the **glomerulus**, originating from the afferent arteriole of the renal artery. The capillary walls are made up of single layer of endothelial cells perforated by pores about 0.1 μm in diameter. The endothelium is closely pressed against the basement membrane which in places is the only membrane between the blood and the cavity of the Bowman's capsule (see Fig. 23.6). The blood pressure in the kidneys is higher than in other organs. This high pressure is maintained because in each Bowman's capsule the afferent arteriole has a larger diameter than the efferent arteriole. As a result of this pressure, substances are forced through the endothelial pores of the capillary, across the basement membrane and into the Bowman's capsule by

ultrafiltration. The glomerular filtrate contains substances with a relative molecular mass (RMM) less than 68 000, e.g. glucose, amino acids, vitamins, some hormones, urea, uric acid, creatinine, ions and water. Remaining in the blood, along with some water, are red blood cells, white cells, platelets and plasma proteins which are too large to pass the filter provided by the basement membrane. Further constriction of the efferent arteriole in response to hormonal and nervous signals results in an increased hydrostatic pressure in the glomerulus and substances with an RMM greater than 68 000 may pass into the glomerular filtrate. This filtering process is extremely efficient. The glomerular filtrate passes from the Bowman's capsule along the kidney tubule (nephron). As it does so, the fluid undergoes a number of changes, since the urine excreted has a very different composition from the glomerular filtrate. These differences are brought about primarily by selective reabsorption of substances useful to the body. The urine when compared with the glomerular filtrate will contain, for example, less glucose, amino acids and water and a relatively higher percentage of urea and other nitrogenous waste products.

Ultrafiltration is a passive process and selection of substances passing from the blood into the glomerular filtrate is made entirely according to relative molecular mass. Both passive and active processes are involved in the selective reabsorption of substances from the nephron. The composition is further altered by the active secretion of substances, such as creatinine, from the blood into the tubule. Without selective reabsorption humans would produce about 180 dm³ of urine per day whereas the actual volume produced is approximately 1.5 dm³.

Proximal convoluted tubule
This is the longest region of the nephron. It comprises a single layer of epithelial cells, with numerous microvilli forming a brush border (see Fig. 23.8). The base of each cell is convoluted where it is adjacent to a blood capillary and there are numerous intercellular spaces. Another notable feature of these cells is the presence of large numbers of mitochondria providing the ATP necessary for active transport. These cells are ideally adapted for

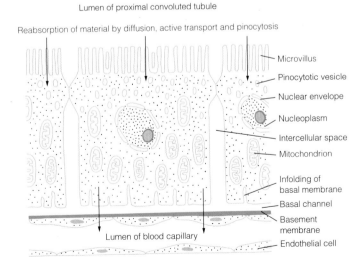

Lumen of proximal convoluted tubule

Reabsorption of material by diffusion, active transport and pinocytosis

- Microvillus
- Pinocytotic vesicle
- Nuclear envelope
- Nucleoplasm
- Intercellular space
- Mitochondrion
- Infolding of basal membrane
- Basal channel
- Basement membrane
- Endothelial cell

Lumen of blood capillary

Fig. 23.8 Detail of cells from the wall of the proximal convoluted tubule

reabsorption and over 80% of the glomerular filtrate is reabsorbed here, including all the food substances and most of the sodium chloride and water. Amino acids, glucose and ions diffuse into the cells of the proximal convoluted tubule and these are actively transported into the intercellular spaces from where they diffuse into the surrounding capillaries. The constant removal of these substances from the cells of the convoluted tubule causes others to enter from the lumen of the tubule by diffusion. The active uptake of sodium accompanied by appropriate anions, e.g. chloride, raises the osmotic pressure in the cells and water enters them by osmosis. About half the urea present in the tubular filtrate also returns to the blood by diffusion. Proteins of small molecular mass which may have been forced out of the blood in the Bowman's capsule are taken up at the base of the microvilli by pinocytosis. As a result of all this activity, the tubular filtrate is isotonic with blood in the surrounding capillaries.

Loop of Henle

It is the presence of the loop of Henle which enables birds and mammals to produce urine which is hypertonic to the blood. The concentration of the urine is directly related to the length of the loop of Henle. It is short in semi-aquatic mammals which have a correspondingly narrow medulla, and extremely long in desert-dwelling mammals such as the desert rat *Dipodomys* which therefore has a wide medulla. *Dipodomys* produces a small volume of urine, ten times more concentrated than that produced in large volumes by a beaver. The loop of Henle is made up of two regions, the descending limb which has narrow walls readily permeable to water and the wider ascending limb with thick walls which are far less permeable to water.

The loop of Henle operates as a **counter-current multiplier** system. So how exactly does this work? Consider Fig. 23.9. Sodium and chloride ions are actively pumped out of the ascending limb creating a high solute concentration in the interstitial region. Normally water would follow, being drawn out osmotically. However the walls of the ascending limb are relatively impermeable to water and so little if any escapes. On the other hand the descending loop is highly permeable to water and so water is drawn from it osmotically. This water is carried away by the blood in the vasa recta. As glomerular filtrate enters the descending loop it progressively loses water and so becomes more concentrated. It reaches its maximum concentration at the tip of the loop because as it moves up the ascending limb, ions are removed making it less concentrated.

It might be thought that as the filtrate in the descending limb becomes more concentrated due to the reabsorption of water, osmosis might cease. However the surrounding fluid also becomes more concentrated, ensuring that an osmotic gradient is maintained right down to the tip of the loop of Henle.

In the same way as water is drawn from the descending limb, so it is too from the collecting duct, which runs alongside the loop in the medulla of the kidney. In this way the urine becomes progressively more concentrated (hypertonic) as it moves out of the nephron. The water which is drawn out passes into the blood of the vasa recta which is both slow-flowing and freely permeable, two factors which aid the uptake of water.

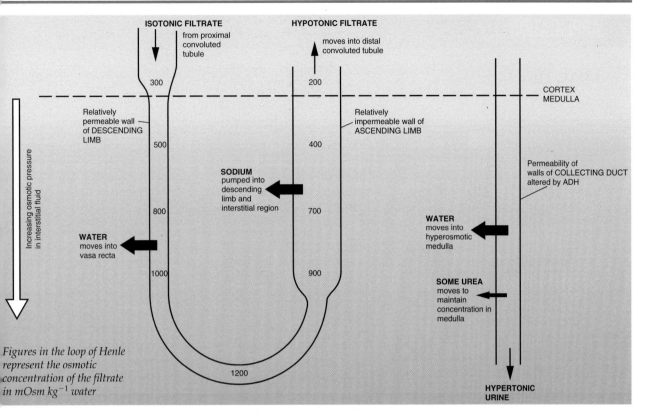

Figures in the loop of Henle represent the osmotic concentration of the filtrate in mOsm kg⁻¹ water

Fig. 23.9 Counter-current multiplier of the loop of Henle

Distal convoluted tubule

The cells in this region are very similar to those of the proximal convoluted tubule, having a brush border and numerous mitochondria. The permeability of their membranes is affected by hormones (see Section 23.5) and so precise control of the salt and water balance of the blood is possible. The distal convoluted tubule also controls the pH of the blood, maintaining it at 7.4. The cells of the tubule combine water and carbon dioxide to form carbonic acid. This then dissociates into hydrogen ions and hydrogen carbonate ions. The absorption of these hydrogen carbonate ions into the blood raises its pH to compensate for the lowering which results from the production of hydrogen ions during metabolic processes. The hydrogen ions from the dissociation of carbonic acid are pumped into the lumen of the distal tubule. In the lumen hydrogenphosphate ions (HPO_4^{2-}) combine with these hydrogen ions to form dihydrogenphosphate ions ($H_2PO_4^-$) which are then excreted in the urine. The acidic effects of the hydrogen ions is therefore buffered by these hydrogenphosphate ions. The events are summarized in Fig. 23.10.

Not only are hydrogen ions (protons) pumped into the tubule, potassium ions (K^+), ammonium ions (NH_4^-) and certain drugs are also pumped in as a means of controlling blood pH or removing unwanted material. This active removal of substances also takes place in the proximal convoluted tubule and is called **tubular secretion**.

Collecting duct

The permeability of the walls of the collecting duct, like the permeability of those of the distal convoluted tubule, is affected by hormones. This hormonal effect, together with the hypertonic

APPLICATION

Dialysis

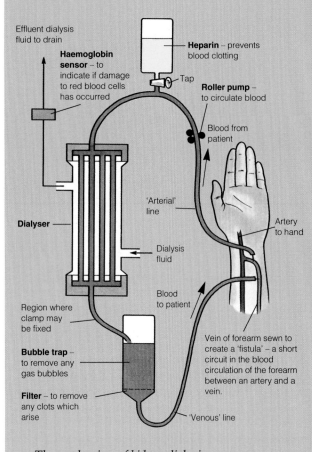

Effluent dialysis
fluid to drain

**Haemoglobin
sensor** – to
indicate if damage
to red blood cells
has occurred

Heparin – prevents
blood clotting

Tap

Roller pump –
to circulate blood

Blood from
patient

'Arterial'
line

Dialyser

Dialysis
fluid

Artery
to hand

Region where
clamp may
be fixed

Blood
to patient

Bubble trap –
to remove any
gas bubbles

Vein of forearm sewn to
create a 'fistula' – a short
circuit in the blood
circulation of the forearm
between an artery and a
vein.

Filter – to remove
any clots which
arise

'Venous' line

The mechanism of kidney dialysis

Kidneys may fail as a result of damage or infection. Upon the loss of one kidney the remaining one will adapt to undertake the work of its partner, but the loss of both is inevitably fatal. Survival depends upon either regular treatment on a kidney machine or a transplant. A kidney machine carries out **dialysis**, a process in which the patient's blood flows on one side of a thin membrane while a solution, called the dialysate, flows in the opposite direction on the other side. This counter-current flow ensures the most efficient exchange of material across the membrane (Section 20.2.4). As the membrane is permeable to small molecules such as urea, this waste product will diffuse from the blood where it is relatively highly concentrated to the dialysate where its concentration is lower. To prevent useful substances like glucose and salts, which are also highly concentrated in blood, diffusing out, the dialysate's composition is the same as that of normal blood. This means that any substance which is in excess, e.g. salts, will also diffuse out until they are in equilibrium with the dialysate. Large molecules such as blood proteins are too large to cross the membrane and there is therefore no risk of them being lost to the dialysate.

A patient's blood needs to pass through the kidney machine many times to ensure the removal of all wastes. Thus, it is necessary for dialysis to take place for up to ten hours every few days. While wastes accumulate in the blood when the patient is away from the machine, adherence to a strict diet ensures they do not build up to a dangerous level before the next treatment.

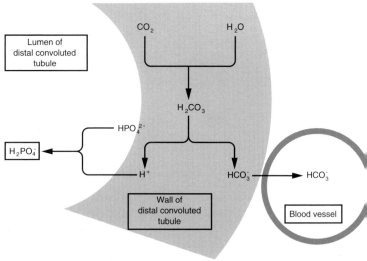

Fig. 23.10 Control of pH

interstitial fluids built up by the loop of Henle in the medulla, determine whether hypotonic or hypertonic urine is released from the kidney.

If the walls of the collecting duct are water-permeable, water leaves the ducts to pass into the hyperosmotic surroundings and concentrated urine is produced. If the ducts are impermeable to water the final urine will be less concentrated. Hormonal control of the permeability of the walls of the collecting duct to water will be considered in the next section.

The mechanism by which urine is concentrated is illustrated in Fig. 23.9.

23.5 Hormonal control of osmoregulation and excretion

If the kidney is to regulate the amount of water and salts present in the body, very precise monitoring systems are required. The two hormones ADH and aldosterone are particularly important in this respect.

23.5.1 Antidiuretic hormone (ADH)

Antidiuretic hormone (ADH) affects the permeability of the distal convoluted tubule and collecting duct.

A rise in blood osmotic pressure may be caused by any one of, or combination of, three factors:

1. Little water is ingested.

2. Much sweating occurs.

3. Large amounts of salt are ingested.

The rise in blood osmotic pressure is detected by **osmoreceptors** in the hypothalamus and results in nerve impulses passing to the posterior pituitary gland which releases ADH. ADH increases the permeability of the distal convoluted tubule and collecting duct to water. This water passes into the hyperosmotic medulla and a more concentrated (hypertonic) urine is released from the kidney.

ADH also increases the permeability of the collecting duct to urea which passes into the medulla, increasing the osmotic concentration and causing more water to be lost from the descending loop of Henle. If the osmotic pressure of the blood falls owing to

1. large volumes of water being ingested

2. little sweating

3. low salt intake

then ADH production is inhibited and the walls of the distal convoluted tubule and collecting duct remain impermeable to water and urea. As a result, less water is reabsorbed and

APPLICATION

Kidney transplant

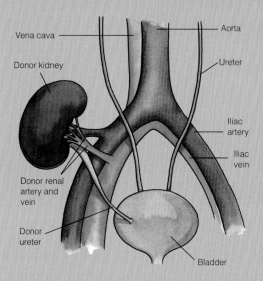

Vena cava

Donor kidney

Donor renal artery and vein

Donor ureter

Aorta

Ureter

Iliac artery

Iliac vein

Bladder

While the kidney machine is an invaluable life-saver, it has many drawbacks. Not only does a considerable time have to be spent connected to the machine, a rigid diet has to be maintained and there is always the risk of anaemia, infection or bone disease. The cost of each treatment is such that there are insufficient machines for all who need them. The preferred solution for many is therefore a **kidney transplant**: a failed kidney must be replaced by a healthy one from a human donor. The donated kidney must be matched so that it is as similar as possible to the failed one since this reduces the chance of it being rejected by the recipient's immune system. A close relative is more likely to have a compatible kidney and therefore live donors are often used. Here a person donates a kidney, safe in the knowledge that he/she can live a normal life with the remaining one. The donor is, however, vulnerable since a disease in the remaining kidney could make dialysis necessary, and there is always a risk in any operation. Against this is the satisfaction of allowing someone to lead a near normal life, free from all the constraints dialysis imposes. However, less than 15% of transplants come from live donors.

The remaining transplants involve the use of healthy kidneys from people who die as a result of other causes. A road accident victim may, for example, provide two functional kidneys. The kidney must be removed within an hour of death, cooled to delay deterioration, and be transplanted within 24 hours. As permission is needed to remove any organ, and as such a request made of a distressed relative is difficult, many people carry donor cards. Their owners sign to say that upon their death they consent to the use of the kidneys (and often other organs) being used for transplants. The card must of course be carried by the owner at all times to avoid any delay in removing organs.

Transplanted kidneys are implanted in the lower abdomen near the groin. The renal vessels of the donor kidney are attached to the iliac vessels of the recipient and the ureter is implanted into the bladder. The failed kidneys are left in place. The survival rate for kidney transplants is high because the operation is relatively straight-forward and because the kidney has a simple vascular supply.

There are around 2000 kidney transplants in the UK each year, with around twice that number of people awaiting a suitable donor.

hypotonic urine is released. Anyone who is unable to produce sufficient levels of ADH will produce large volumes of very dilute urine, whatever their diet. This condition is termed **diabetes insipidus**. ADH is also referred to in Section 26.2. Fig. 23.11 illustrates the regulation of ADH production.

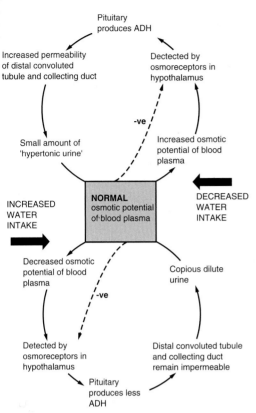

Pituitary
produces ADH

Increased permeability
of distal convoluted
tubule and collecting duct

Dectected by
osmoreceptors in
hypothalamus

−ve

Small amount of
'hypertonic urine'

Increased osmotic
potential of blood
plasma

INCREASED
WATER
INTAKE

NORMAL
osmotic potential
of blood plasma

DECREASED
WATER
INTAKE

Decreased osmotic
potential of blood
plasma

Copious dilute
urine

−ve

Detected by
osmoreceptors in
hypothalamus

Distal convoluted tubule
and collecting duct
remain impermeable

Pituitary
produces less
ADH

Fig. 23.11 Regulation of ADH production

23.5.2 Aldosterone

Aldosterone is the hormone responsible for maintaining a more or less constant sodium level in the plasma and it has a secondary effect on water reabsorption. The control of aldosterone production is very complex.

Any loss of sodium which causes a decrease in blood volume causes a group of secretory cells lying between the afferent arteriole and the distal convoluted tubule to release an enzyme. These cells are known as the **juxtaglomerular complex** and the enzyme they release is **renin**. Renin causes a plasma globulin produced by the liver to form the hormone **angiotensin**. It is this hormone which stimulates the release of aldosterone from the adrenal cortex.

Aldosterone causes sodium ions to be actively taken up from the glomerular filtrate into the capillaries which surround the tubule. This uptake will be accompanied by an osmotically equivalent volume of water, thus restoring the sodium level of the plasma and the volume of the blood.

Further details of aldosterone are given in Section 26.5.

23.6   Questions

1. Describe nitrogenous excretion in animals under the following headings:

    (a) The source of waste nitrogen and the need for its removal. *(4 marks)*

    (b) The removal of waste nitrogen by freshwater animals and by terrestrial insects. *(5 marks)*

    (c) The way in which the structure and function of the mammalian kidney is adapted for the **efficient** removal of waste nitrogen.

(11 marks)

(Total 20 marks)

WJEC June 1993, Paper A1, No. 2

2. Investigations were carried out into the separate effects of sodium chloride and of antidiuretic hormone (ADH) on urine production.
The rate of urine production was measured over a period of 30 minutes. Five minutes after measurements began an intravenous injection was given. On one occasion $10 \, cm^3$ of 2.5% sodium chloride solution was injected; on a separate occasion 1 milliunit of ADH was injected. The results are shown in the table below.

Time/min	Rate of urine production/$cm^3 \, min^{-1}$	
	Sodium chloride injected	ADH injected
0	6.3	5.2
5 injection given	7.0	4.8
10	0.7	2.8
15	0.8	0.5
20	1.2	1.7
25	1.5	3.1
30	1.9	4.0

Adapted from Verney, Proc. Roy. Soc. B (1947)

    (a) Plot the data on graph paper. *(5 marks)*

    (b) Describe and explain the effects of the ADH injection on urine production (i) during the period 5–15 minutes, and (ii) during the period 15–25 minutes. *(4 marks)*

    (c) Discuss the relationship between the two curves on your graph, suggesting reasons for any similarities and differences noted.

(5 marks)

(Total 14 marks)

ULEAC June 1993, Paper III, No. 8

3. The table gives the thickness of the medulla in relation to the rest of the kidney in a number of mammals. The maximum urine concentration for each mammal is also given.

Mammal	Relative thickness of medulla	Maximum urine concentration/ arbitrary units
Beaver	1.0	52
Pig	1.3	110
Human	2.6	140
Rat	5.2	300
Kangaroo rat	7.8	550
Animal X	9.8	940

    (a) Explain the relation between urine concentration and the relative thickness of the medulla. *(3 marks)*

    (b) Suggest the natural habitat of animal **X**.

(1 mark)

(Total 4 marks)

AEB June 1992, Paper I, No. 1

4. The diagram below shows a single nephron from a mammalian kidney together with some associated blood vessels.

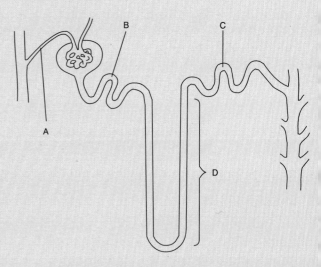

    (a)   (i) Name a substance present in the fluid in A but absent from the fluid in B, and give a reason for its absence from B.

(2 marks)

        (ii) Name a substance present in the fluid in B but absent from the fluid in C, and give a reason for its absence from C.

(2 marks)

    (b) Describe how region D functions as a counter current multiplier. *(3 marks)*

(Total 7 marks)

ULEAC January 1994, Paper III, No. 2

. The diagram is of a kidney tubule showing some of
ne pressure gradients that are involved in the
novement of fluids from blood to the glomerular
ltrate within the Bowman's capsule. The numbers
epresent hydrostatic pressure in kPa.

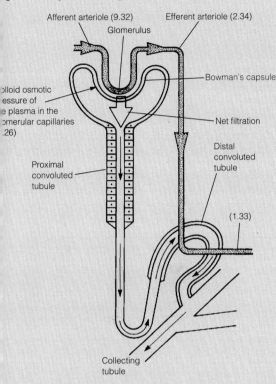

(a) From the figures given, calculate the *filtration
pressure* i.e. the net force moving fluid out of
the glomerulus. Show your working.
(*2 marks*)

Filtration pressure, and therefore the rate at
which fluid passes from the glomerulus into
Bowman's capsule, varies with blood pressure.

(b) If the renal artery was restricted with a
clamp so that the blood pressure in the
afferent artery fell to about 5.3 kPa, what
would happen to the production of urine?
Explain your answer. (*3 marks*)

Caffeine causes vasodilation of the afferent
arterioles.

(c) What effect would caffeine have on the rate
at which fluid passes from the glomerulus
into Bowman's capsule? (*2 marks*)

The table shows the percentage composition of
blood plasma and urine for four substances.

Component	Blood plasma/%	Urine/%
Water	90	90
Plasma proteins	8	0
Glucose	0.1	0
Urea	0.03	2

(d) Explain why
(i) there are no plasma proteins in urine;
(ii) there is no glucose in urine;
(iii) urea concentration is greater in urine
than in blood plasma. (*6 marks*)
(e) The relative length of the loop of Henle
differs greatly between different species of
mammals. State the type of environment in
which you might expect to find species with
relatively long loops, giving reasons for your
answer. (*3 marks*)
(*Total 16 marks*)

UCLES (Modular) June 1992, (Transport, regulation
and control), No. 4

6. (a) Name **two** physical features of body fluids
which are regulated by the mammalian
kidney. (*2 marks*)
(b) The graph below shows the urine production
of a healthy man at half hour intervals
during a period of four hours following
drinking 1 litre of fluid. On day one (solid
line) the fluid was distilled water and on day
two (broken line) it was a salt solution
having a water potential equal to that of
human blood plasma.

Response to the drinking of 1 litre of fluid at time = 0.0 h

Urine accumulated after each half hour/cm^3

● - -● salt solution
○──○ water

(i) Judging by the volume of urine
produced in the half hour period before
the fluid was drunk, estimate what the
total output of urine over a period of 2
hours would have been if no fluid had
been taken.
(ii) Using a similar method, estimate how
much **additional** urine was produced
in the 2 h following drinking.
(1) water
(2) salt solution. (*3 marks*)
(iii) What can you conclude **from these
observations** concerning the factors
controlling urine production in
humans? (*2 marks*)

(iv) In the light of the patterns of urine production in the graphs, describe the solute concentration of the urine relative to that of normal urine (e.g. higher or lower) at the following times. In each case give a brief explanation of your answer.

(1) 1 hour after drinking water
(2) 1 hour after drinking salt solution
(3) 3 hours after drinking water

(5 marks)

(Total 12 marks)

NISEAC June 1991, Paper II, No. 6

7. The drawing has been made from an electronmicrograph of part of the glomerulus and renal capsule (Bowman's capsule) of a nephron.

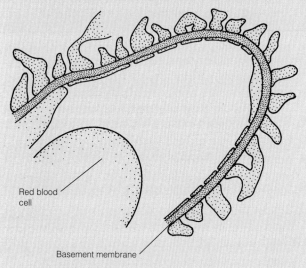

Red blood cell

Basement membrane

(a) Add an arrow to the drawing to show the route by which filtered substances pass into the renal capsule. (1 mark)

(b) Explain how the arrangement of structures shown in the drawing allows effective filtration. (2 marks)

(c) The table contains data concerned with the filtration of some substances in the renal capsule.

Substance	Radius of molecule /nm	Concentration of substance in renal capsule / Concentration of substance in plasma
Urea	0.16	1.0
Glucose	0.36	1.0
Albumin	3.55	0.01

Explain how this information supports the hypothesis that filtration is linked to molecular size. (2 marks)

(Total 5 marks)

AEB November 1993, Paper I, No. 4

8. Urine is routinely analyzed in medical diagnosis since its composition can be affected by certain disease states.

Three of the diseases that can affect urine are:

Nephrosis, in which the basement membrane in the Bowman's capsules becomes permeable to large molecules which would not normally leave the plasma;

Obstructive jaundice, in which the bile duct becomes blocked, sometimes by a 'stone' of cholesterol, so that constituents of bile may 'spill' into the blood and subsequently enter the urine;

Diabetes insipidus ('water diabetes'), in which little or no ADH (antidiuretic hormone) is produced.

Some tests were carried out on samples of urine from different individuals. The results are shown in the table.

Urine sample	Result after testing with			
	Benedict's reagent	Emulsion test	Powdered sulphur	Biuret reagent
P	negative	negative	floated	violet
Q	negative	negative	sank	blue
R	positive	negative	floated	blue
S	negative	negative	floated	blue

(a) (i) Powdered sulphur sinks rapidly in water if emulsifiers are present. What disease might cause emulsifiers to be present in urine?
(ii) Which sample of urine came from an individual with this disease?
(iii) Suggest the identity of the emulsifiers, and their normal function in the body. (3 marks)

(b) (i) Which urine sample came from someone with nephrosis?
(ii) Explain the reasons for your choice. (3 marks)

(c) (i) Which remaining urine sample shows evidence of coming from someone with a disease? Explain your answer.
(ii) Explain why the composition of urine is changed by this disease. (3 marks)

(d) (i) How and why is the composition of the urine affected in someone with diabetes insipidus? Explain your answer.
(ii) Suggest, as precisely as possible, the part of the body directly affected in this disease. (3 marks)

(Total 12 marks)

NEAB June 1993, Paper IIA, No. 2

. There follows a passage illustrating application of
biological principles in medicine. Study the passage
and answer the questions that follow it.

The Artificial Kidney

The artificial kidney is used to replace the
diseased kidney in eliminating the excess ions
and nitrogenous wastes which would otherwise
accumulate in the blood as a result of kidney
5 failure. Total failure of both kidneys would cause
severe problems (possibly fatal) within about ten
days, mainly due to the accumulation in the
blood of excess K^+ ions which adversely affect
heart function. The machine may also be used on
10 a temporary basis after an accident when a
dramatic fall in blood pressure may occur,
resulting in reduced rates of kidney functioning.

In one type of artificial kidney, the patient's
blood is passed (from the radial artery) through a
15 system of fine, coiled tubes bounded by thin
membranes. The other side of the membrane is
bathed by a warm dialysis fluid into which waste
products can pass from the blood. The fine tubes
converge into a tubing which then conducts the
20 blood back into the patient's body through a vein.

The fine tubes of the artificial kidney are made
of differentially permeable cellophane. This
material is used because it has much the same
characteristics as the endothelium of blood
25 capillaries: it is highly permeable to most small
solutes but relatively impermeable to protein. The
dialysis fluid which bathes the cellophane tubes is
a carefully regulated salt and sugar solution with
ionic concentrations similar to those of normal
30 blood plasma.

The function of the artificial kidney can be
explained by the principle of diffusion. Since the
cellophane membrane is permeable to most small
solutes, the concentrations of solutes in the blood,
35 as it flows through the tubes, tend to equal those
in the dialysis fluid. However, if there is an above
normal level of a certain solute in the blood, this
solute will diffuse out of the blood into the
surrounding fluid, which has a lower
40 concentration of the solute. In this way, waste
products and other substances in excess will leave
the blood and pass into the dialysis fluid since
their concentrations in the fluid are very low or
non-existent. The diffusion gradient for any
45 particular substance can be controlled by
dissolving more or less of the substance in the
external solution. The pH of the bathing fluid is
adjusted so that excess H^+ ions are lost from the
patient's blood so preventing it from becoming

50 too acid. Excess water can also be removed by
artificially raising the blood pressure in the
cellophane tubing. To prevent waste solutes from
building up in the fluid and interfering with
diffusion, the fluid is continually being replaced
55 by fresh supply.

Great care must be taken to maintain the
sterility of the artificial kidney. The total amount
of blood in it at any one moment is roughly
$400–500 \ cm^3$. Usually heparin is added to the
60 blood as it enters the artificial kidney to prevent
coagulation; an anti-heparin is added to the blood
before returning it to the patient's body in order
to allow normal blood clotting to take place. An
artificial kidney can clear urea from the blood at a
65 rate of about $200 \ cm^3$ of plasma per minute: this is
almost three times as fast as the clearance rate of
both kidneys working together. But the artificial
kidney can be used for only 12 hours every 3 or 4
days, because of the danger to the blood's clotting
70 mechanisms due to the addition of heparin.

(a) Explain why a fall in blood pressure reduces
the rate of kidney functioning. (line 11)
(1 mark)

(b) What is the chief significance of the phrase
'blood is passed through a system of fine,
coiled tubes'? (line 14)   *(1 mark)*

(c) Explain, in molecular terms, the word
'dialysis'. (line 17)   *(2 marks)*

(d) Suggest a reason why K^+ ions adversely
affect heart function. (line 8)   *(1 mark)*

(e) Name two nitrogenous waste products that
might pass from the blood.   *(1 mark)*

(f) What is meant by the 'endothelium of blood
capillaries'? (lines 24 and 25)   *(1 mark)*

(g) Explain why the dialysis fluid is a salt
solution with similar ionic concentration to
the blood.   *(2 marks)*

(h) Explain the significance of the cellophane
tubes being 'impermeable to protein'. (line
26)   *(1 mark)*

(i) Which phrase explains why maximum
concentration gradients will always be
maintained?   *(1 mark)*

(j) Briefly suggest two possible mechanisms for
the action of heparin. (line 59)   *(2 marks)*

(k) How is the blood's glucose level maintained?
(1 mark)

(l) Which sentence implies that an imposed
pressure will facilitate the diffusion of water
molecules through the membrane?   *(1 mark)*

Oxford June 1990, Paper I, No. 12

See *Further questions*, pages 614–20, question 3.

Immunofluorescent light micrograph of neurones and asterocytes in mammalian spinal cord (*opposite*)

Part V

COORDINATION, RESPONSE AND CONTROL

24   How control systems developed

The ability to respond to stimuli is a characteristic of all living organisms. While many stimuli originate from outside, it is also necessary to respond to internal changes. In a single-celled organism, such responses are relatively simple. No part of it is far from the medium in which it lives and so it can respond directly to environmental changes. The inside of a single cell does not vary considerably from one part to another and so there are few internal differences to respond to.

With the development of multicellular organisms came the differentiation of cells which specialized in particular functions. With specialization in one function came the loss of the ability to perform others. This division of labour, whereby different groups of cells each carried out their own function, made the cells dependent upon one another. Cells specializing in reproduction, for example, depend on other cells to obtain oxygen for their respiration, yet others to provide glucose and others to remove waste products. These different functional systems must be coordinated if they are to perform efficiently. If for example, an animal needs to exert itself in order to capture its food, the muscular activity involved must itself be coordinated. The locomotory organs need to operate smoothly and efficiently with each muscle contracting at exactly the correct time. In addition, more oxygen and glucose will be required and an increased amount of carbon dioxide will need to be removed from the tissues. If breathing is increased so that the oxygen concentration of the blood rises, it is essential that the heart increases its output accordingly. Without coordination between the two systems an increased effort by one could be neutralized by the other. No bodily system can work in isolation, but all must be integrated in a coordinated fashion.

There are two forms of integration in most multicellular animals: nervous and hormonal. The nervous system permits rapid communication between one part of an organism and another, in much the same way as a telephone system does in human society. The hormonal system provides a slower form of communication and can be likened to the postal system. Both systems need to work together. A predator, for example, may be detected by the sense organs, which belong to the nervous system, but in turn cause the production of adrenaline, a hormone. While the nervous system coordinates the animal's locomotion as it makes its escape, the adrenaline ensures that an increased breathing rate and heart beat supply adequate oxygen and glucose to allow the muscles to operate efficiently. The link between these two coordinating systems is achieved by the **hypothalamus**. It is here that the nervous and hormonal systems interact.

All organisms, plant and animal, must respond to environmental changes if they are to survive. To detect these changes requires sense organs. Those detecting external changes are located on the surface of the body and act as a vital link between the internal and external environments. Many other sense cells are located internally to provide information on a constantly changing internal environment. In responding to stimuli, an organism usually modifies some aspect of its functioning. It may need to produce enzymes in response to the presence of food, become sexually aroused in response to certain behaviour by a member of the opposite sex or move away from an unpleasant stimulus. In most cases the organ affecting the change is some distance away from the sense cell detecting the stimulus. A rapid means of communication between the sense cell and the effector organ is essential. In animals the nerves perform this function.

The stimuli received by many sense organs, e.g. the eyes, are very complex and require widely differing responses. The sight of a female of the same species may elicit a totally different response from the sight of a male of the same species. Each response involves different effector organs. The sense organs must therefore be connected by nerves to all effector organs, in much the same way as a telephone subscriber is connected to all other subscribers. One method is to have an individual nerve running from the sense organ to all effectors. Clearly this is only possible in very simple organisms where the sense organs respond to a limited number of stimuli and the number of effectors is small. The nerve nets of cnidarians work in this way.

Large, complex organisms require a different system, because the number of sense organs and effectors is so great that individual links between all of them is not feasible. Imagine having a separate telephone cable leading from a house to every other subscriber's house in Britain, let alone the world. Animals developed a **central nervous system** to which every effector and sense organ has at least one nerve connection. The central nervous system (brain and spinal cord) acts like a switchboard in connecting each incoming stimulus to the appropriate effector. It works in much the same way as a telephone exchange, where a single cable from a home allows a subscriber to be connected to any other simply by making the correct connections at various exchanges.

Where then should the brain be located? The development of locomotion in animals usually resulted in a particular part of the animal leading the way. This anterior region was much more likely to encounter environmental changes first, e.g. changes in light intensity, temperature, pH, etc. It was obvious that most sense organs should be located on this anterior portion. There would be little point locating them in the posterior region, as a harmful substance would not be detected until it had already caused damage to the anterior of the animal. Most sensory information therefore originated at the front. To allow a rapid response, the 'brain' was located in this region. This led, in many animals, to the formation of a distinct head, **cephalization**, concerned primarily with detection and interpretation of stimuli. It was still essential for the brain to receive stimuli from the rest of the animal and to communicate with effector organs throughout the body. An elongated portion of the CNS therefore extends the length of most animals. In vertebrates this is the spinal cord.

With increasingly complex stimuli being received, the brain developed greater powers of interpretation. In particular it developed the ability to store information about previous experiences in order to assist it in deciding on the appropriate response to a future situation. With this ability to learn came the capacity to use previous experience and even to make responses to situations never previously encountered. Thus intelligence developed.

The hormone or endocrine system is concerned with longer-term changes, especially in response to the internal environment. It is an advantage to maintain a relatively constant internal environment. Not only can chemical reactions take place at a predictable rate, but the organism also acquires a degree of independence from the environment. It is no longer restricted to certain regions of the earth but can increase its geographical range. It does not have to restrict its activities to particular periods of the day, or seasons, when conditions are suitable. The maintenance of a constant internal environment is called **homeostasis** and is largely controlled by hormones. In the same way that the brain coordinates the nervous system, the activities of the endocrine system are controlled by the **pituitary gland**.

Responding to changes in the internal and external environments is no less important to survival in plants. Because they lack contractile tissue and do not move from place to place independently, they have no need for very rapid responses. There is therefore no nervous system or anything equivalent to it. Plant responses are hormonal. Their movements are as a result of growth, rather than contractions, and as a consequence are much slower than those of animals.

25 Homeostasis

We have seen that matter tends to assume its lowest energy state. It tends to change from an ordered state to a disordered one, i.e. tends towards high entropy. The survival of biological organisms depends on their ability to overcome this tendency to disorderliness. They must remain stable. This need for constancy was recognized in the nineteenth century by Claude Bernard. He contrasted the constancy of the fluid which surrounds all cells (*milieu interieur*) with the ever changing external environment (*milieu exterieur*). Bernard concluded: *'La fixité du milieu interieur est la condition de la vie libre.'* (The constancy of the internal environment is the condition of the free life.)

By bathing cells in a fluid, the tissue fluid, whose composition remains constant, the chemical reactions within these cells can take place at a predictable rate. Not only are the cells able to survive but they can also function efficiently. The whole organism thus becomes more independent of its environment.

The term **homeostasis** (*homoio* = 'same'; *stasis* = 'standing') was not coined until 1932. It is used to describe all the mechanisms by which a constant environment is maintained. Some examples of homeostatic control have already been discussed, for example osmoregulation in Chapter 23. Further examples are examined in this chapter and the following one.

25.1 Principles of homeostasis

Before examining the detailed operation of homeostatic systems, it is necessary to look at the fundamental principles common to them. Organisms are examples of open systems, since there is exchange of materials between themselves and the environment. Not least, they require a constant input of energy to maintain themselves in a stable condition against the natural tendency to disorder. The maintenance of this stability requires control systems capable of detecting any deviation from the usual and making the necessary adjustments to return it to its normal condition.

25.1.1 Control mechanisms and feedback

Cybernetics (*cybernos* = 'steersman') is the science of control systems, i.e. self-regulating systems which operate by means of feedback mechanisms.

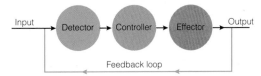

Fig. 25.1 Principal components of a typical control system

The essential components of a control system are:

1. **Reference point** – the set level at which the system operates.

2. **Detector** – signals the extent of any deviation from the reference point.

3. **Controller** – coordinates the information from various detectors and sends out instructions which will correct the deviation.

4. **Effector** – brings about the necessary change needed to return the system to the reference point.

5. **Feedback loop** – informs the detector of any change in the system as a result of action by the effector.

The relationship between these components is given in Fig. 25.1.

An everyday example of such a control system occurs in the regulation of a central heating system in a home, where the various components are:

1. **Reference point** – temperature determined by the occupant, e.g. 20 °C, and set on the thermostat.

2. **Detector** – the thermostat which constantly monitors the temperature of the room in which it is situated.

3. **Controller** – the programmer which can be set to turn the heating on and off at set times. It is connected to the boiler, hot water cylinder, circulation pump and thermostat.

4. **Effector** – the boiler, circulation pump, radiators and associated pipework.

5. **Feedback loop** – the movement of air within the room.

If the temperature of the room falls below 20 °C, the thermostat sends an electrical message to the programmer. The programmer coordinates this information with that in its own programme, i.e. whether the heating is set to operate at this particular time of day. If it is set to operate, it sends appropriate electrical messages which turn on the boiler and the circulation pump. Hot water flows around the central heating system to the radiator in the room. The heat from this radiator warms the air in the room which circulates until it reaches the thermostat. Once the temperature of this air reaches 20 °C, the thermostat ceases to send information to the programmer which then turns off the circulation pump. As the feedback causes the system to be turned off, it is called **negative feedback**. It is possible to have positive feedback systems. Although these are rare in living organisms one example is described in Section 27.1.3.

A similar system operates to control body temperature in birds and mammals. Temperature detectors in the skin provide information on changes in the external temperature which is conveyed to the hypothalamus of the brain, which acts as the controller. This initiates appropriate corrective responses in effectors, such as the skin and blood vessels, in order to maintain the body temperature constant. Details of these processes are given in the following section.

25.2   Temperature control

In extremes of temperature such as those found in polar regions it is the endothermic birds and mammals which are most successful.

The temperature of environments inhabited by living organisms ranges from 90 °C in hot springs to − 40 °C in the Arctic. Most organisms, however, live in the narrow range of temperature 10–30 °C. To survive, most animals need to exert some control over their body temperature. This regulation of body temperature is called **thermoregulation**. In all organisms heat may be gained in two main ways.

1. **Metabolism of food.**

2. **Absorption of solar energy** – This may be absorbed directly or indirectly from

   (a) heat reflected from objects;
   (b) heat convected from the warming of the ground;
   (c) heat conducted from the ground.

Heat may be lost in four main ways:

1. **Evaporation of water**, e.g. during sweating.

2. **Conduction from the body** to the ground or other objects.

3. **Convection from the body** to the air or water.

4. **Radiation from the body** to the air, water or ground.

25.2.1 Ectothermy and endothermy

The majority of animals obtain most of their heat from sources outside the body. These are termed **ectotherms**. The body temperature of these animals frequently fluctuates in line with environmental temperature. Animals whose temperature varies in this way are called **poikilotherms** (*Poikilos* = 'various'; *thermo* = 'heat'). While most ectotherms have a body temperature approaching that of the environment, it is rarely equal to it. During exercise, for example, the metabolic heat produced may raise the body temperature above that of the environment. Moist-bodied ectotherms frequently have temperatures a little below that of the environment owing to the cooling effect of evaporation. Ectotherms do attempt to regulate their temperatures within broad limits. The methods used are largely behavioural.

Mammals and birds maintain constant body temperatures irrespective of the environmental temperature. As their heat is derived internally, by metabolic activities, they are called **endotherms**. As the temperature of the body remains the same, they are sometimes called **homoiotherms** (*homoio* = 'same'; *thermo* = 'heat'). The body temperature of endotherms is usually in the range 35–44 °C.

The higher the body temperature the higher the metabolic rate of the animal. This is especially important for birds where the energy demands of flight make a high metabolic rate an advantage. Most birds therefore have body temperatures in the range of 40–44 °C. The problem with a high body temperature is that the environment is usually cooler, and heat is therefore continually lost to it. The higher the body temperature, the greater the gradient between internal and external temperatures and so the more heat is lost. Mammals, with less dependence on a very high metabolic rate than birds, therefore have body temperatures in the range 35–40 °C to minimize their total heat loss. The body temperatures of all endotherms are something of a compromise, balancing the advantage of a high temperature, in increasing metabolic activity, and the disadvantage of increased heat loss due to a greater temperature gradient between the internal and external environments. The evolutionary advantage of being endothermic is that it gives much more environmental independence. It is no coincidence that the most successful animals in the extremes of temperatures found in deserts and at the poles are mammals and birds. The independence that a constant high body temperature brings allows these groups to extend their geographical range considerably.

25.2.2 Structure of the skin

Most heat exchange occurs through the skin, as it is the barrier between the internal and external environments. It is in mammals that the skin plays the most important rôle in thermoregulation. Mammalian skin has two main layers, an outer epidermis and an inner dermis.

The epidermis
This comprises three regions:

1. The Malpighian layer (germinative layer) – The deepest layer made up of actively dividing cells. The pigment **melanin**, which determines the skin colour, is produced here. It absorbs ultra-violet light and so helps to protect the tissues beneath from its damaging effects. The Malpighian layer has numerous infoldings which extend deep into the dermis, producing sweat glands, sebaceous glands and hair follicles. There are no blood vessels in the epidermis and so the cells of this layer obtain food and oxygen by diffusion and active transport from capillaries in the dermis.

2. Stratum granulosum (granular layer) – This is made up of living cells which have been produced by the Malpighian layer. As they are pushed towards the skin surface by new cells produced beneath, they accumulate the fibrous protein, **keratin**, lose their nuclei and die.

3. Stratum corneum (cornified layer) – This is the surface layer of the skin and comprises flattened, dead cells impregnated with keratin. It forms a tough, resistant, waterproof layer which is constantly replaced as it is worn away. It is thickest where there is greatest wear, i.e. on the palms of the hand and the soles of the feet. Through this layer extend sweat ducts and hair.

The dermis
The dermis is largely made up of connective tissue consisting of collagen and elastic fibres. It possesses:

1. Blood capillaries – These supply both the epidermis and dermis with food and oxygen. Special networks supply the sweat glands and hair follicles. They play an important rôle in thermoregulation.

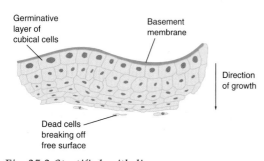

Fig. 25.2 Stratified epithelium

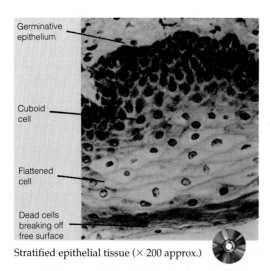

Stratified epithelial tissue (× 200 approx.)

Did you know?

Skin cells form most rapidly in the four hours after midnight when the body's metabolism has slowed down and there is energy available.

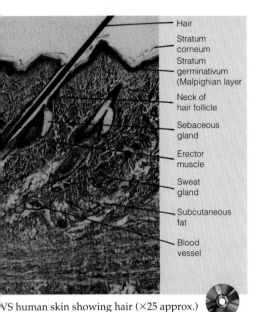

VS human skin showing hair (×25 approx.)

Labels in figure:
Hair
Stratum corneum
Stratum germinativum (Malpighian layer
Neck of hair follicle
Sebaceous gland
Erector muscle
Sweat gland
Subcutaneous fat
Blood vessel

2. Hair follicles – These are formed by inpushings of the Malpighian layer. Cells at the base multiply to produce a long cylindrical hair, the cells of which become impregnated with keratin and die. The more melanin in the hair, the darker its colour. Attached to it is a small bundle of smooth muscle, contraction of which causes the hair to become erect.

3. Sebaceous glands – Situated at the side of the hair follicle these produce an oily secretion called **sebum** which waterproofs the hair and epidermis. It also keeps the epidermis supple and protects against bacteria.

4. Sweat glands – These are coiled tubes made up of cells which absorb fluid from surrounding capillaries and secrete it into the tube, from where it passes to the skin surface via the sweat duct. Sweat has a variable composition, consisting mainly of water, dissolved in which are mineral salts and urea. Evaporation of sweat from the skin surface helps to cool the body.

5. Sensory nerve endings – There is a variety of different sensory cells concerned with providing information on the external environment. These include:

    (a) touch receptors (Meissner's corpuscles);
    (b) pressure receptors (Pacinian corpuscles);
    (c) pain receptors;
    (d) temperature receptors.

6. Subcutaneous fat – Beneath the dermis is a layer of fat (adipose) tissue. This acts both as a long-term food reserve and as an insulating layer.

The structure of human skin is shown in Fig. 25.3 on the next page.

25.2.3 Maintenance of a constant body temperature in warm environments

Endothermic organisms which live permanently in warm climates have developed a range of adaptations to help them maintain a constant body temperature. These adaptations may be anatomical, physiological or behavioural, and include the following.

1. Vasodilation – Blood in the network of capillaries in the skin may take three alternative routes. It can pass through capillaries close to the skin surface, through others deeper in the dermis or it may pass beneath the layer of subcutaneous fat. In warm climates, superficial arterioles dilate in order to allow blood close to the skin surface. Heat from this blood is rapidly conducted through the epidermis to the skin surface from where it is radiated away from the body (Fig. 25.4).

2. Sweating – The evaporation of each gram of water requires 2.5 kJ of energy. Being furless, humans have sweat glands over the whole body, making them efficient at cooling by this means. Animals with fur generally have sweat glands confined to areas of the skin where fur is absent, e.g. pads of the feet in dogs. Sweating beneath a covering of thick fur is inefficient as the fur

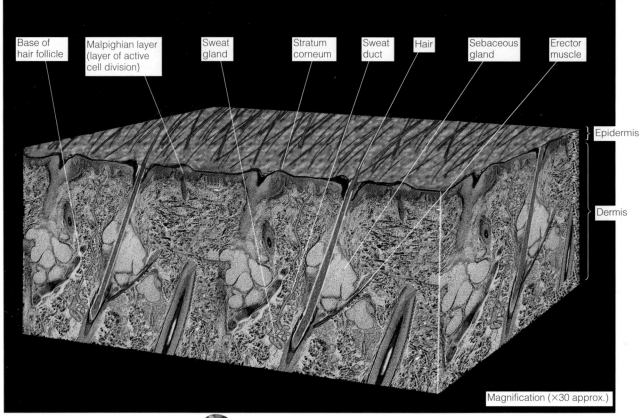

| Base of hair follicle | Malpighian layer (layer of active cell division) | Sweat gland | Stratum corneum | Sweat duct | Hair | Sebaceous gland | Erector muscle |

Epidermis

Dermis

Magnification (×30 approx.)

Fig. 25.3 VS through human skin

prevents air movements which would otherwise evaporate the sweat. Birds lack sweat glands altogether as their covering of feathers makes its evaporation almost impossible. Humans, on the other hand, may produce up to $1000\,cm^3\,hr^{-1}$ of sweat.

3. Panting and licking – Where animals have few or no sweat glands, cooling by evaporation of water nonetheless takes place from the mouth and nose. Panting in dogs may result in the breathing rate increasing from 30 to 300 breaths min^{-1}. This results in excessive removal of carbon dioxide from the blood which is partly offset by a reduction in the depth of breathing. Even so, dogs are able to tolerate a depletion of carbon dioxide which would prove fatal to other organisms. Panting is common in birds. Licking, while not as effective as sweating, may help cool the body. It has been reported in kangaroos, cats and rabbits.

4. Insulation – A layer of fur or fat may help prevent heat gain when external temperatures exceed those of the body. In warm climates the fur is usually light in colour to help reflect the sun's radiation. At high environmental temperatures the hair erector muscles are relaxed and the elasticity of the skin causes the fur to lie closer to its surface. The thickness of insulatory warm air trapped is thus reduced (Fig. 25.5) and body heat is more readily dissipated.

5. Large surface-area to volume ratio – Animals in warm climates frequently have large extremities, such as ears, when compared to related species from cold climates. Fennec foxes have

rctic fox *(top)* and fennec fox *(bottom)* showing
ifferences in lengths of ears

much longer ears than their European counterparts which in turn have longer ears than the Arctic fox. Being well supplied with blood vessels and covered with relatively short hair, ears make especially good radiators of heat.

6. Variation in body temperature – Some desert animals allow their body temperatures to fluctuate within a specific range. In camels this range is 34–41 °C. By allowing their body temperature to rise during the day, they reduce the temperature gradient between the body and the environment and so reduce heat gain. In addition, it delays the onset of sweating and so helps conserve water.

7. Behavioural mechanisms – Many desert animals avoid the period of greatest heat stress by being nocturnal. Some hibernate during the hottest months. This summer hibernation is known as **aestivation**. Other animals avoid the sun by sheltering under rocks or burrowing beneath the surface.

25.2.4 Maintenance of a constant body temperature in cold environments

Endothermic animals living in cold environments show adaptations to the climate. These include:

1. Vasoconstriction – In cold conditions, the superficial arterioles contract, so reducing the quantity of blood reaching the skin surface. Blood largely passes beneath the insulating layer of subcutaneous fat and so loses little heat to the outside. Both vasodilation and vasoconstriction are illustrated in Fig. 25.4.

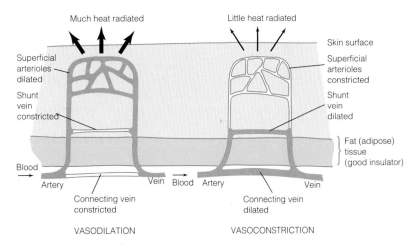

Fig. 25.4 Vasodilation and vasoconstriction

2. Shivering – At low environmental temperatures, the skeletal muscle of the body may undergo rhythmic, involuntary contractions which produce metabolic heat. This shivering may be preceded by asynchronous twitching of groups of muscle.

3. Insulation – Insulation is an effective means of reducing heat loss from the body. It may be achieved by an external covering of fur or feathers and/or an internal layer of subcutaneous fat. The thickness of the fur is related to the environmental temperature, with animals in cold regions having denser and thicker fur. One problem with effective insulation is that it prevents the rapid heat loss necessary during strenuous exercise.

PROJECT

A large organism has a relatively large surface-area to volume ratio and this reduces the rate at which heat is lost

1. Translate this statement into a hypothesis which could be tested.

2. Test your hypothesis by comparing heat loss from round-bottomed flasks of different sizes, using temperature sensors and data-logging equipment if possible.

For this reason the fur on the underside of the body may be thinner to facilitate heat loss. In birds, specialized down feathers provide particularly efficient insulation. Both fur and feathers function by trapping warm air next to the body. In cold conditions, the hair erector muscles contract to pull up the hairs and so increase the thickness of the layer of air trapped, improving insulation (Fig. 25.5).

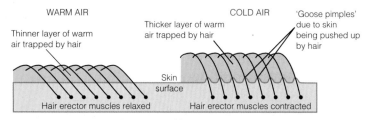

Fig. 25.5 Lowering and raising of hair in controlling heat loss

Subcutaneous fat alone is only half as effective as fur but makes a useful additional contribution. In aquatic mammals (whales, dolphins, seals) fur would be ineffective and has therefore all but disappeared. To compensate for this loss the subcutaneous fat is extremely thick and forms an effective insulating layer, known as **blubber**.

4. **Small surface-area to volume ratio** – Animals in colder climates have a tendency to be more compact, with smaller extremities, than related species in warm climates. In this way heat loss is reduced.

5. **Variations between superficial and core temperature** – The extremities of animals in cold regions are maintained at lower temperatures than the core body temperature. This reduces the temperature gradient between them and the environment. This is especially important in order to reduce heat loss from the feet which are in contact with the cold ground. The reduction of heat loss from these extremities is achieved by **counter-current heat exchangers** found in the limbs of certain birds and mammals. Blood in veins returning from the limbs passes alongside blood in arteries. Heat from the warm blood entering the limb is transferred to cold blood in the vein returning from it. The limb is thereby kept at a lower temperature and cold blood is prevented from entering the core of the body. This system is illustrated in Fig. 25.6.

6. **Increased metabolic rate** – In addition to an increase in heat produced by muscles during shivering, the liver may also increase its metabolic rate during cold conditions. Low temperatures induce increased activity of the adrenal, thyroid and pituitary glands. All these produce hormones which help to increase the body's metabolic rate and so produce additional heat. This requires increased consumption of food, arctic animals consuming more food per gram of body weight than their tropical relatives. Rats kept at 3 °C take in 50% more food than rats kept at 20 °C.

7. **Behavioural mechanisms** – In cold regions, animals are usually active during the day (diurnal). Huddling of groups of individuals is a common way of reducing heat loss. Reproductive behaviour is adapted to allow the young to be

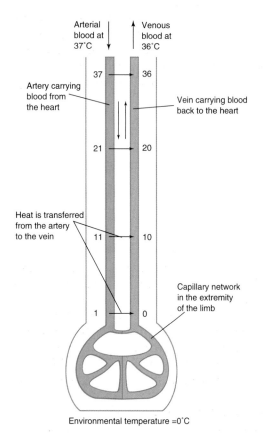

Fig. 25.6 Counter-current heat exchange system (rete mirabilis)

born at a time when food is available by the time they are weaned. Delayed fertilization and implantation of the embryo may occur (Sections 12.7.4–6).

8. Hibernation – One special behavioural mechanism utilized by endothermic animals in cold climates is hibernation. During times of greatest cold and hence shortest supply of food, mammals like squirrels and dormice may undergo a period of long sleep. During this time the metabolic rate is reduced 20–100 times below that of normal with a consequent reduction in food and oxygen consumption. The hibernation may last several months, during which time fat reserves accumulated during the summer are used. These reserves take the form of **brown adipose tissue** which is easily metabolized at low temperatures. Breathing and heart beat become slow and irregular. The body temperature falls close to that of the environment, being 1–4 °C higher. To prevent the body freezing at temperatures of − 4 °C and below, well insulated and sheltered nests are essential, often underground. For this reason hibernation is virtually impossible in areas of permafrost.

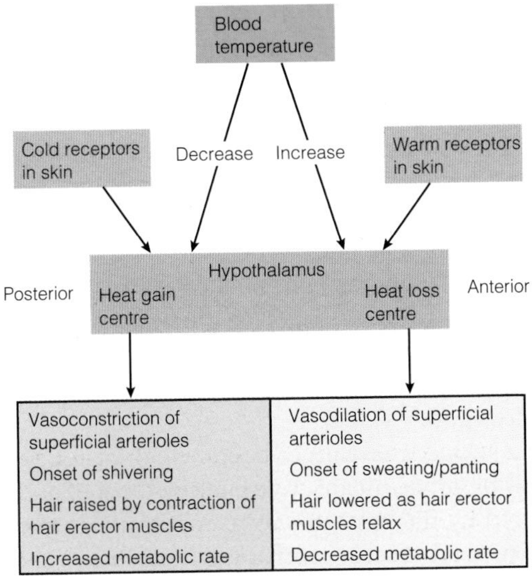

25.2.5 Rôle of the hypothalamus in the control of body temperature

Control of body temperature is effected by the **hypothalamus**, a small body at the base of the brain. Within the hypothalamus is the thermoregulatory centre which has two parts: a heat gain and a heat loss centre. The hypothalamus monitors the temperature of blood passing through it and in addition receives nervous information from receptors in the skin about external temperature changes. Any reduction in blood temperature will bring about changes which conserve heat. A rise in blood temperature has the opposite effect. These effects are summarized in Fig. 25.7.

Fig. 25.7 Summary of body temperature control by the hypothalamus

25.3   Control of blood sugar

All metabolizing cells require a supply of glucose in order to continue functioning. The nervous system is especially sensitive to any reduction in the normal glucose level of 90 mg glucose in 100 cm^3 blood. A rise in blood sugar level can be equally dangerous. The supply of carbohydrate in mammals fluctuates because they do not eat continuously throughout the day and the quantity of carbohydrate varies from meal to meal. There may be long periods when no carbohydrate is absorbed from the intestines. Cells, however, metabolize continuously and need a constant supply of glucose to sustain them. A system which maintains a constant glucose level in the blood, despite intermittent supplies from the intestine, is essential. The liver plays a key rôle in glucose homeostasis. It can add glucose to the blood in two ways:

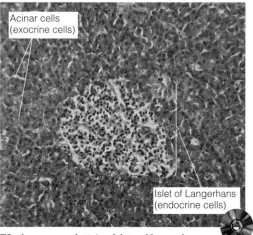

Acinar cells
(exocrine cells)

Islet of Langerhans
(endocrine cells)

TS of pancreas, showing Islets of Langerhans
(×200 approx.)

(i) by the breakdown of glycogen (**glycogenolysis**);

(ii) by converting protein into glucose (**gluconeogenesis**).

It can remove glucose from the blood by converting it into glycogen (**glycogenesis**) which it then stores. A normal liver stores around 75 g of glycogen, sufficient to maintain the body's supply of glucose for about twelve hours. The interconversion of glucose and glycogen is largely under the control of two hormones, produced by the pancreas. In addition to being an exocrine gland producing pancreatic juice, the pancreas is also an endocrine gland. Throughout the pancreas are groups of histologically different cells known as the **islets of Langerhans**. The cells within them are of two types: α-cells, which produce the hormone **glucagon**, and β-cells, which produce the hormone **insulin**. Both hormones are discharged directly into the blood. Some hours after a meal, the glucose formed as a result of carbohydrate breakdown is absorbed by the intestines. The blood capillaries from the intestine unite to form the hepatic portal vein which carries this glucose-rich blood to the liver. Insulin from the pancreas causes excess glucose to become converted to glucose-6-phosphate and ultimately glycogen which the liver stores. The same process can occur in many body cells, especially muscle. Some time later, when the level of glucose in the hepatic portal vein has fallen below normal, the liver reconverts some of its stored glycogen to glucose, to help maintain the glucose level of the blood. This change involves a phosphorylase enzyme in the liver which is activated by the pancreatic hormone glucagon.

Should the glycogen supply in the liver become exhausted, glucose may be formed by other means. Once a low level of blood glucose is detected by the hypothalamus it stimulates the pituitary gland to produce adrenocorticotrophic hormones (ACTH) which cause the adrenal glands to release the glucocorticoid hormones, e.g. cortisol. These cause the liver to convert amino acids and glycerol into glucose. In times of stress, another hormone from the adrenal glands, adrenaline, causes the breakdown of glycogen in the liver and so helps to raise the blood sugar level. Further details of these effects are given in Section 27.5.

The control of blood sugar level is summarized in Fig. 25.8.

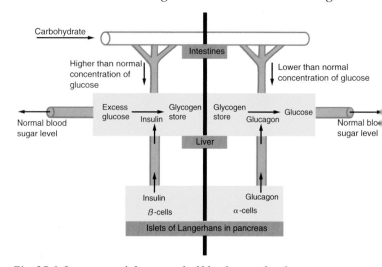

Fig. 25.8 Summary of the control of blood sugar level

25.4   Control of respiratory gases

Metabolizing cells all require a continuous supply of oxygen sufficient to satisfy their respiratory needs. The homeostatic control of both oxygen and carbon dioxide levels in the blood is achieved by the breathing (respiratory) centres in the medulla oblongata of the brain. Details of the mechanisms of control are given in Section 20.3

25.5   Control of blood pressure

Having achieved homeostatic control of the composition of respiratory gases, sugar and other metabolites in the blood, it is clearly essential for a mammal to control the distribution of this blood. It is essential that blood pressure be kept above a certain minimum level if the supply of essential materials to cells is to be maintained. This is achieved by controlling the rate at which the heart pumps blood and by the vasoconstriction or vasodilation of blood vessels. The mechanisms by which these are controlled are detailed in Sections 21.5.4 and 21.5.5 respectively.

25.6   Cellular homeostasis

To some extent each cell of the body is an independent unit capable of selecting which materials enter and leave it. Each cell exerts homeostatic control of the level of each metabolite within it. As the synthesis of substances within a cell is carried out by enzymes, it follows that this control is achieved by controlling enzyme production as follows:

1. If the level of a particular metabolite falls, it stimulates the relevant operon whose operator gene switches on its corresponding structural genes (Section 7.7).

2. Each structural gene forms mRNA by transcription.

3. The various mRNAs are then translated into the particular proteins (enzymes) by the ribosomes in conjunction with tRNA.

4. These enzymes enter the cell's metabolic centre where they synthesize the formation of the specific metabolite.

5. The metabolite enters the cell's metabolic pool where its concentration rises.

6. When the metabolite's level returns to normal it causes the operator gene to switch off (negative feedback).

7. mRNA production ceases as in due course does the production of the metabolite.

These events are summarized in Fig. 25.9

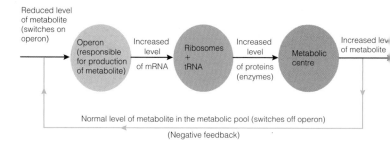

Fig. 25.9 Summary of cellular homeostasis

25.7 The liver

The liver, weighing as it does up to 1.5 kg, makes up 3–5% of the body weight. It probably originated as a digestive organ but its functions are now much more diverse, many being concerned with homeostasis.

25.7.1 Structure of the liver

In an adult human, the liver is typically 28 cm × 16 cm × 9 cm although its exact size varies considerably according to the quantity of blood stored within it. It is found immediately below the diaphragm, to which it is attached. Blood is supplied to the liver by two vessels: the hepatic artery which carries 30% of the liver's total blood supply brings oxygenated blood from the aorta, whereas the hepatic portal vein supplies 70% of the liver's blood and is rich in soluble digested food from the intestines. A single vessel, the hepatic vein, drains blood from the liver. In addition, the bile duct carries bile produced in the liver to the duodenum. The relationship of these structures is given in Fig. 25.10.

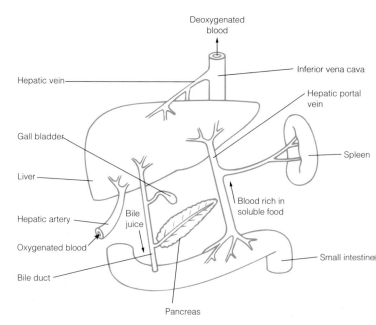

Fig. 25.10 Blood system associated with the mammalian liver

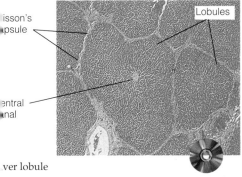

isson's
psule

entral
nal

ver lobule

The branches of the hepatic artery and those of the hepatic portal vein combine within the liver to form common venules which lead into a series of channels called **sinusoids**. These are lined with liver cells or **hepatocytes**. The sinusoids eventually drain into a branch of the hepatic vein called the **central vein**. Between the hepatocytes are fine tubes called **canaliculi** in which bile is secreted. The canaliculi combine to form bile ducts which drain into the gall bladder where the bile is stored before being periodically released into the duodenum. The structure of the liver is shown in Fig. 25.11.

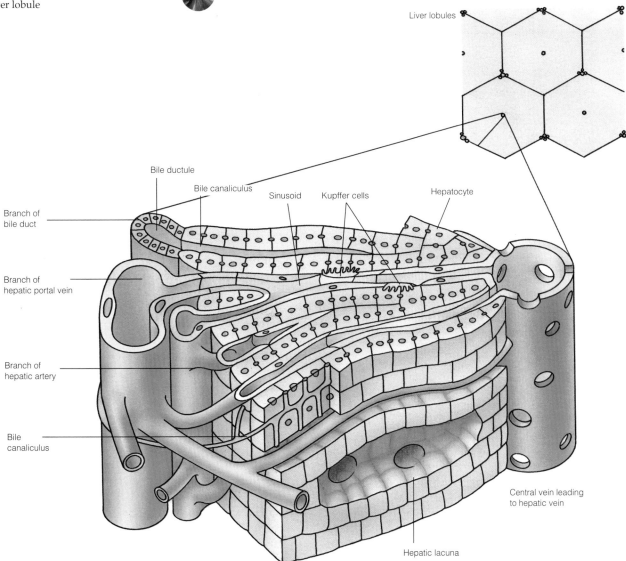

ig. 25.11 Structure of the mammalian liver

The functional unit of the liver is the **acinus**. As blood from the hepatic portal vein and hepatic artery mixes it passes along the sinusoids which are lined with hepatocytes. Materials are exchanged between these cells and the blood. To facilitate this exchange, the hepatocytes have microvilli to increase their surface-area. They also possess a large nucleus, many mitochondria, lysosomes and glycogen granules – all indicate a highly metabolic rôle for these cells. The canaliculi are also lined with microvilli and these appear to remove bile from the hepatocytes by active transport.

503

The sinusoids are lined with flattened **endothelial cells**. Their structure is similar to that in many other organs except for the presence of pores up to 10 nm in diameter. In addition, there are specialized cells lining the sinusoid. These are **Kupffer** cells. They are highly phagocytic and form part of the **reticulo-endothelial system**. They are ideally situated for ingesting any foreign organisms or particulate matter which enter the body from the intestines. They also engulf damaged and worn out blood cells, producing the bile pigment bilirubin as a by-product of this process. The bilirubin is passed into the canaliculi for excretion in the bile.

25.7.2 Functions of the liver

The liver is the body's chemical workshop and has an estimated 500 individual functions. Some of these have been grouped under the following twelve headings:

1. **Carbohydrate metabolism** – The liver's major rôle in the metabolism of carbohydrates is in converting excess glucose absorbed from the intestine into glycogen. This stored glycogen can later be reconverted to glucose when the blood sugar level falls. This interconversion is under the control of the hormones insulin and glucagon produced by the islets of Langerhans in the pancreas (Section 25.3).

2. **Lipid metabolism** – Lipids entering the liver may either be broken down or modified for transport to storage areas elsewhere in the body. Once the glycogen store in the liver is full, excess carbohydrate will be converted to fat by the liver. Excess cholesterol in the blood is excreted into the bile by the liver, which conversely can synthesize cholesterol when that absorbed by the intestines is inadequate for the body's need. The removal of excess cholesterol is essential as its accumulation may cause atherosclerosis (narrowing of the arteries) leading to thrombosis. If in considerable excess its presence in bile may lead to the formation of **gall stones** which can block the bile duct.

3. **Protein metabolism** – Proteins are not stored by the body and so excess amino acids are broken down in the liver by a process called **deamination**. As the name suggests, this is the removal of the amino group ($-NH_2$) to form ammonia (NH_3) which in mammals is converted to the less toxic urea ($CO(NH_2)_2$). This occurs in the ornithine cycle, the main stages of which are shown in Fig. 25.12.

  Transamination reactions whereby one amino acid is converted to another are also performed by the liver. All non-essential amino acids may be synthesized in this way, should they be temporarily deficient in the diet.

4. **Synthesis of plasma proteins** – The liver is responsible for the production of vital proteins found in blood plasma. These include albumins and globulins as well as the clotting factors prothrombin and fibrinogen.

5. **Production of bile** – The liver produces bile salts and adds to them the bile pigment bilirubin from the breakdown of red

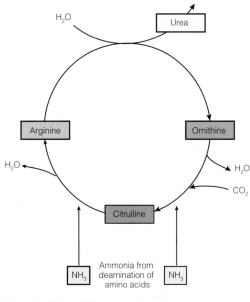

Fig. 25.12 Ornithine (urea) cycle

APPLICATION

Cirrhosis of the liver

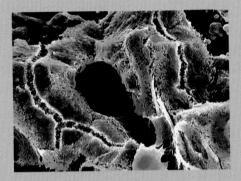

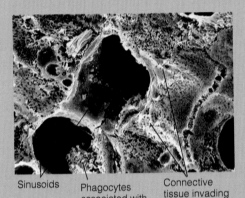

Sinusoids   Phagocytes associated with tissue damage   Connective tissue invading liver

False-colour scanning EMs of cell structure within (top) a healthy liver and (above) one affected by cirrhosis

Worldwide the main cause of the severe liver disease cirrhosis, is infection with the hepatitis B virus but in the western world the commonest cause is alcohol abuse. The chance of developing cirrhosis is related to the amount of alcohol consumed daily. The risks are significantly increased in men with an intake greater than seven units (56 g absolute alcohol) per day and in women exceeding five units (40 g) per day. There is also some evidence that some people are genetically predisposed to develop cirrhosis. Although ethanol itself is toxic the main liver damage is thought to occur as a result of ethanal produced during ethanol metabolism. Ethanal binds to proteins, altering liver structure and function. It inhibits protein secretion from hepatocytes, decreases the activity of many enzymes, encourages cell destruction and stimulates collagen production. The extra collagen produced is responsible for the fibrous scars typical of cirrhosis and has a profound effect on liver function. The collagen deposits in the liver sinusoids depriving the hepatocytes of nutrients. The hepatocytes are therefore unable to produce essential proteins like albumin. The low serum albumin level causes fluid to leak into the tissues and accumulate causing swelling. There is a decreased production of clotting factors leading to a tendency to bleed easily. The liver is unable to carry out its usual functions of dealing with ammonia and other nitrogenous compounds and their accumulation progressively poisons the brain causing confusion and eventually coma.

Many of the severe and irreversible effects of liver damage are preventable if alcohol consumption is reduced and vaccinations against hepatitis viruses are given.

blood cells. With sodium chloride and sodium hydrogen carbonate, cholesterol and water this forms the green-yellow fluid known as bile. Up to 1 dm^3 of bile may be produced daily. It is temporarily stored in the gall bladder before being discharged into the duodenum. The bile pigments are purely excretory. The remaining contents have digestive functions and are described in Section 15.4.4.

6. Storage of vitamins – The liver will store a number of vitamins which can later be released if deficient in the diet. It stores mainly the fat-soluble vitamins A, D, E and K, although the water-soluble vitamins B and C are also stored. The functions of these vitamins are given in Chapter 2, Table 2.1.

7. Storage of minerals – The liver stores minerals, e.g. iron, potassium, copper and zinc, the functions of which are dealt with in Chapter 2, Table 2.1. It is the liver's stores of these minerals, along with vitamins, which makes it such a nutritious food.

8. Formation and breakdown of red blood cells – The fetus relies solely on the liver for the production of red blood cells. In an adult this rôle is transferred to the bone marrow. The adult liver, however, continues to break down red blood cells at the end of their 120-day life span. The Kupffer cells lining the sinusoids carry out this breakdown, producing the bile pigment bilirubin which is excreted in the bile. The iron is either stored in the liver or used in the formation of new red blood cells by the bone marrow. The liver produces **haematinic principle**, a substance needed in the formation of red blood cells. Vitamin B is necessary for the production of this principle, and its deficiency results in pernicious anaemia.

9. Storage of blood – The liver, with its vast complex of blood vessels, forms a large store of blood with a capacity of up to 1500 cm^3. In the event of haemorrhage, constriction of these vessels forces blood into the general circulation to replace that lost and so helps to maintain blood pressure. In stressful situations adrenaline also causes constriction of these vessels, creating a rise in blood pressure.

10. Hormone breakdown – To varying degrees, the liver breaks down all hormones. Some, such as testosterone, are rapidly broken down whereas others, like insulin, are destroyed more slowly.

11. Detoxification – The liver is ideally situated to remove or render harmless, toxic material absorbed by the intestines. Foreign organisms or material are ingested by the Kupffer cells while toxic chemicals are made safe by chemical conversions within hepatocytes. Alcohol and nicotine are two substances dealt with in this way.

12. Production of heat – The liver, with its considerable metabolic activity, can be used to produce heat in order to combat a fall in body temperature. This reaction, triggered by the hypothalamus, is in response to adrenaline, thyroxine and nervous stimulation. Whether the liver's activities produce excess heat under ordinary circumstances is a matter of some debate.

25.8   Questions

(a) Explain what is meant by the term
homeostasis.   (2 marks)
(b) Describe the homeostatic mechanisms
involved in the control of (i) water, and
(ii) sodium ions in body fluids.
(11 marks/10 marks)
(Total 23 marks)

*UCLES (Modular) March 1992, (Transport,
regulation and control), No. 1*

With reference to each of the following, describe
how negative feedback controls operate in the human
body.
(a) Blood glucose levels.
(b) Maintenance of body temperature.
(c) Osmoregulation.   (Total 15 marks)

SEB (revised) May 1991, Higher Grade Paper I, No. 15B

Describe the roles of the liver in
(a) carbohydrate metabolism, and   (10 marks)
(b) amino acid and protein metabolism.
(13 marks)
(Total 23 marks)

*UCLES (Modular) March 1992, (Transport,
regulation and control), No. 2*

The bar chart shows the production efficiency of a
number of organisms. The production efficiency may
be calculated from:

$$\frac{\text{Amount of new food tissue produced}}{\text{Amount of ingested food that is digested and absorbed}} \times 100\%$$

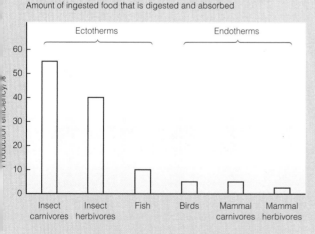

(a) What happens to the ingested food that is
digested and absorbed but is not used to
produce new tissue?   (1 mark)
(b) Suggest **one** explanation for the difference in
production efficiency between ectotherms
and endotherms.   (2 marks)

(c) Describe and explain the effect of body size
on production efficiency in a group of
endotherms such as the mammals.   (3 marks)
(Total 6 marks)

AEB November 1992, Paper I, No. 8

5. The diagram shows the blood supply to and from
the liver.

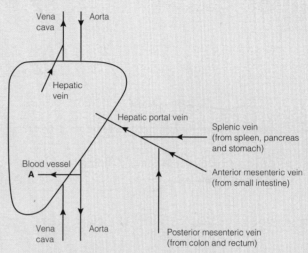

(a) In which blood vessel would you expect to
find the highest level of:
(i)   glucose, 1 hour after eating;
(ii) glucose, 10 hours after eating;
(iii) carbon dioxide;
(iv) insulin?   (4 marks)
(b) (i) What is the name of blood vessel **A**?
(ii) Name **one** substance that is required by
the liver and is mainly supplied by this
blood vessel.   (2 marks)
(Total 6 marks)

AEB June 1991, Paper I, No. 11

6. VS mammalian skin

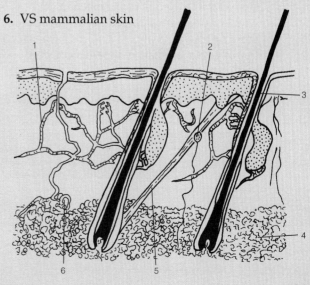

(a) Name the structures labelled 1–6 on the diagram of the skin and state the function each plays in the temperature control mechanism of a mammal. (9 marks)

(b) In 1961 Benzinger, experimenting on human volunteers, obtained the results shown below, after his subjects had ingested ice.

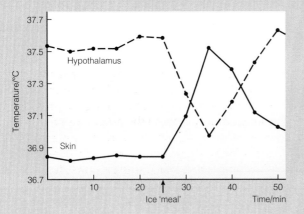

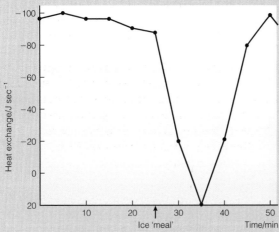

Based on Simpkins & Williams, *Biology of the Cell: Mammal and Flowering Plant* (Mills & Boon, 1980)

Use your knowledge of the mechanism of temperature control to give an explanation of these results. (5 marks)

(c) Endothermic animals maintain a constant body temperature.
   (i) State **four** advantages of endothermy. (4 marks)
   (ii) State **two** disadvantages of endothermy. (2 marks)
(Total 20 marks)

O&CSEB June 1991, Paper I, No. 9

7. The Galapagos marine iguana is a large lizard which lives on rocky shores.
On land, it maintains a temperature of approximately 37 °C by behavioural means. It feeds underwater on marine algae, at sea temperatures between 22 °C and 27 °C.
In an investigation, the heart rate of marine iguanas was measured at different body temperatures, firstly while warming up from 22 °C to 37 °C and secondly while cooling from 37 °C to 22 °C.
The results of the investigation are summarized in the table below.

Body temperature in °C	Heart rate in beats per minute	
	While warming up	While cooling
22	46	28
25	56	41
28	66	40
31	74	40
34	83	42
37	69	66

(a) (i) Plot the heart rates during warming up and during cooling on a single pair of axes, on graph paper. (5 marks)
   (ii) Describe the changes in heart rate shown in the graphs while warming up and while cooling. (4 marks)
   (iii) Comment on the significance of the changes in heart rate while warming up, in relation to the iguana's feeding behaviour. (2 marks)
   (iv) Comment on the significance of the changes in heart rate during cooling, in relation to the iguana's feeding behaviour. (2 marks)

(b) The rocky shores inhabited by the iguanas commonly reach temperatures of 45 °C or more.
   Suggest examples of behaviour which could enable the iguanas to reach and maintain body temperatures of about 37 °C in this habitat. (4 marks)

(c) What are the advantages to animals of a constant body temperature? (3 marks)
(Total 20 marks)

ULEAC January 1990, Paper I, No. 1

See *Further questions*, pages 614–20, questions 4, 6.

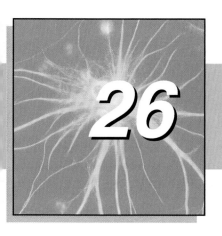

26 The endocrine system

Animals possess two principal coordinating systems, the **nervous system** and the **endocrine system**. The nervous system gives rapid control and details of its functioning are provided in Chapter 27. The endocrine system on the other hand regulates long-term changes. The two systems interact in a dynamic way in order to maintain the constancy of the animal's internal environment, while permitting changes in response to a varying external environment. Both systems secrete chemicals, the nervous system as a transmitter between neurones and the endocrine system as its sole means of communication between various organs and tissues in the body. It is worth noting that adrenaline may act both as a hormone and as a nervous transmitter.

This chapter discusses the principles of the endocrine system, the nature of hormones and the activities of specific endocrine glands. Because of their close association with particular organ systems, the activities of certain endocrine glands are dealt with elsewhere in this book. Reproductive hormones are described in Section 12.6 and digestive hormones in Section 15.5.

26.1   Principles of endocrine control

In animals, two types of gland are recognized: **exocrine glands**, which convey their secretions to the site of action by special ducts, and **endocrine glands**, which lack ducts and transport their secretions instead by the blood. For this reason, the term **ductless glands** is often applied to endocrine glands. The glands may be discrete organs or cells within other organs.

The secretions of these glands are called **hormones**. Derived from the Greek word *hormon*, which means 'to excite', hormones often inhibit actions as well as excite them. All hormones are effective in small quantities. Most act on specific organs, called **target organs**, although some have diffuse effects on all body cells.

Most, but not all, endocrine glands work under the influence of a single master gland, the **pituitary**. In this way the actions of individual glands can be coordinated. Such coordination is essential as hormones work not in isolation, but interacting with each other. Most organs are influenced by a number of different hormones. If the pituitary is considered to be the master of the endocrine system then the **hypothalmus** can be thought of as the manager. It not only assists in directing the activities of

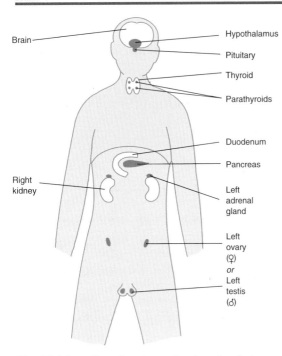

Brain
Hypothalamus
Pituitary
Thyroid
Parathyroids
Duodenum
Pancreas
Right kidney
Left adrenal gland
Left ovary (♀) or Left testis (♂)

Fig. 26.1 Location of major endocrine glands in humans

endocrine glands, it also acts as the all-important link between the endocrine and nervous systems.

The positions of the major endocrine organs in humans are shown in Fig. 26.1.

26.1.1 Chemistry of hormones

A hormone is a chemical messenger produced by an endocrine gland. It is discharged into the bloodstream which carries it around the body to its target organs. Its effects are unrelated to its energy content. Hormones do not belong to any one chemical group. They are generally molecules of medium size, being large enough to have their own specific composition but sufficiently small for them to circulate freely in the body. Some hormones are amines, whereas others are polypeptides and proteins. A few are steroids which are derived from lipids. A summary of the chemical nature of commonly-occurring hormones is given in Table 26.1. The largest hormones are proteins which may possess up to 300 amino acids. The smallest possess less than ten amino acids. The same hormone differs little from one species to another, although its effects may vary in different animals.

26.1.2 Nature of hormone action

Hormones exert their influence by acting on molecular reactions in cells. They achieve this by one or more of the following cell processes:

1. Transcription of genetic information (e.g. oestrogen).

2. Protein synthesis (e.g. growth hormone).

3. Enzyme activity (e.g. adrenaline).

4. Exchange of materials across the cell membrane (e.g. insulin).

While a hormone is transported to all cells by the blood, it only affects specific ones. The explanation is that only target cells possess special chemicals called **receptor molecules** on their surface. These receptors are specific to certain hormones. Both the receptor and the hormone have complementary molecular shapes which fit one another in a 'lock and key' manner, much in the way that enzymes and substrates combine (Section 3.1.2).

There appears to be an alternative means by which this receptor–hormone complex influences the cell. The complex may induce the production of a second messenger (cyclic AMP) which activates enzymes within the cell. The polypeptide and protein hormones act in this way (see Section 26.5.2). Steroid hormones, being lipid derivatives, can pass through the cell membrane. The whole complex enters the cell where it exerts its influence. These mechanisms are illustrated in Fig. 26.2. In some cases the complex simply alters the permeability of the cell membrane. Insulin operates in this manner by increasing the permeability of cell membranes to glucose.

TABLE 26.1 **The chemical nature of commonly occurring hormones**

Chemical group	Hormones
Polypeptides (less than 100 amino acids)	Oxytocin Vasopressin Insulin Glucagon
Protein	Prolactin Follicle stimulating hormone Luteinizing hormone Thyroid stimulating hormone Adrenocorticotrophic hormone Growth hormone
Amines (derivatives of amino acids)	Adrenaline Noradrenaline Thyroxine
Steroids (derivatives of lipids)	Oestrogen Progesterone Testosterone Cortisone Aldosterone

(a) Use of second messenger, e.g. protein and polypeptide hormones such as adrenaline

Hormone approaches receptor site.

Hormone fuses to receptor site, and in doing so activates adenylate cyclase inside the membrane.

The activated adenylate cyclase converts ADP to cyclic AMP which acts as a second messenger that activates other enzymes.

) Steroid hormone mechanism of action

Hormone approaches receptor site.

Hormone combines with the receptor to form a complex.

Hormone–receptor complex passes across the membrane into the cell where it switches on genes on the DNA which produce specific enzymes.

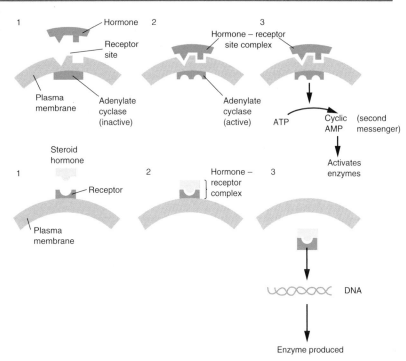

g. 26.2 Mechanism of hormone action

26.2 The pituitary gland

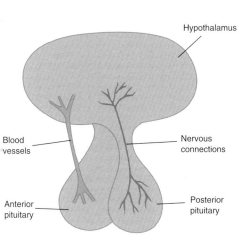

g. 26.3 Structure of the pituitary gland and s relationship to the hypothalamus

Situated at the base of the brain and immediately above the roof of your mouth is the pituitary. Despite its small size (about the dimensions of a pea) it controls, either directly or indirectly, many of your features – your size, sex, metabolism and the rate at which you age. It depends to a large extent upon information it receives from the nearby hypothalamus, to determine which hormones it produces and when. It is sometimes referred to as the master gland because it orchestrates the production of hormones from other endocrine glands. In fact it works not in isolation, but in a complex interrelationship with the brain whose messages are filtered and refined by the hypothalamus. The pituitary gland itself has two distinct portions. The **anterior pituitary** is a region of glandular tissue which communicates with the hypothalamus by means of tiny blood vessels. The **posterior pituitary** is of nervous origin and is in effect an outgrowth of the hypothalamus. Communication with the hypothalamus is by nerves rather than blood vessels. The structure is shown in Fig. 26.3.

26.2.1 The anterior pituitary

This portion of the pituitary gland produces six hormones. Most have other endocrine glands as their target organs. These hormones, called **trophic hormones**, stimulate the activity of their respective endocrine glands. The only non-trophic hormone is growth hormone which, rather than influencing other endocrine glands, affects body tissues in general. The production of all these hormones is determined by small peptide molecules produced by the hypothalamus and passed to the pituitary via small connecting blood vessels. The functions of the hormones produced are given in Table 26.2.

TABLE 26.2   Functions of hormones secreted by the pituitary

Hormone	Abbreviation	Function
Anterior pituitary Thyroid stimulating hormone (thyrotrophic hormone, thyrotrophin)	TSH	1. Stimulates the growth of the thyroid gland 2. Stimulates the thyroid gland to produce its hormones, e.g. thyroxine
Adrenocorticotrophic hormone	ACTH	1. Regulates the growth of the adrenal cortex 2. Stimulates the adrenal cortex to produce its hormones, e.g. cortisone
Follicle stimulating hormone	FSH	1. Initiates cyclic changes in the ovaries, e.g. development of the Graafian follicles 2. Initiates sperm formation in the testes
Luteinizing hormone (interstitial cell stimulating hormone)	LH (ICSH)	1. Causes release of the ovum from the ovary and consequent development of the follicle into the corpus luteum 2. Stimulates secretion of testosterone from interstitial cells in the testes
Prolactin (luteotrophic hormone, luteotrophin)	LTH	1. Maintains progesterone production from the corpus luteum 2. Induces milk production in pregnant females
Growth hormone	GH	1. Promotes growth of skeleton and muscles 2. Controls protein synthesis and general body metabolism
Posterior pituitary Antidiuretic hormone (vasopressin)	ADH	1. Reduces the quantity of water lost from the kidney as urine 2. Raises the blood pressure by constricting arterioles
Oxytocin		1. Induces parturition (birth) by causing uterine contractions 2. Induces lactation (secretion of milk from the nipple)

26.2.2 The posterior pituitary

This portion of the pituitary gland stores two hormones: **antidiuretic hormone (ADH)** or **vasopressin** and **oxytocin**. Both have remarkably similar chemical structures, differing in just one of their nine amino acids. Despite this they exert very different influences as detailed in Table 26.2.

26.2.3 Release of pituitary hormones

The release of pituitary hormones is controlled by a combination of cells within the pituitary itself, neurosecretory cells in the hypothalamus and the endocrine secretions of the target organs of the pituitary hormones. Hormones are released rhythmically – each one having a different periodicity. Gonadotrophins have a monthly pattern of release while luteinizing hormone is produced in pulses about every 2 hours.

26.3   The hypothalamus

Lying at the base of the brain to which it is attached by numerous nerves, the hypothalamus in humans weighs a mere 4 g. Despite its small size it performs many vital functions.

1. It regulates activities such as thirst, sleep and temperature control.

2. It monitors the level of hormones and other chemicals in the blood passing through it.

3. It controls the functioning of the anterior pituitary gland.

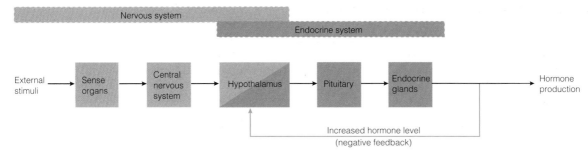

Fig. 26.4 Rôle of the hypothalamus as the link between nervous and endocrine systems

4. It produces antidiuretic hormone and oxytocin which are stored in the posterior pituitary gland.

The hypothalamus is the link between the nervous and endocrine systems as illustrated in Fig. 26.4. By monitoring the level of hormones in the blood, the hypothalamus is able to exercise homeostatic control of them. For example, the control of thyroxine production by the thyroid gland is achieved by this means:

1. The hypothalamus produces **thyrotrophin releasing factor (TRF)** which passes to the pituitary along blood vessels.

2. TRF stimulates the anterior pituitary gland to produce **thyroid stimulating hormone (TSH)**.

3. TSH stimulates the thyroid gland to produce thyroxine.

4. As the level of thyroxine builds up in the blood it suppresses TRF production from the hypothalamus and TSH production by the anterior pituitary gland. By this form of negative feedback the level of thyroxine in the blood is maintained at a constant level.

Fig. 26.5 summarizes these effects.

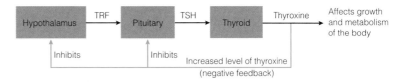

Fig. 26.5 Homeostatic control of thyroxine production

26.4   The thyroid gland

Found in the neck close to the larynx, the thyroid gland, which weighs around 25 g and whose structure is shown in Fig. 26.6 produces three hormones: **triiodothyronine (T_3), thyroxine (T_4)** and **calcitonin**.

Triiodothyronine and thyroxine are very similar chemically and functionally. They regulate the growth and development of cells. In this respect they are especially important in young mammals. In addition, these hormones increase the rate at which glucose is oxidized by cells. One consequence of this is the production of heat and these hormones are therefore produced when an organism is exposed to severe cold. Emotional stress and hunger may elicit a similar production of these hormones.

APPLICATION

Abnormalities of the thyroid

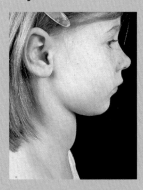

A goitre

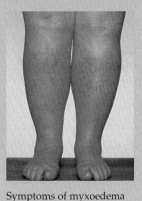

Symptoms of myxoedema

The two main thyroid abnormalities are underactivity (**hypothyroidism**) and overactivity (**hyperthyroidism**).

Underactivity (hypothyroidism) has a more marked effect on immature mammals as mental and physical retardation occur in addition to general sluggishness. The condition is known as **cretinism**. In adults, where mental, physical and sexual development is already complete, underactivity of the thyroid causes mental and physical sluggishness as well as a reduced metabolic rate. The latter leads to a reduced heart and ventilation rate, a lowered body temperature and obesity. The condition is known as **myxoedema**. As a result of underactivity, a swelling called a **goitre** may arise in the throat. The cause of underactivity is often the result of an insufficient supply of thyroid stimulating hormone (TSH). The symptoms of underactivity can be eliminated by taking thyroxine orally.

Overactivity (hyperthyroidism) leads to an increased metabolic rate. This results in an increased heart and ventilation rate and a raised body temperature. Nervousness, restlessness and irritability are other symptoms. A goitre may also be apparent. Extreme cases may result in such overactivity that heart failure occurs. This condition is called **thyrotoxicosis**. One main cause of overactivity is a blood protein which stimulates the thyroid to increase its production of triiodothyronine and thyroxine. Controlling overactivity is achieved by the surgical removal of part of the thyroid gland or the destruction of part of the gland by some means, e.g. administration of radioactive iodine.

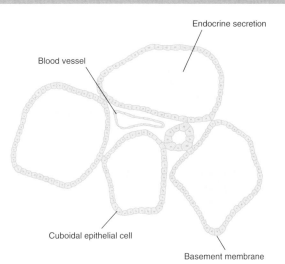

Endocrine secretion

Blood vessel

Cuboidal epithelial cell

Basement membrane

Fig. 26.6 TS Thyroid gland of monkey

The overall effect is to control the metabolic rate of cells and in so doing the hormones work in close conjunction with insulin, adrenaline and cortisone.

Both triiodothyronine and thyroxine are derivatives of the amino acid tyrosine and both contain iodine. Thyroxine possesses four iodine molecules while triiodothyronine has only three. In times of iodine shortage the latter is produced in preference to the former in order to make maximum use of limited iodine. If the iodine supply is severely reduced, the thyroid is unable to make adequate supplies of these hormones and underactivity of the thyroid results.

Calcitonin

Calcitonin is concerned with calcium metabolism. Calcium, in addition to being a major constituent of bones and teeth, is essential for blood clotting and the normal functioning of muscles and nerves. In conjunction with parathormone from the parathyroid gland, calcitonin controls the level of calcium ions (Ca^{2+}) in the blood. A peptide of 32 amino acids, calcitonin is produced in response to high levels of Ca^{2+} in the blood and it causes a reduction in the Ca^{2+} concentration.

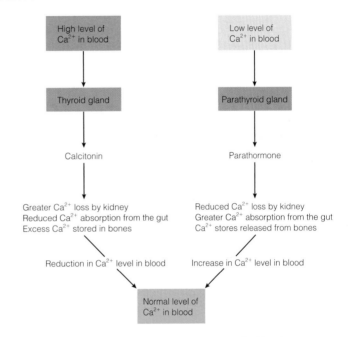

Fig. 26.7 *Summary of the control of calcium in the blood*

26.5 The adrenal glands

Situated above each kidney in humans is a collection of cells weighing about 5 g. These are the adrenal glands. They have two separate and independent parts:

1. The adrenal cortex – This consists of the outer region of the glands.

2. The adrenal medulla – This consists of the inner region of the glands.

26.5.1 The adrenal cortex

Making up around 80% of the adrenal gland, the cortex produces a number of hormones which have relatively slow, long-lasting effects on body metabolism, kidney function, salt balance and blood pressure. All the hormones produced are **steroids** formed from **cholesterol**. Being lipid-soluble they are able to pass across cell membranes along with their receptor molecule (Section 26.1.2). Hormones from the adrenal cortex are collectively called **corticoids** and fall into two groups:

1. Glucocorticoids which are concerned with glucose metabolism.

2. Mineralocorticoids which are concerned with mineral metabolism.

Glucocorticoid hormones

This group of hormones includes **cortisol** which is produced in response to stress. In stressful situations like shock, pain, emotional distress, extreme cold or infection, the hypothalamus induces the anterior pituitary gland to produce adrenocorticotrophic hormone (ACTH). This in turn causes the adrenal cortex to increase its production of glucocorticoids, including cortisol. Where stress is prolonged, the size of the adrenal glands increases. The glucocorticoid hormones combat stress in a number of ways:

1. Raising the blood sugar level, partly by inhibiting insulin and partly by the formation of glucose from fats and proteins.

2. Increasing the rate of glycogen formation in the liver.

3. Increasing the uptake of amino acids by the liver. These may either be deaminated to form more glucose or used in enzyme synthesis.

Mineralocorticoid hormones

This group of hormones includes **aldosterone** which regulates water retention by controlling the distribution of sodium and other minerals in the tissues. Aldosterone cannot increase the total sodium in the body, but it can conserve that already present. This it achieves by increasing the reabsorption of sodium (Na^+) and chloride (Cl^-) ions by the kidney, at the expense of potassium ions which are lost in urine. Control of aldosterone production is complex. In response to a low level of sodium ions in the blood, or a reduction in the total volume of blood, special cells in the kidney produce **renin** which in turn activates a plasma protein called **angiotensin**. It is angiotensin which stimulates production of aldosterone from the adrenal cortex. This causes the kidney to conserve both water and sodium ions. Angiotensin also affects centres in the brain creating a sensation of thirst, in response to which the organism seeks and drinks water, thus helping to restore the blood volume to normal.

26.5.2 The adrenal medulla

The central portion of the adrenal gland is called the **adrenal medulla**. It produces two hormones, **adrenaline** (epinephrine) and **noradrenaline** (norepinephrine). Both are important in preparing the body for action. The cells producing them are modified neurones, and noradrenaline is produced by the neurones of the sympathetic nervous system. These hormones therefore link the nervous and endocrine systems. They are sometimes called the 'flight or fight hormones' as they prepare

TABLE 26.3 **Effects of the hormones adrenaline and noradrenaline on the body and the purpose of these responses**

Effect	Purpose
Bronchioles dilated	Air is more easily inhaled into the lungs. More oxygen is therefore made available for the production of energy by glucose oxidation
Smooth muscle of the gut relaxed	The diaphragm can be lowered further, increasing the amount of air inhaled at each breath, making more oxygen available for the oxidation of glucose
Glycogen in the liver converted to glucose	Increases blood sugar level, making more glucose available for oxidation
Heart rate increased	Increase the rate at which oxygen and glucose are distributed to the tissues
Volume of blood pumped at each beat increased	
Blood pressure increased	
Blood diverted from digestive and reproductive systems to muscles, lungs and liver	Blood rich in glucose and oxygen is diverted from tissues which have less urgent need of it to those more immediately involved in producing energy
Peristalsis and digestion inhibited	Reduction of these processes allows blood to be diverted to muscle and other tissues directly involved in exertion
Sensory perception increased	Heightened sensitivity produces a more rapid reaction to external stimuli
Mental awareness increased	Allows more rapid response to stimuli received
Pupils of the eyes dilated	Increases range of vision and allows increased perception of visual stimuli
Hair erector muscles contract	Hair stands upright. In many mammals this gives the impression of increased size and may be sufficient to frighten away an enemy

Imbalance of adrenal hormones

Where the production of glucocorticoids is deficient, a condition known as **Addison's disease** occurs. Symptoms include a low blood sugar level, reduced blood pressure and fatigue. The condition of the body deteriorates when stresses such as extreme temperatures and infection are experienced. Over-production of glucocorticoids causes **Cushing's syndrome** where there is high blood sugar level mainly due to excessive breakdown of protein. This breakdown causes wasting of tissues, especially muscle. There is high blood pressure and symptoms of diabetes.

Over-production of aldosterone, often as a result of a tumour, leads to excessive sodium retention by tissues; high blood pressure and headaches then arise. The retention of sodium leads to a consequent fall in potassium levels leading to muscular weakness. Under-production of aldosterone leads to a fall in the level of sodium in the tissues. In extreme cases this is fatal.

an organism to either flee from or face an enemy or stressful situation. The effects of both hormones are to prepare the body for exertion and to heighten its responses to stimuli. These effects and their purposes are summarized in Table 26.3.

At a cellular level, adrenaline acts as the first messenger and combines with receptors on the membrane of liver and muscle cells. The hormone–receptor complex on the outer face of the membrane activates the enzyme **adenylate cyclase** which is on the inner face of the membrane. The adenylate cyclase causes the conversion of ATP to cyclic AMP. The cyclic AMP acts as a **second messenger** (intracellular mediator) in that it moves within the cell to activate enzymes such as those involved in glycogen breakdown. The process is a complex series of enzyme reactions in a chain reaction known as a **cascade effect**. The cascade effect amplifies the response. Fig. 26.2 illustrates the mechanism of adrenaline action.

In one respect adrenaline and noradrenaline differ. Whereas adrenaline dilates blood vessels, noradrenaline constricts them. This difference explains the constriction of blood vessels around the gut while those supplying muscles, lungs and liver are dilated. It appears that receptors on some blood vessels are sensitive to noradrenaline and so constrict while others are sensitive to adrenaline and so dilate.

26.6   The pancreas

The structure of the pancreas and its rôle as an exocrine gland are considered in Section 15.4.4. At intervals within the exocrine cells are the **islets of Langerhans** which are part of the endocrine system. Cells known as α-cells produce the hormone **glucagon** whereas β-cells secrete the hormone **insulin**. The two operate

APPLICATION

Diabetes

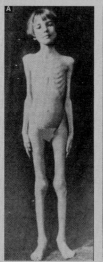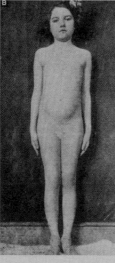

Improvement of health of diabetic child in response to insulin treatment

There are two medical conditions known as 'diabetes'. **Diabetes insipidus** is rare and results from a deficiency of ADH, the hormone which controls urine production. **Diabetes mellitus**, or 'sugar diabetes' is much more common and occurs in two forms: insulin-dependent, or type 1 diabetes, which usually develops before the age of 20 and non-insulin-dependent, or type 2 diabetes, which often comes on in later life and may be associated with obesity. Insulin-dependent diabetes usually results from a massive loss of insulin-secreting β-cells from the pancreas. The lack of insulin means that the uptake of glucose from the blood after a meal is not promoted. Glucagon continues to convert glycogen to glucose and proteins and lipids are broken down to release even more glucose. This leads untreated sufferers to have a thin, wasted appearance.

Excess blood glucose is excreted via the kidneys and is associated with a considerable loss of water resulting in thirst. The fatty acids from lipid breakdown form ketone bodies which may build up in the blood, lowering its pH and leading to coma. Diabetes was a fatal disease until in 1921 Banting and Best succeeded in isolating insulin from the pancreas of pigs and cows. Insulin is a small protein comprising a total of 51 amino acids in two chains. The sequence of amino acids in insulin was determined by Sanger in the 1950s and more recently the gene for human insulin has been isolated and inserted into the DNA of *E. coli* so that the bacteria can make 'human' insulin on a large scale.

Diabetics must regulate their carbohydrate intake, test their blood glucose level and have regular injections of insulin if they are to lead a normal life and avoid some of the longer-term complications of the disease. Such complications include atherosclerosis and degeneration of the kidney, nerves and retina. Insulin cannot be taken orally because, as a protein, it would be broken down by digestive enzymes.

About 90% of diabetics are non-insulin-dependent diabetics. There seems to be some evidence for a genetic predisposition to this form of diabetes and it may be associated with obesity. Some of these patients seem to have receptors which are less sensitive to insulin than normal but the condition can be controlled by limiting the intake of carbohydrates. Other mature-onset diabetics who secrete low amounts of insulin may be helped by drugs such as sulphonylureas which can be used to stimulate insulin production.

antagonistically, with glucagon stimulating the breakdown of glycogen to glucose while insulin initiates the conversion of glucose to glycogen. The mechanism by which these two control the level of sugar in blood is described in Section 25.3.

A summary of the major endocrine glands and the functions of the hormones they produce is given in Table 26.4.

TABLE 26.4   The major endocrine glands and the effects of the hormones they produce

Endocrine gland	Hormone produced	Effect
Pituitary	The pituitary hormones and their various effects are given in Table 26.2	
Thyroid	Triiodothyronine (T_3) Thyroxine (T_4)	Regulate growth and development of cells by affecting metabolism
	Calcitonin	Lowers calcium level of blood
Parathyroid	Parathormone	Raises blood calcium level while lowering that of phosphate
Adrenal cortex	Glucocorticoid hormones, e.g. cortisol	Helps body resist stress by raising blood sugar level and blood pressure
	Mineralocorticoid hormones, e.g. aldosterone	Increases reabsorption of sodium by kidney tubules
Adrenal medulla	Adrenaline Noradrenaline	Prepare body for activity in emergency or stressful situations
Pancreas (islets of Langerhans)	Insulin (from β-cells)	Lowers blood sugar level by stimulating conversion of glucose to glycogen
	Glucagon (from α-cells)	Raises blood sugar level by stimulating conversion of glycogen to glucose
Stomach wall	Gastrin	Initiates secretion of gastric juice
Duodenum	Secretin	Stimulates production of bile by the liver and mineral salts by the pancreas
	Cholecystokinin-pancreozymin	Causes contraction of the gall bladder and stimulates the pancreas to produce enzymes
Kidney	Renin	Activates the plasma protein angiotensin
Testis	Testosterone	Produces male secondary sex characteristics
Ovary	Oestrogen (from follicle cells)	Produces female secondary sex characteristics
	Progesterone (from corpus luteum)	Inhibits ovulation and generally maintains pregnancy
Placenta	Chorionic gonadotrophin	Maintains the presence of the corpus luteum in the ovary

26.7   Other hormone-like substances

Silk moths (*Bombyx mori*) mating

There exist in organisms a number of substances which function in a very similar way to hormones.

26.7.1 Pheromones

Pheromones also form part of the chemical coordination system of certain organisms. Unlike hormones, they operate not within an individual but between members of a species. For this reason they have been called 'social hormones'.

The female silk moth (*Bombyx mori*) produces a scent which attracts male silk moths. So sensitive is the male moth to the pheromone that it can be attracted to a female who is many kilometres away. Commercial variants of such pheromones have been used to control moth populations by luring individuals to their death.

Ants, termites and bees all produce chemicals which aid others in their social groups to locate a food source. In the fire-ant, a volatile liquid is spotted on the ground at intervals when an individual returns to the nest having found a food supply. Other ants in the group follow the scent to the food and they too deposit pheromone on their return, thus reinforcing the trail. Ants also produce an 'alarm pheromone' which warns of danger. In low concentrations it attracts other ants to their aid. These ants also release pheromone, thus increasing its concentration. At very high levels other ants become repelled. This is presumably a mechanism to avoid the whole colony being killed when the danger is beyond their control.

Worker bees lick a pheromone produced by the mandibular glands of the queen which she spreads over her body. The workers transfer the chemical among themselves. It prevents maturation of their ovaries, maintaining their sterility, and so controlling the size of the colony.

26.7.2 Prostaglandins

A group of specialized lipids which act as chemical messengers are the prostaglandins. They differ from hormones in that they are not produced at a discrete site nor do they act at positions far from their origins. Originally discovered in semen, it now appears they are produced throughout the body. Not surprisingly, prostaglandins have diverse effects. These include:

1. Initiation of uterine contractions (this perhaps explains their presence in semen. They would presumably cause the uterus to contract helping to carry the sperm into the oviducts).

2. Regulation of smooth muscle tone.

3. Assistance in blood clotting.

4. Promotion of inflammation in response to injury and infection.

5. Control of the secretion of certain hormones.

26.7.3 Endorphins (natural opiates)

It has been known for some time that many drugs, including **morphine** which is derived from the opium poppy, relieve pain by becoming bound to specific receptor sites on cells in the human brain. The question was therefore posed, why should the brain possess such sites? Do they exist in order to bind natural chemicals rather than externally administered ones? In due course natural opiates were discovered and named endorphins ('inside morphine').

Endorphins operate in ways similar to both hormones and neurotransmitters. Their effects include:

1. Relief of pain.

2. Reduction in thyroxine activity.

3. Lowered ventilation and cardiac rate.

4. Water conservation.

5. Influence on hibernation.

Endorphins are peptides which include some amino acid sequences found in several hormones. Further work is necessary to determine the exact rôle of these substances.

26.8   Questions

1. Discuss the principles and functions of chemical coordination in living organisms.   (*Total 30 marks*)

ULEAC January 1994, Paper III, No. 9(b)

2. Thyroxine is a hormone involved in the control of metabolic activity.

  (*a*)   (i) Explain how thyroxine affects metabolic activity.
        (ii) How is thyroxine transported around the body?   (*4 marks*)

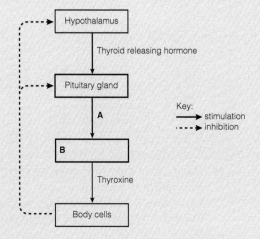

  (*b*) The figure shows some of the factors regulating the secretion of thyroxine.
     (i) Identify **A** and **B**.
    (ii) From which region of the pituitary gland is substance **A** released?
   (iii) The regulation of thyroxine secretion is an example of *negative feedback*. Explain the meaning of this term.   (*5 marks*)

A daily dose of 1 mg of thyroxine increases the adult metabolic requirement from 10 500 kJ per day to 14 700 kJ per day. If the equivalent of 10 500 kJ are consumed in the diet, then body energy reserves will be used.

  (*c*) Suggest why a daily dose of 1 mg of thyroxine would not result in continued weight loss.   (*3 marks*)
     (*Total 12 marks*)

UCLES (Modular) June 1992, (Transport, regulation and control), No. 1

3.   (*a*) Various activities in the body are controlled by hormones. Complete the following table by writing the name of a hormone involved in the activity described and the name of the gland which produces it.   (*20 marks*)

Function	Hormone	Gland
1. decreases blood sugar levels		
2. controls the basal metabolic rate		
3. controls secretion of gastric juice		
4. increases rate of heart beat		
5. controls retention of sodium ions in kidney		
6. alters permeability of distal convoluted tubule		
7. controls growth of the ovarian follicle		
8. controls uterine growth and development		
9. involved in ovulation and maintenance of the corpus luteum		
10. controls male secondary characteristics		

  (*b*) State **two** advantages in the body of controlling physiological processes by the action of hormones compared with nervous control.   (*2 marks*)
  (*c*) A small patch of tissue in a mammal is known to be secretory in function. Outline an investigation to discover whether the secretion acts as a hormone.   (*3 marks*)
     (*Total 25 marks*)

O&CSEB June 1991, Paper I, No. 4

4. This question is about a veterinary investigation of a sick cat, thought to be suffering from a malfunction of the thyroid gland.

  (*a*) Give **one** symptom that might have led the vet to suspect a thyroid hormone deficiency.   (*1 mark*)

A sample of tissue was taken from the thyroid gland of the sick cat and a stained section prepared. This was compared with a similar section from a healthy cat. Drawings of parts of these sections are shown at the top of the next page.

  (*b*) Suggest a function for the lumen of the follicle.   (*1 mark*)
  (*c*)   (i) Name the tubular structure labelled **X**.
        (ii) Give **one** reason why there should be many such tubular structures (**X**) in thyroid tissue.   (*2 marks*)

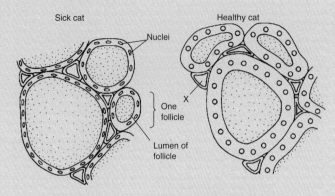

Sick cat — Nuclei — One follicle — X — Lumen of follicle — Healthy cat

(d) Describe **two** ways, apparent from the drawings, in which the thyroid of the sick cat differed from that of the healthy cat.
(2 marks)

(e) The investigator thought that the cat might be suffering as a result of a lack of iodine in its diet. Why should this element be needed for healthy thyroid function? *(1 mark)*

The investigator arranged for both cats to receive radioactive iodine compounds and then measured the levels of radioactivity in their blood. These levels were never high enough to cause harmful effects. The results are shown in the graph.

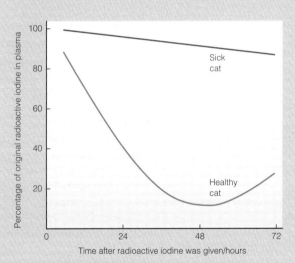

(f) (i) Do these results support the hypothesis that a dietary deficiency of iodine was the cause of the illness? Explain your answer.
(ii) Suggest why the curve for the healthy cat rose after about 48 hours.
(iii) Suggest a treatment which might enable the sick cat to recover. *(4 marks)*
(Total 11 marks)

NEAB June 1990, Paper IIA, No. 3

5. Suggest explanations for the following aspects of diabetes mellitus and its treatment.
(a) Glucose may be present in the urine of a diabetic. *(2 marks)*
(b) A diabetic cannot take insulin orally in tablet form; it must be injected. *(2 marks)*
(c) A person whose diabetes is treated by diet and not by insulin injection must not eat pure sugars but should obtain carbohydrate from bread, potatoes, etc. *(2 marks)*
(Total 6 marks)

AEB June 1993, Paper I, No. 5

6. The diagram below outlines the interaction of the pituitary gland and the testes.

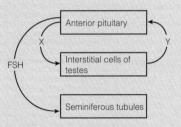

Anterior pituitary — X — Y — FSH — Interstitial cells of testes — Seminiferous tubules

(a) Name hormones X and Y. *(2 marks)*
(b) What effect does FSH have on the seminiferous tubules? *(1 mark)*
(c) How does the secretion of hormones X and Y illustrate the principle of negative feedback? *(2 marks)*
(Total 5 marks)

SEB Human Biology Specimen 1992, Higher Grade Paper II, No. 2

7. Briefly explain the reasons for the following medical practices.
(a) Thyroxine may be administered to young children who show symptoms of cretinism. *(3 marks)*
(b) Patients suffering from diabetes are treated with insulin. *(3 marks)*
(c) In order to avoid unwanted pregnancies, a human female may be prescribed a contraceptive pill that contains oestrogen (oestradiol) or a combination of this hormone and progesterone. *(4 marks)*
(Total 10 marks)

Oxford Summer 1994, (AS) Paper 2 Option E, No. 19

27   The nervous system

The ability to respond to stimuli is a fundamental characteristic of living organisms. While all cells of multicellular organisms are able to perceive stimuli, those of the nervous system are specifically adapted to this purpose.

The nervous system performs three functions:

1. To collect information about the internal and external environment.

2. To process and integrate the information, often in relation to previous experience.

3. To act upon the information, usually by coordinating the organism's activities.

One remarkable feature of the way in which these functions are performed is the speed with which the information is transmitted from one part of the body to another. In contrast to the endocrine system (Chapter 26), the nervous system responds virtually instantaneously to a stimulus. The cells which transmit nerve impulses are called **neurones**.

The nervous system may be sub-divided into a number of parts. The collecting of information from the internal and external environment is carried out by **receptors**. Along with the neurones which transmit this information, the receptors form the **sensory system**. The processing and integration of this information is performed by the **central nervous system (CNS)**. The final function whereby information is transmitted to **effectors**, which act upon it, is carried out by the **effector (motor) system**, which has two parts. The portion which activates involuntary responses is known as the **autonomic nervous system** whereas that activating voluntary responses is termed the **somatic system**. The sensory and effector (motor) neurones are sometimes collectively called the **peripheral nervous system (PNS)**.

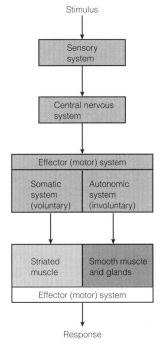

Fig. 27.1 Interrelationships of the various components of the nervous system

27.1   Nervous tissue and the nerve impulse

Nervous tissue comprises closely packed nerve cells or **neurones** with little intercellular space. The neurones are bound together by connective tissue.

All neurones have a cell body containing a nucleus. This cell body has a number of processes called **dendrites** which transmit impulses to the cell body. Impulses leave via the **axon** which

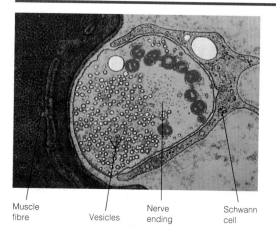

Muscle
fibre   Vesicles   Nerve
ending   Schwann
cell

False-colour transmission EM of neuromuscular
junction (×18 500 approx.)

may be several metres in length. Some axons are covered by a
fatty **myelin sheath** formed by **Schwann cells**.

Nerve fibres may be bundled together and wrapped in
connective tissue to form **nerves**. Nerves may be **sensory**,
comprising sensory neurones, **effector (motor)**, comprising
effector (motor) neurones, or **mixed**, with both types present.

27.1.1 Resting potential

In its normal state, the membrane of a neurone is negatively
charged internally with respect to the outside. The potential
difference varies somewhat depending on the neurone but lies in
the range 50–90 mV, most usually around 70 mV. This is known
as the **resting potential** and in this condition the membrane is
said to be **polarized**. The resting potential is the result of the
distribution of four ions: potassium (K^+), sodium (Na^+), chloride
(Cl^-) and organic anions (COO^-). Initially the concentration of

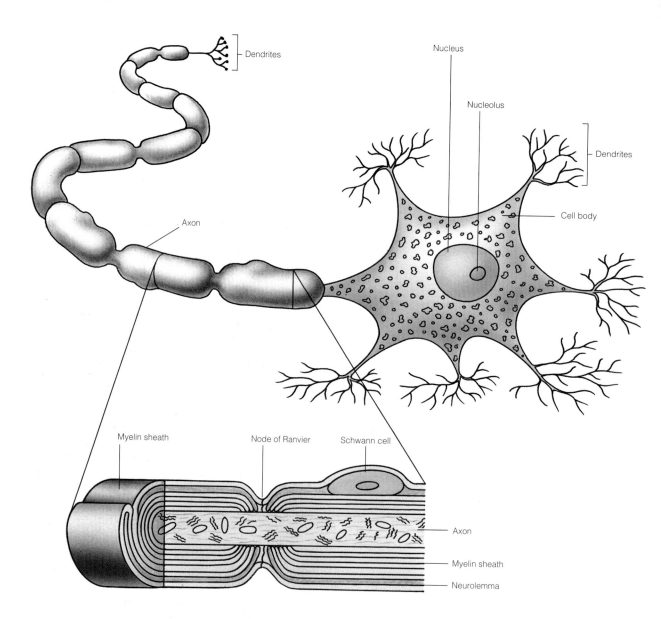

Fig. 27.2 Effector (motor) neurone

APPLICATION

Multiple sclerosis and the myelin sheath

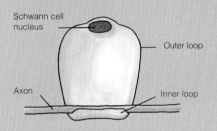

The myelin sheath is produced by the Schwann cell and is rolled around the nerve axon for insulation.

Fig. 1   Schwann cell and axon

On neurones a specialized cell, the **Schwann cell**, wraps itself around the nerve axon to form the numerous concentric layers which comprise the myelin sheath (Fig. 1). This sheath consists of 70% lipid and 30% proteins in the usual bilipid structure; it provides insulation and allows the rapid conduction of electrical signals.

In multiple sclerosis (MS) gradual degradation of the myelin sheath takes place leaving areas of bare, demyelinated axons which cannot conduct impulses (Fig. 2). These regions, known as **plaques**, are approximately 2–10 mm in size and most commonly affect the optic nerve, cerebellum, cervical spinal cord and the area around the ventricles of the brain. Peripheral nerves are unaffected. The most well known symptoms of MS are a weakness of the limbs, 'pins and needles' and numbness, but damage to the optic nerve also results in blurring of the vision and pains in the eyes.

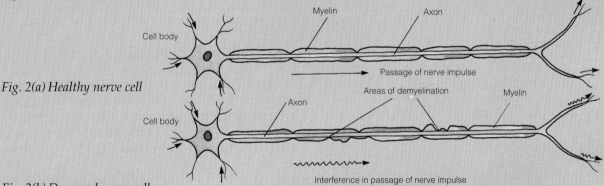

Fig. 2(a) Healthy nerve cell

Fig. 2(b) Damaged nerve cell in multiple sclerosis

MS has two main patterns of disease development. In some people the condition is progressive and unrelenting, leading to severe crippling, but in most there are periods of relapse followed by spells of remission which may last several years. Although in the later stages there is often progressive disability, only about 1 in 10 MS sufferers end up in a wheelchair.

Multiple sclerosis is one of the commonest diseases of the central nervous system in Europe and yet its cause is still unknown. It affects more women than men, in a ratio of 3 : 2 and there is a high prevalence in countries with a temperate climate; it is uncommon in tropical countries. It has been suggested that possibly diet, lifestyle and physical environment have some influence but no theory has yet been proved. There are approximately 80 000 people with MS in the United Kingdom and 250 000 in the USA. The range of onset of the disease varies from 12 years old to 50 years old but the average age of diagnosis is late 20s to mid 30s.

As yet multiple sclerosis cannot be cured and treatment is very limited. Steroids and ACTH (adrenocorticotrophic hormone) are prescribed during relapse to reduce inflammation and promote a remission but otherwise treatment is supportive in the form of pain relievers, muscle relaxants, occupational therapy and physiotherapy, as well as counselling for the individuals and their families.

(a) Unipolar neurone
e.g. arthropod motor neurone

(b) Bipolar neurone
e.g. from mammalian retina

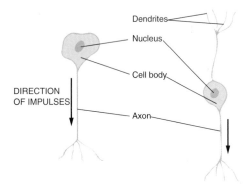

Dendrites

Nucleus

Cell body

DIRECTION
OF IMPULSES

Axon

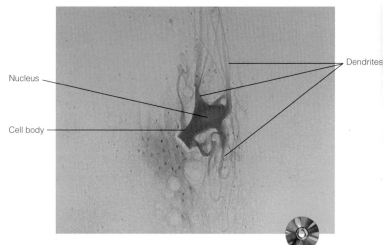

Nucleus

Cell body

Dendrites

Multipolar neurone (×200 approx.)

(c) Branched unipolar neurone
e.g. vertebrate sensory neurone

(d) Multipolar neurone
e.g. from the mammalian spinal cord

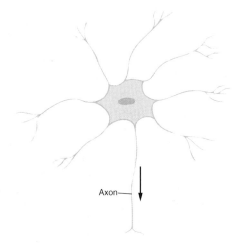

Axon

Fig. 27.3 Types of neurone

potassium (K^+) and organic anions (COO^-) is higher inside the neurone, while the concentration of sodium (Na^+) and chloride (Cl^-) is higher outside. The membrane, however, is considerably more permeable to potassium ions (K^+) than any of the others. As the potassium ion (K^+) concentration inside the neurone is twenty times greater than that outside, potassium ions (K^+) rapidly diffuse out. This outward movement of positive ions means that the inside becomes slightly negative relative to the outside. As more potassium ions move out, the less able they are to do so. In time an equilibrium is reached whereby the rate at which they leave is exactly balanced by the rate of entry. It is therefore the electrochemical gradient of potassium ions which largely creates the resting potential.

The differences in concentration of ions across the membrane are maintained by the active transport of the ions against the concentration gradients. The mechanisms by which these ions are transported are called pumps. As sodium ions (Na^+) are moved in this way they are often referred to as **sodium pumps**. However, as potassium ions are also actively transported they are more accurately **cation pumps**. These cation pumps exchange sodium and potassium ions by actively transporting in potassium ions and removing sodium ions. Being active, this transport requires ATP.

27.1.2 Action potential

By appropriate stimulation, the charge on a neurone can be reversed. As a result, the negative charge inside the membrane of $-70\,mV$ changes to a positive charge of around $+40\,mV$. This is known as the **action potential** and in this condition the membrane is said to be **depolarized**. Within about 2 milliseconds (two thousandths of a second) the same portion of the membrane returns to resting potential ($-70\,mV$ inside). This is known as **repolarization**. These changes are illustrated graphically in Fig. 27.4.

Provided the stimulus exceeds a certain value, called the **threshold value**, an action potential results. Above the threshold value the size of the action potential remains constant, regardless of the size of the stimulus. In other words, the action potential is

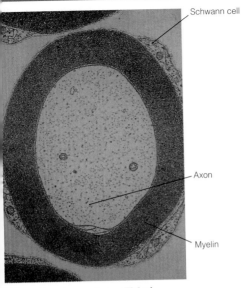

False-colour transmission EM of
myelinated nerve fibre

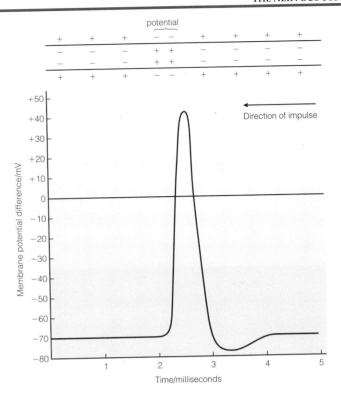

Fig. 27.4 The action potential

either generated, in which case it is always the same, or it is not.
This is called an **all or nothing** response. The size of the action
potential does not decrease as it is transmitted along the neurone
but always remains the same.

27.1.3 Ion movement

The action potential is the result of a sudden increase in the
permeability of the membrane to sodium. This allows a sudden
influx of sodium ions because there is a high concentration
outside which has been maintained by the sodium pump. The
influx of sodium ions begins to depolarize the membrane and
this depolarization in turn increases the membrane's
permeability to sodium, leading to greater influx and further
depolarization. This runaway influx of sodium ions is an
example of **positive feedback**. When sufficient sodium ions have
entered to create a positive charge inside the membrane, the
permeability of the membrane to sodium starts to decrease.

At the same time as the sodium ions begin to move inward, so
potassium ions start to move in the opposite direction along a
diffusion gradient. This outward movement of potassium is,
however, much less rapid than the inward movement of sodium.
It nevertheless continues until the membrane is repolarized. The
changes are summarized in Fig. 27.5.

So why is the movement of ions so rapid? We saw in Section
4.2.2 that some protein molecules span membranes and these
have a fine pore or channel through the middle of them. In a
neurone membrane some of these channels allow sodium ions
(Na^+) to pass through while others permit the movement of
potassium (K^+). In the resting state these channels are closed, but
when the membrane is depolarized by a stimulus they open.

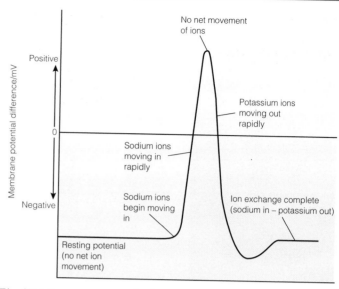

Fig. 27.5 Ion movements during an action potential

They act rather like an entry with a closed gate which can be unlatched. For this reason they are sometimes called **voltage-gated channels**. The gates to the sodium channel open more quickly than those to the potassium channel. This explains why sodium ions entering the neurone cause depolarization followed by the potassium ions leaving which cause repolarization.

27.1.4 Refractory period

Following an action potential, the outward movement of potassium ions quickly restores the resting potential. However, for about one millisecond after an action potential the inward movement of sodium is prevented in that region of the neurone. This means that a further action potential cannot be generated for at least one millisecond. This is called the **refractory period**.

The refractory period is important for two reasons:

1. It means the action potential can only be propagated in the region which is not refractory, i.e. in a forward direction. The action potential is thus prevented from spreading out in both directions until it occupies the whole neurone.

2. By the end of the refractory period the action potential has passed further down the nerve. A second action potential will thus be separated from the first one by the refractory period which therefore sets an upper limit to the frequency of impulses along a neurone.

The refractory period can be divided into two portions:

1. The **absolute refractory period** which lasts around 1 ms during which no new impulses can be propagated however intense the stimulus.

2. The **relative refractory period** which lasts around 5 ms during which new impulses can only be propagated if the stimulus is more intense than the normal threshold level.

Fig. 27.6 illustrates the refractory period in graph form, while Fig. 27.7 demonstrates how the refractory period determines the frequency of impulses along a neurone.

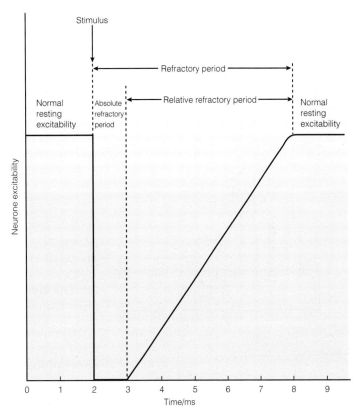

Fig. 27.6 Graph illustrating neurone excitability before and after a nerve impulse

Where the stimulus is at the threshold value the excitability of the neurone must return to normal before a new action potential can be formed. In the time interval shown, this allows just two action potentials to pass i.e. a low frequency of impulses. Where the stimulus exceeds the threshold value a new action potential can be created before neurone excitability returns to normal. In the time interval shown this allows six action potentials to pass, i.e. a high frequency of impulses.

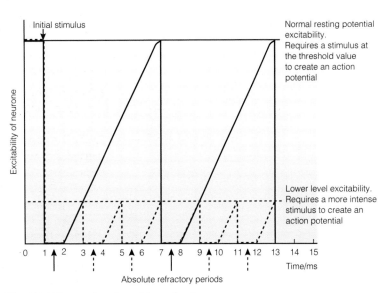

Fig. 27.7 Determination of impulse frequency

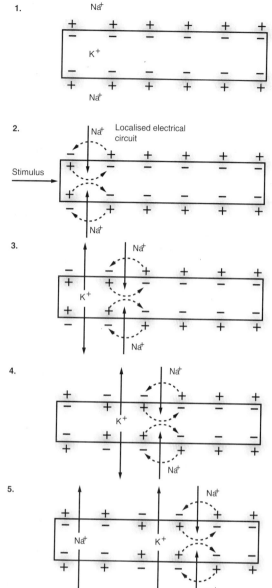

Fig. 27.8 Transmission of an impulse along an unmyelinated neurone

1. At resting potential there is a high concentration of sodium ions outside and a high concentration of potassium ions inside it.
2. When the neurone is stimulated sodium ions rush into the axon along a concentration gradient. This causes depolarization of the membrane.
3. Localized electrical circuits are established which cause further influx of sodium ions and so progression of the impulse. Behind the impulse, potassium ions begin to leave the axon along a concentration gradient.
4. As the impulse progresses, the outflux of potassium ions causes the neurone to become repolarized behind the impulse.
5. After the impulse has passed and the neurone is repolarized sodium is once again actively expelled in order to increase the external concentration and so allow the passage of another impulse.

27.1.5 Transmission of the nerve impulse

Once an action potential has been set up, it moves rapidly from one end of the neurone to the other. This is the nerve impulse and is described in Fig. 27.8.

According to the precise nature of a neurone, transmission speeds vary from 0.5 metres msec^{-1} to over 100 metres msec^{-1}. Two factors are important in determining the speed of conduction:

(a) The diameter of the axon: the greater the diameter the faster the speed of transmission.

(b) The myelin sheath: myelinated neurones conduct impulses faster than non-myelinated ones.

The myelin sheath, which is produced by the **Schwann cells**, is not continuous along the axon, but is absent at points called **nodes of Ranvier** which arise every millimetre or so along the neurone's length. As the fatty myelin acts as an electrical insulator, an action potential cannot form in the part of the axon covered with myelin. They can, however, form at the nodes. The action potentials therefore jump from node to node (**saltatory conduction**), increasing the speed with which they are transmitted (Fig. 27.9).

The insulating myelin causes ion exchange to occur at the nodes of Ranvier. The impulse therefore jumps from node to node.

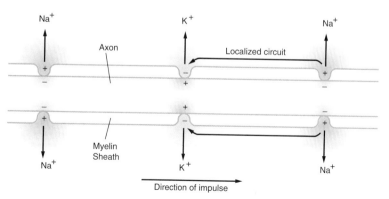

Fig. 27.9 Transmission of an impulse along a myelinated neurone

27.2 The synapse

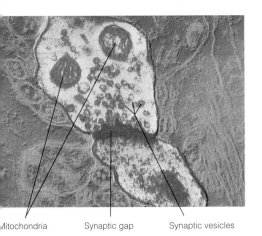

Mitochondria   Synaptic gap   Synaptic vesicles

False-colour transmission EM of synapse
(×34 000 approx.)

The word synapse (*syn* = 'with'; *apsis* = 'knot') means 'to clasp'. It is the point where the axon of one neurone clasps or joins the dendrite or cell body of another. The gap between the two is around 20 nm in width. The synapse must in some way pass information across itself from one neurone to the next. This is achieved in the vast majority of synapses by **chemical** transmission, although at some synapses the transmission is **electrical**. Chemicals which transmit messages across the synapse are called **neurotransmitter substances**. The two main ones in the peripheral nervous system are **acetylcholine** and **noradrenaline** although others include dopamine and serotonin. Neurones using acetylcholine as a neurotransmitter are termed **cholinergic neurones** whereas those using noradrenaline are called **adrenergic neurones**. Amino acids, e.g. L-glutamate are thought to be the most widely used neurotransmitters in the brain. In all over 40 substances with a wide variety of chemical structures act as neurotransmitters.

Did you know?

An average motor neurone may have as many as 15 000 synapses.

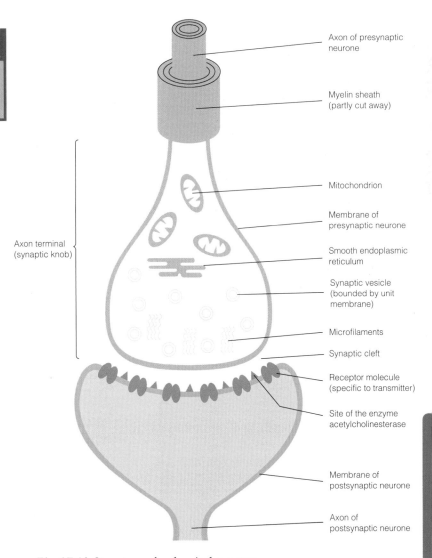

Axon of presynaptic neurone

Myelin sheath (partly cut away)

Axon terminal (synaptic knob)

Mitochondrion

Membrane of presynaptic neurone

Smooth endoplasmic reticulum

Synaptic vesicle (bounded by unit membrane)

Microfilaments

Synaptic cleft

Receptor molecule (specific to transmitter)

Site of the enzyme acetylcholinesterase

Membrane of postsynaptic neurone

Axon of postsynaptic neurone

Fig. 27.10 Structure of a chemical synapse

27.2.1 Structure of the synapse

As it is the most frequently occurring type, we shall deal here with the structure of the chemical synapse.

At the synapse the nerve axon is expanded to form a bulbous ending called the **axon terminal (bouton terminale)** or **synaptic knob**. This contains many mitochondria, microfilaments and structures called **synaptic vesicles**. The vesicles contain a neurotransmitter substance such as acetylcholine or noradrenaline. The neurone immediately before the synapse is known as the **presynaptic neurone** and is bounded by the presynaptic membrane. The neurone after the synapse is the **postsynaptic neurone** and is bounded by the postsynaptic membrane. Between the two is a narrow gap, 20 nm wide, called the **synaptic cleft**. The postsynaptic membrane possesses a number of large protein molecules known as **receptor molecules**. The structure of the synapse is illustrated in Fig. 27.10.

27.2.2 Synaptic transmission

When a nerve impulse arrives at the synaptic knob it alters the permeability of the presynaptic membrane to calcium, which therefore enters. This causes the synaptic vesicles to fuse with the membrane and discharge their neurotransmitter substance which, for the purposes of this account, will be taken to be acetylcholine. The empty vesicles move back into the cytoplasm where they are later refilled with acetylcholine.

The acetylcholine diffuses across the synaptic cleft, a process which takes 0.5 ms. Upon reaching the postsynaptic membrane it fuses with the receptor molecules. In **excitatory synapses** this opens ion channels on the postsynaptic membrane allowing sodium ions to enter and thus creating a new potential known as the **excitatory postsynaptic potential** in the postsynaptic neurone. These events are detailed in Fig. 27.11.

Once acetylcholine has depolarized the postsynaptic neurone, it is hydrolysed by the enzyme **acetylcholinesterase** which is found on the postsynaptic membrane. This breakdown of acetylcholine is essential to prevent successive impulses merging at the synapse. The resulting choline and ethanoic acid (acetyl) diffuse across the synaptic cleft and are actively transported into the synaptic knob of the presynaptic neurone into which they diffuse. Here they are coupled together again and stored inside synaptic vesicles ready for further use. This recoupling requires energy which is provided by the numerous mitochondria found in the synaptic knob.

The excitatory postsynaptic potentials build up as more neurotransmitter substance arrives until sufficient depolarization occurs to exceed the threshold value and so generate an action potential in the post-synaptic neurone. This additive effect is known as **temporal summation**. All events so far described relate to an **excitatory synapse**, but not all synapses operate in this way. Some, known as **inhibitory synapses**, respond to the neurotransmitter by opening potassium ion channels and leaving the sodium ion channels closed. Potassium therefore moves out causing the postsynaptic membrane to become more polarized. It is thus more difficult for the threshold value to be exceeded and therefore less likely that a new action potential will be created.

Did you know?

The fugu fish is considered a culinary delicacy in Japan. However it contains a highly toxic chemical which kills by blocking the sodium ion channels in nerve membranes. As a result only a few chefs are licenced to prepare this dish – even so some deaths have resulted from eating this fish.

The arrival of the impulses at the synaptic knob alters its permeability allowing calcium ions to enter.

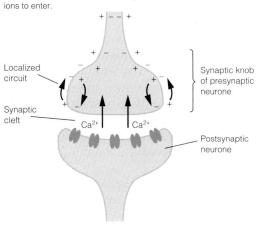

Localized circuit

Synaptic cleft

Ca²⁺   Ca²⁺

Synaptic knob of presynaptic neurone

Postsynaptic neurone

4 The influx of sodium ions generates a new impulse in the postsynaptic neurone.

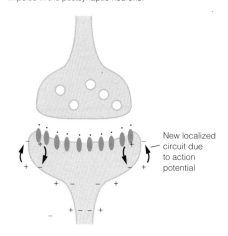

New localized circuit due to action potential

The influx of calcium ions causes the synaptic vesicle to fuse with the presynaptic membrane so releasing acetylcholine into the synaptic cleft.

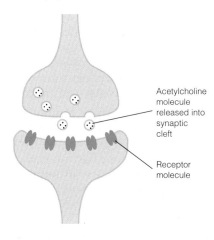

Acetylcholine molecule released into synaptic cleft

Receptor molecule

5 Acetylcholinesterase on the postsynaptic membrane hydrolyses acetylcholine into choline and ethanoic acid (acetyl). These two components then diffuse back across the synaptic cleft into the presynaptic neurone.

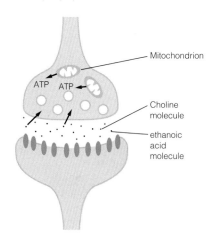

Mitochondrion

ATP   ATP

Choline molecule

ethanoic acid molecule

Acetylcholine fuses with receptor molecules on the postsynaptic membrane. This causes ion channels to open allowing sodium ions to rush in.

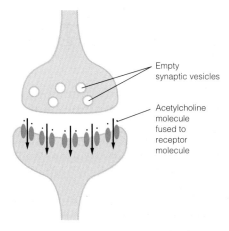

Empty synaptic vesicles

Acetylcholine molecule fused to receptor molecule

6 ATP released by the mitochondria is used to recombine choline and ethanoic acid (acetyl) molecules to form acetylcholine. This is stored in synaptic vesicles for future use.

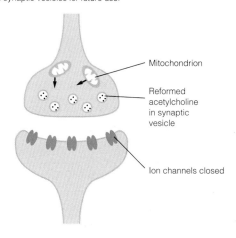

Mitochondrion

Reformed acetylcholine in synaptic vesicle

Ion channels closed

Fig. 27.11 Sequence of diagrams to illustrate synaptic transmission (only relevant detail is included in each drawing)

APPLICATION

Drugs and the synapse

There are many different neurotransmitters responsible for information exchange across synapses. Most of the psychoactive drugs available in society today, such as **ecstasy**, **cannabis** and **cocaine**, cause their effects by interfering with the synaptic transmission of one of these messengers. Those drugs which amplify the process of synaptic transmission are called **excitatory** or **agonistic** drugs, while those which inhibit synaptic transmission are known as **inhibitory** or **antagonistic** drugs.

Narcotics such as **heroin** and **morphine** mimic the actions of the neurotransmitters known as endorphins, binding to their specific receptors and blocking sensations of pain. **Nicotine** similarly mimics natural transmitters.

The presence of **caffeine** in the body raises cell metabolism, leading to the release of more neurotransmitters. **Amphetamines** cause increased release of noradrenaline by interfering with storage mechanisms. This leads to excessive activation of neurones and extra information is transmitted around the brain. A user will feel highly aroused but may also suffer damage to organs including the heart.

Some drugs affect the body by interacting with natural neurotransmitters. The **benzodiazepine tranquillizers**, such as Valium, increase the effect of the inhibitory transmitter GABA in the brain resulting in less transfer of information between neurones. In contrast phencyclidine, the active ingredient in **magic mushrooms**, interacts with excitatory transmitters in the brain and inappropriate information is passed between neurones leading to hallucinations.

It is important to remember that neurotransmitters are rapidly absorbed or broken down. **Cocaine** causes noradrenaline to 'linger' in the synapse producing effects similar to those from amphetamines.

£2m Cocaine Seizure

CRACK BARONS TARGET UK

Drugs 'cocktail' caused teenager's death hears court

27.2.3 Functions of synapses

Synapses have a number of functions:

1. Transmit information between neurones – The main function of synapses is to convey information between neurones. It is from this basic function that the others arise.

2. Pass impulses in one direction only – As the neurotransmitter substance can only be released from one side of a synapse, it ensures that nerve impulses only pass in one direction along a given pathway.

3. Act as junctions – Neurones may converge at a synapse. In this way a number of impulses passing along different neurones may between them release sufficient neurotransmitter to generate a new action potential in a single postsynaptic neurone whereas individually they would not. This is known as **spatial summation**. In this way responses to a single stimulus may be coordinated.

4. Filter out low level stimuli – Background stimuli at a constantly low level, e.g. the drone of machinery, produce a low frequency of impulses and so cause the release of only small amounts of neurotransmitter at the synapse. This is insufficient to create a new impulse in the postsynaptic neurone and so these impulses are carried no further than the synapse. Such low level stimuli are of little importance and the absence of a response to them is rarely, if ever, harmful. Any change in the level of the stimulus will be responded to in the usual way.

5. Allow adaptation to intense stimulation – In response to a powerful stimulus, the high frequency of impulses in the presynaptic neurone causes considerable release of neurotransmitter into the synaptic cleft. Continued high-level stimulation may result in the rate of release of neurotransmitter exceeding the rate at which it can be reformed. In these circumstances the release of neurotransmitter ceases and hence also any response to the stimulus. The synapse is said to be **fatigued**. The purpose of such a response is to prevent overstimulation which might otherwise damage an effector.

27.3   The reflex arc

A **reflex** is an automatic response which follows a sensory stimulus. It is not under conscious control and is therefore involuntary. The pathway of neurones involved in a reflex action is known as a **reflex arc**. The simplest forms of reflex in vertebrates include those concerned with muscle tone. An example of this is the **knee jerk reflex** which may be separated into six parts:

1. Stimulus – A blow to the tendon situated below the patella (knee cap). This tendon is connected to the muscles that extend the leg, and hitting it causes these muscles to become stretched.

2. Receptor – Specialized sensory structures, called **muscle spindles**, situated in the muscle detect the stretching and produce a nervous signal.

3. Sensory neurone – The signal from the muscle spindles is conveyed as a nervous impulse along a sensory neurone to the spinal cord.

4. Effector (motor) neurone – The sensory neurone forms a synapse inside the spinal cord with a second neurone called an effector neurone. This effector neurone conveys a nervous impulse back to the muscle responsible for extending the leg.

5. Effector – This is the muscle responsible for extending the leg. When the impulse from the effector neurone is received, the muscle contracts.

6. Response – The lower leg jerks upwards as a consequence of the muscle contraction.

This reflex arc has only one synapse, that between the sensory and effector neurone in the spinal cord. Such reflex arcs are therefore termed **monosynaptic**. These reflexes do not involve any neurones connected to the brain which therefore plays no part in the response. As the reaction is routine and predictable, not requiring any analysis, it would be wasteful of the brain's capacity to burden it with the millions of such responses that are required each day. Any reflex arc which is localized within the spinal cord and does not involve the brain is called a **spinal reflex**.

Reflexes involving two or more synapses are termed **polysynaptic**. Typical polysynaptic spinal reflexes include the withdrawal of parts of the body from painful stimuli, e.g. removal of the hand or foot from a hot or sharp object. Due to the response involved, such an action is called a **withdrawal reflex**. Fig. 27.12 illustrates a withdrawal reflex where the hand is placed on a sharp object. This reflex involves an additional stage, namely the **connector (intermediate, internuncial** or **relay) neurone**, within the spinal cord.

These simple reflexes are important in making involuntary responses to various changes in both the internal and external environment. In this way homeostatic control of things like body posture may be maintained. Control of breathing, blood pressure and other systems are likewise effected through a series of reflex responses. Another example is the reflex constriction or dilation of the iris diaphragm of the eye in

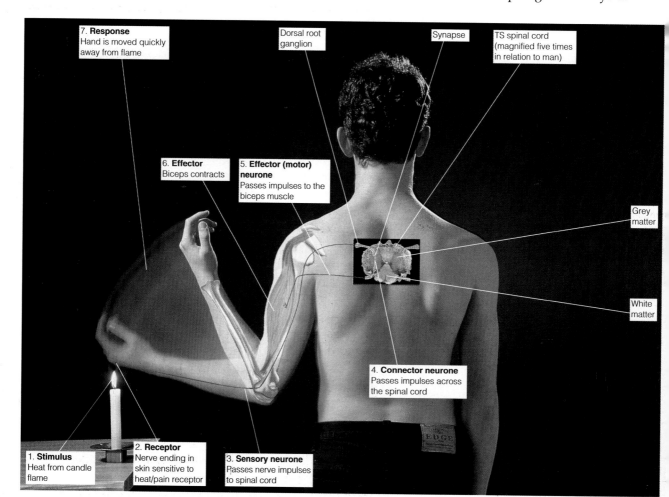

Fig. 27.12 Reflex arc involved in withdrawal from an unpleasant stimulus

response to changes in light intensity. Details of this are given in Section 27.6.1.

Brain reflexes have neurone connections with the brain and are usually far more complex, involving multiple responses to a stimulus. While reflexes are themselves involuntary, they may be modified in the light of previous experience. These are called **conditioned reflexes** and are discussed in Section 27.7.3.

27.4 The autonomic nervous system

The autonomic (*auto* = 'self'; *nomo* = 'govern') nervous system controls the involuntary activities of smooth muscle and certain glands. It forms a part of the peripheral nervous system and can be sub-divided into two parts: the **sympathetic nervous system** and the **parasympathetic nervous system**. Both systems comprise effector neurones, which connect the central nervous system to their effector organs. Each pathway consists of a **preganglionic neurone** and a **postganglionic neurone**. In the sympathetic system the synapses between the two are located near the spinal cord whereas in the parasympathetic system they are found near to, or within, the effector organ. This, and other differences, are illustrated in Fig. 27.13.

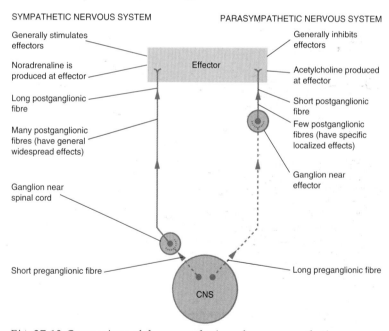

Fig. 27.13 *Comparison of the sympathetic and parasympathetic nervous systems*

The effects of the sympathetic and parasympathetic nervous systems normally oppose one another, i.e. they are **antagonistic**. If one system contracts a muscle, the other usually relaxes it. The balance between the two systems accurately regulates the involuntary activities of glands and organs. It is possible to control consciously certain activities of the autonomic nervous system through training. Control of the anal and bladder sphincters are examples of this.

Table 27.1 lists some of the effects of the sympathetic and parasympathetic nervous systems.

TABLE 27.1   Comparison of some effects of sympathetic and parasympathetic nervous systems

Sympathetic nervous system	Parasympathetic nervous system
Increases cardiac output	Decreases cardiac output
Increases blood pressure	Decreases blood pressure
Dilates bronchioles	Constricts bronchioles
Increases ventilation rate	Decreases ventilation rate
Dilates pupils of the eyes	Constricts pupils of the eyes
Contracts anal and bladder sphincters	Relaxes anal and bladder sphincters
Contracts erector pili muscles, so raising hair	No comparable effect
Increases sweat production	No comparable effect
No comparable effect	Increases secretion of tears

27.5   The central nervous system

The central nervous system (CNS) acts as the coordinator of the nervous system. It comprises a long, approximately cylindrical structure – the **spinal cord** – and its anterior expansion – the **brain**

27.5.1 The spinal cord

The spinal cord is a dorsal cylinder of nervous tissue running within the vertebrae which therefore protect it. It possesses a thick membranous wall and has a small canal, the **spinal canal**, running through the centre. The central area is made up of nerve cell bodies, synapses and unmyelinated connector neurones. This is called **grey matter** on account of its appearance. Around the grey matter is a region largely composed of longitudinal axons which connect different parts of the body. The myelin sheath around these axons give this region a lighter appearance, hence it is called **white matter**.

At intervals along the length of the spinal cord there extend spinal nerves. There are thirty-one pairs of these nerves in humans. They separate into two close to the spinal cord. The uppermost (dorsal) of these is called the **dorsal root**, while the lower (ventral) one is called the **ventral root**. The dorsal root carries only sensory neurones while the ventral root possesses only effector ones; a sort of spinal nerve one-way system. The cell bodies of the sensory neurones occur within the dorsal root, forming a swelling called the **dorsal root ganglion**. The structure of the spinal cord is illustrated in Fig 27.14.

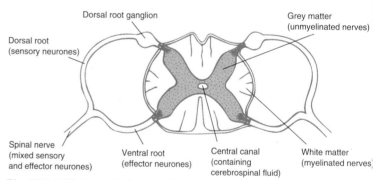

Fig. 27.14 TS through the spinal cord

27.5.2 Structure of the brain

As an elaboration of the anterior region of the spinal cord, the brain has a basically similar structure. Both grey and white matter are present, as is the spinal canal although it is expanded to form larger cavities called **ventricles**. Broadly speaking, the brain has three regions: the **forebrain**, **midbrain** and **hindbrain**.

In common with the entire central nervous system, the brain is surrounded by protective membranes called **meninges**. There are three in all and the space between the inner two is filled with **cerebro-spinal fluid**, which also fills the ventricles referred to above. The cerebro-spinal fluid supplies the neurones in the brain with respiratory gases and nutrients and removes wastes. To achieve this, it must first exchange these materials with the blood. This it does within the ventricles which are richly supplied with capillaries. Having exchanged materials, the fluid must be

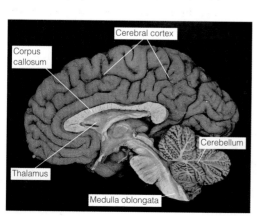

Longitudinal section through a human brain

circulated throughout the CNS in order that it may be distributed to the neurones. This function is performed by cilia found on the epithelial lining of the ventricles and central canal of the spinal cord. The structure of the brain is illustrated in Fig. 27.15.

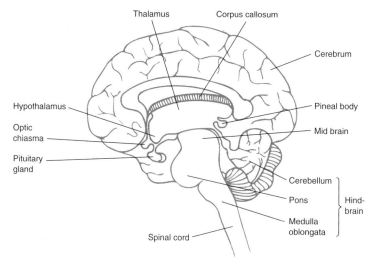

Fig. 27.15 VS through the centre of the human brain

27.5.3 Functions of the brain

The hindbrain

Medulla oblongata

This region of the brain contains many important centres of the autonomic nervous system. These centres control reflex activities, like ventilation rate (Section 20.3), heart rate (Section 21.5.3) and blood pressure (Section 21.5.4). Other activities controlled by the medulla are swallowing, coughing and the production of saliva.

Cerebellum

This is a large and complex association area concerned with the control of muscular movement and body posture. It receives sensory information relating to the tone of muscles and tendons as well as from the organs of balance in the ears. Its rôle is not to initiate movement but to coordinate it. Any damage to the cerebellum not surprisingly results in jerky and uncoordinated movement.

The midbrain

The midbrain acts as an important link between the hindbrain and the forebrain. In addition it houses both visual and auditory reflex centres. The reflexes they control include the movement of the head to fix on an object or locate a sound.

The forebrain

The thalamus

Lying as it does at the middle of the brain, the thalamus forms an important relay centre, connecting other regions of the brain. It assists in the integration of sensory information. Much of the sensory input received by the brain must be compared to previously stored information before it can be made sense of. It is the thalamus which conveys the information received to the appropriate areas of the cerebrum. Pain and pleasure appear to be perceived by the thalamus.

The hypothalamus

This is the main controlling region for the autonomic nervous system. It has two centres, one for the sympathetic nervous system and the other for the parasympathetic nervous system. At the same time it controls such complex patterns of behaviour as feeding, sleeping and aggression. Another of its rôles is to monitor the composition of the blood, in particular the plasma solute concentration, and not surprisingly therefore, it has a very rich supply of blood vessels. It is also an endocrine gland and details of this function are given in Section 26.3.

The cerebrum

In vertebrates, the size of the cerebrum relative to the body increases from fish, through amphibians and reptiles, to mammals. Even with mammals there is a graded increase in this relative size, with humans having far and away the largest. In addition, it is highly convoluted in humans, considerably increasing its surface-area and hence its capacity for complex activity.

The cerebrum is divided into left and right halves known as **cerebral hemispheres**. The two halves are joined by the **corpus callosum**. In general terms, the cerebrum performs the functions of receiving sensory information, interpreting it with respect to that stored from previous experiences and transmitting impulses along motor neurones to allow effectors to make appropriate responses. In this way, the cerebrum coordinates all the body's voluntary activities as well as some involuntary ones. In addition, it carries out complex activities like learning, reasoning and memory.

The outer 3 mm of the cerebral hemispheres is known as the **cerebral cortex** and in humans this covers an especially large area. Within this area the functions are localized, a fact verified in two ways. Firstly, if an electrode is used to stimulate a particular region of the cortex, the patient's response indicates the part of the body controlled by that region. For example, if a sensation in the hand is felt, then it is assumed that the area receives sensory information from the hand. Equally, if the hand moves, this must be the effector centre for the hand. The second method involves patients who have suffered brain damage by accidental means. If the injured person is unable to move his arm, then the damaged portion is assumed to be the effector centre for that arm.

The association areas of the cerebral cortex help an individual to interpret the information received in the light of previous experience. The **visual association area**, for example, allows objects to be recognized, and the **auditory association area** performs the same function for sounds. In humans, there are similar areas which permit understanding of speech and the written word. Yet another centre, the **speech effector centre**, coordinates the movement of the lips and tongue as well as breathing, in order to allow a person to speak coherently.

Despite the methods outlined above that are used to investigate the functions of the brain, certain areas at the front of the cerebral cortex produce neither sensation nor response when stimulated. These are aptly termed **silent areas**. It is possible that they determine certain aspects of personality as their surgical removal has been known to relieve anxiety. The patients, while tranquil, become rather irresponsible and careless, however. The major localized regions of the cerebral cortex are outlined in Fig. 27.16.

Did you know?

Public funding for medical research on Alzheimer's disease has gone down by a fifth.

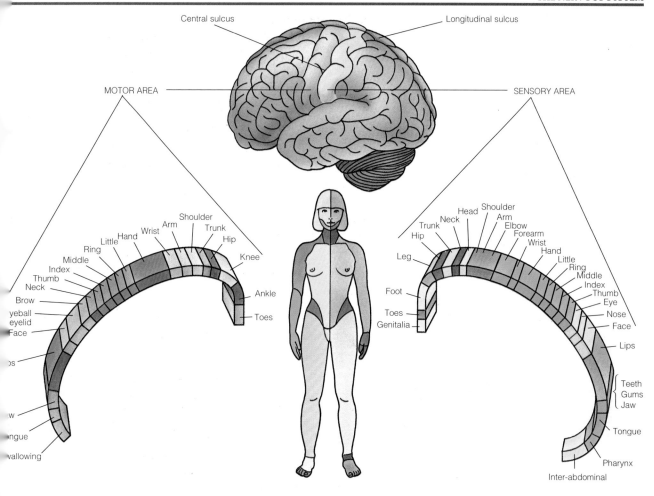

Central sulcus   Longitudinal sulcus

MOTOR AREA   SENSORY AREA

Fig. 27.16 Map of the major regions of the cerebral cortex and their functions

Endocrine system	Nervous system
Communication is by chemical messengers – hormones	Communication is by nervous impulses
Transmission is by the blood system	Transmission is by nerve fibres
Target organ receives message	Effector (muscle or gland) receives message
Transmission is relatively slow	Transmission is very rapid
Effects are widespread	Effects are localized
Response is slow	Response is rapid
Response is often long-lasting	Response is short-lived
Effect may be permanent and irreversible	Effect is temporary and reversible

Diffusely situated throughout the brain stem is a system called the **reticular activating system**. It is used to stimulate the cerebral cortex and so rouse the body from sleep. The system is therefore responsible for maintaining wakefulness. The reticular activating system also appears to monitor impulses reaching and leaving the brain. It stimulates some and inhibits others. By doing so it is likely that the system concentrates the brain's activity upon the issues of most importance at any one time. For example, if searching avidly for a lost contact lens, the visual sense may be enhanced. On the other hand, if straining to hear a distant voice, the system may shift the emphasis to increase auditory awareness.

27.5.4 Comparison of endocrine and nervous systems

Both endocrine and nervous systems are concerned with coordination and in performing this function they inevitably operate together. At the same time the systems operate independently and therefore display differences. A comparison of the two systems is given in Table 27.2.

APPLICATION

Alzheimer's disease

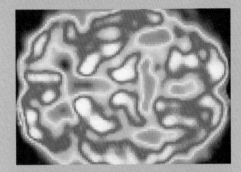

Alzheimer's disease now afflicts more than 600 000 people in the United Kingdom and every day on average a further 42 people develop it. The disease is usually associated with old age, affecting 5% of people over 65 and 20% of those over 80. Sadly it can also affect those as young as 40.

Alzheimer's is a disease which leads to dementia, a term used to describe all illnesses which cause a progressive loss of mental function.

People with dementia have reduced abilities to think and reason; they may not remember people or events, who or where they are. They have difficulty communicating and become increasingly dependent on carers for their every need. Although there are some treatments available which may slow down the progress of the disease, there are no cures. No-one even knows exactly what causes Alzheimer's disease.

There is evidence that patients with Alzheimer's have significantly diminished acetylcholine transferase activity in their brains. This enzyme synthesizes acetylcholine by transferring the acetyl group from acetyl CoA to choline:

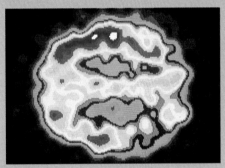

Colour-coded scans of (top) a normal brain and (above) the brain of an Alzheimer's patient, where the pattern of activity is less symmetrical

$$CH_3C \overset{O}{\underset{}{\parallel}} - S - CoA + HO - CH_2CH_2 - \overset{CH_3}{\underset{CH_3}{\overset{|}{N}^+}} - CH_3 \longrightarrow$$

$$CH_3C \overset{O}{\underset{}{\parallel}} - O - CH_2CH_2 - \overset{CH_3}{\underset{CH_3}{\overset{|}{N}^+}} - CH_3 + CoA - SH$$

The cause of the disease may also lie in a defect in the neurotransmitter receptors.

Analyses of diseased human brains demonstrate abnormalities in the neurones of various regions, such as the cerebral cortex (the location of many complex mental processes) and an area which plays a vital rôle in memory. In particular the diseased neurones contain accumulations of filaments known as neurofibrillar tangles and areas of the brain are replaced by extracellular deposits called plaques.

These plaques contain aggregates of protein called amyloid which derives from a normal membrane protein (APP – amyloid precursor protein) found in neurones and other cells. Abnormal catabolism of APP results in the production of insoluble fragments which accumulate outside the neurones and form the plaques. Mutations of the APP gene have been shown to be associated with the early onset of Alzheimer's.

Lysosomes may also play some rôle in the degeneration of neurones. APP and similar hydrophobic proteins are difficult to degrade and their accumulation within cells may cause lysosomes to burst and release their hydrolytic enzymes into the cytoplasm of the neurone. This leads to neuronal death and an aggregation of the proteins in the extracellular space.

By the year 2020 about 40% of the population of Britain, North America and Japan will be over 65 years old. This makes research into the dementing illnesses associated with ageing vital to ensure quality of life for sufferers and their carers.

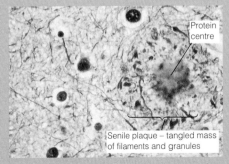

Protein centre

Senile plaque – tangled mass of filaments and granules

LM of brain tissue of an Alzheimer's patient showing a plaque (×250 approx.)

27.6 Sensory perception

All organisms experience changes in both their internal and external environments. Their survival depends upon responding in an appropriate way to these changes, and they have therefore developed elaborate means of detecting stimuli. To some degree, all cells are sensitive to stimuli, but some have become highly specialized to detect a particular form of energy. These are **receptor** cells. In general terms, these receptors convert whichever form of energy it is that they respond to into a nervous impulse, i.e. they act as **biological transducers**.

In its simplest form, a sensory receptor comprises a single neurone in which a single dendrite receives the stimulus and creates an action potential, which it then conveys along its axon to the remainder of the nervous system. This is called a **primary sense cell**, of which the cones of the vertebrate retina are an example. Sometimes the function of receiving the stimulus is performed by a cell outside the nervous system which then passes a chemical or electrical message to a neurone which creates an action potential. This is called a **secondary sense cell**, of which the taste cells on a human tongue are an example.

Whichever form a sensory cell takes, it is often found in groups, usually in conjunction with other tissues, and together they form **sense organs**. At one time it was usual to classify receptors according to their positions. Hence **exterioceptors** collected information from the external environment, **interoceptors** collected it from the internal environment, and **proprioceptors** provided information on the relative position and movements of muscles. It is now more usual to base classification upon the form of stimulus energy. This gives five categories:

1. **Mechanoreceptors** – detect movements, pressures and tensions, e.g. sound.

2. **Chemoreceptors** – detect chemical stimuli, e.g. taste and smell.

3. **Thermoreceptors** – detect temperature changes.

4. **Electroreceptors** – detect electrical fields (mainly in fish).

5. **Photoreceptors** – detect light and some other forms of electromagnetic radiation.

27.6.1 The mammalian eye

That part of the electromagnetic spectrum which can be detected by the mammalian eye lies in the range 400–700 nm. The eye acts like a television camera in producing an ever-changing image of the visual field at which it is directed.

Each eye is a spherical structure located in a bony socket of the skull called the **orbit**. It may be rotated within its orbit by **rectus muscles** which attach it to the skull. The external covering of the eye is the **sclera**. It contains many collagen fibres and helps to maintain the shape of the eyeball. The sclera is transparent over the anterior portion of the eyeball where it is called the **cornea**. It is the cornea that carries out most refraction

Did you know?

A polar bear can smell a seal through ice up to one metre thick.

of light entering the eye. A thin transparent layer of living cells, the **conjunctiva** overlies and protects much of the cornea. The conjunctiva is an extension of the epithelium of the eyelid. Tears from **lachrymal glands** both lubricate and nourish the conjunctiva and cornea.

Inside the sclera lies a layer of pigmented cells, the **choroid**, which prevents internal reflection of light. It is rich in blood capillaries which supply the innermost layer, the **retina**. This contains the light-sensitive **rods and cones** which convert the light waves they receive into nerve impulses which pass along neurones to the **optic nerve** and hence to the brain. There is an especially light-sensitive spot on the retina which contains only cones. This is the **fovea centralis**. The amount of light entering the eye is controlled by the **iris**. This is a heavily pigmented diaphragm of circular and radial muscle whose contractions alter the diameter of the aperture at its centre, called the **pupil**, through which light enters. Just behind the pupil lies the transparent, biconvex **lens**. It controls the final focusing of light onto the retina. It is flexible and elastic, capable of having its shape altered by the **ciliary muscles** which surround it. These are arranged circularly and radially and work antagonistically to focus incoming light on the retina by altering the lens' shape and hence its focal length. The region in front of the lens is called the **anterior chamber** and contains a transparent liquid called **aqueous humour**. Behind the lens is the much larger **posterior chamber** which contains the transparent jelly-like **vitreous humour** which helps to maintain the eyeball's shape. The structure of the eye is illustrated in Fig. 27.17.

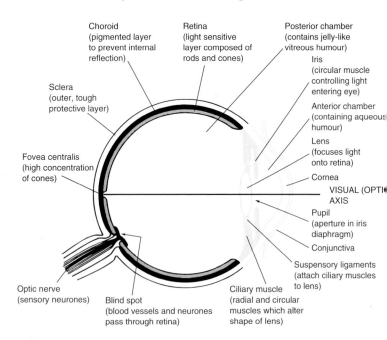

Fig. 27.17 VS through the human eye

Control of the amount of light entering the eye
Controlling the amount of light entering the eye is of importance because if too little light reaches the retina the cones may not be stimulated at all. Alternatively, if the quantity of light is too great the retinal cells may be overstimulated, causing dazzling. Control is exercised by the iris diaphragm as outlined in Fig. 27.1

	BRIGHT LIGHT	DIM LIGHT
	More photoreceptor cells in the retina are stimulated by an increase in light intensity	Fewer photoreceptor cells are stimulated due to decrease in light intensity
	Greater number of impulses pass along sensory neurones to the brain	Fewer impulses pass along sensory neurones to the brain
	Brain sends impulses along parasympathetic nervous system to the iris diaphragm	Brain sends impulses along the sympathetic nervous system to the iris diaphragm
	In the iris diaphragm, circular muscle contracts and radial muscle relaxes	In the iris diaphragm, circular muscle relaxes and radial muscle contracts
	Pupil constricts	Pupil dilates
	Less light enters the eye	More light enters the eye
	Anterior view of iris and pupil	Anterior view of iris and pupil

Pupil constricted — Radial muscle relaxed — Circular muscle contracted

Pupil dilated — Radial muscle contracted — Circular muscle relaxed

Fig. 27.18 Mechanism of control of light entering the eye

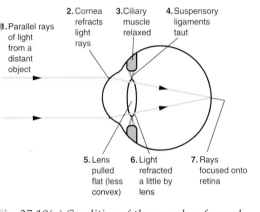

2. Cornea refracts light rays
3. Ciliary muscle relaxed
4. Suspensory ligaments taut
1. Parallel rays of light from a distant object
5. Lens pulled flat (less convex)
6. Light refracted a little by lens
7. Rays focused onto retina

Fig. 27.19(a) Condition of the eye when focused on a distant object

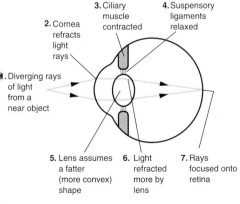

3. Ciliary muscle contracted
4. Suspensory ligaments relaxed
2. Cornea refracts light rays
1. Diverging rays of light from a near object
5. Lens assumes a fatter (more convex) shape
6. Light refracted more by lens
7. Rays focused onto retina

Fig. 27.19(b) Condition of the eye when focused on a near object

Focusing of light rays onto the retina

Light rays entering the eye must be **refracted** (bent) in order to focus them onto the retina and so give a clear image. Most refraction is achieved by the cornea. However, the degree of refraction needed to focus light rays onto the retina varies according to the distance from the eye of the object being viewed. Light rays from objects close to the eye need more refraction to focus them on the retina than do more distant ones. The cornea is unable to make these adjustments and so the lens has become adapted to this purpose. Being elastic, it can be made to change shape by the ciliary muscle which encircles it. The muscle fibres are arranged circularly and the lens is supported by **suspensory ligaments** (Fig. 27.17). When the circular ciliary muscle contracts, the tension on the suspensory ligaments is reduced and the natural elasticity of the lens causes it to assume a fatter (more convex) shape. In this position it increases the degree of refraction of light. When the circular ciliary muscle is relaxed, the suspensory ligaments are stretched taut, thus pulling the lens outwards and making it thinner (less convex). In this position it decreases the degree of light refraction. By changing its shape in this manner the lens can focus light rays from near and distant objects on the retina. The process is called **accommodation**. How the eye accommodates for distant and near objects is shown in Figs. 27.19(a) and (b).

The retina

The retina possesses the **photoreceptor** cells. These are of two types, **rods** and **cones**. Both act as transducers in that they convert light energy into the electrical energy of a nerve impulse. Both cell types are partly embedded in the pigmented epithelial cells of the choroid. The microscopic structure of the retina is shown in Figs. 27.20 and 27.21. In cats and some other nocturnal mammals, there exists a reflecting layer, called the **tapetum**, behind the retina. This reflects light back into the eye and so affords further opportunities for rod cells to absorb it. This greatly improves vision in dim light and is why cats' eyes give bright reflections when light shines into them – it is reflected light from the tapetum which is seen.

As Figs. 27.20 and 27.21 show, the basic structure of rods and cones is similar. However, there are both structural and functional differences and these are detailed in Table 27.3.

Each rod possesses up to a thousand vesicles in its outer segment. These contain the photosensitive pigment **rhodopsin** or **visual purple**. Rhodopsin is made up of the protein **opsin** and a derivative of vitamin A, **retinal**. Retinal normally exists in its *cis* isomer form, but light causes it to become converted to its *trans* isomer form. This change initiates reactions which lead to the splitting of rhodopsin into opsin and retinal – a process known as **bleaching**. This splitting in turn leads to the creation of a generator potential in the rod cell which, if sufficiently large, generates an action potential along the neurones leading from the cell to the brain.

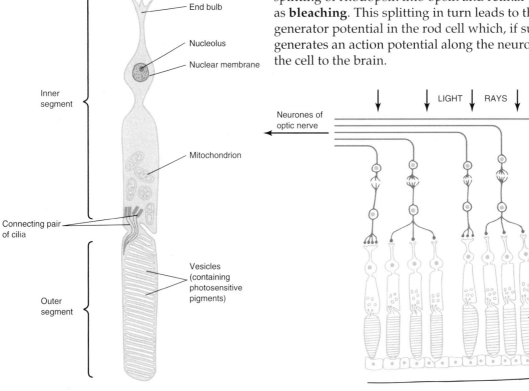

Fig. 27.20 Structure of a single rod cell   Fig. 27.21 Microscopic structure of the retina

TABLE 27.3   **Differences between rods and cones**

Rods	Cones
Outer segment is rod-shaped	Outer segment is cone-shaped
Occur in greater numbers in the retina – being 20 times more common than cones	Fewer are found in the retina – being one twentieth as common as rods
Distributed more or less evenly over the retina	Much more concentrated in and around the fovea centralis
None found at the fovea centralis	Greatest concentration occurs at the fovea centralis
Give poor visual acuity because many rods share a single neurone connection to the brain	Give good visual acuity because each cone has its own neurone connection to the brain
Sensitive to low-intensity light, therefore mostly used for night vision	Sensitive to high-intensity light, therefore mostly used for day vision
Do not discriminate between light of different wavelengths, i.e. not sensitive to colour	Discriminate between light of different wavelengths, i.e. sensitive to colour
Contain the visual pigment rhodopsin which has a single form	Contain the visual pigment iodopsin which occurs in three forms

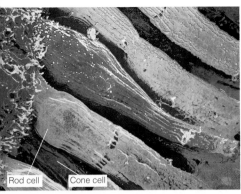

Rod cell | Cone cell

False-colour scanning EM of rod and cone cells

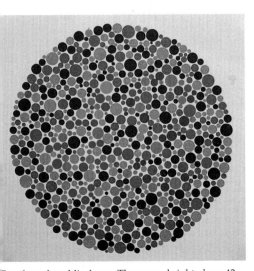

Test for colour blindness. The normal sighted see 42 while those with red–green colour blindness see either just a 4 or a 2 depending on the type of deficiency

Before the rod cell can be activated again in the same way, the opsin and retinal must first be resynthesized into rhodopsin. This resynthesis is carried out by the mitochondria found in the inner segment of the rod cell, which provide ATP for the process. Resynthesis takes longer than the splitting of rhodopsin but is more rapid the lower the light intensity. A similar process occurs in cone cells except that the pigment here is **iodopsin**. This is less sensitive to light and so a greater intensity is required to cause its breakdown and so initiate a nerve impulse.

Colour vision

It is thought that there are three forms of iodopsin, each responding to light of a different wavelength. Each form of iodopsin occurs in a different cone and the relative stimulation of each type is interpreted by the brain as a particular colour. This system is known as the **trichromatic theory** because there are three distinct types of cone, each responding to three different colours of light: blue, green and red. Other colours are perceived by combined stimulation of these three. Equal stimulation of red and green cones, for example, is perceived as yellow. Fig. 27.22 shows the extent of stimulation of each type of cone at different wavelengths of light.

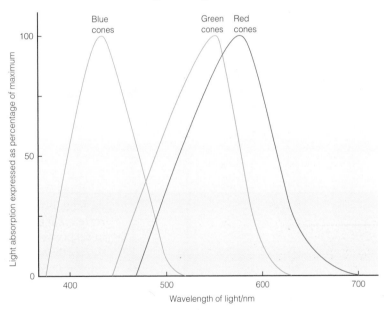

Fig. 27.22 Absorption spectra for the three types of cone occurring in the retina as proposed in the trichromatic theory of colour vision

Colour blindness can be explained in terms of some deficiency in one or more cone type. Deficiencies in the red and green cones, for example, give rise to the relatively common **red–green colour blindness**. The condition is due to a defect on a gene which is linked to the X chromosome (Section 9.4.2). Alternative theories to the trichromatic theory have been put forward in recent years but these have yet to achieve general acclaim.

27.6.2 Rôle of the brain in vision

All light, of whatever colour, intensity or pattern, which enters the eye is transformed into nerve impulses, all of which are the same. It is up to the brain to visualize these impulses into meaningful shapes. This it does by analyzing and interpreting the frequency of impulses and the origins of the neurones which carry them e.g. from a red cone, or green cone. The visual association areas in the cerebral cortex of the brain are especially important here as they match the incoming information with images already stored in these areas. Hence a corgi would be recognized as a dog even if an individual's previous experience of the species was restricted to labradors.

27.7   Behaviour

In order to survive, organisms must respond appropriately to changes in their environment. Broadly speaking, behaviour is the response of an organism to these changes and it involves both endocrine and nervous systems. Behaviour has a genetic basis and is unique to each species. It is often adapted, however, in the light of previous experience. The study of behaviour is called **ethology**.

Although both plants and animals exhibit behaviour, that of animals is considerably more complex. This section deals only with animal behaviour; plant behaviour is covered in Chapter 29.

27.7.1 Reflexes, kineses and taxes

Reflexes are the simplest form of behavioural response. Section 27.3 describes a simple reflex response to a stimulus. These reflexes are involuntary responses which follow an inherited pattern of behaviour. How then do these responses improve an animal's chance of survival? The **withdrawal reflex** illustrates their importance. If the hand is placed on a hot object, the reflex response causes it to be immediately withdrawn. In this way damage is avoided. Invertebrate animals have a similar **escape reflex**. An earthworm, for example, withdraws down its burrow in response to vibrations of the ground. By behaving in this way the earthworm is less likely to be captured by a predator and so improves its chances of survival.

A more complex form of behaviour is a **kinesis**. This is a form of orientation behaviour. The response is non-directional, i.e. the animal does not move towards or away from the stimulus. Instead it simply moves faster, and changes direction more frequently when subjected to an unpleasant stimulus. The greater the intensity of the stimulus the faster it moves and the more often it changes direction. This known as **orthokinesis**. It is exhibited by woodlice which prefer damp situations. If a woodlouse finds itself in a dry area, it simply moves faster and keeps changing its direction. It thereby increases the speed with which it is likely to move out of the dry area. If this brings it into a more moist situation its movement slows, indeed it may cease altogether. It therefore spends a much longer period of time in the more humid conditions. Naturally it cannot remain stationary indefinitely, and factors such as hunger will ultimately cause it to move in search of food.

A **taxis** involves the movement of a whole organism in response to a stimulus, where the direction of the movement is

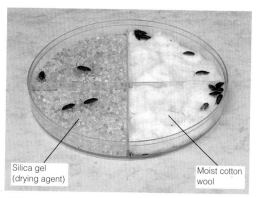

Silica gel (drying agent)

Moist cotton wool

Choice chamber showing the preference of woodlice for a moist environment

related to the direction of the stimulus, usually towards it (positive +), or away from it (negative −). Tactic responses are classified according to the nature of the stimulus, e.g. light – **phototaxis**; chemicals – **chemotaxis**. The flatworm, *Planaria*, shows a positively chemotactic response to the presence of food. By moving its head from side to side it can detect the relative intensity of chemical stimulation experienced on each side of the head. This allows it to locate the source of food. Other examples of taxes are given in Section 29.1.2.

27.7.2 Innate behaviour

Innate or **instinctive** behaviour is inherited and is highly specific. It is normally an inborn pattern of behaviour which cannot be altered. In practice almost all instincts can be modified to some degree in response to experiences. However, innate behaviour is relatively inflexible when compared to learning. Much instinctive behaviour is highly complex and consists of a chain of actions, the completion of each stage in the chain acting as the stimulus for the commencement of the next stage.

This pattern of events is illustrated by a species of digger wasp. As its parents die long before each wasp hatches, there is no opportunity for it to learn from its parents and it must rely on inherited patterns of behaviour. When the time comes to lay eggs, a female digger wasp digs a nest hole and in it constructs small cells, each of which it provisions with a paralysed caterpillar to act as a food source for its offspring. Having done so, it lays a single egg on the roof of each cell and seals it. If upon sealing the cell a caterpillar is placed where it is visible to the wasp, the wasp opens the cell, adds the caterpillar, lays another egg and seals the cell. It will repeat this many times, to the point where the cell is so crammed with caterpillars that there is insufficient air or space for the eggs to develop. It is obvious that the wasp has no appreciation of the purpose of its actions. If it had it would surely not pursue its pattern of behaviour to the point of jeopardizing the survival of its eggs. Instead the wasp reacts automatically, responding to the sight of a caterpillar, which acts as a **sign stimulus**, by carrying out the next stage in its inherited pattern of events. Removal of the roof of the cell where the egg is normally laid causes the wasp some agitation. However, it continues to lay the egg as if the cell roof was still intact, rather than adapt its behaviour to lay it elsewhere. This illustrates the relative inflexibility of innate behaviour.

The characteristics of innate behaviour are:

1. It is inherited and not acquired, although some modifications may result from experience.

2. It is similar among all members of a species and there are no individual differences other than those between males and females of the species.

3. It is unintelligent and often accompanied by no appreciation of the purposes it serves.

4. It often comprises a chain of reflexes. Completion of each link in the chain provides the stimulus for the commencement of the next link.

27.7.3 Learned behaviour

Learned behaviour is behaviour which is acquired and modified in response to experience. As such it takes time to refine and so is of greatest benefit to animals with relatively long life spans. The chief advantage of learning over innate behaviour is its adaptability; learned behaviour can be modified to meet changing circumstances. The simplest form of learned behaviour is **habituation**. This involves learning to ignore stimuli because they are followed by neither reward nor punishment. A snail crawling across a board can be made to withdraw into its shell by hitting the board firmly. Repetition of this action results in the snail taking no notice of the stimulus, i.e. it has learnt to ignore the stimulus as it is neither beneficial nor harmful to it.

Associative learning involves the association of two or more stimuli. One form of associative learning is the **conditioned reflex** exemplified by the classic experiments performed on dogs by the Russian physiologist I. P. Pavlov:

1. He allowed dogs to hear the ticking of a metronome and observed no change in the quantity of saliva produced.

2. He presented the dogs with the taste of powdered meat and measured the quantity of saliva produced.

3. He presented the powdered meat and the noise of the metronome simultaneously on 5 to 6 occasions.

4. He presented the noise of a ticking metronome *only* and observed that the dogs salivated in response to it whereas previously they had not done so (Stage **1**).

5. Repetition of Stage **4** leads to a reduction in the quantity of saliva produced until the stimulus fails to produce any response.

This association of one stimulus with another is the basis upon which birds reject certain caterpillars. If a caterpillar is distasteful it is in its interests to advertise itself by being brightly patterned. Having associated a particular pattern with the unpleasant memory of eating such a caterpillar, the bird ignores it as a potential source of food. This often leads to mimicry by otherwise edible insects, which thereby avoid being eaten.

The features of a conditioned reflex are:

1. It is the association of two stimuli presented together.

2. It is a temporary condition.

3. The response is involuntary.

4. It is reinforced by repetition.

5. Removal of the cerebral cortex causes loss of the response.

A second form of associative learning is **operant conditioning (trial and error learning)**. This form of learning, studied by Skinner, differs from the conditioned reflex in the way it becomes established. Animals learn by trial and error. If mistakes are followed by an unpleasant stimulus while correct responses are followed by a pleasant one, the animal learns a particular pattern of behaviour.

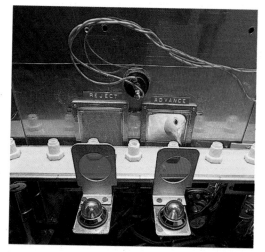

Pigeon choosing in a Skinner box

The features of operant learning are:

1. The associative stimulus *follows* the action, i.e. it does not need to be simultaneous with it.

2. Repetition improves the response.

3. The action is involuntary.

4. While temporary, the association is less easily removed than in a conditioned reflex.

5. Removal of the cerebral cortex does *not* cause loss of the response.

Latent learning arises when an animal stores information while exploring its environment, and uses it at some later time. A rat placed in a maze with no reward as a stimulus will later complete the maze, when a reward is present, more rapidly than a rat which has never been in the maze.

The highest form of learning is **insight** or **intelligent behaviour**. It involves the recall of previous experiences and their adaptation to help solve a new problem. The rapidity with which a solution is achieved excludes any possibility of trial and error. Chimpanzees will acquire bananas fixed to the roof of their cage by piling up boxes upon which they climb to reach them. In the same way sticks may be joined together to form a long pole which is used to obtain bananas which are out of reach outside the cage. Chimpanzees may even chew the ends of the sticks so they can be made to fit one another.

Imprinting is a simple but specialized form of learning. Unlike other forms of learning, imprinted behaviour is fixed and not easily adapted. Newly hatched geese will follow the first thing they see. Ordinarily this would be their mother and the significance of this behaviour is therefore obvious. However, as shown by the Austrian behaviourist, Konrad Lorenz, they will follow humans or other objects should these be seen first. This principle is often used in training circus animals. If a trainer becomes imprinted in an animal's mind, it becomes much easier to train.

27.7.4 Reproductive behaviour

For successful reproduction it is essential that an individual finds a mate who is of the opposite sex, sexually mature and prepared to copulate. **Courtship** is largely geared to achieving this end, and some details are given in Section 12.7.1. The preparedness of an individual to copulate is often controlled by hormones which can be affected by a number of factors (Section 12.6). In female mammals for example there may be behavioural patterns associated with the oestrous cycle. Parental care (Section 12.7.11) is a further example of reproductive behaviour, in this case designed to improve the probability of the young surviving to independence and sexual maturity.

27.7.5 Social behaviour

Even individuals of solitary species interact, if only to bring their gametes together. At the opposite end of the spectrum, the individuals of some species are completely dependent on one

PROJECT

Students often express the view that they can study better if there is background music going on at the same time

Find out if they are right or not.

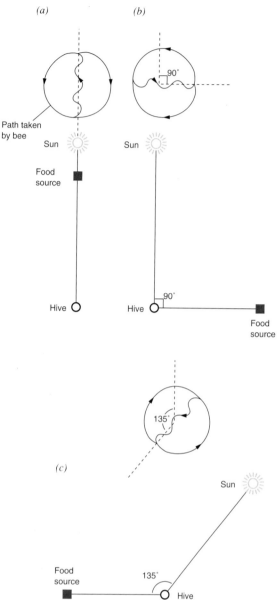

(a)   *(b)*

Path taken by bee

(c)

Fig. 27.23 *Waggle dance of the honey-bee*

another for their survival. It is vital that these social groupings adapt their behaviour so that it is directed towards the interests of the group rather than the individual.

The advantages of a social group include greater opportunities for locating food and better protection against predators. To achieve this there needs to be efficient communication between individuals and an appreciation of each individual's rôle within the community. Frequently there is a **social hierarchy** or **pecking order**, with each individual having its own fixed status within the group. Examples occur in domestic animals such as chickens and cattle. In much the same way there is a dominance hierarchy in human organizations and institutions.

Social behaviour may be illustrated by a honey-bee colony. Here there is a **caste system** where there are distinct types each with a specified rôle in the group. The **queen** is the single fertile female; the remaining females, the **workers**, are sterile as a result of licking a pheromone off the body of the queen (see Section 26.7.1). In this way the caste system is maintained. All males, called **drones**, are fertile. It is particularly in foraging for food that the honey-bees demonstrate complex social cooperation. This form of social behaviour was studied in detail by the German zoologist, Karl von Frisch. He discovered that worker bees returning from foraging missions reveal the location of food sources to other workers using a special **bee dance**. These dances communicate the direction of the food source from the hive and its distance away.

There are two forms of the dance: if the food source is within 100 m of the hive, a **round dance** is performed; if greater than 100 m, a **waggle dance** is carried out. In both dances, the speed at which it is performed is inversely proportional to the distance away from the hive that the food lies. The round dance does not indicate the direction of the food from the hive, but as it lies within 100 m the other workers easily locate it after a brief search.

The waggle dance is more complex. The worker bee moves in a figure-of-eight pattern, waggling its abdomen as it does so. The waggle part of the dance occurs as it moves along the line between the two loops of the figure-of-eight. The number of waggles gives some indication of the quantity of food discovered. The angle of this waggle relative to the vertical is the same as the angle relative to the sun which the other bees should take on leaving the hive, if they are to locate food. Examples to illustrate this are shown in Fig. 27.23.

The obvious question arises – how do bees locate food sources on cloudy days? They seem able to determine the plane of polarized light, which penetrates the cloud cover, and so are able to locate the sun. To navigate accurately the bees must be able to make allowances for movements in the position of the sun according to the time of day. Honey-bees achieve this with the aid of an internal biological clock, which permits them to have a continually changing picture of the sun's movement during the day. Further information on the nature of the food source may be provided by samples of the nectar and/or pollen brought back to the hive by the foraging worker bee.

These complex bee dances serve to illustrate the importance of social behaviour in communicating valuable information quickly and accurately between individuals in the colony.

27.8   Questions

. Write an essay on how living organisms detect
changes in their environment.

(*Total 24 marks*)

AEB November 1992, Paper II, No. 5A

. The graph shows the changes in the permeability
of an axon to sodium ions and to potassium ions
during an action potential.

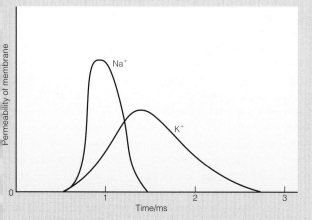

(a) Describe how sodium ions enter the axon
during the passage of the action potential.

(*2 marks*)

(b) Explain how the events shown in the graph:
   (i)   lead to the inside of the axon becoming
         positive with respect to the outside
         during the first stage of an action
         potential;   (*1 mark*)
   (ii) restore the resting potential.   (*1 mark*)

(c) The table shows the speed of conduction of a
nerve impulse along axons from three
different species of animal.

Species	Diameter of axon/μm	Speed of conduction/m s^{-1}
A	7	1.2
B	500	33.0
C	15	90.0

Which **one** of these three species is most
likely to possess myelinated nerve fibres?
Give a reason for your answer.   (*2 marks*)

(*Total 6 marks*)

AEB November 1993, Paper I, No. 5

. The figure at the top of the next column shows the
structures associated with a synapse.
   (a)   (i)   Label structures **A** to **E**.
         (ii) Draw an arrow on the diagram to show
               the direction of a nerve impulse.

(*3 marks*)

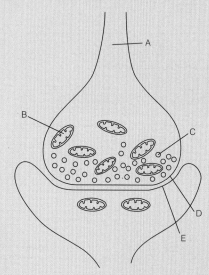

(b) In nerve impulse transmission, what are the
functions of the structures labelled **B** and **C**?

(*2 marks*)

(c) Describe how an impulse passes across a
synapse.   (*5 marks*)

(d)   (i)   What is the term used to describe drugs
            such as heroin which relieve pain and
            produce sleep?
      (ii) Briefly describe the effects of heroin on
            the nervous system.   (*5 marks*)

(*Total 15 marks*)

UCLES (Modular) March 1992,
(Human health and disease), No. 1

4. The diagram below shows the trace obtained on a
cathode ray oscilloscope when an isolated axon of a
neurone is stimulated.

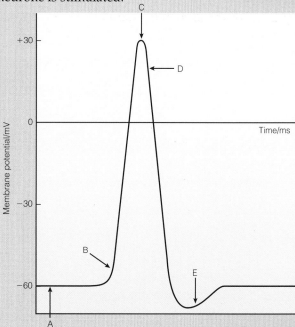

The following statements refer to events at specific points, A, B, C, D and E, on the trace. Match the statements 1 to 5 with points A to E. Each letter should be used once only.

Statement	Letter
1 Region corresponding to relative refractory period	
2 Rapid decrease in the permeability of the axon membrane to sodium ions	
3 No net movement of ions and external surface of the axon membrane is negatively charged	
4 Rapid increase in the number of sodium ions inside the axon	
5 Sodium permeability of the axon membrane insufficient to trigger an action potential	

(*Total 5 marks*)

ULEAC January 1993, Paper III, No. 3

5. The diagram below shows a complete motor neurone from a rabbit, drawn to scale. The section enclosed in the box is shown at a higher magnification.

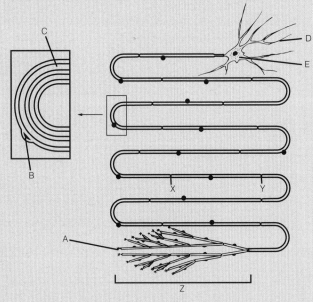

(*a*) Name the parts labelled A, B, C, D and E.
(*5 marks*)
(*b*) X and Y point to constrictions in structure C.
   (i) Explain the nature and significance of these constrictions. (*3 marks*)
   (ii) The magnification of the diagram is approximately ×60.
     Calculate the actual distance between X and Y. Show your working. (*2 marks*)
(*c*) (i) Suggest why the region labelled Z is branched. (*2 marks*)

(ii) Name a substance released from A when the neurone discharges, and state *one* function of this substance. (*2 marks*)
(*Total 14 marks*)

ULEAC June 1991, Paper I, No. 1

6. The diagram below shows the structure of a nerve synapse.

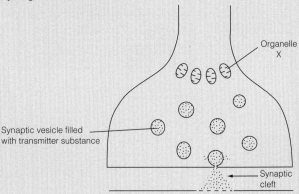

(*a*) Name one transmitter substance. (*1 mark*)
(*b*) Name organelle X and suggest a reason for its presence. (*2 marks*)
(*c*) Some nerve fibres are myelinated. What is the advantage of this myelination? (*1 mark*)
(*d*) In which part of the brain are each of the following functions regulated?
   (i) Speech
   (ii) Muscle coordination
   (iii) Temperature control
   (iv) Ventilation of the lungs (*2 marks*)
(*e*) Describe the effect of the sympathetic nervous system on each of the following:
   (i) Heart rate (*1 mark*)
   (ii) Blood vessels on the surface of the skin (*2 marks*)
(*Total 9 marks*)

SEB Human Biology Specimen 1992
Higher Grade, Paper II, No.

7. Summary

In this question you are given some information concerning the action of a range of chemicals on synapses. You are then asked to provide information about how synapses work, and your interpretation of where and how these chemicals have their action. Further information is given and you are asked to predict the action of a chemical and suggest possible medical uses for it. Finally you are asked some questions concerning nervous conduction.

Many drugs have the action of mimicking or blocking the action of acetylcholine, a widespread neurotransmitter. The table below shows a range of drugs and some of their more common effects on the mammalian nervous system.

Substance	Site of action	Most common effect
botulinum toxin	presynaptic membrane	prevents release of acetylcholine
nicotine	postsynaptic membrane	mimics action of acetylcholine
strychnine	postsynaptic membrane	inhibits the breakdown of acetylcholine
curare	postsynaptic membrane of neuromuscular junction	blocks action of acetylcholine

(a) (i) Describe the events occurring between the arrival of an action potential at the presynaptic membrane of a cholinergic synapse (one in which the transmitter substance is acetylcholine) and the generation of an action potential in the postsynaptic cell. *(5 marks)*

   (ii) By means of a large, clearly labelled diagram of a synapse, indicate the precise site of action of the following substances (using the appropriate numbers).
1. botulinum toxin
2. nicotine
3. strychnine
4. curare *(4 marks)*

   (iii) By means of notes below your diagram, suggest the effect you would expect each substance to have on the postsynaptic generation of an action potential. *(4 marks)*

(c) The drug atropine acts predominantly by depressing the activity of the parasympathetic nervous system. Predict the effect of this drug on
(i) heart rate
(ii) rate of salivation
(iii) pupil diameter
Suggest why atropine could have useful applications in dental surgery and eye examinations. *(5 marks)*

(d) (i) State your understanding of the 'all or nothing' law as applied to the electrical stimulation of a single neurone.

   (ii) Briefly describe **one** other characteristic feature of nerve impulse transmission in a myelinated axon.
(2 marks)
(Total 20 marks)

NISEAC June 1991, Paper III, No. 1X

8. The table compares some effects of the sympathetic and parasympathetic systems.

Feature	Sympathetic	Parasympathetic
Pupil of eye	Dilates	Constricts
Salivary gland	Inhibits secretion of saliva	Stimulates secretion of saliva
Lungs	Dilates bronchi and bronchioles	Constricts bronchi and bronchioles
Arterioles to gut and smooth muscle	Constricts	No effect
Arterioles to brain	Dilates	No effect
Heart rate		
Stroke volume		

(a) Complete the table by filling in the spaces to suggest the effects of these two systems on heart rate and stroke volume. *(2 marks)*

Use information in the table and your own knowledge to answer the following questions.

(b) In giving dental treatment, it is important that any local anaesthetic stays close to the site of its injection.
Explain why dental anaesthetics usually contain adrenaline. *(2 marks)*

(c) Many people suffer from motion sickness when travelling in cars.
A number of drugs are used to control this and some work by inhibiting the parasympathetic system. Explain why side-effects of such drugs may include:
(i) dryness of the mouth;
(ii) blurred vision. *(2 marks)*
(Total 6 marks)

AEB November 1992, Paper I, No. 4

9. Complete the required information about the mammalian eye.
(a) Rays of light entering the mammalian eye are refracted and focused by three methods (or structures). Name the parts of the eye which help to form a sharp image on the retina. *(3 marks)*

(b) Complete the following diagram with ruled pencil lines and a sketch of the image to indicate how such an object will be focused on the retina. *(2 marks)*

(c) There are two chief types of light sensitive cells (here called X and Y) in the healthy human retina but they are not evenly distributed. An investigation was made of the distribution of sensory cell types X and Y across an area of retina, shown as PQ on the diagram below.

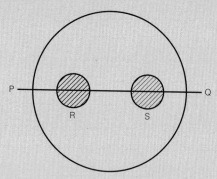

The relative numbers of cell types X and Y were reasonably constant except in positions **R** and **S**. The following graph indicates the relative frequency of sensory cell type X along axis PQ:

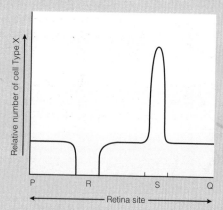

Give the name of
  (i) cell type X;
  (ii) cell type Y;
  (iii) region **R**;
  (iv) region **S**. (4 marks)
(d) Using a pencil, clearly superimpose on the graph printed above, a curve to indicate the relative frequency of cell type Y along the axis PQ. (2 marks)
(e) The diagrams at the top of the next column represent the state of the iris in bright and dim light.

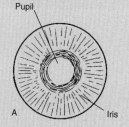
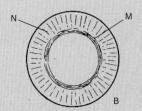

  (i) Name **M** and **N**. (1 mark)
  (ii) Briefly explain what causes the iris in B to have a different appearance from that in A. (2 marks)
  (Total 14 marks)

Oxford June 1991, Paper I, No. ?

10. The diagram below shows a single rod from a mammalian retina.

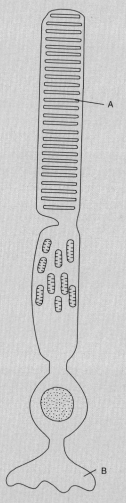

(a) Name the parts labelled A and B and give *one* function of each. (4 marks)
(b) Draw an arrow next to the diagram to indicate the direction in which light passes through this cell. (1 mark)
(c) State *two* ways in which vision using cones differs from vision using rods. (2 marks)
  (Total 7 marks)

ULEAC January 1993, Paper III, No. 4

See *Further questions*, pages 614–20, question 2.

28 Muscular movement and support

28.1 Structure of skeletal muscle

An individual muscle is made up of hundreds of **muscle fibres**. These fibres are cylindrical in shape with a diameter of around $50\,\mu m$. They vary in length from a few millimetres to several centimetres. Each fibre has many nuclei and a distinctive pattern of bands or cross striations. It is bounded by a membrane – the **sarcolemma**. The fibres are composed of numerous **myofibrils** arranged parallel to one another. Each repeating unit of cross striations is called a **sarcomere** and in mammals has a length of $2.5–3.0\,\mu m$. The cytoplasm of the myofibril is known as **sarcoplasm** and possesses a system of membranes called the **sarcoplasmic reticulum**.

The myofibril has alternating dark and light bands known as the **anisotropic** and **isotropic** bands respectively. Confusion between their names can be avoided by reference to the following:

D   L
Anisotropic band **I**sotropic band
R   G
K   H
                     T

Each isotropic (light) band possesses a central line called the **Z line** and the distance between adjacent Z lines is a **sarcomere**. Each anisotropic (dark) band has at its centre a lighter region called the **H zone**, which may itself have a central dark line – the **M line**. This pattern of bands is the result of the arrangement of the two types of protein found in a myofibril. **Myosin** is made up of thick filaments and **actin** of thin ones. Where the two types overlap, the appearance of the muscle fibre is much darker. Anisotropic bands are therefore made up of both actin and myosin filaments whereas the isotropic band is made up solely of actin filaments. These arrangements and the overall structure of skeletal muscle are shown in Fig. 28.1.

Myosin filaments are approximately 10 nm in diameter and $2.5\,\mu m$ long. They consist of a long rod-shaped fibre and a bulbous head which projects to the side of the fibre. These heads are of major significance in the contraction of muscle (Section 28.2.1). Actin filaments are thinner and slightly shorter than those of myosin being approximately 5 nm in diameter and

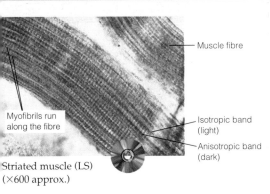

Muscle fibre

Myofibrils run along the fibre

Isotropic band (light)

Anisotropic band (dark)

Striated muscle (LS) (×600 approx.)

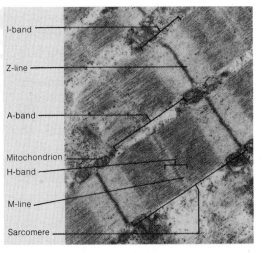

I-band

Z-line

A-band

Mitochondrion

H-band

M-line

Sarcomere

Striated muscle (EM)

557

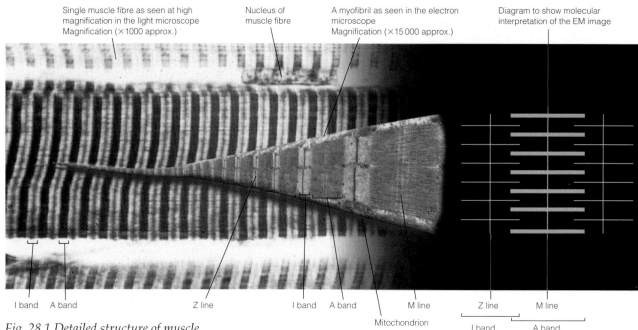

Single muscle fibre as seen at high magnification in the light microscope Magnification (×1000 approx.)

Nucleus of muscle fibre

A myofibril as seen in the electron microscope Magnification (×15 000 approx.)

Diagram to show molecular interpretation of the EM image

I band   A band   Z line   I band   A band   M line   Z line   M line

Mitochondrion

I band   A band

Fig. 28.1 Detailed structure of muscle

Did you know?

Skeletal muscle is the body's most abundant tissue accounting for 23% of the body weight in women and 40% in men.

PROJECT

If you are right-handed presumably you use your right hand more (left-handed people are of course the other way around).

1. Compare muscle strength in left and right hands.

2. Is there any relationship between muscle strength in the hand and handedness?

$2.0\,\mu\text{m}$ long. The filaments comprise two different strands of actin molecules twisted around one another. Associated with these filaments are two other proteins: **tropomyosin**, which forms a fibrous strand around the actin filament, and **troponin**, a globular protein vital to contraction of muscle fibre.

28.1.1 The neuromuscular junction

Skeletal muscle will not contract of its own accord but must be stimulated to do so by an impulse from an effector nerve. The point where the effector nerve meets a skeletal muscle is called the **neuromuscular junction** or **end plate**. If there were only one junction of this type it would take time for a wave of contraction to travel across the muscle and so not all the fibres would contract simultaneously and the movement would be slow. As rapid contraction is frequently essential for survival, animals have evolved a system whereby there are many end plates spread throughout a muscle. These simultaneously stimulate a group of fibres known as an **effector (motor) unit**; contraction of the muscle is thus rapid and powerful. This arrangement also gives control over the force generated by a muscle as not all the units need be stimulated at one time. If only slight force is needed only a few units will be stimulated. The structure of an end plate is shown in Fig. 28.2.

When a nerve impulse is received at the end plate, synaptic vesicles fuse with the end plate membrane and release their acetylcholine. The transmitter diffuses across to the sarcolemma where it alters its permeability to sodium ions which now rapidly enter, depolarizing the membrane. Provided the threshold value is exceeded, an action potential is fired in the muscle fibre and the effector (motor) unit served by the end plate contracts. Breakdown of the acetylcholine by acetylcholinesterase ensures that the muscle is not over-stimulated and the sarcolemma becomes repolarized. This sequence of events is much the same as the mechanism of synaptic transmission (Section 27.2.2).

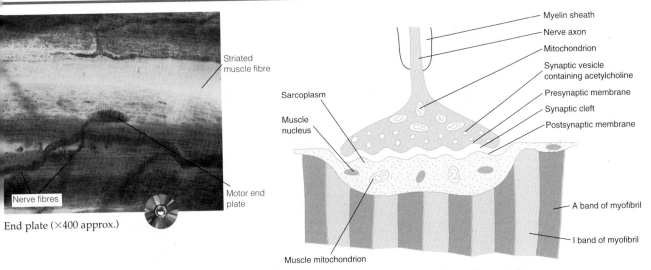

Fig. 28.2 Neuromuscular junction – the end plate

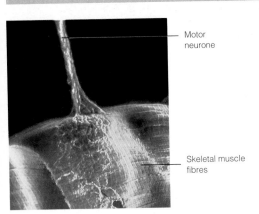

End plate (×400 approx.)

False-colour scanning EM of a neuromuscular junction (×3600 approx.)

Did you know?

There are more than 600 voluntary muscles in the human body.

TABLE 28.1 **Antagonistic pairs of muscles and the movements they perform**

Muscle action	Opposing muscle action
Flexor – bends a limb	**Extensor** – straightens a limb
Abductor – moves a limb laterally away from the body	**Adductor** – moves a limb from a lateral position in towards the body
Protractor – moves a limb forwards	**Retractor** – moves a limb backwards

28.2 Muscular contraction

Muscle cells have the ability to contract when stimulated and are therefore able to exert a force in one direction. Before a muscle can be contracted a second time, it must relax and be extended by the action of another muscle. This means that muscles operate in pairs with each member of the pair acting in the opposite direction to the other. These muscle pairs are termed **antagonistic**. Muscles may be classified according to the type of movement they bring about. For example, a **flexor** muscle bends a limb, whereas an **extensor** straightens it. Flexors and extensors therefore form an antagonistic pair. These and other types of skeletal muscle are listed in Table 28.1.

Skeletal muscle, as the name suggests, is attached to bone. This attachment is by means of **tendons** which are connective tissue made up largely of **collagen fibres**. Collagen is relatively inelastic and extremely tough. When a muscle is contracted, the tendons do not stretch and so the force is entirely transmitted to the bone. Each muscle is attached to a bone at both ends. One attachment, called the **origin**, is fixed to a rigid part of the skeleton, while the other, the **insertion**, is attached to a moveable part. There may be a number of points of insertion and/or origin. For example, two antagonistic muscles which bend the arm about the elbow are the **biceps** (flexor) and **triceps** (extensor). The biceps has two origins both on the scapula (shoulder blade) and a single insertion on the radius. The triceps has three origins, two on the humerus and one on the scapula and a single insertion on the ulna. These arrangements are illustrated in Fig. 28.3.

28.2.1 Mechanism of muscular contraction – the sliding filament theory

Much of our present knowledge about the mechanism of muscular contraction has its origins in the work of H. E. Huxley and J. Hanson in 1954. They compared the appearance of striated muscle when contracted and relaxed, observing that the length

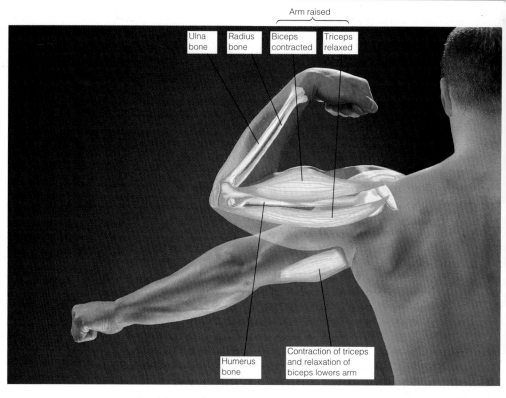

Fig. 28.3 Antagonistic muscles of the forearm

of the anisotropic band (A band) remained unaltered. They concluded that the filaments of actin and myosin must in some way slide past one another – the **sliding filament theory**. It appears that the actin filaments, and hence the Z lines to which they are attached, are pulled towards each other, sliding as they do over the myosin filaments. No shortening of either type of filament occurs. These changes are illustrated in Fig. 28.4. From this diagram it should be clear that upon contraction the following can be observed in the appearance of a muscle fibre:

1. The isotropic band (I band) becomes shorter.

2. The anisotropic band (A band) does not change in length.

3. The Z lines become closer together, i.e. the sarcomere shortens.

4. The H zone shortens.

How exactly do the actin and myosin filaments slide past one another? The explanation seems to be related to cross bridges between the two types of filament, which can be observed in photoelectronmicrographs of muscle fibres. The bulbous heads along the myosin filaments form these bridges, and they appear to carry out a type of 'rowing' action along the actin filaments. The sequence of events involved in these movements is shown in Fig. 28.5.

Each myosin filament has a number of these bulbous heads and each progressively moves the actin filament along, as it becomes attached and reattached. This process is similar to the way in which a ratchet operates and for this reason it is often termed a **ratchet mechanism**. The result of this process is the contraction of a muscle or **twitch**. Having contracted, the muscle then relaxes. In this condition the myosin heads are drawn back towards the myosin filament and the fibrous tropomyosin blocks the

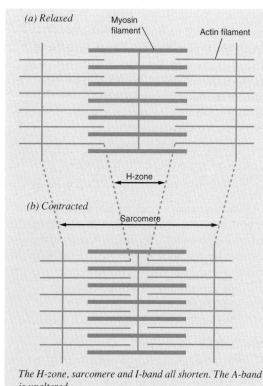

The H-zone, sarcomere and I-band all shorten. The A-band is unaltered

Fig. 28.4 Changes in appearance of a sarcomere during muscle contraction

) The head of the myosin molecule is 'cocked' ready to
attach to the actin filament

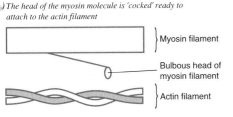

Myosin filament

Bulbous head of
myosin filament

Actin filament

) Myosin head attaches to a monomer unit on the actin molecule

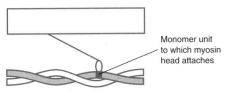

Monomer unit
to which myosin
head attaches

) The myosin changes position in order to attain a lower
energy state. In doing so it slides the actin filament past
the stationary myosin filament

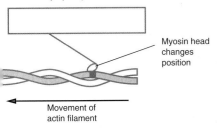

Myosin head
changes
position

Movement of
actin filament

) The myosin head detaches from the actin filament as a
result of an ATP molecule fixing to the myosin head

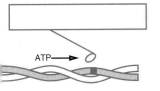

ATP→

) The ATP provides the energy to cause the myosin head
to be 'cocked' again. The hydrolysis of the ATP gives
rise to ADP + P

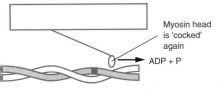

Myosin head
is 'cocked'
again

ADP + P

) The 'cocked' head of the myosin filament reattaches
further along the actin filament and the cycle of events is
repeated

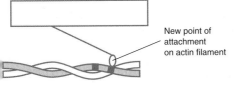

New point of
attachment
on actin filament

*Fig. 28.5 Sliding filament mechanism of muscle
contraction*

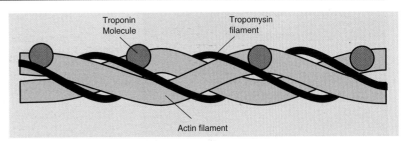

Troponin
Molecule

Tropomyosin
filament

Actin filament

Fig. 28.6 Relationship of tropomyosin and troponin to the actin filament

attachment sites of the actin filament (Fig. 28.6). This prevents
linking of myosin to actin and so stops further muscle contraction.
As the separation of the myosin and actin requires the binding of
ATP to the myosin head, and as this can only be produced in a
living organism, the muscles at death remain contracted. This
results in a stiffening of the body known as **rigor mortis**.

Before a relaxed muscle may be contracted again, the actin
filament must be unblocked by somehow moving the
tropomyosin so that the myosin heads may once again bind with
it. When a muscle is stimulated the wave of depolarization created
spreads, not only along the sarcolemma but also throughout a
series of small tubes known as the **T-system**. The T-system is in
contact with the sarcoplasmic reticulum, both of which are rich in
calcium ions (Ca^{2+}). On depolarization they both release these ions.
The calcium so released binds to part of a protein molecule called
troponin. This causes the troponin to change shape and in so doing
move the tropomyosin molecule away from the actin filament.
This unblocks the actin and so allows the myosin head to become
attached to it, causing the muscle to contract. The calcium ions are
then actively pumped back into the sarcoplasmic reticulum and
T-system ready for use in initiating further muscle contractions.

The energy for muscle contraction is provided by ATP which
is formed by oxidative phosphorylation during the respiratory
breakdown of glucose. The supply of glucose is provided from
the store of glycogen found in muscles. The resynthesis of ATP
after its hydrolysis requires a substance called **phosphocreatine**.

28.2.2 Summary of muscle contraction

The events described fit the observed facts of muscle contraction.
Further research may reveal additional detail, but the basic
mechanism is unlikely to be modified. In view of the complexity
of the process, a summary of the main stages is given below:

1. Impulse reaches the neuromuscular junction (end plate).

2. Synaptic vesicles fuse with the end-plate membrane and
release a transmitter (e.g. acetylcholine).

3. Acetylcholine depolarizes the sarcolemma.

4. Acetylcholine is hydrolysed by acetylcholinesterase.

5. Provided the threshold value is exceeded, an action potential
(wave of depolarization) is created in the muscle fibre.

6. Calcium ions (Ca^{2+}) are released from the T-system and
sarcoplasmic reticulum.

7. Calcium ions bind to troponin, changing its shape.

8. Troponin displaces tropomyosin which has been blocking the
actin filament.

9. The myosin heads now become attached to the actin filament

10. The myosin head changes position, causing the actin filaments to slide past the stationary myosin ones.

11. An ATP molecule becomes fixed to the myosin head, causing it to become detached from the actin.

12. Hydrolysis of ATP provides energy for the myosin head to be 'cocked'.

13. The myosin head becomes reattached further along the actin filament.

14. The muscle contracts by means of this ratchet mechanism.

15. The following changes in the muscle fibre occur:
    (a) I band shortens;
    (b) Z lines move closer together (i.e. sarcomere shortens);
    (c) H zone shortens.

16. Calcium ions are actively absorbed back into the T-system.

17. Troponin reverts to its original shape, allowing tropomyosin to again block the actin filament.

18. Phosphocreatine is used to regenerate ATP.

28.3   Skeleton and support

Organisms originally evolved in water which gave support. Nevertheless some skeletal system was still necessary for most aquatic organisms, either as a rigid framework for the attachment of muscles or for protection. When organisms colonized land, where air provides little support, a skeleton was also necessary to support them against the pull of gravity. Skeletons therefore fulfil three main functions: support, locomotion and protection.

28.3.1 Types of skeleton

Skeletons take a wide variety of forms but they are generally classified into three main types: **hydrostatic**, **exoskeleton** and **endoskeleton**.

Hydrostatic skeleton
As liquids are incompressible they can form a resilient structure against which muscles are able to contract. This type of skeleton is typical of soft-bodied organisms such as annelids where liquid is secreted and trapped within body cavities. Around the liquid, muscles are arranged segmentally. They are not attached to any part of the body but simply contract against one another. The antagonistic arrangement involves circular muscles, whose contraction makes the body longer and thinner, and longitudinal muscles, whose contraction makes it shorter and thicker. In an earthworm (*Lumbricus terrestris*), chaetae on each segment anchor it to the substrate, so that alternate contractions of the two types of muscle bring about movement. Not all circular or longitudinal muscles contract together; instead, waves of contraction pass along the body.

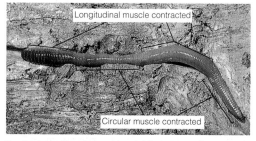
Earthworm (*Lumbricus terrestris*) – waves of contraction of the two types of muscle bring about movement

Exoskeleton

An exoskeleton is a more or less complete external covering which provides protection for internal organs and a rigid attachment for muscles. It is characteristic of arthropods where it comprises a three-layered cuticle secreted by epidermal cells beneath it. The outer layer, the **epicuticle**, is thin and waxy and so forms a waterproof covering. Beneath this is a rigid layer of **chitin** (Section 2.5.4) impregnated with tanned proteins. This is the **exocuticle**. The inner layer, or **endocuticle**, is a more flexible layer of chitin. The exoskeleton may be impregnated with salts, e.g. calcium carbonate, which give it additional strength. This is frequently the case in crustacean exoskeletons.

To permit uninhibited movement, the inflexible parts of the exoskeleton are separated by flexible regions where the rigid exocuticle is absent. Openings to glands and the digestive, respiratory and reproductive systems puncture the exoskeleton as do sensory hairs. Since the exoskeleton cannot expand, it imposes a limit on growth. To overcome this, it is periodically shed by a process known as moulting or **ecdysis** (Section 12.8.5).

Endoskeleton

An endoskeleton forms an internal framework within an organism. Endoskeletons occur in certain protozoans, molluscs (cephalophods) and vertebrates. In the vertebrates, the skeleton is cellular although the bulk of it is a non-cellular matrix secreted by the cells. In the Chondrichthyes (sharks and rays), cartilage forms the entire skeleton. This has the advantage of combining rigidity with a degree of flexibility. Most vertebrates have a skeleton made up of bone which provides a strong, rigid framework. This, however, needs special articulating points known as **joints** if movement is to be possible.

28.3.2 Structure of bone

As bones are usually observed in their dried state, they often give the impression of dead, immutable structures. In fact they are living tissue which is very plastic; capable of moulding itself to meet the mechanical requirements demanded of it. The matrix of compact bone is made up of collagen together with inorganic substances such as calcium, magnesium and phosphorus. These components are arranged in concentric circles, called **lamellae**, around an **Haversian canal** containing an artery, a vein, lymph vessels and nerve fibres. Bone cells, or **osteocytes**, are found in spaces in the lamellae known as **lacunae** and fine channels called **canaliculi** link lacunae. The system of lamellae around one Haversian canal is called an **Haversian system**. Bone is an extremely important and strong skeletal material. It is not static. Its various inorganic components may be deposited or absorbed at different times to meet new stresses put upon the tissue.

When a bone, e.g. the femur, is examined in detail, it is found to have a complex internal structure. There is a hollow shaft, the **diaphysis**, which contains **marrow**, a tissue producing various kinds of blood cell. At each end is an expanded head, the **epiphysis**, which articulates with other bones or to which tendons are attached. While the diaphysis and epiphysis are composed of **hard (compact) bone**, the remainder of the structure is made up of **spongy (cancellous) bone**. This has a honeycomb appearance

Did you know?

Astronauts suffer a loss of bone because, in the absence of gravity, they cannot load their skeletons enough to maintain the structure properly.

APPLICATION

Osteoporosis and HRT

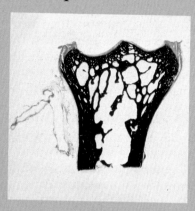

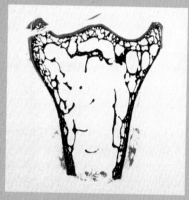

Normal bone (top) and (above) bone showing effects of osteoporosis

Bone is made up of an organic matrix and minerals. The matrix is mainly collagen fibres and the minerals calcium and phosphorus in the form of hydroxyaptite crystals. The surface of bone is covered in cells which continuously remodel it by removing old bone and forming new bone. If the cells remove more bone than is replaced there will be a loss of bone volume which leads to a weakening of the skeleton. This condition is known as osteoporosis.

In the United Kingdom 50% of post-menopausal women will have hip, vertebral or forearm fractures that can be attributed to osteoporosis. Worldwide about 1.7 million people, mostly women, suffer hip fractures as a result of thinning bones – a figure expected to rise to 6 million by 2050.

Long bones consist of an outer layer of cortical bone around the medulla. The latter comprises fat and haemopoietic cells (blood-forming cells) criss-crossed by struts of trabecular bone.

In post-menopausal women the volume of cortical and trabecular bone declines leading to osteoporosis. This has been linked to a fall in their oestrogen production which alters the balance between bone resorption and bone formation.

HRT (hormone replacement therapy) was introduced as a short-term treatment to mitigate the immediate symptoms of the menopause, like hot flushes and night sweats, but the medical indications for HRT have expanded. Some American studies indicate that oestrogen therapy can reduce the chance of fractures resulting from brittle bones, or osteoporosis, as well as halving a woman's risk of suffering a heart attack or stroke. Oestrogen supplements undoubtedly benefit bone density in older women but the hormones may have to be taken for 5 years before they offer any protection against fractures and their effect may decline rapidly when the supplement is stopped. However, oestrogen cannot be the only factor influencing osteoporosis since fracture rates in the western world for both men and women have trebled in the last 30 years.

Some researchers believe that diet plays an important rôle in the prevention of osteoporosis, advocating a diet containing legumes (rich in natural oestrogens) and foods high in vitamin D, like oily fish. Physical exercise also looks promising as an aid in the prevention of osteoporosis, helping to strengthen bones in the elderly. There is even a suggestion that middle-age spread can be good for you! Fat tissue converts a hormone from the adrenal gland into an oestrogenic one so severe dieting in post-menopausal women may be bad for the bones.

Recent research suggests that genes are responsible for more than half the variation in bone density in different people. A group of Australian scientists believes 75% of this genetic influence is the result of a single gene – that for a vitamin D receptor. If the Australians prove to be right, a simple blood test should be enough to identify people at higher risk of brittle bones. Teenagers at risk could be encouraged to change their diet and get more exercise and post-menopausal women might be encouraged to take oestrogen replacement therapy.

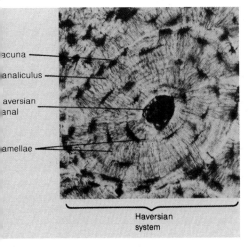

acuna
analiculus
aversian
anal
amellae

Haversian
system

ompact bone (TS) (×400 approx.)

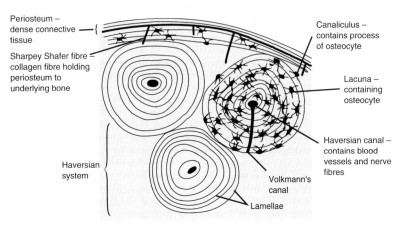

Periosteum – dense connective tissue

Sharpey Shafer fibre = collagen fibre holding periosteum to underlying bone

Haversian system

Canaliculus – contains process of osteocyte

Lacuna – containing osteocyte

Haversian canal – contains blood vessels and nerve fibres

Volkmann's canal

Lamellae

Fig. 28.7 Compact bone (TS)

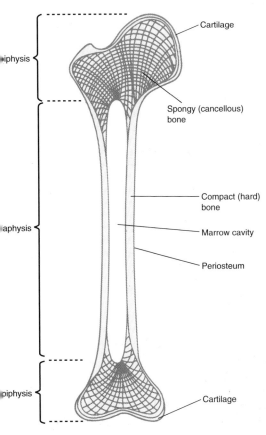

Cartilage

iphysis

Spongy (cancellous) bone

Compact (hard) bone

aphysis

Marrow cavity

Periosteum

piphysis

Cartilage

ig. 28.8 Vertical section through the femur

and provides strength with a minimum of additional mass. A tough, fibrous membrane, the **periosteum**, surrounds the bone. The structure of the femur is shown in Fig. 28.8.

Bone has a number of functions:

1. Providing a framework for supporting the body.

2. Providing a means of attachment for muscles which then operate the bones as a system of levers for locomotion.

3. Protecting delicate parts of the body, e.g. the rib cage protects the heart and lungs; the cranium protects the brain.

4. Acting as a reservoir for calcium and phosphorus salts, helping to maintain a constant level in the bloodstream (Section 26.4).

5. Producing red blood cells and certain white cells, e.g. granulocytes.

28.3.3 Joints

The skeleton has to fulfil two conflicting functions. On the one hand, it needs to be rigid in order to provide support and attachment for muscles; on the other hand, it needs to be flexible in order to permit movement. In the case of bony endoskeletons this paradox is overcome by having a series of flexible joints between the individual bones of the skeleton. The various types of joint found in a mammalian skeleton can be classified into three groups according to the degree of movement possible:

1. Immoveable (suture) joints – No movement is possible between the bones.

2. Partly moveable (gliding) joints – Only a little movement is possible between individual bones.

3. Moveable (synovial) joints – There is considerable freedom of movement between bones; the actual amount depends upon the precise nature of the joint.

The different joints found in mammals are described in Table 28.2.

In moveable joints, the ends of the bones are covered with a layer of **cartilage**. This prevents damage to the articulating

565

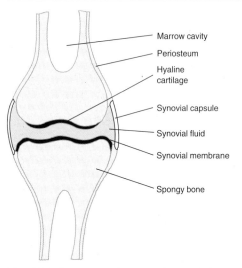

Marrow cavity
Periosteum
Hyaline
cartilage
Synovial capsule
Synovial fluid
Synovial membrane
Spongy bone

Fig. 28.9 Structure of a typical synovial joint

TABLE 28.2   **Joints of the mammalian skeleton**

Name of joint	Type of joint	Example
Suture	Immoveable	Between the bones of the cranium Between the sacrum and ilia of the pelvic girdle
Gliding	Partly moveable	Between adjacent vertebrae In the wrist and ankle
Pivot	Partly moveable	Between the axis and atlas vertebrae
Hinge	Synovial	In the elbow and knee In the fingers and toes
Ball and socket	Synovial	At the shoulder and hip

surfaces of bones as a result of friction between them. The joint is surrounded by a fibrous covering called the **synovial capsule**. The inner lining of this capsule is known as the **synovial membrane**. It secretes a mucus-containing lubricant fluid called **synovial fluid** which also provides nutrients for the cartilage at the ends of the bones. The structure of a typical synovial joint is illustrated in Fig. 28.9.

The bones of a joint are held together by means of strong, but elastic, **ligaments**.

28.4   Questions

(a) Describe the sequence of events which occurs when resting striated muscle is stimulated, contracts and then relaxes.   (14 marks)

(b) Indicate the relationship between ATP and cellular respiration in muscle contraction.   (9 marks)
(Total 23 marks)

UCLES June 1991, Paper II, No. 8

Isolated muscle was used in a series of experiments discover the sequence of reactions involved in a articular respiratory pathway. The tissue was aintained in a suitable bathing medium and rovided with various substances thought to be termediates in the pathway. The rate of respiration ter each substance had been added was measured the fall in oxygen concentration over a set time.

(a) (i) Suggest **one** advantage of using isolated muscle tissue instead of a whole organism.

(ii) Could a suspension of mitochondria have been used instead? Explain your answer.   (8 marks)

(b) (i) State **one** assumption made in using oxygen uptake as the measure of respiration rate.

(ii) Name **two** factors, other than oxygen uptake, which could have been used instead to indicate rate of respiration.

(iii) Describe **two** components of the bathing medium, other than water, which would be needed to keep the muscle tissue alive throughout the experiment.   (5 marks)

was found that the initial rapid rate of respiration in olated muscle preparations could be restored by lding small amounts of succinate or fumarate. The fects of these additions are shown in Table 1.

BLE 1

Substance added	Change in level of substance in muscle		
	Succinate	Fumarate	Citrate
uccinate	–	increase	increase
umarate	increase	–	increase

vo different reaction pathways were suggested to plain the observed results.

ypothesis I: succinate ⟶ fumarate ⟶ citrate

(Succinate is converted to fumarate, which is later converted to citrate.)

Hypothesis II: fumarate ⟶ succinate ⟶ citrate

(Fumarate is converted to succinate, which is later converted to citrate.)

In an attempt to eliminate one of these hypotheses, a chemical was added which inhibits the enzyme catalyzing the reversible reaction between succinate and fumarate.

(c) (i) Complete Table 2 to show the effects you predict (fall, rise or no change) in *citrate level* when mixtures of succinate and the inhibitor, or fumarate and the inhibitor, are added to muscle that has reached a slow rate of respiration.

TABLE 2

Hypothesis assumed to be correct	Change in level of citrate in muscle	
	when succinate plus inhibitor added	when fumarate plus inhibitor added
I		
II		

(ii) Suggest **one** chemical effect (other than a reduction in the rate of respiration) that could be expected if the living muscle were supplied with the inhibitor on its own.   (5 marks)
(Total 18 marks)

NEAB June 1991, Paper IIA, No. 6

3. The diagram below shows some of the structures in the hind limb of a rabbit.

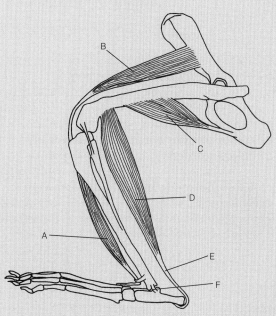

(a) Which of the muscles labelled A, B, C or D are responsible for the following?
   (i) Straightening the knee   (*1 mark*)
   (ii) Lifting the heel   (*1 mark*)
(b) For each of structures E and F, state *one* function and the property which enables this function to occur.   (*4 marks*)
(c) State *two* ways in which friction is reduced at joints.   (*2 marks*)
(*Total 8 marks*)

ULEAC June 1991, Paper I, No. 13

4. The drawing below shows a transverse section through compact bone.

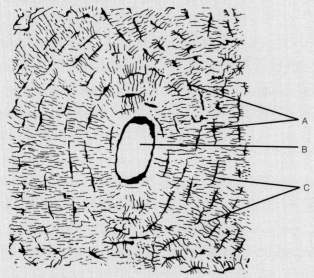

(a) Name parts A, B and C.   (*3 marks*)
(b) Explain how the non-living part (matrix) of bone is adapted for its function.   (*3 marks*)
(c) State why deficiency of vitamin D may result in deformed bones.   (*1 mark*)
(*Total 7 marks*)

ULEAC June 1990, Paper I, No. 6

5. The diagram at the top of the next column shows a human elbow joint in longitudinal section.
(a) Name the parts labelled A, B and C and give *one* function of each.   (*6 marks*)
(b) Indicate by an arrow on the diagram an appropriate position for the insertion of a muscle to extend the arm at this joint.   (*1 mark*)
(c) Name the types of bone tissue represented by D and E.   (*2 marks*)
(d) Comment on the structure of bone tissue D in relation to its functions.   (*3 marks*)
(*Total 12 marks*)

ULEAC January 1990, Paper I, No. 6

See *Further questions*, pages 614–20, questions 2, 13.

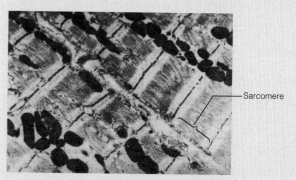

6. The electron micrograph below shows muscle tissue in longitudinal section.

Magnification ×13 500 (approx.)

(a) Label the following structures on the electron micrograph.
   A   dark band
   C   mitochondrion
   I   light band
   Z   Z line   (*4 marks*)
(b) A sarcomere has been labelled on the electron micrograph.
   (i) Measure and record the length of *three* sarcomeres and calculate the mean length. Show your working.   (*2 marks*)
   (ii) Given that the magnification is ×20 000, use your mean length to calculate the actual mean length of a sarcomere. Show your working.   (*2 marks*)
(c) Describe the role of ATP in muscle contraction.   (*2 marks*)
(*Total 10 marks*)

ULEAC June 1992, Paper III, No. 5

29 Control systems in plants

The ordered growth and development of plants shows that, like animals, they are capable of coordinating their activities. Unlike animals, they possess no nervous system and so plant coordination is achieved almost entirely by hormones. These hormones are similar to those of animals in being organic substances which in low concentrations cause changes in other parts of the organism. Unlike animals, plant hormones almost always affect some aspect of growth. This growth may lead to movements of plant parts, although these are relatively slow responses compared to those of animals.

29.1 Plant responses

Plants do not possess any contractile tissue with which to move. Their survival may, however, depend on their ability to move towards certain stimuli such as light and water. These movements are usually performed by growth responses.

TABLE 29.1 **Examples of tropic responses**

Stimulus	Name of response	Examples
Light	Phototropism	In almost all plants, shoots bend towards a directional light source (i.e. are positively phototropic), some roots bend away (i.e. are negatively phototropic) while leaves position themselves at right angles (i.e. are diaphototropic)
Gravity	Geotropism	In almost all plants, shoots bend away from gravity (i.e. are negatively geotropic) and roots bend towards it (i.e. are positively geotropic). The leaves of dicotyledonous plants position themselves at right angles (i.e. are diageotropic)
Water	Hydrotropism	Almost all plant roots bend towards moisture (i.e. are positively hydrotropic) while stems and leaves show no response
Chemicals	Chemotropism	Some fungal hyphae grow away from the products of their metabolism (i.e. are negatively chemotropic). Pollen tubes grow towards chemicals produced at the micropyle (i.e. are positively chemotropic)
Touch	Thigmotropism	The tendrils of peas (*Pisum*) twine around supports. The shoots of beans (*Phaseolus*) spiral around supports
Air	Aerotropism	Pollen tubes grow away from air (i.e. are negatively aerotropic)

Phototropism – the cress seedlings in the pot on the left have been grown in all-round light while those on the right have been subject to unilateral light

29.1.1 Tropisms

A tropism is a growth movement of part of a plant in response to a directional stimulus. The direction of the response is related to that of the stimulus and in almost all cases the plant part moves towards or away from it. Each response is named according to the nature of the stimulus, e.g. a response to light is termed phototropism. The direction of the response is described as positive, if movement is towards the stimulus, or negative if it is away from it. Some examples of tropisms are given in Table 29.1

29.1.2 Taxes

A taxis is the movement of a freely motile organism, or a freely motile part of an organism, in response to a directional stimulus. The direction of the response is related to that of the stimulus, being towards it (positive) or away from it (negative). They occur in both plants and animals. As with tropisms, the type of stimulus determines the name of a tactic response. Examples are given in Table 29.2.

TABLE 29.2   **Examples of tactic responses**

Stimulus	Name of response	Examples	
		Plants	Animals
Light	Phototaxis	*Euglena* swims towards light provided it is not too intense (i.e. is positively phototactic)	Earthworms (*Lumbricus*) and woodlice (*Oniscus*) move away from light (i.e. are negatively phototactic)
Temperature	Thermotaxis	The green alga, *Chlamydomonas*, will swim to regions of optimum temperature. Motile bacteria behave in a similar way	Blowfly larvae and many other small animals move away from extremes of temperature
Chemicals	Chemotaxis	Antherozooids (sperm) of mosses, liverworts and ferns, are attracted to chemicals produced by the archegonium (i.e. they are positively chemotactic)	Many show negative chemotactic responses to specific chemicals – a fact exploited in the use of insect repellents

29.2   Plant hormones

Plant hormones, or **plant growth substances** as they are often called, are chemical substances produced in plants which accelerate, inhibit or otherwise modify growth. Growth at the apices of plants occurs in three stages: cell division, cell elongation and cell differentiation. Plant hormones may affect any, or all, of these processes. There are five groups of growth substances generally recognized: **auxins, gibberellins, cytokinins, abscisic acid (inhibitor)** and **ethene** (ethylene).

29.2.1 Auxins

Auxins are a group of chemical substances of which **indoleacetic acid (IAA)** is the most common. They have been isolated from a large number of plants. Charles Darwin was one of the earliest to investigate the response of plant shoots to light (phototropism) which eventually led to the discovery and isolation of auxins. The historical record of these developments is traced in Fig. 29.1.

The auxin indoleacetic acid, a derivative of the amino acid tryptophane, is largely produced at the apices of shoots and roots. Fig. 29.2 on page 575 outlines its chemical structure.

The transport of auxin occurs in one direction, namely away from the tip, i.e. its movement is **polar**. Short distance movement from cell to cell occurs by diffusion, but long distance transport is possible via phloem.

The rôle of auxin in producing a phototropic response is shown in Fig. 29.1. It appears that unilateral light causes a redistribution

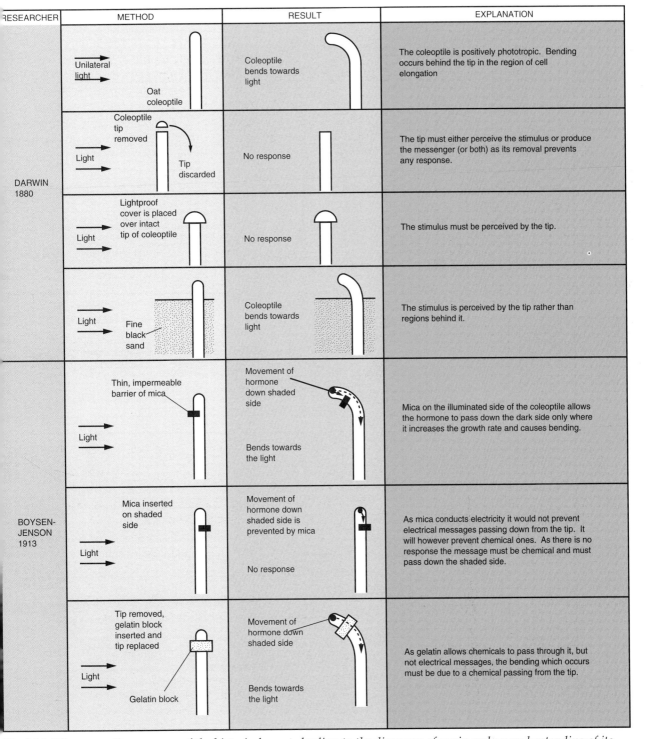

Fig. 29.1 Diagrammatic summary of the historical events leading to the discovery of auxin and an understanding of its mechanism of action (cont.)

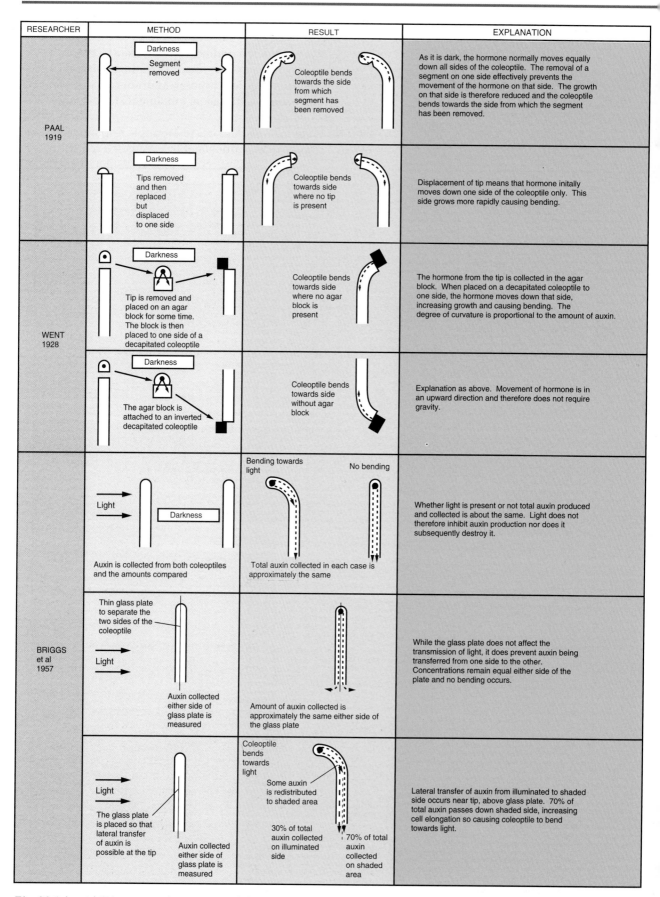

Fig. 29.1 (cont.) Diagrammatic summary of the historical events leading to the discovery of auxin and an understanding of its mechanism of action

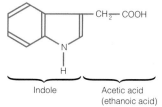

Fig. 29.2 *Structure of indoleacetic acid (IAA)*

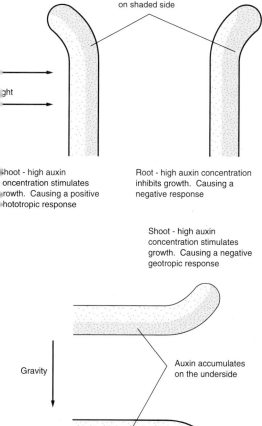

Auxin accumulates
on shaded side

Shoot - high auxin
concentration stimulates
growth. Causing a positive
phototropic response

Root - high auxin concentration
inhibits growth. Causing a
negative response

Shoot - high auxin
concentration stimulates
growth. Causing a negative
geotropic response

Gravity

Auxin accumulates
on the underside

Root - high auxin concentration
inhibits growth. Causing a
positive geotropic response

Fig. 29.4 *Mechanism of auxin action in phototropic and geotropic responses of shoots and roots*

of auxin so that a greater amount travels down the shaded side. As one effect of auxin is to cause cell elongation, the cells on the shaded side elongate more than those on the illuminated side. The shoot therefore bends towards the light. Exactly how unilateral light effects a redistribution is not clear.

The redistribution theory may seem plausible but does not immediately explain why many roots are negatively phototropic. If the same arguments are used, it follows that a root exposed to unilateral light should accumulate auxin on its shaded side. This side should elongate more rapidly and the root bend towards the light. Why then does it bend in the opposite direction? Experiments have revealed that roots are more sensitive to auxin than stems, i.e. they respond to lower concentrations of auxins. As the concentration of auxin increases, the growth of roots, far from being stimulated, becomes inhibited. At a concentration of auxin of one part per million, for example, the growth of roots is reduced while that of stems is considerably increased (see Fig. 29.3). If the effect of unilateral light is to redistribute the auxin so that a concentration of one part per million passes down the shaded side, then in a stem, growth will increase on that side and it will bend towards the light, while in a root, growth will be inhibited on that side and it will bend away from the light.

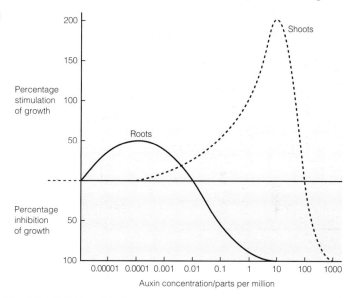

Fig. 29.3 *Relationship between growth and auxin concentration in roots and shoots*

Similar redistribution of auxins could account for the responses of shoots and roots to gravity. It can be shown experimentally that a higher concentration of auxin occurs on the underside of horizontal roots and shoots. The concentrations found inhibit root growth, causing it to bend downwards, i.e. a positive geotropic response. The same concentrations stimulate stem growth, causing it to bend upwards, i.e. a negative geotropic response. However the differences in concentration of auxin are small – smaller in fact than those of another plant hormone – gibberellin. Clearly the process is complex, probably involving a number of hormones. Indeed some theories suggest other factors such as pH or water potential differences are responsible. One mechanism to explain these differences is the **statolith theory**. This suggests that starch grains (or statoliths) fall within the cytoplasm of a cell under the influence of gravity. Their accumulation on the lower side of the cell is responsible for many of the physiological differences between the upper and lower sides of a cell. Both phototropic and geotropic responses of stems and roots are summarized in Fig. 29.4. Phototropic and geotropic responses are due to auxins promoting cell elongation. Auxins, however, have other influences in plants. They stimulate cell division, help to maintain the structure of cell walls and, in high concentrations, inhibit growth. Their ability to stimulate cell division is most easily observed by treating the cut end of plant parts with auxins. Often, an area of disorganized and largely undifferentiated tissue results. The large swellings so produced are called **calluses**. By a similar means, auxins may stimulate fruit development without fertilization (**parthenocarpy**). The maintenance of the structure of cell walls by auxins inhibits **abscission**. This is the separation of leaves, flowers and fruits from a plant as the result of the middle lamellae between cell walls, at the base of a petiole, pedicel or peduncle, weakening to such an extent that small mechanical disturbances, e.g. wind, cause them to fall. Auxin prevents the formation of this abscission layer and so inhibits abscission.

The inhibition of growth by high concentrations of auxin results in **apical dominance**. This is where the bud at the apex of a shoot produces auxin in sufficient concentration to inhibit growth of the lateral buds further down the shoot. These lateral buds remain dormant unless the apical bud is removed, in which case one or more of them develops into side branches. This is the principle behind the pruning of many plants as a means of producing a more bushy form of growth. The effects of auxins are summarized at the end of this section, in Table 29.3.

29.2.2 Gibberellins

The name **gibberellins** was derived from the fungus *Gibberella* (since renamed *Fusarium*). This fungus was shown by Japanese scientists in the 1920s to be the cause of 'foolish seedling disease', a disorder which resulted in rice seedlings growing considerably taller than their healthy counterparts. An extract from this fungus produced an increase in growth when applied to other plants. A group of active substances was finally isolated from the extract; these were called gibberellins. The number of gibberellins now isolated exceeds 50, but all have a similar chemical make-up.

The main influence of gibberellins on plants is to promote cell elongation and so increase growth. Unlike auxins, however, gibberellins can stimulate growth in dwarf varieties, thus restoring them to normal size. These dwarf varieties are thought to result from a genetic mutation which prevents them producing gibberellins naturally. They therefore require an external supply to make them grow to normal size. While the main effect of gibberellins is to cause elongation of the stem, they also influence cell division and differentiation to some extent. Their varied effects sometimes complement those of auxins, e.g. in promoting growth, but at other times they have antagonistic effects, e.g. while auxins promote the growth of adventitious roots, gibberellins inhibit their formation. Gibberellins play a rôle in breaking dormancy in seeds. They mobilize food reserves

Effect of gibberellins on plant growth

in readiness for germination by stimulating the synthesis of enzymes such as α-amylase. It is likely that gibberellins operate at a fundamental level within a cell, possibly by switching on and off genes. Their varied effects are summarized in Table 29.3.

29.2.3 Cytokinins

Plant cells grown in synthetic nutrient media were found to remain alive but did not undergo cell division. Division only occurred if malt extract or coconut milk was added to the culture. It was later found that autoclaved samples of DNA also induced cell division and from these **kinetin**, a substance similar to adenine, was isolated. Partly as a consequence of their effects in promoting cell division (**cytokinesis**), substances like kinetin were termed **cytokinins**. A number of cytokinins are produced naturally by many plants; all are derivatives of adenine. This suggests that they operate through some rôle in nucleic acid metabolism, most probably being involved in tRNA synthesis. They are found largely in actively dividing tissues, especially fruits and seeds, where they promote cell division in the presence of auxins.

One interesting effect of cytokinins is their ability to delay **senescence** (ageing) in leaves. In the presence of cytokinins, leaves removed from a plant remain green and active, rather than turning yellow and dying. Further effects are summarized in Table 29.3.

29.2.4 Abscisic acid

Abscisic acid inhibits growth and so works antagonistically to auxins, gibberellins and cytokinins. Its main effect, as its name suggests, is on abscission (leaf and fruit fall). The process results from a balance between the production of auxin and abscisic acid. As a fruit ripens, the level of auxin (which inhibits abscission) falls, while that of abscisic acid (which promotes abscission) increases. This leads to the formation of an abscission layer which causes the fruit to fall. Other influences of abscisic acid are listed in Table 29.3.

29.2.5 Ethene

Unlike the other hormones, **ethene** (ethylene) has a relatively simple chemical structure. It is produced as a metabolic by-product of most plant organs, especially fruits. Its main effect is in stimulating the ripening of fruits, but it also influences many auxin-induced responses. Other effects are summarized in Table 29.3.

PROJECT

There are a number of hormone rooting powders on the market. How effective are they?

29.3 Control of flowering

Plants flower at different times of the year, daffodils and snowdrops appearing in early spring, roses in summer and chrysanthemums in autumn. The seasonal differences in the time of flowering are related to two main climatic factors: day length and temperature.

TABLE 29.3   **A summary of the effects of plant hormones**

Hormone	Effects		Examples
Auxins	Promote cell elongation	Phototropic responses	Shoots bend towards light
		Geotropic responses	Roots grow down into the soil
	Stimulate cell division	Promotes development of roots	Hormone root powders are used to help strike cuttings
		Stimulates cambial activity	Callus development at the site of wounds
		Stimulates development of fruits	Natural fruit setting may be improved by spraying with synthetic auxins or they may develop parthenocarpically, e.g. in apples
	Maintain cell wall structure	Inhibits leaf abscission	If the supply of auxin from leaves exceeds that from the stem, the leaf remains intact
		Inhibits fruit abscission	If the supply of auxin from the fruit exceeds that from the stem, the fruit remains intact
	Inhibit growth in high concentrations	Apical dominance	Lateral buds remain dormant under the influence of auxin from the apical bud
		Disruption of growth	Synthetic auxins, e.g. 2,4-D and 2,4,5-T are used as selective weedkillers
Gibberellins	Reverse of genetic dwarfism		Dwarf varieties of peas and maize grow to normal size when gibberellin is applied
	Promote cell elongation		Increases the length of internodes
	Break dormancy of buds		Dormancy of many buds, e.g. birch (*Betula*), is broken by the addition of gibberellin
	Break dormancy of seeds		Ash and cereal seed dormancy may be broken as a result of mobilizing food reserves, leading to germination
	Stimulate fruit development		Cherry and peach fruits develop more readily after application of gibberellins
	Remove need for cold period in vernalization		Carrots can be induced to flower without first being subjected to a period of cold
	Affect flowering		Promotes flowering in some long-day plants and inhibits it in some short-day ones
Cytokinins	Promote cell division		Increase growth rate in many plants, e.g. sunflower (*Helianthus*)
	Delay leaf senescence		Maintain leaf for some time once detached from plant
	Stimulate bud development		Promote development of buds on cuttings of African violet (*Saintpaulia*) and protonemata of some mosses
	Break dormancy		Break dormancy in both seeds and buds
Abscisic acid	Inhibits growth		Retards growth in most plant parts
	Promotes abscission		Causes the formation of an abscission layer in the petioles and pedicels of leaves, flowers and fruits
	Induces dormancy		Promotes dormancy in the seeds and buds of many plants, e.g. birch (*Betula*) and sycamore (*Acer*)
	Closes stomata		Promotes stomatal closure under conditions of water stress
Ethene	Ripens fruit		Most citrus fruits ripen more rapidly in the presence of ethene
	Breaks dormancy		Ends dormancy of buds in some plants
	Induces flowering		Promotes flowering in pineapples

APPLICATION

Commercial applications of synthetic growth regulators

As the main function of plant hormones is to control growth, it is hardly surprising that they, or rather their synthetic derivatives, have been extensively used in crop production.

Synthetic auxins such as 2,4-dichlorophenoxyacetic acid (2,4-D) and 2,4,5-trichlorophenoxyacetic acid (2,4,5-T) are used as **selective weedkillers**. When sprayed on crops, they have a more significant effect on broad-leaved (dicotyledonous) plants than on narrow-leaved (monocotyledonous) ones. They so completely disrupt the growth of broad-leaved plants that they die, while narrow-leaved ones at most suffer a temporary reduction in growth. As cereal crops are narrow-leaved and most of their competing weeds are broad-leaved, the application of these hormone weedkillers is of much commercial value. They are also extensively used domestically for controlling weeds in lawns. Other details of these weedkillers are given in Section 18.6.2. Another synthetic auxin, naphthaleneacetic acid (NNA), is used to increase fruit yields. If sprayed on trees, it helps the fruit to set naturally, or in some species causes them to set without the initial stimulus of fertilization (parthenocarpy). This usually results in seedless fruits which may be a commercial advantage. Gibberellins extracted from fungal cultures are used commercially in the same way.

Auxins are the active constituent of rooting powders. The development of roots is initiated when the ends of cuttings are dipped in these compounds. As we have seen, cytokinins will delay leaf senescence. They are therefore sometimes used commercially to keep the leaves of crops, like lettuce, fresh and free from yellowing after they have been picked. Both gibberellins and cytokinins are sometimes applied to seeds to help break dormancy and so initiate rapid germination. The longer a seed remains ungerminated in the soil, the more vulnerable it is to being eaten, e.g. by birds.

Abscisic acid may be sprayed on fruit crops to induce the fruits to fall so they can be harvested together. Ethene is applied to tomatoes and citrus fruits in order to stimulate ripening.

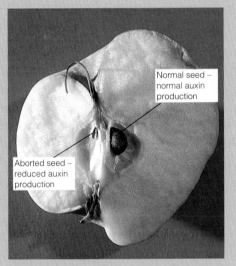

Normal seed – normal auxin production

Aborted seed – reduced auxin production

Apple sectioned to show effects of auxin production

29.3.1 The phytochrome system

Many plant processes are influenced by light. Before a plant can respond to variations in light intensity, duration or wavelength it must first detect these changes. Some form of photoreceptor is necessary. In Section 14.2.3 we discussed the relationships between the absorption spectrum for chlorophyll and the action spectrum for photosynthesis. A similar relationship can be established between a pigment called **phytochrome** and a number of light-induced plant responses (See Fig. 29.5).

Phytochrome, isolated in 1960, exists in two interconvertible forms:

1. **Phytochrome 660 (P_{660})** – This absorbs red light (peak absorption at a wavelength of 660 nm).

2. **Phytochrome 730 (P_{730})** – This absorbs light in the far-red region of the spectrum (peak absorption at 730 nm).

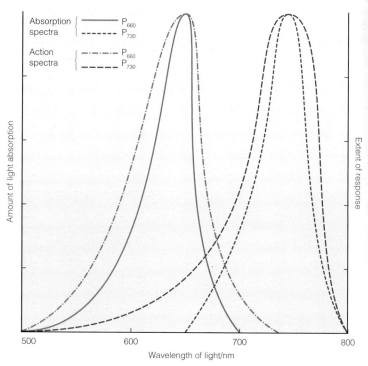

Fig. 29.5 Absorption and action spectra for phytochromes and the responses they control

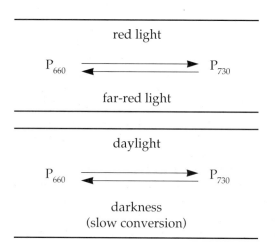

Even a short exposure to the appropriate light wavelength causes the conversion of one form into the other, as shown opposite. These conversions may also be brought about by daylight and darkness. During daylight, P_{660} is converted to P_{730}, while in the dark, a rather slower conversion of P_{730} to P_{660} occurs.

Phytochrome comprises a protein and a pigment. It is distributed throughout the plant in minute quantities, being most concentrated in growing tips. The actions of the two forms are usually antagonistic, i.e. where P_{660} induces a response, P_{730} inhibits it. The various effects of the two forms are listed in Table 29.4.

29.3.2 Photoperiodism

One major influence on the timing of flowering is the length of the day or **photoperiod**. The effects of the photoperiod on flowering differ from species to species but plants fall into three basic categories:

1. Long-day plants (LDP) – These only flower when the period of daylight exceeds a critical minimum length. Examples of long-day plants include radish, clover, barley and petunia.

2. Short-day plants (SDP) – These only flower when the period of daylight is shorter than a critical maximum length. Examples of short-day plants include chrysanthemum, poinsettia, cocklebur and tobacco.

3. Day-neutral plants – These plants flower regardless of the length of daylight. Examples of day-neutral plants include cucumber, begonia, violet and carrot.

Intermediate varieties exist. For example, **short-long-day plants** only flower after a sequence of short days is followed by long ones. These plants will flower naturally in mid-summer when

TABLE 29.4 Summary of the effects of red light and far-red light

Red light effects	Far-red light effects
Phytochrome 660 changes to phytochrome 730	Phytochrome 730 changes to phytochrome 660
Stimulates germination of some seeds, e.g. lettuce (*Lactuca*)	Inhibits germination of some seeds, e.g. lettuce (*Lactuca*)
Induces formation of anthocyanins (plant pigments)	Inhibits formation of anthocyanins
Stimulates flowering in long-day plants	Inhibits flowering in long-day plants
Inhibits flowering in short-day plants	Stimulates flowering in short-day plants
Elongation of internodes is inhibited	Elongation of internodes is promoted
Induces increase in leaf area	Prevents increase in leaf area
Causes epicotyl (plumule) hook to unbend	Maintains epicotyl (plumule) hook bent

the days are long following the shorter ones of spring. **Long-short-day plants** flower after a sequence of long days is followed by short ones. These will flower naturally in the autumn after the long days of summer are followed by the shorter ones of early autumn.

It is rather unfortunate that historically plants were categorized as short-day or long-day, as it is the length of the dark period which is crucial in determining flowering. Short-day plants require a long dark period, whereas long-day plants require a short dark period. This fact was established by a series of experiments summarized in Fig. 29.6.

Interrupting a long dark period with red light is as effective as daylight in stopping short-day plants flowering. Far-red light, however, has no effect and short-day plants flower as if the dark period had been continuous. These and other experiments suggest that phytochrome is the photoreceptor detecting different light wavelengths and ultimately determining whether or not a plant flowers.

Although flowers are formed at the apex, experiments confirm that the light stimulus is detected by the leaves. In some cases only a single leaf needs to be subjected to the appropriate stimulus to induce flowering. A message must therefore pass from the leaves to the apex. As plants coordinate by chemical means, this message is assumed to be a hormone and has been called **florigen**, even though it has not yet been isolated. It is thought to be transported within phloem. Understanding of the mechanism by which phytochrome initiates flowering is poor. The process is clearly complex and, because they can have similar effects to red light, gibberellins may be involved. One possible mechanism is summarized in Fig. 29.7. As long-day plants are known to flower after a short exposure to red light, it is possible that red light (wavelength 660 nm) is absorbed by phytochrome 660 which is rapidly converted to phytochrome 730 which then induces flowering. Short-day plants by contrast flower in response to phytochrome 660. This is formed by absorption of far-red light (wavelength 730 nm) by phytochrome 730 which is then converted to phytochrome 660. This rapid conversion by far-red light can also take place, although far more slowly, in darkness. Hence a long dark period induces flowering in these short-day plants. One postulated mechanism by which

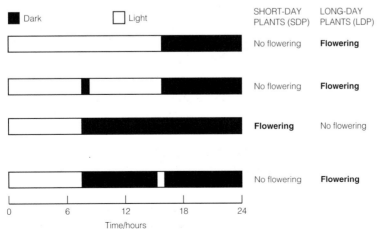

Fig. 29.6 Flowering related to the length of dark period

It is thought that the rate of closing of flowers is related to the frequency of the light

Design and carry out experiments which would 'shed some light' on this statement.

florigen is produced suggests that there are two inactive forms of the hormone, one of which is converted to the hormone by P_{660}, the other by P_{730}. These events are summarized in Fig. 29.7

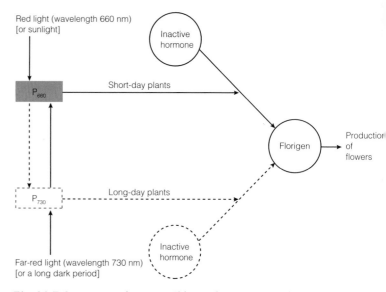

Fig. 29.7 Summary of one possible mechanism to explain the rôle of the phytochrome system in flowering

29.4 Questions

. A plant shoot was kept in the dark for three hours.
: then had a light shone on it from one side only.
he graph shows the growth of the shoot before and
fter the light was shone on it.

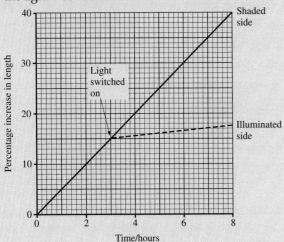

Time/hours

(a) What is the rate of growth of :
   (i) the shoot before the light was shone on it;
   (ii) the illuminated side of the shoot in the
   two-hour period immediately after the
   light was shone on it?   (2 marks)
(b) Use the evidence in this graph to suggest one
   way in which auxins act to produce bending
   of the shoot towards the light.   (1 mark)
(b) An early hypothesis suggested that auxins
   were only produced when the shoot tip was
   illuminated and then slowly diffused to the
   site of action. Explain how the information in
   the graph suggest that this hypothesis is
   incorrect.   (2 marks)
                                  (Total 5 marks)

   NEAB June 1995, Module BY04, No. 5

2. The growth of roots and shoots of plants is affected
   by various plant hormones. The diagram shows the
   effect of IAA (auxin) concentrations on the growth of
   roots and shoots. Results have been expressed as
   percentage stimulation or inhibition compared to a
   control.
   (a) (i) A logarithmic scale has been used for
      plotting the IAA concentrations.
      Explain the reason for using such a
      scale.   (2 marks)
      (ii) Suggest, with reasons, a suitable control
      for this experiment.   (2 marks)
   (b) Using the information in the diagram,
      comment on the growth of shoots and roots
      at:
      (i) low concentrations [10^{-6} to 10^{-3} ppm]
      of IAA;   (4 marks)

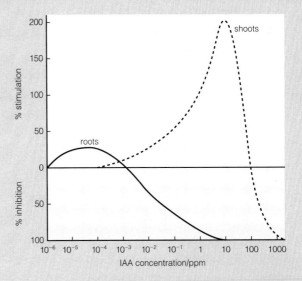

IAA concentration/ppm

   (ii) high concentrations [10 to 1000 ppm] of
      IAA.   (4 marks)

House plants grown on window sills need turning
regularly to prevent them bending towards the light.

   (c) (i) What name is given to this response?
                                    (1 mark)
      (ii) Describe how the growth of plant
      shoots towards the light is brought
      about.   (6 marks)
                                  (Total 19 marks)

   UCLES (Modular) June 1993,
   (Transport, regulation and control), No. 3

3. If a pea seedling is placed horizontally its root soon
   begins to grow downwards. The table below shows
   the geotropic responses of several batches of
   horizontally placed pea seedlings after different
   surgical treatments had been applied to their root
   caps as indicated by the drawings.

Treatment		Response (Curvature/°)
A None (root cap left intact)		Down 63°
B All root cap removed		Down 1°
C Lower half of root cap removed		Up 4°
D Upper half of root cap removed		Down 31°

(a) (i) Does this geotropic response seem to be brought about by a growth stimulator or by a growth inhibitor? Explain your reasoning.

(ii) Account for the different results obtained:

in C compared with A,

in D compared with A. *(3 marks)*

(b) Draw in (or describe) alongside each treatment the results you would expect. Pieces of thin metal foil were inserted between the tip and the region of cell elongation as shown below.

(3 marks)

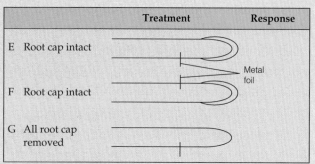

	Treatment	Response
E	Root cap intact	
		Metal foil
F	Root cap intact	
G	All root cap removed	

(c) (i) Describe a control that could be set up to eliminate the possibility that a stimulus other than gravity could be causing these responses.

(ii) Name **two** conditions that would need to be kept constant throughout the experiments. *(3 marks)*

(d) Amyloplasts (starch grains) in the root cap cells are thought to be involved in this response, as they can be observed to fall to the bottom of cells in about the same time as is required to trigger the growth response.

Suggest **two different** ways in which falling amyloplasts might influence the concentrations of plant hormone on upper and lower sides of the root. *(2 marks)*

(Total 11 marks)

NEAB June 1993, Part IIA, No. 5

4. Read through the passage below on the growth of roots under the influence of the Earth's gravitational force. Use your own knowledge and information in the passage to answer the questions which follow.

Once a seed germinates, its immediate problem is to recognize which way is 'up' for the shoot and 'down' for the root. Unless the shoot reaches the light and starts to photosynthesize before the
5 store of food in the seed is used up, the seedling dies.

The most direct pathway to light must be taken. Yet if the seed is buried there is no hint

from the light itself. A reliable signal can be
10 recognized from the Earth's gravitational force directed towards its centre; the root must respond by growing towards the Earth's centre and the shoot by growing away from it.

How can the gravitational force be detected?
15 Newton's clue was the falling apple; perhaps cells contain particles which fall when under the influence of gravity. There is good evidence that the cells which detect the gravitational stimulus contain statoliths, and these fall fairly freely
20 under the influence of gravity, so they tend to accumulate on the bottom of the cell.

We come now to the problem of how falling statoliths can affect the growth of the root towards the Earth's centre. There are two
25 questions here. Firstly, how is their position within the cell recognized; secondly, how is this translated into a signal which affects growth?

It seems probable that the statoliths cause local disturbances inside the cells and the information
30 is then transmitted to the site of response. This indicates the involvement of a hormone, and a persuasive case was made by early workers for the involvement of auxin.

Adapted from Biology Study Guide 1,
Nuffield Advanced Science

(a) (i) What is the term used to describe the response of the root to the Earth's gravitational force (line 10)? *(1 mark)*

(ii) What are root statoliths made of (line 19)? *(1 mark)*

(iii) Give the location of the statoliths in the root. *(1 mark)*

(iv) Suggest a reason why 'destarching' a root might affect its ability to respond. *(1 mark)*

(b) Describe a simple experiment you could set up to demonstrate the effects of the Earth's gravitational force on the growth of a young root. *(4 marks)*

(c) (i) Suggest why statoliths cause local disturbances in the cells and other organelles do not (lines 28–29). *(1 mark)*

(ii) State the site of response in the root (line 30). *(1 mark)*

(iii) Suggest an explanation for the involvement of auxin in causing the root to grow downwards (line 33).

(4 marks)

(Total 14 marks)

ULEAC June 1993, Paper III, No. 7

5. A pea seedling with a straight plumule and radicle was illuminated from one side for a period of 8 hours. Measurements were made of the angle of curvature of

he radicle and the plumule at 2-hourly intervals
during the 8-hour period.

The results are shown in the table below.

Length of exposure to unilateral illumination /hours	Angle of curvature	
	Plumule	Radicle
0	0°	0°
2	+7°	−3°
4	+14°	−4°
6	+18°	−8°
8	+23°	−11°

Adapted from P W Freeland, Problems in Theoretical Biology

(a) (i) Plot these data in suitable graphical form
on graph paper. (5 marks)
(ii) What is the angle of curvature of the
plumule after 3 hours' illumination?
(1 mark)

(b) (i) Compare the responses made by the
plumule and the radicle to the
unilateral illumination. (2 marks)
(ii) Describe the mechanism by which the
stimulus results in the response of the
plumule. (4 marks)

(c) Suggest how the responses of the plumule
and the radicle are important in the
successful growth of the seedling. (2 marks)
(Total 14 marks)

ULEAC June 1992, Paper III, No. 7

6. An experiment was carried out to investigate
germination and growth of seeds of a dwarf variety of
pea. Four groups of seeds were used and treated as
described below.

All batches of seeds were germinated at 26 °C, two
groups being kept in total darkness and two groups
in continuous weak red light. After five days, one
batch in the dark and one batch in red light were each
sprayed with an aqueous solution of gibberellic acid
(GA). The other batches were left untreated to act as
controls. Two days after this treatment random
samples of 20 plants were removed from each batch
and the mean lengths of each of the first, second, and
third internodes were found. The results are shown in
the bar graph at the top of the next column.

(a) Define the term *internode*. (1 mark)
(b) What type (class) of substance is *gibberellic
acid*? (1 mark)
(c) From the results given, what can you deduce
about the effect on the extension of growth in
the shoots of:
(i) red light, (2 marks)
(ii) gibberellic acid in the absence of light,
(1 mark)
(iii) gibberellic acid in the presence of red
light? (2 marks)

See *Further questions*, pages 614–20, questions 5, 11, 15.

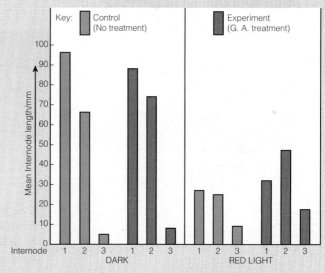

(d) (i) Suggest a logical hypothesis for these
effects. (2 marks)
(ii) Suggest an experiment that might be
carried out to test your hypothesis.
(2 marks)
(Total 11 marks)

Oxford June 1992, Paper I, No. 10

7. In an investigation into photoperiodism, groups of
plants of the same species were exposed to cycles
made up of varying dark periods alternating with a
light period of either 4 or 16 hours. The mean number
of flowers produced by each group of plants is shown
in the graph.

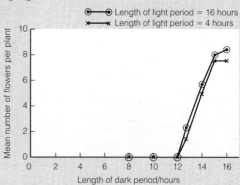

(a) (i) Which of the following terms
short day
long day
day neutral
best identifies this species? (1 mark)
(ii) Give evidence from the graph to
support your answer. (2 marks)

(b) Describe how you would show that the
flowering stimulus is detected by the leaves
and not by the stem apex where the flowers
are produced. (3 marks)
(Total 6 marks)

AEB June 1993, Paper I, No. 7

30 Biotechnology

Biotechnology is the application of scientific and engineering principles to the production of materials by biological agents. Given its recent wide publicity, one could be forgiven for thinking it was a new branch of science. While there is no doubt that recent technological and biochemical advances have led to considerable developments in biotechnology, its origins go back a long way. Food for human consumption has always been vulnerable to spoilage by microorganisms. Ancient civilizations probably found that normally detrimental microbial contamination occasionally conferred some benefit: improved flavour or better preservation for instance. In this way beers would have been developed from 'spoilt' grain and wine from 'spoilt' fruit. Contamination of the alcohol by a different agent led to the production of vinegar which was then used to preserve food. Cheese, butter and yoghurt all resulted from various microbial contaminations of milk.

The modern biotechnology industry had its origins in the First World War. A naval blockade deprived Germany of the supply of vegetable fats necessary for the production of glycerol from which explosives were made. They turned to the fermentation of plant material by yeast as an alternative source. At the same time, the British were using *Clostridium acetobutylicum* to produce acetone and butanol as part of their war effort. In a similar way, the Second World War prompted the mass production of the antibiotic penicillin (discovered by Alexander Fleming in 1929) using *Penicillium notatum*.

Many other chemicals were produced thereafter by use of fermentation techniques, but it was in the 1980s that biotechnology underwent major expansion. This was almost entirely due to the development of **recombinant DNA technology** (Section 7.8.1).

30.1 Growth of microorganisms

Microorganisms (microbes) are found in every ecological niche – from deepest ocean to the limits of the stratosphere, from hot springs to the frozen poles. They are small, easily dispersed and quickly multiply given a suitable environment. They grow on a wide diversity of substrates making them ideal subjects for commercial application. Each species has its own optimum conditions within which it grows best.

30.1.1 Factors affecting growth

The factors which affect the growth of microorganisms are equally applicable to the growth of plant and animal cell cultures.

Nutrients
Growth depends upon both the types of nutrients available and their concentration. Cells are largely made up of the four elements: carbon, hydrogen, oxygen and nitrogen with smaller, but significant, quantities of phosphorus and sulphur. Accounting as they do for 90% of the cell's dry mass, all six are essential for growth.

Needed in smaller quantities, but no less important, are the metallic elements: calcium, potassium, magnesium and iron, sometimes known as **macro-nutrients**. Required in smaller amounts still are the **micro-nutrients (trace elements)**: manganese, cobalt, zinc, copper and molybdenum – indeed not all may be essential to some species. A further group of chemicals, loosely termed **growth factors**, are also needed. These fall into three categories:

1. Vitamins

2. Amino acids

3. Purines and pyrimidines

Up to a point, the more concentrated a nutrient the greater the rate of growth, but as other factors become limiting the addition of further nutrients has no beneficial effect. More details of the nutrients used in culturing cells are given in Section 30.1.3.

Temperature
As all growth is governed by enzymes and these operate only within a relatively narrow range of temperature, cells are similarly affected by it. If the temperature falls too low the rate of enzyme-catalyzed reactions becomes too slow to sustain growth; if too high the denaturation of enzymes causes death. Most cells grow best within the range 20–45 °C although some species can grow at temperatures as low as − 5 °C, while others do so at 90 °C. Three groups are recognized according to their preferred temperature range:

1. **Psychrophiles** (e.g. *Bacillus globisporus*) – These have optimum growth temperatures below 20 °C, many continuing to grow at temperatures down to 0 °C.

2. **Mesophiles** (e.g. *Escherichia coli*) – These have optimum growth temperatures in the range 20–40 °C.

3. **Thermophiles** (e.g. the alga *Cyanidium caldarium*) – These have optimum growth temperatures in excess of 45 °C, a few surviving in temperatures as high as 90 °C. These cells have enzymes which are unusual in not being denatured at high temperatures.

pH
Microorganisms are able to tolerate a wider range of pH than plant and animal cells, some species growing in an environment as acid as pH 2.5, others in one as alkaline as pH 9. Microorganisms preferring acid conditions, e.g. *Thiobacillus thiooxidans*, are termed **acidophiles**.

Oxygen

Many microorganisms are aerobic, requiring molecular oxygen for growth at all times: these are termed **obligate aerobes**. Some, while growing better in the presence of oxygen, can nevertheless survive in its absence; these are called **facultative anaerobes**. Others find oxygen toxic and do not grow well in its presence: these are the **obligate anaerobes**. Some of this group, while tolerating oxygen, nevertheless grow better when its concentration is very low. These are termed **microaerophiles**.

Osmotic factors

All microorganisms require water for growth. In most cases this is absorbed osmotically from the environment, although pinocytosis is used in certain protozoa and all groups produce a little water as a product of aerobic respiration. To ensure absorption, the water potential of the external environment must be less negative (higher) than the cell contents. For this reason most microorganisms cannot grow in environments with a high solute concentration – a fact made use of in preserving foods, e.g. salting of meat and fish, bottling of jam and fruit in sugar. A few called **halophiles**, can survive, however, in conditions of high salt concentration.

Pressure

Although pressure is not a major factor affecting growth in most microorganisms, a few species inhabiting the ocean depths can grow under immense pressure. Some of these **barophiles** cannot grow in surface waters where the pressure is too low for their survival.

Light

Photosynthetic microorganisms require an adequate supply of light to sustain growth.

Water

In common with all organisms, microorganisms require water for a variety of functions. In addition, photosynthetic microorganisms use it as a source of hydrogen to reduce carbon dioxide. Chemoautotrophs may use alternative inorganic hydrogen sources, e.g. hydrogen sulphide, for this purpose.

30.1.2 Growth patterns

The growth of a culture of individual cells, e.g. of unicellular yeasts, protozoa or bacteria is, in effect, the growth of a population and follows the same pattern as that described in Section 12.8.2. The typical bacterial growth curve for a batch culture, illustrated in Fig. 30.1, consists of four phases:

1. **The lag phase** – This is the period after **inoculation** (the addition of cells to the nutrient medium) during which the growth rate increases towards its maximum. Growth during this time is slow initially as the bacteria adapt to produce the necessary enzymes needed to utilize the nutrient medium. The rate of cell division gradually increases during this phase.

2. **The exponential (logarithmic) phase** – During this period, with nutrients in good supply and few waste products being

---

PROJECT

We are advised to keep sugar-reduced jam in the fridge once it has been opened. Investigate the keeping qualities of various preserves.

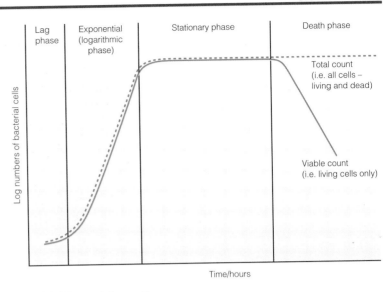

Fig. 30.1 Bacterial growth curve

produced, the rate of cell division is at its maximum, cells sometimes dividing as frequently as every ten minutes. The culture is in a state of **balanced growth** with the doubling time, cell protein content and cell size all remaining constant.

3. **The stationary phase** – As nutrients are used up and toxic waste products accumulate, the rate of growth slows. The changed composition of the medium results in the production of cells of various sizes with a different chemical make-up. They are in a state of **unbalanced growth**. For the total count, this section of the graph is horizontal because the rate at which new cells are produced is equal to the rate at which dead ones are broken down.

4. **The death phase** – While the total number of cells remains constant, the number of living ones diminishes as an ever-increasing number die from a lack of the nutrients necessary to produce cellular energy, or because of poisoning by their own toxic wastes.

30.1.3 Culture media

The correct balance of nutrients, an appropriate pH and a suitable medium are essential to microbial growth.

The medium itself may be liquid (broth) or solid. Solid media are usually based upon **agar**, a seaweed extract which is metabolically inert and which dissolves in hot water but solidifies upon cooling. The more agar, the more solid is the resulting jelly. Any medium composed to satisfy the demands of a single species is called a **minimal medium**. One which provides the nutrients for a small group of microorganisms with similar requirements, e.g. acidophiles, is known as a **narrow spectrum medium**, whereas a medium for general purposes, designed to grow as wide a range of microorganisms as possible is called a **broad spectrum medium**. It is possible to select for the growth of specific types of organisms by use of **selective media** which permit growth of only a single species. The composition of a typical broad spectrum medium is given in Table 30.1.

TABLE 30.1 **A typical broad spectrum medium**

Ingredient	Quantity	Source of
Water	$1.0\,dm^3$	Metabolic and osmotic water, hydrogen and oxygen
Glucose	5 g	Carbohydrate energy source
Yeast extract	5 g	Organic nitrogen, organic growth factors
Dipotassium hydrogen phosphate	1 g	Potassium and phosphorus
Magnesium sulphate	250 mg	Magnesium and sulphur
Iron (II) sulphate	10 mg	Iron and sulphur
Calcium chloride	10 mg	Calcium and chloride
Cobalt, copper, manganese, molybdenum and zinc salts	trace	Metallic ion trace elements

30.1.4 Aseptic conditions

Both in the laboratory and on an industrial scale, pure cultures of a single type of microorganism need to be grown free from contamination with others.

A number of techniques are used to sterilize equipment, instruments, media and other materials. Heat, either passing through a flame, dry heating in an oven, or using water in an autoclave (a type of pressure cooker) is effective in sterilizing instruments, small vessels and culture media. Where heating may affect the media, it may be filtered free of microorganisms using especially fine filters. Ultra-violet light can be used on equipment or even whole rooms. Certain other equipment may be sterilized using a suitable disinfectant, e.g. hypochlorite.

30.2   Industrial fermenters and fermentation

Much of modern biotechnology involves the large-scale production of substances by growing specific microorganisms in a large container known as a fermenter. Fermentation should strictly refer to a biological process which occurs in the absence of oxygen. However, the word is taken to include aerobic processes – indeed the supply of adequate oxygen is a major design feature of the modern fermenter.

30.2.1 Batch versus continuous cultivation

In **batch cultivation** the necessary nutrient medium and the appropriate microorganisms are added to the fermenter and the process allowed to proceed. During the fermentation air is added if it is needed and waste gases are removed. Growth is allowed to continue up to a specific point at which the fermenter is emptied and the product extracted. The fermenter is then cleaned and sterilized in readiness for the next batch.

With **continuous cultivation**, once the fermenter is set up, the used medium and products are continuously removed. The raw materials are also added throughout and the process can therefore continue, sometimes for many weeks.

The continuous process has the advantage of being quicker because it removes the need to empty, clean and refill the fermenter as regularly and hence ensures an almost continuous yield. In addition, by adjusting the nutrients added, the rate of growth can be maintained at the constant level which provides the maximum yield of product. Continuous cultivation is, however, only suited to the production of biomass or metabolites which are associated with growth. **Secondary metabolites**, like antibiotics, which are produced when growth is past its maximum, need to be manufactured by the batch process. The organisms used to produce antibiotics are in any case too unstable for growth by continuous fermentation. In addition, continuous fermentation requires sophisticated monitoring technology and highly trained staff to operate efficiently.

PROJECT

Yeast can respire using different sugars. How would you test the effectiveness of a range of monosaccharides and disaccharides as respiratory substrates?

30.2.2 Fermenter design

The basic design of a **stirred-tank fermenter** is shown in Fig. 30.2. It consists of a large stainless steel vessel with a capacity of up to 500 000 dm^3 around which is a jacket of circulating water used to control the temperature within the fermenter. An agitator, comprising a series of flat blades which can be rotated, is incorporated. This ensures that the contents are thoroughly mixed, thus bringing nutrients into contact with the microorganisms and preventing the cells settling out at the bottom.

Where oxygen is required, air is forced in at the bottom of the tank through a ring containing many small holes – a process known as **sparging**. To assist aeration, increased turbulence may be achieved by adding baffles to the walls of the fermentation vessel. A series of openings, or **ports**, through which materials can be introduced or withdrawn, is provided. The **harvest line** is used to extract culture medium. An outlet to remove air and

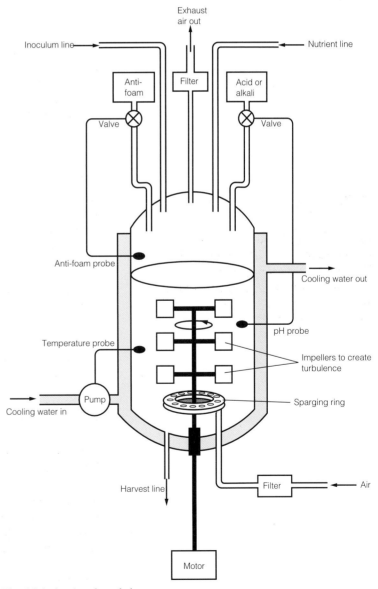

Fermenter for cloned protein

Fig. 30.2 A stirred-tank fermenter

waste gases is needed, as well as one to allow small samples of the culture medium to be removed for analysis. Inlet tubes permit nutrients to be provided and, as the pH changes during fermentation, allow acid or base to be added to maintain the optimum pH. With air being forced into the medium, chemicals often need to be added to reduce foaming. Finally, it is essential to have an **inoculation port** through which the initial inoculum of cells can be introduced once the required conditions in the fermenter are achieved. **Probes**, which constantly register the temperature and pH within the vessel, are used to indicate when adjustments to these factors are necessary.

The stirred-tank fermenter is a well-tried and tested design used extensively in the fermentation industry. It is, however, relatively costly to run, largely as a consequence of the energy needed to drive the agitators and introduce the compressed air. Alternatives have therefore been designed where the air forced into the vessel to provide oxygen is used to circulate the contents, thus making an agitator unnecessary.

One such design, the **pressure-cycle fermenter**, is of two types. In the **air-lift type** the air is introduced centrally at the bottom making the medium less dense. It therefore rises through a central column in the vessel to the top where it escapes. The now more dense medium descends around the sides of the vessel to complete the cycle (Fig. 30.3a). Higher pressure at the bottom of the vessel increases solubility of oxygen, while lower pressure at the top decreases solubility of carbon dioxide, which as a consequence comes out of solution. In the **deep-shaft type**, the principle is similar but the air is introduced at the top (Fig. 30.3b). This has the advantage of giving a more even delivery of oxygen to the microorganisms.

The **tower-** or **bubble-column** (a variety of the air-lift fermenter) has horizontal rather than vertical divisions (Fig. 30.3c). This allows conditions in each section to be maintained at different levels if necessary. A microorganism being carried up from the bottom may therefore pass from a high pH and low temperature to a lower pH and higher temperature to suit each phase of its growth.

While all three types can be used continuously only the air-lift fermenter is used for batch processing. All types must be taller than the conventional stirred-tank vessel to operate effectively.

30.2.3 The operation and control of fermenters

There are two main problems associated with setting up a large-scale fermentation process. Firstly, the inoculum containing the desired strain of microorganism has to be obtained in sufficient quantity; if too little is added to the fermenter, the lag phase is unacceptably long, making the process uneconomic. A small-scale fermentation is set up in a vessel as small as $10\,cm^3$, using frozen culture stock. This is then added to flasks, containing the appropriate nutrient medium, of increasing capacity, e.g. $300\,cm^3$, $3000\,cm^3$, $30\,000\,cm^3$, etc., until the final capacity of the end fermentation vessel is reached. This is known as the **fermenter train**. The problem is that the operational conditions that give the optimum yield in a $300\,cm^3$ fermentation flask are often very different from those for a $300\,000\,cm^3$ vessel.

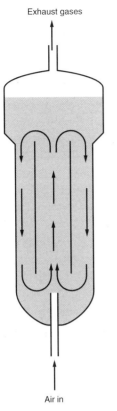

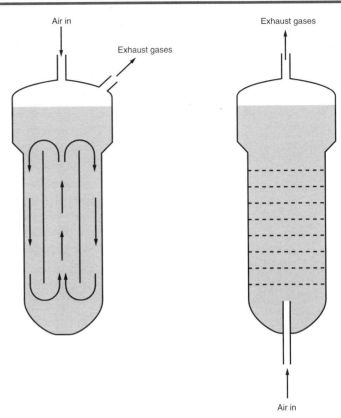

Fig. 30.3(a) Air-lift fermenter   (b) *Deep-shaft fermenter*   (c) *Bubble-column fermenter*

The other main problem arises from the fact that the microorganisms used in fermenters have been genetically selected for the properties (e.g. a high product yield) which make them suitable for use in a large-scale fermenter. They are often enfeebled mutant strains which have resulted from deliberately induced mutanogenesis. As efficient production depends upon rapid growth, i.e. many generations of the microorganism in a short period, there is a tendency for the strain to mutate naturally, often reverting to the parent type which has less desirable properties. One way around the problem is to prevent the microorganism producing the desired product until the final fermentation. This reduces the selection pressure which might alter the gene responsible for the product, but is not always feasible. The use of genetically engineered stock (Section 7.8) has the advantage that it is much easier to express ('switch on') the desired gene at the appropriate time by use of chemical triggers called **promoters**.

Once a sufficient quantity of the inoculum has been produced, usually between 1% and 10% of the total medium, it is added to the fermenter only after the medium within it is of the correct composition and at the desired temperature and pH. The problem now is to maintain all conditions throughout the fermentation. To achieve this the levels of various nutrients, oxygen, pH and temperature are constantly monitored using probes and the information is fed to a computer for analysis, along with information on the composition of the exhaust gases. The necessary corrective changes can then be made. These may be complex – for example, an increase in oxygen uptake could be countered by reducing the air supply, the pressure within the vessel, the nutrient levels or the agitator speed. Which one is selected has implications for other factors and the choice therefore needs to be made advisedly. While the temperature of the medium may need to be increased initially by piping steam through it, once fermentation is under way the heat generated by the microorganisms necessitates continuous cooling by the water jacket around the vessel.

The processes involved in recovering the product are called **downstream processing**. This often involves separation of the cells from the medium which may be achieved in a number of ways:

1. Settlement – The cells may readily settle once agitation and sparging cease. The process can be accelerated by the addition of **flocculating agents**, many of which work by neutralizing the charges on the cells which otherwise keep them in suspension by electrostatic repulsion.

2. Centrifugation – The contents of the fermenter are spun at high speed in a centrifuge causing the cells to settle out. Continuous centrifugation is now possible.

3. Ultrafiltration – The fermenter contents are forced through filters with a pore size less than $0.5\mu m$ which thereby traps cells allowing only liquid through. Some extracellular protein may also be retained.

Where the desired products are the entire cells themselves, these need only be washed, dried and compacted to complete the process. Where the product is contained within the cells, these must be disrupted by some means, the cell debris removed (e.g. by centrifugation or ultrafiltration), and the desired chemical recovered using precipitation, chromatographic or solvent extraction techniques. Where the product lies in the fermentation liquor rather than the cells, this is separated from unwanted enzymes and metabolites, again by precipitation, chromatography or solvent extraction as appropriate.

30.2.4 Sterilization during and after fermentation

The need for aseptic conditions and the basic mechanisms for achieving them were discussed in Section 30.1.4. In industrial fermentation this presents many practical problems considering that not only a very large vessel needs to be sterilized, but also all associated pipework and probes as well as the nutrients, air-supply, and other agents added during the process.

The equipment is designed so that any nooks and crannies which might harbour microorganisms are minimized. The vessel is highly polished, for example, and all components are designed to allow easy access by sterilizing agents. Having been thoroughly washed all equipment is steam sterilized.

The initial nutrient medium may be sterilized in the fermenter by heating it; any added later may be sterilized by heating *en route* to the vessel, as can all other liquid additions. Concentrated acids and alkalis used to adjust the pH may be so inhospitable to contaminating organisms as not to warrant sterilization.

Filtration is used to remove potential contaminants from the air supply although bacteriophages are small enough to pass through – with disastrous consequences. Heat treatment of incoming air can alleviate this problem. The exhaust air is also sterilized to prevent potentially harmful microorganisms being introduced into the atmosphere.

APPLICATION

Commercial uses of enzymes

Immobilised microbial cell pellets in a packed reactor column used to carry out biotransformations

For many years humans have used enzymes to produce bread, cheese, wine, beer and yoghurts but now they have more diverse commercial applications and the sale of enzymes is a multi-billion pound industry.

The commonest sources of these enzymes are bacteria, yeasts and other fungi. The enzymes may be released from the microbial cell, extracted from it when it is broken or the whole cell may be used. Enzymes or whole cells are usually used in an immobilized form so that the enzyme can be reused many times.

Enzymes are very specific and can be used to produce pure products, often more cheaply than by traditional methods. They may also require less energy and result in less pollution.

The specificity of enzymes enables them to be used for sensitive analytical techniques, such as detecting glucose or cholesterol in blood and the enzyme thermolysin is used to produce aspartame, a sweetener sold as *Canderel* and *Nutrasweet*, much more cheaply than by chemical processes.

There is a great demand by the food industry for high fructose syrups which are as sweet as sucrose and which contain glucose and fructose in approximately equal amounts. These syrups are now produced from starch (itself often a waste product of the food industry) using a combination of four enzymes: α-amylase, glucoamylase, pullulanase and glucose isomerase. The first three catalyze the conversion of starch to glucose and then the glucose isomerase converts the glucose to a 50 : 50 mixture of glucose and fructose.

On a commercial scale enzymes are now providing products for pharmaceutical, agrochemical, food, cosmetic and analytical uses. They are used to manufacture semi-synthetic antibiotics, to degrade wastes and in genetic engineering.

PROJECT

When enzymes are immobilized by entrapment they appear to be less prone to denaturing at high temperatures

1. From this statement suggest a testable hypothesis.

2. Test your hypothesis by comparing the activity of an enzyme under normal and immobilized conditions.

30.2.5 Immobilization of cells and enzymes

One problem with the fermentation processes described so far is that at some point the cell culture is removed and discarded. This is fine when the cells are the desired product, but if it is a metabolite they produce which is required, their removal takes away the manufacturing source which then needs to be replaced. Any mechanism for immobilizing the microorganism and/or the enzymes they produce, improves the economics of the process. The idea is not a new one – vinegar manufacture and some stages of sewage treatment have used the technique for a century or more.

There are three basic methods of immobilization:

1. Entrapment – Cells or enzyme molecules are trapped in a suitable meshwork of inert material, e.g. collagen, cellulose, carrageenan, agar, gelatin, polystyrene.

2. Binding – Cells or enzyme molecules become physically attached to the surface of a suitable material, e.g. sand or gravel.

3. Cross-linking – Cells or enzyme molecules are chemically bonded to a suitable chemical matrix, e.g. glutaraldehyde.

However immobilized, the cells or enzymes are made into small beads which are then either packed into columns, or kept in the nutrient medium. The nutrient can be continually added and the product removed without frequent removal of the microorganisms/enzymes. The process cannot be continued indefinitely. Impurities may accumulate preventing further enzyme action or contamination may occur.

30.3 Biotechnology and food production

Until Louis Pasteur showed in 1857 that wine fermentation was the consequence of microbial activity, no one had been aware of the rôle that microorganisms played in the manufacture of some foods. With Pasteur's discovery came further development of the use of microorganisms in food production, an expansion which continues today.

30.3.1 Baking

The use of yeasts in food production is the oldest, and most extensive contribution made by any group of microorganisms. In bread-making cereal grain is crushed to form flour thus exposing the stored starch. Water is added (making a dough) to activate the natural enzymes, e.g. amylases, in the flour which then hydrolyse the starch via maltose into glucose. The yeast, *Saccharomyces cerevisiae*, is added which uses the glucose as a respiratory substrate, producing carbon dioxide. This carbon dioxide forms small bubbles which become trapped in the dough; upon baking in an oven these expand giving the bread a light texture. Dough is often kneaded – a process which traps air within it. This not only helps to lighten the bread directly but also provides a source of oxygen so that the yeast can respire aerobically producing a greater quantity of carbon dioxide. Some anaerobic respiration nevertheless takes place and the alcohol produced is evaporated during baking.

30.3.2 Beer and wine production

Fermentation by yeasts produces alcohol according to the equation:

$$C_6H_{12}O_6 \longrightarrow 2C_2H_5OH + 2CO_2$$

hexose sugar → ethanol + carbon dioxide

Some alcohol produced in this way is for industrial use, but much goes to make beverages like beers and wines which may then be distilled to form spirits. The variety of such beverages is immense and depends largely on the source of the sugar and the type of yeast used to ferment it. Various additives further increase the diversity of alcoholic drinks.

To make wine, the sugar fermented is glucose obtained directly from grapes, whereas beers are made by fermenting glucose obtained from cereal grain (usually barley) which results

rewing beer

Did you know?

In 1992 purchases of beer in Britain were valued at over £13 million, about 3.5% of consumers' expenditure.

from the breakdown of starch in the grain. The yeast used in wine production is often *Saccharomyces ellipsoideus* as this variety can tolerate the higher alcohol levels encountered in wines. Even more tolerant to high alcohol levels are *S. fermentati* and *S. beticus* and these are primarily used in making sherry. Beers are of two basic types – top fermenting varieties of yeast such as *S. cerevisiae* produce a typical British 'bitter' while *S. carlsbergensis*, a bottom fermenting variety, is used to make lager. In beer production, the barley grain is first malted by soaking it in water for two to three days. The grain is then spread on concrete floors and allowed to germinate (about ten days), during which time the natural amylases and maltases in the grain start to convert starch to glucose. This process is stopped by drying the grain and storing it – a process called **kilning** or **roasting**. The higher the temperature during this process, the darker the resulting beer. The germinated grain is often crushed during this stage. The dried, crushed germinated grain is now added to water and heated to the desired temperature – **mashing**. During mashing the remaining starch is converted to sugar to produce a liquid called **wort**. Yeast is added to the wort to convert it to alcohol, as well as hops and other additives, e.g. caramel, which are used to give each beer its characteristic flavour and colour.

The fermentation itself takes place in large deep tanks of around 500 000 dm^3 capacity. No air is introduced as anaerobic respiration is the aim. Although traditionally a batch process, beer production can also be carried out using continuous fermentation. This is more economic and it also allows the carbon dioxide to be collected – a valuable by-product when converted to dry-ice.

The beer is finally separated from the yeast and clarified, and carbon dioxide is added. Sometimes the beer is pasteurized to extend its shelf-life. A good traditional beer, however, retains some yeast in the enclosed barrel which produces the carbon dioxide naturally. Such beers are, for obvious reasons, termed 'live' beers.

30.3.3 Dairy products

Microorganisms have long been exploited in the dairy industry as a means of preserving milk. From this a large number of different products have been manufactured which fall into three main categories: cheese, yoghurt and butter.

Cheese manufacture
An ancient process, cheese-making has altered little over the years. A group of bacteria known as **lactic acid bacteria** are used to ferment the lactose in milk to lactic acid according to the equation:

$$C_{12}H_{22}O_{11} + H_2O \longrightarrow 4CH_3CHOHCOOH$$

lactose   water   lactic acid

Most commercially used lactic acid bacteria are species of two genera – *Lactobacillus* and *Streptococcus*.

Cheese production begins with the pasteurization of raw milk which is then cooled to around 30 °C before a starter culture of the required lactic acid bacteria is added. The resultant fall in pH due to their activity causes the milk to separate into a solid **curd**

Cheese making

595

and a liquid **whey** in a process called **curdling**. The addition of **rennet** at this stage encourages the casein in the milk to coagulate aiding curd formation. Originally extracted from the stomachs of calves slaughtered for food, rennet has now largely been replaced by **chymosin**, a similar enzyme produced by genetically engineered *Escherichia coli*. The whey is drained off and may be used to feed animals. The curd is heated in the range 32–42 °C and some salt added before being pressed into moulds for a period of time which varies according to cheese type.

The ripening of the cheese allows flavour to develop as a result of the action of other milk enzymes or deliberately introduced microorganisms. In blue cheese, for example, *Penicillium* spp. are added. The duration of ripening varies, with Caerphilly taking just a fortnight in contrast to a year required for mature Cheddar. Whereas hard cheeses ripen owing to the activity of lactic acid bacteria throughout the cheese, in soft cheeses it is fungi growing on the surface which are responsible.

Yoghurt manufacture

Yoghurt is made from pasteurized milk with much of the fat removed, and its production also depends on lactic acid bacteria, in particular *Lactobacillus bulgaricus* and *Streptococcus thermophilus*. These are added to the milk in equal quantity and incubated at around 45 °C for five hours during which time the pH falls to around 4.0. Cooling prevents further fermentation and fruit or flavourings can then be added as required.

Butter manufacture

Not essentially a process requiring microorganisms, butter production is nevertheless frequently assisted by the addition to cream of *Streptococcus lactis* and *Leuconostoc cremoris* which help to sour it, give flavour and aid the separation of the butterfat. Churning of this butterfat produces the final product.

30.3.4 Single cell protein (SCP)

Single cell protein comprises the cells, or their products, of microorganisms which are grown for animal, including human, consumption. High in protein, the product also contains fats, carbohydrates, vitamins and minerals making it a useful food. The raw materials for SCP production have included petroleum chemicals, alcohols, sugars and a variety of agricultural and industrial wastes. The microorganisms used to ferment these have been equally diverse – bacteria, algae, yeasts and filamentous fungi. The success of various manufacturing processes has varied. The use of a waste product to produce food seems highly attractive and economical, but the demand it creates for the raw material ceases to make it a waste, its price rises and the process can become uneconomical. Excess food production in some parts of the world has meant the selling off of butter and grain 'mountains' and therefore reduced the need for alternative sources of food such as SCP. It has not therefore proved the success originally anticipated and many countries have ceased production altogether.

The world's largest continuously operating fermenter (600 tonnes) owned by ICI, produces a single cell protein called **Pruteen**. It comprises 80% protein and has a high vitamin content. The process uses methanol, a waste product of some of ICI's other activities, making the raw material relatively cheap. This is acted upon by the aerobic bacterium *Methylophilus methylotrophus* in a pressure-cycle fermenter with a capacity of $1500\,m^3$ to produce the odourless, tasteless, cream-coloured Pruteen which is used as an animal feed.

Similar projects produce a protein based on the fungus *Fusarium graminearum* which can be grown on flour waste. The product, **mycoprotein**, is intended for human consumption and, being high in protein and fibre, but low in cholesterol, is a healthy addition to the diet.

30.3.5 Other foods produced with the aid of biotechnology

Vinegar
One of the oldest products of biotechnology is vinegar; originally produced from wine, much is now the product of fermenting malt. The alcohol produced by *Saccharomyces* spp. (Section 30.3.2) is converted to acetic acid by aerobic bacteria belonging to the genus *Acetobacter* according to the overall equation:

$$C_2H_5OH + O_2 \longrightarrow CH_3COOH + H_2O$$

ethanol   oxygen   acetic acid   water
                                  (ethanoic acid)

In the fermenter, the bacteria are immobilized on the surface of wood shavings or other inert material, while the ethanol is sprinkled over them. Air is forced upwards through the shavings and the vinegar is drawn off at the bottom.

Sauerkraut
Cabbage leaves are shredded, mixed with salt and pressed in containers to exclude air. Lactic acid bacteria are added to ferment the leaves, reducing their pH to around 5.0. This prevents other microorganisms growing and hence acts as a form of preservation, as well as imparting flavour to the cabbage.

Olive and cucumber preservation
Both olives and cucumbers can be preserved by storing them in barrels containing brine. In these anaerobic conditions lactic acid bacteria thrive, reducing the pH to around 4.0 and thus preventing other microbial activity.

Preparation of coffee and cocoa beans
Both coffee and cocoa beans are surrounded by a mucilage coat which must be removed before further processing can take place. The beans are heaped together to allow naturally occurring microflora, including yeasts and bacteria, to ferment the mucilage and also prevent the beans germinating by killing the embryo in the high temperatures, around 50 °C, which are created.

Soy sauce
The fermentation of soya beans and wheat by *Aspergillus oryzae* and other microorganisms is used in the manufacture of soy sauce.

APPLICATION

Enzymes and fruit juice production

Pectin is a substance which helps to hold plant cell walls together. As a fruit ripens the plant produces proteolytic enzymes which convert the insoluble protopectin of the unripe fruit into more soluble forms, causing the fruit to soften. When fruits are mashed and pressed to form juices these more soluble forms of pectin enter the juice, making it cloudy and causing the colour and flavour to deteriorate. They also increase the viscosity of the juice itself, making it difficult to obtain optimal yields. For fruit juice manufacturers the addition of industrial pectinases between mashing and pressing causes complete depectinization so that a good quality, clear juice is obtained which retains its stability when concentrated. Producers may also use other enzymes such as starch-splitting enzymes to reduce cloudiness, especially with apples, cellulases to improve juice yield and colours, and arabanase to reduce the haze caused by the polysaccharide araban passing from the cell walls into the extracted juice.

PROJECT

The production of fruit juices has been greatly improved by the addition of pectinase

1. Compare the yields of fruit juice from apples with and without the addition of pectinase.

2. Determine the optimum, and thus the most economical, concentration of pectinase.

30.3.6 Enzymes associated with the food industry

Many enzymes used in the food industry are produced by microorganisms. Some examples are given in Table 30.2.

TABLE 30.2 **Some enzymes produced by microorganisms used in the food industry**

Enzyme	Examples of microorganisms involved	Application
α-amylases	*Aspergillus oryzae*	Breakdown of starch in beer production Improving of flour Preparation of glucose syrup Thickening of canned sauces
β-glucanase	*Bacillus subtilis*	Beer production
Glucose isomerase	*Bacillus coagulans*	Sweetener for soft drinks Cake fillings
Lactase	*Kluyveromyces* sp.	Lactose removal from whey Sweetener for milk drinks
Lipase	*Candida* sp.	Flavour development in cheese
Pectinase	*Aspergillus* sp.	Clearing of wines and fruit juices
Protease	*Bacillus subtilis*	Meat tenderizers
Pullulanase	*Klebsiella aerogenes*	Soft ice cream manufacture
Sucrase	*Saccharomyces* sp.	Confectionery production

30.4 Biotechnology and pharmaceuticals

Since the antibiotic penicillin was first produced on a large scale in the 1940s, there has been a considerable expansion in the use of biotechnology to produce a range of antibiotics, hormones and other pharmaceuticals.

30.4.1 Antibiotics

Antibiotics are chemical substances produced by microorganisms which are effective in dilute solution in preventing the spread of other microorganisms. Most inhibit growth rather than kill the microorganisms on which they act. Some, like **penicillin**, are only effective on relatively few pathogens – **narrow spectrum antibiotics**, while others, e.g. **chloramphenicol**, will inhibit the growth of a wide variety of pathogens – **broad spectrum antibiotics**. Although around 5000 antibiotics have been discovered only 100 have proved medically and commercially viable.

Antibiotics are made when growth of the producer organism is slowing down rather than when it is at its maximum. They are therefore secondary metabolites and their production takes longer than for primary metabolites. It also means that continuous fermentation techniques are unsuitable and only batch fermentation can be employed.

In penicillin manufacture, a stirred-tank fermenter is inoculated with a culture of *Penicillium notatum* or *Penicillium chrysogenum* and the fungus is grown under optimum conditions: 24 °C, good oxygen supply, slightly alkaline pH. Penicillin production typically commences after about 30 hours, reaching a maximum at around four days. Production ceases after about six days, at which point the contents of the fermenter are drained off. As the antibiotic is an extracellular product, the fungal mycelium is filtered off, washed and discarded. The liquid filtrate containing penicillin is chemically extracted and

Penicillin fermentation

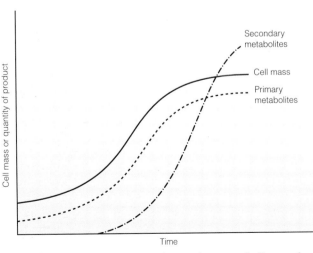

Fig. 30.4 Comparison of primary and secondary metabolite production

APPLICATION

Biosensors

Biosensors have been devised which enable a quick and accurate measurement of glucose to be made. This is important industrially but also medically, enabling diabetics to monitor their own blood sugar levels. The method used relies on the specificity of the enzyme glucose oxidase, allowing glucose to be assayed in the presence of other sugars.

Glucose oxidase catalyses the conversion of glucose to hydrogen peroxide (H_2O_2).

$$\beta\text{-D-Glucose} + O_2 \xrightarrow{\text{Glucose oxidase}} \text{Gluconic acid} + H_2O_2$$

This is coupled to a reaction catalysed by peroxidase and utilising the hydrogen peroxide.

$$\underset{\substack{\text{Colourless} \\ \text{hydrogen donor}}}{DH_2} + H_2O_2 \xrightarrow{\text{Peroxidase}} 2H_2O + \underset{\substack{\text{Coloured} \\ \text{compound}}}{D}$$

The glucose oxidase, peroxidase and the colourless hydrogen donor can be immobilized on a cellulose fibre pad. This forms the basis of 'Clinistix', the glucose dipsticks used by diabetics.

More recently the term biosensor has been used to describe the association of a biomolecule, such as an enzyme, with a transducer which produces an electrical signal in response to a charge in the substrate.

Biosensors are not just used for the quantitative detection of glucose. They have also been devised for pregnancy testing kits (see page 416), to check the freshness of food (there being a correlation between deterioration and the number of microorganisms present) and to detect the presence of pollutants in the environment, as well as for the estimation of many substances, such as urea, ketones and cholesterol, in blood and urine.

purified using solvents to leave a crystalline salt. After sterilization the fermenter is available for the next batch.

The development of resistance to antibiotics by pathogens means there is a continuing need to find new types. The emphasis has moved from searching for new natural antibiotics to the development of new strains using genetic engineering. However, this is not easy. Being secondary metabolites, antibiotics are the product of a long metabolic pathway involving numerous genes. Manipulation of these is complex and difficult. Programmes for enhancing the production rate of existing strains and using random mutation and selection methods to develop new ones, are continually developed. Table 30.3 lists some other antibiotics and their producer organisms.

APPLICATION

BST

The hormone bovine somatotrophin (BST) controls lactation in cows by increasing the number of cells in the mammary glands. When the genetically engineered version is injected into the animals, their milk production can increase by up to 20%. The use of BST was approved by the US Food and Drug Administration (FDA) in November 1993 who declared milk produced from cows given the artificial hormone to be indistinguishable from other milk. Its use is banned in Europe.

Opponents are concerned because the incidence of mastitis among cows given the hormone is 79% higher than normal. They fear that this will lead farmers to use more antibiotics which could stay in the milk and eventually reduce the effectiveness of antibiotics against bacteria that affect humans. Opposition groups have threatened to boycott the milk but the FDA has ruled that it need not carry special labels and neither can other milk simply be declared 'hormone-free'. The introduction of such genetically engineered products without the implementation of strict labelling regulations concerns a number of 'pure food' campaign groups.

TABLE 30.3 **Some antibiotics and their producer organism**

Antibiotic	Producer organism	Type of organism
Penicillin	*Penicillium notatum*	Fungus
Griseofulvin	*Penicillium griseofulvum*	Fungus
Streptomycin	*Streptomyces griseus*	Actinomycete
Chloramphenicol	*Streptomyces venezuelae*	Actinomycete
Tetracycline	*Streptomyces aureofaciens*	Actinomycete
Colistin	*Bacillus colistinus*	Bacterium
Polymyxin B	*Bacillus polymyxa*	Bacterium

30.4.2 Hormones

With the advent of recombinant DNA technology it is now possible to use microorganisms to produce a wide range of hormones which previously had to be extracted from animal tissues. Hormone manufacture using fermentation techniques is relatively straightforward, but the high degree of purity of the final product which is essential, makes downstream processing a complex process. Ion-exchange, chromatography and protein engineering are some of the techniques employed to provide a high level of purity.

Hormones produced in this way include insulin, used in the treatment of diabetes, and human growth factor which prevents pituitary dwarfism. A number of steroids are also manufactured including cortisone and the sex hormones, testosterone and oestradiol. Others with possible commercial and medical value are relaxin which aids childbirth, and erythropoietin for the treatment of anaemia. Bovine somatotrophin (BST), a hormone administered to cows to increase milk yield, is also in current production.

30.5 Biotechnology and fuel production

The rise in oil prices in the early 1970s led to research into alternative means of producing fuel. With only a finite supply of oil available, work continues in this field, accelerated to some extent by the harmful consequences of burning traditional fossil fuels. One method already tried is the fermentation of waste to yield **gasohol** (alcohol) or **biogas** (methane). It may be that in years to come, these fuels will be formed from crops specifically grown for the purpose.

30.5.1 Gasohol production

The 1970s rises in oil price hit oil-importing developing countries, such as Brazil, especially hard, and prompted the initiation of the **Brazilian National Alcohol (or Gasohol) Programme**. The concept was simple – namely to use yeasts to ferment Brazil's plentiful supply of sugar cane into alcohol and so create a relatively cheap, renewable home-produced fuel.

The programme began in 1975 and incorporated research into improving sugar cane production as well as fermenter technology. By 1985 sugar cane production had increased by a third, fermentation conversion by 10% and the fermentation time had been reduced by three quarters. Over 400 distilleries now yield more than $1.2 \times 10^{10} \, dm^3$ of alcohol annually and all Brazilian cars have been converted to using the fuel, either entirely or mixed with petrol. (Some petrol is added to all alcohol fuels as a disincentive to people drinking them.) It is hoped that by the year 2000 all the country's energy needs will be supplied in this way. What makes the programme so successful in Brazil is that the sugar cane is not only a source of the fermentation substrate, but also a fuel for the distilleries. Once the sugar is extracted from the sugar cane, the fibrous waste, called **bagasse**, can be dried and burnt as a power source for the distillery.

The actual process entails a number of stages:

1. Growing and cropping sugar cane.

2. Extraction of sugars by crushing and washing the cane.

3. The crystallizing out of the sucrose (for sale) leaving a syrup of glucose and fructose called **molasses**.

4. Fermentation of the molasses by *Saccharomyces cerevisiae* to yield dilute alcohol.

5. Distillation of the dilute alcohol to give pure ethanol, using the waste bagasse as a power source.

The special circumstances in Brazil have doubtless contributed to its success but schemes in some other countries, e.g. Kenya, have been abandoned as uneconomic. Nevertheless, the potential for solving both the problem of diminishing oil supplies and disposal of waste at the same time has its attractions. Waste straw, sawdust, vegetable matter, paper and its associated waste, and other carbohydrates are all possible respiratory substrates although many require enzyme treatment to convert them into glucose before yeast can act upon them.

30.5.2 Biogas production

The capacity of naturally occurring microorganisms to decompose wastes can be exploited to produce another useful fuel, methane (biogas). It has the advantage over alcohol (gasohol) production of not requiring complex distillation equipment – indeed, the process is very simple. A container

PROJECT

Substances are sold which claim to accelerate the production of compost. Investigate these claims by studying the production of biogas by decomposing vegetable matter.

Biogas generator, Senegal

known as a **digester** is filled with appropriate waste (domestic rubbish, sewage or agricultural waste can be used) to which is added a mixture of many bacterial species. The anaerobic fermentation of these yields methane which is collected ready for use for cooking, lighting or heating. Small domestic biogas fermenters are common in China and India.

30.6 Biotechnology and waste disposal

In the previous section we saw how microorganisms could be used not only to dispose of wastes but also to yield a useful by-product at the same time. These are not the only ways in which microorganisms are used to dispose of unwanted material; the disposal of sewage, the decomposition of plastics and the breakdown of oil are further examples.

30.6.1 Sewage disposal

The basic details of sewage treatment have been discussed in Section 18.5.1 and therefore this section will confine itself to the microbial aspects of the process, in particular biological filtration and sludge digestion.

In a typical sewage filter bed a large variety of microflora are immobilized on layers of coke or stones. Openings above, below and in the sides of the filter, and the spaces between the 'clinker', maintain aerobic conditions. As the effluent trickles down the action of the microorganisms changes its composition. As a result, the species become stratified vertically in the filter bed. On the surface are numerous mobile protozoa as well as fungi of the genus *Fusarium*. In the upper layers occur *Zoogloea* spp., while *Nitrosomonas* spp., *Nitrobacter* spp. and stalked protozoa are found lower down. Between them, these and other species found in the filter oxidize the many organic substances in the sewage, to largely inorganic chemicals. Sludge digestion can be carried out either anaerobically or aerobically. In anaerobic digestion the sewage sludge is drained into large digester tanks where obligate anaerobes like *Clostridium* sp. ferment the protein, polysaccharides and lipids of the sludge into acetic acid, carbon dioxide and hydrogen. These products are then acted upon by *Methanobacterium* spp. which form methane (biogas) a useful fuel, which can be used to power the sewage treatment plant.

Aerobic digestion utilizes a technique known as **activated sludge**. Here the liquid from the primary settlement process is pumped to tanks where it is sparged with air by pumping it through diffusers at the base. Aerobic microflora, including *Zoogloea* spp., *Nitrobacter* spp., *Nitrosomonas* spp., *Pseudomonas* spp., *Beggiatoa* spp. and others, oxidize the organic components. The liquid is moved to other tanks where the remaining solids settle out in lumps called **flocs** – a process facilitated by bacteria, e.g. *Zoogloea ramigera*. This settled material can be returned to the aeration tanks for further treatment.

wage farm showing filter beds

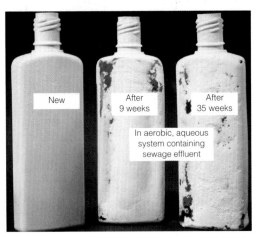

New   After 9 weeks   After 35 weeks

In aerobic, aqueous system containing sewage effluent

Biodegradable plastic bottles

Any solid sludge not decomposed by either of these methods is led into storage lagoons where slow decomposition by anaerobic microorganisms over many years reduces it to a largely inorganic residue. It may otherwise be used for the production of fertilizer.

30.6.2 Biodegradable plastics

One of the commercially attractive features of plastics, polythene, polystyrene and related materials is their durability; they do not corrode or decay easily and are resistant to most forms of chemical attack or other forms of degradation apart from burning. It is this very feature, however, which creates the environmental problem of its disposal once it has ceased to be of use. Incineration produces unpleasant, and sometimes dangerous, gases and hence there is a need to use a more acceptable and safer means of breaking it down. Polythene and polyester polyurethanes of low molecular mass have been developed which can be degraded by microorganisms such as the fungus *Cladosporium resinae*. In general, the more flexible plastics are broken down more easily than rigid ones. Such breakdown is desirable where the material has a short useful life e.g. packaging, but not where its life-expectancy is longer, e.g. furniture and utensils. Research is therefore being carried out into finding new forms of biodegradable plastics for the packaging industry. These products are based on a storage chemical, **polyhydroxybutyrate**, of many microorganisms. Not only does this have the advantage of being biodegradable but it can also be manufactured by microorganisms e.g. *Alcaligenes eutrophus* using biotechnological methods. Another biodegradable packaging product is **Pullulan**, a commercially produced polysaccharide made by *Aureobasidium pullulans*.

30.6.3 Disposal of oil

Oil pollution of oceans, seas and rivers has unfortunately become all too common in recent years. The safe disposal, not only of oil spillages, but also of the 'spent' oil from vehicles and machinery is highly desirable. Like the plastics made from it, oil is very resistant to microbial degradation largely because it contains little if any water. The same fungus *Cladosporium resinae* which degrades some plastics is equally effective in breaking down paraffin-based oils, and some forms of *Pseudomonas* spp., developed using recombinant DNA technology, are used commercially in cleaning up oil spills. These same organisms along with species of the fungi *Aspergillus* and *Cephalosporium* cause problems in fuel systems because their growth degrades the oil, and the fungal hyphae block filters and pipes. The problem is most acute where water, e.g. from condensation, contaminates the oil.

One means of dealing with oil pollution in water is to use emulsifiers to cause the oil to mix with the water and so both disperse it and speed up its microbial breakdown. One such emulsifier, a polysaccharide called **Emulsan**, is produced commercially by the bacterium *Acinetobacter calcoaceticus*.

30.6.4 Disposal of industrial wastes

The economics of industrial processes are such that waste disposal is often an unwanted expense and all too often in the past has led to materials simply being dumped at the nearest convenient point with little regard for the environmental consequences. An increased awareness of ecological issues and tighter legislation have prompted safer waste disposal. To offset the additional costs this entails, many methods incorporate an element of recycling which yields financial benefits.

In Section 30.5 we saw how certain wastes could be used to produce fuel. The use of domestic and agricultural wastes in producing biogas (methane) is equally applicable to many forms of organic industrial waste such as that from paper, cotton and wood mills. Brewery waste can be converted to citric acid, used extensively in the food industry, by the fungus *Aspergillus niger*. Animal feed is manufactured by another fungus *Paecilomyces* spp. from paper-mill waste, or by yeasts of the genera *Candida* and *Endomycopsis* from the waste of a potato-processing plant.

The viability of such schemes often depends on the ease with which the wastes can be converted into a form that micro-organisms can utilize as a respiratory substrate. Genetically engineered microorganisms are making this task increasingly easy.

30.6.5 Recovery of valuable material from low-level sources including wastes

Few elements exist naturally in an uncontaminated form. Separating the desired material from its contaminants is frequently difficult and a point is reached where the level of the required material is so low that the cost of its extraction is no longer economic. The use of microorganisms in recovering such substances from low-level sources is not new (the Romans used the technique) but the range of materials obtained has expanded enormously, not least because of its value in removing environmentally harmful chemicals.

The bacterium *Thiobacillus thiooxidans* is used to extract copper and uranium from waste ore which does not lend itself to metal extraction by conventional means. The ore is dumped and sprinkled with dilute acid to encourage the growth of *Thiobacillus* spp. (an acidophile) which oxidizes the copper sulphide according to the equation

$$CuS + 2Fe_2(SO_4)_3 \longrightarrow 2FeSO_4 + CuSO_4 + S$$

copper (II)   iron (III)   iron (II)   copper (II)   sulphur
sulphide   sulphate   sulphate   sulphate

The soluble metal sulphates so produced are washed out and collected – **metal leaching**. The pure metal can be extracted from the leachate by chemical means, e.g. by use of scrap iron.

$$CuSO_4 + Fe \longrightarrow Cu + FeSO_4$$

copper (II)   iron   copper   iron (II)
sulphate   sulphate

It is hoped that *Thiobacillus* spp. can also be used to remove the sulphur from high sulphur coal, thus making it more environmentally acceptable by reducing the sulphur dioxide it produces on combustion. The reverse of the above process by which insoluble metals are made soluble, can be used to remove toxic metal ions from waste. The bacterium *Desulfovibrio desulfuricans* will make metal sulphates into insoluble sulphides which then precipitate out. The alga *Scenedesmus* will absorb metal ions against a concentration gradient. Both species can therefore be used to remove harmful metal ions from industrial and mining waste before it is discharged. The use of denitrifying bacteria as a means of removing excess nitrates (the result of fertilizer use – see Section 18.5.3) from drinking water is being investigated.

The recovery of some oil also entails the use of microorganisms. With almost half of the world's underground oil being either trapped in rock or too viscous to recover by conventional means, microorganisms are used to reduce its viscosity. The process, known as **enhanced oil recovery**, allows the more fluid oil to drain out of the rock and be pumped to the surface. **Xanthan**, a polysaccharide gum produced by the bacterium *Xanthamonas campestris* is used to thicken water and improve its ability to drive the oil out of the rock.

30.7 Other products of the biotechnology industry

In addition to all the biotechnological applications of microorganisms already discussed, there is an assortment of other products made in this way. Some of these are given in Table 30.4.

TABLE 30.4 **A range of other commercial substances produced by microorganisms**

Product	Producer organism	Function of product
Protease	*Aspergillus oryzae*	Detergent additive
	Bacillus spp.	Removal of hair from animal hides
Butanol and acetone	*Clostridium acetobutylicom*	Solvents
Indigo	*Escherichia coli*	Textile dye
Xanthan gum	*Xanthomonas campestris*	Thickener used in food, paints and cosmetics
Cellulases	*Trichoderma* spp.	Brightener in washing powders
Cyanocobalamin (vitamin B_{12})	*Propioni bacterium shermanii*	Food supplement
Gellan	*Pseudomonas* spp.	Food thickener
Glutamate	*Corynebacterium glutamicum*	Flavour enhancer
Streptokinase	*Streptomyces* spp.	Treatment of thrombosis
Interferon	*Escherichia coli*	Treatment of viral infection
Ergot alkaloids	*Claviceps purpurea*	Vasoconstricter used to treat migraine and in childbirth
Cyclosporin	*Cotypocladium inflatum*	Immunosuppressant drug

APPLICATION

Biological washing powders

The first biological washing powder was manufactured as long ago as 1913 using an extract from the pancreas which contained trypsin. This had very limited success in the alkaline liquids produced by the washing soda containing it. Since then the detergent industry has become the largest single outlet for industrial enzymes. Proteases remove protein stains such as blood, grass, egg and human sweat; lipases digest oily and fatty stains; amylases remove residues of starch substances. Some washing powders also contain a cellulase complex which modifies the fluffy microfibrils which develop when cotton and cotton-mix garments have been washed several times. This has the effect of brightening the colours and softening the fabric as well as removing some dirt particles.

Most of the enzymes are produced extracellularly by bacteria such as *Bacillus subtilis* grown in large-scale fermenters. The bacteria have been genetically engineered to produce enzymes which are stable at a high pH in the presence of phosphates and other detergent ingredients as well as remaining active at temperatures of 60 °C. Subtilisin, a protease, has also had an amino acid residue replaced with an alternative to make the enzyme more resistant to oxidation. During the 1970s the industry suffered a setback when many of the enzymes were shown to produce allergic reactions. This has been solved by producing encapsulated enzymes which are dust-free.

False-colour scanning EM of *Escherichia coli* (×8000 approx)

In addition to their products, microorganisms may be utilized directly to human benefit. The nitrogen-fixing bacterium *Rhizobium* spp. lives symbiotically in nodules on the roots of certain plants where it forms nitrogenous compounds of use to its host. The addition of this bacterium to the soil, along with the seeds of the plants utilizing them, ensures inoculation of the crop and a resultant better yield in areas where natural inoculation is unlikely.

The bacterium *Bacillus thuringiensis* produces a protein which is highly toxic to insects. By contaminating the natural food of an insect pest with the bacterium, some control can be effected. The ecological consequences of using such 'natural' pesticides need further investigation but it seems likely that they will be less harmful than their artificial counterparts.

A recent, and controversial, development is the use of the bacterium *Pseudomonas syringae* to form artificial snow at winter holiday resorts. The bacterium, which is sprayed with water on to a fan, has surface properties which act as nuclei for the formation of ice crystals. The fact that this process could cause considerable frost-damage if used on food crops illustrates the risks that attend many biotechnological advances if used wrongly, either by accident or with malice.

607

30.8   Cell, tissue and organ culture

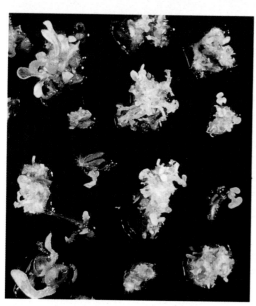

Anther calluses on agar

It is not only microorganisms that can be utilized to produce useful substances. Although more difficult to establish, plant and animal cells can also be grown in vitro in order to manufacture a variety of products. The culture requirements for eukaryotic cells are far more sophisticated than for prokaryotic cells.

30.8.1 Plant cell, tissue and organ culture

The presence of a semi-rigid cell wall around plant cells effectively reduces their ability to divide and grow. For this reason plants retain groups of immature cells which form the only actively growing tissues. These tissues are called **meristems**. If a part of a plant containing a high proportion of meristematic tissue, e.g. bud, root tip or germinating seed, is removed and grown aseptically on a nutrient medium, an undifferentiated mass known as a **callus** frequently develops. Plant hormones – such as auxins, gibberellins and cytokinins (Section 29.2) – a nitrogen and carbohydrate source, as well as vitamins, trace elements and growth regulators need to be present in the nutrient medium. Undifferentiated callus produced in this way is an example of **plant tissue culture**. If the callus is suspended in a liquid nutrient medium and broken up mechanically into individual cells it forms a **plant cell culture**. These can be maintained indefinitely if sub-cultured giving rise to a **cell-line**. If pectinases and cellulases are added to these cultures, the plant cell walls can be removed and the resulting protoplasts can be fused with the protoplasts of different cells to give hybrid cells which can later be grown on into new plant varieties.

Plant organ culture is achieved by taking apical shoot tips, sterilizing them and growing them on a nutrient medium containing cytokinin. A cluster of shoots develops, each of which may be grown on into new clusters. This form of micropropagation results in a large quantity of genetically identical individuals.

30.8.2 Applications of plant cell, tissue and organ culture

There are two main applications of these forms of culture. The first is the generation of plants for agricultural or horticultural use. Vast numbers of plants can be grown in sterile controlled conditions ensuring a much greater survival rate than would be the case if seeds were planted outdoors. These plants can all be identical if required thus ensuring a more uniform crop which incorporates desired characteristics. A particular advantage of this form of propagation is that the stock plants, raised as they are in sterile conditions, are completely pathogen-free when planted out, by which time they have developed defence mechanisms against many diseases.

The second application is the manufacture of useful chemicals by plant cultures. To date only one product, shikonin, a dye used in the silk industry and in the treatment of burns, has been produced commercially, but Table 30.5 lists some of the possible applications of plant cell culture once the processes become economically viable.

Micropropagation

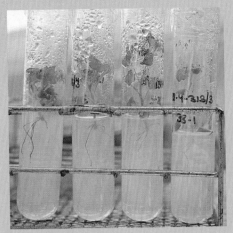

Micropropagation of plants in growing medium

Plant cells are totipotent, that is any differentiated plant cell has the potential to give rise to all the different cells of an adult plant. This is the basis of the tissue culture techniques which form the basis of micropropagation.

Stock plants are kept as pathogen- and pest-free as possible and small pieces are cut (excised) from them. These explants may be pieces of stem tissue, with nodes, flower buds, leaves or tiny sections of shoot tip meristems. The surfaces of the explants are sterilized by using solutions such as dilute sodium hypochlorite and, in aseptic conditions, are transferred to a culture vessel containing nutrients and growth regulators. The medium is usually solidified using a gelling agent, such as agar, to provide a firm matrix for the growing shoots. The vessels are then incubated for 3–9 weeks at 15–30 °C with light for 10–14 hours per day. The new shoots that develop are removed from the explant and subcultured on a new medium. This process is repeated every few weeks so that a few explants can give rise to millions of plants within one year. The composition of medium used varies for different varieties. Tissue culture plants must be acclimatized in special greenhouses until they reach marketable size.

Micropropagation is now widely used as an alternative to conventional propagation of many horticultural species (for pot plants, cut flowers and ornamental bulbs) and some agricultural crops such as potatoes and sugar beet. Tissue culture allows the rapid production of large numbers of genetically identical plants, can be used for plant species which are difficult to propagate by traditional methods, helps to eliminate plant diseases, overcomes seasonal restrictions and enables cold storage of large numbers of plants in a small space.

However, there are also some problems. Plants propagated in this way may be genetically unstable or infertile, their chromosomes being structurally altered or their chromosome numbers being unusual. When oil palms produced by micropropagation were introduced to Malaysia in the 1970s they turned out to be sterile. This problem has now been overcome and most oil palms are the result of micropropagation techniques. It is a technology which requires sterile conditions and a well trained labour force and so costs are higher than with traditional propagation methods.

30.8.3 Animal cell culture

Only in recent years has it proved possible to culture vertebrate cells on any scale. The process begins by treating the appropriate tissue with a proteolytic enzyme like trypsin in order to separate the cells. These are transferred to an appropriate nutrient medium where they attach themselves to the bottom of the container and divide mitotically to give a monolayer of cells.

TABLE 30.5 **Some useful plant chemicals which might be produced using plant cell culture**

Product	Producer plant	Use of product
Atropine	Deadly nightshade (*Atropa belladonna*)	Ophthalmic use – dilation of pupil Treatment of certain heart conditions
Codeine	Opium poppy (*Papaver somniferum*)	Pain killer
Digoxin	Foxglove (*Digitalis* spp.)	Treatment of cardio-vascular complaints
Jasmine	Jasmine (*Jasminium* spp.)	Perfume
Menthol	Peppermint (*Mentha piperita*)	Food flavouring
Quinine	Chinchona tree (*Chinchona ledgeriana*)	Drug used to treat malaria Bitter flavouring in drinks
Pyrethrin	Chrysanthemum (*Chrysanthemum* spp.)	Insecticide

This is referred to as the **primary culture**. Cells from these can be used to establish secondary cultures but their life span is limited, division often ceasing after 50–100 divisions.

It is possible to make these cultures continue to divide indefinitely by the addition of chemicals or viruses which induce the formation of cancer cells. These cell lines are said to be transformed and are **neoplastic**, i.e they can induce cancer if transplanted into a related species.

30.8.4 Applications of animal cell cultures

Of the many applications of animal cell cultures, the production of viral vaccines is the oldest. Viruses are grown on culture cells, frequently monkey kidney cells or those from chick embryos, but increasingly, human cells. The culture medium is harvested and the viruses extracted by filtration. These are then treated to kill them, or, if of the attenuated type (see Section 21.3.3), stored at low temperature ready for use. Poliomyelitis, measles, German measles and rabies vaccines can be produced in this way.

Pharmaceutical products can be harvested from cell lines which over-produce particular products. These include interferon from certain human blood cells, human growth factor and the clotting factors used to treat haemophiliacs. This alternative supply has the added advantage of not transferring AIDS to the recipient, a risk, albeit very small, of the current practice of extracting these factors from donated blood.

Animal cell cultures are used in recombinant DNA technology where appropriate virus vectors can be cloned in them. Perhaps the most important use, however, is in the production of monoclonal antibodies – a topic dealt with in Section 21.3.2.

30.9   Questions

1. (a) (i) Explain, with examples, why and how some microorganisms have proved useful as new sources of protein food. *(12 marks)*

(ii) Describe the important biological principles involved in the culture, extraction and purification of such novel food. *(18 marks)*

(b) Explain the biological role of micro-organisms in any two of:
   (i) sewage (effluent) treatment,
   (ii) brewing,
   (iii) production of dairy foods,
   (iv) production of antibodies,
   (v) any other industrial process.
   (2 × 15 marks)
   (Total 60 marks)

Oxford June 1991, Part II, No. 6

2. 'Pruteen' is the brand name of a single-cell protein produced by microorganisms in an industrial fermenter. It is sold as a supplement to animal feeds. The list below gives some information about the industrial manufacture of *Pruteen*.

Inoculum	A bacterium, *Methylophilus methylotrophus*
Substrate	Methanol (an alcohol), plus ammonia and trace elements
Fermenter	Steam-sterilized 1500 m³ air-lift continuous culture vessel
Cell concentration	Maintained at 3% by volume
Temperature	35–42 °C
pH	6.5–7.0
Doubling time	5 hours
Production capacity	7×10^4 tonne year^{-1}

(a) State the functions of (i) ammonia and (ii) trace elements in the growth medium. *(2 marks)*

(b) Explain why temperature and pH need to be maintained within the ranges indicated. *(2 marks)*

(c) Explain why continuous culture is suitable for the industrial production of single-cell protein, while certain other microbial products (such as antibiotics) have to be produced by batch culture. *(4 marks)*

(d) Describe and explain the procedures that would be necessary to maintain the cell concentration in the fermenter at approximately 3% by volume. *(3 marks)*

(e) Name *two* other sources of industrially produced microbial protein. *(2 marks)*
(Total 13 marks)

ULEAC January 1994, Paper 4A, No. 2

3. The diagram below shows the sequence of processes involved in the brewing of beer.

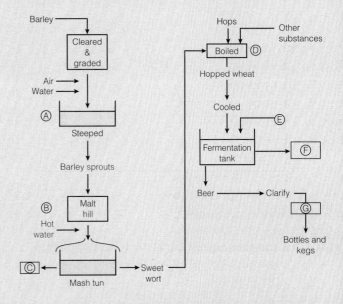

(a) Describe the changes occurring in the barley grains in the vessel labelled A. *(2 marks)*

(b) The malt mill (B) dries and crushes the barley sprouts. State *one* effect this has on the sequence of processes. *(1 mark)*

(c) The nutrient-rich liquor is called sweet wort. Suggest *one* use for the residues which have been separated from the liquor and removed at C. *(1 mark)*

(d) Suggest *one* substance the brewer might add to the mash tun in order to get a low-carbohydrate or 'lite' beer. *(1 mark)*

(e) Suggest *one* reason why the sweet wort is boiled at D. *(1 mark)*

(f) (i) Identify E which is added to the fermentation tank. *(1 mark)*

(ii) Describe exactly what is happening in the fermentation tank to produce beer. *(3 marks)*

(g) (i) Give the economic importance of *one* product removed at F. *(1 mark)*

(ii) State *one* further process which the beer undergoes at G and explain why this is necessary. *(2 marks)*
(Total 13 marks)

ULEAC June 1992, Paper 4A, No. 2

4. The graph on the following page shows changes in the number of cells and in the amount of DNA in a culture of yeast over a period of 5 hours. Temperature and pH were maintained at a constant level throughout the investigation.

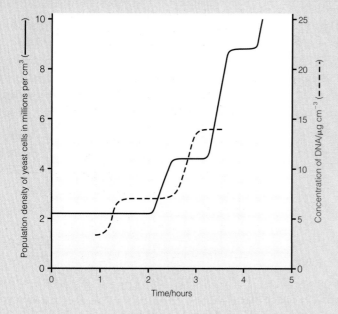

(a) Outline a method by which the population density of yeast cells in the culture could be determined. (4 marks)

(b) (i) What is the time taken for the population of yeast cells at Time 0 to double? (1 mark)

(ii) How does the doubling time for subsequent generations differ from that of the initial population? (1 mark)

(iii) Suggest an explanation for your answer to (b) (ii). (2 marks)

(c) Describe and explain the relation between the two curves shown on the graph. (3 marks)

Yeast and other microorganisms can be used to produce useful substances. The table shows some aspects of fermentation in a culture of the bacterium *Corynebacterium glutamicum* in producing the amino acid lysine which is used to supplement cereal protein.

Time/hours	Cell dry mass/g	Lysine/g dm^{-3}	Residual sugar/%
0	10.0	0.7	20.0
10	12.7	1.1	19.2
20	15.7	9.3	18.2
30	20.1	17.3	13.1
40	24.2	30.2	10.2
50	25.7	40.7	5.4
60	26.0	44.0	2.4
70	26.0	42.1	2.8

(d) Plot the data **in the first three columns** of the table as a graph. Use a single pair of axes and join the points with straight lines. (5 marks)

(e) (i) Calculate the rate of growth of the bacterial population during the period 20 to 40 hours and during the period 50 to 70 hours. Show your working. (2 marks)

(ii) Using evidence from these data, suggest **two** possible explanations for the difference in rates of growth of the bacterial population in these two periods. (2 marks)

The diagram shows part of the biochemical pathway by which lysine is synthesized by *C. glutamicum*. Threonine and lysine both inhibit the enzyme which is necessary to convert aspartate to aspartyl phosphate.

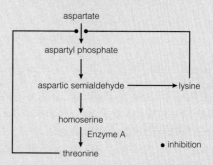

(f) (i) Use the material in this diagram to explain what is meant by negative feedback. (2 marks)

(ii) Suggest how a strain of *C. glutamicum* deficient in Enzyme A can give an increased yield of lysine. (2 marks)

(Total 24 marks)

AEB June 1993, Paper II, No. 4

5. The graph shows the growth of a population of yeast cells over a two-week period. The figures were obtained by counting the number of cells in a sample of the culture.

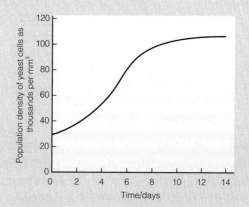

(a) Give **two** explanations to account for the shape of this growth curve after day 12.
(1 mark)

(b) Sketch on the graph a curve showing the *rate* of population growth over the two-week period.
(2 marks)

(c) In a second method of estimating the same yeast-cell population, the cloudiness of the medium was measured.

    (i) Briefly suggest how the cloudiness of the medium might have been measured.
(1 mark)

    (ii) Suggest **one** advantage of measuring the growth of the population in this way.
(1 mark)

    (iii) Suggest **one** disadvantage compared with the density method.
(1 mark)
(Total 6 marks)

AEB June 1992, Paper I, No. 9

6. The production of antibiotics is one example of industrial fermentation.

    (a) Define the term *antibiotic*.
(2 marks)

    (b) The graph summarizes the main events which take place during the production of the antibiotic penicillin by the fungus *Penicillium notatum*.

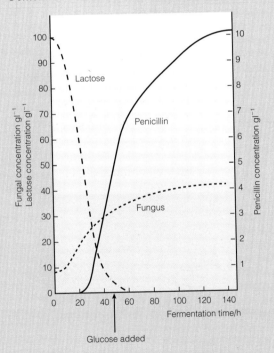

Ref: Adapted from Fermentation Biotechnology by O.P. Ward (1989)

    (i) Describe **three** differences between the pattern of fungal growth and the pattern of penicillin production.
(3 marks)

    (ii) Calculate the rate of penicillin production per hour during the period 32 to 40 hours. Show your working.
(3 marks)

(c) (i) Penicillin production was maintained by adding glucose after 48 h. Explain why this **particular** source of energy for ATP was added instead of more lactose.
(2 marks)

    (ii) Suggest **two** conditions which should be controlled in order to achieve the maximum rate of penicillin production.
(2 marks)

(d) (i) Before cheeses are stored they are often covered with a coating containing an antibiotic. Suggest the purpose of this procedure.
(2 marks)

    (ii) Suggest why penicillin would not be chosen as the antibiotic in this coating.
(2 marks)
(Total 16 marks)

WJEC June 1993, Paper A2, No. 4

7. The use of enzymes in industry is increasing largely as a result of improved methods of producing and stabilizing enzymes. Glucose isomerase is an enzyme commonly used by industry in immobilized form.

    (a) State **two** reasons why it may be advantageous to use an enzyme to catalyse a particular reaction rather than use an inorganic catalyst.
(2 marks)

    (b) Outline **two** methods by which enzymes can be immobilized for use in industry.
(4 marks)

    (c) State **two** advantages which the use of immobilized enzyme systems has over the use of enzyme solutions.
(2 marks)
(Total 8 marks)

UCLES (Modular) June 1993, (Biochemistry), No. 3

Further Questions

1. (a) Relate the structure of a leaf to its activities of photosynthesis and transpiration. *(12 marks)*
   (b) Explain how variations of (i) light and (ii) temperature can affect these functions.
   (8 marks)
   (Total 20 marks)
   NEAB June 1991, Paper IB, No. 1

2. (a) Explain how a series of impulses are initiated and transmitted along a neurone and across a synapse. *(12 marks)*
   (b) Draw a labelled diagram to illustrate the structure of skeletal muscle as seen under an electron microscope. Explain the events which occur when an impulse reaches the end of a motor neurone and suggest how the muscle shortens following the arrival of an impulse. *(8 marks)*
   (Total 20 marks)
   NEAB June 1991, Paper IB, No. 3

3. Write an essay on the maintenance of water and solute balance in plants and animals.
   (Total 30 marks)
   ULEAC June 1993, Paper III, No. 9b

4. (a) State the exact position of the pancreas in the body of a **named** mammal. *(2 marks)*
   (b) Describe the functions of the pancreatic secretions. (Reference to glucagon is **not** required.) *(12 marks)*
   (c) Explain how the release of pancreatic secretions is controlled. *(4 marks)*
   (Total 18 marks)
   UCLES June 1990, Paper I, No. 3

5. (a) State what is meant by a *short day plant*.
   (2 marks)
   (b) Outline a possible sequence of events by which flowering might be initiated in a short day plant. *(7 marks)*
   (c) Describe the features of **named** flowering plants that favour
   (i) cross pollination and (ii) self pollination.
   (9 marks)
   (Total 18 marks)
   UCLES June 1990, Paper I, No. 4

6. Give an account of the control of blood circulation and regulation of breathing in mammals.
   (Total 17 marks)
   SEB May 1990, Higher Grade, Paper II, No. 15B

7. State the significant chemical features and then explain the biological importance of the following molecules:
   (a) phospholipid, *(6 marks)*
   (b) ATP, *(6 marks)*
   (c) haemoglobin. *(6 marks)*
   (Total 18 marks)
   UCLES June 1990, Paper I, No. 1

8. (a) Explain the interactions of plants and animals in maintaining a constant composition of the atmosphere. *(13 marks)*
   (b) Describe ways in which this balance may be upset. *(10 marks)*
   (Total 23 marks)
   UCLES (Modular) June 1992, (Energy in living organisms), No. 1

9. Threonine is an amino acid found in milk protein. Describe how this amino acid could become part of an enzyme molecule in one of your body cells.
   Your answer should include information about:
   (a) protein digestion; *(4 marks)*
   (b) amino acid transport; *(4 marks)*
   (c) protein synthesis. *(7 marks)*
   (Total 15 marks)
   SEB Human Biology Specimen 1992, Higher Grade, Paper II, No. 12A

10. (a) In an experiment to trace the path of carbon in photosynthesizing plants, carbon dioxide containing the radioactive isotope of carbon (^{14}C) was fed to individual leaves of two bean plants – an upper leaf on plant A and a lower leaf on plant B.

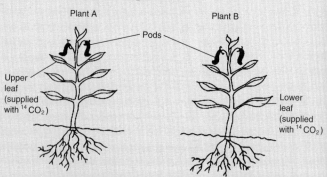

After several hours, various structures of both plants were analyzed and the amount of ^{14}C in each was calculated as a percentage of the total ^{14}C content of each plant. The results of the analysis are tabulated overleaf.

Plant structure	Amount of radioactive carbon present	
	Plant A (Upper leaf treated)	Plant B (Lower leaf treated)
Treated leaf	47.0	51.5
Pods	40.0	24.5
Shoot tip	5.0	2.5
Roots	3.0	16.5
Other leaves	1.0	1.0
Stem	4.0	4.0

(i) The ^{14}C is incorporated into carbon compounds manufactured in the process of photosynthesis. What is the main destination of these carbon compounds in both plants? *(1 mark)*

(ii) Explain why there is very little transport of compounds containing ^{14}C to the 'other leaves'. *(1 mark)*

(iii) What evidence is there, in the data, that the position of a leaf on the plant is related to the destination of the food manufactured in that leaf? *(2 marks)*

(b) The diagrams below represent stages in cell differentiation leading to the formation of phloem tissue in a woody stem.

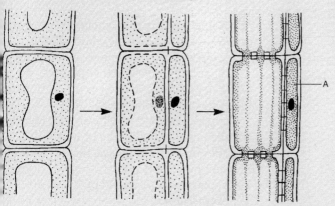

(i) State **three** changes, illustrated by the diagrams, which occur during the formation of phloem tissue. *(2 marks)*

(ii) 1. Name cell A. *(1 mark)*
2. Cell A is found to have many mitochondria. How is this feature related to the function of the cell? *(1 mark)*
(Total 8 marks)

SEB May 1991, Higher Grade, Paper II, No. 3

11. The soil in many tropical rain forests is very poor in nutrients such as nitrate and calcium ions. In some rain forests there are herbaceous plants which, as well as having roots in the soil, have some roots that grow upwards over the surface of tree trunks.

(a) Give **two** external features by which you could distinguish a stem of one of these plants from an upward-growing root. *(2 marks)*

(b) (i) Give **two** stimuli to which these upward growing roots might be responding. *(1 mark)*

(ii) What type of behavioural response are these roots displaying? *(1 mark)*

(iii) Explain how this behavioural response might be an advantage to the plant in nutrient-poor soil. *(1 mark)*
(Total 5 marks)

AEB November 1993, Paper I, No. 3

12. The diagram is of a vertical section through a dicotyledonous leaf.

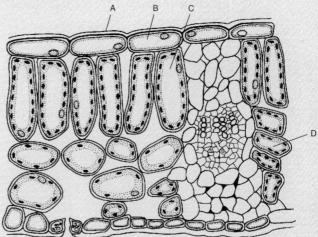

(a) Name the parts labelled **A** to **D**. *(2 marks)*

(b) Describe **three** features shared by leaves and human lungs that make them efficient gas exchange surfaces. *(3 marks)*

(c) Describe **two** ways in which leaves and lungs **differ** as gas exchange surfaces. *(2 marks)*
(Total 7 marks)

UCLES (Modular) March 1992, (Energy in living organisms), No. 2

13. The diagrams show transverse sections of supporting tissues from an herbaceous plant and a mammal.

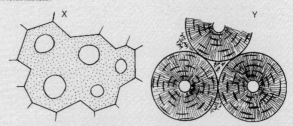

(a) (i) Identify each of the tissues.

(ii) For each tissue, name an organic compound which provides strength.

(iii) State **two** structural similarities and **one** structural difference between these tissues. *(7 marks)*

(b) (i) The plan shows the distribution of tissues in part of a transverse section of a young stem of *Helianthus*. Shade the main regions where tissue X would be found.

(ii) Name the main supporting tissue found in the stem of a woody perennial plant.

(iii) Briefly describe how additional supporting tissue is formed as the woody stem increases in girth. (*4 marks*)

(c) (i) The drawing shows part of a mammalian femur cut longitudinally. Shade the regions where tissue Y would be found.

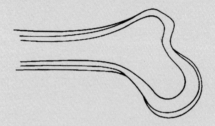

(ii) Explain how the arrangement of tissue Y in the femur confers advantages in relation to the support and locomotion of the mammal.

(iii) Name an additional supporting tissue found in the femur. (*4 marks*)

(*Total 15 marks*)

WJEC June 1992, Paper A2, No. 4

14. For what purpose would you use each of the following pieces of apparatus?
(a) kymograph
(b) colorimeter
(c) pitfall trap
(d) quadrat
(e) haemocytometer

(*Total 5 marks*)

NISEAC June 1993, Paper I, No. 11

15. Comment on the biological significance of the following observations.
(a) Deficiency of molybdate may retard the growth of flowering plants. (*3 marks*)
(b) Some flowers, such as tulip and crocus, close at night and open in the morning. (*3 marks*)

(c) The plant parasite dodder has no leaves, and loses its roots when mature. (*3 marks*)
(d) The stomata of some desert plants open at night and close during the day. (*3 marks*)

(*Total 12 marks*)

ULEAC January 1990, Paper I, No. 5

16. (a) The diagram below shows the nucleus, rough endoplasmic reticulum and Golgi apparatus in a body cell.

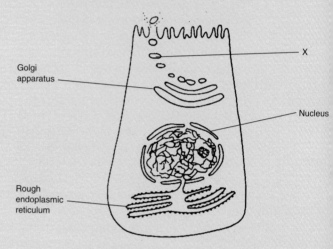

(i) Name the structure labelled X. (*1 mark*)
(ii) What type of organic compound is transported by the rough endoplasmic reticulum to the Golgi apparatus? (*1 mark*)
(iii) Give one function of the Golgi apparatus. (*1 mark*)
(b) Describe one function of the protein molecules in the fluid mosaic model of the cell membrane. (*1 mark*)
(c) (i) Explain what is meant by the term *phagocytosis*. (*1 mark*)
(ii) State one way in which pinocytosis differs from phagocytosis. (*1 mark*)
(d) The graph indicates the gain or loss of water by cells after they had been placed in 0.04% salt solution or in 4.0% salt solution.

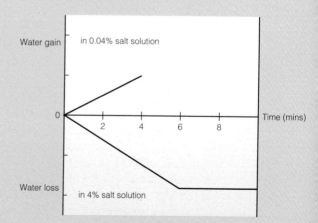

616

(i) Describe the appearance of the cells at 6 minutes in 4.0% salt solution. *(1 mark)*

(ii) Suggest a reason for the lack of data after 4 minutes for the cells in the 0.04% salt solution. *(1 mark)*

(e) The graphs below show the results from experiments to investigate the effect of temperature and oxygen concentration on the rate of uptake of an ion by a body cell. In graph B, the amount of sugar metabolized is also given.

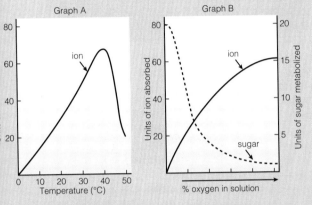

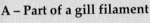

(i) From graph A, by which method is the ion transported across the cell membrane? *(1 mark)*

(ii) With reference to the information in graph B, give two reasons for the answer given in (i). *(2 marks)*

(Total 11 marks)

SEB Human Biology Specimen 1992, Higher Grade, Paper II, No. 16

17. Diagrams **A**, **B** and **C** show part of a gill filament from a fish.

A – Part of a gill filament

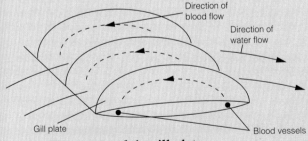

B – Capillary network in gill plate

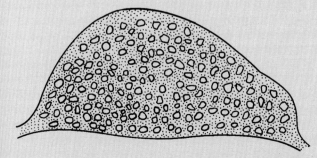

C – Transverse section through gill plate

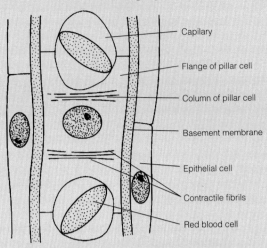

(a) Identify the type of epithelium shown in Diagram **C**. *(1 mark)*

(b) (i) Describe and explain how a high oxygen concentration gradient is maintained across the surface of the gill plate. *(3 marks)*

(ii) Describe how a large surface is provided for gas exchange between the water and the blood. *(3 marks)*

(c) Although fish do not maintain a constant body temperature, they have mechanisms which conserve some of the heat generated by metabolism. In fish living in cold water, heat loss from the gill surface may be a problem. It has been suggested that the columns of the pillar cells are contractile. Suggest how this property of the pillar cells may enable an inactive fish to conserve heat. *(2 marks)*

(d) The data shown in **Table 1** refer to the gills of a number of species of fish.

Table 1

Species	Gill area/ mm²g⁻¹	Thickness/μm		
		Epithelium	Basement membrane	Pillar cell flange
Mackerel	1158	0.17	0.07	0.03
Bonito	595	0.01	0.08	0.02
Dab	188	–	–	–
Skate	–	0.50	0.13	0.03

Mackerel and bonito are both active species living in the open sea. Dab and skate are bottom-dwelling fish that spend much of their time inactive on the sea bed. Explain how the information in **Table 1** relates to the way of life of these fish. *(4 marks)*

(e) Where, precisely, in a cell do the various stages of aerobic respiration take place? *(3 marks)*

(f) **Table 2** lists the amounts of some of the substances used and **Table 3**, some of the substances produced during glycolysis.

Table 2

Substance used	Number of molecules or ions
Glucose	1
ATP	2
ADP	4
NAD	2

Table 3

Substance produced	Number of molecules or ions
ATP	4
ADP	2
reduced NAD	2
pyruvate	2

Copy diagram **D** and use the information in **Tables 2** and **3** to complete the summary of the biochemical pathway of glycolysis. (*3 marks*)

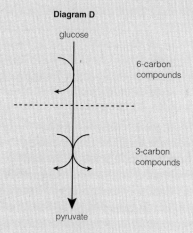

Diagram D

glucose

6-carbon compounds

3-carbon compounds

pyruvate

(g) The equation represents the conversion of pyruvate to lactate during anaerobic respiration in muscle.

$$\text{pyruvate} + \text{reduced NAD} \longrightarrow \text{lactate} + \text{NAD}$$

(i) Explain why this reaction is important in allowing the production of ATP during anaerobic respiration to continue. (*2 marks*)

(ii) What happens to the lactate after the period of anaerobic respiration? (*3 marks*)

(*Total 24 marks*)

AEB November 1993, Paper II, No. 1

18. The photomicrograph at the top of the next column shows part of a mammalian liver cell.

(a) (i) Calculate the actual width of the mitochondrion labelled A, using the scale under the photomicrograph. Show your working. (*2 marks*)

Magnification × 44000

(ii) Name *two* other types of mammalian cell in which you would expect to find relatively large numbers of mitochondria. (*2 marks*)

(iii) What is the function of the cristae in the mitochondria? (*1 mark*)

(b) The table below lists three cellular components (whole cell debris, mitochondria only and residual cytoplasm) and shows their ability to produce carbon dioxide and lactate when incubated with glucose in the presence of oxygen.

Cellular component	Product	
	CO_2	Lactate
Whole cell debris	✓	✗
Mitochondria only	✗	✗
Residual cytoplasm	✗	✓

Key
✓ = Present
✗ = Absent

(i) Comment on the inability of mitochondria to produce carbon dioxide. (*2 marks*)

(ii) Which cellular component contains the enzymes which catalyse the conversion of pyruvate into lactate? (*1 mark*)

(iii) Cyanide is a respiratory poison but it has no effect on the production of lactate from glucose in the residual cytoplasm. Suggest a reason for this. (*2 marks*)

(*Total 10 marks*)

ULEAC January 1993, Paper 1, No. 6

19. Each of the following diagrams contains **three** deliberate errors in the accuracy of the drawing and/or in the labelling. Identify **each** of the errors.

(a) Section of mammalian stomach with associated glands. *(3 marks)*

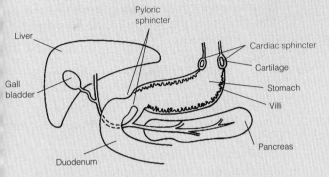

Liver
Gall bladder
Pyloric sphincter
Cardiac sphincter
Cartilage
Stomach
Villi
Pancreas
Duodenum

(b) Section of human testis. *(3 marks)*

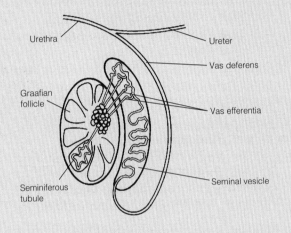

Urethra
Graafian follicle
Seminiferous tubule
Ureter
Vas deferens
Vas efferentia
Seminal vesicle

(c) Diagram to show Went's experiment on the effect of unilateral light on the distribution of auxin. *(3 marks)*

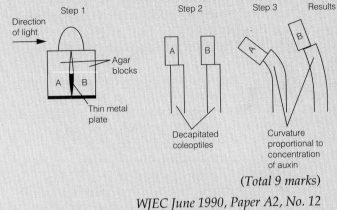

Direction of light
Step 1
Agar blocks
A B
Thin metal plate
Step 2
A B
Decapitated coleoptiles
Step 3
A B
Results
Curvature proportional to concentration of auxin

(Total 9 marks)

WJEC June 1990, Paper A2, No. 12

20. The drawing shows the jaws, dentition and jaw muscles of a sheep, a typical herbivorous mammal.

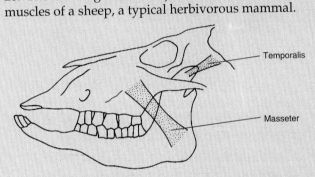

Temporalis
Masseter

(a) (i) Describe how *two* features visible in the drawing adapt the dentition to a herbivorous diet. *(2 marks)*

(ii) Explain how the jaw muscles are involved in the particular jaw action found in a herbivorous mammal like a sheep. *(2 marks)*

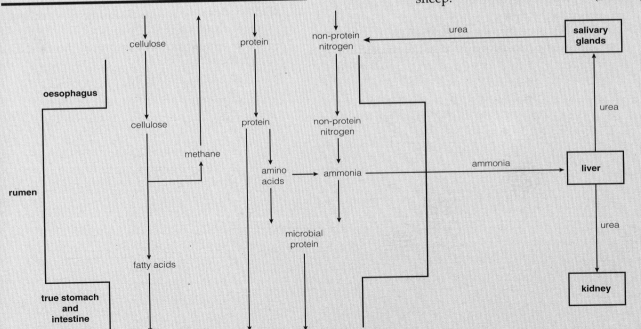

oesophagus

rumen

true stomach and intestine

cellulose
cellulose
methane
fatty acids

protein
protein
amino acids
microbial protein

non-protein nitrogen
non-protein nitrogen
ammonia

urea
salivary glands
urea
liver
ammonia
urea
kidney

A cow obtains most of its nutritional requirements from fermentation by mutualistic (symbiotic) microorganisms in its rumen. The diagram at the bottom of page 619 summarises the biochemical processes carried out by these microorganisms.

(b) With the aid of information in the diagram, explain why:
 (i) the relation between the cow and the organisms living in its rumen may be described as mutualistic; (2 marks)
 (ii) it is possible for cattle to survive on a diet that contains no protein for a considerable period of time; (2 marks)
 (iii) ruminants such as the cow are less efficient than non-ruminant animals at converting energy in their food into energy in their tissues. (2 marks)

(c) (i) What is likely to be the main respiratory substrate of the cow? (1 mark)
 (ii) How does the cow obtain ATP from this respiratory substrate? Details of biochemical pathways are **not** required. (2 marks)

Fermentation in the rumen is sometimes likened to the process in an industrial fermenter.

An industrial fermenter may be used for the continuous production of substances such as penicillin by the fungus *Penicillium*. Sterile medium is continuously added and the penicillin harvested. Temperature and pH are carefully controlled.

(d) (i) Suggest *two* important differences between the fermentation process in an industrial fermenter and fermentation in the rumen. (2 marks)
 (ii) How are constant conditions of temperature achieved in the rumen? (1 mark)

Many plants possess defence mechanisms against invertebrate herbivores. Some forms of clover are cyanogenic. They produce the extremely poisonous chemical, hydrogen cyanide, when their tissues are damaged.

The production of hydrogen cyanide takes place in two steps, each of which is under the control of a separate gene.

$$\text{substrate} \xrightarrow[\text{Enzyme 1}]{\text{Step 1}} \text{cyanogenic glucoside} \xrightarrow[\text{Enzyme 2}]{\text{Step 2}} \text{hydrogen cyanide}$$

Step 1 requires the presence of the dominant allele A and step 2, the presence of the dominant allele L.

(e) Which clover phenotype (cyanogenic or non-cyanogenic) is associated with each of the following genotypes? Explain your answers.
 (i) *AaLL* (2 marks)
 (ii) *aaLl* (2 marks)

(f) Explain how natural selection could lead to a high frequency of these dominant alleles in a clover population. (4 marks)

(Total 24 marks)

AEB June 1994, Common Paper 2, No. 1

21. During a course of lessons on physiology, a teacher was asked the following questions:
 (a) If a double circulation is so efficient, why didn't earthworms evolve one?
 (b) Why don't plants have nervous systems?
 (c) How is it that a field mouse has to eat for twelve hours out of every day, a lion can starve for a week and a polar bear can survive the whole arctic winter without food?
 (d) Why do the leaves of deciduous trees change colour and fall in the Autumn?
 How would you have answered these questions?

WJEC June 1994, Special Paper, No. 4

Index

Main entries are indicated by **bold** type